AGRICULTURAL ENGINEERING
VOLUME 3

PROCEEDINGS OF THE ELEVENTH INTERNATIONAL CONGRESS ON
AGRICULTURAL ENGINEERING / DUBLIN / 4-8 SEPTEMBER 1989

Agricultural Engineering

Edited by
VINCENT A.DODD & PATRICK M.GRACE
Agricultural and Food Engineering Department, University College Dublin

VOLUME 3
Agricultural mechanisation

CRC Press
Taylor & Francis Group
Boca Raton London New York

CRC Press is an imprint of the
Taylor & Francis Group, an **informa** business

Land and Water Use, Dodd & Grace (eds), © 1989 Balkema, Rotterdam. ISBN 90 6191 980 0

Contents

3 Agricultural mechanisation

3.1 *Soil cultivation and seed-bed preparation*

3.2 *Harvesting and handling of agricultural products*

3.3 *Application of plant nutrients and pesticides*

3 Agricultural mechanisation
3.1 Soil cultivation and seed-bed preparation

Land and Water Use, Dodd & Grace (eds), © 1989 Balkema, Rotterdam. ISBN 90 6191 980 0

Keynote paper:

Optimising soil cultivation and seedbed preparation parameters referring to soil conditions

O.S.Marchenko

All-Union Research Institute for Mechanisation of Agriculture (VIM), Moscow, USSR

ABSTRACT: The theme involves all the types of field operations on tillage and seedbed preparation to create the optimum conditions for plant growth, the problems of preventing soil overcompacting by machinery and tractors, different tillage systems and their possibility to ensure the biological demands of plants, tendencies of development of machinery technologies using the systems of minimum and zero tillage, gaining the reduction of labour requirements, the reduction of fuel consumption and means, the results of studies on new tillage tools, combined machines and units.

1 INTRODUCTION

The author has studied 42 reports which according to their content can be divided as follows.

1. New approaches to choosing the optimal systems and parameters of tillage.

2. New working tools of tillage machines.

3. Quality of seedbed preparation.

4. Interaction of tractor tyres with soil, tractor ballasting, small tractors and animal draft as applied to tillage.

The authors of all the reports worry about the condition of soil resources, the increased tendency of reducing the soil fertility and decreasing the humus content in it, development of wind and water erosion, salt content and droughtiness of soil.

To a great extent these phenomena are connected with the lack of well-founded systems of farming and, in particular, wrong choice of technologies and comlexes of machines and tools for tillage in specific soil and climate conditions. So, for example, multiple runs of massive machine and tractor units in field and

soil compaction connected with it, inadequate efficiency of agrotechnical methods of decompacting deep subsoil, water conservation in soil at controlling its water -air regime at tilling and preparing a seedbed during sowing and cultivation do not give the possibility to ensure the optimal conditions for plant growth and high yields.

The results of investigations presented by many authors in their reports are indicative of new approaches to the choice of the optimal parameters of tillage and effective working tools of tillers in accordance with specific soil conditions.

2 NEW METHODS OF CHOOSING THE OPTYMAL SYSTEMS AND PARAMETERS OF TILLAGE

One of the trends of increasing tillage efficiency is differentiated in depth tillage in accordance with a type of soil and cultivated crop, that aids also in significant reducing labour and energy requirements.

However, the usage of conventional implements for tilling

in a reduced scale, for example, in depth of tillage is not always suitable. So Cavalli and Sartori (Italy) indicate that conventional ploughs at tilling soil with a great amount of plant residues cannot carry out shallow cultivation in heavy and dry soils because of difficulties in maintaining constant tillage depth. Moreover, it has been demonstrated that in clay soils the action of p.t.o. driven implements used for shallow cultivation immediately before or contemporary to drilling causes an over pulverization of soil that may lead to crust formation.

The authors suggest a combined machine for shallow cultivation of heavy soils: the passive shanks of the machine perform loosening of deep subsoil and ensure good stability, and p.t.o. drive discs till the upper soil layer and effectively incorporate crop residues into the soil.

Another trend of improving tillage system ensuring the reduction of energy requirements and protection of the environment from degradation is to use a chisel. Panaro and Patruno (Italy) highlight the results of testing some soil tilling techniques using chesel tools in comparison with the conventional deep ploughing.

The main aim of deep soil loosening is to cause maximum disruption of compacted layers within the soil profile. Although not a new concept, it has become commonplace in recent years in connection with continuous tillage practicices without grass in the rotation and also in connection with increasing machinery mass and their effect on the soil.

Larney and Fortune (Ireland) studied the effect of deep loosening on sugar performance. 12 experimental sites were laid down; 4 sites showed significant root and sugar yield increase, 6 showed no significant differences, while 2 gave significant decrease in yield on deep-loosened plots. The authors consider that negative and negligible yield responses to deep loosening are largely explained by recompaction of autum loosened soil during seedbed preparation the following spring. Deep loosening decreased the percentage of forked roots on all the sites where there was a yield response to the operation.

The distinguishing features of deep ploughing in comparison with shallow cultivation in the conditions of horticulture were cited by Torregrosa Mira and Gracia Lopez (Spain). They studied deep ploughing (40-50 cm), tillage by a moldboard plow (30 cm depth), superficial ploughing (10-13 cm deep) using rotary tiller and disk harrows. No significant difference in production have been observed throughout the experiment using any of the tilling systems. However energy requirements were different: tilling with disk harrows - 275 MJ/ha, the moldboard plow - 1550 MJ/ha.

The scale of using the systems of minimum tillage at cultivating row crops including a ridge technology is being enlarged. Hummel, Wax and Siemens (USA) give the results of the comparison of three ridge and one conventional systems of tillage for producing corn and soybeans in the central U.S. cornbelt. In so doing besides the agrotechnical indices they assessed alternative sets of machines, their cost and requirement. A PC-based farm machinery selection program was used to determine the optimal machinery set for 4 farm sizes.

More and more attention is paid to control of the soil compation by mobile energetic means and machine and tractor units. The main ways of reducing the soil compaction besides its decompaction by chisel tools are as follows: decreasing a number of machinery runs in fields , combination of technological operations, reduction of specific pressure of machinery on the soil, performance of field operations in the best agrotechnical period at permissible soil moisture and so on.

Campbell and McGregor (UK) publish the data concerning crop responses to various trafffic treatments for a range of soils, crops and climates in both Europe and North America.

The authors compared conventional and zero traffic systems for winter barley grown for 4 seasons in either a ploughed or direct drilled clay loam. The results show that the zero traffic system gave more uniform crops, smaller seasonal variations in yield, better root penetration, reduced energy requirement and possibility of direct drilling without the soil compaction.

Taylor (USA) presented the results of the research on the interaction of soil-machine and soil-plant. He considered the opposite tendencies of these interactions: farm tractor tyres need dry, highly compacted soil to provide flotation, mobility and good tractive efficiency; agricultural crops need good moisture and low compaction levels. Controlled traffic is suggested as a concept that allows optimization of soil conditions for both crops and tyres. The author considers that a crop production system employing permanent traffic lanes has a great potential.

Vermeulen, Boone, Arts and Tijink (Netherlands) gave us an agricultural engineering approach to investigate crop responses to soil loading treatments; they used the method of simulation of currently practiced as well as possible future loading treatments by field traffic to establish the relationship "wheel-soil-plant".

The report of Kayombo (Tanzania) deals with soil compaction and soil erosion at mechanical land clearing in the tropics ; the author considers that compaction problems could be materially reduced by the avoidance of wet season traffic, reduced topsoil movement and a reduction of vehicular traffic. The practice shows that no-tillage system with crop residue mulch is an effective soil compation control method, provided it can be adopted as a package of all cultural practices that make it work.

Tebrugge (FRG) points out the significance of organic inclusions in the soil for reduction of soil compaction. According to his point of view when the

contact intensity of soil cultivation tools is reduced, power and energy requirements can be cut considerably and, as a result, damage to the soil caused by pressure and slip minimized to a large extent. Homogeneous mixing-in of organic remains leads to major improvements with regard to sludging and erosion protection. Conversion of these substances is accelerated, the biological activity is increased and, as a result of these interrelations the natural fertility of the soil and the crumb and structural atability are improved.

Bourarach (Morocco) and Eichhorn (FRG) studied the influence of first precipitations as related to primary tillage for different tillage systems. It was concluded that the period of primary tillage has a significant influence on yield and profit on a hectare basis. The tillage system had no influence on yield and profit except in dry years.

Braide (Nigeria) studied the effectiveness of usage of tracklaying tractors at state and private farms (more 100 ha). He established, that power requirements at tilling irrigated lands were twice as much as without irrigation.

Blackwell, Jayawardane, Butler, Casegrain and Smart (New Zealand) studied the methods of amelioratting acid and sodic subsoils by means of gypsum/lime slotting to increase soil infiltration and crop yields and to avoid surface crusting. Regular gypsum application and its consequence make it uneconomic. Deep tilling can markedly increase infiltration rates and crop production on these soils, but these effects tend to decrease significantly in the following seasons , apparently due to repacking of the soil under flood irrigation and tractor wheel compaction.

3 NEW TILLAGE TOOLS

It is known that the tillage by moldboard ploughs is a very energy consuming operation, the effectiveness of which is defined by plough parameters according to soil conditions, the power index and traction forces.

The optimal utilization of the tractor's engine capacity at the minimum fuel consumption and the high quality of ploughing is difficult to achieve under variable operating conditions using traditional ploughs.

Jori and Soos (Hungary) examined the American idea to change the working width of the plough bodies on 3 types of soil (sandy, clay loam and clay). The established that exchaning the working width within a certain range (300-500 mm) did not cause any change of ploughing quality and reduced energy requirement. By using ploughs with variable working width we can reduce overall energy consumption under changes of soil conditions as well as tractor power.

The conventional moldboard ploughs are the main implements for tillage which ensure slice cutting, its crumbling, lifting and turning to create the necessary conditions for plant growth. Therefore scientists continue to pay a great attention to improving the geometry and investigation into the power characteristics of the moldboard ploughs.

So Girma (Sweden) measured three orthgonal forces and the resulting moments acting about the orthogonal axes on each plough component (moldboard, share point, share wing and landside) and on plough bodies as a whole. Force distribution on the plough components were determined. Predictive equations were developed from experimental data to determine the relationship between forces and plouging depth and speed in the concrete soil conditions.

Salokhe, Gee-Clough, and Mufti (Thailand) conducted comparative experiments on a Thai-made slit enamel coated plough and an uncoated similar moldboard plough in a laboratory soil bin using clay soil. It was found that enamel coating reduced the plough draft up to 26% depending on the soil moisture content, working speed and the soil sticking.

Kuczewski and Majewski (Poland) developed a mathematical model for selection of optimal work parameters of a tractor-plough combination at reduced engine speed basing on three decision variables: tractor transmission gear, engine speed and working width of plough. As a criterion the fuel consumption per hectare ploughed was taken. The obtained results point out that the most efficiency of the tractor work is at the reduced engine speed in relation to nominal one.

Orlov and Ischeinov (USSR) consider different arrangements of combined mobile and stationary machines and aggregates with active (driven from p.t.o.) working tools and passive working tools. They developed a mathematical model and gave the results of calculating energetic characteristics of the agricultural aggregate including the tractor K-700 and the plough ПТК-9-35 as well as other machine and tractor aggregates, which confirm the efficiency of the developed method for choosing the models of the agricultural machines which satisfy to the requirement of energy consumption reduction.

Fielke (Australia) studied the effect of a chisel plough share wing's width, lift height, rake angle and sweep angle on its draft and vertical force at a range of speed in varying soil types. The tests were conducted on a sandy loam soil and a red brown soil . The results showed that the geometry of the wings can be optimised for minimum draft force and maximum vertical down forces.

Tine (Italy) established that soil pressure along the tine was influenced by tine length, soil water content and speed.

Peruzzi, Consorti and Ciolo (Italy) established the advantages of a mounted implement with vibratory tilling tools in comparison with the drawn vibratiller from the energetic and work productivity point of view.

Venter (South Africa) developed a mathematical model for the energy requirements of a vibratory tillage tool which was used for determining the parameters of a subsoiler with the working depth 500-600 mm to till close to citrus trees with the aim of better aeration and introduction of the necessary fertilizers, thereby extanding the economic productive life of these trees.

Haibo Guo, Burkhardt, Wilkinson (USA) and Makoto Hoki, Tatsuo Tanoue (Japan) presented the results of a computer simulation of the motion of a point on a rotating disk blade of the powered disk tiller. Of some interest is the roller chain drive which has a power loss less than 2%.

Rodgers, Mulqueen and Kyne (Ireland), Spoor and Hann (England) and Centeno (Brazil) presented the results of investigations of the modified mole plough shanks ensuring more intensive influence of the soil to develop a greater number of fissures and to improve the work of a mole plough. The efficiency of any mole drain system depends on fissuring and fissure connections with a mole channel. Increasing expander diameter tends to close the fissure/mole channel connection.

Billot (France), Manor (Israel), Clark and Radcliffe (USA) describe the possibility to take quantitative and qualitative characteristics with the help of vertical (Billot) and horizontal penetrometers which can be used to select the implements and periods best suited to the tillage operations to be carried out on a peculiar field.

4 QUALITY OF SEEDBED PREPARATION

Thoroughly prepared seedbed is required for a uniform crop emergence, plant growth and high yield under different soil and climate conditions and for any crop.

Henriksson (Sweden) considers that when spring sowing is followed by dry weather throughly prepared seedbed is of great importance because in spring the potential of evaporation is often high and the precipitation is low. A careful seedbed prepa-ration is required to ensure a uniform crop emergence at the limited soil moisture content, especially on clay soils. A suitable placement of the seeds is obtained if the drill coulters are pressed against a firm, level seedbed bottom at a suitable depth. The harrows should form such a seedbed bottom as well as a fine tilth to prevent evaporation during germination.

The filed experiments in cereals and sugar beets showed that difference in construction of the harrows such as weight, design of levelling bars, and tines, section width and tine spacing influenced the roughness of the seedbed bottom, the aggregate size distribution and the plant emergence and yield. Repeated harrowings were normally needed to obtain a good seedbed. The harrow construction may influence the optimal number of harrowings.

The other important factor, especially for row crops, is an uniform distribution of seeds in a row, especially at using minimum and zero tillage systems and at the presence of great quantity of crop residues on the soil surface or in the upper soil layer. Therefore scientists pay a great attention to the precision planters which ensure seed incorporation at a required depth on keeping crop residues on the surface.

Heyns (South Africa), presented the results of testing precision planters for row crops. Ji Chungian and Yu Feng (China) developed a device to test the uniformity of seed metering mechanisms. The device can measure seed spacings for the precision seeding of soybean and corn with an error less than 3 mm.

Balsari, Bocchi and Tano (Italy) cited the test results of rice sowing on dry soil by new sowing techniques. They established that when sowing was done in the dry soil the swoing operation time was by 20% lower. The pneumatic fertilizer spreader with the same good performace either in dry or in submerged soil permitted an inincrease of the yield by 27% compared with the yield reached with the traditional spinning ferti-

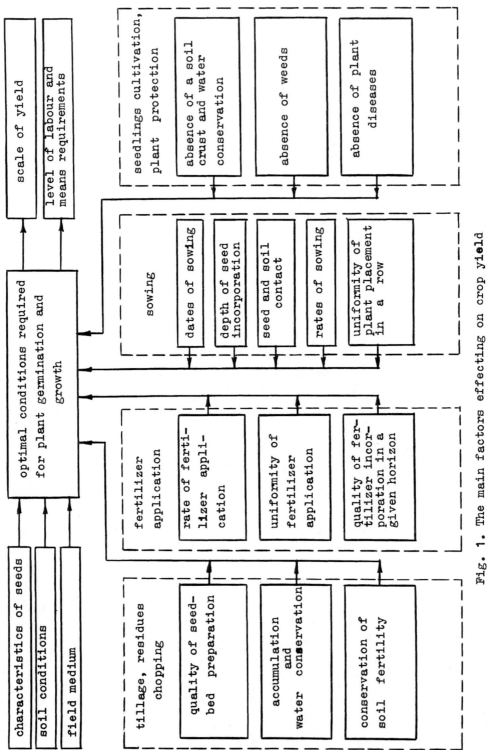

Fig. 1. The main factors effecting on crop yield

lizer distributor.

Odigboh and Akubuo (Nigeria) describe the parameters of a two-row automatic seed yam planter.

Suggs, Peel, Seaboch and Eddington (USA) describe plants growing in a grid of paper cells which unfolds into a continuous strand of plant containing cells; plants are guided to the furrow by a special equipment.

5   INTERACTION OF TRACTOR TYRES WITH SOIL, TRACTOR BALLASTING, SMALL TRACTORS AND ANIMAL DRAFT AS APPLIED TO TILLAGE

The problem of interaction of tractor tyres with soil is paid more and more attention in connection with soil compaction and its negative effect on plants. Armbruster (FRG) paid special attention to investigation of the tractive performance of special grassland tyres, wide tyres with normal and reduced lug height, and low-pressure tyres in comparison with the tractive performance of conventional tractor drive tyres. He measured the difference in yield of the overrun area and the area between the tracks. Wide tyres with low inflation pressure and profiles with wide and flat lugs and overlapping patterns are most suitable.

Plessis (South Africa) gives the side force characteristics of tractor tyres, which on a fairly smooth asphalt surface were in close comparison to those obtained with truck tyres under similar conditions.

Bashford (Morocco) and Esch (USA) presented the data on the tractive performance of a two-wheel drive agricultural tractor on 4 different ballasting modes. They showed that ballasting recommendations based on the ratio of tractor static load to rated PTO power did not reflect the ratio of dynamic load on the drive axles to input power of the driving axles that are occuring under operating conditions.

Venter (South Africa) considers that there is a definite need for an inexpensive, unsophisticated yet reliable agricultural tractor and availability of spare parts for developing nations. A three-

wheel design where traction is obtained from a single rear wheel with sufficient capacity is under development.

Phongsupasamit (Thailand) and Sakai (Japan) consider the problems of creating a hand-tractor plough for the conditions of Thailand.

Mattila (Finland) points out that for the Nordic countries where the average farm size is rather small besides the conventional operations a farm tractor is used for various other purposes, including the work in forestry. This sets special demands on the tractor.

Svensson (Sweden) points out, that maintenance costs for farm machinery of individual farms increase in connection with machinery cost rise, their ageing, including the effect of bad weather, machinery damage and reduction of efficiency.

Betker and Kutzbah (FRG) presented the data concerning the efficiency of animal draft usage for tillage in a semi-desert region of West Africa, where the hand labour is still widely used at tilling and weed controlling.

6   TENDENCIES OF DEVELOPMENT OF TILLAGE TECHNOLOGIES AND IMPLEMENTS

Rational methods of tillage are directed at creating optimal conditions for plant growth and development at which the prepared for sowing soil has necessary for particular soil conditions density, porosity, has a firm seedbed and structural soil elements, ensuring moisture conservation from evaporation and a good nontact of seeds with soil, uniform germination, good plant development and high yields. In addition to the above increase of labour productivity at tilling and minimization of energy requirements and means should be taken into account.

The main factors effecting crop yield are shown at the Fig. 1.

Tendencies of the development tillage technologies are shown on the example of row crop cultivation (Fig. 2).

As we see (Fig. 2), this development follows the two main directions: minimization of tillage

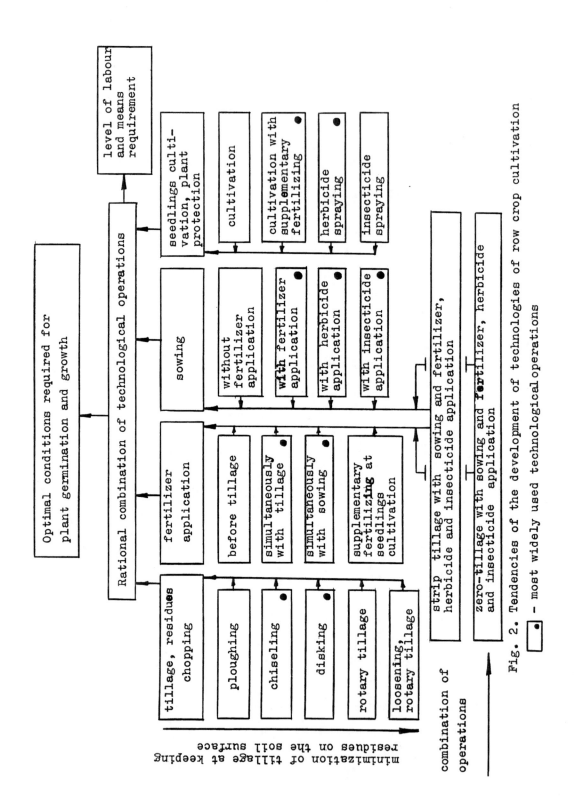

Fig. 2. Tendencies of the development of technologies of row crop cultivation

● - most widely used technological operations

using the strip or zero tillage on keeping chopped plant residues on the surface or in the upper layer where there is some mulch; combination of technological operations such as tillage, sowing, application of fertilizers, herbicides and insecticides and so on.

Minimization of tillage is directed at soil and water conservations. It is reduced tillage, mulching, ridge, strip and zero tillage, when sowing is done into untilled soil.

In connection with the extension of the usage of minimum tillage chisel implements, heavy disk harrows, rotary tillers, combined machines and aggregates are being modified. Partial or complete replacement of some tillage operations by herbicide application for weed control gives great opportunities for combining tillage and sowing, raising the quality of seedbed preparation uniformity of fertilizer and herbicide application, quality of sowing and seed incorporation, reducing the negative effect on soil.

Conventional moldboard tillage is too being improved with the aim to raise tillage quality; this connected with the usage of reversible ploughs for flat tillage, increase of working width of a plough bottom, use of changeable bottoms for tilling soil of different types (digger, semidigger, helicoidal, slat, breaking bottoms and so on).

The usage of chisel tillage gives the possibility to carry out moldboard tillage not every year (one time every 2 - 3 years).

REFERENCES

Armbruster, K. Untersuchungen an Reifen zum boden- und Pflanzenschonenden Befahren von Grünland.

Balsari, P., Bocchi, S., Tano, F., First test results of rice sowing on dry soil by pneumatic fertilizer spreader.

Bashford, L.L., Esch, J.H. Tractive performance of agricultural tractors relative to ballasting.

Betker, J., Kutzbach, H.D. The working performance of animaldrawn implements for soil tillage in Western Africa.

Billot, J.F. Penetrometrie, choix des outils et dates de travail du sol.

Blackwell, J., Jayawardane, N.S., Butler, R., Casegrain, J.C., Smart, R.E. Gypsum/lime slotting technique and machinery for ameliorating acid and sodic subsoils.

Bourarach, E.H., Eichhorn, H. La data du travail primaire du sol en agriculture pluviale au Maroc.

Braide, F.G., Field capacitive performance of track type tractors in plow operation under tropical conditions.

Campbell, D.J., McGregor, M.J. The economics of zero traffic systems for winter barley in Scotland.

Cavalli, R., Sartori, L. Qualitative evaluation of a combined implement for minimum tillage.

Chungian, J., Feng, Yu. An application of solid state cameras to testing the uniformity of seed feeding.

Fielke, J.M. The influence of chisel plough share wing geometry on tillage forces.

Gebresenbet, Girma. Measurements and prediction of forces on plough bodies. I. Measurement of forces and soil dynamic parameters.

Guo Haibo, Burnhardt, T.H., Wilkinson, R.H., Hoki Makoto, Tanoue Tatsuo. Disk trajectory simulation of a powered disk tiller.

Henricksson, L., Effects of different harrows on seedbed quality and crop yeld.

Heynes, A.J. Techniques for the evaluation of precision planters

Hummel, J.W., Wax, L.M., Siemens, J.C. Ridge tillage systems in the central U.S. Cornbelt.

Jori, J.I., Soos, S. Energy-saving ploughing with variable width ploughs.

Kayombo, B. Soil compaction in the tropics and some means of alleviating it.

Kuczewskci, J., Majewski, Z. Selecticn of work parameters of a tractor-plough combination at reduced tractor engine speed.

Larney, J.F., Fortune, R.A. Deep soil loosening effects on sugar beet performance.

Manor, G., Clark, R.L., Radcliffe, D.E. Horizontal penetrometer for trench walls.

Mattila, T. Multipurpose use of a farm tractor in nordic countries.

Odigboh, E.U., Akubuo, C.O. A two-row automatic seed yam planter prototype.

Peruzzi, A., Consorti, S.B., Ciolo, S. Mounted and drawn vibra-tillers and proposal for a program for tillage data processing.

Phongsupsamit Surin, Sakai Jun. Studies on engineering design theories of land-tractor ploughs.

Plessis, H.L.M., The combined lateral and longitudinal forces on a 18.4-35/15-35 tractors tyre.

Rodgers, M. Mulqueen J., Kyne, M. An experimental investigation of the effects of ribbed shanks on the soil displacements induced by by a mole plough.

Orlov, N.M., Ischeinov, V.Ya. Formalization of the automated calculations of agricultural aggregate energetic characteristics.

Salokhe, V.M., Gee-Clough, D., Mufti A.I. Performance evaluation of an enamel coated mouldboard plough.

Spoor, G., Hann, M.J., Centeno, A.S. Influence of mole plough leg and expander geometry on soil disturbance.

Suggs, C.W., Peel, H.B., Seaboch, T.R., Eddington, D.L. Self feeding transplanters.

Svensson, J.E.T. Maintenance costs for farm machinery.

Svensson, J.E.T. Stoppage costs in crop farming.

Taylor, J.H. Controlled traffic research: An international report.

Tebrugge, F. Wechselwirkungen von Bodenbearbeitungssystemen im 10-jährigen Versuch aus technische Leistungsparameter, Bodenstruktur, biologische Aktivitat, Pflanzenkraheiten und Ertrag.

Tine, G. Soil pressure distribution on rigid tines.

Torregrosa Mira A., Gracia Lopez C. Influence of preliminary soil cultural practicies in horticultural crops in Valencia.

Venter, G. Application of a vibratory tillage tool for the reclamation of ageing citrus orchards

Venter, G. Development of a low cost agricultural tractor for developing nations.

Vermeulen, G.D., Boone, F.R., Arts, W.B., Tijink, F.G.Y. An agricultural engineering approach to investigate crop responses to soil loading treatments.

Land and Water Use, Dodd & Grace (eds), © 1989 Balkema, Rotterdam. ISBN 90 6191 980 0

The working performance of animal-drawn implements for soil tillage in Western Africa

J.Betker & H.D.Kutzbach
Institute for Agricultural Engineering, University of Hohenheim, Stuttgart, FR Germany

ABSTRACT: In the semi arid-zone of Western Africa mechanization of soil tillage by the use of draught animals is an important step to increase agricultural production. An experiment has been realized on four different soils in Niger - sandy, sandy clay loam, silt loam and silty clay loam - to analyze the working performance of animal-drawn soil tillage implements (plough, scarifier, ridging plough, strip tiller). Results for draught force requirements and working depths are indicating the need to adapt the actual design to local edaphic conditions.

ZUSAMMENFASSUNG: In der semi-ariden Zone Westafrikas stellt die Mechanisierung der Bodenbearbeitung durch die Nutzung von Zugtieren einen wichtigen Beitrag zur Steigerung der landwirtschaftlichen Produktion dar. In einer im Niger durchgeführten Versuchsreihe wurden verschiedene tiergezogene Geräte zur Bodenbearbeitung - Pflug, Grubber, Häufel-pflug und Tiefenlockerer - in ihrem Arbeitsverhalten auf unterschiedlichen Böden unter-sucht. Ergebnisse über benötigte Zugkräfte und erzielte Arbeitstiefen zeigen, daß die untersuchten Modelle einer weiteren konstruktiven Anpassung an die wechselnden edaphi-schen Bedingungen bedürfen.

RESUME   Dans la zone semi-arid de l'Afrique de l'Ouest la mécanisation de la pré-paration du sol avec des animaux de trait est un pas important pour augmenter la production agricole. Une série d'essais était réaliséeau Niger sur des sols différent - sable, limon argileux sableux, limon fin et limon fin argileux - pour analyser les caractéristiques du travail des matériels aratoire (charrue, scarifieur, buttoir et labour en bande). Les résultats concernant la force de traction et la profondeur du travail montrent la nécessité d'adapter leurs paramètres constructifs aux conditions locaux du sol.

1 OBJECTIVES

In the Sahelien zone of Western Africa animal traction (AT) is a relatively recent technology that has been promoted on a large scale by national governments, research centers and developing projects during the last decades (Munzinger 1981). Draught animals (DA) are increasingly used for transport and field work (Spencer 1988). While seeding and harvesting is still done by hand soil tillage and weeding have been identified as major labor bottle-necks (Hecht, Midohoe 1982) and therefore mechanization efforts are focused on these operations.

The equipment designed for animal traction in this zone - where its number i.e. for Niger is estimated to be about 26,000 (Le Moigne et al. 1987) - is criticised by several authors to be too heavy for light soils (Ashburner and Mamane 1988, Fall 1988) but only average figures about draught forces etc. are available (Anonymus 1978).

Therefore a series of experiments was designed to analyze the working perfor-mance of these implements for soil tillage under various edaphic conditions. Measuring draught force requirements, working depth and width with precise methods it is possible to get detailed informations about the appropriateness of the individual design of the tested equipment and to identify needs for modifications.

2 METHODS AND MATERIALS

Experiments were run on four soil types (tab.1), sandy (A), sandy clay loam (B), clay loam (C) and crusted silty clay (D).

Table 1. Soil types

soil	sand (%)	silt (%)	clay (%)
A	93.8	2.9	3.3
B	68.7	7.4	23.9
C	26.4	49.5	24.1
D	17.3	52.3	30.4

The penetration resistance was low for all humid soils (F < 50 N) and high on dry soils (fig. 1).

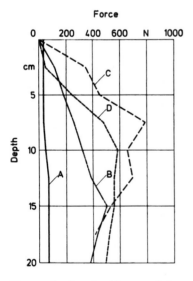

Figure 1. Penetration resistance for dry soils

Experiments were carried out at the end of the dry season and after the first heavy rainfalls (> 15 mm; for soil D only on dry soil). Moisture content (MC) in dry soils was MC < .5 % for all dry soils and varied under humid conditions (tab. 2)

Table 2. Moisture content (MC) in humid soils

soil	A	B	C
MC (%)	5.07	5.75	11.95
S.E.	±0.50		

Under these edaphic conditions the performance of four different animal drawn implements was examined (tab. 3)

Table 3. Characteristics of tested implements

Implement	Characteristics
Plough	DIN 11119, SK 16 cylindric moulboard, cutting length 252 mm 11.5 kg
scarifier	3 tines (1 in front, 2 in rear) reversible tines (50 mm width, 200 mm radius, 58° cutting angle) 6.0 kg
ridger	cylindric moulboard with reversible tine and rectangular wings 10.1 kg
strip tiller	length of support 650 mm Tine: gousefeet-shaped 195 mm width, 65 mm length, 110° cutting angle 6.2 kg
frame	type "Arara" 810 mm length rectangular central tube (50 x 80 x 5 mm)

Experiments were run on flat plots of 25m with millet as last year's crop. Number of replications was four. The mechanical parameters of traction force, inclination of the pulling chain, speed and working depth were recorded continuously in the field (fig.2). For data analysis these analog signals have been converted to digital values at a frequency of 500 Hz.

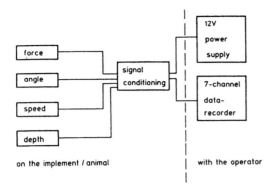

on the implement / animal | with the operator

Figure 2. Portable data acquisition system for measuring mechanical parameters

For the determination of working depth a linear potentiometer was designed measuring the distance from frame to surface (fig. 3); its accuracy is about ±3.0 %.

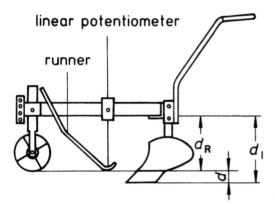

linear potentiometer

runner

Figure 3. Measuring device for working depth.

The actual working depth can then be calculated by (1)

$$d = d_R - d_l \qquad (1)$$

Draught force F is calculated by multiplying traction force F_t with the cosine of the angle of inclination α (2)

$$F = F_t \cdot \cos\alpha \qquad (2)$$

The curve of force is classified in histograms to calculate average draught force (3)

$$F = \frac{1}{n}\sum_{i=1}^{128} n_i\, F_i \qquad (3)$$

To quantify the distribution of force the standard deviation S.D. is determined (4)

$$S.D. = \left[\frac{1}{n-1}\left[\sum_{i=1}^{128} n_i\, F_i - \frac{1}{n}\left(\sum_{i=1}^{128} n_i\, F_i\right)^2\right]\right]^{0.5} \qquad (4)$$

When standard deviation of different treatments is compared this value is related to the average draught force (5)

$$S.D._R = \frac{S.D.}{F}\cdot 100 \qquad (5)$$

To get specific pulling resistance F_R draught force F is related to the cross section (depth d multiplied by width w) (6)

$$F_R = \frac{F}{d\cdot w} \qquad (6)$$

Varying factors that occur when working with different animals (i.e. speed, height of the yoke) were eliminated by pulling the implements by a car with a constant speed of s = 0.9 - 1.0 m/s and a pulling angle of α = 16 - 20°.

The operator was advised not to push or press the handles; the depth gauge wheel was dismounted to avoid misadjustments. The frame was kept in horizontal position to the surface by adapting the hitching point of the pulling chain.

3 RESULTS

Results show the draught force requirements, working depth and relative forces for each combination of implement and soil conditions (tab. 4, see annex).

Draught forces for <u>ploughing</u> vary from F = 852 N (Soil D, dry) to F = 2.07 kN (Soil A, humid), relative forces are raising from F_R = 1.7 N/cm² on sandy, dry soils to F_R = 3.7 N/cm² on heavily crusted soils. For higher moisture content and therefore lower penetration resistance of soil B and C a decrease of F_R can be observed on these soils. On sandy soils, both dry and humid, working depth is high and therefore F also is elevated.
For the distribution of draught forces lower values of SD_R are found in humid soils (i.e. for soil B: 25.4% in dry conditions and 13.7% in humid conditions, fig. 4).

<u>Scarifying</u> requires average draught forces of F = 464 N (soil B) to F = 1.70 kN (soil A) with working depth of d = 3 cm to d = 10 cm. On crusted soils (B - D) tillage effect remains low due to small working

works at sufficient depth (d = 17.2 cm (A), d = 11.3 cm (B)). Relative force F_R is significantly lower on sandy soil than on soil with higher clay content (fig. 5).

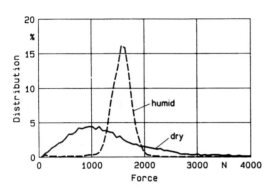

Fig. 4. Histogram of draught forces for ploughing on soil B

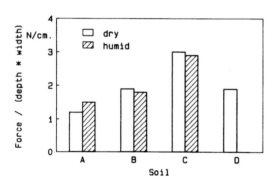

Figure 5. Relative draught forces for scarifying

Regarding data for relative standard deviation scarifying is less tiring for draught animals on humid than on dry soils (fig. 6)

Using a ridging plough on heavy, dry soils (type B,C,D) results in low working depth (i.e. d = 4.0 cm on soil C, fig. 7) and in poorly formed ridges. On humid soils A and B ridges are properly formed and line spacing up to 1m (i.e. for millet) is possible, but draught forces are rather high (F = 1.68 kN (A) and F = 1.24 kN (B), fig. 8).

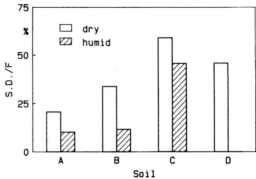

Figure 6. Relative standard deviation for scarifying

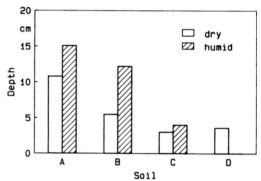

Figure 7. Working depth for ridging

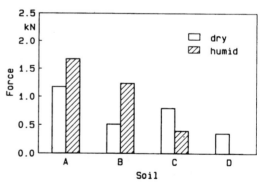

Figure 8. Draught forces for ridging

Due to its special design the strip tiller reaches high working depth on all soils (fig. 9). Except for soil A, humid, draught force does not exceed F = 1.2 kN (fig. 10).

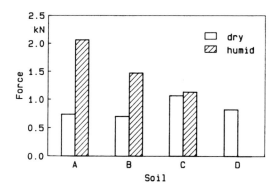

Figure 9. Draught forces for strip tillage

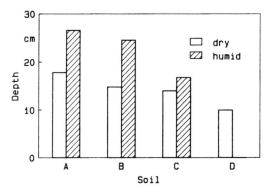

Figure 10. Working depth for strip tillage

Although the strip tiller is working only with one tine no problems of stability have been observed because of the high working depth.

4 DISCUSSION AND CONCLUSIONS

Results are showing the important influence of edaphic factors on the working performance of animal-drawn implements for soil tillage.
The analysis of the results shows:
- draught forces for ploughing are exceeding F = 1.4 kN under most soil conditions. The plough is to heavy for lighter soils where it works very deep.
- the scarifier works well under most of the tested conditions combining a loosing effect of the soil at reasonable working depth with easy handling. A large working width allows high working rates per hour.
- the ridger is an appropriate implement for soil tillage on light soils but is unsuitable for primary tillage on heavy or crusted soils because of low working depth.

- on crusted, heavy soils the strip tiller is useful tool to open the soil deeply without exceeding the capacity of a pair of oxen. It can be used in the dry season to open hard soils to increase infiltration of the first rainfalls.
The experiment has analyzed the status quo of animal-drawn implements. Results show the need to diversify the range of available implements by adapting them to different edaphic conditions. In the long term only appropriate equipment will be accepted by the farmer and thereby will increase the level of mechanization in the region.

5 Acknowledgments

The authors like to thank the 'Projet Machinisme Agricole', the 'Projet Integré Keita', and ICRISAT (International Crop Research Institute for the Semi-Arid Tropics) Sahelian Center, all Niger, for their support during experimental work and Fa. Frank, FRG, for making prototype equipment available.

6 References

Anonymus 1978. Manual on the employment of draught animals in agriculture. Rome, Italy: FAO, CEEMAT.
Ashburner, J.E. and Mamane, Y. 1988. Le developpement de materiel de traction animal adapté pour la production agricole dans les zones pluviales du Niger. In: La traction animale pour le developpment agricole, 7-12 July 1988, Saly, Senegal.
Fall, A. 1988. Adoption et principales contraintes a la diffusion de traction animale en Basse Casamance. In: La traction animale pour le developpment agricole, 7-12 July 1988, Saly, Senegal.
Hecht, H. and Midohoe, K. 1982. Arbeitszeitstudien und Mechanisierungsmodelle Institute of Agricultural Engineering, University of Gießen, FRG.
Munzinger, P. 1981. Animal Traction in Africa. Eschborn, FRG: GATE Publicationes.
Le Moigne et al. 1987. Projet Régional de Machinisme Agricole dans le Pays de l'Union Monetaire Ouest Africaine. (In French) CIRAD, Anthony, France.
Spencer, D. 1988. Farming systems in West Africa from an animal traction perspective. Pages 91-96 in: Animal Power in Farming Systems. Proceedings of the West Africa Animal Traction Workshop, 19-25 Sep 1986, Freetown, Sierra Leone. Eschborn, FRG: GATE Publicationes.

7 ANNEX

Table 4. Data for ploughing, scarifying, ridging and strip tillage on different soils (d-dry, h-humid)

Plough

Soil	A		B		C		D	mean	± S.E.
Moisture	d	h	d	h	d	h	d		
F (N)	1536	2067	1449	1557	1442	922	852	1404	47
SD (N)	294	225	368	214	987	422	380	412	37
SD_R (%)	19.1	10.9	25.4	13.7	68.7	45.7	45.0	32.7	2.6
d (cm)	16.8	22.7	12.0	15.7	7.5	8.0	8.2	13.0	0.4
w (cm)	53.7	55.0	52.0	49.0	52.3	45.0	52.2	51.3	-
F_R (N/cm^2)	1.7	1.7	2.3	2.0	3.7	2.6	2.0	2.3	0.16

Scarifier

Soil	A		B		C		D	mean	± S.E.
Moisture	d	h	d	h	d	h	d		
F (N)	784	1698	464	1337	743	619	494	877	24.0
SD (N)	162	172	157	156	438	283	225	228	20.4
SD_R (%)	20.7	10.2	33.8	11.7	59.0	45.8	45.9	32.4	2.91
d (cm)	10.2	17.2	4.0	11.3	4.0	4.4	4.2	7.9	0.30
w (cm)	62.7	65.0	63.0	65.0	63.0	62.0	63.0	63.4	-
F_R (N/cm^2)	1.2	1.5	1.9	1.8	3.0	2.3	1.9	1.9	0.07

Ridging plough

Soil	A		B		C		D	mean	± S.E.
Moisture	d	h	d	h	d	h	d		
F (N)	1173	1675	515	1240	796	397	355	879	33.9
SD (N)	163	271	197	205	467	145	244	242	31.6
SD_R (%)	14.1	16.1	38.3	16.5	58.6	36.5	66.7	35.3	3.18
d (cm)	10.8	15.1	5.5	12.2	3.0	4.0	3.6	7.8	0.43
w (cm)	74.6	82.0	46.0	76.0	51.0	53.0	51.4	62.0	-
F_R (N/cm^2)	1.5	1.4	2.0	1.4	5.2	1.9	2.0	2.2	0.12

Strip tiller

Soil	A		B		C		D	mean	± S.E.
Moisture	d	h	d	h	d	h	d		
F (N)	738	2062	701	1473	1067	1135	822	1143	49.9
SD (N)	181	415	273	197	718	529	396	387	25.4
SD_R (%)	24.5	20.3	39.0	13.4	67.6	47.0	48.3	37.2	3.12
d (cm)	17.8	26.6	14.8	24.6	14.0	16.8	10.0	17.8	0.73
w (cm)	9.0	9.0	9.0	9.0	9.0	9.0	9.0	9.0	-
F_R (N/cm^2)	4.6	8.6	5.4	6.7	8.5	7.5	9.2	7.2	0.31

Land and Water Use, Dodd & Grace (eds), © 1989 Balkema, Rotterdam. ISBN 90 6191 980 0

La date du travail primaire du sol en agriculture pluviale au Maroc

E.H.Bourarach
Institut Agronomique et Vétérinaire Hassan II, Rabat, Maroc

H.Eichhorn
Institut für Landtechnik, Giessen, RFA

Zusammenfassung: Im Vergleich verschiedener Bodenbearbeitungsmethoden mit Primärbodenbearbeitung vor bzw. nach dem Beginn der Regenzeit hat es sich gezeigt, daß der Zeitpunkt der Primärbodenbearbeitung einem signifikanten Einfluß auf den Kornertrag und der Gewinn je ha hat. Die Geräteauswahl hat dagegen keinen signifikanten Einfluß außer in Trockenjahren.

Résumé: En comparant différentes méthodes de travaux du sol avec travail primaire avant ou après les premières pluies, il a été trouvé que la date du travail primaire à une influence significative sur le rendement et sur le marge nette à l'hectare. L'effet du train d'outils n'est significatif qu'en année à faible pluviosité.

Summary: The influence of first precipitations as related to primary tillage was studied for different tillage systems. It was concluded that the period of primary tillage has a significant influence on yield and profit on a hectare basis. The tillage systems had no influence on yield and profit except in dry years.

1 INTRODUCTION

Le travail du sol est l'un des actes culturaux le plus décesif et le plus coûteux. Il conditionne la réussite des opérations qui le suivent à des degrés variables suivant le milieu et le matériel végétal. Shelton (1979) a montré qu'au Nebraska 42.7 % du gasoil utilisé dans les opérations culturales est requis par les travaux du sol et que 28.4 % revient aux seuls travaux primaires. Il est donc nécessaire d'adopter les techniques culturales qui réduisent la part du travail primaire voire celles qui la suppriment. L'adoption de techniques de conservation du sol adaptées permet d'économiser de l'énergie, du temps et de réduire le coût tout en conservant le potentiel de fertilité du sol (Tebrügge et al., 1985; Collins et al., 1980). En plus de l'aspect économique, le choix d'une méthode de travail du sol est dicté par le système de production de l'exploitation, la disponibilité du matériel et son adaptation aux conditions du milieu (Bentaya et al., 1983, Projet RAB 84-011, 1987). Dans les zones arides et semi-arides à agriculture pluviale, le travail primaire en sec permet d'économiser le temps pour l'installation de la culture et de profiter au maximum de l'eau de pluie disponible. Mais le travail en sec a des retombées négatives sur la tenue à l'usure des outils et sur la consommation d'énergie et eventuellement sur la qualité du travail. Au contraire, le travail en humides c'est à dire après la

chute des premières pluies, sollicite moins les outils et requiert
moins d'énergie. Mais, à matériel égal, il permet d'emblaver une superficie moindre ou occasionne un retard dans l'installation de la culture, ce qui se traduit par la perte d'une partie de l'eau disponible. L'irrigation d'appoint permet de se libérer en partie de ce facteur limitant qui est l'eau mais pose d'autre problèmes d'ordre financier et de maîtrise de la technique (Bouaziz et al., 1985).
L'objectif de ce travail est de comparer les effets des travaux primaires en sec et en humide sur le rendement de céréales d'automne pour différents itinéraires techniques, tout en tenant compte des charges, des recettes et des limitations de chacun d'eux.

2 MATERIELS ET METHODES

L'essai a été réalisé sur une parcelle située à Aïn El Aouda (Maroc), ses coordonnées Lambert sont X= 6° 47', Y= 33° 49' et Z= 230 m. Le climat est à caractère méditerranéen avec un été chaud et sec de plus de 3 mois, et un hiver relativement tempéré et humide. La pluviosité annuelle est irrégulière et mal répartie, 250 à 730 mm, avec une moyenne de 400 mm sur 7 années. Le sol est limono-argileux, contenant 30 % de concrétions ferrugineuses. Le terrain présente une légère pente inférieure à 1 %.
Le dispositif expérimental adopté est celui en blocs aléatoires complets avec parcelles divisées. Il consiste en 3 blocs chacun divisé en 2 grandes parcelles recevant le facteur "date du travail primaire" (avant ou après chute des premières pluies) et chaque grande parcelle est subdivisée en 7 petites parcelles de 9 x 70 m recevant les différentes

Tab. 1: Méthodes de travail du sol utilisées dans l'essai.

symbole	train d'outils
S-CSHC* / H-CSHC*	charrue à soc + combinaison de herses
S-CDHC / H-CDHC	charrue à disque + combinaison de herses
S-CDCC / H-CDCC	charrue à disque + pulvériseur à disques
S-CHRT / H-CHRT	cultivateur à dents lourd + cultivateur rotatif
S-CHHR / H-CHHR	cultivateur à dents lourd + cultivateur rotatif
S-CHCC / H-CHCC	cultivateur à dents lourd + pulvériseur à disques
S-2CC / H-2CC	pulvériseur à à disques + pulvé- seur à disques

* S- et H- indiquent que le travail primaire est réalisé réspectivement avant et après la chute des premières pluies.

méthodes de travail du sol décrites au Tab. 1.
Nous considérons comme travail primaire le travail sur chaumes à l'aide de la charrue à soc (SC) et à disque (CD), du cultivateur à dent lourd (CH) et du pulvériseur à disques (CC). Les caractéristiques techniques des différents outils utilisés dans l'essai sont consignées au Tab. 2.
Au cours des essais nous avons respecté la rotation pratiquée dans l'exploitation: légumineuse/céréales/céréales. Ainsi ont été cultivés succes- sivement le seigle (1984-85), le lupin (1985-86) et l'orge (1986-87).

Tab. 2: Les caractéristiques des outils utilisés.

désignation	caractéristiques
Charrue à soc (CS)	2 socs, 16''
Charrue à disques(CD)	3 disques, ∅=660 mm, 0.75 m
Cultivateur à dents lourd (CH)	5 dents, 1.8 m
Pulvériseur à disques dis-symetrique(CC)	20 disques, 2.3m ∅= 610 mm
Cultivateur rotatif (RT)	2 m, équipé de dents
Herse rota-tive (HR)	2 m
Combinaison de herses(HC)	3.3 m

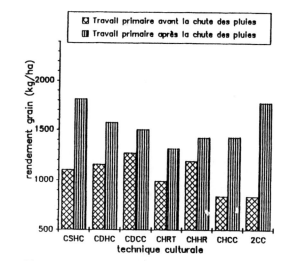

Fig. 1: Le rendement grain du seigle en fonction des techniques culturales (année 1984-85)

Le rendement grain est calculé à partir de la récolte de quatre placettes de 1.2 x 25 m. Les besoins en temps de travaux et en énergie sont mesurés sur des parcelles de longueur variant de 100 à 300 m. La méthode des temps élémentaires (Auernhammer, 1976) est utilisée pour l'analyse des temps de travaux.

3 RESULTATS ET DISCUSSION

Les rendements grain moyens obtenus pour le seigle et l'orge sont présentés à la Fig. 1 et 2. Il ressort de ces 2 années que le rendement des traitements à travail primaire en humide (H, seront appelés traitements en "humide") est supérieur à celui des traitements à travail primaire en sec (S, seront appelés traitements en "sec"), à l'exception des traitements S-CHRT et H-CHRT (cultivateur à dents + cultivateur rotatif) pendant l'année 1986-87 où la tendance est inversée. Ce dernier résultat pourrait s'expliquer par l'humidité défavorable du sol pendant le passage du cultivateur rotatif

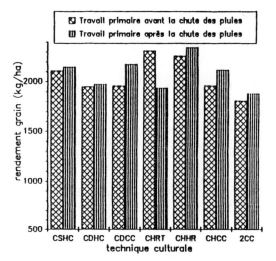

Fig. 2: Le rendement grain de l'orge en fonction des techniques culturales (année 1986-87)

et ne sera pas considéré dans le reste de l'interprétation. L'analyse statistique montre que l'effet "date du travail primaire" est hautement significatif pour les 2 années d'essai alors que l'effet "outil" n'est significatif que pendant l'année à faibles précipitations (1984-85). Pendant la campagne 1986-87 où les rendements au niveau

Tab. 3: Le temps d'opération des outils utilisés (parcelle: 0.8 ha, 1:1).

| outil | temps d'opérations(h/ha) | |
	en"sec"	en"humide"
CS/chaumes	4.3	3.8
CD/chaumes	4.7	4.2
CH/chaumes	1.4	1.3
CC/chaumes	1.4	1.3
CC/CS,CD ou CH	1.7	1.5
CC/CC	1.4	1.4
RT/CH	1.8	1.7
HR/CH	1.8	1.7

* CS/chaumes: indique que la charrue à soc est appliquée sur des chaumes

national étaient les plus élevés des 7 dernières années, les différences entre les traitements en "sec" et en "humide" se sont estompées. Les conditions favorables au développement de la plante ont masqué l'influence du travail du sol. Pendant la première année l'erreur-standard estimée du rendement est faible et à peu près identique pour les différents traitements (± 70 kg/ha), à l'exception des traitements au cultivateur lourd en sec où des difficultés de pénétration de l'outil ont été enregistrées. Pendant l'année 1986-87 l'erreur-standard estimée est variable, elle est particulièrement élevée pour les traitements en "humide" (± 100 kg/ha), ceci est dû à un manque de précipitations après le semis qui a affecté l'homogénéité de la levée et par conséquent celle du rendement. Le Tab. 3 montre le temps d'opération des différents outils. Celui-ci est déterminé pour une parcelle carrée de 0.8 ha, qui est la superficie moyenne par parcelle au niveau national en 1973-74 (Direction des Statistiques, 1974). La

mesure des temps élémentaires a été faite dans des conditions optimales réelles. Ensuite, un modèle a été développé pour calculer les besoins en temps à l'hectare en tenant compte de la superficie et de l'élancement de la parcelle, de la largeur de travail effective de l'outil et du mode de travail (en planche ou à plat). Compte tenu du manque de données fiables sur l'éloignement des parcelles et l'état des chemins d'accès à celles-ci, nous n'avons pas considéré le temps de transport dans le calcul du temps total. La vitesse optimale est estimée sur la base d'un glissement de 10 à 12 % ou à défaut c'est celle autorisée par le micro-relief du sol qui est prise. Les temps requis pour l'installation de céréales (Fig. 3) englobent ceux des travaux du sol, d'épandage d'engrais de fond et du semis. Il ressort de ces résultats que les traitements en "sec" requièrent plus de temps à l'hectare que les traitements en humides ceci s'explique par les conditions sévères en sec (des valeurs extrêmes de 110 kg/dm² sont enregistrées). Aussi, le micro-relief du sol en sec et particulièrement après un travail primaire à la charrue ou au cultivateur impose une vitesse d'avancement réduite. La vitesse du pulvériseur à disques sur chaumes en humide est de 6.5 km/h alors qu'après un labour en sec elle est de 4 km/h). Les traitements utilisant le pulvériseur à disques ou le cultivateur lourd consomment moins de temps à l'hectare, alors que ceux utilisant la charrue sont plus lents (4.6 h/ha pour S-CHCC contre 7.9 h/ha pour S-CDCC). Cette différence s'explique par la largeur et la vitesse d'avancement relativement faibles des charrues. Le pulvériseur à disques communément utilisé au Maroc (plus de 50 % de la superficie

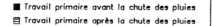

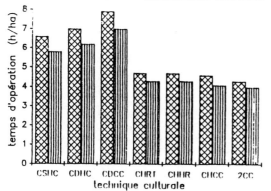

Fig. 3: Le besoin en temps des techniques culturales utilisées pour l'installation des céréales
d'automne (parcelle de 0.8 ha, LG:L = 1:1)

Tab. 4: La capacité théorique
annuelle (ha/an) par itinéraire
technique

itinéraire	superficie (ha/an)	
	en"sec"	en"humide"
CSHC	50	30
CDHC	70	30
CDCC	70	30
CHRT	60	60
CHHR	80	80
CHCC	100	80
2CC	100	80

Tab. 5: La marge nette (%) par
technique culturale exprimée
par rapport au traitement
S-2CC (1984-85)

technique	marge nette (%)	
	en"sec"	en"humide"
CSHC	70	977
CDHC	154	564
CDCC	285	547
CHRT	-103	362
CHHR	42	378
CHCC	-81	797
2CC	0	1112

Tab. 6: La marge nette (%) par
technique culturale exprimée
par rapport au traitement
S-2CC (1986-87)

technique	marge nette (%)	
	en"sec"	en"humide"
CSHC	13	18
CDHC	0	1
CDCC	0	17
CHRT	-3	–
CHHR	18	28
CHCC	10	26
2CC	0	6

emblavée est travaillée à
l'aide de cet outil; Kaddouri
et Benhania, 1986) requiert le
moins de temps d'installation,
4.2 h/ha en moyenne. C'est
l'une des raisons de son
adoption massive par les
agriculteurs en plus de son
faible coût d'usure (Bourarach,
1988), de sa facilité de
réglages et de sa rapidité
d'exécution et de son prix
relativement faible.
La superficie susceptible
d'être emblavée dans les délais
offerts par les conditions du
milieu et par la culture sont
d'importance. En effet, les
traitements en "sec"
bénéficient de plusieurs
semaines en été pour
l'exécution des travaux
primaires alors que pour les
traitements en "humide"les
travaux sont concentrés juste
après les premières pluies.
Ainsi, les premiers permettent
de réaliser de 50 à 100 ha
alors les seconds ne permettent
d'emblaver que 30 à 80 ha (Tab.
4). Cependant, le travail
primaire en humide concilie
entre le système de production
végétale et animale tels qu'ils
sont pratiqués actuellement.

Les dépenses occasionnées par l'installation de la culture sont déterminées à partir du calcul des frais fixes et des frais de fonctionnement. Les dépenses de protection et d'entretien de la culture sont supposées fixes et identiques pour tous les traitements.
Le Tab. 5 présente la marge nette en 1984-85 (année à pluviosité défaillante) exprimée en pourcentage relatif au traitement témoin S-2CC pris comme référence. La marge nette en 1986-87 (année à bonne pluviosité) est représentée au Tab. 6.
Pendant l'année pluvieuse les marges nettes sont élevées et les différences entre les traitements sont faibles (-3 à 28 %). Les traitements en "humide" sont les plus rentables, ceci est dû à la réalisation d'une meilleure structure du sol. Au contraire, en année difficile (Tab. 5) les marges nettes sont faibles et différentiées (-103 à 1112 %), ici les facteurs "outil" et "date du travail primaire" jouent leur plein rôle. En "sec" les traitements utilisant la charrue à disque enregistrent les meilleures marges alors qu'en "humide" ce sont les traitements utilisant le pulvériseur à disque et la charrue à soc qui donnent les meilleures marges nettes. Les traitements CDHC , CDCC et CHCC procurent des marges relativement élevée par rapport au temoin (respectivement 564,547 et 797 %).
Les performances du pulvériseur à disques montrent que contrairement à l'idée répandue mais non justifiée de son inadaptation, cet outil se comporte bien en humide en enregistrant une marge nette de 1112 % par rapport au temoin S-2CC. Il est à remarquer également que malgré leur prix d'acquisition élevé, la herse rotative et la combinaison de herses associées respectivement

au cultivateur lourd et aux charrues procurent des gains élevés.

4 CONCLUSION

Les deux années d'essai ne permettent pas de procéder à une généralisation des résultats, mais nous pouvons tirer les conclusions suivantes:
- Au niveau de l'exploitation, le choix de la date du travail primaire ne doit pas être guidé par la superficie totale susceptible d'être emblavée seule mais également par le rapport des charges et des recettes escomptées;
- Il existe des itinéraires techniques qui permettent de minimiser les risques en année à pluviosité défaillante, tout en restant concurentiels pendant l'année à bonne pluviosité;
- Dans les conditions de notre essai, il semble que:
  * Le pulvériseur à disques utilisé en humidité de sol optimale donne des rendements équivalents à ceux des traitements utilisant la charrue.
  * Les outils animés par la prise de force permettent de dégager une marge nette appréciable. Mais leurs conditions d'utilisation sont plus difficile et les conséquences d'une mauvaise utilisation sont plus graves qu'avec un outil non animé.

5 REFERENCES BIBLIOGRAPHIQUES

Auernhammer,H. 1976. Eine integrierte Methode zur Arbeitszeitanalyse. KTBL-Schrift Nr. 203. Darmstadt-Kranichstein: KTBL.
Bentaya, D. & P.Pascon & L. Zegdouni 1983. L'agriculture en situation aléatoire, Chaouia 1977-82. Rabat: IAV Hassan II

Bouaziz, A. & H.Lamrani & M. Anechoum 1985. Réflexion sur le potentiel céréalier: Approches méthodologiques d'évaluation. Hommes Terre et Eaux. Vol. 15 (60)

Bourarach, H. 1988. Untersuchungen über den Verschleiß von Bodenbearbeitungswerkzeugen in Marokko. Neu-Ulm: VDI-Tagung

Collins, N.E. & T.H.Williams & L.J.Kemble 1980. Measured machine energy requirements for grain production systems. ASAE National Energy symposium. Vol.2

Directions des Statistiques. 1974. Recensement Agricole 1973-74, Données Prioritaires. Vol.2. Rabat: D.S.

Kaddouri, L. & K.Benahnia. 1986 Situation actuelle des travaux du sol, séminaire travail du sol. Rabat: MARA

Projet RAB 84-011 1987. L'adoption des techniques améliorées sur céréales. Rabat: IAV Hassan II

Schelton, D.P. 1979. Nebraska on-farm fuel use survey. ASAE- Technical paper N°: MC 79-204

Tebrügge, F. & J.Griebel & W.Henke 1985. Bodenbearbeitung und Bestelltechnik heute energie-, arbeits-, kostensparend und bodenschonend. Landtechnik 2, Februar 1985

Land and Water Use, Dodd & Grace (eds), © 1989 Balkema, Rotterdam. ISBN 90 6191 980 0

The influence of chisel plough share wing geometry on tillage forces

J.M.Fielke
South Australian Institute of Technology, Adelaide, Australia

ABSTRACT: This paper looks at the effect of a chisel plough share wing's width, lift height, rake angle and sweep angle on its draft and vertical force at a range of speeds in varying soil types. Tests were conducted in the SAIT Tillage Test Track which contains a sandy loam soil and in two field soils, one a sandy loam and the other being a red brown earth. The results showed that the geometry of the wings can be optimised for minimum draft force and maximum vertical down force.

RESUME: Cet article présente l'effet de la largeur des ailerons du soc d'une charrue de type ciseau, l'élévation verticale, l'angle par rapport à l'horizontale ainsi que l'angle d'attaque des ailerons, sur la force de résistance et la force verticale dans le sol à divers vitesses et dans différents types de sols. Les essais ont été conduits sur la piste d'essai de SAIT qui contient une terre sabloneuse, et dans deux champs, un à sol sabloneux, et le deuxième une terre rouge-marron. Les résultats ont montré que la géométrie des ailerons peut être optimisée pour une force de résistance et une force verticale minimales vers le bas.

AUSZUG: In diesem Dokument wird die Wirkung der Breite, Hebehöhe, Neigungswinkel und Bewegungswinkel eines Meisselpflugscharflügels auf seine Zug- und Senkrechtkraft bei verschiedenen Geschwindigkeiten in unterschiedlichen Erdsorten betrachtet. Es wurden Versuch auf der SAIT Pflugteststrecke durchgefuhrt, die aus einem sandigem Lehmboden und zwei Feldböden, einer davon aus sandigem Lehm und der andere aus roter brauner Erde, besteht. Die Ergebnisse zeigten, dass die Geometrie der Flügel fur minimale Zugkraft und maximale Senkrechtkraft optimiert werden kann.

1 INTRODUCTION

With the increase in tillage speed from the horse drawn 3km/h to speeds of 8km/h or more, large increases in productivity per man hour have been achieved but with a more aggressive tillage action. Associated with this aggressive action are high power requirements and fuel consumptions. With large fuel usages for tillage, methods of reducing fuel usage are worth investigating.

Little is known of share geometry and its effects on tillage, especially in relation to increasing speed . The aim of this paper was to examine the draft and vertical forces of various chisel plough share wing geometries at a range of speeds. The tests were conducted under controlled conditions in the South Australian Institute of Technology (SAIT) Tillage Test Track which contains a sandy loam soil and under field conditions in a sandy loam and a red brown earth.

2 DEFINITION OF SHARE GEOMETRY

To examine the influence of chisel plough share geometry on tillage forces, shares of three major manufacturers were examined. It was observed that they could be simplified to consist of two inclined flat plates mounted on a vertical shank. To full describe these flat plate wings four geometric variables as shown in Figure 1 were defined

Width: the width of the share wings perpendicular to the direction of travel.

Lift Height: the vertical height that the share wing lifts the soil.

Rake Angle: the angle the share wing lifts the soil, measured in the direction of travel.

Sweep Angle: the angle enclosed by the two cutting edges of the share wings.

For the experiment, which was based on the work of Sirohi & Reaves 1969, each geometric variable was varied individually about the current geometric shape with the nominal share geometries studied shown in Table 1.

Table 1. Nominal share geometries.

	Width (mm)	Lift Height (mm)	Rake Angle (°)	Sweep Angle (°)
Varying Width	200	32	10	70
	300	32	10	70
	400	32	10	70
	500	32	10	70
	600	32	10	70
Varying Lift Height	400	12	10	70
	400	22	10	70
	400	32	10	70
	400	42	10	70
	400	52	10	70
Varying Rake Angle	400	32	6	70
	400	32	7.5	70
	400	32	10	70
	400	32	15	70
	400	32	25	70
	400	32	45	70
Varying Sweep Angle	400	32	10	50
	400	32	10	70
	400	32	10	100
	400	32	10	130
	400	32	10	180

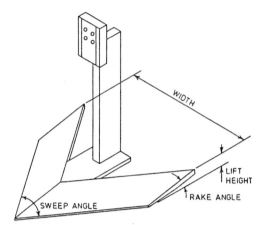

Figure 1 Definition of share geometry

3 EQUIPMENT

3.1 Share construction

The shares were built by the SAIT School of Mechanical Engineering workshop with the wings made from cold drawn 12mm thick S1040 steel. The shares were constructed with a constant 3mm thick cutting edge and a constant 5° underside relief (measured in the direction of travel). The share tip was placed 295mm in front of the 25 x 75mm vertical shank. The share was designed to be mounted with a zero set (pitch) angle to give a constant tillage depth across the width of the tool.

3.2 Force measurement

To measure the draft and vertical forces of each of the geometries in the Tillage Test Track the individual share under evaluation was mounted on a extended octagonal ring transducer as described in Godwin 1975. For the field tests the draft and vertical forces of the individual shares were measured using a two force frame described in Riley & Fielke 1988. The signals from the transducers were sent to two channels of the SAIT purpose built four channel strain bridge and amplifier and then to the SAIT purpose built integrator which uses voltage to frequency conversion principles. The integration period was varied to give a constant distance travelled over which the signals were integrated.

4 SITES

4.1 Tillage Test Track

The Tillage Test Track is a unique outside continuous soil bin described in Fielke & Pendry 1986 and consists of two straights 50m in length which are joined by two curves 50m in diameter giving 250m per lap. The test soil is 2.5m wide and 0.3m deep and is a sandy loam soil.

An 80kW tractor towing two trolleys each capable of tillage testing and soil reconditioning, travels around the track at speeds up to 15 km/h.

For the experiment the Tillage Test Track rig was set up with the front trolley reconditioning the total width of soil by ploughing, grading out undulations and rolling. The rear trolley was set up to measure the tillage tool's draft and vertical forces along with the speed and depth of tillage. The rear grader and roller were removed to allow the furrow created to remain for evaluation.

4.2 Field testing

Field testing was conducted using a sandy loam field soil at Avon, South Australia, and a red brown earth at Hoyleton, South Australia. A comparison of the soil properties at these 3 sites are shown in Table 2.

Table 2. Soil properties for the 3 test sites.

Site	Moisture Content (%)	Bulk Density (kg/m³)	Cohesive Strength (kPa)	Friction Angle (°)
Tillage Test Track (Sandy Loam)	8	1900	6	45
Avon, S.A. (Sandy Loam)	5	1370	4	30
Hoyleton, S.A. (Red Brown Earth)	26	1400	13	25

5 PROCEDURE

5.1 Tillage Test Track tests

The Tillage Test Track tests were conducted in November 1986. The experiment was set up using a split-split block design with the experiment broken down into the four geometric variations and also into speeds. Each geometric variation was tested at three speeds of 5, 10 and 12.5 km/h with the tests being replicated 3 times at a nominal depth of 70mm. Each speed and replication combination was conducted on a separate half day.

Prior to a day of testing the Tillage Test Track soil was wet up to a nominal 8% moisture content using a towed tank with a boom over the track. For the tests the required share was used to till the soil for four laps giving 1 km of travel. The draft and vertical forces were measured while travelling along the two 50m straights for the four laps.

5.2 Field tests

The field tests were conducted in May 1988. The experiment at each of these sites was again split into the four geometric variations. Within each geometric variation the shares were tested at speeds of 5, 10 and 15 km/h using a 50mm depth in a completely randomised order with three replications being conducted. The force was integrated over a 50m distance in a 70m run to allow for speeding up and stopping at the ends of the runs.

During the field testing several of the shares bent upwards at the shank mount, explaining the missing values from the field results.

6 DISCUSSION

6.1 Width effects on tillage forces

The results of draft and vertical force versus wing width for the three sites are shown in Figure 2.

The results show a linear increase in draft force with width for all of the sites with the draft force per unit width (slope) increasing with

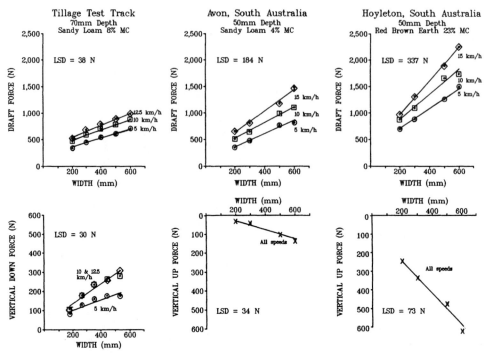

Figure 2 Draft and vertical force versus width
(32mm lift height, 10° rake angle and 70° sweep angle)

speed as shown in Figure 3. This increase in draft force per unit width with speed was shown in Fielke (1988) to be a result of increased soil throwing, friction and soil pulverisation.

For the Tillage Test Track tests where the share was held level in a vertical plane, the vertical force was shown in Fielke 1988 to be close to the vertical down force calculated for the combination of wing upward soil throwing and soil weight above the wing.

In contrast, the field results showed a linearly increasing vertical up force with width. This upwards force may be explained by the field undulations causing the share to vibrate vertically and to compress the soil under the cutting edge resulting in a vertical up force, increased draft force due to underside friction and increased wear on the underside of the cutting edge. In addition, speed was observed to have little effect on the vertical force for the field tests.

Figure 4 of draft force efficiency (cross sectional area of soil cut

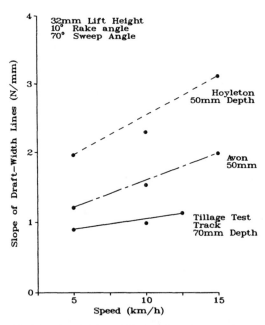

Figure 3 Slope of draft-width lines versus speed (32mm lift height, 10° rake angle and 70° sweep angle)

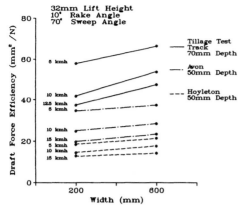

Figure 4 Draft force efficiency
versus width (32mm lift height,
10° rake angle and 70° sweep angle)

divided by the draft force which can
be considered as the volume of soil
cut per unit energy) versus wing
width showed that for all of the
sites the draft force efficiency
increased with increased share width
and decreased with increased speed.
Typically, the increase in share
width from 200 to 600mm resulted in
a 17% increase in the draft force
efficiency.

6.2 Lift height effects on tillage forces

The lift height effects on the draft
and vertical forces for the three
sites are shown in Figure 5. The
results show the minimal effect lift
height had on the draft force with
each 10mm increase in lift height
increasing the draft force by 2% in
the sandy loam soil. Fielke 1988
showed that this could be accounted
for by the increased top surface
friction. The red brown earth tests
showed the lift height to have no
significant effect on the draft
force, which can be explained by the
higher forces masking the small lift
height effects.

The vertical force results showed
it to increase in a downwards
direction with increased lift height
for all of the sites which can be
explained by the increase in weight
of soil above the wing.

6.3 Rake angle effects on tillage forces

Figure 6 shows the rake angle
effects on the draft and vertical
forces for the three sites. The

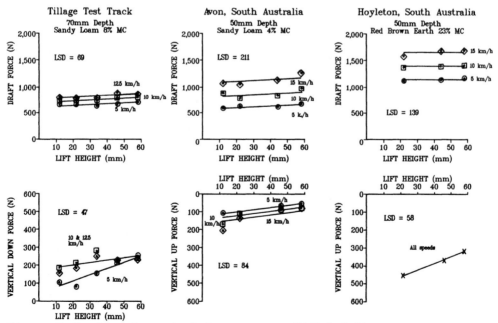

Figure 5 Draft and vertical force versus lift height
(400mm width, 10° rake angle and 70° sweep angle)

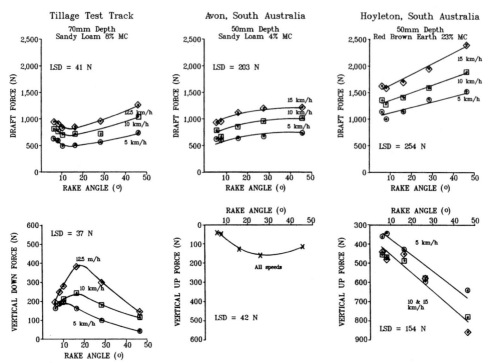

Figure 6 Draft and vertical force versus rake angle
(400mm width, 32mm lift height, and 70° sweep angle)

Tillage Test Track tests showed there existed a range of optimum rake angles of between 10 and 20° where the draft force was close to its minimum and the vertical down force was close to its maximum. A minimum in draft force for a 15° rake angle was shown in Fielke 1988 to be predicted by the Universal Earthmoving Equation (as presented in McKyes 1985) for a sandy loam soil.

The shallower field results showed the draft force to increase and the vertical down force to decrease with increased rake angle as reported by researchers such as Payne & Tanner 1959.

6.4 Sweep angle effects on tillage forces

The influence of sweep angle on the draft and vertical forces for the three sites are shown in Figure 7.

The results for the sandy loam soils showed that as the sweep angle increased the draft force increased. This can be explained by the increased bulldozing of the soil by the wing. For the Tillage Test Track, where bulldozing was noted not to occur at the slow 5 km/h speed, sweep angle had no significant effect on the draft force. The vertical force was also observed to follow closely the weight of soil above the wing.

In contrast, for the red brown earth where the soil adhered to the top and underside of the wing, the draft force decreased and the vertical down force increased with increasing sweep angle. An explanation for this is possibly that the vertical up force is related to the cutting edge length with the higher vertical up forces, being associated with the longer cutting edges. The draft force decrease with increasing sweep angle can likewise be explained by the longer cutting edges of the small sweep angle shares increasing the friction and hence the draft force.

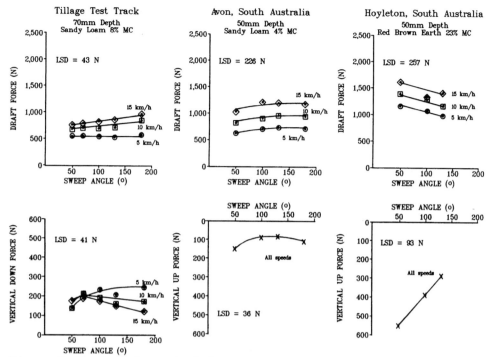

Figure 7 Draft and vertical force versus sweep angle
(400mm width, 32mm lift height, and 10° rake angle)

7 CONCLUSIONS

The results of the tests indicated that:

- For the three sites the draft force efficiency (volume of soil cut per unit energy) was observed to increase as the share width increased.

- The lift height had minimal effect on the draft force while the vertical down force increased as a function of the weight of the soil above the wing.

- A range of optimum rake angles of between 10 and 20° existed for minimum draft force and maximum vertical down force for the Tillage Test Track while for the field tests increasing rake angle increased the draft force and decreased the vertical down force.

- Increasing sweep angle increased the draft force in the sandy loam soils due to bulldozing by the wing and the vertical force was a function of the weight of soil above the wing while for the sticky red brown earth the draft force decreased and the vertical down force increased with increasing sweep angle.

8 ACKNOWLEDGEMENTS

The author wishes to acknowledge the financial support of the Australian National Energy Research Development and Demonstration Council (NERDDC), South Australian Barley Industry Research Committee and the South Australian Wheat Industry Research Committee. Special thanks are due to Messrs. R. Manley and K. Jericho for the use of their properties for the field tests.

9 REFERENCES

Fielke, J.M., 1988. The Influence of the Geometry of Chisel Plough Share Wings on Tillage Forces in Sandy Loam Soil, Master of Engineering Science Thesis, University of Melbourne.

Fielke, J.M. & Pendry, S.D., 1986.
SAIT Tillage Test Track,Conference
on Agricultural Engineering, 1986,
Adelaide, The Institution of
Engineers, Australia, p.90-95.

Godwin, R.J., 1975. An Extended
Octagonal Ring Transducer for Use
in Tillage Studies, Journal of
Agricultural Engineering Research,
Vol. 20, p.347-352.

McKyes, E., 1985. Soil Cutting and
Tillage, Developments in
Agricultural Engineering 7,
Elsevier.

Payne, P.C.J. & Tanner, D.W., 1959.
The Relationship between Rake
Angle and the Performance of
Simple Cultivation Implements,
Journal of Agricultural Engin-
eering Research, Vol. 4, p.312-
325.

Riley, T.W. & Fielke, J.M., 1988.
Use of a Two Force Dynamometer For
Direct Drilling Point Evaluation,
Conference on Agricultural Engineering, 1988, Sydney, The Institution of Engineers, Australia,
p.83-84.

Sirohi, B.S. & Reaves, C.A., 1969.
Similitude Techniques Applied to
Performance Studies of Cultivator
Sweeps, ASAE Transactions, p.786-
789.

Land and Water Use, Dodd & Grace (eds), © 1989 Balkema, Rotterdam. ISBN 90 6191 980 0

Measurement and prediction of forces on plough bodies – 1. Measurement of forces and soil dynamic parameters

G.Girma
Department of Agricultural Engineering, Swedish University of Agricultural Sciences, Uppsala, Sweden

ABSTRACT: The present study reports measurements of three orthogonal forces and the resulting moments acting about the orthogonal axes on each plough component (mouldboard, share point, share wing and landside) and on plough bodies as a whole, separately and simultaneously using five modified extended octagonal ring type transducers. Force measurements were made at the rate of 1 kHz under field conditions using a tractor with an onboard computer in three different soils. Ploughing speed and depth were measured at the same rate as the forces using a fifth wheel and a pull-back mechanism equipped with a potentiometer, respectively. Sufficient data on soil mechanical properties (cohesion, soil internal friction, adhesion and soil-metal friction) were gathered at each ploughing depth increment parallel with force measurement using a torsional shear device, mounted on another instrumented tractor. Soil moisture content and bulk density were measured for each series of tests. Force distributions on the plough components were determined. Predictive equations were developed from experimental data to determine the relationship between forces and ploughing depth and speed. The preliminary multiple linear regression analysis of force and soil parameters, speed and depth showed that most of the parameters were significant with resulting R-squared value of 0.92.

ZUSAMMENFASSUNG: Es wurden die drei orthogonalen kräfte und die resultierenden momente an allen pflugteilen (streichblech,scharspitze,scharblech und landseite) sowie am Plugkörper als ganzem gemessen. Die messungen erfolgten mittels fünf modifizierten oktagonalen Ringdruckgebern, ohne dass die normale Ausrichtung der Pflugteile verändert werden musste. Die messfrequenz betrug 1 kHz. Es wurde unter feldbedingungen in drei verschiedenen bodentypen gearbeitet. geschwindigkeit und furchentiefe wurden mittels eines fünften rades und einer potentiometerversehenen pull-back anordnung ebenfalls mit 1 kHz messfrequenz registriert. Werte der bodenmechanik (kohäsion, interne bodenfriktion, adhesion sowie die friktion zwischen boden und metall) wurden für jede furchentiefe parallel zur kraftmessung mittels eines radial arbeitenden Schergerätes an einem zweiten traktor ermittelt. Bodenfeuchtigkeit und - dichte wurden in jede tesstserie bestimmt. Es wurde die kraftverteilung an den Pflugteilen bestimmt. Aus den ergbenissen wurden schätzgleichungen für die beziehung zwischen kräften und Pflugschwindigkeit und -tiefe abgeleitet. Die werte der kraftmessung, der bodenparameter, der arbeitsgeschwindigkeit under der funrchentiefe wurden einer regressionsanalyse unterzogen. Die mehrzahl der variablen war signifikant (R^2 = 0.92).

RESUMÉ: La présente étude concerne les mesures des trois forces orthogonales et des moments quifen re'sultent autour du axes orthogonaux au niveau de chaque composante de la charrue (moule "mouldboard", point d'appui "share point", aile "share wing et landside") et de l'ensemble de la charrue separément et simultanément en utilisant cinq types de dynamomètres á anneau modifié et allongé. Les mesures in situ des forces sont effectuées à une fréquence de 1 kHz, en utilisant un tracteur et un micro-ordinateus pour trois differents types de sol. La vitesse et la profondeur du labourage sont mesurées à la même frequence que les forces, en utilisant respectivement, une cinquiéme roue, un mecanisme "pull-back" equipé d'un potentiométre. Les donneés suffisantes relatives aux propriétres mécaniques du sol (cohésion, friction interne du sol, adhésion et friction metallique du sol) sour collectreés à chaque labouraqe et incrementteés par la profondeur, parrallelement aux mesures des forces utilisant un instrument à cisaillement et torsion monté sur un tracteur. L'humiddité et la densite du sol sout mesurées à chaque serie de test. La distribution des forces dans les differentes composantes de la charrue est determinee. Des équations sont developpées à partir des donneés expérimentales afin de déterminer la relation entre les forces et la profondeur ou la vitesse de labourage. L'analyse preliminaire des forces, des parametres du sol, de la vitesse et de la profondeur à l'aide de la regression linéaire multiple à montré que la plupart des paramétres sont significatifs (R^2 = 0.92).

1 INTRODUCTION

The conventional mouldboard plough remains the most important agricultural implement to cut, loosen, lift and invert soil in order to promote plant growth. This necessitates the carrying out of rigorous force analysis to achieve efficient design and qualitative field operations with minimum energy requirements. Important and dedicated research works (Nichols and Kummer , 1932; Doner and Nichls, 1934; Gao et al., 1986) have been done to describe mouldboard surface and classify forces produced during

performance and relating them to the dynamic properties of soil. O'Callaghan and McCoy(1965) have conducted detailed analysis of forces acting by a mouldboard on a soil slice and calculated with the aid of a digital computer. The soil path was determined experimentally and polynomial curves were fitted to the obtained data.

A number of plough draught prediction models (Larson et al., 1968; Wang et al., 1972; Gee-Clough et al., 1978) have been developed on the basis of dimensional analysis. Oskoui and Witney (1982) and Oskoui et al.(1982) proposed a model of specific draught of a plough, adapting the formula developed by Garyachkin, taking into account the effect of a mouldboard tail angle and assuming a cone index as a measure of soil strength. However, little emphasis was placed on determining a relation between forces and plough geometry and pertinent soil dynamic parameters which could be used to evaluate designs and quality of field operations. Clyde (1961) conducted important research to find the best location for the center of resistance and measured forces on plough bodies. Getzlaff (1951) measured three orthogonal forces and moments acting on plough bodies using hydraulic based dynamometers. However most of the previous force measurement works focused on the determination of forces acting on plough bodies as a whole, which does not provide the knowledge needed to understand variations of force components on each plough component and to justify the proposed plough force theories.

The objectives of the present study reported in this paper were to:

1. Measure three orthogonal forces and the resulting moments acting on each plough component (share point, share wing, mouldboard and landside) and on plough body as a whole simultaneously and independently of each other, to investigate variations of component forces on plough body,

2. Measure pertinent soil dynamic parameters which were assumed to influence the resulting forces during field performance,

3. Conduct preliminary studies to develop a model which permits description of forces in terms of plough and soil paramters.

It should be noted that since tests for the effect of plough geometries have not yet been completed the scope of the present paper is restricted to perform multiple regression at constant plough parameters.

2 EQUIPMENT AND METHOD

Three experimental sites of different soil types (sandy loam, silty loam and clay) were chosen to carry out field tests. Five modified extended octagonal ring type dynamometers (Fig. 1) described by Gebresenbet (1988) have been used to measure three orthogonal forces and moments exerted on each plough component and the plough body as a whole. The dynamometers were integrated into the plough frame without affecting the usual positions of the components in order to record forces simultaneously and independently of each other.

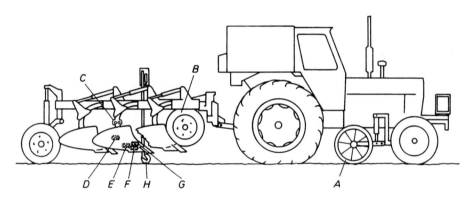

Fig. 1. Semimounted plough equipped with dynamometers; A, fifth wheel; B, wheel; C, D, E, F, G, Extended octagonal ring dynamometers; H, depth measuring device.

The measurements were conducted using a tractor with an onboard computer. A computer program was made to record samples in 32 channels at a sampling rate of 1 kHz. One extra channel was used to observe the effect of various error sources. The offset levels of the channels were adjusted to the offset level observed during calibration and these values were recorded when the plough was ideally free from external load except the body forces. This was repeated for each test series.

Ploughing depth and speed were measured simultaneously at the same rate as the forces. Depth measurement were made using a device (Fig. 1, denoted by H) operating on the basis of pull-back mechanism equiped with a potentiometer. The ploughing depth was varied from 100 mm to 300 mm with increments of 20 mm at a constant speed to test for the effect of depth upon the forces. Ploughing speed was measured using a fifth wheel (Fig. 1, denoted by A), and varied from 0.11 to 2 m/s with increments of 0.3 m/s to test for speed effect and an extra wheel (Fig. 1, denoted by B) was mounted on the plough frame to control depth variations.

For the measurement of mechanical properties of soil, the torsional shearing device described by Olsen (1984) has been used. The device was mounted on an instrumented tractor and intended to measure cohesion, angle of soil internal friction, adhesion and angle of soil-metal friction. An annulus with grousers was attached to the head of the device to measure cohesion and angle of internal friction. Normal force and torque were applied by hydraulic means to press the annulus into the levelled soil surface and to rotate the annulus respectively. Both the normal and shear stresses at soil failure were recorded. The test was repeated several times varying the normal force and the obtained shear stresses were plotted versus the normal stresses. The slope and intercept define (Gill et al., 1972) the angle of soil internal friction and cohesion respectively, and are represented by Coulomb's equation:

$$\tau = c + \sigma \tan \phi \qquad [1]$$

where τ = shear stress,

c= cohesion,

σ= normal stress,

φ= angle of soil internal friction

Adhesion and angle of soil-metal friction were measured following the same procedure replacing the former annulus by an annulus without grousers. Then the slope and intercept obtained from tests represents the angle of soil-metal friction and adhesion, respectively.

3 RESULT AND DISCUSSION

3.1 FORCE VARIATIONS

The output signals from the dynamometers were recorded at a rate of 1 kHz as mentioned above so that forces on all plough components may be measured simultaneously, to study variations of horizontal, vertical and lateral forces. Ploughing speed and depth varied systematically. At each small increment of depth, the speed varied from 0.11 to 1.5 m/s in clay soils and 0.11 yo 2 m/s in sandy soils. For each depth 3 to 8 different speeds values were used. This can be clearly observed in Fig. 6 that for the test numbers 1 to 6 the depth was constant while speed varied from 0.16 to 1.25 m/s, and then for test numbers 7 to 10 the depth was increased by 10 mm and 4 different speeds were used. This method of sampling may enable to observe the combined effect of depth and speed up on force components. The results are presented in graphic form, plotting the mean of 300 readings versus test number.

Fig. 2, 3 and 4 illustrate the distribution of horizontal, vertical and lateral forces at various speed and depth. Table 1 reports the contribution of plough components to the horizontal and vertical forces measured in two fields at a constant speed (0.6 m/s) and depth (190 mm).

Assuming that sliding speed has a significant effect on the forces as reported by Stafford and Tannre (1983), the tractor on which the plough was mounted was driven at approximately 0.11 m/s in order to maintain the sliding speed of the annulus. The mean sliding speed of the annulus was 0.1 m/s.

Samples for the determination of moisture content and dry bulk density were taken during each force test series using cylinders of 72 mm diameter and 50 mm height, and drying them in an oven at 105 °C until a constant weight was maintained.

Table 1. Distributions of forces on ploough components

plough component	Force distribution, %		Soil
	Horizontal	Vertical	
Share point	70	44	sandy
	63	32	clay
Share wing	9	32	sandy
	8	27	clay
Mouldboard	17	21	sandy
	23	38	clay
Landside	4	3	sandy
	6	3	clay

The results of several tests confirmed that 63 to 72 % of the sum of horizontal forces due to the plough components was resulted from share point while 16 to 26 %, 6 to 9 % and 5 to 10 % were resulted from the mouldboard, sharewing and the landside respectively for the soils and plough geometries used in the present work.

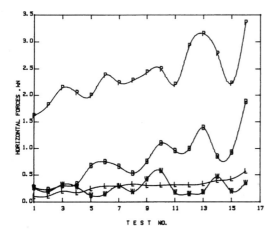

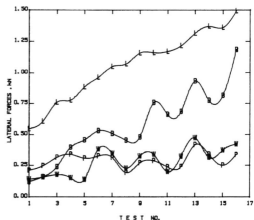

P—P, Share point; B—B, Mouldboard; W—W, Share wing; L—L landside

Fig. 2. Horizontal forces on plough components.

Fig. 3. Lateral forces on plough components

1541

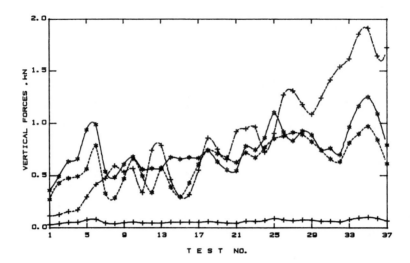

——, Share point; +——+, Mouldboard; *——*, Share wing; +——+, Landside

Fig. 4. Vertical forces on plow components.

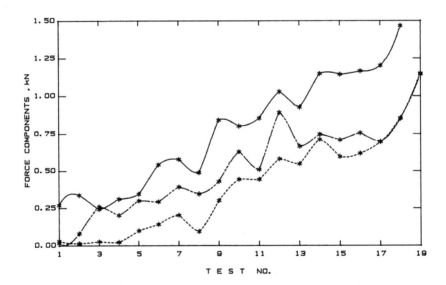

——, Horizontal force; *——*, Vertical force; *——*, Laterale force

Fig. 5. Force components on the mouldboard.

The variations of the results were due to the effect of soil conditions, ploughing speed and depth since particularly forces on the mould board are relatively more sensitive to depth and speed. At the lower values of depth the vertical and lateral forces are higher for the share point and share wing than the mouldboard (Fig.3 and 4).

As the depth increases, the rate of increase of the vertical force on the mouldboard is very rapid (Fig.4). The lateral force acting on the mouldboard varies greatly with speed and is the dominant (Fig. 5) of all force components while it is relatively low on the share point (Fig.6). The increase of lateral force acting on the landside with depth is apparent in clay soils. However, in sandy soils the force is less affected by depth. It should be noted that the sum of the side forces generated by the tilling components (mouldboard, share wing and share point) were almost equal to that of on the landside with 0.5 to 5 % error for clay soils while for sandy soils the range of error is significant.

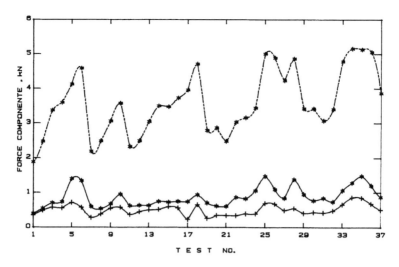

←··→, Horizontal force; ←—→, Vertical force; +——+, Lateral force

Fig. 6. Force components on the share point.

A summary of the above results is presented in Table 2. The possible explanation of this variation may be that the side reaction forces were not only caused by the landside, but also by the plough beam, since the 'ploughing wall'of the sandy soils can easily be deformed.

Table 2.

Number of tests	Lateral forces, kN			Soil
	Mould board, share piont, and share wing	Landside	Plough beam	
10	0.94	0.66	0.29	sandy
6	1.12	0.84	0.27	sandy
31	2.83	2.70	0.12	clay
37	2.57	2.53	0.04	clay

From Table 2 it can be observed that the sum of the reaction forces from the plough beam dynamometer and the land-side are equal to the sum of the forces from mouldboard, share wing and share point.

Variations of moments about the horizontal, vertical and lateral axes are mostly influenced by depth variation. All moments above the depth value of about 170 mm are substantially pronounced on the mouldboard.

3.2 EFFECT OF A DISK COULTER

Forces on the disk coulter were not measured directly in the test. Its effect was investigated by sampling the forces on the plough components when the coulter was performing its normal operation and also when dismounting

the coulter. At the later moment, the horizontal and vertical components on the share point and mouldboard were significantly increased. The observed results of the horizontal force are presented in Table 3.

Table 3.

Horizontal force, kN			
Coulter under normal operation		Coulter dismounted	
Share point	Mouldboard	Sharepoint	Mouldboard
3.25	0.91	4.43	1.28
2.26	0.62	3.11	0.88
2.57	0.93	3.54	1.30
2.55	0.86	3.59	1.25
2.14	0.74	2.88	1.03
2.45	0.83	3.38	1.21

3.3 EFFECT OF DEPTH UPON FORCES

In general, determination of the best relationship between plough forces and depth under field conditions is difficult due to the wide variations of soil properties and conditions. However, attempts have been made by keeping the ploughing speed constant (0.8 m/s). To demonstrate the depth effect, the measured values from several runs of the horizontal force acting on the plough bodies were plotted versus depth (Fig. 7).

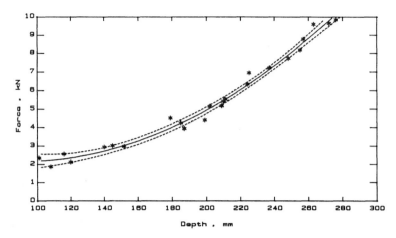

Fig. 7. Effect of depth upon forces

The curve obtained show a rapid increase of force at greater values of depth especially below the plough pan (200 mm). To describe the experimental data a polynomial curve equation was fitted. The predicted curve is well correlated with the actual measurement points and the final equation obtained after regression was:

$$F = 4.07 - 0.04D + 2.3 \cdot 10^{-4}D^2 \qquad [2]$$

where F is the total horizontal force acting on the plough body (kN) and D is the ploughing depth (mm). It should be noted that on all the plough components the vertical forces were most sensitive to depth, and this effect is very apparent on the mouldboard.

3.4 EFFECT OF SPEED UPON FORCES

Several runs were conducted to determine the functional relationship between force and speed at a constant depth and the variability of soil properties and conditions were assumed to be constant. The test was performed at a moisture content of 22 % (w/w) on a clay soil. The collected data showed that force increases considerably with speed at higher speeds (Fig. 8) and it was therefore reasonable to fit a polynomial curve of the second degree to the measured data.

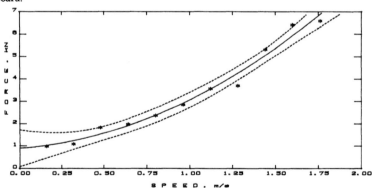

Fig. 8. Effect of speed upon forces

A regression analysis yields a correlation coefficient of 0.98 at a 95 % confidence limit, and the final equation obtained has the form:

$$F = 0.9 + 0.55V + 1.62V^2 \qquad [3]$$

where F is the horizontal force acting on the share point (kN) and V is ploughing speed (m/s). It was noticed that the second degree term was highly significant. However, for sandy soils the exponential curve fits best to the collected data since at higher speeds the force maintain an asymptotic value.

3.5 SOIL PARAMETERS AND FORCES

Investigation of data analysis revealed that most of the soil properties measured depend entirely on soil moisture content. Cohesion and soil internal friction increase with increasing moisture content (Fig. 9) and then the curve tends to decline with further increases of moisture content. For clay soil, adhesion inreases with increased moisture content, while the increase of soil-metal friction was not so significant as adhesion for the 18-28%(w/w) range of moisture content. It was noticed that the change of dry bulk density with moisture content was not significant.

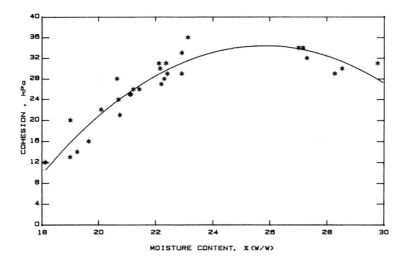

Fig. 9. Effect of moisture content on cohesion.

A multiple linear regression analysis was conducted to study the relationship between plough forces and the pertinent soil parameters, speed and depth. The result showed that most of the variables were significant with an R^2 value of 0.92 and (P<1%). The least significant parameter was the bulk density. However, since the study of the influence of plough geometry parameters to which the forces are most sensitive has not yet been completed, the result of force prediction is limited to the preliminary investigation.

4 CONCLUSION

Results of measurements of force components acting on plough components revealed that the share point is the part which is most subjected to the horizontal soil reaction force (63 to 72 %) of the sum of the horizontal forces acting on all plough components). At greater depth, the vertical and lateral forces are predominant on the mouldboard. The proportions of horizontal, vertical and lateral forces vary significantly on each plough component depending mostly on the depth and speed of operation.

The predictive equations developed from experimental results confirmed that ploughing depth and speed have a quadratic relation with forces. It was noticed that soil parameters measured in the present study depend entirely on moisture content over the moisture content range of 16-28 % (w/w). The correlation coefficients obtained as a result of regression analysis of these parameters with moisture content were very high. The preliminary multiple linear regression analysis of forces and the pertinent soil parameters, depth and speed showed that most of the variables are significant with a correlation coefficient of 0.92.

Acknowledgement

The author expresses his grateful thanks to Professor N. Möller and Dr. H.J. Olsen for useful discussions. The author is also greatly indebted to Mr. Örjan Bergwall for his help in typing the paper.

REFERENCES

Clyde, A.W. 1961. Force Measurements Applied to Tillage Tools. Transactions of the ASAE, Vol 4, No.2: 153-154,157.

Doner, R.D. and Nichols,M.L. 1934. The Dynamic Properties of Soil V. Dynamics of Soil on Plow Mouldboard Surfaces Related to Scouring. Journal of ASAE, Vol 15, No.1:9-13.

Gao Qiong; Pitt,R.E. ; Ruina,A. 1986. A Model to Predict Soil Forces on the Plough Mouldboard. Journal of Agric. Engng Res 35: 141-155.

Gebresenet,G. 1988. Multicomponent Dynamometer to Measure Forces on Plough Bodies. AgEng paper No. 88.090

Gee-Clough, D., McAllister,M., Pearson,G. and Evernden,D.W. 1978. The Empirical Prediction of Tractor - Implement Field Performance. Journal of Terramechanics, Vol 15, No.2: 81-94.

Getzlaff, V.G. 1951. Messung der Kraftkomponenten an einem Pflugkörper. Grdlgn d Landtechn. Heft 1:16-24.

Gill, W.R. and Vanden Berg,G.E. 1968. Soil Dynamics in Tillage Tractions. Agricultural Handbook No. 316,Agricultural Research Service,U.S. Dept. of Agriculture.

Larson, L.W., Lovely,W.G. and Bockhop,C.W. 1968. Predicting Draft Forces using Model moldboard plows in Agricultural Soils. Transaction of the ASAE, Vol 11, No.5:665-668.

Nichols, M.L. and Kummer,T.H. 1932. The Dynamic Properties of Soil IV. A Method of Analysis of Plow Moldboard Design Based upon Dynamic Properties of Soil.Agricultural Engineering 13(11)279-285.

O'Callaghan, J.R.; and McCoy,J.D. 1965. The Handling of Soil by Mouldboard Ploughs. Journal of Agric. Engng Res 10: 23-35.

Olsen, H.J. 1984. A torsional shearing device for field test. Soil Tillage. Res., 4:599-611.

Oskoui, K.E., Rackham,D.H. and Witney,B.D. 1982. The Determination of Plough Draught. Part II. The Measurement and Prediction of Plough Draught for two Mouldboard Shapes in three Soil Series. Journal of Terramechanics, Vol 19, No.3: 153-164.

Oskoui, K.E. and Witney,B.D. 1982. The Determination of Plough Draught - Part I. Prediction from Soil and Meteorological Data with Cone Index as the Soil Strength Parameter. Journal of Terramechanics, Vol. 19, No.2: 97-106.

Stafford, I.V. and Tanner,D.W. 1983. Effect of Rate on Soil Shear Strength and Soil-Metal Friction II. Soil-Metal Friction. Soil Tillage Res., 3:321-330.

Wang,J, Kwang,L. and Liang,T. 1972. Predicting Tillage Tool Draft using Four Soil Parameters. Transactions of ASAE, Vol 15, No.1:19-23.

Land and Water Use, Dodd & Grace (eds), © 1989 Balkema, Rotterdam. ISBN 90 6191 980 0

Disk trajectory simulation of a powered disk tiller

H.Guo, T.H.Burkhardt & R.H.Wilkinson
Michigan State University, East Lansing, Mich., USA

M.Hoki
Mie University, Tsu, Japan

T.Tanoue
Toyosha Company Ltd, Osaka, Japan

ABSTRACT: A joint research effort was initiated at Mie University and Michigan State University to study the field performance of the powered disk tiller. This paper presents the results of a computer simulation of the motion of a point on a rotating disk blade of the powered disk tiller. The simulation results showed that the tiller worked in the slipping condition. Given the gang angles of 26° through 34° and the ground speeds of 2 through 8 km/hr, the disk slippages were between 52.79% and 89.11%. The absolute velocity of a point on the edge of a rotating disk blade ranged between 1.098V and 2.835V and was always greater than the implement ground speed (V). The direction and magnitude of the absolute velocity were affected by changes in ground speed and gang angles.

1 INTRODUCTION

The powered disk tiller is a PTO-driven tillage implement which was introduced in the Japanese marketplace in April, 1983 by Toyosha Company, Ltd. Power is transferred from the PTO to the disk blades through a gear box followed by a roller chain drive which is enclosed in an oil bath. The disk blades operate with a rotation direction the same as that of the tractor wheels.

The powered disk tiller has an efficient system for transferring power from the PTO to the disk blades. Transfer of power through the gear box and the roller chain drive has a power loss less than two percent and is much more efficient than the hydrostatic drive of powered disk implements used in other studies. A joint research effort was initiated at Mie University and Michigan State University to study the field performance of the powered disk tiller. Results of field tests for two different models of the powered disk tiller were reported by Tembo et al. (1986) and Hoki et al. (1988).

This paper presents the results of a computer simulation of the motion of a point on a rotating disk blade of the powered disk tiller. This simulation was done for a tiller with a tilt angle (β) = 0°, a disk diameter (D) = 710 mm, a disk radius of curvature (r) = 680 mm and a PTO rotation speed (n) = 540 rpm. Simulation results are presented for 5 gang angles and for 4 ground speeds.

2 REFERENCE SYSTEMS

Nartov (1985) defined three different coordinate systems to study the kinematics of a disk blade (Figure 1). The origin coincided with the center of the circular cutting edge of the disk. OX was in the direction of the assembly movement. OY was in the lateral direction, and OZ was in the vertical direction. By rotating the OXYZ coordinate system +α degrees about the Z axis, the $OX_1Y_1Z_1$ coordinate system resulted, and then rotating this $OX_1Y_1Z_1$ system -β degrees about the X_1 axis, the $OX_2Y_2Z_2$ system resulted. OX_2 was in the horizontal direction perpendicular to the axis of rotation. OY_2 was along the disk axis, and OZ_2 was in the vertical direction perpendicular to the axis of rotation (Figure 2). Since the powered disk tiller used in this study had a tilt angle of β = 0°, the $OX_1Y_1Z_1$ coordinate system was the same as the $OX_2Y_2Z_2$ system.

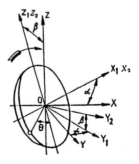

Fig. 1 Reference coordinate systems

The gang angle (α) was the angle between a horizontal diameter of the disk face and the travel direction of the assembly (Figure 3). As the disk rotated, it had a tendency to go in the X_1 direction which was the disk diameter

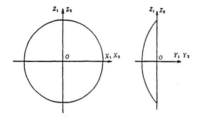

Fig. 2 $OX_2Y_2Z_2$ and $OX_1Y_1Z_1$ coordinate systems when $\beta = 0$

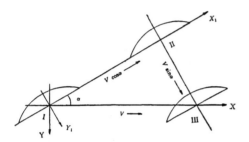

Fig. 3 Disk movement

direction. But the entire implement was hitched to the tractor which traveled in the X direction. The disk movement can be visualized as a two step process (Figure 3). The first step is motion along line I-II to position II, and the second step is motion along line II-III to position III.

3 SLIPPING DISK MOTION

The movement of the disk blade was complicated. The blade rotated about its own axis while the implement moved along with the tractor at a certain speed. The blade movement was the combination of the angular rotation powered by the PTO shaft and the translation movement pulled by the tractor.

In the $OX_1Y_1Z_1$ coordinate system along the line I-II in Figure 3, the disk had an angular velocity (ω) and a forward translation speed (V_1) in the X_1 direction given by:

$$V_1 = V \cos\alpha, \qquad (1)$$

where V is the assembly ground speed (km/hr). The tangential peripheral velocity (V_1') on the disk edge is given by:

$$V_1' = R\omega, \qquad (2)$$

where R is the disk radius (m) and ω is the disk angular velocity (rad/s). The disk motion was a combination of translatory motion and rotary motion for any point on a

disk edge. There were three possible combinations. If the forward speed was equal to the linear tangential speed ($V_1 = R\omega$), the disk rotation was pure rolling. If the forward speed was greater than the tangential speed ($V_1 > R\omega$), the disk was skidding. If the forward speed was less than the tangential speed ($V_1 < R\omega$), the disk was slipping. Trajectories of a single point on the edge of a rotating disk for these conditions are given in Figure 4. When the PTO speed is 540 rpm, the disk shaft speed is 113.8 rpm and the disk angular velocity is 11.92 rad/s. The tangential speed of the disk cutting edge is 4.23 m/s at this disk angular velocity. Table 1 gives possible V_1 values for different ground speeds and different gang angles. These ground speeds were used for the field test phase of this research. Since the tabular values were all less than 4.23 m/s, the powered disk tiller worked in the slipping condition for all of these ground speed and gang angle combinations.

Table 1. Possible speeds in X_1 direction

GROUND SPEED (m/s)	0.556	1.111	1.667	2.222
GANG ANGLE (degrees)		V_1 SPEED (m/s)		
26	0.499	0.999	1.498	1.997
28	0.491	0.981	1.472	1.962
30	0.481	0.962	1.443	1.925
32	0.471	0.942	1.413	1.885
34	0.461	0.921	1.382	1.842

4 TRAJECTORIES OF MOTION FOR A POINT ON THE EDGE OF A ROTATING DISK

An interactive program, MOTION, written in FORTRAN was used to simulate motion trajectories for a point on a disk edge given different gang angles and ground speeds. Point coordinate equations for any point on a disk in the OXYZ coordinate system given by Nartov (1985), were as follows:

$$X = Vt + \Delta R \cos\beta\sin\alpha + p\cos\theta\sin\beta\sin\alpha \\ - p\sin\theta\cos\alpha, \qquad (3)$$

$$Y = \Delta R \cos\beta\cos\alpha + p\cos\theta\sin\beta\cos\alpha \\ + p\sin\theta\sin\alpha, \qquad (4)$$

$$Z = \Delta R \sin\beta - p\cos\theta\cos\beta, \qquad (5)$$

$$\Delta R = (r^2 - (D/2)^2)^{0.5} - (r^2 - p^2)^{0.5},$$

where t is the time (s), r is the disk curvature (m), p is the simulating point's radial distance from the rotation axis (m), and θ is the simulating angle ($\theta = \omega t$) rotated from the bottom of the disk clockwise (rad) as shown in Figure 5.

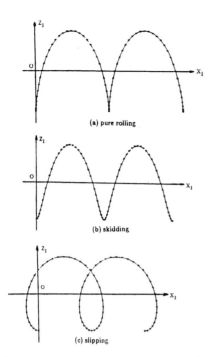

(a) pure rolling

(b) skidding

(c) slipping

Fig. 4 Disk trajectories in three conditions

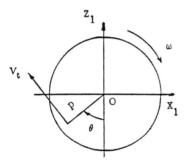

Fig. 5 Illustration of p and θ

The representation of the disk trajectories was in the $OX_1Y_1Z_1$ coordinate system (Figures 1 and 3) where the disk's axis of rotation was in the Y_1 direction. The coordinate point equations in the $OX_1Y_1Z_1$ system were obtained by rotating the OXYZ system +α degrees about the Z axis, that is, multiplying the vector [XYZ] by the matrix operator (Rogers, 1976):

$$\begin{bmatrix} \cos\alpha & \sin\alpha & 0 \\ -\sin\alpha & \cos\alpha & 0 \\ 0 & 0 & 1 \end{bmatrix}$$

The resulting equations for β = 0 are:

$$X_1 = Vt\cos\alpha - p\sin\theta, \qquad (6)$$

$$Y_1 = Vt\sin\alpha + \Delta R, \qquad (7)$$

$$Z_1 = -p\cos\theta. \qquad (8)$$

Some simulation results are presented in Figures 6 through 8 where the vertical movement is Z_1 of Equation 8 and the horizontal movement is X_1 of Equation 6. All of results are in the slipping condition. The more the disk point trajectory overlaps, the more the disk slips. The percentage of slippage varies with gang angle and ground speed. The percentage of slippage was defined (Zhou, 1980) as:

$$SLIP(\%) = (S_t - S)(100)/S_t, \qquad (9)$$

wnere S_t is the theoretical distance the disk moved, and S is the actual distance the disk moved. As a disk with a diameter of 0.71m was rotated one revolution, the distance it moved theoretically in the X_1 direction was 2.23m. Since the disk implement was hitched to the tractor, it could only move along at the ground speed (V) of the tractor. The distance that the implement moved at this speed was the actual distance. This distance was simulated by using the program MOTION. By using Equation 9, disk slippages were calculated. Part of the simulation results are listed in Table 2. Disk trajectories with various slippages are shown in Figures 6 through 8. For a single revolution, the actual distance varied with changes in gang angles and ground speeds, while the theoretical distance remained constant. A comparison of Figures 6 and 7 illustrates the effect of changing ground speed on slippage for constant PTO speed and constant gang angle. A

Table 2. Slippage simulation results

Gang angle degrees	Ground Speed km/hr	Actual distance m	Theoretical distance m	Slip %
26	2	0.263	2.231	88.20
	4	0.527		76.39
	6	0.790		64.59
	8	1.053		52.79
30	2	0.254	2.231	88.63
	4	0.507		77.25
	6	0.761		65.88
	8	1.015		54.51
34	2	0.243	2.231	89.11
	4	0.486		78.22
	6	0.729		67.34
	8	0.971		56.45

comparison of Figures 6 and 8 illustrates that changing gang angle has very little effect on slippage for constant PTO speed and constant ground speed. Table 2 shows that with a gang angle of $26°$ to $34°$ and ground speed of 2 to 8 km/hr, the disk slippages varied from 52.79% to 89.11%.

5 VELOCITY DISTRIBUTION SIMULATIONS

Any point on the rotating disk had an angular velocity (ω) and radius (p). Its linear tangential speed (V_t in Figure 5) was expressed as:

$$V_t = p\omega. \tag{10}$$

By employing the reference systems described earlier, velocity components for a point on a rotating disk in the $OX_1Y_1Z_1$ system produced by ω were:

$$V_{x1} = -p\omega\cos\theta, \tag{11}$$

$$V_{y1} = 0, \tag{12}$$

$$V_{z1} = p\omega\sin\theta \tag{13}$$

Rotating the vector $[V_{x1}\ V_{y1}\ V_{z1}]$ $-\alpha$ degrees about the Z_1 axis (Figure 9), that is, multiplying the velocity vector in the $OX_1Y_1Z_1$ system by a matrix operator, (Rogers and Adams, 1976) $[V_{x1}\ V_{y1}\ V_{z1}]$ resulted:

$$[V_x\ V_y\ V_z] =$$
$$[V_{x1}\ V_{y1}\ V_{z1}] \begin{bmatrix} \cos\alpha & -\sin\alpha & 0 \\ \sin\alpha & \cos\alpha & 0 \\ 0 & 0 & 1 \end{bmatrix} \tag{14}$$

Replacing V_{x1} V_{y1} and V_{z1} by using Equations 11 through 14, Equation 14 became:

$$V_x = -p\omega\cos\alpha\cos\theta, \tag{15}$$

$$V_y = p\omega\sin\alpha\cos\theta, \tag{16}$$

$$V_z = p\omega\sin\theta \tag{17}$$

The above equations resulted only from the disk blade rotations. The assembly ground speed (V) also contributed to the velocity vector $[V_x\ V_y\ V_z]$. Since V was in the X direction, it only increased the magnitude of V_x. When both linear speed and rotation velocity were considered, that is, add V into Equation 15, the equation became:

$$V_x = V - p\omega\cos\alpha\cos\theta. \tag{18}$$

Equations 16, 17 and 18 had similar forms as those Nartov (1985) derived.

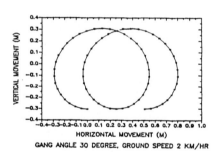

Fig. 6 Disk trajectory with slippage of 88.63%

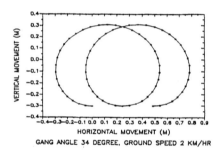

GANG ANGLE 34 DEGREE, GROUND SPEED 2 KM/HR

Fig. 8 Disk trajectory with slippage of 89.11%

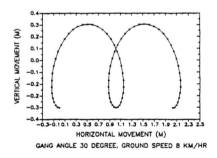

GANG ANGLE 30 DEGREE, GROUND SPEED 8 KM/HR

Fig. 7 Disk trajectory with slippage of 54.51%

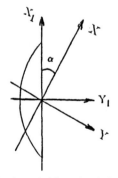

Fig. 9 Coordinate rotation about Z_1

Equations 16, 17 and 18 described the disk velocities in three dimensions. Using these equations, computer simulations were made with the program MOTION. A computer program PLOT 3.VXYZ was used to plot the simulation results. Velocity distributions in both XZ and YZ planes were ellipses and in the XY plane the distribution was a straight line. The ellipse in the YZ plane was symmetric about the origin, and the ellipse in the XZ plane was symmetric about the X axis.

Velocity distributions were affected by changes of the gang angle and changes of ground speed. The change of gang angle did not affect V_z, and changing the ground speed had no effect on either V_y or V_z. For all of the tests, V_z had the same distribution in any one disk rotation (Figure 13). V_y had the same distribution for all ground speeds at a given gang angle. Increasing the gang angle increased the velocity in the Y direction for positive V_y and decreased the velocity for negative V_y. V_y remained at zero for $\theta = 90^\circ$ and $\theta = 270^\circ$ as demonstrated in Figure 10. Increasing the gang angle at a given ground speed slightly decreased the velocity in the X direction for positive V_x and slightly increased the velocity for negative V_x. V_x remained at zero for $\theta = 90^\circ$ and 270° (Figure 11). The velocity in the X direction was also affected by the ground speed, when the ground speed was increased, V_x was increased (Figure 12).

Figure 13 gives the general distributions of V_x V_y and V_z in two revolutions of the disk blade. The working part of the disk was located on the lower part of the disk. In

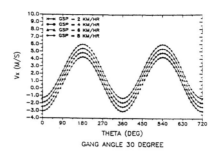

Fig. 12 Effect of ground speed on V_z

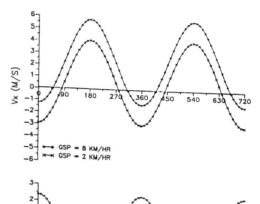

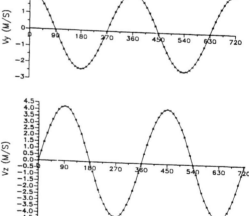

Fig. 13 General velocity distribution in degrees of rotations

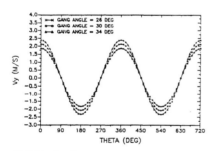

Fig. 10 Effect of gang angle on V_y

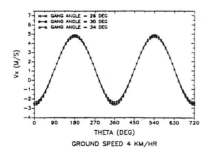

GROUND SPEED 4 KM/HR

Fig. 11 Effect of gang angle on V_x

Figure 13, the working part of the disk was between 0° to 90° and 270° to 360°. The lower front part of the disk, at an angle of 270° to 360°, cut and moved the soil, and the lower rear part at an angle of 0° to 90° moved and threw the soil. In Figure 10, it was demonstrated that V_y was always positive for the lower part of the blade. V_x was negative and became slightly positive as ground speed increased, and V_z was negative for the lower front part and positive for the lower rear part of the disk blade (Figure 13). Therefore, the lower front part of the disk cut the soil and then moved it backward, to the right and somewhat lower. The lower rear part of the disk moved the cut soil upward, to the right and backward.

Given Equations 16, 17 and 18, the magnitude of the absolute velocity of a disk point in three dimensional space is:

$$V_{abs} = (V_x^2 + V_y^2 + V_z^2)^{0.5} . \tag{19}$$

Its direction can be determined from:

$$\cos\lambda = V_x/V_{abs} , \tag{20}$$

$$\cos\gamma = V_y/V_{abs}, \tag{21}$$

$$\cos\delta = V_z/V_{abs}, \tag{22}$$

where λ is the angle (rad) between vector V_{abs} and the X axis, γ is the angle between V_{abs} and the Y axis, and δ is the angle between vector V_{abs} and the Z axis.

The simulation using the above equations and the program MOTION, showed that the magnitude of the absolute velocity varied from 2.4 m/s up to 6.3 m/s. λ was in the range of 20° to 150°, γ was between 40° and 120°, and δ varied in the largest range from 5° to 180°. These results also showed that variation of gang

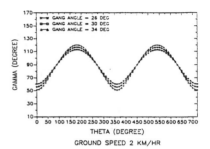

Fig. 16 Effect of gang angle on γ

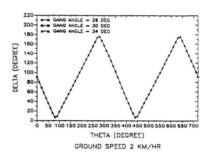

Fig. 17 Effect of gang angle on δ

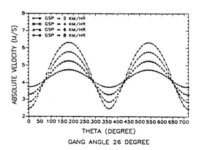

Fig. 18 Effect of ground speed on magnitude of V_{abs}

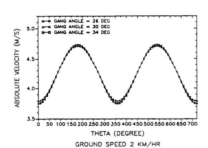

Fig. 14 Effect of gang angle on magnitude of V_{abs}

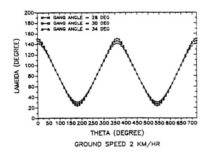

Fig. 15 Effect of gang angle on λ

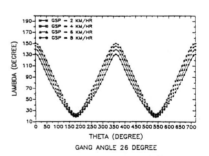

Fig. 19 Effect of ground speed on λ

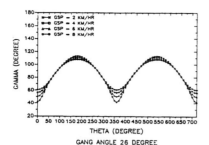

Fig. 20 Effect of ground speed on γ

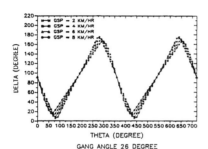

Fig. 21 Effect of ground speed on δ

angle and ground speed had an influence on the vector of absolute velocity.

From the simulation result plots in Figures 14 through 17, the gang angle did not have much effect on the magnitude and the direction of the absolute velocity. The change of gang angle influenced the γ values slightly more than for the others.

The ground speed had more effect on both the magnitude and direction. Figure 18 shows that as the ground speed increased, the magnitude varied widely. When the ground speed was at 2 km/hr, the magnitude of the absolute velocity varied from 3.76 to 4.72 m/s. As the ground speed increased up to 8 km/hr, the magnitude varied from 2.56 to 6.25 m/s, for a gang angle of 26°. The λ value shifted upward when the ground speed went higher (Figure 19). The ground speed also affected γ and δ as indicated in Figures 20 and 21.

Nartov (1985) reported that the maximum amplitude of the absolute velocity equaled 1.95V for an unpowered disk implement. In this study of the power disk, the absolute velocity, with the minimum amplitude of 1.098V and the maximum of 2.835V, was always greater than the ground speed given the range of gang angles between 26° and 34° and ground speeds between 2 km/hr and 8km/hr.

6 CONCLUSIONS

A computer simulation of the motion of a point on the edge of a rotating disk blade on a powered disk tiller provided the following results:

1. The powered disk tiller worked in the slipping condition. Given the gang angles of 26° through 34° and the ground speeds of 2 through 8 km/hr, the disk slippage was between 52.79% and 89.11%.
2. The disk velocity distributions were ellipses in the XZ and YZ planes and a straight line in the XY plane. The distributions were affected by changes in ground speed and changes in gang angle.
3. The absolute velocity of a point on the edge of a rotating disk blade ranged between 1.098V and 2.835V, and was always greater than the implement's ground speed (V). The absolute velocity was affected by changes in ground speed and changes in gang angle.

REFERENCES

Draper, N.R. & H.Smith 1981. Applied regression analysis. Second edition. John Wiley & Sons Inc.

Finnery, J.B. 1982. The role of discs in primary cultivation. Agricultural Engineer 37(1):14-19.

Gill, W.R., A.C.Bailey & C.A.Reaves 1982. Harrow disk curvature - influence on soil penetration. Trans. of ASAE 25(5):1173-1180.

Gordon, E.D. 1941. Physical reactions of soil on plow disks. Agricultural Engineer 22(6):205-208.

Harrison, H.P. 1977. Soil reacting forces for disks from field measurements. Trans. of ASAE 20(5):836-838.

Hoki, M., T.H. Burkhardt, R.H. Wilkinson & T.Tanoue 1988. Study of PTO driven powered disk tiller. Trans. of ASAE 31(5):1355-1360.

Klenin, N.I., I.F.Popov and V.A. Sakun 1985. Agricultural machines, theory of operation, computation of controlling parameters and condition of operation. Amerind publishing Co.PVT Ltd.

Nartov, P.S. 1966. Method of calculation of basic parameters and kinematic of disk working tools of forest soil working implements. NAL translation No. 25690.

Nartov, P.S. 1967. Basics of parameters of working tools of forest disk implements. NAL translation No. 25681.

Nartov, P.S. 1967. Study of working tools of disk plows. NAL translation No. 23846.

Nartov, P.S. 1985. Disk soil-working implements. Vornoezh University Press, Voronezh.

Neter, J., W.Wasserman & M.H. Kutner 1985. Applied linear statistical models. Second edition. Richard D. Irwin Inc.

Rogers, D.F. and J.A.Adams 1976. Mathematical elements for computer graphics. McGraw-Hill Book Company.

Tembo, S., T.H.Burkhardt, R.H. Wilkinson & M.Hoki 1986. Performance of the PTO power disk tiller. ASAE paper no. 86-1010. ASAE, St. Joseph, MI 49085.

Young, P.E. 1976. A machine to increase productivity of a tillage operation. Trans. of ASAE 19(6):1055-1061.

Zhou, Yiming 1980. The tractor mechanics (in Chinese). Beijing Agricultural Engineering China.

Land and Water Use, Dodd & Grace (eds), © 1989 Balkema, Rotterdam. ISBN 90 6191 980 0

Disc configuration and performance

M.J.Hann, R.J.Godwin, T.A.Green & A.Al-Ghazal
Silsoe College, Cranfield Institute of Technology, Silsoe, Bedford, UK

ABSTRACT: Disc performance was evaluated, under controlled conditions in terms of the soil forces, area of disturbance, specific resistance and penetration rate for a range of sweep and tilt angles, disc edge profiles, interactions with deeper secondary tines and positively driving the disc. The degree of soil inversion reduced as the top of the disc is tilted backwards. Disc penetration is best at sweep angles of 35° and low tilt angles. Tilting the disc backwards retards penetration but has the advantage of reducing the draught force. Cutting scallops from the edge of the disc improves the penetration rate. Deeper secondary tines increase both the area of soil disturbance and draught force but not in the same proportion, such that the performance is greatly improved. Under field conditions positively rotating the disc significantly reduces the draught force at the expense of an additional power requirement.

1 INTRODUCTION

The use of simple discs for soil manipulation is widespread in both temperate and tropical agriculture. The disc has the advantages of being relatively simple to manufacture, robust and by the varying geometry can be a most versatile tool. This versatility enables a range of different operations to be performed ranging from primary tillage (with greater or lesser soil inversion depending upon the setting of the disc), disintegration of dry clods, ridging, lifting of root crops and burial of surface trash. The major limitation of discs are the potential compaction caused by the underside of the disc, slow penetration into hard dry soils and the relatively high draught and specific resistance (draught/area of soil disturbed) figures.

The objective of this paper is to examine these problems in more detail and to produce recommendations for disc design, manufacture and operation from studying the effect of:

1. disc geometry on soil disturbance and re-arrangement,
2. disc geometry on the rate of penetration into soils,
3. the attachment of deeper secondary tines following the disc upon soil disturbance, draught force, specific resistance,
4. positively driving the discs on the resulting soil forces and power requirements.

Figure 1 shows the basic geometrical configuration of agricultural discs, which for any given radius, diameter and edge sharpening is influenced by the attitude of the disc to the:

1. direction of travel, ie sweep angle, and
2. vertical, ie tilt angle.

Earlier studies of discs with zero tilt angle, Harrison and Thivavarnvongs (1976) and Godwin et al (1985) have shown the effect of the interaction between sweep and clearance angle. This interaction is such that if the sweep angle is larger than the clearance angle no scrubbing occurs on the convex face of the disc and as a result the soil forces acting upon the disc are significantly reduced. The clearance angle of the disc is influenced by the compromise between the need for a large disc radius for small clearance angles and the small radius required for soil inversion.

With the exception of various combination tillage tools where discs have to be attached to a multistage tool bar together with chisel or spring tines, Watts and Patterson (1984) no detailed studies have considered the interaction between discs and tines, as shown in Fig 2. The benefit of the interaction would be most beneficial if the tine followed the disc and at a greater depth of work. This would have the added advantage of removing the smeared layer or hard pan caused by the disc.

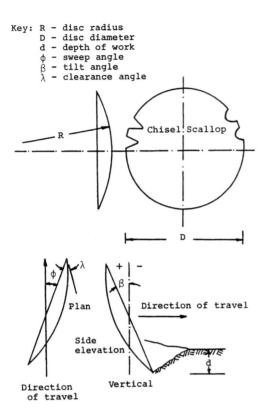

Key: R – disc radius
     D – disc diameter
     d – depth of work
     φ – sweep angle
     β – tilt angle
     λ – clearance angle

Fig 1. Basic disc geometry.

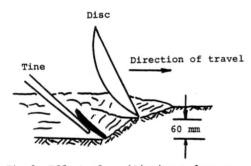

Fig 2. Effect of positioning a deeper chisel tine following a shallower disc.

Under normal circumstances the agricultural disc is free to rotate when in work, being passively driven by the forces of the soil disc reaction. Earlier studies by Getzlaff and Soehne (1959) showed that there could be advantages in positively driving the disc. This advantage could be like many in powered and vibratory tillage where the draught force is very significantly reduced with the total power requirement remaining relatively unaffected or increasing marginally. This can be an effective operational advantage as it reduces the need for tractive pull and hence lighter

tractors with smaller ground drive systems may be used. The benefit of powering a disc rather than vibrating a tillage tine is that the motion is continuous and hence the peak accelerations do not occur at 10-20 cycles/sec with all of the design, manufacturing, maintenance and operational problems that ensue.

2 EXPERIMENTAL METHOD

The majority of the experimental work was conducted under controlled soil bin conditions, using a sandy loam soil (sand 72%; silt 10%; clay 18%) compacted to a bulk unit weight of 14 kN/m^3 at a gravimetric moisture content of 11% dry base.
    The discs were either set to run at a fixed depth or mounted on a parallel linkage which enabled them to float and penetrate to a final equilibrium depth. The discs were attached to a trunion which enabled both sweep and tilt angles to be varied. This in turn was attached to instrumentation to measure the draught, vertical and side forces as shown in Godwin et al (1985).
    For the studies on soil disturbance a secondary soil box (2.4 m long; 1.2 m wide and 0.21 m deep) was used in which chalk dust tracers (220 mm long; 20 mm wide and 3 mm deep) were positioned 80 mm apart horizontally with a different colour positioned 20 mm vertically above. Five different colours were used in all, see Fig 3, with the final colour on the soil surface. Once the secondary box was located in the soil bin its ends were removed to allow the disc to pass through the soil at the required setting. To enable the soil flow to be fully understood the disc was made from transparent material which enabled the process to be recorded on video film from the rear of the disc.
    To enable the interaction studies to be conducted a 50 mm wide, 25° rake angle plane tine was attached to operate 60 mm below the disc, as shown in Fig 2.
    A hydraulic motor was used to rotate the disc via a chain drive and torque transducer at a range of rotational speeds to give driven peripheral speed/free rotational peripheral speed ratios from 1 to 8 in both the forward and reverse directions. The tilt and sweep angles of the disc were kept constant at 15° and 35° respectively. These studies were also conducted in field conditions to confirm the laboratory results and provide realistic figures for subsequent design purposes. The field soil was of similar texture, moisture content and bulk unit weight to that used in the laboratory, but was of higher shear strength due to the age-cementing process which cannot be easily simulated in laboratory conditions.
    Details of the discs geometry and forward speed for each of the studies is given in Table 1.

Table 1. Disc geometry and forward speed.

Study	Diameter (D) mm	Radius (R) mm	Depth (d) mm	Forward speed (V)
Soil disturbance	415	540	80	0.5 m/sec
Penetration rate	610	480 560 880	Variable 0- >180	0.3 m/sec
Tine interaction	610	880	100	1.0 m/sec
Driven discs	610	560	150	0.75 m/sec

3 EFFECT OF DISC GEOMETRY ON SOIL DISTURBANCE AND RE-ARRANGEMENT (AL-GHAZAL, 1989)

Following careful soil excavation, analysis of the positions of the chalk markers, see Fig 3, and the accompanying video recordings show the following effects of changes in sweep and tilt angle over the ranges of 20° to 50° and 0° to 35° respectively.

1. At zero tilt angles inversion of the soil occurs at all sweep angles, however the amount of inversion increases as the sweep angle is increased. This is due to the change from an initial rolling-throwing action to a lateral mass transfer.

2. As tilt angles increase from zero the initial rolling action is reduced and the soil flows without inversion, this is especially evident at sweep angles in excess of 30°.

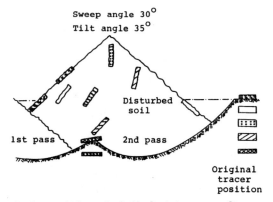

Sweep angle 30°
Tilt angle 35°

Disturbed soil

1st pass   2nd pass

Original tracer position

Fig 3. Position of chalk dust tracers after passage of disc; showing little inversion.

4 DISC PENETRATION

The penetration performance of discs has been investigated by measuring the equilibrium depth they achieve when in a free float mode on a parallel linkage, Kokabi (1987).

The effects of varying the tilt and sweep angles of a standard (560 mm radius of curvature) disc on the depth of penetration are shown in Figure 4. For

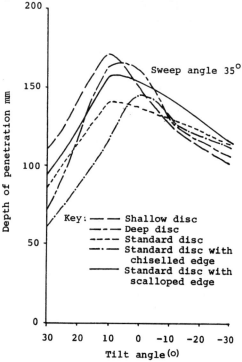

Sweep angle 35°

Key: — — Shallow disc
     —--— Deep disc
     ----- Standard disc
     —·— Standard disc with chiselled edge
     ——— Standard disc with scalloped edge

Fig 4. Relationship between tilt angle and depth of penetration of discs with various concavities and edge geometries.

all tilt angles the equilibrium depth increases with increased sweep angle to a maximum at approximately 35°. The greatest equilibrium depth is achieved at small tilt angles, increasing this angle backwards retards disc penetration more than increased forward tilt.

The effects of changing disc concavity and edge geometry on penetration performance are presented in Figure 5. All disc arrangements have a maximum penetration depth at low tilt angles.

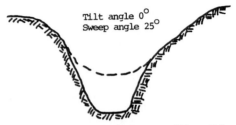

Fig 6. Soil disturbance profile with (solid line) and without (broken line) deeper tine.

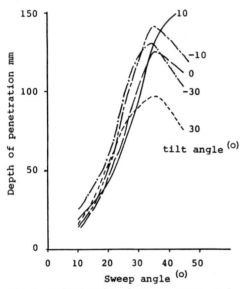

Fig 5. Relationship between sweep and tilt angle and the depth of penetration of a standard disc.

When tilted forwards all concavities perform similarly, however shallower discs are superior when tilted backwards.

The standard disc performance is improved when its edge is scalloped, however this improvement is lost if the edge is a chisel shape. This can be explained by considering the difficulty of producing a v-shape which is both sharp and self cleaning.

5 EFFECTS OF THE ATTACHMENT OF A DEEP SECONDARY TINE, AND CHANGES IN SWEEP AND DISC ANGLE ON PERFORMANCE

Figure 6 shows the cross section of the disturbed profile for a disc with and without a deeper secondary tine. The secondary tine breaks a significantly greater area of soil when set 60 mm below the disc.

The effects of the attachment of the secondary tine on the draught force, area of soil disturbances and specific resistance for a range of disc sweep and tilt angles is shown in Fig 7. This shows that the attachment of the tine increases both draught and area of disturbance but not in the same proportion so that the

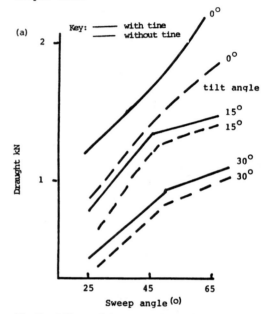

Fig 7a. Effect of deep secondary tine on draught force for a range of sweep and tilt angles.

specific resistance falls for all angular combinations. The secondary tine had the advantage of reducing the magnitude of the upward vertical force by between 10 and 30% depending upon disc geometry.

The effect of sweep angle on draught show a similar effect to that shown by Gill et al (1978) and Godwin et al (1985) but Fig 7 adds to this by showing the dramatic reduction in draught by increasing the tilt angle of the disc. The increase in tilt angle also increases the area of disturbance as the disc is effectively wider at any depth of work. The combination of these two effects significantly reduces the specific resistance at all sweep angles.

Despite the reduction in upward vertical force, field tests with secondary tines attached to a disc plough, showed no advantage in the attachment of the tines on either the working depth or rate of penetration. In order to achieve this a more sophisticated linkage would need to be developed.

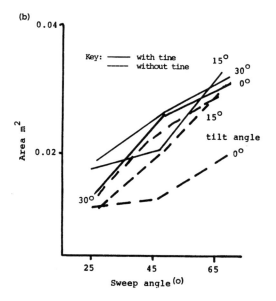

(b)

Key: ——— with tine
      ——— without tine

tilt angle

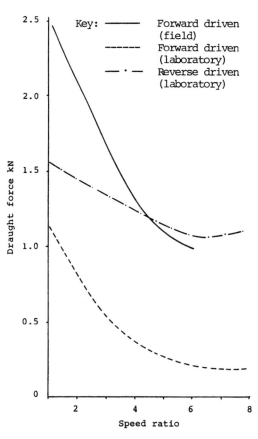

Key: —————— Forward driven
                (field)
      - - - - - Forward driven
                (laboratory)
      — · — · — Reverse driven
                (laboratory)

Fig 8a. Relationship between the speed ratio and draught force.

(c)

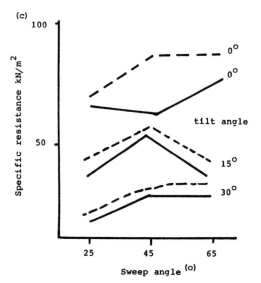

tilt angle

Fig 7b and c. Effect of deep secondary tine on area of disturbance and specific resistance for a range of sweep and tilt angles.

6 DRIVEN DISCS

The results of the study by Green (1989) into the effects on performance of driven discs are presented in Figure 8.

The investigations show that when the disc is driven forwards draught reduces significantly with increasing speed ratio. This reduction can be explained by considering the change in soil failure mechanism. Freely rotating discs produce a passive soil failure, a driven disc however causes soil failure by a positive cutting action with lower draught. Also the soil travelling over the disc is subject to the benefit of the mechanics of reverse friction where the disc moves upwards faster than the soil which has been shown to produce substantial draught reductions, Hann (1983).

A similar draught speed relationship is achieved when the disc drive is reversed however the draught reduction is significantly smaller. This can be explained by considering the greater effective path length of the soil over the disc producing larger sliding frictional forces. The high draught forces from the field work are a reflection of the greater soil shear strength.

Vertical forces are largely unaffected by driving the disc forwards. There is however a change in direction of the force which may have some adverse effects on penetration. Driving the disc in reverse, however restores the downward force and increases its magnitude.

The soil bin trials showed side force to be relatively unaffected by powering the disc. However in the stronger field soils side force increased with speed ratio. This may be due to the change in soil failure mechanism discussed above changing the direction and magnitude of the resultant forces.

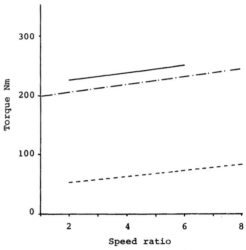

Fig 8c. Relationship between the speed ratio and torque.

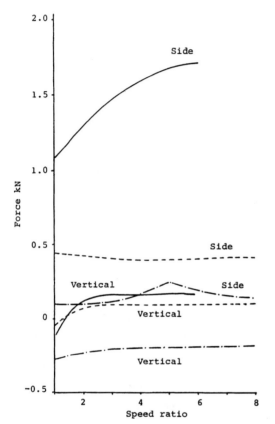

Fig 8b. Relationship between the speed ratio and vertical and side force.

As the torque requirement increases marginally with disc speed, the power to drive the disc is almost directly proportional to speed. This is also reflected in the total power requirement of the disc. Although the results from the soil bin investigation show that the forward driven disc had minimum power requirement at a speed ratio of 3:1 the field results show that driving the disc requires an increase in power. Assuming a speed ratio of 3:1 a 30% total power increase would reduce the draught force by 50%. Which could be very beneficial when as with most tractors, it is tractive pull rather than total power which is the limiting factor.

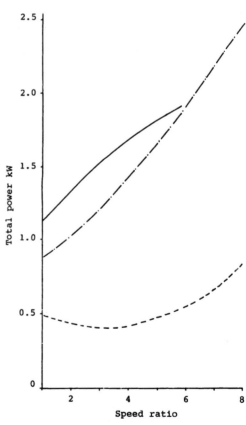

Fig 8d. Relationship between the speed ratio and total power requirement of the disc.

7 CONCLUSIONS

1. To obtain soil inversion discs should be operated with small tilt angles.
2. To minimise the inversion effect discs should be operated with large tilt and sweep angles.
3. Disc penetration is most effective at sweep angles of 35°, tilting the top of the disc backwards retards penetration more than tilting forwards.
4. Discs with a large radius of curvature penetrate to a greater depth when tilted back than discs with a small radius of curvature.
5. Scallop edge discs penetrated more effectively than standard discs.
6. Deeper secondary tines increase the area of disturbance at a greater rate than the increase in draught force and hence, improves performance by significantly reducing the specific resistance.
7. Increasing the tilt angle of the disc reduces the draught force, increases area of disturbance and improves performance by reducing the specific resistance.
8. Deeper secondary tines reduce the upward vertical force and could assist in penetration, although addition of these tines to a disc plough did not improve either the rate of penetration or the ultimate depth of work.
9. Significant draught reductions can be achieved when driving the disc both in and opposing the normal direction of rotation. Under field conditions, however, the total power requirement is not reduced.
10. Driving the disc in opposition to the normal direction of rotation increases the downward vertical force and should aid penetration.

ACKNOWLEDGEMENTS: The authors would like to acknowledge the technical assistance provided by Messrs A.J. Reynolds, R. Newland, G.C. Soane and Mrs. G. Lovelace and the support in terms of equipment and specialist discs from Victor Ward Engineering and W.A. Tyzack plc.

REFERENCES

Al-Ghazal, A. 1989. An investigation into the mechanics of agricultural discs. Unpublished PhD Thesis (in press), Silsoe College, Cranfield Institute of Technology, Silsoe, Bedford, UK.

Getzlaff, G. and W. Soehne. 1959. Forces and power requirements of freely rotating and driven plough discs on hard, dry, clayey loam. Grundlagen der Landtechnik 11: 40-52, illus. Natl. Inst. Agr. Engin. Eng. Translation 106.

Gill, W.R., C.A. Reaves and A.C. Bailey. 1978. The effect of geometric parameters on disc forces. A.S.A.E. Paper 78-1535, St. Joseph, Michigan, A.S.A.E.

Godwin, R.J., D.A.T. Seig and M. Allot. 1985. The development and evaluation of a force prediction model for agricultural discs. Proc. Int. Conf on Soil Dynamics Vol. 2, 250-265, Auburn, Alabama.

Green, T.A. 1989. An investigation into driven discs. Un published MPhil Thesis, Silsoe College, Cranfield Institute of Technology, Silsoe, Bedford, UK.

Hann, M.J. 1983. An investigation into the mechanics of tines with a power driven leading face. Unpublished MSc Thesis, Silsoe College, Cranfield Institute of Technology, Silsoe, Bedford, UK.

Harrison, H.P. and T. Thivavarnvongs. 1976. Soil reacting forces from laboratory measurements with discs. Can. Agric. Eng. 18: 49-53.

Katukula, B.M. 1987. Effects of soil loosening tines working behind disc ploughs. Unpublished MSc Thesis, Silsoe College, Cranfield Institute of Technology, Silsoe, Bedford, UK.

Kokabi, M. 1988. Design of a disc implement. Unpublished MSc Thesis, Silsoe College, Cranfield Institute of Technology, Silsoe, Bedford, UK.

Watts, C.W. and D.E. Paterson. 1984. The development and assessment of high speed shallow cultivation equipment for autumn cereals. J. agric. Engng. Res., 29, 115-122.

Land and Water Use, Dodd & Grace (eds), © 1989 Balkema, Rotterdam. ISBN 90 6191 980 0

Performance characteristics of the PTO powered disk tiller

M.Hoki
Mie University, Tsu, Japan

T.H.Burkhardt
Michigan State University, East Lansing, Mich., USA

T.Tanoue
Toyosha Company Ltd, Osaka, Japan

ABSTRACT: Field tests were conducted to evaluate a PTO powered disk tiller. The research identified some interesting characteristics of power requirement, specific work and side force under varying ground speeds of the tractor. It was found that in comparison with the rotary tiller the energy efficiency of the PTO powered disk tiller was better specifically for high ground speed. The side force acting on a disk blade was measured to find that it depended upon the ground speed under specific soil and operating conditions. Also the soil reaction acting horizontally to the rear furrow wheel was measured to show that the rear wheel was functioning to provide a horizontal force which counteracted the side force acting on the disk tiller from the soil being inverted. Some of the advantages of the PTO powered disk tiller over the conventional tillage equipment is discussed.

1 INTRODUCTION

The PTO driven disk tiller which has been made commercially available in recent years has some features of deep tillage and better soil turning at relatively small power requirements. Its characteristics are considerably different compared to the PTO driven rotary tillers, yet their advantages and limitations compared to the conventional tillage equipment have not been fully investigated.

In the past some research on the driven disks was made by Getzlaff(1953, 1959) and Sohne(1963). The former indicated that the equipment of this type required considerably high power and resulted in more complex drive mechanisms which is not desirable. The latter mentioned that in spite of the advantages of the driven disk it was not as advantageous from the points of high power requirement for driving and pulling the disks. In recent years Young (1976) pointed out the advantage of a driven disk and tried make a more detailed evaluation, yet was unable to reach definite conclusions from the point of practical applications. Abernathy(1976) reported that although a driven disk had various advantages its overall performance from the practical points varied depending upon the conditions, which have not yet lead to definite conclusions. In the early years of 1980s PTO driven disk tillers were made commercially available in Japan. By the

year of 1989 around 30,000 units of varying types and sizes of the driven disk tillers have been produced and distributed for use in many agricultural areas in Japan. Kawamura(1985) reported studies on a disk tiller of two gangs which was normally used for paddy field tillage. Hoki(1985) and Tembo(1986) also reported on the PTO power and total power requirements for PTO driven disk tillers of single gang which were used for upland tillage. There is yet remarkably little data available on the performance of driven disk tillers.

The purposes of the study were:

1. To investigate energy efficiency of the driven disk tiller by power requirement and specific work.

2. To find characteristics of the side force acting on the disk as the ground speed was incresed.

3. To measure the horizontal reaction acting on the rear wheel and investigate its effect in counteracting the side force.

2 RESEARCH EQUIPMENT AND METHODS

2.1 Driven disk tiller used

The equipment used for the experiment had four driven disks with single gang and is mounted through three point linkages on the tractor. The dimensions and specifications of the disk and the tractor are presented in Table 1 and 2 respectively.

Table 1. Specifications of the PTO driven
disk tiller

Number of disks	4
Diameter of disks	610 mm
Gang angle	27°, 31°
Working width	1000 mm
Overall width	1420 mm
Weight	270 kg
Mounting system	3-point hitch

Table 2. Specifications of the tractor

Drive system	Four wheel drive
Length	3240 mm
Wheel base	1800 mm
tread	1280 mm
Weight	1655 kg
Engine displacement	1498 cc
Engine output	24 kw/2600 rpm

2.2 Measuring systems

The torque of the PTO shaft was measured
by applying strain gages to the surface of
the circular section shaft near the universal joint connected to the tractor PTO.
In order to measure the revolving speed of
the PTO shaft a reflection type photo
sensor unit was affixed close to the PTO
shaft. The light reflected by a reflector
plate attached to the revolving PTO shaft
generated a pulse of current for every
revolution which was recorded on a data
recorder.
 The side force acting on a single disk
in parallel to the shaft was measured by
sensing compressive strain on a steel
collar spacer between the right end disk
and the second disk. The cylindrical
spacers are acting as a covering for the
hexagon shaft which turned the disks and
therefore the spacers only supported the
compressive force from the disk. In other
words the side force was supported by the
spacer and the drive torque was transmitted to the disk only through the hexagon

shaft.
 The rear wheel reaction which was acting
horizontally to the rear wheel at a right
angle to the travel direction of the tractor
was measured by sensing bending strain on
the arm base of the rear wheel.
 For measuring tilling depth an ultrasonic
system was employed. An ultrasonic trans-
mitter-receiver unit was attached at the
center of the frame of the disk tiller.
The ultrasonic wave of 40 KHz was directed
to the untilled ground surface and reflected
back to the receiver. The time required for
the wave to go and back from the ground was
measured and recorded which was converted
to the tilling depth.
 All the signals from the sensors were
lead to a data recorder mounted on the
tractor. A block diagram of the measuring
system is shown in Fig. 1. The system was

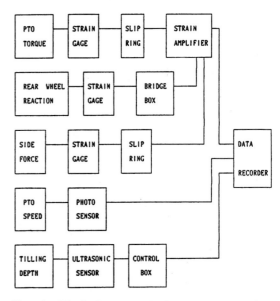

Fig. 1 Block diagram of the measuring and
recording system

AC powered from a DC-AC inverter using the
tractor battery.

2.3 Field conditions

The tests were conducted in a field at a
Mie University Experiment Station. The
field was covered with feed oats previously
and was harvested leaving about 10 cm
residue on the ground. The soil was sandy
loam type containing 20-30 % moisture at
20 cm depth. The soil hardness measured
with a cone penetrometer of 30° cone angle

and 7 cm^2 base area was 4.0–6.0 kg/cm^2 at
20 cm depth.

2.4 Experimental method

During the field tests the engine speed of
the tractor was fixed at 2600 rpm. The PTO
speed therefore was set to 540 rpm which
resulted in the disk speed to be 117 rpm
with some variation with loading conditions.
The ground speed was set to be 0.3, 0.7 and
1.2 m/s. But actual speed was calculated
from the time and distance measured with a
stop watch and a tape measure for the
tractor to move during 2 revolutions of a
tractor rear wheel. The angle of the rear
furrow wheel was set to be 15° to the
travel direction and 10° to the vertical
line. The gang angle of the disk tiller
was fixed to 31°.

3 RESULTS AND DISCUSSION

The data recorded were PTO torque, disk
side force, rear furrow wheel reaction,
tilling depth and PTO speed. The torque,
side force, furrow wheel reaction data were
converted into numerical data and averaged
by a laboratory computer. The sampling was
done for 7 seconds at the rate of 100 Hz
to produce 700 data points for each.

3.1 Power requirements

Fig. 2 shows the PTO torque requirements.
Increasing ground speed increased PTO
torque requirements. But the rate of
increase was kept relatively low. With the
increase of ground speed around 3 times
from 0.3 m/s to 1.2 m/s, the torque requirement increased from 100 Nm to 130 Nm only
by 30 % at the 21 cm tilling depth. In
other words the rate of increase of PTO
torque with the increase of ground speed
was not high which was greatly different
from conventional rotary tillers. While
the practical ground speed for the rotary
tiller is considered to be about 0.5 m/s
at 15 cm tilling depth(Ishihara 1983), the
powered disk can be said to travel at the
speed more than two times with the tilling
depth of 21 cm. The deep tillage and high
ground speed is one of the advantages of
the powered disk tiller.
  Fig. 3 shows the power requirements for
the disk tiller. The PTO speed was set to
540 rpm which varied slightly during tests
depending upon field conditions. The power
requirements were shown to be increased
only from 5.8 Kw to 7.7 Kw by 30 % when the
ground speed increased by 4 times from 0.3
m/s to 1.2 m/s under the tilling depth of

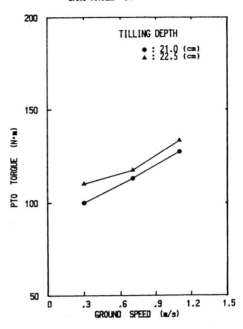

Fig. 2 PTO torque versus ground speed

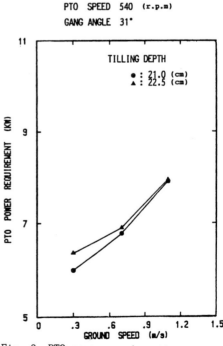

Fig. 3 PTO power requirement versus
ground speed

22.5 cm.

Fig. 4 shows the specific work for the disk tiller used. The specific work is the

PTO SPEED 540 (r.p.m)

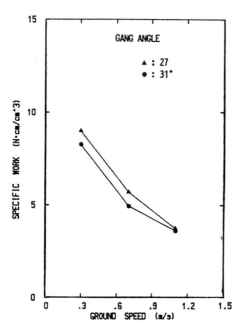

Fig. 4 Specific work versus ground speed

PTO SPEED   540 (r.p.m.)
TILLING DEPTH   ● : 21 (cm)
                   ■ : 23 (cm)

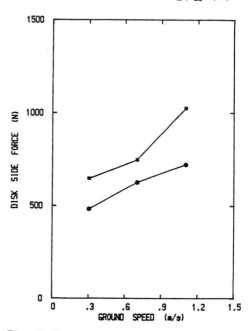

Fig. 5 Disk side force versus ground speed

energy required to till the soil of unit volume or the efficiency of energy consumption of the implement. The specific work of 8 Ncm/cm^3 at the ground speed of 0.3 m/s was decreased to 3.5 Ncm/cm^3 at 1.2 m/s. Compared to the conventional rotary tiller of which specific work ranged from 10-18 Ncm/cm^3(Kawamura 1980), the specific work for the disk tiller was considered much less especially for higher ground speed.

3.2 Disk side force and rear wheel reaction

The disk tiller inverts soil to the right side and as a result each disk receives reaction from the soil to the left side. When the reaction or the side force becomes excessive the disk tiller is swung to the left and as a result the front wheel is swung to the right making steering difficult. Fig. 5 shows a relationship between ground speed and side force. The side force increased with the increase of the ground speed. Also it was large for the case of deep tillage. Since each data point indicates the value of side force per single disk the actual side force act-

ing to the disk tiller(4 disks) receives 4 times more than these. Therefore the disk tiller working at the ground speed of 1.2 m/s tilling 23 cm deep for example receives about 1000 N x 4 = 4000 N of the side force acting in parallel with the tilling shaft. The actual side force being perpendicular to the travel direction can be obtained by correcting with gang angle of 31°. Therefore it becomes to be 4000 N x cos 31° = 3430 N.

Fig. 6 shows the rear wheel reaction in relation to the ground speed at the gang angles of 27° and 31°. The rear wheel angles of V-10° and T-15° in the figure indicate the angles of the wheel to the vertical and the traveling directions respectively. When the gang angle was changed to 27°, the actual angle to the traveling direction of the rear wheel which was attached to the gang frame increased by 4° to become 19°. As shown in the figure the rear wheel reaction was larger for the case of 19°(gang angle 27°) than the case of 15° of rear wheel angle. For the gang angle of 31° the maximum reaction obtained at the ground speed of 1.2 m/s was about 40 % of the expected maximum

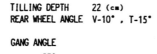

TILLING DEPTH 22 (cm)
REAR WHEEL ANGLE V-10°, T-15°

GANG ANGLE
● : 27°
■ : 31°

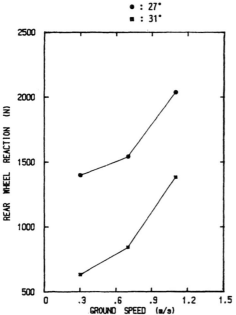

Fig. 6 Rear wheel reaction versus ground speed

tiller increased with the increase of the ground speed. This force must be kept within certain range not to affect tractor steering adversely.

4. The rear wheel reaction increased with the increase of ground speed. The reaction plays an important role to compensate some of the side force. The rest of the side force is supported by the three point linkage.

REFERENCES

Abernathy, G. H. 1986. Draft requirements of self-powered disk. ASAE Paper No.76-1021. ASAE, St. Joseph, MI 49085.
Gatzlaff, G. 1953. The forces acting on a power driven plow disc. Grundlagen der Landtechnik, 3(5): 36-41(N.I.A.E. Translation 6).
Gatzlaff, G. and W. Sohne 1959. Forces and power requirements of freely rotating and driven plow discs on hard, dry and clayey loam. Grundlagen der Landtechnik, 9(11): 40-52(N.I.A.E. Translation 106).
Hoki, M., T. H. Burkhardt, R. H. Wilkinson and T. Tanoue 1988. Study of PTO driven powered disk tiller. Trans. of ASAE 31 (5): 1355-1360.
Ishihara, A. 1983. New agricultural machinery, P. 117-118. Asakura Publishing Co., Tokyo.
Kawamura, N. 1980. Farm machinery, P. 65-73. Bun-eido Publishing Co., Tokyo.
Kawamura, N. 1985. Soil dynamics and its application to tillage machineries. International Conferences on Soil Dynamics Proceedings, June 17-19, 1985, Auburn, USA.
Sohne, W. 1963. Aspects of tillage. Canadian Journal of Agricultural Engineering, 5(1): 2-3, 8.
Tembo, S., T. H. Burkhardt, R. H. Wilkinson and M. Hoki 1986. Performance of the PTO power disk tiller. ASAE Paper No. 86-1010. ASAE, St. Joseph, MI 49085.
Young, P. E. 1976. A machine to increase productivity of tillage operations. Trans. of ASAE 19(5): 1055-1061.

side force of 3430 N shown above. Thus the difference of about 2000 N was considered to be supported by the three point linkage of the tractor. It is understood however that the rear wheel reaction is still effective to compensate for a considerable part of the side force thus improving tractor implement balancing. For the exact analysis of balancing more detailed data for the individual forces including specific points of action are needed and therefore is left for the future study.

4 CONCLUSIONS

The performance of the PTO driven disk tiller was evaluated and the following statements can be made in conclusion.

1. The increase of torque requirements were small with the increase of ground speed permitting high speed tillage under large tilling depth.

2. By increasing ground speed the specific work of the disk tiller can be reduced. This is one of the advantages over the conventional rotary tiller.

3. The side force acting to the disk

Land and Water Use, Dodd & Grace (eds), © 1989 Balkema, Rotterdam. ISBN 90 6191 980 0

Effects of different harrows on seedbed quality and crop yield

L.Henriksson
Swedish University of Agricultural Sciences, Uppsala, Sweden

ABSTRACT: Under Swedish conditions spring sowing is often followed by dry weather, and autumn sowing is done shortly before winter. Therefore, thoroughly prepared seedbeds are required for a uniform crop emergence, especially on clay soils. A suitable placement of the seeds is obtained if the drill coulters are pressed against a firm, level seedbed bottom at a suitable depth. The harrows should form such a seedbed bottom, as well as a fine tilth to prevent evaporation during germination.

Field experiments in cereals and sugarbeets showed that differences in construction of the harrows such as weight, design of levelling bars and tines, section width and tine spacing influenced the roughness of the seedbed bottom, the aggregate size distribution and the plant emergence and yield. Repeated harrowings were normally needed to obtain a good seedbed. The harrow construction may influence the optimal number of harrowings.

RESUME: En Suède les semis de printemps sont souvent suivies d´une période de sécheresse et les semis d´automne sont faites immédiatement avant l´hiver. C´est pourquoi des lits de semences minutieusement préparées sont nécessaires pour le développement d´une culture uniforme, particulièrement dans les terres argileuses. Un bon placement des grains est obtenu si les socs du semoir sont pressés contre une base du lit ferme et aplanie à un profondeur suffisant. Les herses doivent former une telle base du lit, aussi bien qu´une terre fine pour protéger contre évaporation pendant la germination.

Les essais des céréales et des betteraves ont montré que différences en construction de herses comme poids, formation des lames de nivellement et des dents, la largeur de la section et la distance entre les dents ont influencé la rugosité de la base du lit, l´étendue de les proportions entre la terre fine et les mottes, la levée simultanée et le rendement. Hersages répétés étaient normalement nécessaires pour obtenir un bon lit de semence. La construction de la herse peut influencer le nombre optimal des hersages.

ZUSAMMANFASSUNG: Unter schwedischen Verhältnissen ist die Frühjahrssaat oft von Trockenheit gefolgt und zur Keimung der Herbstsaat steht nur eine kurze Zeit vor dem Winter zur Verfügung, Sorgfältige Saatbeetbereitung ist deshalb eine Grundvoraussetzung für einen gleichmässigen Feldaufgang insbesondere auf Tonböden. Eine gute Placierung der Samen wird erhielt wenn die Drillscharen gegen eine feste, unbearbeitete, ebene Fläche auf geeigneter Tiefe gedrückt werden. Die ebene Fläche wird durch Eggen zubereitet sowie auch eine feinkrümlige Deckschicht, die gegen Wasserverdunstung schützt.

Feldversuche mit Getreide und Zuckerrüben haben gezeigt, dass Verschiedenheiten in der Konstruktion der Eggen wie Gewicht Form von Schleppschienen und Zinken, Breite von Sektionen und Streichabständen die Unebenheiten der Fläche, den Feinheitsgrad des Saatbeetes, den Feldaufgang sowie den Ertrag beeinflussen. Mehrmaliges Eggen ist normaler weise erforderlich um ein gutes Saatbeet zu erhalten. Die Konstruktion der Egge beeinflusst die optimale Anzahl der Feldbefahrungen.

1 INTRODUCTION

A careful seedbed preparation is important under Swedish conditions. In spring the potential evaporation is often high and the precipitation low, with water stress in the seedbed as a result. In autumn, fast germination is needed after the sowing while the soil is still warm. Later, the temperature will soon become too low as winter approaches. Soils with high clay contents are common and increase the requirements placed on the seedbeds and on the implements used to obtain them.

Spring-sown crops ought to be sown as soon as possible in order to utilize a short growing season. The starting-point in spring is a situation where the soil is saturated with water and a certain amount must evaporate before the tillage can start. On clay soils with a low capillary conductivity a sharp limit between the dry and the moist soil is formed. According to model experiments made by Håkansson and von Polgár (1984), and data received from a sampling investigation of 300 spring-sown fields made by Kritz (1983), it is possible even under dry weather conditions to obtain rapid and uniform germination. This can be achieved if the seeds are placed on a moderately compacted seedbed bottom with >5% (w/w) plant-available water and, in the case of cereals, they are covered with a 40-50 mm thick layer of mainly fine aggregates to minimize further evaporation. For proper placement of the seeds, the drill coulters are pressed against the seedbed bottom. Their work is improved by a uniform thickness of the seedbed, a level seedbed bottom and the absence of big clods.

Sugarbeets need shallower sowing depths (Sperlingsson 1981), which means that the need to protect against evaporation from the seedbed layer is greater. This can be compensated by a larger amount of fine aggregates, and by growing sugarbeets on soils with lower clay content, where the moisture content gradient by depth is flatter. In most cases, the sowing depth of sugarbeet drills is controlled from the soil surface. By displacing the top layer of dry, often cloddy soil in front of the drill to both sides of the row, the seeds can be placed in moist soil at a correct depth, even if the harrowing depth exceeds an optimal sowing depth.

Crust formation does not cause poor plant emergence as often as drought, and special measures are seldom considered. A rapid germination diminishes the risk of a crust developing before emergence. Breaking the crust or resowing may sometimes be necessary.

The demands on the seedbeds of winter cereals are in principle the same as those of the spring cereals. Rapid germination is important as the approach of winter soon stops plant development. Low temperatures sometimes delay the germination, and in northern parts of the areas growing winter cereals the winter sometimes starts before emergence has taken place. Germination can also be delayed if the soil profile is dry

after the previous crop. Normally there is sufficient water for good germination, even if we can never get the same good quality of the seedbeds on clay soils in autumn as in spring, when we can benefit from the frost action.

With the objective of seedbed preparation in mind, the harrows have to loosen and to make a fine tilth of the top layer, and to form a level seedbed bottom at a proper depth with regard to the crop and to the moisture conditions. The difficulty to make a shallow seedbed will increase if the soil surface is uneven. Levelling of the surface by a harrow after the ploughing in autumn is an advantage on well-structured soils but should be avoided on soils that easily slake and form a crust, as this makes the seedbed preparation more difficult (Henriksson, 1974).

A fine tilth is important to prevent evaporation during the germination period. Although some water losses cannot be avoided during repeated harrowings, the final effect of the improved seedbed quality is better germination. Earlier experiments (Henriksson, 1974) have shown that more harrowings are needed under dry weather conditions and on heavy soils.

This paper presents field experiments with recently developed types of harrows to study the influence of their different tools on seedbed qualities and yield. The experiments in sugarbeets were done in cooperation with the Swedish Sugar Company, and made up the final part of the work to develop a harrow especially for sugarbeets but suitable also for other crops.

2 Methods and materials

2.1 Experiments in cereals

At the experimental site at Ultuna, Uppsala, 13 field experiments in spring-sown barley, wheat or oats were carried out during 1980-85 and 5 in winter wheat during 1981-83. The clay content on the sites varied from 34% to 54% with a median value of 45%, and the organic matter content varied from 2.2% to 8.4% with a median value of 3.3%.

The spring cereals were sown in late April or in the first half of May and the winter wheat in September. During May and June, months critical for the establishment of the spring sown crops, the mean precipitation during the experimental period was 21 and 71 mm, respectively; corresponding normal values being 32 and 44 mm. The potential evapotranspiration was 86 and 95 mm, which is almost normal. During September and October, the germination period for winter wheat, the mean temperatures were 11.7° and 5.8° C, the precipitation 47 and 82 mm and the evapotranspiration 34 and 10 mm respectively. These values are nearly normal except for an excess of 33 mm of rain in October.

All the sites were mouldboard ploughed in autumn. In cereals a splitplot design was used with different harrows as main plots and different numbers of harrowings as subplots. In spring cereals, 1, 2 and 3 harrowings were tested. In winter wheat the numbers were varied due to the soil conditions. Low intensity tillage consisted of 1-3 harrowings, normal intensity of 2-4 and high intensity of 3-5. Repeated harrowings were done in diagonal directions which was made possible by headlands between the main plots. Driving speed was 5-7 km h⁻¹. The length of the plot was normally 25 m and the width equaled that of the harrow (3-5.5 m). Each treatment was replicated 3 or 4 times.

2.2 Experiments in sugarbeets

An attempt to develop a better harrow for sugarbeets started in 1980 in collaboration between the Division of Agricultural Engineering at the Swedish Sugar Company (Sockerbolaget), the Danish company Kongskilde A/S and the Division of Soil Tillage of the Swedish University of Agricultural Sciences. The project ended in 1983-84 with the testing of a new harrow, the Germinator, in 7 field experiments in the surroundings of Lund in southernmost Sweden. The Sugar Company did the ordinary field work whereas the Soil Tillage Division was responsible for the measurements of seedbed quality.

The soils used in these experiments are moraines with some flint stones, a clay content of 15-20% and an organic matter content of about 2%. The normal precipitation in this area in April, May and June is 37, 42 and 54 mm, respectively. In 1983 the sugarbeets were sown in the middle of May, in poor soil conditions after a rainy period, which was followed by a dry period. In 1984, the sowing could be done in the middle of April in good conditions. The results of the different harrows were assessed after 2 harrowings in spring on sites that were mouldboard ploughed and levelled in the autumn with a harrow.

2.3 Harrows used in the experiment

The different types of harrows used were

developed or improved during the experimental period and, apart from the Bastant SD, were not included in each experiment. The basic designs of the harrows are shown in Figures 1-4.

Fig.1. Wiberg Bastant SD.

Fig. 2. Kverneland levelling harrow.

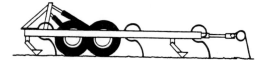

Fig. 3. Väderstad NZD.

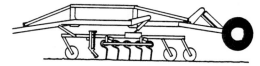

Fig. 4. Kongskilde Germinator.

Bastant SD (Fig.1) has s-tines in three rows and a tine spacing of 100 mm. It has a trail-harrow with two rows of long-finger tines. The harrow has a rigid levelling board in front and another spring-loaded board between the last two rows of s-tines. The weight is about 250 kg m⁻¹ width. A 3.5 m wide centre-section was used in all of the experiments.

The Kverneland levelling harrow (Fig.2) has two levelling boards in front equipped with scrapers underneath, oblique-angled to the direction of travel, which loosen the soil. The three rows of s-tines behind with a spacing of 100 mm should be adjusted not to work deeper than the scrapers. The Kverneland harrow is equipped with crumbler rollers. In the first two years a mounted and then a trailed model with a working width of 3.5 m and a weight of roughly 350 kg m⁻¹ was used.

The Väderstad NZD (Fig.3) harrow mainly used in the experiments was a 3.2 m centre-section with wheels in a tandem axle arrangement and a weight of approximately 300 kg m^{-1}. The s-tines are mounted in four rows with a spacing of 80 mm. The harrow has two spring-loaded levelling boards and a trail-harrow with a single row of long-finger tines. In the first experiments a model with one levelling board and a tine spacing of 68 mm was used.

Kongskilde Germinator (Fig.4) consists of 1 m wide sections. Each section is carried by one crumbler roller in the front and one at the rear. A levelling board is placed after the front crumbler. The sections have spring tines especially designed for the Germinator with a rake angle of nearly 90° and a spacing of 50 mm. When tilling, the hydraulic system is designed to divide the weight of the frame equally amongst all sections regardless of changes in the soil surface level. Trail harrows with working widths of 3 or 5 m and a weight of 300 kg m^{-1} were used. In the experiments in sugar-beets the Germinator was compared with a normal, mounted skid harrow with s-tines in three rows with a spacing of 100 mm and a weight of 170 kg m^{-1}.

2.4 Measurements after the tillage

Seedbed qualities were measured in some of the experiments. Harrowing depth was determined by pressing down a steel frame (area 0.25 m^2), collecting the cultivated, loose soil within it, measuring the volume in a bucket and calculating the depth assuming no change in bulk density (Kritz, 1983).

When the loose soil was removed, the roughness of the seedbed bottom within the frame was subjectively graded on a 1-5 scale, where 1 is a very smooth bottom and 5 a very rough.

After air drying the soil samples, the aggregate size distribution was determined in a rotary sieve with mesh sizes of 4, 8, 16 and 32 mm.

About two weeks after the first seedlings had emerged, the number of plants was counted on 2x0.25 m^2 per plot.

3 Results and discussion

In Table 1 the results from the experiments in cereals are summarized. All harrows are compared with Bastant SD which was used in all the experiments. In group A, compari-

Table 1. Effects of different harrows and number of harrowings on seedbed qualities, plant emergence and crop yield. Ne = Number of experiments, Hd = Harrowing depth mm, Sbr = Seedbed bottom roughness 1-5, Ag = Aggregate <4 mm %, Np = Number of plants per m^2, Gy = Grain yield kg ha^{-1} and Gm = Grain moisture content at harvest %.

Treatment	Ne	Hd	Sbr	Ag	Np	Gy	Gm
1. SPRING CEREALS							
A Bastant SD	12	44	2.9	46	436	4590	25.1
Kverneland		44	2.2*	48	452	4700	25.0
B Bastant SD	9	42	3.0	46	428	4600	25.2
Väderstad NZD		42	3.0	46	412	4660	25.5
Germinator		43	2.2*	48*	436	4800*	24.6
C 1 harrowing	13	40	2.8	41	417	4450	25.3
2 harrowings		40	2.2***	44*	434	4630**	24.4
3 harrowings		40	1.9***	46**	441	4680*	24.2
2. WINTER WHEAT							
D Bastant SD	5	39	2.3	17	355	5440	18.9
Kverneland		43	2.5	20	388	5310	19.1
E Low intensity	5	44	3.3	16	340	5360	19.1
Normal "		42	2.2	19**	374	5390	19.0
High "		41	2.1	21**	384	5410	19.8

*, **, *** Differences significant at 0.05-, 0.01- and 0.001-level respectively compared to the first treatment in the same group.

sons with Kverneland are made. Group B includes the experiments with Bastant SD, Väderstad NZD and the Germinator. Differences between harrows are based on the results of the main plots, i.e. mean values of the three different harrowing intensities. In group C, the effects of an increased number of harrowings are shown regardless of the type of harrow. Groups D and E show the corresponding results in winter wheat. The results in sugarbeets are given in Table 2.

3.1 Experiments in cereals

The Kverneland levelling harrow is characterized by its high weight and the oblique scrapers that cover the whole width. They can create a level seedbed bottom, especially in the spring. In autumn, big clods may be broken out of the furrow slice, and form considerable roughness in the bottom. The percentage of fine aggregates and the number of plants is somewhat higher after the tillage with the Kverneland harrow, but the differences as well as those of the yields are not significant. Studies in Norway (Berntsen, 1987) have shown similar results. The harrow has a rather high power requirement, stones can get caught between the scrapers and wet soil can stick to them. The harrow is best suited to heavy soils.

Väderstad NZD has a narrower tine spacing than Bastant SD, 80 mm instead of 100 mm. In some of the experiments, where the ploughed surface was even and soft, the Väderstad harrow made a smoother seedbed bottom, but the mean values do not differ. When the soil surface is uneven and hard, this effect of a minor difference in tine spacing between the two harrows is overshadowed by the movements of the harrow and the breaking of clods by single tines. There are no differences in aggregate size, emergence or yield in the comparison with Bastant SD and these two harrows can be considered to give equal working results.

Seedbed preparation with the Germinator has resulted in a more level seedbed bottom and a greater portion of fine aggregates. The narrow sections that follow the soil surface well, the rake angle of the tines and their narrow spacing, together with the crumbling effect of the levelling boards and the rollers, contribute to the effects. The better seedbed quality has resulted in a higher yield. These effects have been observed for all harrowing intensities.

In spring cereals, the number of harrowings has had a great influence on the quality of the seedbed. A smoother bottom, a

finer tilth and finally a higher yield, were obtained by increasing the number of runs. The differences in number of plants and in the moisture content in the grains at harvest are not significant. A lower moisture content is an effect of an even germination resulting in an even ripening of the crop. In practice, the optimal number of runs has to be considered from an economical point of view.

In winter wheat, the effects are similar except on yield. The amount of aggregates <4 mm is much smaller in these seedbeds, but enough for the germination as the moisture content is much higher. At the higher harrowing intensities, the number of plants is greater but not the yield. In winter cereals a low plant number can be compensated to a greater extent by more and bigger spikes than in spring cereals.

3.2 Experiments in sugarbeets

Table 2 shows the results of the experiments in sugarbeets.

Table 2. Effects of an S-tine harrow and Kongskilde Germinator on seedbed qualities and yield of sugarbeets. Hd=Harrowing depth mm, Sbr=Seedbed bottom roughness 1-5, Ag=Aggregate <4 mm %, Np=Number of plants thousands ha^{-1}, Sy=Sugar yield kg ha^{-1}.

Variable	No of comparisons	S-tine harrow	Germinator
Hd	4	31	34
Sbr	4	2.8	1.8
Ag	4	46	50
Np	7	82.7	88.0*
Sy	5	8890	9160*

* Difference significant at 0.05-level

A shallower harrowing depth was used, 31-34 mm instead of 39-44 mm in the cereals. This is in accordance with the concepts in Sweden as regards good seedbeds for these crops. At shallow working depths there is a high requirement placed on the implement to follow the soil surface and to till the soil not only in the higher parts but also in the lower. Calculations of the coefficient of the variation of the harrowing depth showed a lower value for the Germinator. This is of importance for a good placement of the seed when the depth of the drill coulter is controlled from the surface. In agreement with the experience with cereals, harrowing with the Germinator in sugarbeets resulted in a more level seedbed bottom, more fine aggregates, more plants ha^{-1} and a higher yield in comparison with an s-tine harrow.

4 Conclusions

Carefully prepared seedbeds are important for good plant emergence and yield, especially on clay soils under dry weather conditions.

The quality of the seedbed is influenced by the design and weight of the harrows. The shape and spacing of the tines are of importance for a proper loosening of the soil and for producing a level seedbed bottom. Clods are mainly broken by pressure between a tool and the soil surface. Levelling boards, crumble rollers and tractor and implement wheels operate in this way.

In most cases, repeated harrowings in different directions are needed to obtain a good seedbed.

5 REFERENCES

Berntsen, R. 1987. Tillaging av såbed med trekredskaper. Norsk landbruksforsking, Supplement No. 2. Ås. Summary in English.

Håkansson, I. & von Polgár, J. 1984. Experiments on the effects of seedbed characteristics on seedling emergence in a dry weather situation. Soil Tillage Res. 4:115–135.

Henriksson, L. 1974. Studier av några jordbearbetningsredskaps arbetssätt och arbetsresultat. Rapporter från jordbearbetningsavdelningen, nr 38. Sveriges Lantbruksuniversitet, Uppsala. Summary in English.

Kritz, G. 1983. Såbäddar för vårstråsäd. En stickprovsundersökning. Rapporter från jordbearbetningsavdelningen, nr 65. Sveriges Lantbruksuniversitet. Uppsala. Summary In English.

Sperlingsson, C. 1981. The influence of the seedbed soil physical environment on seedling growth and establishment. Institut international de recherches betteravieres. 44th winter congress, Proceedings. 59–77. Bruxelles.

Land and Water Use, Dodd & Grace (eds), © 1989 Balkema, Rotterdam. ISBN 90 6191 980 0

Low input ridge tillage systems

R.Reeder

Ohio State University, Columbus, Ohio, USA

ABSTRACT: Results of ridge tillage research in Ohio, Indiana and Iowa and observations from several private farms in Ohio, Indiana and Illinois are presented. The ridge till system of soil cultivation and seedbed preparation has generally improved yields, especially on poorly drained, compacted clay soils. Soil structure and earthworm populations have improved. Although fewer than 2 percent of the corn-soybean farmers in the north central region of the USA use this conservation tillage practice, it has proven to be a good system for controlling soil erosion and reducing the cultural inputs of inorganic fertilizers and pesticides. Compared to tillage systems based on a plow, ridge tillage requires less machinery investment and reduces operating and labor costs. Farmers report overall savings of about $50 per hectare. Recent design improvements in ridge till planters and cultivators are also discussed.

RESUMÉ: Résultats de la recherche du "Ridge Tillage" (= labourage de sillon) en Ohio, Indiana et Iowa et plusieurs observations en Ohio, Indiana et Illinois sont présentées. Le système "Ridge Tillage" -culture du sol et préparation du semis- a generalement amélioré les rendements, specialement pour des sols argileux compactés avec une drainage pouvre. Structure de la terre et les populations des verres de terre sont améliorées. Quoique moins de 2 pourcent des cultivateurs de maïs et de soya dans la region Nord-Central des Etats Unis utilisent la "Rigde Tillage", ce système a prouvé d'etre un bon système pour régler d'erosion de terre et rèduire à des puissances d'alimentations agricoles -les engrais inorganiques et les pesticides. En comparaison, "Ridge Tillage" demande moins investissement de machine que le système du labourage avec la charrue et réduit le coût d'operation et le coût de labour. Cultivateurs reportent épargne globale de $50 par hectare. Améliorations récentes concernant la dessein des planters et des extirpateurs sont aussi discutées.

ZUSAMMENFASSUNG: Resultaten der Untersuchung des "Ridge Tillage" Systems (= Furchenrain-Ackerbau) in Ohio, Indiana und Iowa und Beobachtungen von vieler Bauernhöfe in Ohio, Indiana und Illinois werden präsentiert. Das "Ridge Tillage" System von Bodencultivation und Saatvorbereitung hat im allgemeinem die Ernte erhöhert, im besonderem auf schwach drainierten, compactierten Lehmboden. Der Bodenstructur und die Erdwurm Population hat sich verbessert. Obwohl weiniger als 2 prozent der Bauern von Maisschnaps und Sojabohnen in der Nordcentrale Region der Vereinigten Staten diese Schutzpraxis -"Ridge Tillage"- gebrauchen, beweißt es ein gutes System zu sein für Controllierung der Bodenabwaschung und Verminderung der agrarischer Leistungsaufname wie anorganischen Kunstdünger und pesticiden. Im vergleich mit intensiven Ackerbausystemen, braucht "Ridge Tillage" weiniger Machineninvestirungen und verringert es Operations- und Arbeitskosten. Bauern rapportieren gesammt Ersparnisse von ungefär $50 pro Hektar. Kürzliche Planverbesserungen in "Ridge Tillage" Pflanzmachinen und Cultivatoren werden auch diskutiert.

1 INTRODUCTION

Ridge tillage is a reduced tillage system where the crop, usually corn or soybeans in the USA, is grown on "permanent" ridges. The only tillage tool is a heavy-duty cultivator which provides or aids in weed control and rebuilds the ridge each year in the standing crop. Rows are typically 76 cm apart with one row per ridge. The ridges are 12 to 20 cm higher than row middles so they dry out and warm up faster for earlier planting in spring.

The planter must have a tool in front of the planting unit to remove 2 to 4 cm of soil to leave a smooth residue-free seedbed. In addition to the cultivator and planter, most ridge till farms have a chemical sprayer, fertilizer applicator, a tractor sized to pull the cultivator, and a combine for harvesting. Some or all herbicides, insecticides and fertilizers are applied during planting and/or cultivation with attachments on those implements. Farmers wanting to minimize or eliminate herbicides may use a rotary hoe, a high speed, shallow tillage tool which can uproot small or emerging weeds without undue harm to the small crop.

Wheels must be spaced accurately for successful ridge farming. Tires must never run on the row. With 76-cm row spacing, tractor wheels will be typically 1.52 m on centers and combine wheels 3.04 m. Tires must be no more than about 55 cm wide to prevent mashing the side of the ridge. Large combines require dual wheels which are separated so they run in adjacent row middles.

There is no compaction under the rows and in some row middles. This controlled traffic has several benefits (Taylor, 1983) and is perceived to be the major advantage offered by ridge tillage. In the US, 92% of the ridge tilled land is in the twelve states comprising the North Central region. (The twelve states, listed in descending order of land in ridge tillage, are Nebraska, Minnesota, Iowa, Illinois, Indiana, Kansas, South Dakota, Ohio, Missouri, Wisconsin, North Dakota, and Michigan.) The 880,000 ha of ridge till is only 1.3% of the cropland in the region. By comparison the region has 3 million ha in no-till. Nationally ridge tillage land has increased six years in a row; 1988 data show a 15% increase over 1987 according to the 1988 Survey of Conservation Tillage Practices (Anonymous, 1988).

2 RESEARCH RESULTS AND ECONOMIC ANALYSIS

Researchers have found that ridge tillage compares favorably with other tillage systems on most soils. Reported yields have usually been nearly equal to or above those for systems depending on a moldboard plow, chisel plow or disk for primary tillage. On well drained sloping soils, no-till yields typically equal or exceed ridge till. Because of lower costs for machinery and chemicals, ridge till is usually the best economic choice.

Tillage studies have been conducted since 1982 on a Hoytville silty clay in northwest Ohio and since 1986 on Crosby silt loam and Kokomo silty clay loam in southwestern Ohio by Eckert (1987 and 1989). On the poorly drained lake plain Hoytville soil, corn planted on ridges has produced yields approximately equal to corn planted following intensive (plow) tillage for both continuous corn and corn-soybean rotations, with or without tile drainage (0.13 mg/ha advantage for ridges). Soybeans in rotation with corn have yielded the same for both tillage systems in 76-cm rows. Soybeans planted in 18-cm rows after plowing have yielded about 0.2 mg/ha higher.

For both Crosby and Kokomo soils ridge tillage has not produced corn yields as high as those for moldboard plow, chisel plow, or no-till plots. (0.34 to 1.7 mg/ha advantage for plowing) Soybeans in rotation with corn have yielded about the same in all tillage systems.

I established a 12 ha tillage demonstration for corn in a field consisting of both Crosby and

Kokomo soils at the same southwestern Ohio location in 1987. (Ridges were formed in 1986.) Yields in ridge tillage have equalled moldboard and chisel plow systems.

On the same northwest Ohio research farm above, Fausey (1987 and 1989) initiated a study in 1984 to determine the effects, if any, of ridge tillage on a soil with a compacted layer resulting from an extended history of annual grain production. The use of ridge tillage and controlled traffic was intended to illustrate whether the soil can recover permeability and drainage on its own if it is not abused by random, heavy traffic. Yields for 5 years of ridge till (1984-1988) in a corn/soybean rotation were compared to yields for the previous 3 years of continuous corn in intensive tillage on the same plots. Corn yields doubled and the planting date was 12 days earlier with ridge till.

The Agricultural Research Service of the US Department of Agriculture does erosion and other research on watersheds at Treynor, in southwestern Iowa. The terrain is rolling. The soil is wind-blown loess deposited over glacial till. Typically, the loess is 5 to 30 m thick. Two research watersheds, one farmed with intensive tillage, the other ridge till, are quite similar although they are 5 km apart. Slopes average about 10%. Each watershed has about 40 ha and has been in continuous corn since 1972. Contour farming practices are used. The average sediment yield (soil loss) from 1972 to 1987, measured at the bottom of each field, was 14.2 t/ha/yr for intensive tillage but only 1.8 t/ha/yr for ridge tillage (Alberts, 1989). The estimated gross erosion for the watersheds was 26 and 2.2 t/ha/yr, respectively, a 90% reduction for ridge till. Surface runoff from ridge tillage was 53% less than from the intensively tilled watershed. Crop yields also showed a strong advantage for ridge tillage. The 16-yr mean average was 7.3 mg/ha for plowing and 8.2 mg/ha, or 12% higher, for ridges. The average yield was also calculated for 4 dry years when corn suffered from moisture stress. For those years the intensively tilled corn averaged 4.2 mg/ha and ridged corn yielded 5.8 mg/ha, a 38% increase for ridge till.

In Indiana, researchers at Purdue University have summarized yield relationships for six tillage systems based on twenty years of work on Purdue farms (Doster, et al, 1987). A yield index was prepared for soils in five major groups for continuous corn and for corn and soybeans in rotation. In general, ridge till yield potential ranged from 2% lower than fall plow to 10% higher. Ridge till yield potential was equal to or 2% lower than fall plow on dark, poorly drained silty clay loams to clays, 0-2% slope, (Group I), such as Kokomo. On soils, such as Crosby, which are light (low organic matter), somewhat poorly drained silt loams to silty clay loams, nearly level to 4% slope, (Group II), yield potential with ridges was the same to 2% higher than fall plow. For light, well and moderately well drained sandy loams to silt loams on 3 to 6% slopes, (Group III), ridge till yield potential was 10% higher. For similar soils on slopes greater than 6%, (Group IV), ridge till yield potential was 3 to 9% higher than fall plow. On light, somewhat poorly and poorly drained silt loams, nearly level to gently sloping, overlying very slowly impermeable fragipan-like pans, (Group V), ridge till potential yields were the same to 3% higher.

By comparison, no-till had a yield potential, compared to fall plow, 2 to 10% lower for Group I soils, the same to 5% lower on Group II, 10 to 12% higher on Groups III and IV, and 5 to 10% higher for Group V.

The Purdue researchers also completed an economic analysis for a hypothetical 300 ha farm on each soil type. On a basis of returns per ha, ridge tillage was better than fall plowing for all situations. Specifically, for a corn/soybean rotation on soil Groups I to V, the advantage, in

$/ha, was 4, 5, 11, 7 and 6, respectively. For continuous corn the advantage, in the same order, was 2, 7, 12, 10, and 4. No-till had an economic disadvantage compared to ridge till on soil Groups I and II but had approximately the same hypothetical return for Groups III, IV and V. For their analysis the Purdue researchers assumed that herbicides cost $3/ha more for ridges and $4.50/ha more for no-till compared to intensive tillage systems. In fact, most ridge till farmers reduce their use of herbicides substantially, saving much more than the cost of an extra one or two passes with the cultivator.

3 FARMER EXPERIENCES

Most farmers have at least maintained crop yields after switching to ridge tillage. Production costs are usually reduced compared to intensive tillage systems. There is less investment in tractors and machinery. Fewer hours of use means lower costs for repairs, maintenance and fuel. Less labor is required. Herbicide costs are often reduced by two-thirds by banding the chemicals over the row and using cultivation to control weeds in row middles. (The cost of cultivation will partially offset the reduction in herbicides, but usually results in a net savings.) Fertilizer inputs may be reduced by split applications of nitrogen which provide the nutrient when needed instead of applying the full amount in the spring.

The environment benefits from a reduction in the application of herbicides and nitrogen fertilizer. Crop residue left on the surface minimizes or eliminates runoff which carries chemicals into streams. Split applications of nitrogen at cultivation time allow rates to be reduced if crop growth and soil moisture reserves indicate a reduced potential yield. Excess nitrogen in the soil is often carried by the drainage system into waterways.

A survey of ridge till farmers in Ohio, Indiana and Illinois by R. Donald Moore (1987) revealed a positive picture for ridge tillage. The 118 farmers who responded to the mail survey had an average of 150 ha of corn and 130 ha of soybeans. Eleven had started ridge tilling before 1980; for the remainder, the mean starting year was 1983. Most had previously used either a moldboard plow or chisel plow as their primary tillage tool. Reported yields for the 1985 and 1986 crop years were 10.1 mg/ha for corn and 3.0 mg/ha for soybeans. One farmer said his yields decreased, 42 reported no change, and 72 reported that yields increased an average of 1.1 mg/ha for corn and 0.34 mg/ha for soybeans. Every farmer claimed production costs decreased. The average cost reduction was $59/ha for corn and $50/ha for soybeans. Twenty-four of the farmers indicated their previous soybean system was "solid seeding" (soybeans are planted with a drill in rows 18 cm apart and weeds are controlled with herbicides). No one reported a yield decrease, 10 had no change, and the other 14 claimed increased yields of 0.42 mg/ha.

A farmer in western Ohio, who has used ridge tillage for ten years, experimented with a rotary hoe for weed control instead of herbicides banded over the row in 1988 (McNelly, 1989). Yields were the same for each treatment for both corn and soybeans. Visual observation indicated equal weed control. Cost of herbicides was $18/ha for soybeans and $13/ha for corn. There was no cost figure reported for hoeing but a reasonable estimate would be $1/ha per pass over the field. Each crop was hoed three times.

Since 1988 was a dry year farmers had no difficulty scheduling field operations. In the infrequent seasons when rains prevent use of a rotary hoe or cultivator, post-emergence herbicides can be used for weed control. If the final ridge forming cultivation cannot be completed before the crop grows too tall, this operation can be finished after fall harvest. The disadvantage of ridging after harvest is that much crop residue is covered, making the soil more vulnerable to erosion.

Ridge till farmers have reduced their use of fertilizer without affecting yield or soil test measurements. McNelly applies 30 to 50% as much phosphorous and potassium as he did ten years ago. He has not changed nitrogen rates significantly. A farmer in north central Ohio has also reduced fertilizer use (Rickenbacker, 1989). Typically he applies half of the estimated amount of nitrogen needed for the season prior to planting corn and the other half during one or two cultivations. In 1988 he reduced the amount applied at cultivation to about 10% because of expected low yields, saving about $60/ha.

Soil structure, tilth, and air and water permeability usually improve greatly after intensively tilled fields have been switched to ridge till for 5 to 10 years. A firm soil layer forms in the trafficked row middles; but in row middles with no tire traffic and under the rows the soil is usually well structured to provide an ideal environment for root growth (Fausey, 1989). McNelly, Rickenbacker and most other ridge till farmers have reported increases in earthworm population, sometimes going from virtually no earthworms to over two million per ha.

4 RIDGE TILLAGE DEMONSTRATION

A demonstration of low input ridge tillage was started in 1988 on an exhibition farm operated by Ohio State University in central Ohio. A total of 70 ha of continuous corn and corn/soybean rotation in ridge tillage is expected to provide ample opportunities for researchers to show how reduced inputs affect crop yields, weed growth, earthworm and insect activity, soil physical and chemical properties, and economics over a period of several years. The farm has about 160 ha in intensive tillage and 10 ha no-till. Each September 100,000 farmers come to the farm to attend a "Farm Science Review" which features demonstrations of new harvesting, tillage and planting machinery, educational exhibits, and programs presented by university faculty.

5 MACHINERY INNOVATIONS

Many machine improvements and innovative features have been developed since ridge tillage research began about 30 years ago. New planters space seeds and place them at the desired depth consistently under a variety of soil and crop residue conditions to provide uniform stands. In addition to planters designed for ridge tillage, companies manufacture attachments for no-till planters which remove soil and residue from the top of the ridge ahead of the planting unit. Several designs of "paired wheels" are available to keep the planter centered on top of the ridges. Planters and cultivators are available commercially in 4 row to 16 row widths. Bigger machines fold for transport. Lift-assist wheels can be attached to machines so slightly smaller (and lighter) tractors can handle 3-point hitch implements.

Cultivators have wide sweeps to undercut weeds between rows and angled disc blades remove weeds close to the row without damaging the plants. New cultivator sweep designs and crop shields prevent loose soil from covering the small crop during the first cultivation, even at high speed (10 km/hr). Adjustable ridging wings, or ridgers, form the ridge in the desired shape against the standing crop during the final cultivation. Wide gauge wheel provide accurate depth control. Component design and placement minimize plugging by residue of the previous corn crop. Attachments allow accurate placement of anhydrous ammonia, liquid nitrogen, or spot spraying of weeds during cultivation.

New cultivator guidance systems make cultivation more accurate, faster and less stressful for the tractor operator. They sense the location of the row and keep the cultivator centered automatically.

As a result cultivator components can be placed closer to each row and narrower bands of herbicides may be used.

Improved rotary hoe designs minimize plugging in high residue conditions making rotary hoes an acceptable alternative to banded herbicides on many farms.

REFERENCES

Alberts, Eugene 1989. Personal communication. USDA-ARS, Columbia, Missouri, USA

Anonymous 1988. National Survey of Conservation Tillage Practices. West Lafayette, Indiana, USA: The Conservation Technology Information Center

Doster, D.H., D.R.Griffith, J.V.Mannering and S.D.Parsons 1987. Economic returns from alternative corn and soybean tillage systems in Indiana. West Lafayette: Purdue University.

Eckert, D.J. 1987. Evaluation of ridge planting systems on a poorly drained lake plain soil. Journal of Soil and Water Conservation. 42:208-211.

Eckert, D.J. 1989. Personal communication. Ohio State University, Columbus, Ohio, USA

Fausey, N.R. 1987. Impact of cultural practices on drainage of clay soils. St. Joseph, Michigan, USA. ASAE Proceedings of the Fifth National Drainage Symposium, p.288-292.

Fausey, N.R. 1989. Personal communication. USDA-ARS, Columbus, Ohio.

McNelly, D. 1989. Personal communication. Farmer, Arcanum, Ohio.

Moore, R.D. 1987. Results of Ohio, Indiana and Illinois ridge tillage survey. Columbus: Ohio Cooperative Extension Service.

Rickenbacker, W. 1989. Personal communication. Farmer, Forest, Ohio.

Taylor, J.H. 1983. Benefits of permanent traffic lanes in a controlled traffic crop production system. Soil & Tillage Research 3: 385-395.

Land and Water Use, Dodd & Grace (eds), © 1989 Balkema, Rotterdam. ISBN 90 6191 980 0

Energy-saving ploughing with variable width ploughs

J.I.Jóri & S.Soós
National Institute of Agricultural Engineering, Gödöllő, Hungary

ABSTRACT: Ploughs are difficult to operate profitably. There are a
number of methods to decrease fuel consumption required for ploughing,
one of which is to match the tractor and the plough. The optimum utilisation of the tractor's engine capacity, that would imply the minimaliza-
tion of fuel consumption, is difficult and costly to achieve under
variable operating conditions using traditional methods.
In Hungary, the American idea to change the working width of the plough
bodies has been tested on three types of soil /sandy, clay loam and
clay/ with three types of ploughs /JD-2800, Kverneland-CC, Rába-IH-10-
-735-6TJ/. Both the laboratory and the fields tests have shown that the
working width can be changed within a certain range /300-500 mm/ without any major difference in the quality of work done. Since the increasing of the working width involves concurrent reduction in the
energy requirement, the new solution offers extensive advantages. By
using ploughs with variable working width, agriculturists may accomod-
ate different soil varieties, changes in soil conditions as well as
tractor power and changes in their conditions and, as a result, overall
energy consumption can be reduced.

RÉSUMÉ: L'éxploitation des groupes de machines à labourer est une tâche
difficile. On peut avoir recours à plusieurs sortes de solutions pour la
diminution de la matière combustible nécessaire au labourage, dont
l'une est l'adaptation énergétique de la charrue-tracteur. La charge
optimale de la puissance du moteur du tracteur qui signifie la minimi-
sation de la consommation de combustible entre circonstances d'exploitation diveres, est difficile est coûteuse avec les solutions traditionnelles. La méthode américaine développée pour la modification de la
largeur de coupe variable a été effectuée, dans les essais hongrois sur
trois sortes de sols /sable, argileux et terre a adobes/ avec trois
éspèces de charrue /JD-2800, Kverneland-CC, Rába-IH-10-735-61J/. Les
essais en laboratoire et en champ ont prouvé que la modification de la
largeur de coupe peut être appliquée sans endommagement dans la
qualité du travail dans des limites données /300 a 500 mm/. Etant
dônné que l'augmentation de la largeur de coupe entraine une diminution
en énergie, c'est pourquoi la nouvelle solution résulte des avantages
considérables. Avec les charrues à largeur de coupe variable on peut
s'adapter au changement de l'éspèce et de l'état des sols, au rendement
du tracteur et aux changements d'état, et en résultat, diminuer l'utilisation en énergie.

ZUSAMMENFASSUNG: Der wirtschaftliche Betrieb der Pflugmaschinengruppen
ist eine schwierige Aufgabe. Die Herabstzung des Treibstoffverbrauches
bei der Ackerbestellung kann auf verschiedenen Wegen erreicht werden,
eine der wichtigsten Verfahren ist die energetische Abstimmung zwischen
Ackerschlepper und Pflug. Die optimale Auslastung der Motorleistung des

Schleppers, die eine Minimierung des Treibstoffverbrauches bedeutet,
ist unter veränderlichen Betriebsbedingungen mit den herkömmlichen
Methoden schwerfällig und kostspielig. Die Untersuchung der unter Anderung der Griffbreite des Pfluges entwickelte amerikanische Methode
wurde in Ungarn an drei Bodentypen /Sand, lehmiger Ton und Lehm/ mit
drei verschiedenen Pflügen /JD-2800, Kverneland-CC, Raba-IH-10-735-6TJ/
durchgeführt. Die Untersuchungen im Laboratorium und auf dem Acker
haben bewiesen, dass die Anderung der Griffbreite zwischen bestimmten
Grenzen /300-500 mm/ ohne wesentliche Verschlechterung der Arbeitsqualität anwendbar ist. Da die Erhöhung der Griffbreite mit der Verminderung des Energiebedarfes einhergeht, ergibt das neue Verfahren weittra-
gende Vorteile. Mit Hilfe von Pflügen mit veränderlicher Griffbreite
kann man sich an Anderungen der Bodenqualität und des Bodenzustandes
anpassen, und als Gesamtergebnis den Energieverbrauch herabsetzen.

1 INTRODUCTION

There are a number of factors
that affect the volume of fuel
used for ploughing. Some of them
are controllable while in operation, some of them are not. Considering the machine-lines of
Hungary's large-scale farms, one
our most important choices perhaps is to match the tractor and
the plough, since optimum load in
respect of the tractor engine may
be achieved by using a given working width under certain conditions. If deviation from this in
either direction occurs, capacity
utilisation would deteriorate.
One can realise quite easily, considering the typical size of
arable land in Hungary's large-
scale farms /5,000-10,000 ha/,
that the conditions for ploughing
cannot be taken as constant. The
type of soil changes within the
farm plots /often from light to
heavy soil/, but subject to
moisture content, the tractive
resistance is varied too within
the same plot. Changes in operating conditions call for changes
in the optimum working width as
well. For traditional ploughs,
changes in the working width imply
changes in the number of plough
bodies. This may be achieved by
replacing the whole plough, by
mounting or removing or uplifting
or lowering plough bodies. The
first option is investment-inten-
sive while the other two imply
extra labour and costs. However,
the method developed in the 1970s
in the USA to adjust the cutting
width of the plough bodies
appeared to be an ideal solution
to the problem of changing the

working width, ie. matching the
tractor and the plough. Wide
ranging tests have been initiated to assess the adaptability of
the new method under Hungarian
conditions and to identify the
effects of changing the cutting
widths on work quality and energy-
related aspects. This paper includes the results of our test
series on the three most frequest
soil varieties in Hungary /such
as sandy, clay loam and clay/.

2 TEST METHODS AND CONDITIONS

The series of tests were started
in 1981 on clay loam, which is the
most characteristic type of soil
in Hungary, using the Type JD-2800
ploughs with the widest adjustment
range. Having ascertained that
ploughs with variable working
widths have promising prospects,
control tests were conducted first
in 1986 with Kverneland-CC ploughs
on sandy soil, then in 1988 on
clay soil with Hungarian-made
RABA-IH-10-735-6TJ ploughs.
/1-3 figures/

Before starting the tests, the
major technical parameters of the
ploughs were specified /as in
Table 1/ and the characteristic
curves of the plough bodies were
identified for their pulverizing
and turning effect under laboratory conditions /Figure 4/. In the
course of the field tests the work
quality parameters of the ploughs
were defined at different working
widths and at the highests speed
possible under the given conditions, while the energy-related

parameters were defined at three distinct speeds. The field conditions of the tests are listed in Table 2.

3 TEST RESULTS

Plough performance was characterized in general by the pulverizing and turning effects, although in the case of the JD plough, that may be adjusted within a wider range, furrow positioning /ie. open furrow width and the furrow angle/ was also analysed. On top of that, the variation coefficient of working depth and width data were calculated too, to characterize the running performance of plough. /Tables 3 through 5/

Analysing the consequences of the changes in cutting width it was found that plough performance was acceptable in each and every pattern. Of course, the best indices were received in the range closest to the optimum cutting width, ie. between 350-450 mm. In some cases, certain problems have arisen at the extreme cutting width values. On sandy soils and clay loams, the turning effect left a bit to be desired at a cutting width of 300 mm. On clay loams, the pulverizing effect began to deteriorate at cutting widths above 500 mm and soil fallen in could also be noted. On heavy soils, where the clod-making is usual, the detrimental effect of increasing the cutting width was evident. The rate of clods on the surface with a diameter over 5 cm rose from 25.3 % to 43.1 % as a result of expanding the cutting width from 300 mm to 500 mm.

The energy characteristics of the ploughs were defined in a speed range of 5 to 10 km.h^{-1}. During the tests both the tractive resistance and the fuel consumption were monitored. /Fuel consumption however, was not measured on clay soil./ The effect of changing the cutting width is demonstrated by specific energy characteristics /Figures 5 through 7/. /The curves have been defined under identical speed gears, at the operating speed of 7.0-7.5 km.h^{-1}./

The tractive resistance per soil cross section ploughed, the power requirement as well as the changes in fuel consumption required to plough a unit of land show that by increasing the cutting width the energy requirement is reduced. Of course, the rate of reduction is affected by the soil type, the type of tool and a number of other factors.

4 CONCLUSIONS

With respect to the results of tests, conducted with plows of variable working width on various types of soil, the following conclusions may be drawn:

- the tested plough bodies can be used to vary cutting width within a predetermined range,

- based on work quality considerations, the range to be used under Hungarian conditions is 300 to 500 mm,

- as a result of increasing the cutting width, specific energy consumption decreases /the rate of reduction is 10-30 % subject to the soil conditions/,

- with respect to work quality and energy characteristics, on sandy and loam soil the upper end of the cutting width range /400-500 mm/ was found to be more favourable, while on clay soil the lower end of the range /300-400 mm/ was found to be better,

- by using ploughs with variable working width
  - varied soil varieties, and
  - changes in soil conditions and
  - tractor power and condition may be accomodated,
  - optimum load can be ensured for the tractor at all times, and
  - energy, ie. fuel, can be saved as a result of optimum load on the tractor.

REFERENCES

Buckingham, F. 1975. Update on tillage research. Implement and Tractor. No.7.

Cooksley, J. 1984. Farm verdict on the Kverneland vari-width plough. Arable Farming Ipswich. No.3.

Jóri J.I.-Soós S. 1983. A változtatható fogásszélességü ekék magyarországi tapasztalatai. /Field experiment with vari-width ploughs in Hungary/ Agrotechnicar, Beograd, 9/XIX.

Jóri J.I.-Soós S. 1983. Változtatható fogásszélességü ekék. /Vari-width ploughs/. Müszaki fejlesztési eredmények, Agroinform, Budapest No.205.

Jóri J.I.-Soós S. 1984. Változtatható fogásszélességü ekék./Test with vari-width ploughs/ Jármüvek, Mezögazdasági Gépek Budapest. 31.évf. No.10.

Newhall, M. 1983. "Inventor's Report" On succesful farm inventions, Farm Show. Vol.7. No.6.

Pearce, A. 1982. On-the-move plough adjustment, Power Farming No.8.

Reaves, C.A.-Schafer, R.L. 1975. Force versus width of cut for moldboard bottom. Transactions of the ASAE.

Richardson, C.D.-Patterson.O.E. 1983. Evaluation of Kverneland ploughs . Contract Report of NIAE.

Soós S. 1986. Analysis of the tractor-plough matching related to the width variation of bottom. University of Agricultural Sciences, Gödöllö.M.Sc. Report.

Stroppel, A.-Shafer, W, 1981. Maximierung der Schlagkraft beim Pflügen durch stufenlose Arbeitsbreitenverstellung. Grundlagen der Landtechnik, Düsseldorf. 31/6.

Wainwright, R.P. et.al. 1983. A variable approach-angle Moldboard plow. ASAE No.2.

Table 1   The major technical data of the ploughs

Types		John Deere 2800	Kverneland-CC	RÁBA-IH-10-735-6TJ
– length	mm	9800	9600	9100
– width	mm	3600	3900	3800
– height	mm	1800	1500	1800
Working width	m	2,1-4,2	1,8-3,0	1,8-3,0
– no. of plough bodies,	pcs	7	6	6
– clearance fore and aft.	mm	975	1000	1000
– cutting width	mm	300-600	300-500	300-500
– clearance under frame	mm	870	800	870
– weight of plough	kg	2400	2300	2150

Table 2   Test conditions

Equipment tested	RÁBA-300 + Kverneland-CC		RÁBA-250 + John Deere 2800		RÁBA-250 + RÁBA-IH-10-735-6TJ	
Test location	Bakonyszombathely		Nádudvar		Mosonmagyaróvár	
Date of test	06.09.1986		18.08.1981		08.08.1988	
Test site features	flat, stubbled grain field		flat, stubbled grain field		flat, stubbled grain field	
Sampling depth cm	5-10	20-25	5-10	20-25	5-10	20-25
Soil moisture /W/W/ %	9,4-10,5	10,5-11,7	15,6-17,0	19,8-21,1	21,2-21,4	18,3-20,7
Soil density g.cm⁻³	1,36-1,41	1,42-1,60	1,11-1,24	1,37-1,50	1,31-1,44	1,35-1,54
Physical type of soil	sandy		clay loam		clay	

Table 3 Work quality features of the RÁBA-IH-10-735 plough

Cutting width	mm	300	500
Working speed	km.h^{-1}	8,2	6,4
Average working depth	cm	25,30	25,55
Variation	cm	1,45	1,47
Variation coefficient	%	5,75	5,75
Average working width	cm	179,95	325,10
Variation	cm	4,55	4,78
Variation coefficient	%	2,53	1,47
Ratio of clods over 5 om	%	25,3	43,1
Turning effect	%	100	100

Table 4 Work quality features of the Kverneland plough

Cutting width	mm	300	400	500
Working speed	km.h^{-1}	8,8	9,0	9,0
Average working depth	cm	24,6	23,6	24,8
Variation	cm	1,05	1,19	1,01
Variation coefficient	%	4,28	5,05	4,05
Average working width	cm	208,6	291,45	349,3
Variation	cm	6,15	10,58	4,64
Variation coefficient	%	2,95	3,63	1,33
Furrow angle	deg.	45,2	40,7	39,5
Ratio of clods over 5 cm	%	8,85	9,55	13,0
Turning effect	%	90,0	100	100

Table 5 Work quality features of the JD-2800 plough

Cutting width	mm	305	350	400	450	500	600
Working speed	km.h^{-1}	9,7	9,2	9,2	8,7	7,0	7,0
Average working depth	cm	23,6	23,6	22,7	24,9	24,5	23,3
Variation	cm	1,40	1,87	1,24	0,94	1,17	1,21
Variation coefficient	%	5,05	7,93	5,48	3,79	4,76	5,20
Average working width	cm	207,1	239,5	262,6	316,0	337,4	419,2
Variation	om	5,48	2,79	4,31	3,12	4,25	7,69
Variation coefficient	%	2,64	1,16	1,64	0,99	1,26	1,83
Furrow angle	deg.	35,9	38,6	42,5	41,4	45,2	42,8
Open furrow width	cm	23,7	27,7	31,9	30,5	32,0	28,7
Ratio of clods over 5 cm	%	20,3	11,3	10,6	14,9	27,2	26,5
Turning effect	%	80,5	91,5	100	100	100	100

Fig.1.: John Deere-2800 plough

Fig.2.: Kverneland-CC plough

Fig.3.: RÁBA-IH-10-735-6TJ plough

GEOMETRICAL PARAMETERS OF THE MOULDBOARDS

PULVERIZING CURVES

TURNING CURVES

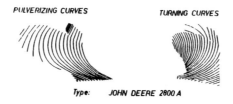

Type: JOHN DEERE 2800 A

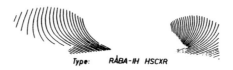

Type: RÁBA-IH HSCXR

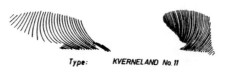

Type: KVERNELAND No.11

FIGURE 4

Specific resistance with different cutting widths

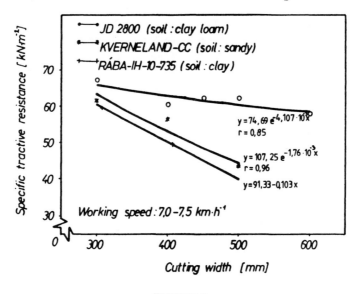

Cutting width [mm]

FIGURE 5

Specific power requirement with different cutting widths

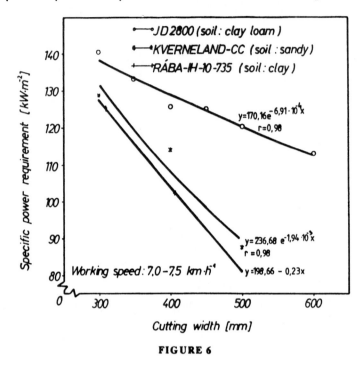

FIGURE 6

Specific fuel consumption with different cutting widths

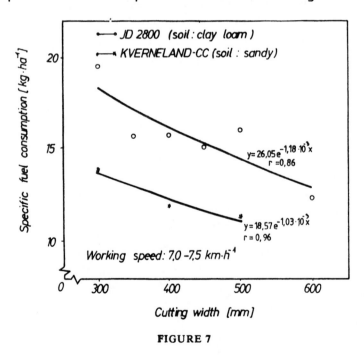

FIGURE 7

Land and Water Use, Dodd & Grace (eds), © 1989 Balkema, Rotterdam. ISBN 90 6191 980 0

Selection of work parameters of a tractor-plough combination at reduced tractor engine speed

J.Kuczewski & Z.Majewski
Institute of Agricultural Engineering, Agricultural University, Warsaw, Poland

ABSTRACT: A mathematical model for selection of work parameters of a tractor-plough combination at reduced engine speed was developed, based on three decision variables: tractor transmission gear, engine speed, and working width of plough. As a criterion the fuel consumption per hectare ploughed was taken. The obtained results point out at usefulness of tractor work at the reduced engine speed in relation to nominal one. The carried out verification of the model proved the theoretical results of analysis.

Es wurde ein mathematisches Modell für die Auswahl der Arbeitsparameter des Aggregates Schlepper-Pflug bei reduzierter Drehzahl des Schleppermotors auf Grund von drei Entscheidungsvariabeln: Schleppergeschwindigkeit, Drehgeschwindigkeit des Motors und Arbeitsbreite des Pfluges ausgearbeitet. Als Kriterium für die Auswahl des Aggregates wurde der Kraftsoffverbrauch je Hektar der Ackerfläche angenommen. Die erhaltenen Ergebnisse beweisen die Zweckmässigkeit des Schleppereinsatzes bei der, in Bezug auf die Nenudrehzahl, reduzierten Motordrehzahl. Die. in Feldversuchen durchgeführte Verifikation des Modells bestätigte die Richtigkeit der theoretischen Ergebenisse der Analyse.

On a établi le modèle mathématique pour la sélection des paramètres de travail de l'agrégat tracteur-charrue à la vitesse réduite du tracteur, basé sur trois variables décisives: vitesse du tracteur, vitesse de rotation du moteur et largeur de travail de la charrue. La consommation du combustible à l'hectare de la surface labourée a été admise comme critère du choix de l'agrégat. Les résultats obtenus démontrent l'utilité du travail du tracteur à la vitesse réduite par rapport à la vitesse nominale. La vérification du modèle réalisée au cours des essais au champ à confirmé que les résultats théoriques de l'analyse sont corrects.

1 INTRODUCTION

The application of powerful tractors to agricultural operations creates two possibilities: work with wide implement at the highest, under given conditions, engine speed, or work at reduced engine speed using the implement of appropriate width. Determination of the engine operating point at its reduced speed calls for application of a general contour characteristic of the engine. Stroppel (1980) analysed work of a tractor at its partial loading, while Schäfer (1983) attempted to formulate dependence between the actual engine loading and specific fuel consumption. Jahns and Steinkampf (1982) and Schimmel and Hulla (1983) also considered the shape of general contour characteristic of the engine, while Lyne, Meiring and Hansen (1985) investigated the possibility of application of a device, informing the operator at which operating point of the engine he is working. The mention sources do not include any complete analytical models, which can be taken as a basis for selection of work parameters of a tractor-plough combination at reduced engine speed under various exploitationsl conditions.

2 BASIC ASSUMPTIONS

The starting point in selection of a combination for ploughing is the range of agronomically acceptable working speeds which are to be used for a given plough body under given field conditions. Among the available tractor transmission gears one can distinguish: those, which can not be used for ploughing because they do not ensure the minimal required working speed, those at which work of the tractor is possible at both the full and recued engine speeds, and those at which reduction of engine speed is essential in order to maintain the maximal admissible speed, if there is enough engine power at the given point. Thus, in order to select the work parameters of a tractor-plough combination, the operator has three possibilities: selection of tractor gear, selection of engine speed and selection of implement's width. The fuel consumption per hectare ploughed is a criterion of correct matching of the combination.

Assuming, that the operator uses certain working speed of ploughing, which is usually the highest possible speed under given working conditions enabling to achieve the highest rate of ploughing, one can determine, for a given implement, the horizontal and vertical load on tractor driving wheels, using the equations:

$$P_k = (k_o + k_v \cdot v^2) a \cdot b (1 + M_2 \cdot M_3 \cdot f_c) + \tag{1}$$
$$(G_c + G_{pl} \cdot b \cdot M_3) g \cdot f_c$$

$$Y_k = \left\{ G_c \cdot g + \left[(k_o + k_v \cdot v^2) a \cdot b \cdot M_2 + \right.\right.$$
$$\left.\left. G_{pl} \cdot b \cdot g \right] \cdot M_3 \right\} \cdot M_1 \tag{2}$$

Then, using the coefficient of adhesion utilization:

$$\varphi = \frac{P_k}{Y_k} \tag{3}$$

one can determine tractor wheel slip according to formula developed by Grecenko (1963):

$$\mho = \frac{c_1 \cdot \varphi - c_2 \cdot \varphi^2}{c_3 - \varphi} \tag{4}$$

Thus, the operating point of the engine under conditions assumed is determined by the coefficients λ_M and λ_n:

$$\lambda_n = \frac{30 \cdot v \cdot i_c}{\pi \cdot r_d (1 - \mho) n_{nom}} \tag{5}$$

$$\lambda_M = \frac{P_k \cdot r_d \cdot n_{nom}}{9.55 \cdot i_c \cdot \rho_m \cdot N_{nom}} \tag{6}$$

In order to calculate the specific fuel consumption at a given point, the general contour characteristic of fuel consumption was described by polynomial regression equation. Basing on analysis of empirical contours of the series of characteristics the following equation was developed:

$$\lambda_g = b_o + b_1 \cdot \lambda_M + b_2 \cdot \lambda_M^3 + b_3 \cdot \lambda_n + b_4 \cdot \lambda_n^2 +$$
$$b_5 \cdot \lambda_M \lambda_n + b_6 \cdot \lambda_M \cdot \lambda_n^2 + b_7 \cdot \lambda_M^3 \cdot \lambda_n \tag{7}$$

As a result, the fuel consumption per hectare can be calculated for the given operational conditions from the expression:

$$Q = \frac{0.0265 \cdot N_{nom} \cdot g_{enom} \cdot \lambda_M \cdot \lambda_n \cdot \lambda_g}{b \cdot v} \tag{8}$$

Basing on statistical analysis of several characteristics there was assumed that the external part of torque curve can be described by cubic polynomial of the form:

$$\lambda_M = a_o + a_1 \lambda_n + a_2 \lambda_n^2 + a_3 \lambda_n^3 \tag{9}$$

while the governor part of the curve by equation:

$$\lambda_M = -\lambda_n \cdot \frac{1}{\xi_n - 1} + \frac{\xi_n}{\xi_n - 1} \tag{10}$$

The available engine power at a given setting can be obtained at intersection of external part of curve and the parallel to governor characteristic, running through a given engine operating point. Introducing to analytical model the admissible value of wheel slip and value of essential margin of power as variables set up arbitrarily by the operator, one can obtained the required limitations for the operating point determined.

Basing on the presented assumptions a computer program was developed, considering, besides the dependences described above, also the variability of tractor rolling resistance and variability of mechanical efficiency of tractor transmission, depending on operating point of the engine.

3 CALCULATION AND RESULTS

The exemplary results of calculations are presented for tractor of power 35 kW and the parameters corresponding to tractor Massey Ferguson MF255. The ploughing was assumed to a depth of 25 cm on medium soil of specific resistance equal to 40 kN/m². As admissible range of working speed 4-8 km/h was taken.

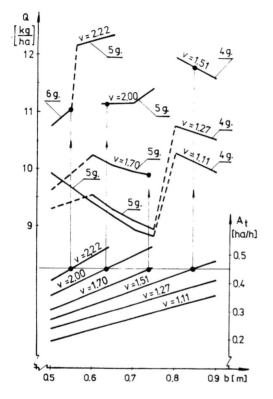

Fig. 1. Results of compputer calculations: Q - fuel consumption, kg/ha, A_t - rate of ploughing, ha/h, b - working width, m, v - working speed, m/s

Considering the obtained results (Fig.2) one can observed that ploughing at the maximal admissible speed (2.22 m/s) is possible using 6th or 5th gear of considered tractor, with application of engine speed reduction, while the fuel consumption amounts to: from about 11 to above 12 kg/ha Ploughing at the minimal requested speed (1.11 m/s) can be performed on 5th or 4th gear, also at

reduced engine speed, and the fuel consumptions range from under 9 to above 10 kg/ha. Assuming that the required rate of ploughing should reach e.g. 0.45 ha/h, one can read from the presented diagram, that for ploughing at the highest admissible speed it is needed to use the plough 0.56 m wide and that the fuel consumption on 5th gear will amount to 12.3 kg/ha, while on 6th gear it will be equal to 11.1 kg/ha. Assuming slightly lower, than the maximal, speed equal to 2 m/s one should apply the plough of width 0.63 m, which would result in fuel consumption equal again to 11.1 kg/ha on 5th gear. Using the speed of 1.7 m/s and the plough width 0.74 m ploughing will be performed on 5th gear also, but the fuel consumption will be lower than 10 kg/ha. At last, decreasing the working speed to 1.5 m/s in order to obtain the ploughing rate assumed, one has to apply plough of width 0.85 m, achieving the fuel consumption above 10 kg/ha on the 4th gear. The working speeds lower than mentioned above do not allow to achieve the required exemplary rate of ploughing equal to 0.45 ha/h.

The presented exemplary analysis points out that the most advantageous operating conditions of a given combination can be obtained by application of the plough of working width close to about 0.75 m and when ploughing is performed on the 5th gear, with the proper engine speed reduction, depending on the requested rate of ploughing. Using the presented method it is possible to select the work parameters of tractor-plough combination at reduced enginee speed.

In order to compare the fuel consumption values obtained in ploughing at nominal engine speed with those at reduced one, the field tests on combinations of a heavy tractor with 4-furrow plough and medium-size tractor with 2-furrow plough were accomplished (Fig. 2). It was found, that at corresponding values of ploughing speed the method of work at reduced engine speed allowed to decrease the fuel consumption per hectare ploughed by about 12-20%, in relation to the method of nominal engine speed.

4 CONCLUSIONS

The present analytical methods enables to select the work parameters of tractor-plough combination with application of three decision factors: selection of tractor transmission gear, selection of engine rotational speed, and selection of most advantageous plough width for a given tractor.

If tractor power is sufficient, the ploughing at reduced engine speed enables to achieve lower fuel consumption per hectare, when compared with work at nominal engine speed.

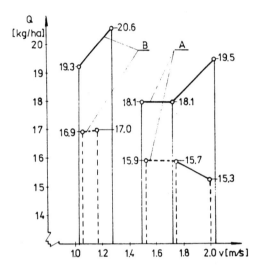

Fig. 2. Results of field tests: A - heavy tractor with 4-furrow plough, B - medium-size tractor with 2-furrow plough

NOTATION

a - working depth, m

a_o - a_3 - coefficients of equation of engine external characteristic

b - working width of plough, m

b_o - b_7 - coefficients of equation of engine general contour characteristic

C_1 - C_3 - coefficients of field surface

f_c - coefficient of rolling resistance

G_c - weight of tractor, kg

G_{p4} - specific weight of plough, kg/m

g - acceleration of gravity, m/s^2

g_{enom} - nominal specific fuel consumption g/kWh

i_c - transmission ratio of tractor

k_o, k_v - constant and variable parts of specific resistance coefficient, kN/m^2, kN.s^2/m^4

M_1 - coefficient of vertical load distribution on tractor wheels

M_2 - coefficient of vertical soil forces

M_3 - coefficient of weight transfer from plough

N_{nom} - nominal power of tractor, kW

n_{nom} - nominal engine speed, rev/min

P_k - force on tractor driving wheels, kN

r_d - radius of tractor driving wheels, m

v - working speed, km/h or m/s

Y_k - vertical load on tractor driving wheels, kN

σ - tractor wheel slip

ξ_n - coefficient of governor insensibility

η_m - mechanical efficiency

λ_g - coefficient of specific fuel consumption

λ_M - coefficient of engine torque utilization

λ_n - coefficient of engine speed utilization

φ - coefficient of adhesion utilization

REFERENCES

Grecenko, A. 1963. Kolove a pasove traktory. Praha, Statni Zemed. Nakladat.

Jahns, G. and H. Steinkampf. 1982. Einflussgrössen auf Flächenleistung und Energieaufwand beim Schleppereinsatz. Grundl. Landtech. 1:20-27.

Lyne, P.W.L., P. Meiring and A.C. Hansen 1985. Optimizing tractor operation. Proc. Conf.on Soil Dynamics.4:756-764.

Schäfer, W. 1983. Teoretische Untersuchungen zur optimalen Kombination von Allradschleppern und gezogenen Geräten zur Bodenbearbeitung. PhD thesis. Univ. Hohenheim, Stuttgart.

Schimmel, J. and H. Hulla 1983. Einsatzoptimierung von Ackerschleppern durch elektronische Fahrerinformation. GrundlLandtech. 1:72-80.

Stroppel, A. 1980. Energie- und Arbeitszeitbedarf für gezogene Geräte der bodenbearbeitung bei unterschiedlicher Schleppermotorauslastung. Grundl. Landtech. 4:135-140.

Land and Water Use, Dodd & Grace (eds), © 1989 Balkema, Rotterdam. ISBN 90 6191 980 0

Deep soil loosening effects on sugar beet performance

F.J.Larney & R.A.Fortune
Agricultural Engineering Department, Teagasc, Oak Park Research Centre, Carlow, Republic of Ireland

ABSTRACT: Twelve experimental sites were laid down to ascertain the effect of autumn deep loosening in alleviating soil compaction and improving sugar beet performance. Four sites showed significant root and sugar yield increases, six showed no significant differences, while two gave significant decreases in yield on deep-loosened plots. Negative and negligible yield responses to deep loosening were largely explained by recompaction of autumn loosened soil by excessive tillage traffic during seedbed preparation the following spring. Deep loosening decreased the percentage of forked roots on all sites where there was a yield response to the operation.

RESUME: 12 sites d'expérimentation ont été mis en oeuvre afin de déterminer l'impact de l'ameublissement en profondeur du sol en automne pour atténuer le compactage du sol et améliorer le rendement en betterave à sucre. Sur 4 des sites on a observé une augmentation considérable du rendement en matière de racines et de sucre. Six des sites n'ont révélé aucune différence et deux des sites ont révélé une diminution importante du rendement sur les parcelles ou un ameublissement en profondeur a été réalisé. Des réactions de rendement négligeables ou négatives ont été en grande partie expliqués par le recompactage de sols ameublis en automne par une circulation excessive de labourage au cours de la préparation des couches de semis au printemps suivants. L'ameublissement en profondeur diminuait le pourcentage de racines fourchues sur tous les sites sur lesquels on a observé une réaction de rendement à l'opération.

AUSZUG: Es wurden zwölf Versuchsparzellen angelegt, um die Auswirkungen einer Bodenlockerung im Herbst auf die Kompaktierung der Erde und eine Verbesserung der Zuckerrübenerträge zu untersuchen. Vier Parzellen zeigten beträchtliche Steigerungen der Wurzel- und Zuckererträge, bei sechs waren keine wesentlichen Unterschiede festzustellen, während bei zwei Parzellen mit tiefer Bodenlockerung betrachtlich niedrigere Erträge zu verzeichnen waren. Negative und geringe Ertragsverbesserungen bei tiefgelockerten Parzellen konnten größtenteils auf die Wiederverhärtung von herbstgelockerter Erde durch übermäßigen Bodenbearbeitungsverkehr bei der Saatbettvorbereitung im folgenden Frühjahr zurückgeführt werden. Eine tiefe Bodenlockerung verringerte den Prozentsatz von gegabelten Wurzeln in allen Parzellen, in denen eine Ertragserhöhung aufgrund des Versuchs festgestellt werden konnte.

1 INTRODUCTION

The basic aim of deep loosening is to cause maximum disruption of compacted layers within the soil profile. Although not a new concept, it has become commonplace in recent years to alleviate soil compaction caused by the combined trends of heavier farm machinery and continuous tillage practices without grass in the rotation. Large increases in crop yield have been reported following deep loosening (Braim et al. 1984; Stone 1982; Unger 1979). Sugar beet is particularly sensitive to soil compaction as severe root forking results which reduces yield potential.

In the light of increasing levels of soil compaction and a deterioration in soil structural properties, a study was conducted between 1981 and 1984 to ascertain the effects of soil physical properties, deep loosening, seedbed cultivations and sowing methods on growth and yield of sugar beet (Larney 1985; Larney 1986; Larney and Fortune 1985; Larney and Fortune 1986; Larney et al. 1988a; Larney et al. 1988b). This paper reports specifically on yield parameter (percentage emergence, root shape, sugar content and root yield) responses of sugar beet to deep loosening.

2 MATERIALS AND METHODS

Deep loosening treatments were imposed on four different sites in each of the three autumns 1981, 1982 and 1983. Each autumn, three test sites (1981: Ashfield, Churchtown and Donabate; 1982: Hollymount, Athy and Oak Park; 1983: Busherstown, Paulstown and Maganey) were located on Grey Brown Podsolic soils (Orthic Haplaqualfs) which had developed soil compaction and weak structure and had produced low sugar beet yields (35-45 t/ha) in the past. The fourth site (Ferns, Fenagh and Pump Field in 1981, 1982 and 1983, respectively) was chosen each autumn on high yielding (55-66 t/ha) Brown Earth soils (Orthic Dystrochrepts). All 12 sites were located in the sugar beet growing area of south-east Ireland, where the mean annual rainfall is approximately 800 mm. The inherent soil physical properties of each site have already been outlined by Larney et al. (1988b).

On 10 of the 12 sites two different types of deep loosener were compared with a non-deep loosened or control treatment with four replications. Deep loosening was carried out to the maximum depth possible which ranged from 350-500 mm. Working depth was governed by soil moisture conditions, soil strength of subsurface compact layers and machine capabilities. Three different machines were used for the deep loosening operation: (i) Armer Salmon, which consisted of three heavy duty C-shaped legs 800 mm apart and which had a maximum working depth of 450-500 mm, (ii) Howard Paraplow, which had four 45 degree slanted legs which caused a lifting and bending action of the soil as it passed over them. Its maximum working depth was 400 mm with a 450 mm spacing between each leg, (iii) McConnel Shakaerator, which comprised of five legs mounted on a frame vibrated by a pto-driven flywheel. The legs were spaced at 500 mm and the machine had a maximum

working depth of 500 mm.

All sites were mouldboard ploughed to a depth of 230 mm as soon as possible after deep loosening and before the soil had re-wetted. Seedbed preparation on deep loosened and non-deep loosened plots followed in April using powered rotary, spring tine or disc harrow cultivation, in accordance with normal farmer practice in Ireland. Sugar beet was sown on 560 mm rows.

Percentage emergence was estimated when the more advanced sugar beet seedlings had reached the four true leaf stage. For yield estimation, a 12 m^2 area from each replicate treatment was hand-harvested. Before bagging a visual assessment of root shape was conducted and the percentage of well-shaped roots was estimated. In the following discussion the growing season year is used (1982, 1983 and 1984) which relates to effects of deep loosening carried out in the autumns of 1981, 1982 and 1983, respectively.

3 RESULTS AND DISCUSSION

3.1 Percentage seedling emergence

In 1982 deep loosened treatments had higher seedling emergence than control treatments on two of the four sites (Ashfield and Ferns) (Table 1). At Ferns the Armer Salmon treatment resulted in significantly higher emergence than the Paraplow which in turn was significantly higher than the control. As the deep loosened treatments were generally wetter than the control treatments at sowing time on this soil, the extra moisture may have increased seedling emergence in the dry period which followed sowing in 1982.

In 1983 deep loosening gave increased emergence at all four sites, being significant at Hollymount and Oak Park (Table 1). This may have been due to the better water infiltration on the deep loosened plots as rainfall was well above normal following sowing in 1983. The control plots suffered from surface water ponding which led to waterlogged conditions. Deep loosening had no significant effect on emergence at any of the four sites in 1984 (Table 1), when rainfall amounts were closer to the long-term average.

3.2 Root shape: severity of forking

In 1982, deep loosening increased the percentage of well-shaped roots at all

Table 1. Sugar beet seedling emergence (%) on deep loosening treatments at 12 study sites, 1982-84

	Control	Paraplow	Armer Salmon	McConnel	F-test
1982					
Ashfield	67.2	68.8	–	–	NS
Churchtown	64.0	61.6	–	–	NS
Donabate	63.4	63.8	–	63.0	NS
Ferns	68.6	72.7	75.0	–	**
1983					
Hollymount	73.0	76.1	74.1	–	*
Athy	83.0	85.8	85.1	–	NS
Oak Park	68.2	74.3	–	71.9	*
Fenagh	72.4	76.8	74.4	–	NS
1984					
Busherstown	69.6	68.2	69.7	–	NS
Paulstown	73.6	–	72.2	70.5	NS
Maganey	61.7	65.6	–	64.7	NS
Pump Field	81.1	81.7	79.7	–	NS

*, ** = Significant at 0.05 and 0.01 levels, respectively;
NS = not significant

sites except Churchtown (Table 2). Beneficial loosening effects had been lost on this soil as structural damage had been incurred during seedbed preparation. Four passes of a rotary cultivator were required for acceptable seedbed conditions, as the topsoil had dried substantially leading to increased clod strength and resistance to fracture. This feature, coupled with the wet and plastic nature of the 200-400 mm soil layer led to soil compaction and cancellation of the benefits of deep loosening. The largest effect of deep loosening occurred at Ferns, where the Armer Salmon treatment led to a significant increase in well-shaped roots. Ferns was the well-structured soil chosen for

Table 2. Well-shaped sugar beet roots (%) on deep loosening treatments at 12 study sites, 1982-84

	Control	Paraplow	Armer Salmon	McConnel	F-test
1982					
Ashfield	83.1	84.0	–	–	NS
Churchtown	70.7	66.6	–	–	NS
Donabate	71.0	74.5	–	77.1	NS
Ferns	78.8	80.6	84.5	–	**
1983					
Hollymount	70.9	73.9	75.1	–	NS
Athy	79.6	77.8	79.0	–	NS
Oak Park	69.8	72.1	–	71.5	NS
Fenagh	82.8	86.0	83.6	–	NS
1984					
Busherstown	63.9	66.5	67.6	–	NS
Paulstown	75.7	–	75.8	77.4	NS
Maganey	62.7	64.3	–	67.8	NS
Pump Field	82.8	82.2	82.2	–	NS

** = Significant at 0.01 level; NS = not significant

comparative purposes in 1982.

In 1983, deep loosening increased the proportion of well-shaped roots at all sites except Athy, however none of these increases was significant (Table 2). The largest increase over the control treatment was on the Armer Salmon treatment at Hollymount. In 1984, deep loosening increased (non-significantly) the numbers of well-shaped roots at Busherstown, Paulstown and Maganey but had no effect at Pump Field (Table 2). The largest increase was on the McConnel Shakaerator plots at Maganey.

These results agreed with those of Larney et al. (1988a) who found that root shape was highly correlated with structural properties of the Ap horizon of the 12 study soils. Therefore, any improvement of soil structural conditions, e.g. by deep loosening, was likely to reduce the prevalence of forking.

3.3 Sugar content

Deep loosening increased percentage sugar content of sugar beet roots on all four sites in 1982, being significant at Ashfield, Churchtown and Ferns (Table 3). The largest increase occurred at Ashfield with deep loosened plots having a 0.67% higher sugar content. At Ferns the Armer Salmon deep loosener increased sugar content over the Paraplow (+ 0.29%) and the control (+ 0.41%) treatments.

In 1983, deep loosening increased sugar content at the Fenagh site only, but this was non-significant (Table 3). At Hollymount there was a significant decrease in sugar content on the Paraplow treatment of 0.18%. In 1984, there was no significant effect of deep loosening on sugar content at any of the four sites (Table 3).

3.4 Root yield

Deep loosening gave a significant increase in yield of 12% at Ferns on the Paraplow treatment and 15% on the Armer Salmon treatment (Table 4). There was an overall increase of 15% on the Paraplow plots at Ashfield but this was non-significant. The Paraplow treatment yielded significantly lower than the control treatment at Churchtown while there were no differences between treatments at Donabate. The yield response to deep loosening was directly related to the number of cultivator passes required to create acceptable seedbeds. At Ferns and Ashfield (where yield increases occurred) three passes were required. At Donabate and Churchtown, four runs were needed and no response occurred. In addition, the Donabate and Churchtown soils were of poorer structure than Ashfield and Ferns and the benefits of deep loosening were

Table 3. Sugar content (%) of sugar beet roots on deep loosening treatments at 12 study sites, 1982-84

	Control	Paraplow	Armer Salmon	McConnel	F-test
1982					
Ashfield	17.09	17.76	–	–	**
Churchtown	15.11	15.51	–	–	**
Donabate	17.39	17.49	–	17.23	NS
Ferns	16.24	16.36	16.65	–	*
1983					
Hollymount	16.19	16.01	16.17	–	*
Athy	16.10	16.07	15.97	–	NS
Oak Park	15.73	15.71	–	15.98	NS
Fenagh	15.33	15.68	15.49	–	NS
1984					
Busherstown	19.46	19.24	19.43	–	NS
Paulstown	19.42	–	19.34	19.40	NS
Maganey	16.89	16.91	–	17.06	NS
Pump Field	15.88	15.60	16.06	–	NS

*,** = Significant at 0.05 and 0.01 levels, respectively;
NS = not significant

Table 4. Root yield (t/ha) on deep loosening treatments at 12 study sites, 1982-1984

	Control	Paraplow	Armer Salmon	McConnel	F-test
1982					
Ashfield	25.1	28.8	-	-	NS
Churchtown	37.5	35.9	-	-	*
Donabate	47.3	47.1	-	46.7	NS
Ferns	46.9	52.5	53.9	-	**
1983					
Hollymount	50.7	55.4	53.4	-	*
Athy	47.6	47.8	47.2	-	NS
Oak Park	39.9	35.9	-	35.9	*
Fenagh	58.3	61.3	60.6	-	NS
1984					
Busherstown	55.4	58.3	57.7	-	*
Paulstown	50.1	-	53.2	50.5	*
Maganey	44.0	44.5	-	45.6	NS
Pump Field	42.7	44.6	45.4	-	NS

*, ** = Significant at 0.05 and 0.01 levels, respectively;
NS = not significant

lost to recompaction of their wet plastic 200-400 mm soil layer during seedbed preparation. The yield response to deep loosening at Ferns was unexpected as the soil was not considered to have physical problems prior to the operation. However, the action of deep loosening may have changed the orientation of shale stones which comprised 40% of the soil mass at loosening depth, thus allowing better root penetration and increased yield.

Deep loosening increased root yield by overall values of 7% on the Paraplow plots and 5% on the Armer Salmon plots at both Hollymount (significant) and Fenagh (non-significant) in 1983 (Table 3). At Athy there was no increase in yield due to deep loosening, while at Oak Park there was a significant 10% decrease in yield on both the Paraplow and McConnel Shakaerator plots. The yield response to deep loosening was again governed by the number of passes needed for seedbed preparation which were: Fenagh, 2; Hollymount, 5; Athy, 6; and Oak Park, 7. The Paraplow resulted in higher yields than the Armer Salmon deep loosener on the three sites where they were compared, but these differences were non-significant.

In 1984, deep loosening increased root yield on all four sites being significant at Busherstown and Paulstown (Table 3). At Busherstown, 5% and 4% yield increases resulted on Paraplow and Armer Salmon plots, respectively. At Paulstown, a 6% increase in yield occurred on the Armer

Salmon treatment, while the McConnel was only slightly higher than the control. Yield increases of 3-6% also occurred on deep loosened treatments at Maganey and Pump Field. Less seedbed cultivation passes (1-3) were required on the 1984 soils and hence the beneficial loosening effects were better preserved than in previous years.

4 SUMMARY AND CONCLUSIONS

1. There was some evidence that deep loosened treatments resulted in higher sugar beet emergence and establishment in the dry early growing season of 1982 (Ferns) and in the wetter than normal early growing season of 1983. This was possibly due to more moist soil conditions in 1982 and better infiltration in 1983 on the deep loosened plots. In the more normal early growing season of 1984 there was no effect of deep loosening on seedling emergence.

2. Deep loosening increased the proportion of well-shaped roots, and hence increased yield potential, at those sites where benefits of the operation were not lost to recompaction during subsequent tillage operations.

3. Deep loosening increased sugar content of the roots at all four sites in 1982, but in 1983 and 1984 trends were less distinct. A climatic interaction with deep loosening may have occurred in

1982, which was not repeated in the following two years.

4. Eight sites showed root yield increases due to deep loosening: Ferns, Hollymount, Busherstown, Paulstown, Ashfield, Fenagh, Maganey and Pump Field, the first four being significant. Positive response soils included the three well-structured comparative sites (Ferns, Fenagh, Pump Field) which indicated that deep loosening can be beneficial even when no apparent soil structural problems are present. Two sites showed no response to deep loosening (Donabate and Athy), while two showed significant yield decreases on deep loosened treatments (Churchtown and Oak Park).

5. The poorly structured soils in this study showed evidence of soil compaction and hence would be expected to benefit from deep loosening. However, because of their poor structural nature, they also required more seedbed passes for sufficient clod breakdown. Hence, while having the greatest need for deep loosening, advantageous effects of the operation were more likely to be lost on these soils. Yield response of sugar beet to deep loosening was directly related to the number of subsequent cultivation passes required for seedbed preparation. Where the number of post-loosening passes was minimised, increased yields of sugar beet were recorded.

REFERENCES

Braim, M.A., K. Chaney and D.R. Hodgson 1984. Preliminary investigation on the response of spring barley (Hordeum sativum) to soil cultivation with the 'Paraplow'. Soil Tillage Res. 4:277-293.

Larney, F.J. 1985. Effects of soil physical properties, deep loosening and seedbed cultivations on yield and growth of sugar beet (Beta vulgaris L.). The National University of Ireland, Dublin. 516 pp.

Larney, F.J. 1986. Drills or flat for sugar beet? Farm & Food Res. 17:24-25.

Larney, F.J. and R.A. Fortune 1985. Is powered rotary cultivation best on all soils? Farm & Food Res. 16:167-169.

Larney, F.J. and R.A. Fortune 1986. Recompaction effects of mouldboard ploughing and seedbed cultivations on four deep loosened soils. Soil Tillage Res. 8:77-87.

Larney, F.J., R.A. Fortune and J.F. Collins 1988a. Soil physical factors influencing sugar beet emergence, root shape and root yield on a range of Irish soils. Proc. 11th Int. Conf. of Int. Soil Tillage Res. Organisation, Edinburgh, Scotland, 11-15 July 1988. 1:79-84.

Larney, F.J., R.A. Fortune and J.F. Collins 1988b. Intrinsic soil physical parameters influencing intensity of cultivation procedures for sugar beet seedbed preparation. Soil Tillage Res. 12:253-267.

Stone, D.A. 1982. The effects of subsoil loosening and deep incorporation of nutrients on yield of broad beans, cabbage, leek, potatoes and red beet. J.agric.Sci., Camb. 98:297-306.

Unger, P.W. 1979. Effects of deep tillage and profile modification on soil properties, root growth, and crop yields in the United States and Canada. Geoderma. 22:275-295.

Land and Water Use, Dodd & Grace (eds), © 1989 Balkema, Rotterdam. ISBN 90 6191 980 0

Economic and technical appraisal of trial results of some innovatory land tilling methods

V.Panaro
Institute of Agricultural Mechanics, University of Bari, Italy

G.Patruno
Institute of Political Economy, University of Bari, Italy

SUMMARY: In the paper the authors highlight the results deriving from testing some soil tilling techniques: rippering by a five-tined chisel with later clod breaking; rippering via a seven-tined chisel with subsequent clod brewking; scarification by a three-tined scarifier with later skim ploughing and clod breaking; rippering via a chisel combined with milling. The said techniques have been compared with conventional deep ploughing. Our research has been primarily aimed at reducing energy inputs and protecting the environment from degratation.

Economic evaluations have been made as well on the basis of experimental results, which have stressed the convenience for operators to introduce such innovations. Mention has also been made of the EEC 1985 regulation 797 tangibly substantiating the opportunity for being granted aids on the introduction of technologically advanced farm implements, with a view to reducing production costs, since it may be inferred on the basis of the results related to a scientific and technical analysis that the above equipment proves to be effective both from the economic and technical standpoint, given its highly competitive user charge vs the one required by traditional ploughs.

1 FOREWORD

In 1985 the EEC issued Regulation 797 for the improvement of agricultural building. As is well-known, the regulation in question is mainly aimed at fostering a continuous agricultural development within the EEC on the basis of a common policy designed to further the evolution of structures and to guarantee everlasting conservation of natural land resources.

The form of action being referred to consists principally in granting aid for farm investment on the conditions laid down by Article 2, with the aim of improving qualitatively and implementing productive reconversion with respect to the market needs.

Surely, in this regard, all of the above turns out to be extremely important; yet, its limitations set in within the regulation itself when proclaiming that granting an aid on making investments can be excluded or restrained, provided that a rise in productivity of goods "which are not marketable" is achieved.

These limitations have often acted as a spur for operators to increase certain crop areas; to this end, suffice it to mention proteaginous plants which have spread in a few years' time to such an extent that marketing difficulties have arisen. This is also the case with the sunflower and the soy-bean, or especially in inland areas, with the so-called small-sized fruits which, after a succession of good harvests, proved disappointing. A further path taken lies in the introduction of new crops - both food (e.g. actinidia, avocado, babako, pink grapefruit) and nonfood ones (e.g. jojoba, kenaf, topinambour).

Now, what is actually being questioned is the market reaction to the massive introduction of such unknown multi-taste farm products which not always are enojoyable. Mention should be made of the kiwi whose overspreading has contributed to lowering prices to the extent that profits have been greatly reduced, whereas, as everybody knows, its installation cost is really sky-rocketing; and this after a few years of high incomes earned by the first growers.

Nonetheless such problems should not command us to desist from pursuing these forms of actions; anyway, a thorough, more effective research and a sequence of tests should be carried out, assessing that "the great diversity of natural causes and of the gravity of structural problems within agriculture do call for clear and distinct solutions which can be easily adapted in time according to the regions examined", as stated in the introduction to the regulation.

On the grounds of the awkward market conditions of a great number of agricultural products, particularly in Southern Italy where 60% of areas are invested in bumper corps, we may deem it necessary to catalyze financial aids on investments which, would allow to reduce production costs, especially if these are a great boost to environment - protecting and energy - saving techniques.

It is within this framework that land tilling techniques have been tested. And this with a view to checking whether the introduction of new techniques, or the restoration of previous ones, entails cost reduction, saving of energy and environmental protection.

As far as protection of the environment is concerned, the problem of erosion has been deeply felt in many areas; hence, the need for putting several forms of reduction in soil processing to the test in order to try to lessen the phenomenon, without any crop yield being impaired or by resorting to reduced losses, or simply to the improvement of the same. In this respect, experiments were performed whose methodology and results obtained both being illustrated below.

2 EXPERIMENTAL METHODOLOGY

2.1 In order to better evaluate the influence exerted on sugar-beet output by various land tilling methods, experimental results obtained on the "Tomacelli" farm situated in the countryside around Metaponto (Mt) in a plot stretching over an overall 5-hectare surface with an average one-per-cent slope, were examined.

The silty soil is presented in summer with great phenomena of withdrawal, giving rise to wide vertical in-depth cracking, with a tendency to reproduce macroporosity spontaneously.

Given a clay soil from the standpoint of mechanical processing, physical properties will not be favourable since the soil, if not adequately waterfilled, proves to be hard to till because of its high cohesion when dry, and its high plasticity and stickiness when wet.

The conventional cultivation method applied practically on a farm (deep ploughing by means of a double-shared plough to a 50-centimetre tilling depth) was compared with newly adopted land tilling techniques. The allotment was subsequently split up into five plots of approximately 1 hectare, with the same geometric features and the same average slope so as to make field surveys as homogeneous and comparable as possible.

On this plot hard wheat had been previously grown, and stubble mulch preliminarily burnt. Sugar-beet had been regarded as a rotational crop. Four theses were confronted with the traditional cultivation method by means of a two-share plough in order to prepare earth for seeding, which specifically concerned sugar-beet.

The said theses were the following:

I) "rippering" by means of a five-tined giant chisel with subsequent clod breaking;

II) "rippering" by means of a seven-tined giant-size chisel with subsequent clod breaking;

III) "scarification" via a three-tined scarifier equipped with a rear spinner, with subsequent surface "labour" and clod breaking;

IV) "rippering" by means of a Rambo chisel with combined milling and rolling.

The main cultivations as for all four theses being confronted affected an average depth of some 50 centimetres, while later surface labour performed in thesis n. 3 after rippering reached a 30-centimetre depth.

As for the fourth thesis worked out via a combine machine, one single tilling was sufficient to prepare the seed-bed; conversely, as for the other theses, we had to employ further implements after the main tillage for clods to be so well-pulverized that they would turn suitable for a good seed-bed.

Such further operations consisted of two clod breakings as to the first three theses compared and the stake; more precisely, clod breaking was performed along the longest side of the field (290 metres), thereafter follow by a cross clod breaking. All plots were subsequently sown by means of an integral six-row precision seeder.

The results stemming from surveys performed on the five plots were examined on the basis of fuel consumption, specific use of tools and labour per hour/hectare, production, the polarimetric degree (observed from a sugar refinery) of the four alternative tilling techniques alongside those pertaining to conventional ploughing.

2.2 The tractor employed was a FIAT four -

wheel drive - type 180/90 DT Turbo of a maximum horsepower of 132 kW (180 CV).

The implements used were as follows:
- an integral two way moldboard turnwrest plough of NARDI's (stake);
- SOGEMA's five tined subsoiler - type 120-CA 3P-5D (1st thesis, Fig. n. 1);
- SOGEMA's seven-tined chisel - type 120-CA 3P-7D (2nd thesis, Fig. n.2);
- SOGEMA's three-toothed ripper with spinner - type 180-TC-3P-3RD (3rd thesis, Fig. n.2 and3);
- a combine machine made up of a five-tined chisel (Rambo-type) with horizontal shaft rotary
hoe RTP 2300/4, plus a Paker roller by Renter (4th thesis, Fig. n. 4);
- a four-section drag disk harrow having six gangs for a total of 24 disks;
- Gaspardo's six-row precision seeder.

The seed employed was a sugar-coated monogerm one of the Maribo Auta Mono variety (autumnal); crop spacing was 11 centimetres on the rows, and 50 centimetres between the rows; seeding was performed at a depth of three centimetres.

Renter's combine machine is fundamentally made up of a chisel, followed by a hoe powered by power-takeoff from the same tractor, coupled-on the same support-with a straight roller having protrusions, which serve to further crush clods and to level the soil.

Such elements can be easily broken down. Obviously, in this case, the feed speed is strictly connected with the operating requirements of the rotary machine.

Furthermore, the different types of soil processing compared in the various theses found short-range application, which made it possible to operate on a soil under the same physical conditions.

The data concerning processing capacity, fuel consumption converted to power input for an immediate assessment of the energy input needed for the wide range of techniques, and the specific output reported in table 1 are expressed as an absolute value and index numbers vs the monitor-serving conventional method (conventional deep ploughing = 100).

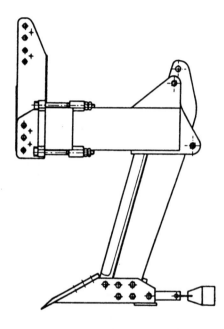

Fig. n. 2 - SO.GE.MA.'s standard subsoiler equiped with mole ball.

3 ECONOMIC ASPECTS

3.1 Besides the technical side, our attention is going to be focused on the results deriving - in economic terms - from the tested alternative land tilling techniques.

At any rate, it should be pointed out that, apart-from the criteria adopted for experimentation-related physical values to be converted to money values, conclusions cannot be generalized to the whole agricultural sector owing to greatly

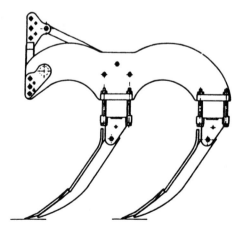

Fig. n. 1 - SO.GE.MA.'s subsoiler chisel.

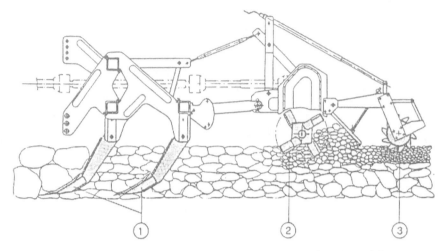

Fig. n. 3 - Side view of RENTER's compound machine formed by:
1) subsoiler chisel with 5 tines;
2) rotary tiller;
3) paker's rotary hoe.

varying operational costs within a wide range of cropping farms, depending on a great number of interacting variables.

As a result, the economic judgement we'll come to may be felt sound within the hypotheses guessed.

Such being the picture, it stands to reason that we are now handling capital projects which can be differently analyzed in view of a shrewd assessment of the same.

There follows mention of the most significant ones.

There are methods which do not take the present value into account, like, for example, the payback period by which the length of time necessary to recover the initial capital employed can be estimated, as well as the rate of return by which an average investment profitability can be calculated, thus assessing - in either case - the capital needed for output achievement and the annual cash flows for the duration of the investment policy.

On the other hand, there are methods which do consider the present value, the most widely used being the net present value by which it is possible to obtain a summation of the present values of projected cash flows paid off at a rate of discount fixed according to the average interest on a long - and mid-term investment basis. Shoould this summation go beyond the initial capital issued, it is worth achieving it.

The ratio between the summation of the present values of cash flows and capital investment at source will result in a cost/benefit analysis wich also denotes a profitability index.

Obviously, if the ratio is greater than the unit, the investment will prove to be acceptable but instead the project will not be taken into account if the index is lower. Indicative of this is the internal rate of return, namely the rate of discount marking the summation of the present values of cash flows equal to the value of the primary capital investment.

Such a method will thus enable us to estimate the real profitability index which - once compared with an adequate market rate of interest - may ascertain whether or not there is chance of the project being carried out.

Great difficulties arise when making such economic evaluations, especially if we consider that in time all the hypotheses guessed tend to vary, and that data are related to a given period, thereby changing unevenly.

The above lays stress on the difficulties specific to economic evaluations, the number of wich ever being on the rise if highly relevant to the agricultural sector whereby some variables (e.g. the climatic one) come into play, with a trend which makes it hard to make whatever forecasting.

TYPE OF SOIL PROCESSING	Tilling depth	Tilling width	Specific employment of means and labour			Fuel consumation and electrical input					Specific output		
	m	m	(h/ha)	v.a.	n.i.	(kg/ha)	v.a.	(MJ/ha)	v.a.	n.i.	(t/ha)	v.a.	n.i.
Deep ploughing	0,50	1,10	3,12			67,9		2875					
Clod breaking	0,15	2,70	0,93			11,4		482					
Cross-clod breaking	0,15	2,70	0,43	100,00		5,2		221		100,00	92,00		100,00
Seeding	0,03	3,00	0,87			5,0		213					
(stake)													
Rippering	0,50	2,40	1,72			37,6		1592					
Clod breaking	0,15	2,70	0,73			9,0		382					
Cross-clod breaking	0,15	2,70	0,44	64,51		5,2		221		61,32	96,83		105,25
Seeding	0,03	3,00	0,87			5,0		213					
(1st thesis)													
Rippering	0,50	3,20	2,22			42,0		1779					
Clod breaking	0,15	2,70	0,90			11,1		470					
Cross-clod breaking	0,15	2,70	0,45	79,68		5,4		228		69,23	91,81		99,79
Seeding	0,03	3,00	0,90			5,0		220					
(2nd thesis)													
Scarification	0,50	2,70	0,92			22,3		944					
Skim ploughing	0,30	1,10	2,50			49,7		2104					
Clod breaking	0,15	2,70	0,67			8,3		352					
Cross-clod breaking	0,15	2,70	0,42	100,89		5,1		215		101,03	100,00		108,70
Seeding	0,03	3,00	0,85			5,1		215					
(3rd thesis)													
Rippering +	0,50	2,50	3,33	74,33		67,9		2873		80,35	92,35		100,38
holing and rolling	0,25	2,50											
Seeding	0,03	3,00	0,87			5,0		213					
(4th thesis)													

3.2 In the case in point, we'll bump into further stumbling blocks. In fact, we are bound to evaluate trial results relevant to restricted areas and specific cultivation, and not to one or more cropping farms with a well-defined production system which would allow to analyze both expense and revenue related to a form of long-term investment.

This aspect surely represents one of the main goals of farm management. The above is of major importance if we consider that experimentation is all about aiming for an alternative employment of land till methods; this cultural operation is known to be performed for each crop intended for realization; hence the shorter production cycles of selected crops are, the more frequent capacity ratio of mechanical implements increases, this leading - from the economic standpoint - to an allocation of costs according to a greater number of hours whereby they find a convenient use on the farm.

A more discrimination between the user cost of the various mechanical tools employed will enable us to make sound judgements on alternative techniques.

At this point we deem it useful to remind that the user cost of a machine is accounted for by the summation of fixed - and variable costs. The former are made up of three elements: amortization, maintenance and interest on invested capital, functions of working life, annual capacity ratio and interest rate. Variable costs accounted for mainly by fuel and wages are proportional to the use of machines; they may further be easily determined, these corresponding to real outlays, unlike oncosts to be merely regarded as estimated values.

Evaluations thereof will be only economic to be precise, since we do leave rather huge public financial aids on investments within farm mechanization out of consideration - both for sunk contributions and credit facilities. Obviously, if such hypotheses were to be advanced, judgements would largely vary; anyway, we felt it fit to dwell upon the economic appraisal in order to better grasp the scope of innovations.

3.3 Use was made of the methodology adopted generally by agricultural economists as to user pricing.

In this respect, the new value of machines stems from adding VAT to late - 1987 ruling prices, with the recovery value standing at 20% as to tractors and 5% as to operating machines. It stands to reason that the difference between new- and residual values accounts for the value to be recovered.

In order to calculate amortization, the linear sharing principle was followed according to the renowned formula:

$$q = \frac{nV - rV}{n}$$

(q = yearly quota, nV = new value, rV = recovery value) where n, service life, was estimated to last 10 years as to the tractor, 20 years as to the disk harrow, hoe and roller, 12 years as to other operating machines. Interests allowed were reckoned on the locked-up mean value, the formula of which being as follows:

$$mV = \frac{nV - rV}{2}$$

at a rate of 8%. As regards maintenance, the latter was calculated as percentage value of the new value being as much as 2% for operating machines and 6% for the tractor.

As for this one, insurance as much as 1% of the new value was taken into account.
3.4 On the basis of the above methodology annual constant costs as to tractor can be held to amount to Lit. 21.857.180; therefore, supposing an annual employment of 1.000 hours - which is certainly suitable for so highly-powered a tractor - there will be a time-incidence of Lit. 21.857 to be resorted to in the following evaluations.

As for operating machines, processed data concerning oncosts allocated according to different tilling techniques have been reported in Table 2. In this context, the seeder was not taken into account since its cost has no bearing on the desirable results, given some basically equal technical parametres for all theses.

Now, at least data enable us to analyze the problem. In the case in point, one should consider that the different theses compared have effects on the value of the crop yield. In fact, from experimentation it may be argued that on plots the resulting crop differs both quantitatively and qualitatively (degree of polarization)

with subsequent price differentiation and uneven revenue, ewing to different jointly-acting quantities and prices. In Table 3 report has been made on relevant data per hectare.

For the sake of convenience, it is not enough to compare unit costs of the different alternative solutions; it will be for more indispensable to analyze the value of the crop, or product, except the cost of operation performed by various techniques, as well as tend to the highest value, or crop plusvalue.

There follows pertinent formula:

$$R_i \lesseqgtr R_j$$

where both Ri and Rj stand for the value of the crop obtained upon application of technique i or j.

The crop value stems from the difference QP-CTU where Q accounts for the crop quantity, P for the price, and CTU for the cost of operation in relation to the technique adopted.

Therefore, by using the methodology described so far and working out relevant data suitably, as well as by adopting the hypothesis that operating machines are to be used, on average, 200 hours annually, the results reported in table 4 can be obtained, from which it may be inferred that the technique allowing to achieve the crop plusvalue is the one based on rippering via a five-tined chisel whose specific costs are quite low although the revenue per hectare is not specifically high.

Moreover, on the basis of the data available it may be also inferred that, though not to the same extent, all alternative land tilling techniques vis-à-vis the one of deep ploughing are less exacting and more advantageous.

However superfluous it may be, it should be stressed that all evaluations attempted up to now have been based on the results related to technical and operating experimentation; hence they are to be considered within the range of the data available besides other hypotheses guessed (annual capacity ratio, working life, and so on).

4 CONSIDERATIONS AND CONCLUSIONS

The above economic and technical analysis, whose results have been already reported, enables us to make some operating comments and, at the same time, to stress their limits.

As things stand, we see fit to consider the four theses separately and then compare them with the stake.

TAB. N. 2 - Oncosts of operating machines used in different tillages

Operating machines \ Operations	Deep ploughing	Pentatoothed rippering with autosequent clod breaking	Eptatoothed rippering with later clod breaking	Scarification with subsequent surface labour and clod breaking	Rippering with combined hoeing and rolling
Double-shared plow	1.064.853	- - -	- - -	- - -	- - -
Five-tined or seven-tined chisel	- - -	894.783	894.783	- - -	- - -
Three-toothed scarifier	- - -	- - -	- - -	733.720	- - -
Four-share plow	- - -	- - -	- - -	912.677	- - -
24-disk harrow	676.700	676.700	676.700	676.700	- - -
"Rambo" type chisel	- - -	- - -	- - -	- - -	724.593
Rototiller with a Paker roller	- - -	- - -	- - -	- - -	811.270
TOTAL	1.741.553	1.571.483	1.571.483	2.323.097	1.535.863

TAB. N. 3 - Sugar beets yelds had in the plots according to different tillages, average degree of polarization and pertinent average unit price.

TYPE OF SOIL PROCESSING	Net specific collect output			Degree of polarization			Average total unit price		
	(t/ha) v.a.		n.i.	(%) v.a.		n.i.	(£/t) v.a.		n.i.
Deep ploughing (stake)	92,00		100,00	10,34		100,00	51118,80		100,00
Rippering (1) (1st thesis)	96,83		105,25	11,55		111,70	59978,00		117,33
Rippering (2) (2nd thesis)	91,81		99,79	10,75		103,96	54580,55		106,77
Scarification with skim ploughing (3rd thesis)	100,00		108,70	11,27		108,99	58110,00		113,68
Rippering + holing and roling (4th thesis)	92,35		100,38	10,65		103,00	53695,00		105,04

v.a. = absolute value
n.i. = index number (deep ploughing) = 100
(1) by a five-tined chiesl
(2) by a seven-tined chisel

TAB. N. 4 - Values of crop had according to different tillages

TYPE OF TILLING	REVENUE (a)		IMPUTED COST										CROP VALUES c = a-b	
			FIXED				VARIABLES				TOTAL (b)			
			OPERATING		TRACTORS		FUELS AND LUBRIFICANTS		LABOUR					
	v.a.	n.i.	v.a.	n.i.	v.a.	n.i.	v.a.	n.i.	v.a.	n.i.	v.a.	n.i.	v.a.	n.i.
Deep ploughing (stake)	4.702.930	100,00	23,882	100,00	98,044	100,00	59,356	100,00	44,857	100,00	226,139	100,00	4.476.791	100,00
Rippering (1) (1st thesis)	5.807.670	123,49	12,498	52,33	63,141	64,40	36,607	61,67	28,888	64,40	141,134	62,41	5.666.536	126,58
Rippering (2) (2nd thesis)	5.011.040	106,55	14,507	60,74	78,063	79,62	41,805	70,43	35,715	79,62	169,090	74,77	4.841.950	108,16
Scarification with surface labour (3rd thesis)	5.811.000	123,56	18,517	78,54	98,870	100,84	59,954	101,01	45,235	100,84	222,576	98,42	5.588.424	124,83
Rippering + rolling and holing (4th thesis)	4.958.733	105,44	25,597	107,18	72,857	74,31	47,278	79,65	33,333	74,31	179,065	79,18	4.779.668	106,77

v.a. = absolute value
n.i. = index number (deep ploughing) = 100
(1) by a five-tined chisel
(2) by a seven-tined chisel

The five-tined chisel rippering, followed by clod breaking, has been shown to be the most promising technique. As far as specific employment of cultivation implements and labour is concerned, the said technique actually proved to be most time-saving one, since there has been some 35% time-reduction with respect to the stake, and some 40% decrease in electrical input. The output per hectare itself, which was 5% higher than the stake, has turned out most promising, with a high average degree of polarization of the resulting harvest; so the average total unit price standing at 17% in comparison with the conventional technique. Hence, from the economic standpoint it can be held that this technique has contributed to achieving the crop plusvalue.

The technique based on the use of a seven-tined chisel has turned out less beneficial. And this because of a lower speed of farm implements due to an increase in the tilling width. Thus, this kind of slowing down can be regarded as the cause of a soil loosened to a smaller extent, which would involve a yield per unit slightly lower than the stake. The polarimetric degree, however, has proved to be higher, with an average total unit price and a crop value both increasing by 7 and 8% respectively.

As regards the third thesis, it must be specified that, of those tested, soil scarification, followed by skim ploughing, is the only method of "two-layered processing". Technically speaking, it presents no considerable advantage whatever since both specific use of implements and labour and power input have been as high as those of deep ploughing. Anyway, this highly effective method, which implies a thorougher crushing and an average rise of the ground, has brought about the highest output per unit. Moreover, the crop itself has presented a high degree of polarization; hence economic results have turned out presented a high degree of polarization; hence economic results have turned out satisfactory although to a lesser extent than those achieved upon application of the first trial technique.

As for the last thesis, the resulting crop may be held to be quite low - both quantitatively (yield per unit) and qualitatively (degree of polarization) - although it shows a considerable reduction in processing time and absorbed energy, or power input.

Examining experimentation globally, all four alternative techniques may be claimed to be more beneficial than the conventional one, though not in equal measure.

The most advantageous technique is the one consisting of rippering via a fivetined chisel, followed by a two-layered processing, rippering via a seven-tined subsoiler and, lastly, the combine machine.

Both economic and technical considerations have been made on the basis of only a year's experimentation. Therefore, it is advisable to further promote and substantiate experimentation in the next few years even though the results obtained up to now are nothing other than the fruit of detailed analyses and appraisals. And this with a view to checking whether results may be borne out, in which case methodology would be extended to agricultural entreprises having the same pedoclimatic features.

Upon consideration of such features, tests as above prove to be worth carrying out, thereby affecting different realities because, as is known, Italian agriculture presents a wide range of crops; as a results, solutions suitable for certain producing regions can be of no help to others. Experimental research is therefore aimed primarily to detect methodologies and techniques well-suited for various requirements.

In this respect, it should be stressed that the significance attached to multiannual experimentation is bound to the necessity to dispel all doubts and perplexity already existing in the previous three-year experimentation and (1) still affecting the said techniques - both directly and indirectly - as to physical, chemical, and biological properties of soil. The use of the combine machine, which in last year's experimentation did not bring about any remarkable economic outcome, technically proves to be beneficial in that the seed-bed can be prepared by operating the machine only once, thus avoiding any subsequent soil hardness accompanied eventually by structural degradation. Therefore only a long-term experimentation will give us more detailed indications on the use of the said machinery. Hence the need for interdisciplinary investigations; for analyses of this kind do call for collaboration between researchers showing great professional competence in the relevant fields, viz pathological, microbiological, and agronomic beside the mechanical and economic one.

Research funded both by the Ministry of Education (60%) and the National Council for Scientific Research.

(1) Panaro V.- Patruno G. "Technical analysis on the use of new soil processing methods on cereal farms" being printed in the "Journal of Agricultural Engineering".

Land and Water Use, Dodd & Grace (eds), © 1989 Balkema, Rotterdam. ISBN 90 6191 980 0

Mounted and drawn vibratillers and proposal for a program for tillage data processing

A.Peruzzi, S.B.Consorti & S.Di Ciolo
Agricultural Machinery Institute, Pisa University, Italy

ABSTRACT: This paper describes an experimental test, carried out on a dry clay soil, on the performance of a mounted and drawn vibratiller. The mounted implement has been equipped with a counter-frame for vibration damping. The results show the advantages, from an energetic and work productivity point of view, of the mounted operative machine which also presents an "economical threshold" of vibration use lower than the drawn implement one. A program of computerized calculus, created by the research workers of the Institute of Agricultural Machinery of Pisa University, for a fast processing of experimental data relative to field tillage test, is also described. The authors finally outline possible further development of these studies.

RÉSUMÉ: Cet article décrit une étude comparative entre un scarificateur à vibration traîné et un autre porté, sur sol argileux sec. L'outil porté était équipé d'un contrechassis pour l'amortissement des vibrations. Les résultats montrent qu'l est plus avantageux, en termes d'économie d'énergie et de productivité, d'utilizer un outil porté. Celui-ci a en autre un "seuil économique" d'utilisation plus bas que l'outil traîné. On décrit ensuite un programme sur ordinateur pour l'élaboration rapide des données expérimentales concernant le travail du terrain mis au point par des chercheurs de l'Institut de Mécanique Agraire de l'Université de Pisa. Enfin, on envisage quels peuvent être les développements futurs de ces études.

ZUSAMMENFASSUNG: Dieser Werk beschreibt einen vergleichenden Untersuchung, auf einem trockenen tonhaltigen Boden zwischen die Leistungen eines geschleppten und eines aufgestellten Vibratiller (der zweite mit einem Gegenchassis, um die Vibrationen anzuhalten). Die Ergebnisse zeigen den Vorteil - kraft in Energieesparnis und in der größer Arbeitsproduktivität - des aufgestellten Gerät, das auch eine kleinere "Wirtschaftschwelle" als die geschleppten Gerät bietet. Ein Computerprogramme auch erklärt ist, das die Forscher aus Landwirtshaft Mekanik Institut von der Universität Pisa erfinden haben, um eine shnelle Experimentellendatenverarbeitung von Bodendbearbeitung zu versuchen. Die Autoren untersuchen eigentlich die möglichen weiteren Entwicklungslinien dieser Forschungen.

1 INTRODUCTION

Primary tillage in Italy absorbs the greatest part of the total fuel consumption, used in agriculture, affecting herbaceous crops total energetic input by about 50% (Triolo et al..1987).

There is therefore a clear need of replacing traditional tillage (consisting in a ploughing 0.5-0.6 m deep) with alternative techniques, in order to obtain either a relevant energy saving, or an increase of work productivity and consequently a reduction of running costs without any reduction of crops productive standards.

A good alternative to deep ploughing could

(a)

(b)

Fig.1. Vibratillers at work: a) modified
mounted implement, b) drawn implement.

be rapresented by deep subsoiling, either
as a single technique, or as the first and
basic part of two layer tillage.

The operative machines available for deep
subsoiling equipped with a vibrating system
can be considered very versatile because
they can be used either as static subsoilers
ers or as vibratillers. In fact the use of
vibrations in subsoiling of compact soils
should be advantageous from a mechanical,
energetic and economical point of view.

The use of vibratillers can cause some
troubles due to the effects of vibration
propagation to the tractor (stress crack-
ings, resonance problems, ergonomics and
security problems, etc.).

It is possible to limit these troubles by
using drawn vibratillers which have a
large mass. At the same time operative ma-

chine tires allow a certain damping of vibrations.

Anyway it is preferable to use mounted
vibratillers equipped with an appropriate
dampener able to reduce at least the main
trouble due to vibration propagation to the
tractor.

A study has been carried out on this sub
jet at the Agricultural Machinery Institute
of Pisa University.Tests have been made by
comparing the relative performances of a
drawn and a mounted vibratiller (Fig.1).
The last implement had been equipped with a
damping counter frame connecting the vibra
tor-tines unit to the frame of the operative machine. The two tools have similar operative characteristics as: theoretic working width; tine number, form and relative
distance. They have been used either as sta
tic subsoilers or as vibratillers.

This paper describes the results of the
experimental tests and a short presentation
of a gwbasic program, created by the research workers of the Agricultural Machinery Institute of Pisa University on p.c.Oli-
vetti M24, which is a proposal for data pro
cessing standardization of tillage field
tests.

2. MATERIALS AND METHOD

2.1 Technical characteristics of the imple
     ments

For both tools the working width is 1.5 m.
Each implement has three tines. The tines
are bent in the feed direction, they have
a relative distance of 0.75 m, they are
0.850 m long and 0.025-0.040 m thick.

The two implements have, on the contrary,
different size, different weight and diffe
rent characteristics of the vibrating system. The drawn subsoiler is 2.35 m long,
1.90 m wide and 1.30 m high and its weight
is 1850 daN, while the mounted tool has a
smaller size and weight (it is 0.5 m long,
1.5 m wide and 1.4 m high and its weight is
500 daN).

The vibrating system of the drawn implement consists in a rotating eccentric mass
which has the form of a truncated pyramid;
it weighs 78 daN and its eccentricity is
0.12 m. In the mounted vibratiller the vibrator consists in the two counter-rotating
eccentric masses, each one costituited by

a toroidal sector with central angle of 144°. Each sector weighs 46 daN and has ec centricity of 0.08 m.

2.2 Test methodology

The tests have taken place in the Experimental Research Center "E.Avanzi" of Pisa University on a compact soil. The main physical-mechanical characteristics of the tested soil are given in table 1.

Both the implements were coupled to a 93.55 kW 4WD tractor.Test fields were 130-160 m long and 18-25 m wide.

A single furrow deep ploughing was carried out to have a comparison with the results obtained with the implements described abo ve.

Table 1.Physical-mechanical characteristics of the tested soil.

PARTICLE SIZE ANALYSIS:
- texture (∅ > 2 mm)   absent
- sand   (0.02 < ∅ < 2 mm)   33 %
- silt   (0.002 < ∅ < 0.02 mm) 25 %
- clay   (∅ < 0.002 mm)   42 %

CLASSIFICATION: clay soil

ATTERBERG LIMITS:
- S.P.L.   22.0
- I.P.L.   47.5
- Ip   25.5

AVERAGE MOISTURE CONTENT
IN THE TEST PERIOD:
- layer 0-0.3 m   14.5 %
- layer 0.3-0.6 m   16.5 %

2.3 Data processing

A gwbasic program was created to have a da ta processing system which is fast and in dipendent of the size of worked fields.The flow chart of the program is shown in figu re 2. Experimental data are processed by the program and are referred to an ideal field 333 m long and 30 m wide (so that the field area is nearly one ha). The program allows a comparison among five tillage tests.

Input data requested by the program are shown in table 2. For the quantities quoted

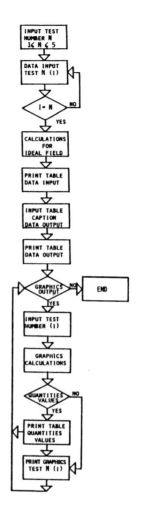

Fig.2. Flow-chart of the program.

with and asterisk, the user has to specify the number of experimental data (up to a maximum of 50).

Output data processed by the program (as shown in table 3) allow a comparison among the different tests indipendently of the size of the worked fields.

On request of the user, the program proceeds then to a graphic presentation of the results on (CO), (ce),(Em) and (R)(see table 3)as a function of the field length from 50 m to 950 m (in steps of 50 m). A quick inspection of the graphic output allows the agrarian operators to adapt the final data to their personal needs.

1611

Table 2. Input data requested by the program.

DATA INPUT	SYMBOL	UNIT
field length	F.LEN	m
field width	F.WID.	m
work effective depth (*)	pe	m
work effective width (*)	be	m
passages number	PASS.N.	-
passage working time (*)	T.E.	s
turn time (*)	T.A.V.	s
"tempus adjuvandi curando"	T.A.C.	s
"tempus morandi"	T.M.	s
total working time	T.T.	h
50 meters working time (*)	T.W.50m	s
50 meters idling time (*)	T.I.50m	s
fuel consumption	FUEL C.	l
drawbar pull	F	kg
idling drawbar pull	Ra	kg
PTO power	Ppto	HP
tractor type (2WD-4WD-TRACK)	T.TYPE	-
tractor power	T.POWER	HP
speed used (1->N)	SPEED	-
speed reduction used (1->N)	REDUCT	-
engine revolution number	En.R.N.	r.p.m.

Table 3. Output data processed by the program.

DATA OUTPUT	SYMBOL	UNIT
tractor type	-	-
tractor power	T.POWER	kW
speed-reduction	SPEED	-
engine revolution n.	EN.R.N.	r.p.m.
operative work width	bo	m
coeff.of width use	bo/be	-
effective work depth	pe	m
worked section	pexbo	m^2
"tempus efficientiae"	TE	h/ha
"tempus adjuv.vert."	TAV	h/ha
"tempus operandi"	TO	h/ha
TE/TO %	-	-
TAV/TO %	-	-
effective work speed	vb	m/s
effective work speed	vb	km/h
operative work speed	vo	m/s
operative work speed	vo	km/h
operative work capacity	CO	ha/h
effective work capacity	CE	ha/h
operative volume of worked soil	VO	m^3/h
fuel consumption per hour	ch	kg/h
fuel consumption per ha	ce	kg/ha
fuel consumption per m^3 of worked soil	cm	g/m^3
drawbar pull	F	daN
drawbar pull per m^2 of worked section	F/pexbo	daN/m^2
drawbar power	Pg	kW
PTO power	Ppto	kW
effective power	Pu	kW
effective power per m of bo and cm of pe	Pu/boxpe	kW/mxcm
effective energy	Eu	kWh/ha
effective energy	Eu	MJ/ha
mechanical energy	Em	kWh/ha
mechanical energy	Em	MJ/ha
driving power	Pm	kW
tractor skidding %	-	-
tractor efficiency %	Rp	-
energetic efficiency	R	-
specific consumption	ce/Em	kg/kWh
eff.energy per 1000 m^3 of worked soil	Eum	$MJ/1000m^3$

3. RESULTS AND DISCUSSION

Test results, shown in table 4, show immediatly that any kind of subsoiling allows relevant energy saving and remarkable increase of work productivity with respect to deep ploughing.

The use of the mounted implement causes a operative time (TO) reduction of 79% (as a static subsoiler) and of 76% (as a vibra tiller), while the drawn tool causes a (TO) reduction of 77% and 69% respectivily with respect to deep ploughing.

There is a similar trend for fuel consumption per ha (ce). Mounted subsoiler causes fuel savings of 79% (static) and of 78% (vibrating) in comparison with deep ploughing, while the drawn implement causes a (ce) reduction of 77% and 74% respectively.

The mean values of fuel consumption per m^3 of worked soil and of effective power per meter of (bo) and centimeter of (pe) are smaller for the two implements used as vibratillers than for the two

Table 4. Results of the tests of the two implements, each used with and without vibrations.

CHARACTERISTICS	SYMBOL	UNIT	(1)	(2)	(3)	(4)	(5)
tractor type	-	-	4WD	4WD	4WD	4WD	4WD
tractor power	-	kW	95.55	95.55	95.55	95.55	95.55
speed	-	-	2 - 2	2 - 2	1 - 2	2 - 2	1 - 2
engine revolution n.	-	r.p.m.	2100	2100	1900	2150	2200
effective work width	be	m	0.57	1.50	1.50	1.50	1.50
operative work width	bo	m	0.58	2.48	2.52	2.49	2.64
coeff.of width use	bo/be	-	1.02	1.65	1.68	1.66	1.76
effective work depth	pe	m	0.55	0.42	0.52	0.48	0.57
worked section	boxpe	m^2	0.32	1.04	1.31	1.20	1.50
"tempus operandi"	TO	h/ha	3.763	0.872	1.171	0.785	0.921
TE/TO %	-	-	94	87	89	92	94
TAV/TO %	-	-	6	13	11	8	6
effective work speed	vb	km/h	4.688	5.123	3.686	5.484	4.403
operative work speed	vo	km/h	4.575	4.618	3.390	5.116	4.117
operative work capacity	CO	ha/h	0.266	1.147	0.854	1.274	1.086
operative volume of worked soil	VO	m^3/h	1461	4817	4442	6114	6190
fuel consumption per h	ch	kg/h	16.92	17.06	14.23	16.79	15.44
fuel consumption per ha	ce	kg/ha	63.66	14.88	16.66	13.18	14.22
fuel consumption per m^3 of worked soil	cm	g/m^3	11.57	3.54	3.20	2.74	2.49
drawbar pull	F	daN	2900	3080	2870	3400	3200
drawbar pull per m^2 of worked section	-	daN/m^2	9079	2953	2190	2845	2129
drawbar power	Pg	kW	37.00	42.95	28.79	50.76	38.35
PTO power	Ppto	kW	0	0	6.62	0	5.15
effective power	Pu	kW	37.00	43.00	35.00	51.00	43.00
effective power per m of (bo) and cm of (pe)	-	kW/mxcm	1.158	0.412	0.270	0.425	0.289
driving power	Pm	kW	70	71	59	70	64
effective energy	Eu	MJ/ha	470	118	134	132	136
effective energy per 1000 m^3 of tilled soil	Eum	$MJ/1000m^3$	85.4	28.0	25.8	27.6	23.8
mechanical energy	Em	MJ/ha	950	222	249	197	212
energetic efficiency %	R	-	49	53	54	67	64
tractor skidding	-	-	25	23	22	12	13

(1) Single furrow plough; (2) Drawn static subsoiler; (3) drawn vibratiller;
(4) mounted static subsoiler; (5) mounted vibratiller.

tools used as static subsoilers. In fact there are average reductions of 76.5% for vibratillers and of 66% for static subsoilers in comparison with deep ploughing.

The mounted operative machine (either as static subsoiler or as vibratiller) results in higher values of energetic efficiency (R) than the drawn implement. This fact seems to be due to the reductions of tractor skidding and of "tempus adjuvandi vertendo"

(TAV) achieved with the mounted tool.

Figure 3 presents a graphic comparison of operative working times (TO), fuel consumption per ha (ce) and effective energy (Eu) in index values (the values of deep ploughing beeing equalized to 100) for the four tests. The plot shows that using the mounted tool there is no much difference between the static and vibrating modes. For this reason it can be stated that the use

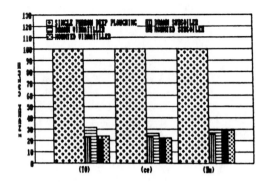

Fig. 3. Operative working time (TO), fuel consumption per ha (ce) and effective energy (Eu) in index value (the values of the deep ploughing being equalized to 100) for to the four tests.

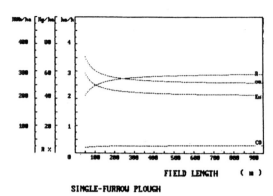

SINGLE-FURROW PLOUGH

Fig. 4. Work capacity (CO), fuel consumption per ha (ce) mechanical energy (Em) and energetic efficiency (R) as a function of field length for deep ploughing.

of vibrations is more convenient for the mounted tool than for the drawn one.

Figures 4,5,6 show the graphiccal presentation of the results on (CO), (ce), (Em) and (R) as a function of the field length, for deep ploughing (fig.4), mounted tool (fig.5) and drawn implement (fig.6) respectively. The plots show that the curves are asymptotic and that above a length of 300-500 m, the results are almost independent of the field length.

4. CONCLUSIONS

Experimental test results prove the superiority of the mounted operative machine (either as static subsoiler or as a vibratiller) for deep subsoiling with respect to the drawn tool. The mechanical, energetic and work productivity performances of the mounted tool are superior to those of the drawn implement, with similar operative characteristics.

For the subsoiling of the test soil the use of vibrations did not appear convenient (from an energetic and work-producti vity point of view) for both drawn and mounted tools. However it must be stressed that vibratillers produce an increase of working depth and a relevant reduction of drawbar pull per m² of worked section.

Moreover vibrations result in a better quality of subsoiling, by inducing more cracks in the tilled soil.

In general, the problems related to the

use of vibrations depend on the soil typ. pology. It would be appropriate to fix limiting values for soil toughness and moisture content beyond which an operative machine can be properly used as a vibratiller or as a static subsoiler.

According to tests results, the mounted implement presents a lower "economical threshold" for the use of vibrations with respect to the drawn tool.

The protracted use of the mounted vibratiller equipped with the vibrations dampener has not cause any kind of trouble. Any way the performances of the dampener used in the test has not been compared with other thecnical soluctions for vibrations damping. In this way, the differences in performance between drawn and mounted vibratillers could be properly evaluated and the best soluction for vibrations damping for the mounted implement could be chosen.

The program created by the research work ers of the Agricultural Machinery Institute of Pisa University must be considered a first approach towards a fast and standardized processing of the experimental data relative to field tillage tests.

Improvements and suggestions are obvious ly needed and welcomed.

REFERENCES

Baraldi G. 1981. I ripuntatori nella lavorazione profonda del terreno. IMA-Il trat torista.3/4:8-9.

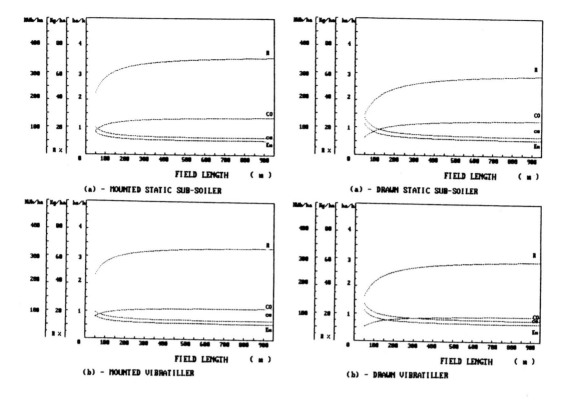

FIELD LENGTH (m)

(a) - MOUNTED STATIC SUB-SOILER

FIELD LENGTH (m)

(a) - DRAWN STATIC SUB-SOILER

FIELD LENGTH (m)

(b) - MOUNTED VIBRATILLER

FIELD LENGTH (m)

(b) - DRAWN VIBRATILLER

Fig. 5. Work capacity (CO), fuel consumption per ha (ce), mechanical energy (Em) and energetic efficiency (R) as a function of field length for the mounted tool: a) without vibrations; b) with vibrations.

Fig. 6. Work capacity (CO), fuel consumption per ha (ce), mechenical energy (Em) and energetic efficiency (R) as a function of field length for drawn tool: a) without vibrations; b) with vibrations.

Baraldi G. & Pezzi F. 1981.Macchine e metodi per la lavorazione principale del terreno. IMA-Il trattorista.19/20:7-23.

Baraldi G. & Pezzi F. 1986. Gli attrezzi a punte nella lavorazione principale del terreno.L'Informatore agrario.6:85-93.

Bolli P. &Scotton M. 1987. Lineamenti di tecnica della meccanizzazione agricola. Bologna:Edagricole.

Cera M. 1976. Meccanizzazione agricola - volume primo -. Padova:Patron.

Colonnelli F.D. & Farina G. 1982. Sub-soiler and chisel-plow performances for primary tillage. Rivista di Ingegneria Agraria.3:211-214.

Dalleinne E. 1979. Les façons en travail du soil. Etudes du CNEEMA.428,438;445, 455.

De Vecchi Pellati N. 1971. Lo scarificatore vibrante in sostituzione dell'aratro

rovesciatore nella lavorazione dei terreni argillosi asciutti. M&MA.7:27-39.

De Zanche C. & Frisio D. 1986. L'assorbimento energetico dei ripuntatori.L'Informatore agrario.26:29-34.

De Zanche C. & Frisio D. 1987.L'impiego di un ripuntatore vibrante nella lavorazione a due strati.L'Informatore agrario. 9:95-101.

Di Ciolo S.,Fiorentini R. & Zoli M. 1972. Lavorazione del terreno con operatrici vibranti e oscillanti. Rivista di Ingegneria Agraria.2:49-57.

Di Ciolo S.& Consorti S.B. 1984. Discissore vibrante: proposta per impedire la propagazione delle vibrazioni alla motrice. Rivista di Ingegneria Agraria.3: 176-182.

Di Ciolo S. & Peruzzi A. 1988. Proposal for data processing standardization for

tillage field tests.Agricoltura Mediter
ranea.3:231-236.

Pellizzi G. 1987. Meccanica e meccanizzazione agricola. Bologna:Edagricole.

Peruzzi A.,Consorti S.B. & Di Ciolo S.1988.
Discissura del terreno con attrezzi vi-
branti portati e trainati. Rivista di
Ingegneria Agraria.3:149-156.

Pezzi F. 1985. Lavorazione del terreno con
discissori a differenti caratteristiche e regolazioni. L'Informatore Agra-
rio.31:25-29.

Terrier H.1987.Le décompactage. Motorisation et Technique Agricole.4:27-36.

Triolo L.,Mariani A. & Tomarchio L. 1987.
L'uso dell'energia nella produzione
agricola vegetale in Italia. Agricoltu
ra ed Innovazione. 2/3:100-109.

Wolf D.,Garner T.H. & Davis J.W. 1981. Til
lage mechanical energy input and soil-
crop response. Transactions of the ASAE
4:1412-1425.

Land and Water Use, Dodd & Grace (eds), © 1989 Balkema, Rotterdam. ISBN 90 6191 980 0

Studies on engineering design theories of hand-tractor ploughs

S.Phongsupasamit
Chulalongkorn University, Thailand

J.Sakai
Kyushu University, Japan

ABSTRACT: The objective of this study is to derive design equations for planning design of hand-tractor ploughs. The theoretical relationship between hand tractors, wheels, ploughs and human engineering in the best and standard ploughing forms in three dimensions are analyzed and expressed in mathematical equations. Under such a situation, whole planning design theories are developed and applied to actual design practices so as to attain ideal ploughing performance. The effective location of ploughs elements which influences ploughing efficiency of hand tractors are primarily discussed. The specialized lug-shaped of iron wheels in stable ploughing without holding on handled is also discussed in this paper.

1 INTRODUCTION

Thailand has developed her own hand tractors as well as small riding tractors mainly for lowland-paddy fields. Meanwhile, there were also hand tractors and riding tractors imported from advanced countries. However, farmers preferred locally-made hand tractors and riding tractors to those imported ones. This is because the retail price of them is quite cheaper than imported tractors. The total number of hand tractors utilized on farms in Thailand was more than 240,000 units in 1983, while 10 to 15 percent of which were imported hand tractors. Such locally-made hand tractors are sometimes inconvenient to be used especially for ploughing. The main reason is due to the lack of basic engineering design know-hows of ploughing with hand tractors.

Fig.1 Thai hand-tractor plough

2 ENGINEERING DESIGN PROBLEMS OF THAI PLOUGHS AND HAND TRACTORS

At the beginning stage of hand tractorization in Thailand, hand tractors and ploughs were mostly developed by small machine repair shops and blacksmiths who were scattered all over the country. Subsequently, those shops became small local manufacturers to produce not only hand tractors and ploughs but also other farm machines. A typical model of Thai hand-tractor with plough is shown in Fig.1

This plough has only a control mechanism for ploughing depth because it was developed on the basis of the Thai conventional plough (see Fig.2). The ploughing depth can be adjusted through the upper screw of the plough. The plough is directly coupled to the hitch bar of the hand tractor by means of a simple pin. And the ploughman has to change the hitch point of the left in order

to plough edges of the field lot, as shown
in Fig.3, because the plough has no swinging
beam mechnism.

Fig.2 Thai animal-draft plough

Fig.3   The hitch bar of Thai hand tractor

When edges of the field lot are ploughed,
the ploughman has to change the hitch point
of plough again to other positions for
appropriate ploughing. And the ploughing
pattern is mostly right turn and moves from
levees side to the center of the field lot
by the continuous ploughing method. Therefore, busy turning and a dead furrow happen
at the center of the field lot after ploughing, and the dead furrow causes trouble in
levelling the soil surface through puddling.
Concerning some effects of cage wheels in
ploughing, the hand tractor is usually fitted with cage wheels (see Fig.3) for ploughing on wet or dry paddy fields. Cage wheels
are very effective to puddling but they are
not always suitable to be used for ploughing.
This is because the hand tractor always inclines in ploughing, and the wheel lug shape
should fit to the soil cutting-lines of the
ploughshare, as shown in Fig.4, in order to
ensure stable ploughing with joyful ploughing.

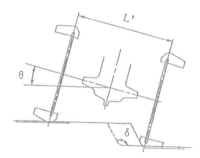

Fig.4   The reasonable shape of lug
for ploughing

However, in the case of the top soil of
paddy fields is very soft because it is
flooded by irrigation water. Thus, the top
soil can be puddled directly by a disk
harrow with cage wheels without ploughing.
The technical problems of hand tractor
ploughing in Thailand cause not only by the
plough and wheels but also the hand tractor.
The Thai locally-made hand tractors in 1970s
had no steering clutches, and heavy diesel
engines of 8 to 10 horsepower were commonly
installed. Furthermore, those hand tractors
had long handle structure. This is because
the long handle is very useful for the
ploughman to support and handle such a heavy
hand tractor, and moreover, the hand tractor
can be easily turned during ploughing by the
continuous ploughing method. Although current hand tractors have steering clutches,
their handles are still long. Therefore,
if the plough point (tip of ploughshare)
locates near the driving wheel axle of the
hand tractor and its handle is too long,
the handle has to be lifted up too high in
order to raise the plough point above the
soil surface as shown in Fig.5. In such a
condition, the ploughman is inconvenient to
work. Thus, the technological development
of hand-tractor ploughing in Japan should
be induced and adapted to Thai agriculture
in order to overcome those technical problems.

Fig.5   Problem of ploughing with the hand
tractor having long handle

3 BASIC DESIGN THEORIES OF HAND-TRACTOR PLOUGHS

3.1 Deviating angle η and stabilizing clearance Δl

In standard ploughing, as shown in Fig.6 projected from the view X in Fig.7, the setting direction of the hand tractor must slightly deviate from the side of ploughed land in order to make the hand tractor always travel straight and stable.

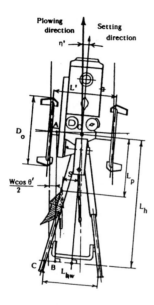

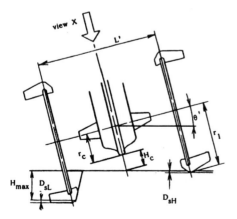

Fig.6 The whole machine in the best and standard ploughing form projected in perpendicular to the wheel axle

Fig.7 Determination of maximum ploughing depth H_{max}

The reason is that the hand tractor has a tendency to travel towards the side of unploughed or ploughed land if it set along the ploughing direction. This is because the soil resistance acting upon the plough and gross-tractions produced by both driving wheels always fluctuate and also are not equal. Therefore, in order to avoid the hand tractor traveling on the ploughed land, it must be set with a deviating angle to the side of the unploughed land. Under such a condition, the traveling action of the tractor to the unploughed land along the setting direction is prevented by the furrow wall against the inside lugs of the driving wheel (Fig.4) on plough-pan surface. As a result, the hand tractor travels straight along the furrow in the ploughing direction.

The authors propose to name the horizontal angle η as " deviating angle ". Also the authors propose to name the important clearance as " stabilizing clearance " of the stabilizer bolts of the plough yoke in hand tractor ploughing (Fig.8).

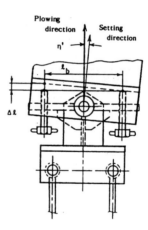

Fig.8 Plough yoke with stabilizing clearance Δl

3.2 Side-swing distance S and maximum side-swing distance S_{max}

The side-swing distance of the plough point S in standard ploughing (Fig.9) is computed from the following equation when the hitch-pin is approximately vertical:

$$S = oc - ob - W \cos \theta$$
$$S = (L' \cos \eta')/2 - L_y \sin \eta' - W \cos \theta \tag{1}$$

1619

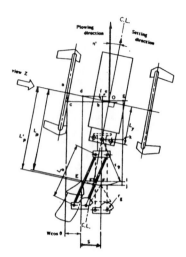

Fig.9　Determination of distance S and L'p

From Eq.(1), the optimum value of S can be decided according to the hand tractor specifications as Ly , L' and ploughing conditions as η' , θ and W.

The ploughshare has to swung behind each driving wheel in order to plough the edges of paddy field lot (Fig.10).

$$S_{max} = \frac{L'}{2} + L_0 + \frac{H'}{\tan \delta} \qquad (2)$$

The value of S_{max} is one of parameters for deciding the horizontal distance of swing beams L_b.

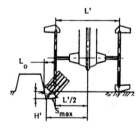

Fig.10　Determination of distance S_{max}

3.3 Inclined-angles θ and θ' of wheel axle

In standard ploughing with the hand-tractor plough, the wheel axle of the machine slightly inclines from the horizontal line.

There are two inclined-angles of the wheel axle: one is projected along the ploughing direction θ (Fig.11); and the other is projected along the machine setting direction θ' (Fig.7).

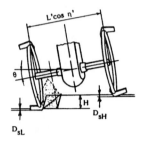

Fig.11　Determination of inclined-angle θ·

The inclined-angles θ and θ' are given by

$$\theta = \sin^{-1} \left(\frac{H + D_{sL} - D_{sH}}{L' \cos \eta'} \right) \qquad (3)$$

and

$$\theta' = \sin^{-1} \left(\frac{H + D_{sL} - D_{sH}}{L'} \right) \qquad (4)$$

3.4 Maximum ploughing depth, H_{max} vs hand tractor specifications

The maximum ploughing depth of hand-tractor ploughs can not be decided only by the plough structures because the lowest point of transmission case of hand tractors must not be pushed on the field surface.

The relation between the maximum ploughing depth H_{max} vs hand tractor specification is determined as follows:

From Fig.7, the outside radius of the wheel r_1 can be decided as following equation.

$$r_1 = \frac{1}{2}(L'\tan\theta') + r_c + H_c + \frac{D_{sH}}{\cos \theta'} \qquad (5)$$

Then,　$H_{max} =$

$$\frac{2L'C \sqrt{(L')^2 + 4C^2 - 4(D_{sH})^2 -} \quad (L')^2(D_{sL} + D_{sH}) + 4C^2(D_{sL} - D_{sH})}{(L')^2 + 4C^2} \qquad (6)$$

1620

where

$$C = r_1 - r_c - H_c$$

Figure 12 shows the calculated values of the maximum ploughing depth H_{max} with various of wheel radius r_1 and wheel space L' when the other parameters are constant. The maximum ploughing depth is 0 when wheel radius approximately equals 21.5 cm. When L' is 70 cm, the maximum ploughing depth of 15 cm correlates to 29.0 cm of wheel radius. However, in a case of soft paddy fields, both driving wheels will sink and travel on the plough-pan surface. Therefore, in such a condition when the ploughing depth is 15 cm, the optimum wheel radius is 35 cm.

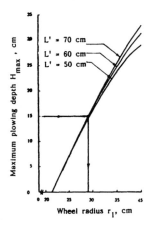

Fig.12 Relation between maximum ploughing depth and wheel radius and wheel space L'

All these dimensions as a case study concern drive-type hand tractors of 7 to 8 PS water-cooled diesel engines. Not all but some of these dimensions should be approximately proportional to engine PS.

3.5 Height of higher hand grip : H_{hL}

The height of the hand grips should be decided in the range of comfortable working manner of ploughman. The height H_{hL} of the higher hand grip in standard ploughing (Fig. 13) is computed from the following equation:

$$H_{hL} = H_{hR} + L_{hw}\left(\frac{H + D_{sL} - D_{sH}}{L'}\right) \quad (7)$$

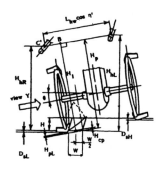

Fig.13 Dertermination of Distance H_{pL} and H_p

The value of H_{hL} depends on six parameters. And the height of the lower hand grip H_{hR} can be assumed to be 75 cm to 85 cm as the height for standard ploughing. Moreover, the width of hand grips should be 60 cm to 65 cm for handling the machine effectively. Figure 14 presents the calculated values of the hand grip height H_{hL} with various values of ploughing depth H, hand grip height H_{hR} and wheel space L'. The height H_{hL} is recommended to be 90 cm to 110 cm.

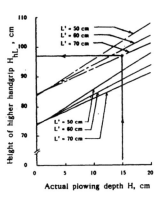

Fig.14 Relation between height of higher hand grip and actual ploughing depth, height of lower hand grip and wheel space

3.6 Effective height of hand grips : H_p

It is accurate and convenient to form side-view planning design of hand-tractor ploughs by projecting along the wheel axle. Then, the relation between the plough point and hand grip H_p has to be determined after the height of higher hand grip H_{hL} was decided. From Fig. 13, H_p is computed from the following expression:

$$H_p = \frac{H_{hR}}{\cos\theta} + \frac{W\sin\theta}{2} \qquad (8)$$

Figure 15 shows the relation between the effective height of hand grips H_p and the lower height of hand grips H_{hR} while the other parameter are constant.

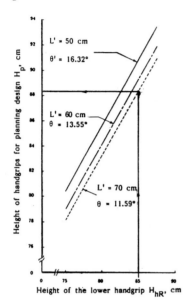

Fig.15 Relation between H_p and H_{hR} with various values of wheel space L' and inclined-angle θ

3.7 Relation between plough point and lowest point of lug : H_{pL}

The theoretical relative location between the plough point and the lowest point of the lug traveling on plough-pan surface (Fig.13) is determined from the following expression:

$$H_{pL} = W\sin\theta - \frac{D_{sL}}{\cos\theta} \qquad (9)$$

Figure 16 presents the relation between the calculated value of H_{pL} and various inclined-angle θ while the ploughing width and sinkage of the lower wheel are constant. For example in normal paddy fields of good irrigation and drainage functions, ploughing

conditions might be as follows; ploughing width is 16 cm, wheel sinkage D_{sL} and D_{sH} are approximately 1 cm and 2 cm, respectively. Therefore, when the wheel space L', ploughing depth H and deviating η' are 70 cm, 15 cm and 5 degrees, the calculated inclined-angle θ from Eq.(3) is 11.6 degrees which is correlated to 2.1 cm of H_{pL}.

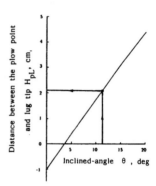

Fig.16 Relation between distance H_{pL} and inclined-angle of machine

3.8 Horizontal distance of plough point to wheel axle : L_p

The relative location of the plough to the hand tractor in the side view, from the view Y in Fig.13, also plays an important role to ploughing performance. Because the location of the rear-end portion of the plough must not be in the range of the ploughman's step in ploughing. And the ploughman has to lift the plough point above the field surface by lifting the hand grips when the hand tractor turns on headlands in continuous return ploughing or in order to inspect the plough during ploughing.

Therefore, the relative distance L_p of the plough point to the wheel axle (Fig.17) has to be decided carefully in consideration of the whole machine structure as well as ploughing conditions.

Figure 17 is a side view of the whole machine projected in paralled to the wheel axle as indicated as view Y in Fig.13. Considering the optimum location of the plough point in standard ploughing, when the plough point is elevated just above the field surface from point P_1 to P_2, and the hand grips have to be raised upward with H_L from point G_1 to G_2. The point O is considered as the origin of X - Y coordinate system. The plough point at P_1 moves to P_2 and the hand grip at G_1 moves to G_2 when the machine is rolated about point O with the rolating angle ϕ .

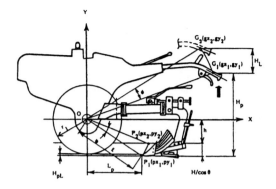

Fig.17 Determination of distance H_L when plough point is lifted above soil surface

The relation between the coordinates P_1 (px$_1$, py$_1$) and P_2 (px$_2$, py$_2$) and also G_1 (gx$_1$, gy$_1$) and G_2 (gx$_2$, gy$_2$) are given by

$$\begin{pmatrix} px_2 \\ py_2 \end{pmatrix} = \begin{pmatrix} \cos\phi & -\sin\phi \\ \sin\phi & \cos\phi \end{pmatrix}\begin{pmatrix} px_1 \\ py_1 \end{pmatrix} = \begin{pmatrix} px_1\cos\phi - py_1\sin\phi \\ px_1\sin\phi + py_1\cos\phi \end{pmatrix} \quad (10)$$

and

$$\begin{pmatrix} gx_2 \\ gy_2 \end{pmatrix} = \begin{pmatrix} \cos\phi & -\sin\phi \\ \sin\phi & \cos\phi \end{pmatrix}\begin{pmatrix} gx_1 \\ gy_1 \end{pmatrix} = \begin{pmatrix} gx_1\cos\phi - gy_1\sin\phi \\ gx_1\sin\phi + gy_1\cos\phi \end{pmatrix} \quad (11)$$

The value of $H/\cos\theta$ is computed from the following expression:

$$\frac{H}{\cos\theta} = py_2 - py_1 \quad (12)$$

Substituting Eq.(10) for py$_2$ into Eq.(12) and after transforming, we get:

$$px_1\sin\phi + py_1\cos\phi = \quad py_1 + \frac{H}{\cos\theta} \quad (13)$$

When the rotating angle ϕ is greater than 0 but less than 90 deg., the values of $\sin\phi$ and $\cos\phi$ from Eq.(13) are given by

$$\sin\phi = \frac{px_1\left(py_1 + \dfrac{H}{\cos\theta}\right) - py_1\sqrt{px_1^2 + py_1^2 - \left(py_1 + \dfrac{H}{\cos\theta}\right)^2}}{px_1^2 + py_1^2} \quad (14)$$

$$\cos\phi = \frac{py_1\left(py_1 + \dfrac{H}{\cos\theta}\right) + px_1\sqrt{px_1^2 + py_1^2 - \left(py_1 + \dfrac{H}{\cos\theta}\right)^2}}{px_1^2 + py_1^2} \quad (15)$$

According to Fig.16, we get:

$$l = \sqrt{px_1^2 + py_1^2}$$

$$h = py_1 + \frac{H}{\cos\theta}$$

Then, the values of $\sin\phi$ and $\cos\phi$ is reformulated by the following equations:

$$\sin\phi = \frac{px_1 h + py_1\sqrt{l^2 - h^2}}{l^2} \quad (16)$$

$$\cos\phi = \frac{py_1 h + px_1\sqrt{l^2 - h^2}}{l^2} \quad (17)$$

From Fig.16, the value of H_L is computed from the relation:

$$H_L = gy_2 - gy_1 \quad (18)$$

Substituting Eq.(11) for gy$_2$ into Eq.(18), we get:

$$H_L = gx_1\sin\phi + gy_1\cos\phi - gy_1 \quad (19)$$

Then, by substituting Eqs.(16) and (17) into Eq.(19), H_L is finally given by

$$H_L = \frac{h(gx_1 px_1 + gy_1 py_1) - (gx_1 py_1 - gy_1 px_1)\sqrt{l^2 - h^2} - gy_1 l^2}{l^2} \quad (20)$$

The value of H_L in Eq.(20) depends on six parameters, including the effective distance of the plough point to the wheel axle L_p or px_1. The values of gy_1 and py_1 can be determined from the values of H_p, r_1 and H_{pL} and gx_1 which is the length of hand-tractor handles. Moreover, the machine inclined-angle θ is computed from the decided parameters as well as the ploughing depth H. Therefore, H_L is computed by assuming the value of px_1. However, the calculated value of H_L must not be large. Otherwise, it is inconvenient for ploughman to lift the hand grips in turning the tractor.

For example:

H_p : 88 cm, H_{pL} :   2 cm •

θ : 11.59 deg., r_1 :   35 cm

H : 15 cm, L_h :   135 cm

Therefore, gx_1, gy_1 and py_1 equal 135 cm, 51 cm and -37 cm respectively. By assuming L_p or px_1 49 cm and substituting all those parameters into Eq.(26), H_L is :

$$H_L = \frac{-102\ 536.1 + 430\ 473.6 - 192\ 268.0}{3770.0}$$

$$= \quad 36.0 \quad cm$$

Hence, in order to lift the plough point above the field surface in turning the machines, the total heights of both lower and higher hand grips are computed from the following expressions:

$$H'_{hR} = H_{nR} + H_L \cos\theta \qquad (21)$$

and

$$H'_{hL} = H_{hL} + H_L \cos\theta \qquad (22)$$

Substituting the value of H_{hR}, H_{hL}, H_L and $\cos\theta$ into Eqs.(21) and (22), we get:

$$H'_{hR} = 85 + 36.0 \cos 11.59 = 120.3 \ cm$$

$$H'_{hL} = 97 + 36.0 \cos 11.59 = 132.3 \ cm$$

These heights are not so high that the ploughman can easily handle the hand tractor.

3.9 Horizontal distance of plough point to wheel axle : L'_p

The plough point is located at point E with side-swing distance S (Fig.9). In order to make the way of planning design easy, the plough point at F should be swung to F' with radius r_p about point O'. At that position, the swing beams are perpendicular to the wheel axle.

From Fig.9, L'_p is determined by the expression:

$$L'_p = L_y + r_p \qquad (23)$$

and since

$$r_p = \frac{1}{\cos\eta'}(L_p - L_y) + L_b - \sqrt{L_b^2 - S^2} - S\tan\eta'$$

Then, the value of L'_p is computed from the following equation:

$$L'_p = L_y + \frac{1}{\cos\eta'}(L_p - L_y) + L_b - \sqrt{L_b^2 - S^2} - S\tan\eta' \quad (24)$$

For example:

L' : 70 cm, L_b : 40 cm, L_y : 30 cm

W : 16 cm, L_p : 49 cm, θ : 11.59°, η' : 5°

Substituting these parameters into Eq.(1), we get side-swing distance S, 16.6 cm. Hence, L'_p is

$$L'_p = 30 + 19.1 + 40 - 36.4 - 1.45$$

$$= 51.3 \quad cm$$

This length is suitable because the length F'D(83.7 cm) is sufficient for designing the plough base, having the position of plough-sole end beyond the range of ploughmen's foot step.

4 SIDE-VIEW OF THE WHOLE PLANNING DESIGN

Figure 18 shows a side-view of the whole planning design, projected from view Z in Fig.9 of a turn-wrest plough for a hand tractor. The plough is designed in consideration of technical relationship between the whole machine involved in ploughing as well as human engineering. Those relationships are expressed in mathematical equations with many parameters of machine specifications, including ploughing conditions, especially on lowland paddy fields.

5 DRAWING OF PLOUGHING WHEEL

Figure 19 shows the whole planning design of the ploughing wheel. The wheel is designed with technical relations between the hand tractor, the plough and the wheels are taken into consideration.

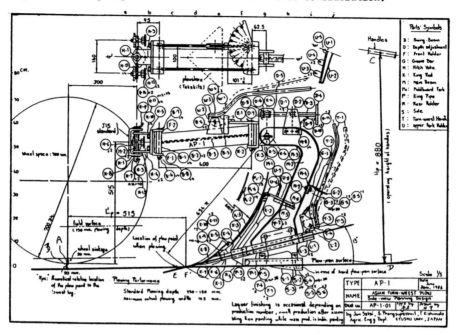

Fig.18 Side-view planning design of Asain turn-wrest plough AP-1

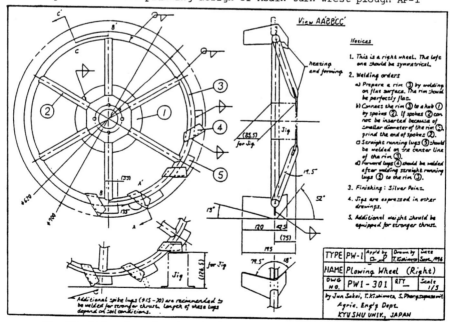

Fig.19 Side-view planning design of right ploughing wheels PW-1

6 CONCLUSIONS

(1) Designers have to consider not only the plough structures but also hand tractors, wheels and human engineering which are involved in ploughing in order to realize an efficient and stable ploughing.

(2) The maximum ploughing depth of hand-tractor ploughs in designing can not be decided by considering only the plough structures because its value depends on six parameters; there are hand-tractor specifications and the others are ploughing conditions.

(3) The operating heights of both left and right hand grips are not equal, and they should be decided in the range of comfortable working of ploughmen.

(4) The plough point is located lower than the lowest point of the lug traveling on the plough-pan surface in standard ploughing when the machines are projected in paralled to the wheel axle.

(5) The plough point has to be located at a proper horizontal distance L_p from the wheel axle in standard ploughing so that the ploughman can raise it above the soil surface by lifting the handles in turning the tractor or inspecting the plough.

(6) In order to lay out " whole planning design " easily, the center line of the plough should be set perpendicularly to the wheel axle of the hand tractor by shifting the plough from standard ploughing position to align with the center line of the hand tractor.

REFERENCES

Sakai, J., Phongsupasamit, S. and Kishimoto, T. 1986. Studies on evolutional steps for ploughing and tractorization in Japan,- The unique mechanization promotion in small-scale farming-. AMA, Vol.17, No.4: 11-19.

Sakai, J., Phongsupasamit, S. and Kishimoto, T. 1987. Study on basic knowledge of ploughing science for Asian lowland farming,- Plough structures related to ploughing principles-, AMA, Vol.18, No.1: 11-18.

Sakai, J., Phongsupasamit, S. and Kishimoto, T. 1987. Studies on basic knowledge of ploughing science for Asian lowland-paddy farming,- Appropriate plough structures for farming requirement-, AMA, Vol.18, No.2:11-17.

Phongsupasamit, S. and Sakai, J. 1987. Technology transfer problems for mechnizing small-scale family farms in Thailand, - Tillage and tractorization-. Proceedings of the International Symposium on Agricultural Mechanization and International Cooperation in High Technology Era, Tokyo: 539-553.

Sakai, J., Phongsupasamit, S. and Kishimoto, T. 1987. Engineering design theories of hand-tractor ploughs, AMA, Vol.18, No.3: 11-20.

Sakai, J., Kishimoto, T. and Phongsupasamit, S. 1987. A study on engineering design theories of iron wheels for ploughing,- relations between the whole machine and design theories for the wheel-, AMA, Vol. 18, No.4:11-18.

Phongsupasamit, S., Sakai, J. and Kishimoto, T. 1988. A study on engineering design theories of hand-tractor plough II,- Detail planning design of turn-wrest plough-, AMA, Vol.19, No.2:9-19.

Land and Water Use, Dodd & Grace (eds), © 1989 Balkema, Rotterdam. ISBN 90 6191 980 0

An experimental investigation of the effects of ribbed shanks on the soil displacements induced by a mole plough

M.Rodgers
Civil Engineering Department, University College, Galway, Ireland

J.Mulqueen
Agriculture and Food Development Authority, Ballinrobe, Co. Mayo, Ireland

M.Kyne
Civil Engineering Department, University College, Galway, Ireland

ABSTRACT: Many Irish soils are classed as being unsatisfactory for moling due to poor fissuring with existing mole plough designs. In this paper, mole plough shanks were modified by fitting ribs on the shanks to see if fissuring, resulting from mole ploughing could be improved. Tests were carried out in the laboratory and in the field with shanks ribbed on one side. A special model mole plough apparatus was designed and developed for the laboratory tests. Fissuring was evaluated by measuring residual soil displacements on the ribbed and smooth sides of the shanks. Three soils namely a clay, a silt and a sand were used in the laboratory and field tests. Results of the laboratory tests showed that soil displacements were greater on the ribbed side of the shank. Results of field tests appeared to be similar but were more difficult to evaluate due to the variability of the soil and the nature of the fissuring.

RESUME: Beaucoup de sols irlandais sont classés comme peu satisfaisants pour le Taupage, à cause de la médiocrité de fissuration avec les plans de Taupage existants. Dans cet article, les lames pour le Taupage ont été modifiées en adaptant des côtes à ces lames pour voir si les fissures provenant du Taupage pouvaient être améliorées. Des tests ont été faits en laboratoire et sur le terrain avec des lames à côtes sur un côté. Un appareil d'un modèle spécial de Taupage a été dessiné et amélioré pour les tests en laboratoire. Les fissures ont été évaluées en mesurant les déplacements des résidus de terre sur les côtes fixées aux lames et sur la partie non tranchantes des lames. Trois sortes de terre, de l'argile, du limon et du sable ont été utilisés pour les test en laboratoire et sur le terrain. Les résultats des tests en laboratoire ont montré que les déplacements de terre étaient plus grands sur les côtes fixées aux lames. Les résultats des tests sur le terrain semblaient être similaires mais étaient plus difficile à évaluer à cause de la variabilité du sol et de la nature de fissuration.

AUSZUG: Viele irische Böden sind mit der herkömmlichen Maulwurfsdränung wegen ihrer unbefriedigenden Rißbildung als nicht zufriedenstellend eingestuft. In der hier berichteten Untersuchung wurden die Pflugschenkel durch Anbringen von Rippen an die Schenkel verändert, um festzustellen, ob das Aufreißen - resultierend aus der Maulwurfsdränung - verbessert werden könnte. Tests wurden im Labor und im Feld mit an den Schenkeln einseitig angebrachten Rippen ausgeführt. Ein Spezialmodell Maulwurfspflug wurde für die Labortests entworfen und entwickelt. Die Rissbildung wurde an der verbleibenden Bodenumwendung an den gerippten und den glatten Seiten des Schenkels gemessen und ausgewertet. Drei Bodenarten, Ton, Schlick und Sand wurden in den Labor- und Feldtests verwendet. Die Ergebnisse der Labortests zeigten, daß die Bodenverschiebung an der gerippten Seite des Schenkels größer war. Ergebnisse des Feldtests erschienen ähnlich, waren aber wegen der Verschiedenheit der Böden und der Art der Rissigkeit schwieriger auszuwerten.

I INTRODUCTION

A mole drain is a trenchless drain which facilitates the removal of surface water and ground water from soils of low permeability. Fissures may develop as the mole drain is formed thereby improving permeability. Water percolates down through the soil and into the mole drain through these fissures. The two main criteria for a

successful mole drainage system are a
stable mole channel and a high degree of
fissuring. Soil, climatic and machine
factors have been identified as having a
large influence on the life and performance
of mole drains. The machine factor is the
one most readily controlled.

The existing mole plough design was developed in the eastern counties of England
for use on a limited range of soils. Many
Irish soils are graded as being unsatisfactory for moling with the existing mole
plough design. This paper describes the
results of modifications to mole ploughs
consisting of attaching vertical ribs to
the mole plough shanks. The modified
ploughs were tested on three Irish soils
in the laboratory and in the field.

The objectives of the study were:

1. To develop a laboratory model mole
plough apparatus whereby the suitability
of a soil for mole ploughing in the field
could be assessed.

2. To examine the effects of modifications to the model plough shank in the laboratory.

3. To carry out experiments to investigate if the results of modifications to the
model plough shank in the laboratory are
valid for similar modifications to commercial mole ploughs in the field.

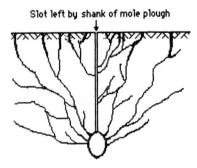

Slot left by shank of mole plough

Figure 1. Illustration of the fissuring
caused by mole drainage.

2 MOLE DRAINAGE

Mole drainage involves the creation of a
continuous cylindrical channel or mole
drain below ground surface with open fissures running from the channel to the surface (Figure 1). Mole drainage is required
wherever an impervious soil underlies a
thin top soil at shallow depth. The purpose of the fissures is to allow surface
water from the topsoil and ground water to
percolate into the mole channel through
which it can flow into an outlet system.
A mole plough is used to create the channel
and fissures. The mole plough consists of
four main components, a torpedo, a shank,
an expander and a beam (Figure 2).

An increase in the bearing capacity and
shear strength of the soil is one of the
main reasons for installing any drainage
system. Drainage of the soil lowers the
watertable and reduces the pore water
pressure of the soil. This results in an
increase in the effective stress of the
soil and hence of its bearing capacity.

2.1 Mole plough components

The torpedo is a steel cylinder of 75 mm
diameter, with a cutting face at its leading end and between 300 mm and 450 mm in
length as shown in Figure 2. The torpedo
serves to form a channel and deforms the
surrounding soil.

A vertical steel shank attaches the torpedo to the beam of the plough. As the
shank is pulled through the soil, it cuts,
displaces and deforms the soil. A wedge
of soil builds up in front of the shank
during the pulling operation.

The expander is usually a steel cylinder
with a maximum diameter of 100 mm. It is
attached to, and trails behind the torpedo
expanding the channel formed by the torpedo.
It serves to strengthen the mole wall.

The beam of the plough is usually mounted
on two wheels which control the working
depth of the torpedo and expander forming
a channel approximately parallel to the
ground surface.

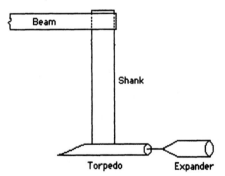

Figure 2. Main components of a mole
plough.

2.2 Depth

Mole drains are commonly pulled at depths

varying from 350 mm to 600 mm. These depths are required so that the mole channel will be able to withstand the applied loads from cattle and farm machinery and to lower the watertable to at least 300 mm below the soil surface. The draught force required to pull the mole plough at these depths can be readily accommodated by crawler tractors and by large wheeled tractors when traction conditions are favourable.

2.3 Soil permeability

The discharge and spacing of mole drains can be calculated using drain spacing formulae such as those of Toksöz and Kirkham (1961). These formulae show that the discharge and spacing of mole drains are very dependent on the permeability of the soil. The permeability of fine grained soils is very dependent on fissuring of the soil (Youngs 1985). During the pulling of mole drains at close spacings, the soil is fissured and the degree of fissuring is related to mole plough design and the nature of the soil. Fissuring results in an increase in the volume occupied by the soil and this volume increase can be measured from soil displacements.

3 LABORATORY APPARATUS

A laboratory model mole plough was built to approximately one third the size of a commercial machine. This model mole plough consisted of a shank, torpedo and expander. The shank was attached to a shaft which was mounted on a frame. This shaft was driven by an electric motor with speed control through a reduction gear box. The soil specimen was extruded into a moling cylinder rigidly fixed to the frame and 600 mm high and 101 mm internal diameter. Two diametrically opposed slots with guide plates were provided in the moling cylinder to allow the model mole plough to enter and leave the soil specimen.

The force required to drive the model mole plough through the soil specimen was measured by a proving ring and was recorded by an X-Y plotter. The displacements induced by the model mole plough were measured using a surface relief meter (Kyne 1988). This enabled the elevation of 78 points of the soil surface of the specimen to be measured.

4 SOIL CHARACTERISTICS

Three different soil types were used in

the tests. These were a predominantly clay soil from Ballinamore, County Leitrim, a predominantly silt soil from Inagh, County Clare and a predominantly sand soil from Mountshannon, County Clare. Some characteristics of these soils are shown in Table 1.

Table 1. Some characteristics of the clay, silt and sand soils.

Parameter	Clay	Silt	Sand
Moisture content (%)	28.0	31.7	15.5
Liquid limit	52.7	53.0	22.0
Plasticity index (%)	25.3	24.7	NP*
Bulk weight (kN/m^3)	18.6	19.8	20.9
Coarse sand (%)	5	3	41
Fine sand (%)	6	8	30
Silt (%)	35	48	16
Clay (%)	52	41	13
Organic matter (%)	0.98	0.78	0.90
Shear vane (kN/m^2) in situ	95	105	–

*NP – Non plastic.

In the clay and silt soils there are indications that the fissures and permeability arising from existing mole plough designs may not be adequate to drain the soil (Robinson, Mulqueen and Burke 1987). In the sand soil, fissuring appears to be adequate. Stability problems of the mole channel can be overcome by using a gravel fill in the channel (Mulqueen 1985, Galvin 1987).

Because of the gravelly nature of the soils derived from glacial deposits, the laboratory tests were carried out on reconstituted samples. The soils were air-dried and passed through a 2 mm sieve. The soils were then reconstituted to moisture contents and bulk unit weights determined from soil classification tests and trial mixes.

5 LABORATORY TESTS AND RESULTS

Preliminary direct shear box tests were carried out on undisturbed specimens of clay and silt soils using three different vertical normal loads. Three types of tests were carried out as follows:

1. An undisturbed soil specimen was sheared in the standard way.

2. A smooth metal plate mounted in the upper chamber was driven over the upper

surface of an undisturbed soil specimen placed in the lower chamber. The plane of contact was coincident with the shearing plane of the apparatus.

3. A ribbed metal plate mounted in the upper chamber was driven over the upper surface of an undisturbed soil specimen placed in the lower chamber. The plane of contact was coincident with the shearing plane of the apparatus. The space between the ribs was filled with soil which was then trimmed flush with the metal surface.

For soil specimens in test type 1, the shear strengths were measured according to the normal method used for the direct shear box test. In the type 2 test, the adhesions were calculated from the force required to drive the smooth metal plate across the soil surface. In the type 3 test, the adhesion derived is a compound value as it reflects soil to soil and soil to metal adhesion. Each test type was repeated six times

Table 2. Maximum stresses (kN/m^2) measured in direct shear box tests on undisturbed clay specimens. Test types 1, 2, and 3 are as described in the text.

Test type	Applied normal stresses (kN/m^2)		
	24.7	38.4	51.9
1	37.3 ± 2.9*	45.9 ± 3.7	55.7 ± 5.3
2	13.5 ± 0.8	23.1 ± 3.0	27.3 ± 6.3
3	23.7 ± 4.1	35.0 ± 5.1	38.4 ± 1.6

*Standard deviation

The results of the tests are shown in Table 2. The specimens tested for smooth metal to soil contact gave the lowest stress values. The specimens tested in the standard manner gave the highest stress values. The specimens tested with the ribbed metal to soil contact gave intermediate values. This indicates that there is greater adhesion between ribbed metal to soil contact than smooth metal to soil contact. This suggested that laboratory and field investigations should be carried out to examine the effect of ribbed and smooth mole plough blades on soil displacements.

5.1 Model mole plough tests

In these tests on reconstituted soils two shanks were used. One shank had two smooth faces and the other shank had one ribbed face and one smooth face (Figure 3). The main objective of the tests was to establish the difference in residual displacement between the ribbed and smooth faces of the blade. Twenty four model mole plough tests were carried out on each of the clay, silt and sand soils. A 101 mm diameter specimen was trimmed to the required length and the specimen was extruded into the moling cylinder. The original surface profile was taken using the displacement measuring device (Kyne 1988). The cylinder was fixed in the mole apparatus and the grooves of the shank were filled with soil and the soil was trimmed flush with the metal surface. The shank was driven through the soil at a constant speed of 150 mm/min. The force was recorded on the X/Y plotter. After the shank had passed through the soil, residual displacement readings of the surface profile were recorded. Samples were taken for moisture content. Three tests on each soil type were carried out using the blade with two smooth sides. Two depths of soil were tested, one at 150 mm and the other at 250 mm. Each soil was tested at a constant predetermined moisture content near the field value (Kyne 1988).

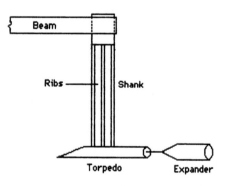

Figure 3. Diagrammatic view of ribs fixed to a mole plough shank.

Table 3 shows comparisons of average displacements on the ribbed (R) and smooth (S) side of the shank at two intensities of compaction on the three soils. C, M and S refer to clay, silt and sand respectively and D and E refer to the two compaction intensities employed and F refers to an all smooth blade. The model mole plough was drawn at a constant depth of 150 mm. The results show that residual displacements on the ribbed side were significantly greater than on the smooth side of the blade. In the case of the all-smooth blade the displacements were similar on both sides.

Table 3. Comparison of average displacements on ribbed and smooth sides of shank.

Test Type	Number of tests	R (mm)	S (mm)	$\sigma_{\bar{d}}$*	Signif.
CD	8	13.6	6.3	0.62	P<0.0001
CE	4	16.2	7.3	1.05	P<0.0I
CDF	3	6.6	7.8	1.64	P<0.5
MD	8	11.5	5.0	0.87	P<0.0001
ME	4	9.0	6.1	0.80	P<0.05
MDF	3	5.7	6.4	1.20	P<0.5
SD	8	14.1	3.4	1.40	P<0.0001
SE	4	7.6	4.9	0.65	P<0.05
SDF	3	5.1	4.7	1.10	P<0.5

* = Standard error of mean difference.

were almost symmetrical on either side of the shank.

6 FIELD TESTS AND RESULTS

The purpose of the field tests was to establish if the results of the laboratory mole plough tests on the three soils were similar in the field. In the field tests, ribs were attached to one side of a commercial mole plough shank. The effects of using different thicknesses of ribs and of attaching an extension plate to the back of the shank were examined. Initially the grass sod was removed from 2 m wide by 1 m long plots. Surface relief measurements were then made before and after drawing of the mole plough by using an electronic recording relief meter on three randomly selected cross sections in each plot. The relief meter consisted of 50 number 10 mm diameter steel measuring rods of equal

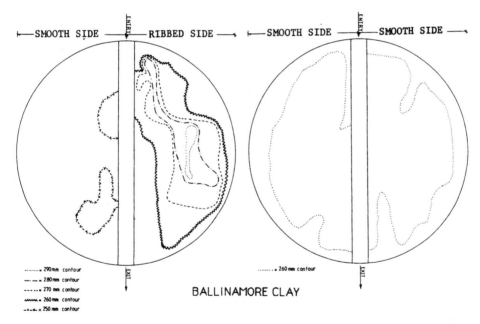

BALLINAMORE CLAY

Figure 4. Contour plots of surface relief after passage of a smooth/ribbed shank (left) and an all-smooth shank (right). The contours on right of entry-exit slot on left circle were caused by the ribbed side of the shank.

Contours of displacements were plotted for each test. Typical plots are shown in Figure 4. These plots clearly indicate much greater displacements on the ribbed side of the model mole plough. In the case of the all-smooth blade the displacements

length and spaced at 30 mm centres. A gamma ray bulk density meter was also used to examine soil densities before and after the field tests. The number of plots examined in the three different soils were as follows: clay - ten, silt - fifteen and

1631

sand - eleven. At least two plots were examined for each shank and typical soil displacements are shown in Figure 5.

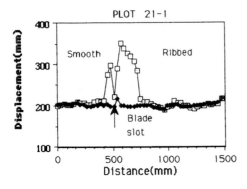

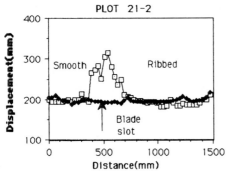

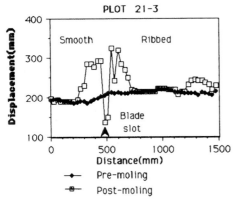

Figure 5. Residual soil displacements in the clay at 3 randomly selected cross sections on plot 21 using 25 mm deep ribs.

Difficulties were encountered in evaluating the results of the field measurements due to:
1. Fracture of the soil into large blocks.
2. Presence of large fissures.
3. Presence of stones.

4. Falling back of displaced soil into the shank slot and fissures, especially in the case of the sand soil.
5. Variability of the three soils which was shown by static cone penetrometer tests. For example, end resistances in the silt soil at the 400 mm depth varied from 0.7 MPa to 3.5 MPa.

7 CONCLUSIONS

The model mole plough was developed and was successfully used to test the effects of modifications to the mole plough shank on the three soil types. This model could be used in further studies on the suitability of soils for mole drainage and to model the effects of modifications to field equipment. The results of the laboratory tests, showed conclusively that soil displacements and fissuring were much greater on the ribbed side of the shank.

Field results also appeared to show that fissuring was much greater on the ribbed side of the shank. Due to the difficulties in evaluating the results of the field tests the effects of modifications to mole plough shanks may be best assessed from drain hydrographs on experimental plots.

REFERENCES

Galvin, L. 1986. Impermeable soils require stable channels and good crack formation for effective drainage. In K.V.H. Smith & D.W. Rycroft (eds.), Hydraulic design in water resources engineering: land drainage, p.413-422. Berlin: Springer-Verlag.
Kyne, M. 1988. An experimental investigation of the effects of ribbed blades on the soil displacements induced by mole ploughs. Unpubl. M. Eng. Sc. thesis. Galway: University College.
Mulqueen, J. 1985. The development of gravel mole drainage. J. agric. Engng. Res. 32: 143-151.
Robinson, M., J. Mulqueen & W. Burke.1987. On flows from a clay soil-seasonal changes and the effect of mole drainage. J. Hydrol. 91: 339-350.
Toksöz, S. & D. Kirkham 1961. Graphical solution and interpretation of a new drain spacing formula. J. Geophys. Res. 66: 509-516.
Youngs, E.G. 1984/1985. An analysis of the effect of the vertical fissuring in mole-drained soils on drain performances. Agr. Water Manage. 9: 301-311.

Land and Water Use, Dodd & Grace (eds), © 1989 Balkema, Rotterdam. ISBN 90 6191 980 0

Performance evaluation of an enamel coated mouldboard plough

V.M.Salokhe, D.Gee-Clough & A.I.Mufti
Asian Institute of Technology, Bangkok, Thailand

ABSTRACT: Experiments were conducted to compare a Thai-made slit type enamel coated plough with an uncoated similar mouldboard plough under similar working conditions. Experiments were conducted at 21%, 31%, 51% and 58% (db) moisture content in a laboratory soil bin using Bangkok clay soil. The effect of speed was also studied. It was found that enamel coating reduced the plough draft up to 26% depending on the soil moisture content and working speed. This reduction was found to be statistically significant.

1 INTRODUCTION

The mouldboard plough is a historic, useful and widely utilized tillage implement for primary tillage. A mouldboard plough cuts the soil slice, transports it over its surface and inverts it. These procedures require high energy input. Ploughing accounts for more traction energy than any other field operation (Kepner et al., 1978). Despite this, it is a very popular tractor as well as animal drawn implement.

The power requirement of tillage tools is an important aspect of design. The designer and manufacturer not only pay attention to the design of a better tool but also give due importance to power requirements. Mouldboard ploughs generally work well as low speed soil inverting implements and improvements can be obtainly mainly by reducing draft and wear. It was estimated that the frictional component of draft contributes about 30% of the total draft for ploughs at 3 kmph speed (O'Callaghan and McCoy, 1965). In the past several efforts were made to reduce this frictional force by coating the mouldboard surface with glass, Teflon and other low friction materials (Koolen and Kuiper, 1983). None of these efforts were very successful due to poor wear resistance, complicated application techniques or cost factors. Recently, while working at reducing the soil adhesion on cage wheel lugs, Salokhe and Gee-Clough (1987, 1988) tried different coating materials. They found

that enamel coating on a cage wheel lug surface reduces the soil adhesion considerably. The enamel coating is a simple and cheap technique and the technology is readily available in most parts of the world. Therefore in this particular investigation it was decided to use enamel coating on a mouldboard plough to study its effect on the draft requirement of the plough.

2 REVIEW OF LITERATURE

The draft requirement of the mouldboard plough is affected by soil type and condition, ploughing speed, plough-bottom shape, depth and width of furrow slice and friction characteristics of the soil-engaging surfaces. Kepner et al. (1982) found that the values of specific draft requirement of the plough increased from 1.4 to 2 N/sq.cm in sandy soils to 10 to 14.4 N/sq.cm in heavy gumbo soils. Reed (1937) found that the magnitude of the draft force is not only affected by the type of soil but also varies with change in travel speed. The same speed effects on ploughs draft were later observed by many research workers (Randolph and Reed, 1938; O'Callaghan and McCoy, 1965; Panwar et al., 1973 and Oskoui et al., 1982).

Many researchers, (Cooper and McCreery, 1961; Fox and Bockhop, 1962) reported that friction is greatly modified by adhesion. Soehne (1960) found that coefficient of friction increased with an increase in the length of the sliding path in different types of soil.

Stafford and Tanner (1983) carried out a study on the effect of strain rate on soil shear strength and soil-metal friction. It was found that, frictional stress was positively related to normal stress by a Coulomb relationship over the speed range investigated. They also found that speed also affected the frictional stress. Nikiforov and Bredun (1965) showed that frictional stress decreases with increasing speed in the speed range 1-5 m/sec. Kummer and Nichols (1938) found that friction between the soil and metal surface had a great influence upon the life and draft of tillage tools because the intensity of this frictional force governs largely the wearing and scouring properties of the implement. Nichols (1931) concluded that roughness of the mouldboard surface over which soil slides influences friction force and in effect increases the total draft of a mouldboard plough. According to Wismer et al. (1968), friction on mouldboard plough surface may represent as much as 30% of the total draft. They found that covering the plough bottom with Teflon reduced the draft by 23%. However the Teflon showed low wearing resistance and the technique was expensive. Fox and Bockhop (1965) also found that Teflon wears much faster than steel.

Various other attempts at coating the plough surface to reduce draft have been reported by Koolen and Kuiper (1983). Salokhe and Gee-Clough (1987, 1988) tried various materials for coating cage wheel lugs to reduce soil adhesion on it. They used lead oxide paint, silicone lubricant oil, gloss paint, gloss paint and varnish, Teflon tape and sheet, ceramic tile and enamel. They found that enamel coating was the best suitable material for cage wheel lugs coating to avoid soil adhesion and consequently cage wheel blocking. The experiments reported here were performed to see if enamel coating could reduce draft and improve scouring in wet, heavy clay soils.

3 METHODOLOGY

Experiments were conducted in a laboratory soil bin using Bangkok clay soil with 47% clay, 38% silt and 15% sand. For soil processing the soil bin carriage facility was used. Before each experiment soil moisture content, bulk density, specific weight and cone index were measured. The average soil properties are listed in Table 1.

Table 1 : Soil properties

Soil moisture content	Specific weight	Cone index	Cohe- sion	Angle of int- ernal friction
% (db)	kN/m^3	kPa	kPa	deg
21	10.4	73	1.1	35
31	11.4	53	3.6	33
51	10.7	9	-	-
58	10.1	7	-	-

Two identical shaped Thai traditional power tiller drawn slat type mouldboard ploughs were used in these experiments. One plough was covered with an enamel coating (mouldboard plough and share). The forces on the uncoated and coated plough were measured. The same test conditions were maintained while testing each plough and each test was replicated thrice.

Fig. 1 : Experimental set up

Fig. 1 shows the complete experimental set up. The horizontal, vertical and side forces were measured using six load cells. These load cells were connected in between the mainframe and subframe through a rose end bearing. The plough was attached to the subframe. It was possible to adjust the width of cut and the lift angle of plough. Data were analysed using a computer statistical package. Experiments were conducted at four different soil moisture levels viz. 21%, 31%, 51% and 58% (db). Two speed levels, 1 kmph and 2 kmph, were selected for testing the effect of speed at 21% and 31% soil moisture levels and 2 kmph and 3 kmph for tests conducted at 51% and 58% soil moisture contents. These values were low to avoid contribution of acceleration forces to the plough draft.

4 RESULTS AND DISCUSSION

In general the variation in the vertical and side forces were not consistent and it was therefore difficult to draw any conclusion. The variations could be attributed to different soil flow conditions on the plough because of the design itself and the absence of landside. A major design defect in the plough was the triangular shape share with pointed edge and a long pipe resting on the soil which created uneven hardpan and caused variations in vertical force. However, there was a definite pattern of variation in the draft due to variation in the coating, travel speed and soil moisture content. Therefore only the results of draft requirement are presented here.

4.1 21% soil moisture content

Results obtained at 21% soil moisture content are shown in Fig. 2. This moisture content was close to the plastic limit of soil. It was observed that at both 1 kmph and 2 kmph the draft required to pull an enamel coated plough was lower than the draft required for the same uncoated plough. The average values of draft for enamel coated and uncoated plough were 1023 N and 1045 N at 1 kmph speed and, 1137 N and 1275 N at 2 kmph speed respectively. An enamel coating on the plough reduced the draft requirement as compared to uncoated plough at both the forward speeds. The draft of the plough was related to soil strength and cohesion of the soil.

The effect of increase in speed was found to increase the draft force of both the ploughs. The decrease in draft of enamel coated plough compared to uncoated plough was found to be 2.11% at 1 kmph speed which increased to 11% at 2 kmph speed. This can be attributed to the difference in soil-material friction on the ploughs. Also increase in plough speed caused more volume of soil movement on both ploughs which resulted in an increase in draft at 2 kmph speed. At this soil moisture content and at both travel speeds it was observed that the

soil failed in the form of small blocks and also in a dry failure manner. Fig. 3 shows small soil blocks formed during the uncoated plough operation at 21% soil moisture content. The variation in the specific draft at this soil moisture content for uncoated plough was found to be 56 to 68 kPa compared to 54 to 60 kPa for enamel coated plough.

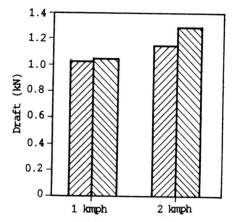

Fig. 2 : Draft requirement at 21% soil moisture content, ▨ coated plough, ▧ uncoated plough

4.2 31% soil moisture content

A similar trend in draft variation for coated and uncoated ploughs was observed at 31% soil moisture content. At this moisture content draft recorded for both the ploughs was higher than that measured at any other moisture content. Fig. 4

Fig. 3 : Soil failure pattern under the uncoated plough at 21% soil moisture content

shows the draft required for enamel coated and uncoated ploughs at 31% soil moisture content and at 1 kmph and 2 kmph forward speeds. The average values of draft for coated and uncoated plough were 1361 N and 1664 N at 1 kmph speed while they were 1685 N and 2253 N at 2 kmph speed respectively. Again the draft required for coated plough was less than the draft required for uncoated plough.

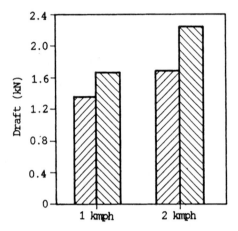

Fig. 4 : Draft requirement at 31% soil moisture content, ▨ coated plough, ▨ uncoated plough

The increase in speed caused similar effects as observed at 21% soil moisture content. A 19-26% reduction in the draft requirement for the uncoated plough was found when coated plough was used. At this moisture content the soil under the action of the mouldboard plough failed by block formation. Fig. 5 shows soil blocks formed over the mouldboard plough at 1 kmph speed. The similar trend was observed at 2 kmph speed however the block size was larger in this case than observed at 1 kmph speed. The specific draft for the coated plough varied from 72 to 90 kPa while for the uncoated plough it varied from 88 to 119 kPa.

4.3 51% soil moisture content

Fig. 6 shows the comparison of draft requirement for the enamel coated and uncoated plough at 2 kmph and 3 kmph forward speeds and 51% soil moisture content. In this case also the draft required for the enamel coated plough was less than the similar uncoated plough at

both speeds. Increase in speed caused an increase in the draft required for the ploughs. The average values of draft

Fig. 5 : Soil failure pattern under the uncoated plough at 31% soil moisture content

recorded for uncoated and coated plough at 2 kmph speed were 903 N and 774 N and at 3 kmph they were 1048 N and 1020 respectively. The decrease in draft of coated plough compared to uncoated plough was 3 to 14% at 51% soil moisture

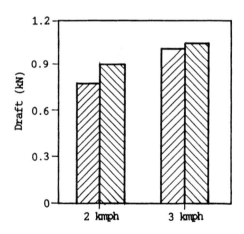

Fig. 6 : Draft requirement at 51% soil moisture content, ▨ coated plough, ▨ uncoated plough

content. At 51% soil moisture content different soil failure patterns were noted under the action of the plough. The soil failed in a flow failure manner. Compared to the draft requirement at 21% and 31% soil moisture content, the draft required at 51% moisture content was less. This was due to reduced soil strength at 51% moisture content. Due to higher moisture content less shearing resistance was offered and soil sticking and accumulating on the plough contributed significantly to the draft. The scouring on the coated plough was found to be much better than on uncoated plough. The specific draft requirements at the moisture content varied from 41 kPa to 54 kPa for coated plough and 48 kPa to 58 kPa for uncoated plough.

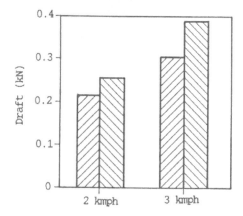

Fig. 7 : Draft requirement at 58% soil moisture content, ▨ coated plough, ▨ uncoated plough

4.4 58% soil moisture content

Similar observations as noted at 51% moisture content were noticed at this moisture content. Increase in soil moisture content caused further reduction in the draft compared to the draft required for both the ploughs at lower moisture content. A flow failure was noticed at this moisture content. The average values of draft for the coated plough were 217 N and 303 N, while they were 255 N and 361 N for uncoated plough at 2 kmph and 3 kmph forward speed respectively (Fig. 7). Thus a decrease in the draft requirement for the coated plough compared to uncoated plough varied from 15 to 19%. The specific draft for the coated plough varid from 11 to 16 kPa while for uncoated plough it varied from 13 to 19 kPa, respectively. Figs. 8 and 9 show soil sticking on coated and uncoated ploughs at 58% soil moisture content. The enamel coated plough showed

excellent scouring compared to the uncoated plough. A large amount of soil accumulated on the uncoated plough.

The statistical analysis results of all these data indicated that the effect of coating on the draft at all moisture contents and travel speeds was significant.

Fig. 8 : Soil scouring on enamel coated plough at 58% soil moisture content

Fig. 9 : Soil scouring on uncoated plough at 58% soil moisture content

5 CONCLUSIONS

An enamel coating on a mouldboard plough surface was found to reduce the draft requirement significantly. As compared to an uncoated plough the draft requirement reduced by up to 26% depending on soil moisture content and speed of travel. Further investigations about the effects of enamel coating on a larger tractor drawn plough are being carried out. Durability tests on the enamel coating are also being carried out.

REFERENCES

Cooper, A.W. and W.F. McCreery. 1961. Plastic surface for tillage tools. Am. Soc. Ag. Engrs. Paper No. 61-649.

Fox, W.R. and C.W. Bockhop 1965. Characteristics of a Teflon-covered simple tillage tool. Trans. Am. Soc. Agr. Engrs. 8.

Kepner, R.A.; Bainer, R. and E.L. Barger 1982. Principles of Farm Machinery, AVI Publ. Connecticut.

Koolen, A.J. and H. Kuiper 1983. Agricultural Soil Mechanics. Springer Verlag, Germany.

Kummer, F.A. and M.L. Nichols 1938. The dynamic properties of soil: a study of the nature of physical forces governing the adhesion between soil and metal surface. Agril. Eng. 19: 73-78.

Nichols, M.L. 1931. The dynamic properties of soil: soil and metal friction. Agril. Eng. 12: 321-324.

Nikiforov, P.E. and M.I. Bredun 1965. The sliding friction of soil on metal and plastic surfaces. Vestn. Skh. Nanki. Moscow 10: 112-116.

O'Callaghan, J.R. and J.G. McCoy 1965. The handling of soil by mouldboard ploughs. J. Agric. Engng. Res. 10.

Oskoui, E.; Rackhan, D. H. and B. D. Witney 1982. The determination of plough draft - I : Prediction from soil and meteorological data with cone index as soil strength parameter. J. Terramech. 19(2):97-106.

Panwar, J. S.; Clark, G. N. and J. A. Weber 1973. Similitude prediction of model forces in artificial soils. Trans. ASAE 16(5):825-830.

Randolph, J.W. and I.F. Reed 1938. Test of tillage tools: II - Effect of several factors of the reactions of 14 inch mouldboard plough. Agril. Eng. 19.

Reed, I.F. 1937. Tests of tillage tools: I - Equipment and procedure for mouldboard ploughs. Agril. Eng. 18: 111-115.

Salokhe, V.M. and D. Gee-Clough 1987. Studies on effect of lug surface coating on soil adhesion of cage wheel lugs. Proc. 9th Int. Conf. of Int. Soc. for Terrain-Vehicle System, Barcelona, Spain. Aug. 31 - Sept. 4. pp. 389-396.

Salokhe, V.M. and D. Gee-Clough 1988. Coating of cage wheel lugs to reduce soil adhesion. J. agril. Engng Res. (in press)

Soehne, W. 1960. Suiting the plow body shape to higher speeds. Grundlagen der Landtechnik 12: 51-52.

Stafford, J. V. and D. W. Tanner 1983. Effect of rate on soil shear strengthand soil-metal friction. II - Soil-metal friction. Soil & Tillage Research 3:321-330.

Wismer, R. D.; Wegscheid, E. L.; Luth, H. L. and B. E. Raning 1968. Energy application in tillage and earth moving. SAE Trans. 77:2486-2494.

Land and Water Use, Dodd & Grace (eds), © 1989 Balkema, Rotterdam. ISBN 90 6191 980 0

Influence of mole plough leg and expander geometry on soil disturbance

G.Spoor & M.J.Hann
Silsoe College, Cranfield Institute of Technology, Silsoe, Bedford, UK

A.S.Centeno
Fundacão de Ciência e Tecnologia, CIENTEC, Porto Alegre, Brazil

ABSTRACT: The effectiveness of any mole drainage system is dependent upon having soil fissure development and fissure/mole channel connections appropriate to field requirements in that particular situation. Changes to mole plough leg and expander geometry have been shown to be capable of influencing the nature of these fissures and connections at the time of mole drain installation. Increases in leg thickness will tend to increase fissure size and spacing, and increases in leg side length and roughness will enlarge the fissures. Increasing expander diameter tends to close the fissure/mole channel connection.

RESUME: L'efficacité de tout système de drainage-taupe dépend du développement d'une fente dans le sol et d'une liaison de celle-ci avec le drain crée par la taupe qui sont appropriés aux besoins du champ dans cette situation spéciale. On a montré que des changements de la géometrie de l'étai de la taupe et de l'étendeuse sont capables d'influencer la nature de ces fentes et des connexions au moment de l'installation du drainage-taupe.
Des augmentations de l'épaisseur de l'étai montrent une tendance à agrandir la dimension des fentes et l'espacement entre celles-ci, de même que des augmentations de la longeur du côté de l'étai et de son aspérité. Des augmentations au diamètre de l'étendeuse tendent à fermer la liaison de la fente avec le drain.

ABSTRACT: Du Wirksamkeit eines jeden Maulwurfsdrainageverfahrens hängt davon ab, ob sich Spalten im Boden und Verbindungen zwischen den Spalten und dem Dränkanal gemäss den spezifischen Anforderungen des Bodens ausbilden.
Es wird gezeigt, dass Änderungen der Geometrie des Schwertes und des Zeihkegels des Drainagepfluges die Ausprägung dieser Spalten und deren Verbindungen zum Dränkanal beim Anlegen einer Maulwurfsdrainage beeinflussen können.
Mit zunehmender Schwertdicke nehmen Spaltweite und - abstand zu, grössere Seitenlänge des Schwertes und rauhere Oberfläche führen ebenfalls zu grösseren Spalten.
Verwendet man einen grösseren Ziehkegel, so werden leicht die Verbindungen zwischen Spalten und Dränkanal unterbrochen.

1 INTRODUCTION

The usual action of a mole plough is to create a mole channel with associated interconnecting leg slot and fissures as shown in Fig 1. The extent of these fissures and the effectiveness of the connection between them, the leg slot and the channel, controls the discharge rate and flow route of water to the mole drains. Fig 2 (after Leeds-Harrison et al 1982) shows the very different discharge hydrographs for moling systems with and without good fissure development and good and bad connections. Fig 3 illustrates schematically the major water flow routes in each case.

The actual flow route required depends upon the circumstances. Fissure flow (Fig 3a) is particularly advantageous for rapid water removal from the surface horizons in soils of very low hydraulic

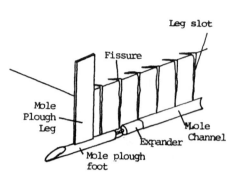

Fig 1: Section showing the mole plough and associated mole channel and fissures

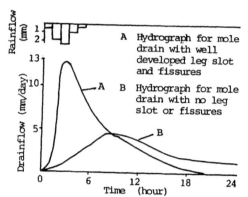

A Hydrograph for mole drain with well developed leg slot and fissures

B Hydrograph for mole drain with no leg slot or fissures

Fig 2: Influence of soil fissure and leg slot development on the mole drain hydrograph

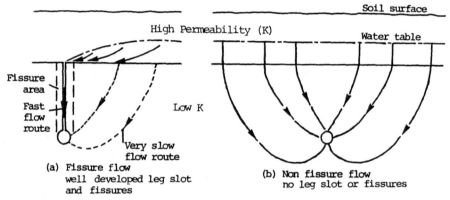

Fig 3: Water flow routes to mole drains

conductivity. Non-fissure flow (Fig 3b), is required for the reclamation of saline soils, to leach salts uniformly from the whole profile, see Spoor and Leeds-Harrison (1986).

In moling practice, the degree of fissure development and the channel connection vary with the soil conditions, particularly moisture content and density. Usually, either a small number of wider cracks are produced or a larger number of small ones. To improve the efficiency of the moling operation, it would be helpful in certain situations if some control could be exercised over fissure development and the connection at the time of installation.

Hopewell (1953) demonstrated the development of larger cracks with a thicker mole plough leg, which increased

water flow rates to the mole channel. Unfortunately this leg modification also reduced mole channel stability. Overy (1952) in a theoretical study, identified links between fissure length and leg thickness, predicting length would increase linearly with increasing leg thickness. He also provided an explanation based on soil mechanics principles for this form of leg fissure development. These studies have been extended further at Silsoe College, where the practicalities of modifying fissure development by changing mole plough geometry, without impairing channel stability, are being assessed. This paper reviews the influence of leg and expander geometry on the nature of the soil fissuring and the fissure/leg slot/mole channel connection.

2 SOIL FISSURE FORMATION

The stages in the development of the vertical soil fissures by the mole plough leg are shown in Fig 4. A vertical frontal fissure is first opened up immediately ahead of the leg, taking the line of least resistance and running approximately parallel to the direction of travel (stage 1). This fissure is widened by sideways soil movement as the leg moves forwards and further vertical lateral fissures (herringbone fissures) are generated at approximately 45° to the direction of travel, just ahead of the leg (stage 2). With further leg movement, the drag resulting from the sliding resistance between the side of the leg and the soil may open up these herringbone fissures further, through a tensile failure (stage 3).

The size of the herringbone leg fissures appears to be closely related to the relative magnitude of the leg/ soil interface sliding resistance and the soil shear strength. The greater the interface sliding resistance, the greater the chance in any soil situation, of larger fissures being formed and of some of the initial smaller fissures being closed. Whilst in any given situation it is impossible to change soil shear strength, it should be possible by modifying leg geometry, to change the interface sliding resistance and hence fissure size.

Potential ways of changing the sliding resistance at the side of the leg include changes to the following:
a. leg thickness (influencing normal stress)
b. leg roughness (influencing angle of soil/interface friction and adhesion)
c. leg side length (influencing adhesive force).

The effect of these leg variables on fissure geometry is now considered.

3 EXPERIMENTAL METHOD

Tests were carried out under both laboratory and field conditions with mole plough legs of varying thickness, side length and side roughness, to determine their influence on fissure dimensions. The laboratory soil comprised an artificial clay/oil mix with stable shear strength properties, and the field test sites all clays, are detailed in Table 1. All soils were plastic in consistence at the time of the tests and the field soils differed in terms of their bulk density and strength properties, see Table 2.

The leg thicknesses and side lengths tested in the laboratory ranged between 7-25 mm and 50-200 mm respectively, the working depth remaining constant at 200 mm. Surface roughness was increased by bonding sand paper to the tine surface. The field test treatments are detailed in Table 3, the rough surface being produced by forming a 50 mm square grid of weld beads on the leg face as shown in Fig 5.

Fissure width, depth and spacing, see Fig 6, were determined by direct measurement along one side of the leg slot, with sampling lengths of 0.5 and 2.0 m in the laboratory and field respectively. The soil shear strength and interface parameters were measured in situ, using a torsional shear box (Payne and Fountaine 1952) and a torsional shear plate.

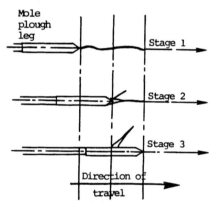

Fig 4: Development of vertical soil fissures by mole plough leg (Plan view)

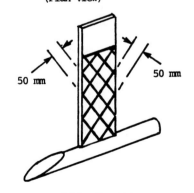

Fig 5: Weld bead pattern on mole plough leg

Table 1 Soil characteristics at moling depth

| Site | Soil series* | Particle size | | | Plastic limit % | Liquid limit % |
		Sand %	Silt %	Clay %		
A	Wicken (Evesham)	16	22	62	37	83
B	Rowsham	23	26	50	20	60
C	Oak	50	17	33	20	56

*King (1969)

Table 2 Soil conditions and strength properties

Site	Moisture content %	Density t m^{-3}	Cohesion kN m^2	Angle of shearing resistance (Degrees)	Adhesion kN m^2	Angle of soil/metal friction (Degrees)
A	38.9	1.20	17.5	25	1.4	34
B	21.7	1.69	26.5	27	6.1	24
C	26.3	1.42	28.9	22	6.0	19

Table 3. Mole plough leg geometries in field tests

Leg	Thickness mm	Side length mm	Surface roughness mm
1	12	200	Smooth
2	25	200	Smooth
3	40	200	Smooth
4	40	250	Smooth
5	40	350	Smooth
6	25	200	Rough

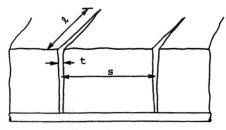

ℓ : fissure depth
s : spacing between
t : fissure width

Fig 6: Characteristics of leg fissures

4 INFLUENCE OF LEG GEOMETRY ON FISSURE DEVELOPMENT

4.1 Laboratory results (Webb 1981)

The laboratory soil bin tests under controlled soil conditions confirmed the prediction of Overy (1952), namely that the depth of the fissures generated, increases linearly with increasing leg thickness; this is shown in Fig 7. Increasing the leg side length and hence side contact area had little influence on fissure size, whereas increasing leg roughness using sand-paper, produced a significant increase in fissure depth at the same leg thickness, see Fig 7. The lack of response to increasing side contact area can be explained on a basis of the low soil/leg adhesion (0.5 kN/m²) prevailing. Increasing side area would due to the low adhesion, have little effect on the resultant adhesive force and hence sliding resistance. The sand-paper bonded leg however, clogged with soil, transforming the leg into a perfectly rough one, thus significantly increasing the sliding resistance.

No significant changes occured in the ratio of fissure width to fissure depth between treatments, suggesting this ratio is largely a function of soil condition. Fissure volume however, increased markedly with increasing leg thickness and side roughness.

The force required to produce a fissure of given size and volume depended upon whether the fissure was formed with a thin rough leg or a thick smooth one. In the laboratory soil, the thicker leg required almost twice the effort of the thin leg.

4.2 Field results (Centeno 1988)

Although the fissuring pattern produced by any given leg often differed markedly between the 3 field sites, the effects of changes in leg geometry were similar, differing only in magnitude. To remove the effects of soil condition, the results presented in this paper are the mean values for all sites. Figs 8 and 9 show the influence of leg thickness, side length and side surface roughness on the average spacing, width and depth of the soil fissures generated in the 3 soils.

The major leg geometry factor influencing fissure spacing appears to be leg thickness, side length and surface roughness having only a minimal effect.

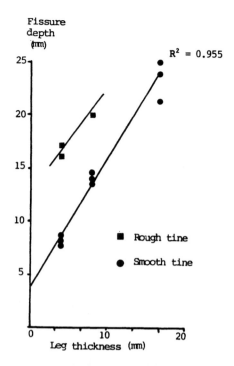

Fig 7: Relationship between fissure depth and leg thickness

Increasing the leg thickness of very thin legs increased spacing markedly, but beyond a certain thickness it had little further effect. Fissure spacings of 32, 20 and 28 mm were measured on sites A, B and C respectively, with the widest leg tested (40 mm wide).

Fissure size, defined in terms of fissure width and depth and hence volume, is influenced very significantly by all three leg variables, particularly near mole channel depth. Increases in both leg thickness and side length, increased both fissure width and depth, by increasing sliding resistance at the side of the leg. Fissure width and depth were always much greater near the surface, than at mole channel depth, see figs 8 and 9.

The weld bead roughened side face in these tests actually reduced the sliding resistance at the side interface, rather than increasing it as expected. Soil did not pack into the spaces between the beads and hence a non-scouring situation, which would have markedly increased sliding resistance, never developed. Instead the sliding area was reduced from that of the complete leg

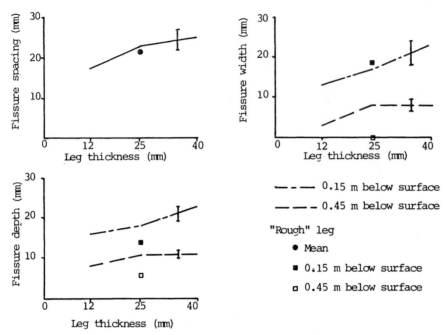

Fig 8: Influence of leg thickness and side roughness on fissure
        dimensions (leg side length 200 mm)

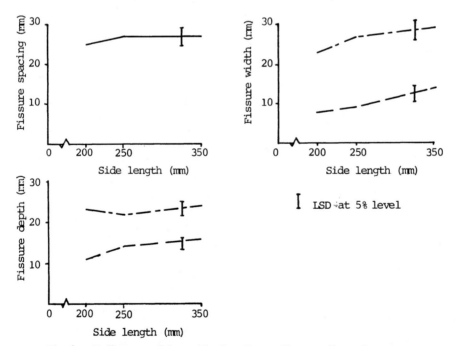

Fig 9: Influence of leg side length on fissure dimensions
        (leg thickness 40 mm)

face to the area of the beads alone, effectively eliminating all adhesive forces and reducing sliding resistance. The very low sliding resistance was insufficient to open up further the fissures generated by the leading leg face and the weld beads tended to rework the soil at the leg slot, sealing off many fissures. Reducing the effective side roughness therefore, significantly reduces fissure size and volume.

Comparing fissure development on the 3 soils, fissure spacing and size for any given leg geometry were always greater in soil A than in soil C and were always least in soil B.

5 INFLUENCE OF EXPANDER GEOMETRY ON FISSURE AND LEG SLOT CLOSURE

Expander action as identified by Spoor and Ford (1987), moves soil in the channel roof area largely forwards and upwards, reworking soil in the leg slot/fissure zone above the channel. Observations on numerous sites suggest that the extent of leg slot and fissure closure by the expander, depends upon expander diameter relative to channel diameter, expander shape, initial slot and fissure size and soil condition. In any given situation, the chances of fissure and slot closure increase with increasing expander diameter and with the use of a barrel shaped rather than a tapered expander. The barrel expander induces a larger forward component of soil movement than the tapered one (Castle and Spoor 1989), causing greater soil reworking and instability in the roof area.

Complete channel roof sealing following fissure and leg slot closure and hence a complete loss of connection, depends largely on soil plasticity. The more plastic the soil, the greater the chance of sealing.

6 CONCLUSIONS

Within limits, depending upon the soil conditions, modifications to mole plough leg and expander geometry can alter the size and spacing of the leg fissures generated and the effectiveness of the connection between these fissures, the leg slot and the mole channel itself. Fissure spacing is particularly sensitive to leg thickness, and fissure width, depth and volume to the soil/leg sliding resistance at the side of the leg.

Increasing leg thickness and sliding resistance, increase fissure spacing and size (particularly near the mole channel) respectively. Sliding resistance can be increased by increasing leg thickness, side length and side surface roughness. The response to change in side length depends upon the value of soil adhesion. The draught force on the leg is more sensitive to changes in leg thickness than to changes in side length and roughness. The change in specific resistance for a given change in fissure size will therefore depend upon the particular leg geometry variable changed.

Increasing mole expander diameter relative to channel diameter, increases the possibility of closing the connection between the leg fissures, leg slot and mole channel. This connection can only be satisfactorily sealed following closure when the soil is in a plastic state.

ACKNOWLEDGEMENTS

The authors wish to acknowledge the contributions made by Professor R J Godwin and Dr P B Leeds-Harrison to the work.

REFERENCES

Castle, D.A. and G. Spoor. 1989. Mole plough design factors influencing the hydraulic connection between mole channels and the permeable backfill placed over pipe drains. Ms submitted to J. Agric. Eng. Res.

Centeno, A.S. 1988. Influence of mole plough leg width and side contact area on soil crack development. M.Sc. Thesis, Silsoe College, Silsoe, Bedford.

Hopewell, H.G. 1953. Tile drainage investigations in New Zealand. III Investigation to determine the effect of increasing the thickness of the mole plough blade or knife. N.Z. J Sci Tech. 35A.(3):237.

King, D.W. 1969. Soils of the Luton and Bedford District: a reconnaissance survey. Soil Survey of England and Wales, Rothamsted, Harpenden.

Leeds-Harrison, P.B., G. Spoor and R.J. Godwin. 1982. Water flow to mole drains. J. Agric. Eng. Res. 27: 81-91.

Overy, B.B. 1952. The application of
soil mechanics to a study of mole
drainage. M.Sc. Thesis, Univ of
Cambridge.

Payne, P.C.J. and E.R. Fountaine. 1952.
A field method of measuring shear
strength of soils. J of Soil Sc.3:
136-144.

Spoor, G. and R.A. Ford. 1987.
Mechanics of mole drainage channel
deterioration. J of Soil Sc. 38:
369-382.

Spoor, G. and P.B. Leeds-Harrison. 1986.
Extension of mole drainage practices.
In K.V.H. Smith and D.W. Rycroft(eds),
Hydraulic design in water resources
engineering : Land drainage p.443-453.
Publ. Springer-Verlag. Proc. 2nd Int.
Conf. Southampton Univ.

Webb, D.R. 1981. Vertical crack
formation with narrow drainage tines.
M.Sc. Thesis 81/262 Silsoe College,
Silsoe, Bedford.

Land and Water Use, Dodd & Grace (eds), © 1989 Balkema, Rotterdam. ISBN 90 6191 980 0

Wechselwirkungen von Bodenbearbeitungssystemen im 10-jährigen Versuch auf technische Leistungsparameter, Bodenstruktur, biologische Aktivität, Pflanzenkrankheiten und Ertrag

F.Tebrügge
Institut für Landtechnik, Justus Liebig Universität, Giessen, Bundesrepublik Deutschland

ÜBERBLICK:

Bei Anwendung reduzierter, konservierender Bodenbearbeitungsverfahren bis hin zur Direktsaat nehmen die Mechanisierungskosten für Schlepper, Gerät und Lohn ab. Der Leistungs- und Energiebedarf sinkt mit abnehmender Eingriffsintensität, wobei Bodenschäden durch Druck und Schlupf erheblich reduziert werden.
Durch die homogene Einmischung der organischen Restsubstanzen kann der Erosionsschutz auf gefährdeten Standorten erheblich verbessert, die Umsetzung dieser Substanzen beschleunigt, die biologische Aktivität gesteigert und durch diese Wechselwirkungen die natürliche Bodenfruchtbarkeit und Gefügestabilität erhöht werden.

When the contact intensity of soil cultivation tools is reduced, power and energy requirements can be cut considerably and, as a result, damage to the soil caused by pressure and slip minimised to a large extent.
Homogeneous mixing-in of organic remains leads to major improvements with regard to sludging and erosion protection at locations which are at risk. Conversion of these substances is accelerated, the biological activity is increased and, as a result of these interrelations, the natural fertility of the soil and the crumb and structural stability is improved.

Avec l'adoption des techniques de travail du sol réduit, depuis le travail de conservation du sol jusqu'au semis direct, le coût de mécanisation du tracteur, de l'outil et des travaux à façons augmentent. Les besoins en puissance et en énergie diminuent avec 12 diminution de l'intensité d'attaque du sol. Ce qui se traduit par une réduction des effets néfastes dûs à la compaction et au glissement.
A travers un enfouissement homogène des résidus végétaux, la protection cote l'érosion dans les zones menacées se trouve améliorée. La dégradation de ces substances accélère l'activité biologique. Les réactions qui en découlent améliorent la fertilité du sol et la stabilité structurale.

1 EINLEITUNG

Über die zweckmäßigste und erfolgversprechendste Bodenbearbeitung wird oft und auch leidenschaftlich diskutiert. Ein Indiz für die Wechselwirkungen zwischen Standort, Pflanze und Technik, aber auch ein Zeichen für den Mangel an langjährigen Vergleichsuntersuchungen auf verschiedenen Standorten. Es genügt nicht, die augenblickliche Wirkung eines Werkzeuges oder Arbeitsgerätes zu bewerten, da sich das Ökosystem Boden erst allmählich auf eine

verändere Bodenbearbeitungsmaßnahme einstellt.
Ein umfassende Bewertung und Quantifizierung der Wechselwirkungen von Bodenbearbeitungssystemen mit unterschiedlicher Eingriffstiefe und -intensität in das Ökosystem Boden, setzt einen engen interdisziplinären Forschungsansatz voraus, wie er an der Universität Giessen durch die Institute: Landtechnik, Zoologie, Mikrobiologie, Phytopathologie, Pflanzenernährung und Bodenkunde realisiert wurde.
Grundlage der gemeinsamen Forschung bilden

Langparzellenanlagen unterschiedlicher Bodenart mit betriebsüblicher Fruchtfolge, die seit nunmehr 10 Jahren konsequent jeweils mit dem entsprechenden System bearbeitet wurden.

2 METHODE UND ERGEBNISSE

Teilt man die Vielzahl der Bodenbearbeitungsgeräte nach der Funktion ihrer Werkzeuge ein, so kristallisieren sich wenige Systeme heraus, die sich durch ihre spezifische Einwirkungsweise klar voneinander abheben. Die Spanne reicht dabei von der krumentiefen Wendung oder Lockerung der Primärwerkzeuge über eine auf die Arbeitstiefe des Sekundärgerätes beschränkte Einwirkungstiefe bis hin zum totalen Verzicht auf jegliche Bearbeitungsmaßnahme (Abb.1).

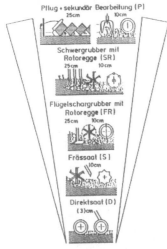

Abb. 1. Bodenbearbeitungssysteme mit abnehmender Eingriffsintensität.

2.1 Landtechnische Parameter

Da die Geräte sowohl Zug- als auch Zapfwellenleistung beanspruchen, die Effektivität, der Wirkungsgrad der Kraftübertragung zwischen beiden Leistungstransformatoren aber unterschiedlich ist, ist es für einen Verfahrensvergleich notwendig, die anteiligen Ansprüche zu untersuchen.
Die aus den resultierenden Zugkräften für Unter- und Oberlenkerbelastung und aus Drehmoment zwischen Schlepperzapfwelle und Gelenkwelle ermittelten Daten, zeigen sowohl hinsichtlich des Gesamtleistungsbedarfes wie im %-Anteil für Zug- und Drehleistungsbedarf deutliche verfahrensbedingte Differenzen (Abb. 2).

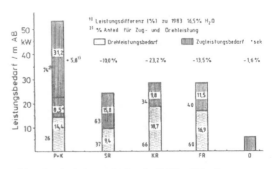

Abb. 2. Leistungsbedarf (kW/m AB) der Bearbeitungssysteme -suL-Boden, 18 % H_2O, 1986.

Bei gleicher Eingriffstiefe der Primärwerkzeuge von 25 cm weisen die Grubbervarianten einen deutlich geringeren Zugleistungsbedarf gegenüber der Pflugvariante auf. Auch hinsichtlich der Kraftkomponentenverteilung sind die nicht-wendenden Verfahren günstiger zu beurteilen, wobei sich auf Grund der Werkzeuggestaltung die %-Anteile für Zug und Drehleistung bei den Systemen SR und FR umkehren. Trotz erheblicher Lockerung im Verfahren P liegt der Energiebedarf, selbst ohne Berücksichtigung des Zugleistungsanteils für die Sekundärbearbeitung (K) auf nahezu dem gleichen Niveau wie bei den übrigen Verfahren.

2.2 Bodenphysikalische Parameter
2.2.1 Einarbeitung pflanzlicher Reststoffe und Influktuation

Die oberflächennahe homogene Einarbeitung pflanzlicher Residuen in den Boden bei den nicht-wendenden Systemen oder das Belassen auf der Oberfläche (Tab. 1) bewirkt nicht nur einen schnelleren Abbau dieser Substanzen durch Mikro- und Makrofauna, sondern verbessert in erheblichem Maß das Vermögen des Bodens kurzfristig hohe Niederschlagsmengen aufnehmen zu können.

Tab. 1. Strohverteilung in den Horizonten (%) und Abbau bis Frühjahr

Bearb.-system	0-10 cm	10-15 cm	15-25 cm	Abbau %
P	17,4	16,3	66,3	26
SR	74,7,	16,5	8,8	44,6
FR	89,1	5,0	5,9	36,2
D	95,9*	2,0	2,2	1,9

:* 91,7% auf der Bódenoberfläche

Wiederholt aufgegebene Niederschlagsmengen
von jeweils 19 mm wurden zeitlich wie men-
genmäßig auf denen durch die Systeme SR,
FR und S bearbeiteten Flächen vom Boden
problemlos verdaut, während bei der Variante P eine unzureichende Influktuation
wegen der bodenverschlämmenden Wirkung bereits durch die ersten 19 mm zu verzeichnen war (Abb. 3).

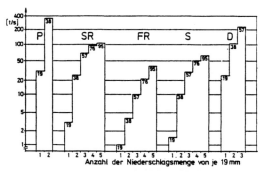

Abb. 3. Wirkung der Bodenbearbeitungssysteme auf die Influktuationskapazi-
tät (mm $H_2O \times 5^{-1}$) eines sL-Bodens.

Daraus kann abgeleitet werden, daß durch
oberflächennahes Einarbeiten von Pflanzen-
reststoffen (Stroh) der Bodenerosion auf
gefährdeten Standorten erheblich entgegen-
gewirkt werden kann.

2.2.2 Porencharakteristik und Tragfähig-
keit

Auf einem seit 9 Jahren mit den unterschiedlichen Systemen bearbeiteten Standort (Parabraunerde) läßt sich ableiten,
daß zwischen nicht- (FR) und wendendem System (P) nur geringfügige Unterschiede im
GPV und in der Porencharakteristik bestehen (Abb. 4).

Eine Belastung des Bodens durch einen
110 kW Schlepper führt bei beiden Varianten zu einer Abnahme des GPV, wovon insbesondere die Poren > 120 µm betroffen
sind. Die D-Variante weist dagegen bis zu
20 cm Bodentiefe ein geringeres GPV, insbesondere durch den geringeren Grobporen-
raum > 120 µm auf.
Im Bereich von 40 cm Tiefe vergrößert sich
der Porenraum bei der Variante D durch die
Zunahme der Porenbereiche > 50 µm und
liegt damit im GPV deutlich über den Ver-
gleichsvarianten. Eine Belastung des Bodens bewirkte hier keine Reduktion des Po-
renvolumens (Gruber 1989).
Daraus kann abgeleitet werden, daß durch
Bodenruhe mit einhergehender Sackungsver-
dichtung die Strukturstabilität des Bodens
und damit die Tragfähigkeit positiv beeinflußt werden. Dies spiegeln sich auch die durch
die Überfahrt verursachten Spurtiefen für
D von 1,3 cm, FR 5,9 cm und P 7,6 cm wieder.

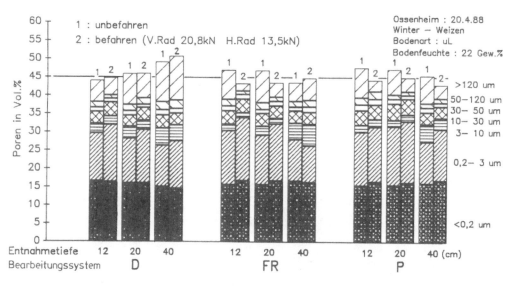

Abb. 4. Porengrößenverteilung vor und nach der Belastung bei verschiedenen
Bearbeitungssystemen.

2.3 Bodenfauna und -flora

Jeder Bodeneingriff wie auch dessen Intensität hat eine mehr oder weniger starke Wirkung auf die Lebensbedingungen für Bodenfauna und -flora. Zwei Parameter die hierfür u.a. herangezogen werden können sind biologische Atmungsaktivität (CO_2) und Biomasse der Lumbriciden. Die ermittelten Daten (Henke 1987) von drei verschiedenen Standorten im 8. Versuchsjahr lassen einen erheblichen Einfluß durch die Bodenbearbeitungssysteme auf diese Parameter erkennen. Der Verzicht auf Bodenwendung bei dennoch krumentiefer Lockerung führt zu einer Zunahme der Lumbriciden-Biomasse, die um den Faktor 3, bei absoluter Bodenruhe sogar um den Faktor 5, über dem wendenden Pflugverfahren liegt (Abb. 5).

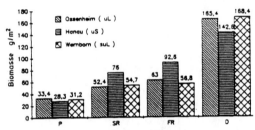

Abb. 5. Entwicklung der Lubriciden-Bio masse (g/m²) unter dem Einfluß der Bearbeitungssysteme (7. Versuchsjahr).

Durch Einmischung der pflanzlichen Reststoffe (siehe 2.2.1) in aerobere Zonen des Bodens werden aber auch im Vergleich zur vergrabenden Wirkung des Pfluges die Lebendsbedingungen der Mikroorganismen gefördert. Die hierauf untersuchten drei Standorte spiegeln dies im Oberkrumenbereich durch eine deutlich höhere Atmungsaktivität wieder, ohne dabei im Unterkrumenbereich signifikant unter der Variante P zu liegen (Abb. 6).

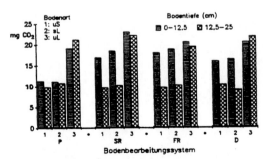

Abb. 6. Biologische Atmungsaktivität (mg CO₂ /50g Boden) unter dem Einfluß der Bearbeitungssysteme.

2.4 Halmbasiserkrankungen

Auf den gleichen Standorten wie zuvor, mit einem Getreideanteil zwischen 71-85 % in der Fruchtfolge wurden die Halmbasiserkrankungen an Winterweizen nach visueller Bonitur laut BBA-Richtlinie 4.5.1.6 aufgenommen und konnten durch nachfolgenden Klewitz-Labortest bestätigt werden. Entgegen der vorherrschenden Meinung, daß bei Anwendung nicht-wendender Bearbeitungssysteme bis hin zur Direktsaat zwangsläufig eine erhöhte Infektionsgefahr durch Schaderreger (P und P/F) für die Folgefrucht einhergeht, erbrachten die Erhebungen (Bräutigam 1989) ein völlig umgekehrtes Bild. Denn mit einer Abnahme der Bearbeitungsintensität nahmen auch die Befallswerte in der Reihenfolge P-SR-FR-S-D ab (Tab. 2).

Tab. 2. Halmbasiserkrankung (%)

Bearb.- system	Bodenart - % Getreide*		
	suL-85%	uL-71%	uS-75%
P	44	32	41
SR	46	20	25
FR	39	16	26
D	14	11	15

*in der Rotation

Ursache ist vermutlich der Aufbau eines phytophagen Potentials, welches maßgeblich durch Humusbestand, Bakterien und gewebezersetzenden Saprohyten beeinflußt wird. Eine Wechselwirkung zur Bearbeitungsintensität scheint vorhanden zu sein.

2.5 Pflanzenertrag

Für den Praktiker steht der nachhaltige Pflanzenertrag im Vordergrund seiner Überlegungen. Tab. 3 gibt die Erträge der Bearbeitungssysteme als Summe von jeweils 8 Jahren für die verschiedenen Standorte und Fruchtfolgen wieder.

Hier sei angemerkt, daß die Pflanzenschutz- und Düngemittelaufwendungen für alle Verfahren, mit Ausnahme der Direktsaat, bei der jeweils eine Vorsaatspritzung mit einem nicht selektiven Herbizid durchgeführt wird, gleich gehalten wurden. Auf den Standorten (suL und uL) mit nur einmal Raps bzw. Erbsen in der 8-jährigen getreidebetonten Fruchtfolge zeigen die beiden nicht-wendenden Systeme SR und FR

eine Ertragsüberlegenheit von bis zu 9 %
gegenüber dem System P. Der um 3 % geringere Ertrag von D auf den suL-Boden beruht
allein auf dem Mindererterag an Erbsen von
4,6 dt gegenüber P.
Auf den Standorten (uS und uL) mit Zuckerrüben in der dreijährigen Fruchtfolge ergeben sich für die Systeme hinsichtlich
der Getreideerträge keine eindeutigen Unterschiede. Betrachtet man hingegen die
durchschnittlichen prozentualen Zuckererträge, so schneiden die nicht-wendenden
Verfahren auf dem Sandboden am besten, das
Verfahren D am schlechtesten ab.
Auf dem uL-Boden kehrt sich das Bild zu
Gunsten der D-Variante um. Denn sowohl im
3ten als auch im 6ten Versuchsjahr lag der
Zuckerertrag beim System D deutlich über
den anderen Systemen, die unter sich betrachtet keine Unterschiede aufwiesen.

Tab. 3. Pflanzenertrag (%) der Bearbeitungssysteme von 8 Versuchsj.

System Boden	P	SR	FR	D
uS-Getreide	98	100	100	100
-Z.-Rüben	104	109	111	79
uL-Getreide	103	102	97	97
-Z.-Rüben	95	95	97	104
suL-Getr.u. Erbsen	99	103	106	96
uL-Getreide u. Raps	95	100	104	-

Ein vertikaler Vergleich über die Standorte hinweg läßt tendenziell eine Überle-
genheit der Systeme SR und FR gegenüber D
und P erkennen, wobei die Direktsaat, mit
Ausnahme des Anbaus von Zuckerrüben auf
dem Sandboden, durchaus vergleichbare Erträge zum konventionellen Pflugsystem er-
brachte.
Unterstellt man ein über die Jahre annähernd gleiches Ertragsniveau, so treten
bei ökonomischer Betrachtung (Tebrügge
1985) die Vorteile der nichtwerdenden Systeme durch Kostenreduktion noch deut-
licher zum Vorschein.

ZUSAMMENFASSUNG

Das wirtschaftliche Ziel des maximalen Ge-
winns ist nicht allein durch die Ertrags-
bildung eines Systems zu erreichen. Es
wird durch stagnierende beziehungsweise
abnehmende Produktpreise sowie Kontingen-
tierung bei steigenden Produktionsmittel-
preisen im zunehmenden Maße eingegrenzt.

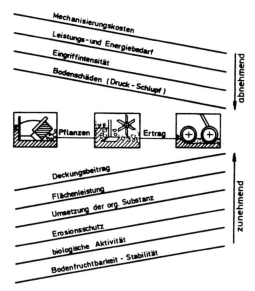

Abb. 6. Wirkungen abnehmender Bodenbe-
arbeitungsintensität

Andere, nicht ökonomisch bewertbare Faktoren mit ihrem Einfluß auf das Ökosystem
Boden werden langfristig die Berwertung
eines Bodenbearbeitungssystems maßgeblich
beeinflussen. Denn bei sachgerechter Anwendung nicht-wendender Systeme unter Ein-
schluß der Direktsaat können bei Erhaltung
des Ertragsniveaus die vielfältigen Funk-
tionen des Ökosystems Boden gefördert und
stabilitsiert und damit Bodenschäden langfristig vermieden werden. (Abb. 6)

LITERATUR

Bräutigam, V. 1989. Wechselwirkungen von
Bodenbearbeitungssystemen auf Fußkrank-
heiten des Getreides. Symposium Universität Gießen, Agrarwissenschaften 17./
18. Mai 1989
Gruber, W. 1989. Befahrbarkeit von Acker-
böden bei unterschiedlichen Bodenbearbeitungsverfahren. Landtechnik 1,
S. 25-28
Henke, W. 1987. Einfluß der Bodenbearbeitung auf die Regenwurmaktivität. Mittl.
d. DBG, 55, S. 885-889
Tebrügge, F., J. Griebel, W. Henke 1985.
Bodenbearbeitung und Bestelltechnik
heute - energie-, arbeits-, kostensparend und bodenschonend. Landtechnik, 2,
S. 73-77
Tebrügge, F. 1986. Einfluß von Bodenbearbeitungsverfahren auf das Bodengefüge.
KTBL-Schrift 308, S. 137-152
Tebrügge, F. 1988. Bodenbearbeitungs-
systeme im mehrjährigen Vergleich.
Landtechnik 9, S. 364-366

Land and Water Use, Dodd & Grace (eds), © 1989 Balkema, Rotterdam. ISBN 90 6191 980 0

Soil pressure distribution on rigid tines

G. Tine
Facoltà d'Agraria, Università di Napoli, Italy

ABSTRACT:Records of local pressures on a rigid tine were taken by means of a transducer,suitably positioned on the frontal face of the implement,which was carried by a truck, moving on a suspended rail.The vertical tine was made to travel in sandy soil at given operation depths and known speed. Tentative results showed that soil pressure along the tine is influenced by: tine length,soil water content and speed.

RESUME':Les valeurs locaux de la pression, exercé par le sol sur la face anterieur d'un outil rigide et étroit, ont été mesuré au moyen d'un convenable transducteur. Le fourchon vertical est porté par un chariot, mouvant avec une vitesse donnée le long d'un rail suspendu. On a utilisé du sol sableux,en différentes conditions d'humidité. Les premiers résultats montrent que l'accroissement de la pression le long du fourchon dépends de la longueur de celui-ci, de sa vitesse et de le contenu d'eau du sol.

ZUSAMMENFASSUNG: Die Werte des Drück des Boden über die vorhergehende Fläche eines schmales Gerät sind durch ein eigen Drückelement gemessen. Die ersten Errungenschaften beweisen die abhängigkeit von der Lage über die Gerät und von der Feuchtigkeit und der Verunstaltung des Boden.

1 INTRODUCTION

Mechanical behaviour of agricultural soil under the action of a cultivator share or tine is of interest to both agronomist and mechanician. Soil structure,i.e. in a broad sense agronomical results, as well as draught, i.e. the source of all mechanical problems,strongly depend on it.
The matter has already been investigated by many scholars both in twodimensional and threedimensional case, (e.g. Collins, Godwin 1977, Hettiaratchi 1980, Payne); draught in particular has been considered from both experimental and analytical point of view (e.g.Hettiaratchi 1966, Johnson, Orlandea, Perumbral).
Various measuring devices have been conceived and used ;in general draft components are measured and, in some instances, components of the torque acting on the structure are too (e.g.Godwin 1975, Godwin 1987, O'Dogherty). Resort to mathematical models was made to forecast draft value (Kuczewski 1981,1982, McKyes). Obviously the prevision is more or less accurate depending on the hypotheses made on deformation and properties of the soil. Particularly for narrow tines mathematical models are based on theoretic hypotheses regarding the distribution of the soil action on the implement face (Plasse, Hettiaratchi 1967, Godwin 1977, McKyes,Perumbral 1983,Grisso 1985). At the best knowledge of the author,direct observations on the pressure distribution have not yet

been made, consequently its experimental determination on the frontal face of a rigid narrow tine was deemed worth of investigation. To attain the objective an appropriate equipment was suitably designed. First tentative tests, whose results are here reported, aimed to check the equipment and to show, if any, pressure dependance on speed, soil deformation and water content. In these tests emphasis was laid more on qualitative than quantitative results.

2 EQUIPMENT

Experiments were carried out using a vertical narrow rigid tine, purposedly designed, which was pulled trough sandy soil in a tank: speed, soil water content and operation depth were separately varied. Only few notes on equipment details follow.

2.1 Tine

Some words on the phylosophy which led to the tine design seem appropriate. It was considered that local interaction between soil and implement, for given soil properties, depends essentially on both: depth, where the interaction is evaluated, and global soil perturbation induced by the implement.
Tine, schematically drawn in fig.1, consists of a main body (A) (continuous line), bearing the measuring system, which is made up of: an estensimetric ring (B), a lever with equal arms (C), and a pressure button (D). The operating part of the tine, 14 cm long, contains (D) and one arm of (C); its normal section is a 2 cm square. The button emerges on the frontal face and soil presses on it, more or less intensively, according to position and soil properties. For a given soil, in fact, pressure presumably depends on the global deformation, (i.e. on tine shape and operation depth,) and on the position along the tine face. To vary the latter and mantain at the same time the operation depth at a given value,

extensions (E) can be connected to the main body (A), as shown in fig.1 by a dashed line. Of course the body A must be suitably raised according to the length of the extension used. The above arrangement was chosen first to use only a transducer for the various measuring positions, second to avoid that the transducer bulk, if close to the sensing area, might alter soil deformation: in any instance it is above the soil surface. The sensing apparatus is a steel ring whose deformations are detected by means of a strain gage electrical bridge. The force, resulting from the soil pressure acting on the base of the pressure button, is transmitted to it by the lever (C). The greatest distance between the sensing area and the tine end is 8 cm, the smallest 2 cm, the longest extension 6 cm long; operation depth was not higher than 10 cm.

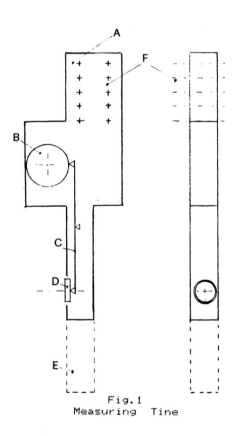

Fig.1
Measuring Tine

2.2 Soil bin

Tests were performed in a soil bin 2.5 m long, 0.8 m wide and 0.5 m deep, filled with a layer of sandy soil 0.4 m thick. Measurements were made with two soil water contents. The bin was equipped with a motored tiller, fitted with nine tines, by which soil was remoulded and conditions brought to a standard before each run.

2.3 Drag system

Tine is fastened to a pulling truck, which moves along a rail suspended, 10 cm high, above the soil surface at half width of the bin. Fastening in the required position, according to the location of the sensing point referred to the tine end, is accomplished by properly chosing in a row of holes, fig.1 (F), provided on the body (A).
Drag on the truck is exerted by a friction device, which automatically engages and, after a run of predetermined length, disengages. A pulling rope, hooked up to the truck, runs horizontally and winds forming one coil round a cylinder; the latter is keyed on the shaft of an electric motor. Next a pulley turns the rope vertical; to give rise to the drag a proper ballast must be hung to the free end. The tension due to the ballast makes the rope to adhere to the cylinder, whence the pulling action. This ceases as soon as the ballast, after a fall of predetermined length, is suitably stopped.

2.4 Recording

Pressure records were taken on a writing recorder whose maximum sensitivity is 0.004 mV/mm. The whole measuring system was calibrated at proper feeding condition of the electrical bridge.

3 TEST PROCEDURES

3.1 Soil conditions

After each test run, before the next one, soil conditions were brought back to a standard by: filling the drill, then tilling, with the joined motored implement, and finally rolling the soil.

3.2 Average speed

Average speed of the pulling truck was teoretically fixed by properly chosing the driving cylinder diameter according to the syncronous speed of the squirrel-cage driving motor. A fly-wheel was keyed on the shaft to obtain both: a higher starting torque and, under load, a lesser speed reduction. Actual average speeds of the tine were measured by recording time taken to cover a measured base.

3.3 Tine adjustments

It has been already pointed out that two adjustments of the tine are needed: one to properly set the pressure button along the frontal face of the transducer, the other one to keep the operation depth. The first requirement is fulfilled by fastening a proper extension to the main body; the second one by raising the whole tine.

3.4 Measurements

Each measurement was repeated at least three times, when values were almost the same, or more when they proved to be somewhat scattering. Measures were recorded on graph; then the average diagram of soil pressure versus position on the tine of the sensing button was plotted.

4 RESULTS

4.1 Tranducer tests and calibration

Transducer was checked to be unaffected by spurious loads: the tine stem was loaded, except on the sensing point, and no bridge unbalance resulted. The transducer was also tested for mechanical shock: to this end the truck was driven at the highest speed, with no soil into the bin, and then

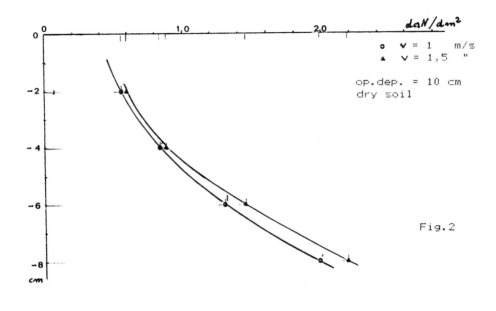

Fig.2

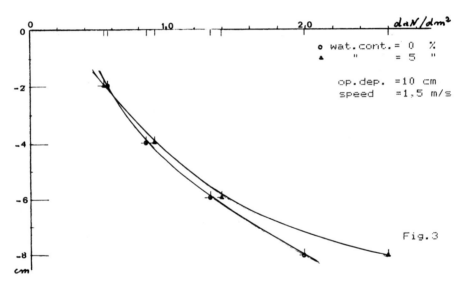

Fig.3

collided against a fixed barrier: only a 2 mm deviation was recorded. Calibration was performed by keeping the tine orizontal, with the frontal face upwards and by loading and unloading the pressure button with increasing weights: deviations were recorded.
Sensitivity was 0.3 daN/dm^2.mm, measuring range 75 daN/dm^2. With approximation of $\pm$0.5 daN/dm^2,

calibration curve resulted a straight line.

4.2 Errors

As pointed out, measuring precision was about $\pm$0.5 daN/dm^2, whereas values of any measure scattered within the range of 2 daN/dm^2. This

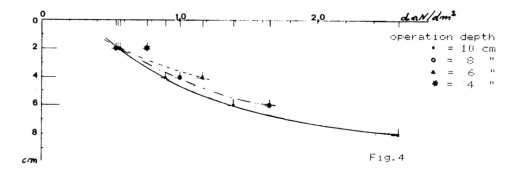

Fig.4

was interpreted as produced by
minor casual soil perturbations
which affected every single
determination. For any single
test,however, mean values lay on a
continuous well defined curve.

4.3 Pressure measurements

At first tests were performed using
dry sandy soil at operation depth
of 10 cm; two speeds were
used,namely: 1 and 1.5 m/s. Fig.2
summarizes the main results of the
above experiments. On the graph
only mean values are marked;
whereas circles refer to the speed
of 1 m/s, triangles correspond to
1.5 m/s. It is straightway to note
how measures taken in the same
position on the tine, but at
different speed, are strictly
comparable.Both curves show that
pressure markedly increases along
the tine with the increasing of
depth.The rate of change of
pressure versus depth increases
too: at 2 cm below the soil surface
(neglecting upheaval) the average
rate is of 4.5 daN/dm^2.cm but at 8
cm it almost redoubles. The curves
however are so close that all
further experiments were made only
at the speed of 1.5 m/s,also
because, at such a level,
literature claims drag to be almost
independent of speed (Payne,
Collins, Stafford).
Tests were repeated with damp soil
at 5% of water content, operating
depth was still 10 cm. Results are
compared with the former ones in
fig.3, where circles refer to dry
soil and triangles to the damp
one.As said no quantitative
analysis was attempted but only

qualitative difference observed: it
is plain that water content does
influence the tine-soil
interaction, in the sense of
increasing soil pressure on the
implement.
To ceck if pressure distribution was
influenced by soil global
deformation, this was changed by
varying the operation depth, i.e.
tine length; soil water content was
5%.
Pressures were taken along the tine.
New values are reported in fig.4,
where every curve is labelled
according to operation depth.
Results clearly show that tine
length, i.e. soil deformation,
strongly influences the pressure
distribution on the implement.
Last it was noticed that any lack of
homogeneity in the soil,due for
instance to compaction or water
content, affected pressure values.

5 CONCLUSIONS

A measuring equipment to check soil
pressure distribution on the frontal
face of a rigid tine was constructed
and tested. All experimental
results showed pressure to increase
along the tine.Tine length and soil
water content appeared to be the
major factors which governed
pressure; in dry soil, in fact, a
50% increase of speed caused only a
sligth rise of values, as,besides,is
claimed in literature (Collins,Payne
Stafford).
Sensitivity and accuracy,in author's
opinion, make the above measuring
system useful for deeper
investigations in studying relations
between pressure on tine and
deformation in soil.

REFERENCES

Collins,E.R;Lalor,W.F.1973.Soil-bin investigation of a deep-working tillage tool.Transactions of the ASAE. 16,(29-33)

Godwin,R.J1975 An extended octagonal ring transducer for use in tillage studies. Journal of Agricultural Engineering Research 20,(347-352)

Godwin,R.J.;Spoor,G. 1977 Soil failure with narrow tines. Journal of Agricultural Engineering Research 22,(213-228)

Godwin,R.J.;Magalhaes,P.S.G.;Miller S.M.;Fry,R.K.1987.Instrumentation to study the force systems vertical dynamic behaviour of soil-engaging implements. Journal of Agricultural Engineering Research 36,(301-310)

Grisso,R.D.;Perumbral,J.V. 1985. Review of models for predicting performance of narrow tillage tool. Transactions of tha ASAE 28, (1062 - 1067)

Hettiaratchi,D.R.P.;Witney,B.D.; Reece,A.R. 1966.The calculation of passive pressure in two-dimensional soil failure. Journal of Agricultural Engineering Research 11,(89-107)

Hettiaratchi,D.R.P.;Reece,A.R. 1967.Symmetrical threedimensional soil failure. Journal of Terramechanics 4, (45 - 67)

Hettiaratchi,D.R.P.;O'Callaghan,J.R. 1980. Mechanical behaviour of agricultural soils. Journal of Agricultural Engineering Research 25,(239-259)

Johnson,C.E.;Jensen,L.L.;Schafer,R.L Bailey,A.C.1980.Some soil tool analogs. Transactions of the ASAE.23,(9-13)

Kuczewski,J.1981. Soil parameters for predicting the draught of model plough bodies. Journal of Agricultural Engineering Research 26,(193-201)

Kuczewski,J.1982.A multiparametric model for predicting the draught of two model plough bodies in four different soils. Soil & Tillage Research.2,(213-223)

McKyes,E;Ali,O.S. 1977.The cutting of soil by narrow blades.Journal of Terramechanics 14,(43 - 58)

McKyes,E.;Desir,F.L.1984.Prediction and field measurements of tillage tool draft forces and efficiency in cohesive soils. Soil & Tillage Research 4,(459-470)

O'Dogherty,M.J. 1986. A tri-axial dynamometer for soil trough measurements of forces on soil working components. Journal of Agricultural Engineering Research 34,(141-147)

Orlandea,N.;Chen,S.H.;Berenyi,T.1983. A study of soil-tool interactions Transactions of the ASAE 26,(1619-1625)

Payne,P.C.J.1956.The relationship between the mechanical properties of soil and the performance of simple cultivation implements. Journal of Agricultural Engineering Research 2,(23-50)

Perumbral,J.V.;Chance,L.C.;Woeste,F. E.;Desai,C.S 1980 A matrix method for force and moment analysis on a tillage tool. Transactions of the ASAE 23,(1072-1075)

Perumbral,J.V.;Grisso,R.D.;Desai,C.S 1983. A soil-tool model based on limit equilibrium analysis. Transactions of te ASAE 26,(991-995)

Plasse,R.;Raghavan,G.S.V.;McKyes,E. 1985. Simulation of narrow blade performance in different soils. Transactions of the ASAE, 28, (1007-1067)

Stafford,J.V.1979. The performance of rigid tine in relation to soil properties and speed. Journal of Agricultural Engineering Research 24,(41-56)

Land and Water Use, Dodd & Grace (eds), © 1989 Balkema, Rotterdam. ISBN 90 6191 980 0

Influence of preliminary soil cultural practices in horticultural crops in Valencia

A.Torregrosa Mira & C.Gracia López
Universidad Politécnica de Valencia, Spain

ABSTRACT: Experiments on preliminary tilling in sowing horticultural crops were initiated in 1985 in twoareas of the Valencia region (Spain). The cropsinvolved were: carrot, sweet corn, potato, spinach, and green beans. Deep ploughing operations (40-50 cm) with a subsoiler and a moldboard plow (30 cm depth) were evaluated. Other sub-experiments were conducted, including superficial ploughing (10-13 cm deep) using rotary tiller and disk harrows. Mesurements of energy consumed in the tilling practices were done, and their quality determined (soil loosening, roughnes and size of the aggregates), as well as their effects on chemical and physical properties of the soil (Cone Index, bulk density and moisture). No significant differences in production have been observed throughout the experiment using any of the tilling systems. Energy consumed operating a subsoiler was far lower (621 MJ/ha) than that consumed with the moldboard plow (1550 MJ/ha). This is also true with tilling with disk harrow (275 MJ/ha) vs the rotary tiller (660 MJ/ha).

1.INTRODUCTION

Soil preparation in sowing horticultural crops is made with a lot of different implements what election is due more to farmer's subjective impressions than to an objective knowledge about the better. However there are big differences in the fuel requirements and work quality between them, but there are not Spanish references about experiences with this kind of cultures, and the foreigner are scarce.

It is possible to have some idea about the problem looking at works carried out with species like corn and soya, very tested in tillage research (Van Doren (1976), Siemens (1978), Camp (1984), Hallauer (1985), Dick (1985), Hargrove (1985), Jenane (1988)). All this authors are interested in the effect of no-tillage techniques, and in general they observe that traditional ploughing obtain good results in dry years, because it is very important to reserve the higher amount of water possible, but in years with a good rainfall the differences between ploughing and no-tillage are minimum.

Wald (1971), Mckibben (1973), Bolton (1976), Smith (1978) and Nave (1980) test different levels of tillage and some times irrigation rates with the same species. They obtain similar results than the first group of researchers, and see the higher the rainfall and the irrigation, the lower the importance of the tillage.

Horticultural crops precise high temperature and a lot of water. Sweet corn is a good example of these cultures. Doss (1980 and 1981), Smittle (1981) and Dickey (1983) tested the influence of pre-liminary tillage in these plants obtaining the best productions with moldboard plough, but the differences with other tools had no signification.

Shotton (1973), Curwen (1976), Orzolek (1978), Schuller (1976), Phatak (1981) using roots and tubercles found that in general productions and/or the quality was better as much energetic was the soil treatment. Some, like Millette (1981), found that a deep work with plough is not necessary all the years, been enough to do this one each two or three years.

2. MATERIALS AND METHODS

The experimental fields were located in Villena and Liria, traditional areas of horticultural productions near the Mediterranean Sea. In Villena the altitude is 300 to 400 m upper the sea level, 400 mm annual rainfall and cold winters. Liria has a good weather, without frozen, and the cultures follow another without interruption. The soils are similar in both sites: clay loam.

The crops tested were: potatoes, carrots, sweet corn, spinach and green beans. All they cultivated with spray irrigation.

Tests were distributed in field in random blocks with elementary units of 325 m2 or higher, and four or eight repetitions by treatment.

The treatments were:
-Moldboard (30 and 35 cm deep).
-Subsoiler (40 to 50 cm deep).
-Subsoiler twice (40-50 cm deep).

This deep works were combined with two shallow treatments of tillage:
-Rotary tiller (11-13 cm).
-Disk harrow (10-12 cm).

In one singular experience also was tested the absence of deep tillage, preparing the soil only with rotary tiller.

For the traction was used a 68 kw p.t.o. power tractor with two driving wheels.

They were measured energetic requirements and work rates. Fuel consumption was measured with a flowmeter Solex PLU-100 with an accuracy of 1 cm3.

Labor quality was evaluated with a profilmeter that took in each lecture 50 points with separation between points 2.5 cm. For the loosening and roughest we used the formulas:

$$e = \frac{b - a}{p} \; 100 \quad ; \quad R = 100 \log \sigma$$

e = loosening %, a and b = average of soil after and before tillage respectively, p = deep of tillage, R = Kuipers index and σ = standard deviation of after tillage data.

Chemical analysis were done before tillage, after tillage and at harvesting. Humidity, bulk density and cone index were measured along the culture. Cone index was taken with a hand penetrometer provided with a cone of 1 cm2 section and 30 degrees. The lectures were continuous from the surface to 40 cm depth.

Nascency was measured annotating the number of plants burnt in three meters long in ten segments of each treatment.

Weeds were well controlled chemically, but in carrot field was an infestation of Cyperus ssp.; in this case the population of weeds was evaluated counting the number of weeds in a square of 50 x 50 cm, taking 15 samples of each elemental plot.

Production was evaluated harvesting whole plots of potatoes, carrots and sweet corn in Villena. In the other crops only was collected an area of 150 m2 by plot.

3 RESULTS

3.1 Energy requirements

Deep tillage was done at 3 and 5 km/h, velocities of work in the region, obtaining respectively work rates of 0.77 and 0,99 ha/h with the subsoiler and 0.26 and 0.33 ha/h with the moldboard plough (table 1). Fuel consumption was similar for both tools (9 and 15.5 l/h for each velocity), the subsoiler was three times more efficient than the plough.

In the shallow tillage disk harrow was mo-re effective (1.4 ha/h) than rotary tiller (0.77 ha/h), with similar fuel consumption. (Table 2).

The energetic requirements (table 3) show that moldboard plough is the most expensive tillage (1.6 MJ/ha), subsoiler and rotary tiller have similar exigencies (0.66 and 0.62 MJ/ha respectively) and disk harrow is the cheaper (0.27 MJ/ha).

Table 1. Field capacity and fuel consumption in deep tillage.

Tractor data	Plough		Subsoiler	
Real velocity (m/s)	0.89	1.14	0.97	1.25
Slipage (%)	16	20	9	13
Width (m)	0.8	0.8	2.2	2.2
Depth (m)	0.36	0.36	0.39	0.39
Work rate (ha/h)	0.26	0.33	0.77	0.99
Fuel consumption (dm3/h)	8.82	15.06	9.42	16.26

Table 2. Field capacity and fuel consumption in shallow tillage.

Tractor data	Disk Harrow		Rotary Till.	
Precedent tillage	Mold.	Subs.	Mold.	Subs.
Real velocity (m/s)	2.42	2.44	0.94	0.94
Slipage (%)	3	1	-4	-3
Width (m)	1.6	1.6	2.3	2.3
Depth (m)	0.12	0.10	0.15	0.13
Work rate (ha/h)	1.38	1.41	0.77	0.77
Fuel consumption (dm3/h)	13.0	12.5	13.0	12.8

Table 3. Energetyc requirements of preliminary soil cultural practices.

Operation	Widh m	Depth m	Fuel l/ha	Field h/ha	Energy MJ/ha
Rotary till.	2.3	0.12	17	1.3	660
Disk Harrow	2.4	0.10	7	0.5	275
Mold. Plough	1.2	0.30	40	2.0	1550
Subsoiler	2.4	0.40	16	1.0	621

+ Tractor 68 kw single drive wheels.

3.2 Tillage quality

The moldboard loosed the soil twice than subsoiler, but the first let the surface rougher (figures 1 and 2). The sallow tillage loosed invaluably the soil, and the roughness depended of the previous deep tillage. Over subsoiled soil was not difficult to work, but over ploughed soil the disk harrow had problems to desegregate the clods, being necessary various passes to let an acceptable surface. (figure 3).

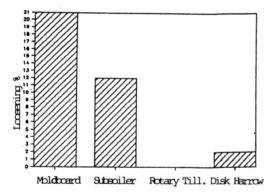

Figure 1. Soil loosening by labours.

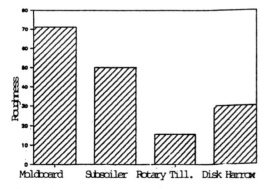

Figure 2. Soil surface roughness afrter tillage.

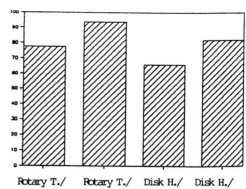

Figure 3. Proportion of soil surface covered by aggregates smaller of 3 cm diameter.

3.3 Physic properties of soil

1. Soils of study presented a theoretic field capacity around 20 %. At the moment of tillage the humidity was 13-15 %, but along the culture, the water content was near field capacity in the 20-40 cm bottom layers, and two or three points less in the 20 cm shallow. We did not find differences in humidity between tillage treatments.

2. Soil bulk density before tillage was 1.5-1.7 g/cm3. After work, the density lets down to 1.0-1.3 g/cm3, and along the culture is increasing, and at the moment of harvesting experiment an sudden raise to return to the initial values.

In the topper 20 cm there was not differences between treatments, but in the bottom 20-40 cm we found:

Ploughing at 35 cm and subsoiling at 40 cm, the first let the soil looser (1.5 g/cm3) than the second (1.6 g/cm3); but the differences were not significant.

Ploughing at 30 cm, subsoiling (40 cm) and double cross subsoiling (40 cm), we found that the two first treatments did not differ between them (1.57 and 1.59 g/cm3 respectively), but both let the soil significantly more compact than the treatment of double cross pass of subsoiler (1.47 g/cm3).

3. Cone index measurements were very correlated with bulk density and showed four clear stages in soil compaction (Figure 4). The comparison between treatments gave the same results than bulk density.

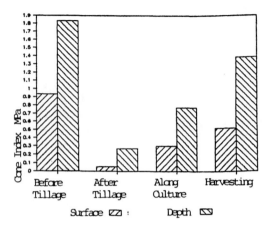

Figure 4. Four typical stages of soil compaction in horticultural crops.

3.4 Yields

Only carrot (Villena-86) and spinach (Villena-87) crops presented big differences between treatments, but these were not statistically significant because that crops suffered cultural accidents that produced losses of blocks in carrot and irregularities in the spinach production.

The other crops did not present relevant differences between deep tillage treatments (table 4).

Table 4. Yields of the tested crops.

Crop (Locality-Year)	Yield (kg/ha) Standard Desviation Plough	Subsoiler	Subs.twice
Sweet corn	10108	10061	
(Villena-86)	2128	1905	
Potatoes	13962	14068	
(Villena-86)	2888	2704	
Carrots	51987	41944	
(Villena-86)	2580	3680	
Spinach	4571	5679	5567
(Villena-87)	1273	1690	856
Sweet corn	23558	22075	21501
(Liria-87)	2083	2674	1665
Green Beans	10128	10095	10096
(Villena-87)	1368	1201	1240
Spinach 1th cut	13571	13844	
(Liria-88)	1908	2871	
Spinach 2nd cut	19991	19789	
(Liria-88)	3840	3372	

As we did not find differences in the response of the cultures to deep tillage, we did one experience to know the necessity of these. We used soils with two degrees of compaction, one 600 kPa and other 1300 kPa average of 40 cm cone index. Half of these were subsoiled and rotary tilled, and half only rotary tilled. The crop tested was green bean.

In the more compacted soil the subsoiler pass was fundamental (table 5), whereas in the loose soil there was not significant differences.

Table 5. Effect of subsoiling and initial compaction conditions of soil on green bean yield.

Labour	Soil 600 kPa	Soil 1300 kPa
Subsoiler +	14500 kg/ha	13367 kg/ha
Rotary Tiller	Std Dv. 756	Std Dv. 555
Rotary tiller	14082 kg/ha	6108 kg/ha
	Std Dv. 890	Std Dv.1513

4 CONCLUSIONS

Crop yields responses were indifferent to tillage treatments used.

But it is necessary to do some kind of deep work to loose soils highly compacted, happening the higher compaction at harvesting.

It has not been observed any improve in yields spending a lot of energy tilling with moldboard or subsoiling twice respect to a single pass of subsoiler.

Disk harrow had to many problems to desegregate the clods in this kind of soils with very compacted surface.

Probably a chisel tillage to 30 cm deep will be enough for this operations, because moldboard ploughing at that deep gave the same production than subsoiling to 40-50 cm, and the turning of the soil over does not seem to affect yields.

REFERENCES

Bolton E., Findlay W., Dirks V. 1976. Effects of depth and time of plowing Brookston clay on yield of corn and tomatoes. Canadian Agricultural Engineering. Vol. 19. No. 1. 45-47.

Camp C., Christenbury G., Doty C. 1984. Tillage effects on crop yield in coastal plain soils. Transactions of the ASAE. 1729-1733.

Curwen D., Weis G., Bubenzer G. 1976. Tillage practices effects on yield and quality of snap beans, potatoes and field corn. ASAE PAPER NO.2555, 9 pp.

Dick W., Van Doren D. 1985. Continuous tillage and rotation combinations effects on corn, soybean, and oat yields. Agron. J. 77. 459-465.

Dickey C. et al. 1983. Yield comparisons between continuos no-till and tillage rotations. Transactions of the ASAE. 1682-1686.

Doss, B., Turner J., Evans C. 1980. Irrigation methods and in-row chiseling for tomato production. J. Amer. Soc. Hort. Sci. 105 (4). 611-620.

Hallauer A., Colvin T. 1985. Corn hybrids responses to four methods of tillage. Agron. J.77. 547-550.

Hargrove W. 1985. Influence of tillage on nutrient uptake and yield of corn. Agron. J. 77: 763-768.

Jenane C., Chaplin J. 1988. Comparison of selected tillage systems with respect to energy input and crop response. International Conference on Agricultural Engineering. Paris. Paper 88.202. 9 pp.

Mckibben J., Whitaker F. 1973. Tilling various soil zones affects corn yields. Transactions of the ASAE. 404-406.

Millette J., Vigier B. 1981. Seedbed preparation for carrot production on organic soil. J. Amer. Soc. Hort. Sci. 106 (4). 491-493.

Nave W., Hummel J. 1980. Tillage for corn and soybeans. ASAE paper núm.80-1013. 22 pp.

Orzolek M., Carroll R. 1978. Yield and secondary root growth of carrots as influenced by tillage system, cultivation and irrigation. J. Amer. Soc. Hort. Sci. 103 (2). 236-243.

Phatak C. et al. 1981. Preplant tillage soil strength and root growth of turnips greens. J. Amer. Soc. Hort. Sci. 106 (2) 137-140.

Schuller R. 1979. Reduced tillage studies in potatoes following corn. ASAE paper núm. 79-1524. 10 pp.

Shotton E. 1973. Aspects of soil preparation for the potato crop. The Agricultural Engineer. Spring 1973. 37-42.

Siemens J., Oschwals W. 1978. Corn-soybean tillage systems: erosion control, effects on crop production, costs. costs. Transactions of the ASAE. 293-302

Smith E., Vaughan D., Brann D. 1978. Under-row ripping with various tillage practices for corn and soy- beans. ASAE paper núm. 78-1512.

Smittle A. 1981. Sweet corn growth, yield and nutrient uptake responses to tillage systems. J. Amer. Soc. Hort. Sci.

106 (1). 49-53.

Torregrosa A. 1988. Labores preparatorias del suelo en los cultivos de huerta: evaluaciones energéticas y agronómicas. Tesis Doctoral Universidad Politécnica Valencia, Spain.

Van Doren D., Tripplet G., Henry J. 1976. Influence oflong term tillage, crop rotation, and soil type combinations on corn yield. Soil Sci.Soc. Am.J. Vol 40. 100-105.

Wald A., Bockhop C., Lovely W. 1971. evaluation of rotary tillage systems for corn production. Transactions of the ASAE. 195-200.

ACKNOWLEDGEMENTS

-Grupo de Campo de Frudesa (SAT núm 1969)
-Comisión Asesora del MEC
-Consellería de Cultura, Educación y Ciencia la Generalitat Valenciana.

Land and Water Use, Dodd & Grace (eds), © 1989 Balkema, Rotterdam. ISBN 90 6191 980 0

Application of a vibratory tillage tool for the reclamation of ageing citrus orchards

G. Venter

Department of Agricultural Engineering, University of Pretoria, RSA

ABSTRACT: A mathematical model for the energy requirements of a vibratory tillage tool was developed. A theoretical analysis of the influence of factors such as forward speed, oscillating frequency, amplitude and direction of oscillation, depth of tillage etc, was developed. The addition of a coil spring with a spring constant of sufficient magnitude to ensure a natural frequency of vibration which was close to the frequency required for a certain tilth helped to obtain certain energy savings.

These principles were further analysed through the development of the necessary software in Turbo Pascal, and the results were applied in the development of a subsoiler with a working depth of 500 - 600 mm. The implement width was designed so that it could be used for tillage close to ageing citrus trees in an effort to stimulate root growth through better aeration and the introduction of the necessary fertilisers, thereby extending the economic productive life of these trees.

The vibrating action was obtained through the use of unbalanced masses which rotated at different speeds. The masses and speeds of rotation were chosen so that the sentrifugal forces could coïncide with the cutting portion of a full cycle where maximum energy is required.

Résumé: Un modèle mathématique établissant l'énérgie nécéssaire pour un outil vibratoire, utilisable pour labourer, a été développé. Une analyse théorique de l'influence des facteurs suivants a été développée: la vitesse de propulsion, la fréquence oscillatoire, l'amplitude et la direction de l'oscillation, la profondeur du labour etc. L'addition d'un ressort à boudin dont la constante de ressort est d'un magnitude suffisante pour assurer une fréquence naturelle de vibration est prévue. Cette fréquence est celle requise pour un certain labourage, et permet d'obtenir certaines économies d'énérgie. Ces principes ont été analysés davantage par le développement du software nécessaire en Turbo Pascal. Les résultats ont été utilisé dans le développement d'un outil pour travailler sous-sol, agissant jusqu'à une profondeur de 500-600 mm. La largeur de l'outil a été concue pour pouvoir travailler tout près des arbres déjà plus anciens; de facon à stimuler la croissance de leurs racines, en retournant d'une manière plus efficaca le sol et permettre d'y introduire des engrais nécessaires, ce qui a permis de prolonger la vie productive et économique de ces arbres. L'action vibratoire a été obtenue en utilisant des masses déséquilibrées, tournant à différentes vitesses. Les masses et la vitesse de rotation ont été choisies pour que les forces centrifuges puissent coincider avec la partie d'un cycle complet correspondant à la section du sol, ce qui exige un maximum d'énérgie.

ZUSAMMENFASSUNG: Ein mathematisches Modell für den Energiebedarf eines vibrierenden Bearbeitungsgeräts wurde entwickelt. Eine theoretische Analyse des Einflusses von Faktoren, wie Vorwärtsgang, Schwingungsfrequenz, Schwingungsweite und Schwingungsrichtung, Bearbeitungstiefe, etc. wurde entwickelt. Eine zusätzliche Bandfeder mit einer Federkonstanz von einer ausreichenden Grösse, um eine natürliche Vibrationsfrequenz, die der für ein bestimmtes Ackerland erforderlichen Frequenz nahe lag, zu garantieren, half um eine gewisse Ersparnis an Energie zu erzielen.

Diese Prinzipien wurden weiter analysiert durch die Entwicklung der benötigten Software in Turbo Pascal, und die Ergebnisse wurden bei der Entwicklung eines Tiefenlockerers mit einer Arbeitstiefe von 500 - 600 mm angewandt. Die Breite des Geräts wurde so geplant, dass es für die Beackerung nahe von alternden Citrus-Bäumen benutzt werden konnte, in dem Bestreben Wurzelwachstum durch bessere Belüftung anzuregen und notwendige Düngemittel einzubringen, um dadurch das wirschaftliche produktive Leben dieser Bäume zu verlängern.

Eine vibrierende Bewegung wurde mit Hilfe von unausgeglichenen Massen, die mit verschiedener Geschwindigkeit rotieren, erzielt. Die Massen und Umlaufgeschwindigkeit wurden so gewählt, dass die Zentrifugalkräfte mit dem Schnitteil eines vollen Zyclus wo die maximum Energie benötigt wird, zusammenfallen können.

1 INTRODUCTION

Citrus orchards have a limited productive life span. Production capacity could start deteriorating as early as 15 years after planting and could reach uneconomical production levels at the age of 20 years if unfavourable conditions could not be rectified.

It was found that one of the main reasons for the deterioration in the economical life span of these orchards was soil compaction by wheeled traffic such as tractors, sprayers, trailers etc. These machines have to move through the orchards on a regular basis, often with full loads.

Preliminary trials with deep tillage showed promising results, but the limited space between the rows of trees resulted in serious damage to the trees where crawler-type tractors were used for cultivation. Heavy draught requirements also resulted in unacceptable performance of conventional tractors, and it was felt that better results might be obtained through the use of a vibratory tillage tool with lower draught requirements.

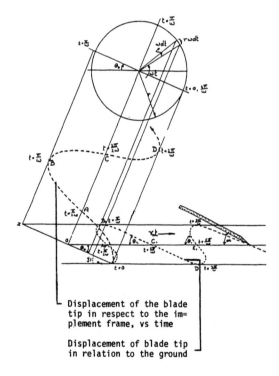

Displacement of the blade tip in respect to the im=plement frame, vs time

Displacement of blade tip in relation to the ground

2 THEORETICAL ENERGY APPROACH

(See figures 1 and 2)

Figure 1. Schematic diagram of the movement of a vibrating blade tip through the soil.

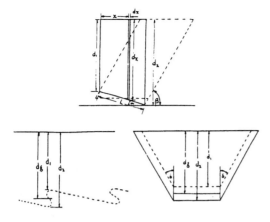

Figure 2. Tillage depths of vibratory blade moving through the soil.

A theoretical analysis of the energy requirements of a vibrating tillage tool shows that the total energy requirements could be expressed as

$$E_t = E_a + E_l + E_c + E_s + E_r$$

where E_t = Total energy requirement
E_a = Energy required to overcome accelleration forces
E_l = Energy required for the lifting action
E_c = Energy required to overcome cohesive and adhesive forces
E_s = Energy to overcome the cutting action
E_r = Energy required to overcome friction

After a certain degree of simplification, the energy equation becomes

$$E_t = \frac{\gamma}{g}\sqrt{1-2k\cos\theta_0+k^2}\; v_x r\omega + \gamma\left[(2r\sin\theta_0 + \frac{r^2\omega^2\sin^2\theta_0}{2g} - \frac{g}{2\omega^2})\right] +$$
$$\frac{kcA_2\sin\theta_0}{\pi A_g \sin^2 8}\sqrt{\frac{\pi^2-4\pi\,k\cos\theta_0+4k^2}{4-8k\cos\theta_0+4k^2}} +$$
$$\frac{bf_s}{2\,\pi A_g}\sqrt{\pi^2+4\pi\,k\cos\theta_0+4k^2} +$$
$$\frac{\mu_1\gamma\omega\,L^2\cos\alpha}{2\,\pi A_g\,v_x}\left[A_1+\frac{L\sin\alpha}{3d_1}(A_1-\frac{bd_1}{2}) + \frac{L^2\sin^2\alpha\,\tan\delta}{12}\right]$$

where γ = soil density g/m³
g = gravitational accelleration (m/sec²)
k = $r\omega/v_x$ = ratio between rotational and forward speed of the implement
r = amplitude of vibration (m)

ω = angular velocity due to vibration alone (rad/sec)
v_x = forward speed of the implement (m/sec)
θ_0 = angle of vibration
c = cohesive force (N/m²)
A_2 = sectional area of tilth with vibratory tool at its deepest position (m²)
A_g = average area of tilth (m²)
β = inclination of the shear plane in front of the tillage tool
b = lateral width of the soil engaging blade (m)
f_s = unit cutting force (N/m)
μ_1 = friction constant (soil on steel)
L = width of the blade (m)
α = inclination of the blade
d_1 = depth of tillage at the rear end of the blade (m)
A_1 = tilth area above the rear end of the blade
δ = shear angle lateral to the direction of travel

This equation could be further simplified if it is expressed as a function of k and v_x viz

$$E_t = \frac{\gamma}{g}\sqrt{1-2k\cos\theta_0+k^2}\; kv_x^2 + \gamma(2r\sin\theta_0 + \frac{k^2\sin^2\theta_0}{2g}v_x^2 - \frac{gr^2}{k^2v_x^2}) +$$
$$\frac{kcA_2\sin\theta_0}{\pi A_g\sin^2 8}\sqrt{\frac{\pi^2-4\pi\,k\cos\theta_0+4k^2}{4-8k\cos\theta_0+4k^2}} + \frac{bf_s}{2\,\pi A_g}\sqrt{\pi^2(4\pi\,k\cos\theta_0+4k^2} +$$
$$\frac{\mu_1\gamma kL^2\cos\alpha}{2\pi A_g\,r}\left[A_1+\frac{L\sin\alpha}{2d_1}(A_1-\frac{bd_1}{2}) + \frac{L^2\sin^2\alpha\,\tan\delta}{12}\right]$$

As L is normally much smaller than 1, the last factor $L^2\sin^2\alpha\tan\delta/12$ becomes negligible and could thus be ignored.

3 FACTORS INFLUENCING TOTAL ENERGY REQUIREMENTS

Close scrutiny of the above energy equation shows that the following factors should be kept in mind, ie.

3.1 The angle of vibration : (θ_0)

The angle of vibration should be kept to a minimum, provided that the right mode of movement through the soil is maintained.

3.2 The blade andle (α)

The inclination of the blade should also

be kept to a minimum, but should be large enough to prevent contact between the bottom of the blade and the untilled soil below the blade.

3.3 Blade length (L)

The length of the blade should be kept to a minimum, and preferably not longer than double the amplitude of vibration.

3.4 Speed ratio (k)

A speed ratio in exess of 1,25 is required to ensure that the bottom of the blade is free from contact with untilled soil below the blade, and to obtain a shearing action under tensile stress during the lifting phase of the blade. A minimum value of 1,5 is recommended.

3.5 Speed of vibration ($r\omega$)

The energy required is a complex function of $r\omega$, and should normally be kept as low as possible, with the proviso that speed ratio requirements should be maintained. It is better to increase the amplitude of vibration (r) rather than the speed of rotation ($r\omega$) to satisfy the right k-value requirements.

3.6 Forward speed (v_x) in relation to the rotational speed ($r\omega$)

Higher v_x values will normally lead to a decrease in the specific energy requirements if $r\omega$ is kept the same. This could be attributed to the coarser tilth due to the smaller number of shear planes that are formed per meter of forward travel if the forward speed is increased in relation to $r\omega$.

4 FACTORS THAT WILL INFLUENCE DRAWBAR PULL

Tillage energy reaches the implement via the drawbar and the P.T.O. drive, the latter of which is responsible for the vibratory action. It could thus be stated that the requirements which are favourable for a reduction in drawbar pull will lead to conditions where high energy rates are transmitted via the P.T.O. shaft.

The following recommendations could thus be made for requirements on reduced draught reduction with a vibratory tillage tool.
1. Increase the amplitude of the vibration (r)

2. Increase the rate of vibration (ω)
3. Decrease the forward speed (v_x) in relation to 1 and 2 above.
4. Use a larger angle of vibration (θ_0)
5. Use a larger blade angle (α)
6. Keep the blade length as short as possible.

5 TURBO PASCALL SOFTWARE

Due to the complicated nature of the energy equation, the necessary software was developed for the design of vibratory equipment. The above veriables were thus used in a Turbo-Pascall program which could be installed in almost any IBM-compatable computer, and this was found to be of tremendous value in the design of this specific implement.

6 DESIGN CONSIDERATIONS

The following criteria had to be considered during the design of this implement:
1. Power Source: An agricultural tractor of 80 kW. (maximum.)
2. Maximum width of implement 5 meters.
3. Maximum depth of tillage : 600 mm
4. Maximum permissable clod size after tillage : 100 mm.
5. Two strips of $\pm$ 1 meter to 1,5 meter wide had to be cultivated close to the trees, leaving an uncultivated central portion for all further traffic.
6. Due to the possible detrimental effects of root damage, it was decided to cultivate every alternative row of trees. The rows skipped will then be cultivated during a later season.
7. Soil inversion was not required.
8. The implement had to be low enough to pass underneath low-hanging branches ($\pm$ 750 mm from the ground) without undue damage to the trees.
9. The implement width had to be adjustable for row spacings between 5 and 7 meters.
10. Total operating cost for this type of tillage had to be much lower than that of crawlar type tractors.

Preliminary research on vibratory tillage tools where cams were used to obtain the right vibratory movement through the soil showed that this design had the following disadvantages, i.e.
(i) The whole drive train was subjected to impact loads when solid ob-

jects such as rocks etc. were encountered. This had to be considered when designing such items as gearboxes, shafts, bearings, chain drives etc.

(ii) Due to the above considerations, the implement tended to become heavy and very expensive.

(iii) The vibrating mechanism and total drive became very complicated, especially with multiple tillage units.

(iv) No provision could be made for easy adjustment of energy input, angle of vibration etc.

7 ROTATING MASSES

A prototype, where preference was given to rotating masses, was subsequently developed and tested. Masses of different sizes were rotated at different speeds and in different directions, thus delivering the maximum impact to the soil tillage tool in the right direction when required.

The vibrations were applied to the shanks and blades of the implement only, (and not to the whole implement) in order to eliminate the damping action of the massive frame, reduce the transfer of vibrations to the tractor, and to get a more effective application of the vibration energy to the soil.

The right angle of vibration was obtained by placing the pivoting points in front of the frame to ensure that the share point would be at the right position ($\pm$ 20 degrees behind a vertical line through the pivoting point.)

Figure 3. General lay-out of the vibratory ripper.

Figures 3 and 4 give an impression of the general lay-out of the machine without the guards which are necessary to prevent contact between the rotating masses and low hanging branches.

Figure 4. Detail of suspended arms and springs (under compression)

Figure 5 demonstrates the radial forces of
(a) A single rotating unbalanced mass. (Circular force distribution)
(b) A large mass in combination with a small mass rotating at double the speed of the larger mass - note the cancellation effect at 180° from maximum.
(c), (d) and (e) A combination of large and smaller masses as was used in the design of the implement under discussion. Note the way in which most of the energy is released during the portion of the cycle where the resultant force is at its maximum.

The actual vibratory force applied to the shank of the implement could be obtained from the following equation, i.e.

$$F = \frac{M_1 r_1 \omega_1^2 \ell_1 + M_2 r_2 \omega_2^2 \ell_2 + M_3 r_3 \omega_3^2 \ell_3 + \ldots)}{\ell}$$

Where M_n = the mass of the specific rotating mass (g)

r_n = distance between the shaft centre and the centre of gravity of the specific rotating mass. (m)

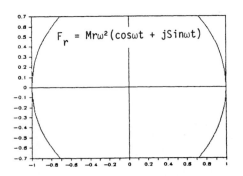

a. Single Mass (Circular force distribution)

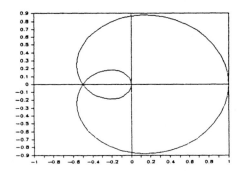

$$F_r = M_1 r_1 \omega_1^2 (\cos\omega_1 t + j\sin\omega_1 t) + M_2 r_2 \omega_2^2 (\cos\omega_2 t + j\sin\omega_2 t)$$

with $M_1 r_1 = 4M_2 r_2$

b. Two rotating masses, same direction $(\omega_2 = 2\omega_1)$

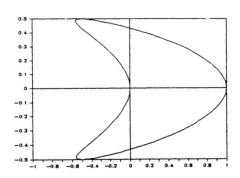

$$F_r = M_1 r_1 \omega_1^2 (\cos\omega_1 t + j\sin\omega_1 t) + M_2 r_2 \omega_2^2 (\cos\omega_2 t + j\sin\omega_2 t) -$$
$$M_3 r_3 \omega_2^2 \{\cos(-\omega_2 t) + j\sin(-\omega_2 t)\}$$

c. Three rotating masses, larger masses rotating in the same direction $(\omega_2 = 2\omega_1)$

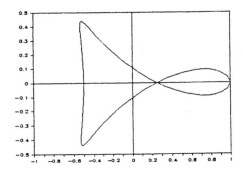

$$F_r = \frac{M_1 r_1 \omega_1^2}{2} \left[2(\cos\omega_1 t + j\sin\omega_1 t) + \cos(-\omega_2 t) + j\sin(-\omega_2 t) + \cos(-2\omega_2 t) + j\sin(-2\omega_2 t) \right]$$

d. Three rotating masses, larger masses rotating in opposite directions (case 1) $(\omega_2 = 2\omega_1)$

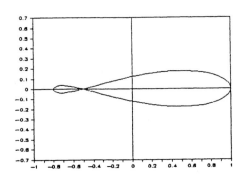

$$Fr = M_1 r_1 \omega_1^2 \left[(\cos\omega_1 t + j\sin\omega_1 t) + 0.8 \{\cos(-\omega_1 t) + j\sin(-\omega_1 t)\} + 0.2(\cos 2\omega_1 t + j\sin 2\omega_1 t) \right]$$

e. Three rotating masses, larger masses rotating in opposite directions (case 2) $(\omega_2 = 2\omega_1)$

Figure 5. Force distribution patterns of rotating masses.

ω_n = rotational speed (angular speed) of the specific rotating mass (r/min)

l_n = distance between the specific shaft centre and the centre of the pivoting point

l = distance between the centre of the shank and the pivoting point.

The direction of the main impact could be adjusted by altering the relative positions between the weights, and thus the angle at which they will come into phase.

The spring action was designed to give a natural free vibration of a frequency close to the required frequency of vibration, thus reducing the energy input into the system itself.

8 PRELIMINARY EXPERIMENTAL RESULTS

Initial tests on a fairly dry clay - loam plot gave very promising results, and after small alterations, the unit was tested under actual field conditions in a 26 year old orchard with a 72 kW SAE four wheel drive agricultural tractor.

Figure 6. Vibrating implement operating in a 26 year old orchard.

Figure 6 shows the implement during these tests. It was established that the total draught was within acceptable levels, that the necessary tillage depth could be reached, and that the tilth was remarkably fine - somewhere between that of an ordinary mouldboard plough and a rotary plough.

The degree of tilth could also be regulated by a change in the gear ratio, thereby altering the forward speed in relation to the vibratory speed.

Figure 7. Excavation of tilled area to show effective depth and fragmentation.

Figure 7 shows the actual depth and area tilled, and gives an idea of the fine tilth and proper fragmentation obtained by this implement.

It was found that root development was limited to the top 150 to 200 mm before cultivation, and though it is still a bit early to report on improvements in production, root development was stimulated to such a degree that the full area tilled was filled with new roots within two months after cultivation.

Further tests will be necessary to establish the effect of this machine on factors such as

* root diseases and the possible propagation of these by the implement

* effect on yield

* long-term effects on crop production, profitabillity and general condition of the trees.

* total economic feasibility of this method in comparison to the present replacement of orchards on a shorter cycle.

* Total costs involved in comparison to crawler-type tractors.

LITERATURE

ALEKSANDRYAN, K.V., 1963. The ap= plication of vibration for the deep cultivation of compacted rocky soils. Na. Inst. of Agric. Eng. Transl. No. 179 by P.J. Kemp, Nat. Inst. of Agric. Eng., Silsoe.

DUBROVSKII, A.A., 1960. Experiment to oscillate a drainage plough. Trudy 27, V.I.M., 241-263, & 214-240. (Opgesom in Agric. & Hort. Eng. Abstr., Vol 15 (1), 8.)

DUBROVSKII, A.A., 1962. The use of oscillation techniques for improving agricultural technological processes. Mechaniz. Elektrif. Sots. Sel. Khoz., Vol. 20 (2), 29-32. (Nat. Inst. of Agric. Eng. Transl. No. 190.)

DZYUBA, V.I., 1961. Effectiveness of using oscillating machine components. Vestn. Sel. Khoz. Nauki., Vol. 6 (3), 83-87. (Opgesom in Agric. & Hort. Eng. Abstr. Vol. 12 (4), 208.)

DZYUBA, V.I., 1963. The effect of oscillation on internal soil friction. Mechaniz. Elektrif. Sots. Sel. Khoz., Vol. 21 (5), 50-51. (Opgesom in Agric. & Hortic, ENg. Abstr., Vol. 16 (2), 74.)

EGGENMüLLER, A., 1958a. Veldversuche mit einem schwingenden Pflugkorper. Grundl. der Landtech., Vol. 10.

EGGENMüLLER, A., 1958c. Schwingende Bodenbearbeitungswerkzeuge. Kinematik und versuche mit einzelssen Modelwerkzeugen. Grundl. der Landtech. Vol. 10.

ENDO. S., 1963. Studies on the oscil= lating subsoiler. Transl. No. 15825 from the Soc. of Agric. Mach., Japan. Amer. Soc. of Agric. Eng. Journal, Vol. 25 (2), 76-82.

FOX, W.R., DEASON, D.L. & WANG, L., 1965. Tillage energy applications. Paper No. 657, presented at the 1965 winter meeting, A.S.A.E., Chicago, III.

GUNN, J.T. & TRAMONTINI, W.N., 1956. Oscillation of tillage implements. Agric. Eng. Vol. 36 (11), 725.

GUPTA, C.P. & PANDYA, A.C., 1967. Beha= viour of soil under Dynamic loading: Its application to tillage implements. Trans. Amer. Soc. of Agric. Eng., Vol. 10 (3).

HENDRICK, J.G., 1962. Transmission of tillage energy by vibration. Unpu= blished Ph.D. Thesis, Michigan State University, East Lansing.

HENDRICK, J.G. & BUCHELE, W.F., 1963. Tillage energy of a vibrating tillage tool. Trans. Amer. Soc. Agric. Eng., Vol. 6 (3), 213-216.

HETTIARATCHI, D.P.R., WITNEY, B., & REECE, A.R., 1966. Passive pressure in two dimensional soil failure. J. Ag= ric. Eng. Res., Vol. 11 (2).

JOHNSON, C.E. & BUCHELE, W.F., 1967. Energy in clod size reduction of vibratory tillage. Proc. 60th Ann. Meeting Amer. Soc. Agric. Eng. & Can. Agric. Eng., Saskatoon, Saskatchewan.

KEMNITZ, L.A., GLENN, P.B., & LLOYD, F.B., 1966. Vibrating ploughs for direct burial of cables, wires, tubing. Soc. of Automotive Eng., Bull. No. 660043, Detroit, Mich.

LARSON, L.W. 1967. The future of vibratory tillage tools. Trans. Amer. Soc. Agric. Eng., Vol. 10 (1), 78-83.

SHKURENKO, N.S., 1958. Experimental data on the effect of oscillation on the cutting resistance of soil. Nat. Inst. of Agric. Eng. Transl. from Trudy Akademii Stroitelstra I Arkhitektury. U.W.S.R. Inst. Osnovanii I Pdzemrykh Soorukzhenii. Vol. 32, Published in J. Agric. Eng. Res. Vol. 5 (2), 226.

TETSUKA, U. & ENDO. S., 1962. Studies on the oscillating subsoiler. A Trial manufacture of an oscillating sub= soiler for small tractors. J. of the Soc. of Agric. Mach., Japan. Vol. 24 (1/80), 21-24. (Opgesom in Agric. Hort. Eng. Abstr., Vol. 13 (4), 239.)

VENTER G. 1969. Energy requirements of Soil Tillage Tools; Agricultural Engineering in South Africa Vol 3 (1) p 18.

VENTER, G. 1970. Energy requirements of a simple. inclined blade. Agricultural Engineering in South Africa Vol. 4 (1) p 27.

VENTER, G. 1971. Energy requirements of a vibrating blade. Agricultural Engineering in South Africa Vol 5 (1) p 77.

VENTER, G. 1972. Application of the theoretical principles of a damped spring-mass system for the development of a vibrating tillage implement.

Land and Water Use, Dodd & Grace (eds), © 1989 Balkema, Rotterdam. ISBN 90 6191 980 0

First test results of rice sowing on dry soil by pneumatic fertilizer spreader

P.Balsari
Institute of Agricultural Mechanics, University of Turin, Italy

S.Bocchi & F.Tano
Institute of Agronomy, University of Milan, Italy

ABSTRACT: One of the most interesting new sowing tecniques for rice growers who will face future possible water shortages seems to be sowing in dry soil conditions. In the 1987, at the Veneria farm (VC), the performance of two fertilizer spreaders (one pneumatic an the other with spinning disk distributor) sowing in submerged and dry soil were evaluated. After this first year of field experiment different results were obteined from the machines working in the different experimental condition. When the sowing was done in the dry soil the sowing operation time was by 20% lower. Moreover the pneumatic fertilizer spreader, with the same good performance either in dry or in submerged soil, permitted an increase of the yeld by 27% compared with the yeld reached with the traditional spinning fertilizer distributor.

RESUME': parmi les différentes techniques conseillées aux producteurs de riz pour faire face aux pénuries possibles d'eau dans la phase initiale de la culture, l'ensemencement en terrain sec semble offrir de bonnes perspectives. En 1987, dans l'entreprise Veneria (VC), deux répandeurs d'engrais ont été testés mécaniquement (un pneumatique et l'autre centrifuge) pour semer le riz, sur terre submergée et sur terre sèche, ensuite les résultats obtenus ont été estimés. A la fin de la première année les résultats ont été différents suivant les machines et les conditions d'essai. En particulier en ce qui concerne les conditions menées sur terrain sec, il a été possible de réduire de 20% le temps nécessaire à l'ensemencement. En outre, le répandeur pneumatique d'engrais, qui a opéré sur terrain sec comme sur terrain immergé, a permi d'augmenter la production de grains d'environ 20% par rapport au répandeur centrifuge d'engrais de type traditionnel

ZUSAMMENFASSUNG: unter den verschiedenen Techniken, die den Reisanbauern zur Begegnung der evtl. Wasserlosigkeitsprobleme beim Anlauf der Produktion empfohlen werden, scheint die Trockensaat interessante Perspektiven zu bieten. 1987 wurden in dem Betrieb Veneria (VC) die maschinellen Leistungen zweier Düngerstreuer (einer pneumatisch und einer zentrifugal) bei der Reisaussaat auf überschwemmtem Boden und auf trockenem Boden, sowie die entsprechenden Agrarresultate bewertet. Am Ende des ersten Jahres konnten zwischen den Maschinen Leistungsunterschiede bei den verschiedenen Bedingungen festgestellt werden. Besonders bei der Trockenbedingung war es möglich, die für die Säharbeiten erforderlichen Zeiten um 20% abzukürzen.
Ferner ermöglichte der pneumatische Düngerstreuer - der sowohl auf trockenem als auf überschwemmtem Boden gleich gut arbeitet - eine Zunahme der Kornproduktion um etwa 27% gegenüber dem traditionellen Zentrifugal-Düngerstreuer.

1.FOREWORD

At our latitudes, rice is generally sowed by broadcasting on a submerged rice-field by means of spinning fertilizer spreaders; the other sowing modalities are now linked to particular situations and therefore little diffused.

Sowing on submerged soil presents remarkable mechanical, agronomical, organization and water problems. More in details, this sowing modality brings about: less mobility and working capacity of the machines, the necessity to equip them with subsidiary grappling devices, lack of reference points to determine the correct distance between the different passages of the sowing machine, relative asphyxia of the environment during rice germination, high water consumption.

Concerning what above said, sowing on dry rice-field seems to be a valid alternative to sowing on submerged rice-field on water-proof soil not requiring active state obstruction and it is the only possible solution on scarce capacity soils.

The dry sowing can be performed spreading on the surface or in lines covering the seed with earth.In the first case, soil submersion must follow immediately (therefore, the seed will have to be treated with surface-active substances to avoid its displacement with subsequent accumulations), whilst it must be delayed to germination occurred in the second case.

Sowing in dry soil covering the seed with earth is generally performed with suitably modified wheat sowing machines. This system allows both avoiding problems characterizing sowing on submerged soil and saving water or, however, to dilute its requirements on a longer time when starting the cultivation if soil is submerged after the emergency
Nevertheless, this system has some inconveniences linked to the increased difficulty in checking weed besides the not easy adjustment of seed fall due to glumes roughness.

On the other hand, even if surface sowing does not offer special water advantages, since submersion is only delayed of a few days, it allows increasing machines working capacity and, in special conditions and with the use of suitable equipment, to check the initial development of weeds.

Finally, it must be remembered that, independently from the presence of water, good soil levelling and seed distribution uniformity throughout the plot of land is extremely important. As known, uniform seed distribution depends upon the construction and operation characteristics of the machines used besides their operation conditions.

In order to estimate both working characteristics of some machines present on the market that are or that could be used for sowing rice as well as the technical and operative advantages of surface sowing on dry soil, the herebelow-described experiments have been carried out.

2. MATERIALS AND METHODS

In 1987, following a preliminary analysis started the previous year, three different rice sowing machines have been compared acording to a randomized block diagram with four repetitions : one pneumatic fertilizer spreader (fig.1) and one seed drill (fig.2) for sowing on dry soil; one pneumatic and one spinning fertilizer spreader (fig.3) for submersion sowing.

The test took place in four contiguous fields made available by the Veneria farm in Vercelli. For the experiment, the Veneria variety has been used due to its characteristic to adapt to test environmental conditions, due to the rapid growing of the germ and due to its good resistance to low initial temperatures.

Each elementary plot, 50 m.long, had a 36 linear m. front corresponding to three passages of each machine which were necessary to allow observations concerning the transversal uniformity of seed distribution.

Sowing took place on 28th and 29th April with 140 kg/ha of rough rice. In the case of dry soil sowing, the seed was previously treated with Siacton surface-active agent of Siapa (250 g.diluted in 7 liters of water per quintal of seed); this allowed flooding the field only some hours later than for submersion sowing. From this time till harvesting, all plots underwent the same treatments.

Fig.1 - Spinning fertilizer spreader used for the experiment.

Fig.2 - Pneumatic fertilizer spreader used for the experiment.

Table 1 - Machines features

	LELY	AMAZONE JET	ACCORD DL
Hopper capacity (dm3)	2000	1200	500
Empty mass (kg.)	650*	450	450
Work width (m.)	15	12	3.6
Type of distribution	spinning	pneumatic	pneumatic
Type of coupling	towed	carried	half-carried

* with iron wheels

During experimentation, some climatic and cultivation variables have been detected (meteorological data and water temperatures in the various fields, the development phases of some sample plants that will be published in a note to come).
Besides the here-below described mechanical observations, at the end of the experimentation ten tufts per plot have been picked in order to determine production components: number of panicle/plant, number of grains/panicle, number of not ripened grains/panicle and weight of 1000, making a distinction between main stem and cutting stem.
Before harvesting, in three different points of each plot, the panicles present in each square meter have been counted. Furthermore, the damages caused by lodging and by animal or vegetal parasites have been estimated.
The harvesting, which took place on 20th October has been performed both manually(two 1 sq.m.samples per plot) and by harvester(one 20 sq.m. sample per plot). The present note refers to the results obtained with the mechanical harvesting performed in the central part of plots, both far from banks and from water inlet openings.
Concerning the more mechanical aspects, it must be pointed out that all machines used for experimentation are mainly including a frame on which one hopper, one adjustment and one distribution system are fixed.
The internal machines organes, exception made for pneumatic sowing machine adjustment system, are activated by means of the power outlet of the tractor with which machines are connected (table 1).
The pneumatic fertilizer spreader used for the tests (produced by the company Amazone, mod.jet 1202-1) has an empty mass of 450 kg.and a 1200 dm3 hopper capacity. The distribution system includes a radial fan whose air flow is subdivided in 16 transport tubes (8 per each side) constituting the distribution bar having a total width of 11.25 m. By gravity, the seed contained in

the hopper is brought in contact with two metering rollers whose angular speed is controlled by means of a continuous adjustment eccentric shaft (fig.4). Each metering roller conveys the desired quantity of material in the eight small hoppers located below and each one of them is located at the beginning of a transport pipe. The material,after reaching the hoppers, falls in the air flow transporting it towards the spreader.
The actual machine working width is 12 m.

Fig.3 - Detail of the beaters and of the pneumatic spreader eccentric cam.

Fig.4 - Pneumatic seed drill.

whilst the recommended height of the bar compared with the surface is 0,8 m. The tests for estimating machines operation characteristics concerned the determination of the following parameters:
- uniformity of transversal distribution;
- quantity adjustment system efficiency;
- damages caused to the seed;
- working capacity.

The transversal distribution uniformity of each machine has been estimated according to O.C.S.E. rules concerning the standard procedures for fertilizer spreaders tests. For this purpose, some cassettes have been realized (500 x 500 x 150 cm.) internally equipped with 8 separation plates (85 cm.high) suitable to reduce possible material rebounds during the distribution phase (fig.5). The material fallen inside each box has been weight by means of an electronic, two decimal weighing machine. With a subsequent processing of the data thus collected, the following has been determined:
- the average value of the quantity of seed fallen in the box;
- the percentage variation near the average;
- the variation coefficient;
- the real distribution quantity (kg/ha) referred to a given advancement speed.

As for the pneumatic sowing machine, seed collection has been performed near each distribution element.

Weighing the material thus collected, the transversal distribution diagram has been derived. Furthermore, with the aim to determine whether the simple seed passage through the mechanical organes of the various sowing machines could cause damages to the seed thus reducing its possibility to germinate, laboratory tests on germination have been carried out.

For this purpose, seed samples have been taken from each outlet device of the two compared sowing machines (Amazone e Accord) randomly extracting sub-samples for analysis on Petri plates at 28 degrees C.

For each outlet, three 100 seeds plates have been prepared and the reference sample was a seed sample directly taken from the hopper. The germinated seeds have been examined for three days also considering the seeds with broken glume or the faulty seeds (plumette only, radicle only, musty seeds, etc.). The germination percentages, suitably transformed, have been subjected to variance analysis.

3. RESULT AND DISCUSSION

The unfavourable climatic trend (temperature decrease and strong wind) remarkably influenced the investment causing a high percentage of faults. This inevitably reflected in an undifferentiated way amongst the various sowing methods which have been investigated, on the investment and mainly on the number of panicles per surface unit.

Fig.5 - Test boxes used for the transversal accuracy of spreading test.

Table 2 - Working time and productivity of labour related to traditional sowing and dry-soil sowing.

Sowing technique	Operation	Work capacity (ha/h)	N.of operators	Labour productivity (hxman/ha)
Traditional	Puddling	1,75	1	0,75
	Sowing	4	3	0,75
Total				1,32
On dry soil	Rolling	5	1	0,20
	Sowing	5,5	2	0,38
Total				0,58

3.1 Yield and yield components

The recovery occurred with the tillering has been partial and anyhow smaller with the Accord pneumatic seed drill used for seeding with dry field. This explains the not very high productions obtained also because the good fertility of the main stem has been accompanied by the less acceptable fertility of the tillering stem.
In fact, the main stem panicles have always borne the double or more of grain/panicle compared with tillering stem.
However, the Amazone sowing machine distinguished itself in both cultivation start conditions which have been experimented, thus allowing to obtain grain productions that, as an average, are approx.27% higher than those obtained with the compared sowing machines (table 2).
This improved production result is partly due to the higher number of grains which matured on main stem, concerning both spinning fertilizer spreader (+27%), in submersion sowing, and Accord (approx.+22%) for dry sowing; the correlation between production and number of grains/panicle resulted highly significant (r=0.77).
It is interesting to note (table 2) that, when sowing in submersed rice-fields, the Amazone reached the same number of panicles/sq.m. than spinning fertilizer spreader, even if with the latter one a higher tillering has been reached. The largest number of main stem panicles, together with the smaller incidence of not ripened grains obtained with Amazone, brought about its better production result.
For dry soil sowing, Amazone and Accord have not reached remarkably different results concerning the degree of tillering; however, the number of not ripened or ripened per panicle has always resulted favourable in Amazone.
The weight of each grain, both for master and for tillering stem, has shown a tendency to decrease where the panicles had a higher number of ears.However, this did not help to recover all what has been lost due to the lower fertility of plants.

3.2 Machines working quality

Concerning transversal distribution uniformity, the experimentation revealed that the spinning fertilizer spreader is characterized by an asymmetric distribution diagram. In fact, with 10 m. distance between one passage and the following one (normally used in the farm), maximum mistakes or approx.22% compared with the average are made, with a variation coefficient of 17% (fig.6).

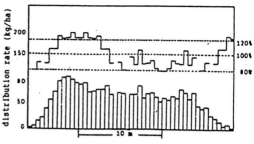

C.V. = 17,02   distribution rate = 149,46 kg/ha

Fig.6 - Diagram of transversal distribution of the spinning distributor.

The transversal distribution uniformity seemed sensibly influenced by environmental conditions, particularly by the wind (drift phenomena) and by soil asperities which caused continuous machine inclination variations compared with water surface.
The lack of a device for work width partialization caused remarkable waste of seed near field borders as well as application rate alterations during sowing stages completion. During the beginning and end phases of distribution, remarkable work quality decreases were detected, due to the anomalous seed distribution during the period of time requested to reach the optimum rotation rate of the distribution disk and viceversa.
The distribution uniformity tests performed with the Amazone Jet 1202-2 pneumatic fertilizer spreader evidenced a trapezoidal distribution diagram requiring limited overlap between the two contiguous passages. With 12 m. work width the maximum error values compared with the average appeared as ranging between + 16,38% and -9,48%, with 6,6% variation coefficient (fig.7). In order

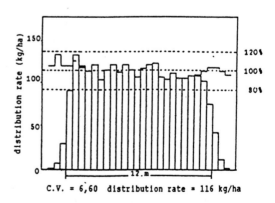

C.V. = 6,60   distribution rate = 116 kg/ha

Fig.7 - Diagram of transversal distribution of the pneumatric fertilizer spreader.

Table 3 - Yeld and yeld components obteined in different conditions

Machine	Yield (dt/ha)	panicle (n/m²)	Tiller/ hill (n)	Grains/ main stem (n)	Grains/ tillering stem (n)	not ripened grains (%) main stem	not ripened grains (%) till. stem	Weight of 1000 kernels main stem	Weight of 1000 Kernels till. stem	H₂O at harvesting (%)
AMAZONE (submersed)	58,64 Aa	205,5 a	2,6 b	174,5 Aa	64,25 a	9,8 a	5,1 a	30,95 a	29,63 a	18,6
SPINNING DISK (submersed)	46,59 Bc	199,7 a	3,8 a	137,0 Bb	70,75 a	12,6 a	10,2 a	34,80 a	30,88 a	19,4
AMAZONE (dry soil)	65,34 Aa	192,7 a	3,4 a	157,0 Aa	88,5 a	7,6 a	7,0 a	30,65 a	30,25 a	19,9
ACCORD (dry soil)	46,02 Bc	174,2 a	3,2 ab	128,7 Bb	69,75 a	14,8 a	8,6 a	32,98 a	32,48 a	19,4

The averages indicated with the same letters are not statistically different (P<0,05 small letters, P<0,01 capital letters)

to keep the correct distance between two contiguous passages, in the case of sowing in dry soil, it is possible to trace the lines with the help of a foam device, which is however already supplied with the machine, whilst, when operating in a submersed soil, it is useful to locate the diffusers of both bar extremities in a high position so as to widen seed track area. Due to its construction characteristics, the pneumatic seed drill does not require overlap between following passages has its distribution diagram is characterized by maximum error values compared with the average ranging between+21,40 and -23,64% and by a variation coefficient of 12,82% (fig.8). This high value of variation coefficient seems due to flow divider not fully satisfactory operation.
The results of laboratory germinability tests are rather interesting.
Analyzing fig.9, containing the data concerning the Amazone pneumatic fertilizer spreader, it is possible to observe that, whilst seed distributor has not damaged rice seeds at the extreme outlets of the distribution system, both right and left side outlets brought about small germinability reductions from 97-98% to sample to 86% of the left intermediate outlets and 92% of right intermediate outlets.
On the contrary, the seed samples drawn from the 14 outlet outlets of the Accord pneumatic seed drill did not reveal neither damages nor decreases of germinability which always remained at higher values than 90% (fig.10).

3.3 Machines work capacity

Concerning machines work capacity it has been revealed that, with dry soil sowing using the pneumatic fertilizer spreader, it

is possible to increase work capacity by 37,5% and to reduce the number of operators necessary to perform the operation. These advantages derive from the higher advancement speed of the machine on dry soil and from the possibility to use elements tracing on soil instead of an operator in order to determine the correct overlap between subsequent passes. Operating in this way, it is possible to reduce by 50% the need of operators to perform sowing operations. In fact, in our test conditions, we passed from 0,75 hours⋅man/ha, in the case of traditional sowing, to 0,38 hours⋅man/ha in the case of sowing on dry soil (table 3).
Therefore, in the case of dry soil sowing by pneumatic fertilizer spreader it has been

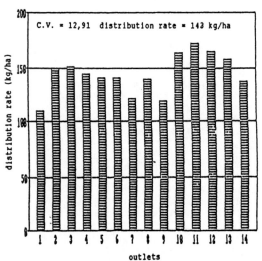

Fig.8 - Diagram of transversal distribution of the seed drill machine.

possible to obtain both cost reduction and a more timely development of sowing operations.

4 CONCLUSION

The use of the Amazone fertilizer spreader has revealed particularly interesting due to the positive consequences it can have on grain production and on work quality and productivity. Compared with the spinning fertilizer spreader, the Amazone machine permitted a more uniform transversal distribution as well as a more homogeneous covering of sectors near headlands, thanks to a smaller track area and to the possibility to divide the distribution bar in sectors.

The better seed distribution reflected in a higher percentage of main stem (more productive than the tillering ones) as well as in a higher number of grains/panicle with favourable influences on grain yield which increased by approx.27% compared with the one obtained with the compared sowing machine.

Furthermore, in dry soil, manpower productivity with sowing by means of the pneumatic fertilizer spreader has increased by approx.60%. The Amazone fertilizer spreader, due to the various technical and operational advantages found, has a higher purchasing price (10 million lire) than the traditional spinning fertilizer spreaders. Nevertheless, if the positive production results obtained in our test conditions were confirmed, the higher purchasing price of the machine could be compensated in one year only operating on a surface of 15 ha only.

REFERENCES

Baldi, G. e Malagoni, R. e Giovannini, G. 1974. Studi sulla possibilita' di coltiva-zione del riso con irrigazione turnata. Il Riso,XXIII,20-31

Baldi, G. e Mazzini F. 1983. Coltivazione del riso senza sommersione: risultati ottenuti con linee selezionate per l'ambiente italiano. Annali dell'Accademia di di Agricoltura di Torino,CXXV,1-16.

Balsari, P. e Provolo, G. e Airoldi, G. 1987. Concimazione minerale con impiego di spandiconcime pneumatico. Macchine e Motori Agricoli,3,101-109.

Crescini, F. 1951. Prove di semina ritardata del riso e possibilita' di ottenere due prodotti in un anno. Conf.Soc.Agr. di Lomb. 17.2.51.

Moletti, M. e Villa, B. e Mazzini, F. e Baldi, G. 1987. Un quinquennio di speri-mentazione sul controllo delle erbe infe-stanti nella coltivazione del riso senza sommersione. Informatore Fitopatologico 7-8,49-62.

Piacco, R. 1953. Prove di irrigazione della risaia. Il Riso, 11,7.

Piacco, R. 1973. Prospettive sulla semina del riso all'asciutto. Atti VIII Convegno Internazionale di Risicoltura, Vercelli.

Poli, P. 1957. Esperimento di coltura di riso con irrigazione a pioggia. Il Riso, VI,2.

Spada, C. 1964. Indagine sulla semina del riso a secco con postarella interrata. Il Riso, 2,15-22.

Tano, F. e Sebastiani, E. e Sparacino, A. e Chiapparini, L. 1977. Ricerche sul diserbo chimico del riso coltivato con irrigazione turnata. Atti Incontro S.I.L.M., 59-72.

Tinarelli, A. 1973. La coltivazione del riso. Edagricole.

a)

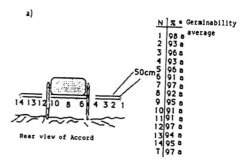

* average values equal to 5% correspond to equal letters.

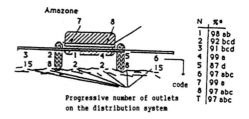

b)

* average values equal to 5% correspond to equal letters.

Fig.9 - Germination test results of the seed passed through the tested machines distributors.

Land and Water Use, Dodd & Grace (eds), © 1989 Balkema, Rotterdam. ISBN 90 6191 980 0

Seed encapsulation system

R.E.Garrett, J.J.Mehlschau & N.E.Smith
University of California, Davis, Calif., USA

ABSTRACT: A system has been developed for encapsulating tomato seeds in an alginate gel. Seeds climb spiral paths in a vibrating bowl. Air jets and seed sensors allow one seed at a time to be admitted to one of two encapsulation nozzles on a self-contained unit. Gel, pumped in discrete drops from an annular nozzle, forms a meniscus which catches the next seed. A puff of air inflates the meniscus slightly as the next pulse of gel traps the seed inside the drop. Capsules are inspected by a digital camera as they fall to a hardening bath. Three successive empty capsules cause a channel to stop and signal the operator. The unit stops when the combined output of acceptable capsules from the two nozzles reaches a pre-set value. Six encapsulating units are monitored by a computer which records date, time, number of capsules, number of rejects, and identity of the encapsulating unit and prints labels.

1 JUSTIFICATION

Encapsulation of seeds is performed for several purposes. Many seeds are difficult to singulate for precision planting. Encapsulation can produce a unit which is more easily handled. The encapsulating medium affects the microenvironment around the seed. The effect may be beneficial or detrimental. Encapsulation also provides an opportunity to combine with the seed one or more adjuvants selected to enhance the growth of the seedling or to curtail the growth of competing organisms in the surrounding environment.

Use of a hydrated alginate gel appears to improve the likelihood of germination of seeds and to improved emergence, survival and vigor of seedlings. It also appears to be a good medium for carrying a wide variety of adjuvants, especially microorganisms; and the gel provides good protection for primed or pre-germinated seeds. The gel, a mixture of sodium alginate in water, can be applied to raw seeds, to pregerminated seeds and even to plant embryos produced through tissue culture techniques. Encapsulation of plant embryos in gel has been proposed as a means of producing artificial seeds.

The costs and benefits of encapsulating seeds obviously need to be evaluated before a decision is made to encapsulate them. From the anticipated rate of production and estimates of the cost of encapsulation equip-

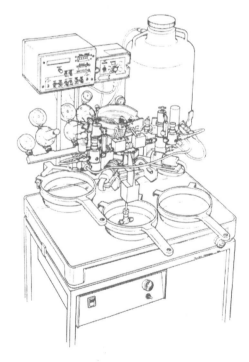

Figure 1. Seeds are fed individually to one of two nozzles where a gel coating is applied. Capsules are inspected by a digital camera as they fall to a hardening bath.

ment, approximate costs of seed encapsulation can be established. Benefits of encapsulating seeds in gel need to be determined by comparing their performance with that of non-coated seeds or seeds coated by other methods. The more costly seeds are to produce, the more important it is to ensure survival of their seedlings.

The development reported here was initiated and supported by Plant Genetics, Inc. They performed cost/benefit analyses in the context of their own operations and in view of their anticipated future needs. The encapsulation system was designed to encapsulate hybrid tomato seeds in sodium alginate gel. During the development phase, no adjuvants were added to the seeds or the gel.

2 THE SYSTEM

One of the units is shown in figure 1. The system, as shown in figure 2, consists of multiple, self-contained units. Each employs two capsule forming nozzles fed from one dual channel seed singulating device. Figure 3 shows a cutaway view of the seed singulating device and capsule forming nozzles. Separate sorting devices inspect the

capsules from each nozzle before the capsules are combined in a single container where they are collected and allowed to harden. Six of the self-contained units are combined with a single packaging unit, and information from the separate units is fed to a single computer which monitors their performance and drives a label printer. The seed singulation, capsule formation, capsule inspection and control components will be described in some detail. The remainder of the system will be discussed only in general terms.

3 SEED SINGULATION

A rotary vibrating feeder (figure 3) having a bowl with two spiral ledges on its inner surface, is used to provide single files of seeds along two separate paths. Near the upper end of each seed path is an air jet used to return excess seeds back into the bowl. When a seed is to be sent to an encapsulation nozzle, the air jet is shut off. Next to the air jet orifice, is a seed drop tube which routes seeds from the vibrator bowl to the encapsulation nozzle by gravity. An infrared sensor, mounted on the edge of the vibrator

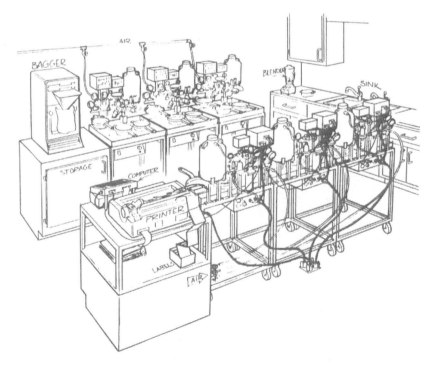

Figure 2. A system of six seed encapsulating units is monitored by a computer. Plastic bags of capsules are labelled to identify the number of capsules, the date and time, and the unit by which they were encapsulated.

bowl and focused at a point near the edge of the seed drop tube, detects seeds as they move into the drop tube. The sensor signal turns the seed return air jet back on in time to prevent more than one seed at a time from falling into the seed delivery tube. Both spiral seed paths are similarly equipped and the output from each seed drop tube is routed to a separate seed encapsulation nozzle.

4 SEED ENCAPSULATION

A coaxial tube nozzle, (figure 3) oriented vertically, is used to encapsulate the seeds. Seeds channelled from the seed drop tube on the vibrating bowl fall through a passage and enter the central tube from the side about midway of its length. Surrounding the lower portion of the central tube is an outer tube with an internal shoulder at its lower end which, together with the central tube, forms an annular orifice. Alginate is supplied through a gallery between the central tube and the outer tube by a positive displacement, adjustable rate pump, creating a hollow cylindrical flow from the nozzle tip. Alginate remaining at the end of the nozzle, following a pump stroke and formation of a capsule, creates a meniscus which stretches

across the lower open end of the central tube. A seed, falling in the central tube, comes to rest on the meniscus, ready for encapsulation. Slightly before alginate is pumped through the nozzle, a small pulse of air is injected into the upper end of the central tube. The air pulse inflates the meniscus slightly before pump flow begins, aiding acceleration of the meniscus and seed with gel flow. A thin-walled cup is generated with the seed in the bottom. The air pulse is terminated midway through the pump cycle; when pump flow stops, inertia of the gel stretches and breaks the thin cylindrical wall. The top of the detached cup closes to encapsulate the seed, and gel remaining attached to the nozzle forms a new meniscus to catch the next seed.

5 CAPSULE INSPECTION

As the newly formed capsule falls toward the hardening bath and capsule collecting device, it passes through an inspecting and sorting zone. An air jet directed transverse to the path of the capsule is turned on at the same time that the gel pump is activated. This air jet will deflect the capsule into a reject container unless its air supply is shut off before the capsule reaches the jet.

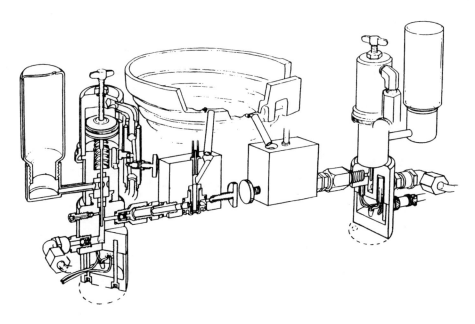

Figure 3. Seeds climb spiral paths in a vibrating bowl. Air jets blow seeds back into the bowl until one is selected for encapsulation. Gel, pumped in discrete drops from a nozzle, forms a meniscus which catches a seed. A puff of air inflates the meniscus slightly, and the next pulse of gel traps the seed inside a drop.

Before reaching the reject air jet, the capsule passes between a light source and a digital camera (figure 2). A line of pixel elements in the digital camera is scanned rapidly and repeatedly by the control system. The light source is of such intensity that the gel material in the capsule appears transparent. If the capsule contains a seed, the seed blocks light to pixels in the line during at least one of the scans. When light is blocked by a seed to two or more consecutive pixels, the reject air jet is turned off in time to allow the capsule to pass freely into the hardening bath and collection container.

6 SYSTEM CONTROL

Spacing and timing of events are quite critical to successful operation. The seed singulation device has its own control system. A request for a seed triggers deactivation of the appropriate seed return air jet. When a seed is detected falling into the corresponding drop tube, the air jet is turned back on. The remaining functions are coordinated by a microprocessor-based control system which independently controls the two separate capsule forming and inspecting channels. Figure 4 shows a typical sequence of control signals.

At the beginning of a cycle, it is assumed that a meniscus exists at the nozzle, that a seed is on the meniscus, and that seed return air is blowing seeds back into the bowl to prevent additional seeds from entering the nozzle. When the control system is switched to the automatic mode, a seed ready signal is already ON indicating that a seed has been delivered to the meniscus and the seed return air is ON. At that time, the seed return air is turned OFF to allow the next seed to advance toward the seed delivery tube, and a capsule form delay period is initiated. (The time required to select and deliver a seed is sufficiently long that the search for the next seed can precede formation of a capsule. This procedure results in minimal cycle time and allows seeds maximal time to settle onto the meniscus.) At the end of the capsule form delay period, the capsule reject air is turned ON and an encapsulation cycle is initiated to encapsulate the seed already on the meniscus. Meniscus inflation air is turned ON; shortly thereafter, the pump is actuated. (Pulses actuating the inflation air and pump are both of fixed duration and overlap each other.) The capsule reject air is turned OFF by the presence of a seed detected in the capsule as it is pumped out of the nozzle. If no seed is detected by the sensor, the capsule reject air will be turned OFF again after a timed period following capsule

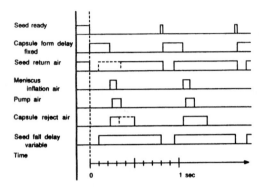

Figure 4. A typical sequence of control signals, during operation in the automatic mode, begins when a seed has been singulated and given time to fall to the meniscus on the nozzle. To minimize the cycle time, the search for the next seed begins before one is encapsulated.

formation to reduce air consumption.

While this capsule forming sequence has been in progress, the system has also been seeking the seed to be used in the next capsule. When that seed is detected falling into the seed drop tube, the seed return air is turned back ON and a seed fall delay is initiated. The seed fall delay is made sufficiently long to ensure that the seed admitted to the drop tube will have time to reach and settle on the meniscus before the next pump action. The end of the seed fall delay period causes the seed ready signal to switch ON again. In the automatic mode, this initiates the next cycle. In the manual mode, a separate signal is required to initiate the next cycle.

In addition to controlling the sequence of events, the control system also counts both the number of times the pump cycles (taken to be the number of capsules formed) and the number of capsules with seeds detected by the capsule inspection component. The difference between these two counts is taken to be the number of blank capsules generated. The control system also counts consecutive blanks. When three consecutive blanks are detected, the control system initiates a clearing cycle. This involves three pump strokes without seeds. After the clearing cycle, the system reverts to normal automatic action. If a second series of three consecutive blanks is detected immediately, the control system shuts down that particular nozzle (the other nozzle is not affected) and signals the operator through an audible tone and a visual readout. The operator must take corrective action (generally clean out the nozzle and ensure that the seeds are entering the nozzle properly) and reset the control system to resume operation.

The control system has provisions by which the operator may set in an identification number for the capsule collection container and the number of capsules to be formed into that container (figure 2). When the sum of the numbers of capsules accepted from both nozzles equals the preset number, the unit stops and signals the operator.

7 THE REST OF THE SYSTEM

Capsules are collected in a screen-bottomed basket setting in a tray of calcium chloride in water (figure 2). An air bubbler in the basket causes capsules to flow to the side of the basket out of the way of incoming capsules. The bubbler also disturbs the surface of the calcium chloride solution and reduces the chance of solution splashing back onto the nozzle.

When one basket is filled, it is set aside for approximately 20 minutes to allow the capsules to harden. A computer, which has been monitoring the capsule-forming units, indicates which baskets are filled and which have set the proper amount of time in the hardening bath. After the hardening period, the basket of capsules is suspended on a bracket above the tray to drain (figure 2). Then the basket is carried to the packaging station and emptied into a plastic bag. The number of the basket is entered into the computer. The computer prints a label identifying the unit which formed the capsules, the number of capsules in the basket, and any other pertinent information desired, such as time, date, and variety of seeds.

The computer communicates via serial interface with the control system on each of the units. Except for this sharing of packaging and computer facilities, each unit is self-contained, needing only supplies of seeds and alginate and sources of 690 kPa (100 psig) dry, filtered compressed air and 120 v, 60 Hz electricity.

8 OBSERVATIONS

Fabrication of the nozzle requires precise machining to provide an annular orifice with proper radial and axial clearances and with a high degree of concentricity. For the units described in this report, designed to encapsulate tomato seeds, the orifice ID was 5.500 mm (0.2170 in) with a radial clearance of 0.038 mm (0.0015 in). The orifice is formed between an internal shoulder on the outer tube and the outer diameter of the inner tube. The shoulder begins at a point, coinciding with the end of the inner tube. Overlaps with the inner tube, ideally zero, should be no more than 0.010 mm (0.0004 in). The orifice clearances and the amount of overlap are major determinants of the velocity and direction of gel flow which, in turn, determine whether a cylindrical flow of gel is generated, a capsule detaches and a new meniscus forms. Parts which perform satisfactorily together should be kept together as matched sets. The ability to produce precise nozzles appears to represent a practical limit to the minimum size of capsules which can be formed. For the present device, the minimum capsule size was approximately 80 ml; larger capsules (on the order of 100 ml) were produced more easily and reliably.

The ability to form a meniscus and the strength of the meniscus depend on the surface tension of the gel material, the diameter of the central tube and the thickness of the meniscus. Surface tension depends primarily on the concentration of the gel. The thickness of the meniscus, in turn, is dependant on viscosity in concert with the velocity of the gel flow through the orifice. Velocity of the gel affects capsule separation and the quantity of gel remaining for formation of the next meniscus. Viscosity, dependent primarily on concentration, is also affected by temperature. (The use of adjuvants which modify the surface tension or viscosity of the gel may also affect the ability to establish a satisfactory meniscus.) The velocity depends on the orifice clearance, the concentricity of the annular orifice, the volume of gel pumped in each stroke and the time interval for the pump stroke. If velocity is too high, a meniscus will not form; if velocity is too low, a drop will not separate and the meniscus becomes thicker.

Velocity is also critically dependent on compressibility of the gel. It is important to ensure that no air is entrained in the gel during mixing and that no air bubbles are trapped in the passages from the pump to the nozzle. An air-operated, positive displacement pump with inlet and outlet check valves was used to meter the gel. Air bleed valves were provided in the pump and in the encapsulation nozzle. Pump stroke and operating speed must be carefully controlled to control the size of a capsule and to ensure separation of the capsule.

Nozzle parts need to be cleaned and assembled carefully to maintain proper clearances and alignment. Care is required in mixing the alginate and in charging and bleeding the system to ensure that no air is trapped in the system. Erratic behavior of the gel during capsule formation is an indication that air is present in the system. Pump stroke speed must be carefully controlled to provide a gel velocity sufficient to cause separation of a capsule from the nozzle. Inflation air pressure must be carefully controlled to provide just the right acceleration of the meniscus and seed. Excess inflation air pressure may rupture the meniscus or interfere with formation of the next meniscus. It may also result in entrapment of air bubbles in the capsule. Inadequate inflation air pressure will result in seeds being left behind in the next meniscus or in failure of capsules to separate.

The cycle time for a nozzle is determined to some extent by the time required for a seed to fall from the singulating device to the meniscus. For the units described here, the passage length was about 90 mm (3.5 in) and seeds reached the meniscus in approximately 0.4 sec. There is, however, another factor which also affects the speed of operation; seeds appear to need some time to settle onto the meniscus before they will move reliably with the meniscus when it is inflated. Efforts to reduce the cycle time for encapsulation frequently resulted in seeds being left behind at the nozzle, either to form a double in the next capsule or to plug the nozzle as other seeds accumulated. This condition seemed to exist even though seeds arrived at the meniscus well ahead of inflation of the meniscus. The result was that a cycle time of about one second was needed even though arrival of the seeds seemed fast enough for a cycle time of about one-half second. This condition places a serious limitation on the capacity of the units. The underlying cause is not understood.

9 APPLICABILITY

The unit described here was developed specifically to encapsulate tomato seeds. The applicability of the concepts employed in this device to other seeds will depend primarily on the size and shape of those seeds. The seed passages from the vibrating bowl and through the encapsulation nozzle of the present units have inside diameters (ID) of 4.8 mm (3/16 in). The seed passages can be made larger to accommodate larger seeds, but at some point seed size and nozzle ID are likely to be limited by the need to form a meniscus sufficiently strong to support the seed.

Seeds must be large enough to be detectable by the seed sensor in the singulating component and by the sensor of the capsule inspection component. Very small seeds would obviously be difficult to detect. Adjuvants which change the transparency of the gel will also affect the ability of the capsule inspection component to detect seeds.

Seeds must flow freely as individuals in response to the vibratory action of the bowl and in response to gravity through the seed drop tube and in the nozzle. Plant embryos may need to be dried or coated to allow them to flow freely. The ability to encapsulate embryos by this process may be limited by the ability of the embryos to survive the treatment.

REFERENCES

Gautz, L. D., S. K. Upadhyaya and R. E. Garrett. 1987. Vibratory Separation of Gel Encapsulated Plant Propagules. Transactions of the ASAE, Vol. 30, No. 33, pp. 652-656.

Gautz, L. D., S. K. Upadhyaya and R. E. Garrett. 1988. Hydro-pneumatic Singulation of Gel Encapsulated Propagules. Accepted for publication in Transactions of the ASAE.

Redenbaugh, M. K. 1986. Analogs of Botanic Seed. U. S. Patent Number 4,562,663.

Redenbaugh, M. K. 1986. Delivery System for Meristematic Tissue. U. S. Patent Number 4,583,320.

Upadhyaya, S. K., L. D. Gautz and R. E. Garrett. 1987. Retrofitting Vegetable Planters to Seed Gel Encapsulated Propagules. Applied Engineering in Agriculture, Vol. 3, No. 2, pp. 211-214.

RESUME: On a développé un système pour mettre en capsules des graines de tomates dans un gel d'alginate. Les graines montent suivant une trajectoire en spirale dans une cuve vibrante. Des injections d'air et des détecteurs de graines permettent l'admission d'une graine à la fois dans une des deux canules d'"encapsulation"* présentes sur une "unit" (unite)** indépendante. Du gel, pompé au goutte à goutte à partir d'une canule, forme un menisque capable de saisir une graine. Une bouffée d'air gonfle légèrement le menisque alors que l'injection suivante de gel bloque la graine à l'intérieur de la goutte. Les capsules sont examinées à l'aide d'un "digital camera" (appareil photographique numérique)** alors qu'elles sont recueillies dans un bain durcissant. Trois rejets successifs provoquent l'arrêt d'une voie et avertissent l'opérateur. L'unité s'arrête quand le rendement combiné des deux canules atteint une valeur préétablie. Six unités d'"encapsulation" sont contrôlées par un ordinateur qui imprime des étiquettes et enregistre date, heure, nombre de capsules, nombre de rejets et identité de l'unité d'"encapsulation."

*Id.: Faute d'un équivalent français, le traducteur a conservé le terme américain.
**Note du traducteur: La parenthèse contient une traduction littérale et possiblement imprécise de l'expression américaine entre guillemets.

ABRIß: Ein System zur Verkapselung von Tomatensamen in ein algisches Gel ist entwickelt worden. Samen steigen auf spiralen Wegen in einer vibrierenden Schüssel hoch. Luftdüsen und Samensensoren gestatten jeweils einem Samen den Eintritt in eine der zwei Kapsulierungsmündungen einer in sich geschlossenen Einheit. Gel, das in einzelnen Tropfen von der Mundung gepumpt wird, bildet einen Meniskus, welcher einen Samen einfängt. Ein Hauch von Luft bläst den Meniskus leicht auf, während das darauffolgende Gel den Samen in dem Tropfen einschließt. Die Kapseln werden von einer digitalen Kamera inspiziert, wenn sie in das Härtebad fallen. Drei aufeinanderfolgende Ausstoßungen bringen den Kanal zu einem Stillstand und signalisieren dem Bedienungspersonal. Die Einheit stoppt, wenn der gesamt Ertrag der zwei Mündungen einen vorbestimmten Wert erreicht. Sechs Kapsulierungseinheiten sind an einen Computer angeschlossen, welcher Etikette druckt und Datum, Zeit, Anzahl der Kapseln, Anzahl der Ausstoßungen, sowie die Identität der Kapsulierungseinheit, notiert.

Land and Water Use, Dodd & Grace (eds), © 1989 Balkema, Rotterdam. ISBN 90 6191 980 0

Techniques for the evaluation of precision planters

A.J.Heyns
Agricultural Mechanisation, Silverton, RSA

ABSTRACT:In an official test programme of single seed planters, extensive tests were carried out on the seed metering mechanisms in the laboratory and in the field. By presenting test results in a modified pie-chart format, results can easily be interpreted and compared. Literature also indicates that too much emphasis could unnecessarily be placed on seed spacing accuracy in tests.

ZUSAMMENFASSUNG: In einem amtlichem Testprogram fur Einzelsaat-Pflanzer, wurden Saatmessmechanismen im Labor und auf dem Acker umfassend gepruft. Eine graphische Darstellung der Ergebnisse ermoglicht eine einfache Auswertung und einen objectiven Vergleich. In der Litteratur wird darauf hingewiesen das des ofteren eine unnotig hohe Wertung auf den Abstand zwischen folgenden Saatkornern gelegt wird.

RESUME: Dans un programme officiel d'essais sur des machines à planter de semences individuelles, des essais approfondis ont été entrepris sur les mechanisms comptant une seule graine à la fois, autant au laboratoire que dans les champs. La présentation des résultats d'essais par un diagramme en cercle modifié, rend facile l'interprétation et la comparaison de ceux-ci. La littérature montre également qu'on pourrait inutilement mettre trop d'accent sur la précision de l'espacement du semis dans des essais.

1 INTRODUCTION

One group of implements that featured strongly in the official test programmes of the South African Directorate: Agricultural Mechanization has been single kernel planters. A total of 21 different models, covering most of the market has been tested up to 1984. These precision type planters are used for planting the main crop in South Africa, maize, as well as for diverse other crops such as soyabeans, groundnuts, grain sorghum, sunflower, edible beans, cotton and in some cases even wheat. During 1984 it was decided to embark on a new test programme for single planters and to specifically attend to the shortcomings identified with previous tests. A major complaint was that planter test reports have tended to concentrate on the laboratory evaluation of seed metering with very little data on in-field performance and with no interpretation of overall test results. Data on the durability of the different machines has also been very limited. A major change in

the type of planter has also been taking place over the previous decade. By 1984 approximately half of the planters sold locally made use of pneumatic seed metering systems with the balance using the more conventional mechanical seed plate or finger pick-up seed metering principle. Another significant change has occurred in the type of seedbed conditions a planter has to be able to work. There has been a definite trend away from conventional tillage practices towards using tine implements in a stubble mulch system with the emphasis on keeping all crop residues on the surface. This consequently places higher demands on the planter in terms of handling of stubble and depth control.

2 LABORATORY TESTS

During the laboratory evaluation of planters, the inherent ability of different planters to handle different seed types and seed sizes was evaluated. Each of the planters was positioned to plant onto a moving belt, adjustable to speeds of 4, 6, 8, 10 and 12 km/h. A synchronized drive through either the planter press wheel or land wheel powered the metering mechanism and oil provided the medium into which was planted. SAE 250 mineral oil with a tackiness additive eliminated any seed bounce on the belt. After each run, seed position was determined by using an optical rotary encoder coupled to a HP 85 desktop computer. A sliding cursor, coupled with a cable and pulley arrangement to the encoder, enabled determining the position of seeds on the stationary belt with an accuracy of 1 mm. Statistical analysis of pilot runs showed that a minimum of 120 seed positions gave a representative sample. Seed damage was also recorded during the tests.

and quick comparisons between machines, speeds and seed types. In figure 1 a typical report sheet for a number of representative planter types is shown.

Figure 1 highlights some of the common characteristics found with the different planter types. The pneumatic metering system gives excellent metering of seeds over the full speed range with few misses or doubles, especially due to a small seed drop height. The mechanical metering systems such as the horizontal seed plate, finger wheel pick-up, inclined seed plate and perforated belt types gives comparable performance at lower speeds or wider spacings but with an increase in speed both metering accuracy and uniformity of spacing deteriorates to differing degrees. With seeds such as kidney beans which are more

Table 1. Example of spacing results according to ISO format

MAIZE 5 FLAT				TARGET SPACING OF 360 mm					SET SPACING (T) OF 362 mm		
1	2	3	4	5	6	7	8	9	10	11	12
SPEED km/h	DAMAGE %	AVERAGE SPACING mm	STANDARD DEVN. mm	COEFF. VAR. %	MISSES %	DOUBLES (0,5T) %	0,5T– 1,5T %	1,5T– 2,5T %	2,5T– 3,5T %	3,5T– 4,5T %	>4,5T %
6	0	334	114	34	2	14	96	2	0	0	0
8	0	351	121	34	3	10	95	3	0	0	0
10	0	383	153	40	11	8	79	11	0	0	0
12	0	426	188	44	14	3	76	7	4	0	0

2.1 Analysis of seed metering results and presentation in report

Data from the tests was analysed using the international standard ISO 7256/1 for single seed drills. This statistical basis for evaluation, defines a "miss" as a measured seed spacing of more than 1,5 times the theoretical spacing. Multiples or doubles are defined as an actual seed spacing of less than 0,5 times the theoretical spacing. It was quickly realized that to present the test results in tables according to the ISO-standard as shown in table 1, would lead to an unwieldy report format and would rather confuse than inform end users of the report. Instead, a much simplified "pie chart" format was developed which is shown in figure 1. As can be seen in table 1, the figures for misses, multiples and correctly spaced seeds do not add up to 100% and therefore the pie chart is only an approximation of the results. This way of presentation, however, condenses a huge amount of data in a form which allows easy

susceptible to damage than maize, pneumatic metering gave less damage than some mechanical metering systems. The perforated rubber belt type metering system, however, produced least seed damage.

2.2 Seed grading and accuracy of seed metering

One of the often cited reasons for the increase in popularity of planters with pneumatic metering, is that of freedom from the tedium of matching seed plate to seeds and the ability to plant ungraded or poorly graded seeds. The question of the "plantability" or "meterability" of different seed grades was addressed during the planter test programme as a separate but related exercise. Using maize seed, the aim was to investigate whether there are differences in plantability of different sizes of seeds, flat or round seed and long or short seeds. The results and conclusions can be summarized as follows:

MAIZE 5 ROUND (360 mm spacing)

| Planter | 4 km/h | 6 km/h | 8 km/h | 10 km/h | 12 km/h |

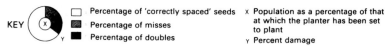

Finger wheel pick-up seed metering — 96, 93, 97, 95

Horizontal seed plate type metering — 94, 92, 92, 94

Pneumatic seed metering — 105, 108, 106, 106

Inclinded plate type metering — 102, 100, 101, 91

Perforated rubber belt metering — 114, 111, 106, 97, 87

Note : No pie-charts are given for the majority of planters at speeds of less than 6 km/h as these speeds were considered impractical under present farming practice for the planting of maize seed.

KEY

☐ Percentage of 'correctly spaced' seeds
▨ Percentage of misses
■ Percentage of doubles

x Population as a percentage of that at which the planter has been set to plant
y Percent damage

1. Pneumatic planters can deal with the whole spectrum of seed sizes and shapes using only one or a very limited number of different seed plates.

2. The finger pick-up mechanism can handle the whole spectrum of commonly available seed sizes and shapes. Round seeds greater than size 2 is, however, preferred.

3. Depending on how well the seed plates are matched to the seed, mechanical plate type planters cope equally well with all sizes. Seeds, as round as possible, are preferred.

2.3 Other laboratory tests

In addition to measurements of the quality of seed metering, different other aspects of the performance of the planters were also evaluated under laboratory conditions. The fertilizer metering equipment were evaluated in terms of minimum and maximum delivery settings, speed sensitivity of application rates over a speed range of 4 to 12 km/h, evenness of application down the row as well as the degree to which fertilizer granules are pulverized in the fertilizer metering system. All the tested planters were equipped with auger type fertilizer meters which can cover a wide range of application rates using different drive ratios as well as low- and high-rate augers. The application rate of the auger type system is normally relatively insensitive to speed variation but the application from point to point down the row, showed wide variations, another inherent characteristic of this type of system. The tendency of a fertilizer bin to block, especially in conditions of high humidity, is related to the degree of pulverization of fertilizer granules taking place in the metering system. By measuring the degree of pulverization using test sieves, it was shown that bins with small clearances between the auger and the walls or floor of the bin, are very prone to blockages by accumulated fertilizer dust, if not cleaned regularly.

Some planters were fitted with granule applicators for herbicides and insecticides. These attachments were also evaluated for range of application rates, evenness of metering, speed sensitivity and in the case of herbicide applicators,the cross distribution pattern of the spreading attachment of fishtail was measured.

3 FIELD EVALUATION

In order to obtain an indication of the performance of the different planters under a reasonable spectrum of field conditions, field evaluations had to be carried out at different locations in the country. With nine different planter models entered for testing, transport of these planters, all of them in a four row format, was a serious logistical problem, especially since test locations were as far as 400 km apart. This problem was overcome by not evaluating each planter as a complete unit, but by using a seeding unit from each model and mounting the units on two separate test rigs. Special care was taken to mount each separate unit in such a manner as to represent the position and characteristics of the unit on its original frame. Two test rig frames, one for the pneumatic metering units and one for the mechanical systems were assembled while an elaborate drive system made it possible to drive each unit at the correct speed for a specific target seed spacing.

The seedbed conditions under which field evaluations were carried out, varied from light windblown sandy soils to heavy black turf. Seedbed preparation practises included conventional ploughed and disced, plough and plant, chisel ploughed and shallow tine cultivation. The amount of stubble from the previous crop on the soil surface also varied according to the type of seedbed preparation.

After a few dummy runs to carry out final depth adjustments and to set attachments such as stubble furrow openers and seed covering scrapers, the test plots were planted at two speeds, namely 6.9 and 8.8 km/h using maize seed. After emergence of the seedlings, the positions of approximately 120 plants were noted. The seedlings were also dug up carefully and measured to evaluate evenness of planting depth.

3.1 Seed spacing results

The results of the in field seed spacing evaluation, was analysed using the ISO-standard and the results reported in the "pie-chart"-format discussed earlier. Figure 2 gives a typical example of field results. As can be expected, there are marked differences between laboratory performance and the performance measured in the field. Accuracy of seed spacing is related to the type of seed metering system as well as the distance the seed drops before reaching the soil. In the field, factors such as vibration can upset seed pick-up as well as the path of seeds during their fall to the soil.

3.2 Depth control results

The results of the seed depth measurements were reported in the form of histograms for each planter and each seedbed condition. As shown in the example in figure 3, the X-axis represents the spread of seed depth values measured around the mean depth in millimeters while the Y-axis denotes the proportion of results occurring at a given depth as a percentage of the total sample. The ideal depth control system should therefore be represented by a single column at the mean point with a height of 100%.

In theory, the smaller the distance between the point where planting depth is gauged and the point of seed drop, the more accurate and responsive depth control becomes. However, in practice various field effects such as uneven soil penetration and the character of the surface irregularities appear to distort this theory. Even on relatively even fields it was found that with a furrow opener set to open the soil too deeply, a marked amount of soil is washed or blown into the furrow depression causing uneven planting depths with an otherwise excellent depth control system.

The ability of a planter to handle plant

Figure 2. Example of typical field seed spacing results in pie-chart format

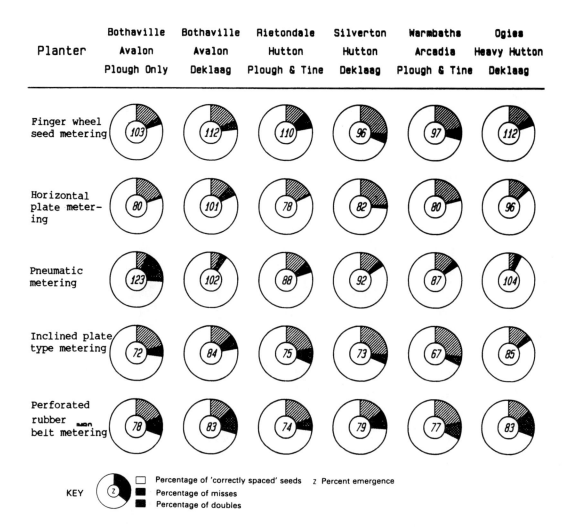

FIELD RESULTS
6.9 km/h

material without blockages depended on the design of the specific planter regarding fertilizer and seed openers, press and depth wheels, seed covering scrapers and the drive arrangement.

4 THE EFFECT OF PLANTING UNIFORMITY ON YIELD

A study of test reports on planters published by different institutions worldwide, shows clearly that in all cases the main emphasis in the reporting of results is placed on seed spacing accuracy. In earlier planter test programmes, this approach has also been followed by the Directorate, but in the latest tests, it was decided to obtain more information on what the effect of uneven spacing really is on the eventual yield of crops such as maize. This research was limited to a literature study only.

Three groups of researchers have done work in this field, albeit not to evaluate planters, but research fields such as crop replant decisions. Johnson and Mulvaney

Figure 3. Example of seed depth results

DEPTH CONTROL
Rietondale   Hutton   Plough & Tine

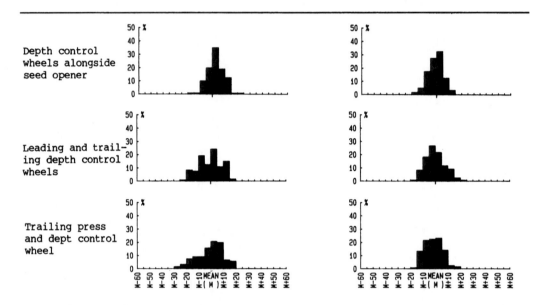

Planter   6.9 km/h   8.8 km/h

Depth control
wheels alongside
seed opener

Leading and trail-
ing depth control
wheels

Trailing press
and dept control
wheel

(1980) did work with maize, Strivers and Swearingen (1980) worked with soyabeans and Robinson, Ford, Lueschen, Rabas, Warnes and Wiersma (1980) investigated the effect of seed spacing on sunflower yields. Although all this research has been carried out in the USA under conditions that can differ vastly from local conditions, it is felt that some basic conclusions can still be made. All the results indicate that a more even spacing of seeds and plants leads to higher eventual yields, but even with very uneven spacing patterns, much more uneven than found with the worst seed metering situation, the decrease in yield is much less than anticipated. It appears that the crop has the ability to compensate to a major extent for variations in plant spacing and also to a lesser extent for a lower plant density.

This would indicate to us that when one compares the laboratory results and also the field test results for different planters, there is little sense in embarking on a hair splitting exercise. As long as the seed distribution is reasonable with no undue drop in seed population, the eventual yield will not be noticeably affected, especially if one consider the potential influence of other factors, such as weather, on eventual yield.

5 DURABILITY INFORMATION

The needs of the farmer regarding information on the durability of machines, is very difficult to satisfy in an official test programme. Few test stations has the infrastructure to do extensive field work with groups of implements under reasonably controlled conditions. Accelerated wear testing often lacks credibility. For this test programme the approach was to interview users of the different planter models. A total of 199 farmers were visited, representing 371 individual planters and from a careful sifting of information, a reasonable picture of the weak points and strengths of each planter could be build up. Collecting durability information in this manner, requires experienced personnel, especially since very few farmers keep formal records of the incidence of breakdowns and cost of repairs.

6 CONCLUSIONS

1. Using the test methods from ISO 7256/1 together with a "pie-chart" presentation for test results resulted in a readable report which should enable test report users to make a logical choice between different machines.

2. Laboratory tests must be supplemented by field evaluation. Field conditions can have a marked effect on planter performance.

3. It would appear that for most crops under dryland conditions, the accuracy of seed spacing is of less importance than previously thought.

REFERENCES

International Organization for Standardization 1984. International Standard 7256/1 Sowing Equipment - Test Methods - Part 1: Single seed drills (precision drills).

Johnson, R.R. & Mulvaney, D.L. 1980. Development of a Model for Use in Maize Replant Decisions. Agronomy Journal, Vol. 72, No.3, p459.

Robinson, R.G., Ford, J.H., Lueschen, W.E., Rabas, D.L., Warnes, D.D. & Wiersma, J.V., 1982. Response of sunflower to Uniformity of Plant Spacing. Agronomy Journal, Vol. 74, No. 2, p363.

Stivers, R.K. & Swearingen, M.L., 1980. Soybean Yield Compensation and Missing Plant Patterns. Agronomy Journal, Vol.72, No. 1, p98.

Land and Water Use, Dodd & Grace (eds), © 1989 Balkema, Rotterdam. ISBN 90 6191 980 0

An application of solid state cameras to testing the uniformity of seed feeding

Ji Chungian & Yu Feng
Jilin University of Technology, Changchun, People's Republic of China

ABSTRACT: A device was developed to test the uniformity of seed metering mechanisms. The device consists of a solid state camera, a microcomputer and interface circuits. The solid state camera composed of lenses and a piece of CCD (Charge-Coupled Device) solid state image sensor was used to obtain the seed presentation which was processed by the computer. The device can measure seed spacings for the precision seeding of soybean and corn. The measuring error of the seed spacing is less than 3mm.

RESUME: Un disposidif est réalisé pour déterminer l'uniformité des semailles du semoir. Ce dispositif est composé d'une caméra à régime solide, d'un microcalculateur et d'un circuit d'interface. La caméra qui est constituée d'un objectif optique et d'une unité sensible à image à régime solide - CCD (Charge-Coupled Device)est prévue pour obtenir l'image des semailles, et ce calculateur est utilisé pour l'acquisition et le traitement des données. Ce dispositif peut mesurer la distance entre grains de soja et de maïs, etc, ensemencés avec précision. Lerreur de la distance entre grains est inférieur à 3mm.

ZUSAMMENFASSUNG: Der Apparat wurde entwickelt, der ist angewandt zur Prüfung der Gleich-mässigkeit von der Saat-Beschickungseinrichtung. Dieser Apparat besteht aus einer festen aggregatzustanden Bildkammer, einer elektronischen Rechenanlage und den Zwischenkreis. Die feste aggregatzustande Bildkammer, das besteht aus der Glaslinse und einer Stück CCD (Charge-coupled Device) festem aggregatzustandem bildem Fühler, ist augewandt zur Erhaltung des Bildes der Saat, die ist bearbeitet durch eine elektronische Rechenmasch-ine. Der Apparat kann abmessen den Abstand der Saat fur die Genauigkeits-Aussat der So-jen und der Maise. Der Fehler der Saat-Abstand ist weniger als 3 mm.

1 INTRODUCTION

With the development of precision seeding at high speeds, more and more new princi-ples of design and new structures are being developed for seed metering mechanisms. The main performance criteria to evaluate these metering mechanisms are the number of seeds dropped in a given time and the uniformity of the seed spacings. These metering mecha-nisms are often tested on test bed to ob-tain their performance data in our labora-tory. In the past, seed spacings were meas-ured with rulers. In order to make the measuring accurate, we have developed a seed spacing measuring device.

The forming of a seed spacing has close relation to the pathways during dropping of seeds. The dropping pathways of two seeds metered sequentially from a feeding device are different, therefore the seed spacing of the two seeds cannot be made

out only by the time interval between two moments when the two seeds reach the soil respectively. If a measuring device needs acquire the seed spacing accurately, it must measure not only the time interval but also the deviation from a stationary reference point around which each seed falls on. We take the advantage of photo-graphic surveying to detect the time inter-vals and the seed longitudinal coordinates with which seed spacings are calculated.

The method to measure seed spacings is summarized in the followings. Now, a solid state camera scans horizontally at regular intervals, delivers serial pulses whose amplitudes are directly proportional to the intensity of the luminous flux on every light-sensitive unit. The pulse amplitude shows whether a seed is detected. The ad-dress of the light-sensitive unit in cor-respondence with the pulse expresses the coordinate of the seed. When a seed   is detected, the interface circuits tell the computer to take down the information about the seed, including the time and the seed position. The computer, according to the information, discriminates seed overlap from a single seed and calculates the seed spacings. When the number of seeds being tested is equal to the number given, the computer makes the statistical evaluation, and gives out the testing results. This method carries out non-contact measuring and real time data processing.

This measuring device, available to precision metering of soybean and corn, can measure the speeds of the conveyer apron accepting seeds, the rotation rate of a seeding wheel, the air pressure in a seed filling cell and the spacings of dropped seeds. The measuring system consists of two revolution transducers, an air pressure sensor, a solid state camer, interface circuits, and a computer. All the signals from the sensors are transmitted to the interface circuits which provide the computer with data, and the computer processes the data.

2 THE SOLID STATE CAMERA

The solid state camera in the system consists of lenses and a piece of CCD. The piece of CCD is an alignment solid state image sensor with 256 light-sensitive units. In addition, it includes a electric charge transfer gate, triphase analogue shift registers, a charge output detector, and an amplifier. The luminous energy of incident light on light-sensitive units generates electric charges the number of which is proportional to the light intensity, then the charges are transfered to the shift registers, driven by three-phase clock, and shifted to the amplifier where they become video signals the amplitudes of which are modulated by light signals. The width of the light-sensitive unit is 15 μm, the width of the light sensitive alignment is 3.84 mm. The lenses can project an image of 45 mm width 350 mm away on the alignment. Since what we want to know is whether a seed is detected, the image processing is a problem of two state values. Therefore, the pulse signals from the solid state image sensor are dierctly transmitted to a voltage comparator where the signals are compared with a threshold, and become the signals compatible with TTL circuits. The measuring scheme and the waveform genarated when a seed is detected are shown in Fig.1.

The pulse amplitudes of the light-sensitive units in receipt of a seed image are less than those of others. If a proper

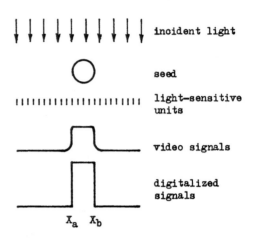

Fig.1 Measuring sheme and waveform

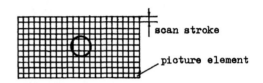

scan stroke

picture element

Fig.2 A seed presentation during sequent scanning

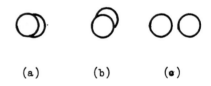

(a)   (b)   (c)

Fig.3 Seed overlap disposition-plans

voltage threshold is set, the digitalized signals can be generated. When the addresses X_a (of light-sensitive unit corresponding to the front edge of a seed) and X_b (of that corresponding to the back edge) are taken down, the width D of a seed can be calculated by the formula

$$D = C(X_b - X_a), \qquad (1)$$

and the seed central coordinate can be acquired from formula

$$XC = C(X_a + X_b)/2, \qquad (2)$$

where C is a constant coefficient.

A seed presentation is disoribed through sequent scanning, as shown in Fig.2.

In most cases of precision drilling,

only one seed is detected at one time, however, unavoidably two seeds are occasionaly detected at the same time, this case is the seed overlap. When this case happens, there are three possible disposition-plans of the seeds, as shown in Fig.3.

The first case (as shown in Fig.3 a)can be discriminated with the measured width of the seeds, the second case (as shown in Fig.3 b) with the number of the scanning lines on the whole image, the third (as shown in Fig.3 c) with the number of the seeds detected on the same scanning line.

3. INTERFACE CIRCUITS

The output signals from the solid state camera must be handled through interface circuits before a computer receives it. According to our measuring scheme, the interface provides for the computer the scanning moment when the information about

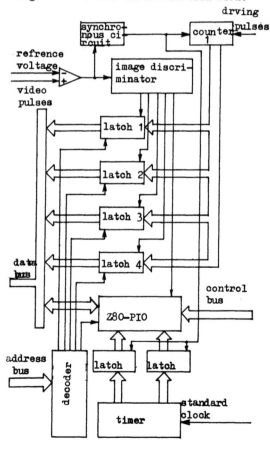

Fig.4 Interface block diagram

a seed is obtained by the camera on the current line and the addresses of light-sensitive units to which an image of a seed correspondes.

Through the interface circuits, the video signals from the camera are first digitalized, then the addresses of light-sensitive units corresponding to the front and back edges of the seed respectively are latched. At the same time, the moment of scanning the current line is also latched. After that, the computer will be announced that the data are ready. Having read in the data, the computer begins to process them, meantime the interface circuits may transfer the information on the next line. The block diagram is shown in Fig. 4.

Through the comparator video signals are transformed to the waveform as shown in Fig.5 b. From left to right, the first minus pulse represents the image information on the first light-sensitive unit, by analogy, the second represents that on the second unit, and so on. When a seed appears an empty space without minus pulse between A and B is formed. Point A represents the front edge of a seed image and B the back edge. The addresses at points A and B are the data to be recorded. For obtainning them counter 1 has been set,which counts clock pulses driving the CCD chip. Because video signals from the CCD transmit in according to the beats of the driving clock pulses, that is, the driving clock pulses correspond to the addresses of light-sensitive units, so the addresses of photosensor units can be determined by counting driving clock pulses. Because the driving clock pulses are continuous, and the video signal output is by line, so sync-circuit is set. The syncronous signal is obtained at point S and makes counter 1 be cleared. Thus, the number in counter 1 is the address of photosensor units. But counter 1 does continous plus numeration, so when the number recorded by counter 1 is the address at point A, it must be latched. The same must be done at point B. Latching signals are delivered from a seed image discriminator as shown in Fig. 5 c.

The seed image discriminator is mainly composed of a pulse width discriminator. By the waveform in Fig. 5 b, it is clear that the pulse width is even where there are no seed images, e.g. between S and A, and a wide pulses is formed where there is a seed image, e.g. between A and B. At points A and B,   the pulse width discriminator generates signals which latch the addresses. The pulse width discriminator is mainly composed of two monostable flip-flop and two NAND gates, as shown in Fig. 6. The pulse width of T1 and of T2 at output pins

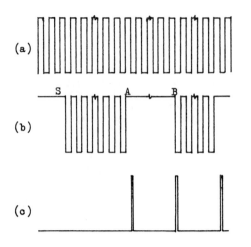

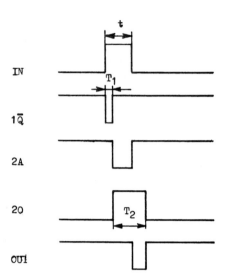

Fig.5 Waveforms (a) driving clock (b) digitalized pulses (c) latching pulses

Fig.7 Waveforms of every point of the pulse width discriminator

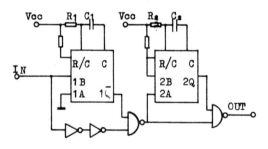

Fig.6 pulse width discriminator

$1\bar{Q}$ and 2Q of the monostable flip-flop is determined by the values of R1, C1 and R2, C2 respectively. Fig. 7 expresses the waveform when the pulse width t at the input pin is fitted to the condition $T1 < t < T_1 + T_2$.

Suppose the width between A and B is t (See Fig. 5). When the R1, C1 and R2, C2 are properly selected, the condition $T1 < t < T1+T2$ is satisfied and the T1 is as small as possible, so long as its width is bigger than that of video pulse's. As shown in Fig. 7, the descent edges at point 2A and the output port are taken as trigger signals which latch the addresses of the front edge and back edge of a seed image respectively. Then the data in latch 1 is the address of the front edge of a seed image and the data in latch 2 is that of the back edge. For providing the scanning moment on every line, a 16 bit counter is set in the interface circuits. The clock pulses are provided by the standard clock pulse from

the computer. The timer numerates continuously. Output pins of the timer are connected to two data latches. The data in the timer are latched with sync-signals on every scanning line. Then the content in the data latches represents the moment on scanning every line.

The communication between the interface and the computer is carried out through a parallel interface Z80-PIO. Its two ports A and B are taken as input ports of scanning moment. It is programmed as IM2. When a seed appears, the signal from the image discriminator transmits to pin $\overline{ASTB}$ of the Z80-PIO, then the level of pin $\overline{INT}$ is effective. It requests the computer to read the data in. After the computer responds to it, the data of the moment on scanning the current line and the adresses of a seed are taken out. Then the computer analyses and processes the data.

4. Data acquisition and data processing

If a seed is detected during scanning, the information about the seed will be transmitted to the memory in the computer. The information about the seed will occupy 4 storage locations. A seed is generally scanned about 6 times, so the whole information will occupy 24 storage locations. In consideration of the finiteness of storage capacity, a method, processing data over data acquisition, is taken. The procedure is shown in Fig. 8. From the data

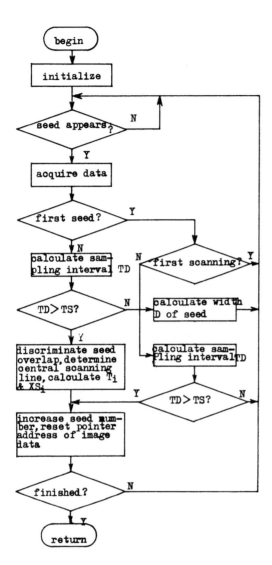

Fig.8 Block diagram of data acquisition and data processing

corresponding to the 128th light-sensitive unit, the direction in which the planter moves is considered positive, so

$$XS_i = C \left[(X_a + X_b)/2 - 128 \right], \qquad (3)$$

where C, X_a, X_b have been defined in formula 2. The seed spacing can be calculated by formula

$$D_i = T_i \cdot V - XS_{i-1} + XS_i, \qquad (4)$$

where V is the operation speed of the planter.

The degree of accuracy of seed spacings calculated in this way depends upon frequency of line scanning. Although the central coordinates of seeds are used in the calculation, the degrees of accuracy of them are quite high comparatively, the error of the coordinates are less than 0.2mm. So its effect can be neglected. However, the scan stroke DS is far larger than 0.2mm. Suppose the line scan period is TS. The scanning line does not always pass the seed centres when the camera scans regularly. Therefore the scanning line which is selected as the central line of a seed deviates ± 0.5 DS from the actual centre of the seed in the worst case. When the time interval T_i is calculated according to the selected scanning lines, its maximum error ΔT_{imax} is not larger than the line scan period TS, ie. $\Delta T_i < TS$. So the seed spacing error ΔD_i is not larger than TS times V, ie. $\Delta D_i < TS \cdot V$, the error ΔD_i enlarges with the increasing of the forward velocity V. The line scan period of the camera is 1 ms. If the forward veliocity V is 8 km/h, the maximum seed spacing error ΔD_{imax} is 2.2mm.

When all data about a seed have been read in, the computer determines which scanning line is taken as the central line of the seed, then subtracts the trace time of the scanning line from that of the anterior seed's, makes out the time interval T_i which is stored in two storage locations. And the deviation XS_i is calculated according to the front edge address X_a and the back edge address X_b.

In the course of information processing, whether there is seed overlap is discriminated. From the data acquired, a picture of a seed in one direction can be drawn. By comparing the picture with the actual contours of seeds, the seed overlap is discriminated.

The number of seeds to be tested in the detection is set beforehand. When finishing with the number of seeds, the computer begings to make statistical evaluation. When a seed spacing is less than half a theoretical seed spacing, a seed overlap is recorded. When a seed spacing is

acquired, seed spacings can be worked out, and seed overlap can be recongnized. When seed spacings are calculated the time interval of seed's reaching the ground must be know. The time interval T_i can be obtained by subtracting the time when the centre of the seed is scanned from that of the anterior seed's. Then the seed centre coordinate deviation XS_i from a reference point when a seed falls on the ground must be calculated. Here, the reference point is the point below the metering device,

larger than one and a half theoretical
seed spacings, a seed missing is recorded. If someone needs all the data of seed
spacings, the data are delivered from a
printer, otherwise, only statistical results
are delivered.

5 CONCLUSIONS

The method, using a solid state camera to
test the uniformity of seed spacings, is
ideal. It is necessary to measure the seed
spacings when the quality of precision metering is to be evaluated. It is not easy
to measure the spacings with any other
sensors.

The CCD solid state image sensor is of
higher performance in anti-interference,
and its signals are easy to process.

The test data are accurate. The errors
of seed spacings mainly depend upon the
frequency of line scanning.

REFERENCES

Gin Jirong 1987. The priciple of photoelectric detection and how to use it. Peking:
National defence industry.
Zhou Mingde 1982. Hardware & software of
microcomputers and how to use them. Peking: Ginghua university.
Zhang Boping 1983. The principle for designing seeding mechanisms. Peking: Engineering industry.

Land and Water Use, Dodd & Grace (eds), © 1989 Balkema, Rotterdam. ISBN 90 6191 980 0

A two-row automatic seed yam planter prototype

E.U.Odigboh & C.O.Akubuo
Agricultural Engineering Department, University of Nigeria, Nsukka, Anambra State, Nigeria

ABSTRACT: The design and development of a two-row automatic seed yam planter are described. The prototype planter is fully tractor mounted to facilitate manoeurability. However, the metering mechanism of the seed yams is ground wheel driven to ensure that spacing within rows is independent of planting or tractor speed. The prototype planter is designed to plant seed yams on one-metre ridges, with agronomically recommended within-row and between-row spacings of one metre. The planter prototype simultaneously makes the ridges, meters and places the seed yams in the ridges to a selected depth of 100 - 150 mm, and then covers up the soil around the planted seed yams. Thus the planting operation is automatic. At an optimum planting speed of 3.5 - 5.5 km/h, the planter gives a very satisfactory performance at a theoretical field capacity of 0.7 - 1.1 ha/h.

RESUME: La réalisation et le développement d'une machine automatique à double-rangée pour planter l'igname sont décrits. Le prototype de la machine est tracté pour faciliter sa manipulation. Le compteur mécanique de l'igname est tiré par terre pour assurer l'indépendance de l'éspace entre les rangées. L'éspace entre les rangées est de'un mètre et la machine est réalise'a'base de cette même distance entre les rangées de ces plantations d'ignames. Le rôle de cette machine est de costruire les arêtes et en même temps d'éspacer les plantations d'une distance d'un mètre. La profondeur est entre ioomm-i50mm. Ainsi fait, les plantations sont ensuite recouvertes de terres par la même machine. Si on considère une vitesse optimale de 3.5 a 5.5 par heure, nous avons découvert que les résultats donnes sont satisfaisant s

ZUSAMMENFASSUNG: Der Entwurf und die weiter Entwirklung von den Zwei-Reihen automaten saatjamwurzel Pflanzer wird beschrieben. Das Muster Pflanzer wird an der Traktor montiert um einfacher Manoever zu leisten. Jedenfalls, das Metrum aggregat fur den Pflanzer wird von eigenen radern angetrieben um selbst unabhangig innerhalb den Grat Reihen als auch von der geschwindigkeiten des Traktoren zu erzielen. Das Muster Pflanzer war Entwolfen fur saatjamwurzel pflanzer mit agronomicshe empfehlung von ein meter abstand in Reihen als auch in quivier der saat im Grat und den Uberdeckung automatish erledigen. Die Tiefe von den Grat wird wahrbar zwischen 100 - 150 mm. Der Pflanzer wird automatish operiert. Mit Optimal Pflanzer geschwindigkeiten von 3.5 - 5.5 km/Stunde, der Pflanzer wird befriedigender Leistun von theoretishe Capacitat von 0.7 - 1.1 ha/stunde erzielen.

1 INTRODUCTION

Nigeria is a major yam growing country where about 1.5 million hectares are annually devoted to yam production, to yield upto 15 million tonnes of yams per annum (Okoli, 1987). Almost the entire annual yam production is consumed as human food, since yam constitutes about 20% of the daily caloric intake of Nigerians (Iwueke et al., 1983).

In Nigeria, yam is produced by the peasant farmers for whom 'yam husbandry' is a way of life and an age-honoured occupation, undertaken even when no profit is possible. Traditionally, yams are planted manually in mounds and sometimes in previously prepared and manured holes or trenches, covered flat, mounded or ridged, depending on locality. Manual planting of yams therefore, entails very heavy back-breaking work which, coupled with the very high labour intensity of all the other phases of its culture, severely limits the development of large-scale yam industries in Nigeria. The average manual labour requirement for producing yams is given at about 45 man-days per tonne of the harvested crop or per 0.01 ha of area planted (Coursey, 1966). To be able to satisfy the growing domestic demand for yam, both as food and as an agro-industrial raw material, requires the urgent development of a requisite mechanization package for all phases of its production. The development of a yam planter is considered a necessary first step in this regard.

Although a great deal of work has been done on yam agribiology, breeding, pests and diseases, as well as its economics (Osuji, 1985) and sociology (Coursey, 1983), very little is reported in the literature on the mechanization of yam production. Vandevenne (1973) reported on stanchion ridgers for mechanizing the planting of yams. Our preliminary efforts in this regard, involving three final-year students' projects (Njoku, 1984; Omirefa, 1985 and Oluka, 1986) did not yield satisfactory results. This paper discusses the subsequent design and development of a two-row, tractor mounted automatic seed yam planter prototype.

2 CULTURAL PRACTICES AFFECTING PLANTER DESIGN

Yams are usually propagated with whole tubers called 'seed' yams which may weigh between 100 and 150 grams. Sometimes too, pieces called 'setts' may be cut from large yam tubers and used as the planting materials.

Many yam varieties of the Dioscorea species are grown in Nigeria. Among the cultivated varieties, there is a lot of variation in the size, length, weight and shape of the plantable seed yams, as shown in Fig. 1. Plantable diameters of seed yams vary between 45 mm and 80 mm while their lengths may vary between 150 mm and 250 mm (Onwueme, 1982). The seed yams are placed in mounds, holes or ridges to depths of 100 - 150 mm, and usually spaced 1 m apart within and between rows. But traditionally, wider or narrower spacings may be used in ill-defined rows, depending on the size of the seed yams. The traditional practice of planting yams in mounds or holes is not amenable to mechanization with commonly available machinery. Yet, there is no proof or experimental evidence that mounds have any agronomical advantage over ridges for yams. In fact, Coursey (1966) noted that there is no significant difference in the yields of yams planted on mounds and ridges under similar conditions. However, it is known that the broader and higher the ridge, the better. A high ridge provides ample depth of loose fertile soil for tuber and root penetration, while a broad ridge is less readily washed away by rains during the course of the season.

Studies on yam planting show that orientation of the yam does not significantly affect its emergence and future growth. Good viable setts will sprout

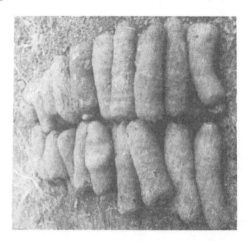

Fig.1   Variations in sizes of seed yams

and emerge irrespective of their orientation (Onwueme, 1982). However it is very necessary to ensure that the head is not placed way down in the soil since this creates a greater distance for the sprout to travel before it can emerge.

3 DESCRIPTION OF PLANTER PROTOTYPE

A pictorial description of the two-row yam planter prototype is given in Figs. 2 - 4 which are photographs of the front view, rear view and side view, respectively. The planter is tractor mounted and consists of pairs of hoppers, metering drums, delivery-chutes and drive wheels which are attached onto a standard commercial 2-row disc ridger which can make 1 m large ridges. The arrangement is such that the planter components can easily be detached to free the ridger for normal ridging operations when the need arises; the cost advantage of this

Fig. 4 Side view of planter prototype

arrangement is, of course, obvious. The design of the planter has taken into consideration the known cultural requirements for yams, as detailed in the following descriptions of the special features of the planter components.

3.1 The hopper

The planter has two hoppers which are 1000 mm wide, 200 mm deep and 700 mm high, one for each plant row. The shape and special features of the hoppers are as shown in Fig. 5. Each hopper is divided into four compartments with free swinging strips of galvanized steel plate. Each of these compartments is further divided with strips of polythene sheets into cells about 80 mm wide. The resulting cellular arrangement of the hopper facilitates the free flow of the seed yams down to the hopper throat. At the bottom of the hopper is located a

Fig. 2 Front view of planter prototype

Fig. 3 Rear view of planter prototype

Fig.5 Hopper open to show design details

component that serves as a load relief platform (see Fig. 6) which takes up the weight of most of the seed yams in the hopper so as to reduce the dead load on the yams in the hopper throat, thereby facilitating the flow of the seed yams in the metering zone. This component also provides for the channelling of the seed yams, one by one, to the hopper throat where they can be picked up by the metering cylinder.

A very important design feature of the hopper is the provision of the metal flap, shown in the sketch of Fig. 6, which acts as a gate to allow only seed yams in one half of each hopper to flow towards the hopper throat at a time. This avoids choking or blocking of the hopper throat which occurs when seed yams flow simultaneously from two directions or from the two sides of the

hopper, as was experienced during the preliminary testing of the planter without this gating arrangement. As illustrated in Fig.6(b),the pressure of the yams in the left-hand-side of the hopper pushes the metal flap gate out of the way after all the seed yams in the right-hand side of the hopper have been metered.

The two hoppers together can hold about 400 seed yams so that, for the agronomically recommended planting density of one seed yam per square metre, about 0.04 ha can be planted up with one filling of the hoppers.

3.2 The metering mechanism

The metering arrangment consists of two 250 mm dia, 200 mm long metering cylinders rolled from 1.5 mm m/s plates. Each metering cylinder has two semi-circular receptacles of 80 mm dia, recessed inwards from the cylinder surface and located diametrically opposite each other. Each metering cylinder is mounted so as to seal off the hopper throat under which it rotates to receive a seed yam every time a receptacle is directly below the hopper throat, as illustrated in Fig. 7. The seed yam in the receptacle is carried round until directly above the delivery chute into which it is released to

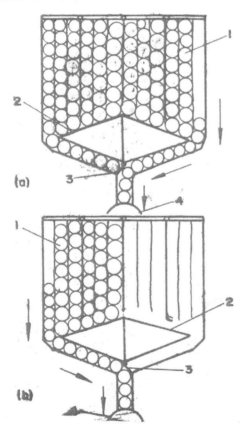

(a)

(b)

Fig.6 Gate regulating flow of seed yams metered from(a) RHS hopper,(b) LHS hopper 1, seed yams; 2, load relief platform; 3, sheet metal gate; 4, metering cylinder.

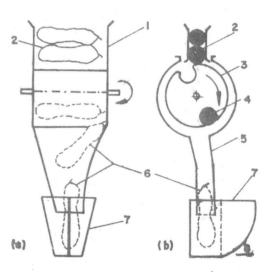

(a)   (b)

Fig.7 The metering mechanism: (a) front view, (b) side view 1, hopper throat; 2, seed yams; 3, metering cylinder; 4, seed yam in receptacle; 5, delivery chute; 6, seed yam descending in chute/ furrower system; 7, furrower; 8, direction of travel.

descend into the furrower. At that point, the other receptacle is in a position to receive a seed yam. Thus, as one receptacle is picking, the other is releasing a seed yam so that two seed yams are metered in one revolution of the metering cylinder. As illustrated in Fig. 7, the design of the delivery chute is similar to that of the cassava planter by Odigboh (1988) such that the seed yam descends in the chute-furrower system tail first, to ensure that the seed yam is always planted with its head raised above its tail to facilitate sprouting.

3.3 The drive train of the metering mechanism

The drive train for the metering mechanism shown sketched in Fig. 8, is modelled after that of the cassava planter by Odigboh (1988). Thus, the metering mechanism is driven by a series of chain drives which are in turn driven by ground-driven wheels. The drive

wheels have a diameter of 635 mm and traction lugs which make the effective diameter such that an average distance of 2 m is covered in one revolution. This ensures that the two seed yams metred per revolution of the drive wheel or the metering cylinder are planted at within-row spacing of 1 m, independent of the tractor foward speed. As illustrated in Fig. 8, the drive wheels are trailed behind the disc ridger. The weight of the drive wheel subassembly, together with the special shock absorber arrangements, ensures that the drive wheels are always pressed against the ground with sufficient pressure for effective traction while at the same time, the drive wheels have sufficient freedom to follow the rough or undulating surface of the field and still remain in contact with the ground. Although the two drive wheels are mounted on separate shafts, they are constrained to move together through a linking chain drive (see Fig. 8). As discussed by Odigboh (1988), the idea is that, if one of the wheels happens to lose contact with the

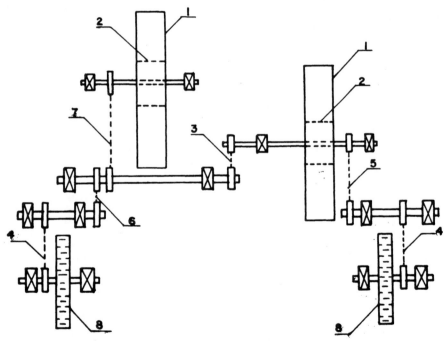

Fig.8 Top view of drive train 1, hopper; 2, metering cylinders; 3, chain drive to synchronize rotation of (2); 4, chain drive from ground wheel shafts to counter shafts; 5, chain drive from RHS counter shaft to RHS metering cylinder; 6, chain drive from LHS counter shaft to synchronizing drive shaft; 7, chain drive from (6) to LHS metering cylinder; 8, ground wheels.

ground for any reason, then the other wheel alone momentarily drives the two metering cylinders.

4 THE PLANTER OPERATION

The hoppers are filled with seed yams, making sure that the head ends of the yams face the front of the hoppers (see Fig. 5). The planter must be raised so that the drive wheels are off the ground to transport to the field or to turn round or about in the field. The planter must be operated in straight lines since there is no provision for differential movements of the drive wheels.

When lowered at the beginning of a row, the disc ridger and the drive wheels contact the soil, ready for planting to start. As the tractor moves forward and the discs are lowered to the appropriate ridging depth, the drive wheels are pressed firmly to the ground against the springs of the pairs of shock absorbers which are compressed to provide appropriate pressures to ensure effective traction of the drive wheels.

Located between each pair of discs of the ridger are the delivery chute-furrower units which shield the metered seed yams from the soil coming from the discs. The ridger makes the ridges whose advancing front covers the bottom, the rear end and the sides of the furrowers. For every half revolution of the drive wheels, each metering cylinder releases a seed yam which descends into the delivery chute, turns the right way up and descends into the furrow in a nearly vertical orientation. The furrower then moves forward to allow the soils of the ridge to quickly close in on the seed yam while the rear portion of the furrower pushes the seed yam to incline it to between 15 and 60° to the horizontal, but with the head end above the tail end, as earlier stated.

5 FIELD TESTS AND RESULTS

Mounted on MF 165 tractor, the prototype planter was subjected to a limited number of tests in a ploughed and harrowed field at average planting speeds of 2.5, 3.5, 4.5 and 5.5 km/h. Seed yams of 70 mm mean diameter and 165 mm mean length were used. At each planting speed, the within-row spacings, are practically independent of the planting speed and do not deviate

appreciably from the design value of 100 cm. This of course indicates that the metering mechanism performs efficiently at all the tested speeds. However, the planting depth seems to vary with the planting speed. This is attributed to the variation in the dimensions of the ridges produced in the very loose soil of the December month when the tests were performed. A planting depth of 10 - 15 cm is recommended for the shallowest part of the planted seed yams and this has been achieved at the higher speeds tested. The planter is yet to undergo more rigorous field tests during the normal yam planting season, to evaluate emergence and other yield performance parameters of the yams planted. However, it was observed in the tests conducted that all the seed yams were inclined at between 45° and 60° to the horizontal, with the heads above the tails, which is the agronomically preferred orientation of planted seed yams.

Table 1 FIELD TEST RESULTS

Planting speed (km/h)	Mean Within-row spacing (cm)	Mean depth of shallowest part of seed yam(cm)	Height and base width of ridges (cm)
2.5	107	8.5	32; 98
3.5	103	9.8	38;102
4.5	98	10.4	42; 96
5.5	101	12.2	45; 95

6 CONCLUSION

On the strength of the field tests conducted, the prototype planter is adjudged to perform very satisfactorily indeed. It makes acceptable sizes of ridges, places the seed yams in the ridges at average within-row spacings of close to 100 cm at the agronomically recommended depths and orientation of planting. At the tested planting speeds of 2.5 to 5.5 km/h, the planter achieves a theoretical field capacity of 0.5 to 1.1 ha/h.

7 REFERENCES

Coursey, D.G. 1966 . Yams - An account of the nature, origins, cultivation and utilization of the useful members of the Discorealese, Longmans, London 40 - 135 pp.

Coursey, D.G. 1983. Yams. Chap. 15 in Handbook of Tropical Foods (Ed. H.T. Chan, Jr.) Marcel Dekker, Inc., New York and Basel.

Iwueke, C.C., Mbata, E.N. and Okereke, H.E. 1983. Rapid Multiplication of seed yam by Minisett Technique. National Root Crops Research Institute, Umudike Advisory Bulletin No. 9.

Njoku, F.O. 1984. Design and testing of a seed yam metering device. B.Eng. Thesis. Department of Agric. Engineering, University of Nigeria, Nsukka.

Odigboh, E.U. 1988. A commercial model two-row automatic cassava planter. Paper presented at ANSTI International Conference on Appropriate Technologies in Agricultural Production and Processing in Africa. Arusha, Tanzania 4 - 7 April, 1988.

Okoli, O.O. 1987. Propagation techniques and prospects for breeding of yams (Dioscorea spp). Paper presented at the 2nd International Symposium on the Yam Tuber. Anambra University of Technology, Enugu, 8 - 10 June, 1987.

Oluka, S.I. 1986. Design of a Two-row Yam Planter. B.Eng. Thesis. Department of Agricultural Engineering, University of Nigeria, Nsukka.

Omirefa, J.O. 1985. Development of a yam metering device. B.Eng. Thesis, Department of Agricultural Engineering, University of Nigeria, Nsukka.

Onwueme, I.C. 1982. The Tropical Tuber Crops: Yams, Cassava, Sweet potato and cocoyams. John Wiley and Sons. London.

Osuji, G. (Ed.) 1985. Advances in Yam Research. Pub. Biochemical Society of Nigeria in collaboration with Anambra State University of Technology, Enugu.

Vandevenne, R. 1973. Mechanization of Yam Cultivation in the Ivory Coast. Proceedings of the thrid symposium of the International Society for Tropical Root Crops. IITA. Ibadan, Nigeria.

ACKNOWLEDGEMENT

The funds for the work reported in this paper were taken from a research grant by the (Nigeria) Federal Ministry of Science and Technology which is hereby gratefully acknowledged.

Land and Water Use, Dodd & Grace (eds), © 1989 Balkema, Rotterdam. ISBN 90 6191 980 0

A punch planter for vegetables on plastic mulch covered beds

L.N.Shaw
University of Florida, Gainesville, Fla., USA

K.H.Kromer
Institut für Landtechnik, Bonn, FR Germany

ABSTRACT: Crop establishment with plastic mulch makes earlier planting possible due to higher soil temperatures and better weed control along with other advantages. Bucket wheel planters were developed for planting through plastic mulch but they are complicated and hard to maintain because of the many moving parts that are working in the soil.

A punch type planter for seeding crops through plastic film mulch has been developed that has no moving parts in the dibble spades. Hollow punches or spades attached to an inclined, ground driven wheel deposit seeds from a seed metering mechanism that is synchronized with the wheel. Cavities for seeds are formed by the lateral movement of the spades in the soil and the dibbles can be left open for easier emergence of the plants or closed to provide better seed to soil contact.

The planter has been compared to bucket wheel planters on both plastic film mulch planting and on bare soil and the crop emergence and yields are equal or better with the simpler planter.

1 INTRODUCTION

The production of high quality fresh vegetables in Florida would not be possible without plastic mulch. The cultural and economic justifications for use of film mulch have been summarized by Everett, et.al (1972) and Fluck et.al (1977). These included good weed control; improved action of soil fumigant for disease, weed, and nematode control; confinement of plant roots to the fumigated soil; reduction of root rot; reduction of soil moisture evapo- ration; prevention of bed erosion; reduction of root damage from mechanical cultivation; energy conservation; and improved cleanliness and quality of the vegetables. Typical mulch-covered beds are spaced 183 cm on center with 152 cm wide polyethylene film used to cover the soil. The edges of the film are buried in the soil leaving an exposed mulch of 81 to 122 cm wide. Generally 1.25 to 1.5 mil. polyethylene film is used.

The beds are formed with disc-type ridging equipment and a bed press which forms a level firm bed on which most of the required fertilizer is distributed. The fertilizer application is followed with a combined operation where a soil fumigant is injected into the soil, irrigation drip tube is laid, and the plastic mulch film is installed with special film laying equipment. A two week interval separates the fumigation and film-laying operation from the planting operation. Tomato or pepper plants are generally established by using the plug mix (Hayslip, 1973) seeding operation or setting seedling plants with a transplanting machine. The holes for the plants or the plug mix are either punched or melted in the plastic mulch during the planting operation.

Over ten years ago vegetable growers were concerned about the residual fer- tilizer left after a crop being a wasted resource and causing ground water pollution. Up to that time only one tomato or pepper crop was grown on the plastic covered beds but growers real- ized that second vegetable crops of squash, melons, beans or corn could utilize the residual fertilizer if these crops could be planted. A punchplanter was developed by the Agricultural Engineering Department at the Univer- sity of Florida from concepts originated by some growers.

Fig.1 Punch planter for plastic mulch

2 ORIGINAL PLANTER DEVELOPMENT

This punch or bucket wheel planter for plastic mulch consisted of a punch wheel, a seed metering device, and a system for delivering the seeds to each punch. The hollow wedge shaped punches which projected out from between the dual rims of the wheel assembly, penetrated the mulch film and the underlying soil and deposited seeds at regular intervals. The seed metering device was driven from the punch wheel. As the wheel turned, seeds were metered into a star shaped manifold which directed them towards each punch as it approached the ground. One or more seeds were delivered to each punch where they remained until they were deposited in the soil. Just as each punch started to withdraw from the soil the seed was released through a spring loaded door on the rear of the wedge. The door was forced open by a lever arm activated by a stationary cam. The cam was designed with a constant acceleration contour and an adjustable mounting allowed changes in the timing of the door openings.

The projections of the punches from between the dual wheel rims established the punch penetration which could be varied from 0 to 3.81 cm. Seed spacing along the row could be varied by changing the number of punches on the wheel and planting intervals of 7.6, 10.2, 15.2, 20.3, 22.9, 30.5, 45.7, 61.0, 91.4, and 182.9 cm were possible. As with most planting equipment a variety of seed cells and cell speeds provided a range of seeding rates and seed types. With nine punches on the wheel, a six cell plate, and a planter wheel to seed plate drive ratio of 0.67:1, a cell was matched to each punch. The roller chain drive provided some timing adjustment between the seed metering device and the planter wheel. A semi-pneumatic press wheel firmed the soil around the seeds in each cavity.

Some operational difficulties resulted when the planter was operated in wet soil or with a high accumulation of plant residue. The moisture caused flat squash seeds to stick inside the punches and some holes were not planted. So planter operation has had to be limited to dry soil conditions.

After initial field trials it was realized that the slit cut in the plastic by the punch was not sufficiently large to allow the seedlings to emerge through the mulch. Additional cutters were welded perpendicular to the faces of the wedges and the resulting punctures were x-shaped. These larger openings were sufficient for plant emergence.

Limited use of this planter in a double cropping system with squash following tomatoes has shown that the yields are about equal to yields on bare ground with fertilizer applied at planting time. In addition to the savings in fertilizer, farmers reduced production costs by eliminating the cultivating and hand hoeing required for weed control.

This planter design was introduced into Germany and it was found that the planting date of maize could be advanced by nearly two weeks because of the soil warming provided by plastic mulch. Extensive studies were carried on at several locations and at least four manufacturers put plastic mulch punch planters on the market.

The original mulch planter design had many moving parts that were difficult to lubricate so a great deal of maintenance was required. The advantages of plastic mulch culture justified continued planter development so cooperative work was initiated between the University of Florida and the Institut fur Landtechnik at the University of Bonn in Bonn, F.R.G. This cooperation has resulted in the development of the revolving spade punch planter.

3 DEVELOPMENT OF THE REVOLVING SPADE PUNCH PLANTER

Several punch and seed delivery concepts were considered and one of the simpler designs is shown in Fig. 2. This con-

Fig.2 Revolving spade planter

cept was to insert a spade into the
soil and to shift the spade to form a
cavity for seed. If spades were
mounted radially on a wheel, and the
wheel moved in such a manner as to cause
some movement of each spade in the
soil, then cavities would be formed.
Cavities could be formed by operating
the wheel with positive or negative slip
or slightly yawing it relative to the
direction of travel. The latter method
of forming holes was selected for simplicity. A cavity having the cross-
section of the end of a very tin ellipse
would be formed with this orientation of
the wheel assembly. The wheel axis
could also be inclined from the horizontal to allow the mounting of the
seed metering device as close to the
hollow wedge-shaped spades as possible
and to allow easier entry of the spades
into the soil. With the entry of a
spade into the soil, it would displace
the soil to one side. The displaced
soil would be compacted on one side of
each cavity but there would be minimal
compaction on the remaining surfaces.
This variation in compaction might
provide the optimum amount of compaction
for good seed to soil contact.
   After the seed is deposited and each
spade withdrawn, some of the displaced soil would fall back into each
cavity and an inclined, oppositely-yawed
press wheel could be used to close the
holes and cover the seeds.
   It was envisioned that seeds dropped
from a seed-metering device could be
synchronized with the action of the
spades; thus the kernels. could be

dropped in immediately behind each spade
and settle to the bottom of the holes
for uniform depth of planting.

3.1 Planter prototype design

For a dynamic analysis of the soil-
opening mechanism and seed drop, prelim-
inary design specifications as follows
were drawn up:
   Spade wheel diameter- 72 cm
   No. of spades   - 12
   Planting depth   - 62 mm (max).
   Planting speed   -   5 km/h
   Wheel inclination   - 30° from
                              vertical
   Yaw angle   -   0° to 10°
   If each seed had to drop 9 cm., the
drop interval would be about 0.1 sec.
because centrifugal force would contribute almost g/2 to the acceleration.
The period of spade contact with the
soil would be about 0.33 sec. at a
travel speed of 5 km/h. Thus with
proper timing of the seed drop, satisfactory deposition of the seed should
be possible.
   Using available parts and materials,
a spade wheel was designed and fabricated. Spades, 60 mm. wide, with a
radial length of 93 mm. were mounted on
a wheel for soil bin and field evaluation. Cavity formation was satisfactory in tilled ground with a yaw angle
of 7° (measured in a horizontal plane)
and acceptable even in hard clay sod
with sufficient weight applied to the
wheel.
   After the field tests of the spade
wheel, the problem of mounting and
driving a seed-metering device was
addressed. Originally a pneumatic
seeder was selected so that a more rapid
seed delivery could be obtained than
with the gravity drop from a plate-type
planter unit but none were easily
adapted to the soil opening wheel. An
inclined spoon wheel seed metering
device that could be driven directly
from the hub of the spade wheel was
selected and adapted. Of concern with
the use of this seeder unit was the
seed-drop distance of over 24 cm. A dynamic analysis of the seed movement,
showed that the drop point on the seed
metering unit had to be advanced by
40°. Provision was made in the metering unit drive coupling to adjust the
seed drop timing. The spoon wheels
were modified so seeds would only be
picked up when there was a spade to
receive them. The seed metering unit
was mounted in the planter frame and
coupled to the spade wheel with a

flexible coupling. A seed delivery chute with a clear plastic side was attached to the seeder unit to direct the seed to each spade. When a test run was made on a planter test stand, using a strobe light to study the seed dynamics, most of the seeds followed the theoretical trajectory. But some missed the spades. Rectangular metal seed tubes were added to the top of each spade to intercept the seeds from the seed chute. The addition of the seed tubes did not change the action of the spades in the soil but improved the seed delivery to the spades. After some adjustments in the seeder-to-wheel timing the performance was judged to be quite satisfactory on the test stand at planter travel speeds of 3.5 to 6.5 km/h.

A pneumatic gage and press wheel was mounted on the rear of the planter frame. The wheel mount was designed so that it could be adjusted in and out and inclined to effectively press the soil back over the seeds.

3.2 Field trials of the spade planter

The planter unit was first field tested on some rather rough clay soil that was almost too wet to be tilled, yet the performance was quite good. The maximum speed under these field conditions was about 4.5 km/h; otherwise the seed wasn't placed at a uniform depth. The addition of a longer seed chute to direct the seeds into the spade seed tubes was needed because a few seeds missed the tubes when the planter bounced on rough ground. Also, it was observed that more weight was needed to get adequate penetration into hard soil.

After the longer seed delivery chute was added and the seed plate to wheel timing adjusted, further field trials were conducted. Trials were conducted on well-tilled soil with and without plastic mulch as well as on established sod. The performance of this spade-wheel planter on unmulched soil were at least equal that of a bucket planter. In addition the scraper on the revolving spade planter allowed it to operated under wetter soil conditions.

For comparison tests on plastic mulch, the revolving spade planter was mounted on a tool bar with a bucket wheel planter. It had been anticipated that there would be a lateral thrust on the plastic mulch from the action of the spade wheel, but it was found that a properly positioned

Fig. 3 Revolving spade planter and bucket wheel planter on plastic mulch

and adjusted gage wheel compensated for this action and the mulch was not pulled out of place. Performance of the two planters working together was judged to be nearly equal. Only very limited trials have been conducted with this spade planter on plastic mulch because this design is being evaluated as a no-till or minimum till planter. More extensive test work in planned for this planter in the U.S on plastic mulch.

4 DISCUSSION AND SUMMARY

A simple design for a punch planter has been developed and evaluated in field tests. Further development work is underway in Europe, and research on this device is planned in the U.S. Limited tests have demonstrated that this design is equal in performance to more complicated punch planters and the simplicity of the design should make it cheaper to manufacture and easier to maintain.

REFERENCES

Everett, P.H., H.H.Bryan, and N.C. Hayslip 1972. Growing fresh market tomatoes for machine harvest. Proc. Fla. State Hort. Soc. 85:152-156.

Fluck, R.C., L.N.Shaw, and P.H. Everett 1977. Energy analysis of the use of full-bed plastic mulch on vegetables. Proc. Fla. State Hort. Soc. 90:382-385.

Hayslip, N.C. 1973 Plug-mix seeding developments in Florida. Proc. Fla. State Hort. Soc. 86:179-185.

Land and Water Use, Dodd & Grace (eds), © 1989 Balkema, Rotterdam. ISBN 90 6191 980 0

Self feeding transplanters

C.W.Suggs, H.B.Peel & T.R.Seaboch
North Carolina State University, Raleigh, N.C., USA

D.L.Eddington
University of Missouri, Columbia, Mo., USA

ABSTRACT: Two self-feeding transplanters are being developed. One of the machines uses plants grown in a grid of paper cells which unfolds into a continuous strand of plant containing cells. The strand passes between a pair of feed rollers which spaces the plants and finally between a pair of acceleration rollers which snap the perforated strand and propel the plant into a chute which guides it to the furrow. Special features include plant straightening equipment and a chute with a controlled exit angle to compensate for transplanter speed. The other machine uses bare root plants grown in rows in a cold frame. Plant are machine dug and stitched/taped into a strand which is fed through the transplanter with computer control of the sequence of events. Field tests with planting speeds as high as 140 plants per minute have been successful with the first (paper cell) machine. The bare root machine has an operating speed of approximately 40-60 plants/minute.

Zwei selbstregulierende Umpflanzungsmaschinen warden entwickelt. Eine der Maschinen setzt Pflanzen, die in einem Netz von Papierzellen gezogen sind, das sich zu einer Kette von Pflanzenenbestückten Zellen ausfaltet. Die Kette lauft durch zwei Zuführrollen, die den Abstand regulieren, und danach durch Beschleunigungsrollen, die Kette an der Perforierung zerreißen. Dann rutschen die Pflanzen an eine Schütte in die offene Furche. Besondere Eigenschaften schließen eine Vorrichtung zum Aufrechtstellen der Pflanzen und einen kontrollierbaren Rutschwinkel ein, der sich der Geschwindigkeit der Maschine anpaßt. Die zweite Maschine setzt Pflanzen mit entbloßten Wurzeln, die in Reihen im Frühbett gezogen warden sind. Die Pflanzen werden mit Maschine ausgegraben und in eine Kette genäht/geklebt, die in die Umpflanzungsmaschine eingeführt wird. Die Reihenfolge der Vorgänge ist durch einen Computer kontrolliert. Experimente mit Pflanzgeschwindigkeiten bis zu 140 Pflanzen pro Minute sind mit der ersten (Papierzellen) Maschine erfolgreich ausgeführt worden. Die zweite Maschine erreichte eine Arbeitsgeschwindigkeit von etwa 40-60 Pflanzen pro Minute.

1 INTRODUCTION

Although field seeding is easier, many crops are transplanted for a number of compelling reasons. Some crops, especially smaller seeded plants such as tobacco, will not survive under field conditions. Other crops appear to be stimulated by the transplanting shock.

Transplanting also allows land to be used more efficiently because large land areas are not needed until the seedlings are of transplant size. Transplanted crops can mature in short season areas where seeded crops cannot. Because the in-field season is reduced it also makes multicropping possible in areas where it would not otherwise be possible and allows for the production of an early "preseason" crop. With transplanting and multicropping, photosynthetic conversion is increased because more of the land area is covered with plant leaves more of the time.

Transplanted crops, because of the shorter field season, typically require less irrigation and fewer or lower rate applications of herbicides and pesticides. For many crops, sugar beets for example, yield is increased because the effective growing season is longer.

The main disadvantages or deterrents to transplanting are labor, machine costs and the production of transplants. The development of automatic feeding transplanters would significantly reduce labor requirements and possibly could reduce unit machine cost if the transplanters have high production rates and could be used to transplant large areas.

Although transplanting methods and machines have been improved dramatically, as compared to the hand methods used several decades ago, transplanting still requires a significant labor input. At the present time the machines provide mechanical power to open and close the furrow and to convey the plant from the feeding location into the furrow. Plant feeding which consists of singulation, selection, alignment and transfer is the operation which has resisted mechanization and must still be done by hand. Automation of this operation has been difficult because of plant fragility and nonuniformity with respect to size, shape, and location.

Transplanting labor for tobacco (Pugh 1983) averaged 1.12 h of labor per 1000 plants transplanted or 15 plants per minute. Since this figure includes turn time and time to reload the machine with water and plants, actual transplanting rates, while in the row, were found to be approximately twice the overall rate. Labor to pull seedling from field plant beds, a method used for tobacco and many horticultural crops, requires about the same amount of time as the feeding operation. The total time to pull and feed plants is, therefore approximately, 2.24 h per 1000 plants. Total transplanting labor, then, for widely spaced crops like tomatoes and tobacco fall between 25 and 36 h/ha (10,000-16,000 plants/ha). For closely spaced crops the total could easily be 70 to 100 h/ha depending on the plant spacing.

2 PREVIOUS WORK

In recognition of the fact that seed are easier to singulate than plants, most of the approaches to automatic feeding of seedlings into transplanters involve the production of plants in trays or other orderly arrays from which plants can be located mechanically (Suggs et al 1987). Typically, plant pulling is also eliminated as the tray or other seedling support structure is taken to the field and indexed on the transplanter where it becomes a supply magazine from which the machine automatically removes the plants as they are placed into the furrow. In this approach the soil mass remains on the roots (intact root) in contrast to field grown plants which lose most of the mass from the roots when they are pulled from the soil (bare root).

Automatic feeding is based either on plants being held on a belt, chain, tape or similar device which holds the plants in a single strand or a tray which holds plant containing cells in a rectangular grid. With these systems it is possible to predict the position of the next plant. Attachment of the plant to a uniform element circumvents the problems associated with the nonuniform size and shape and the unpredictable location of mass grown seedlings. Thus the problem of plant singulation and selection can be reduced to seed singulation by growing the plants in one of these structural arrangements.

Plant strands are usually folded or coiled to conserve space and expedite handling and plants may be grown in cells along the strand or transferred from other sources to the strand (Roth 1973, Brewer 1978, Moden & Hauser 1982, Hauser 1985, Boa 1984). In most cases more labor is eliminated if the seedlings can be grown directly in these strands or lines of cells instead of being transferred from some other growth areas. Transplanting is thus simplified from the singulation of plants to the singulation of cells and consisted of feeding the strand through a device which meters cells and cuts or breaks the strand between plants.

Although various materials have been used in the fabrication of the strands, paper is cheap and does not have to be removed from the plant roots (Brewer 1978, Smith et al 1984, Eddington 1988a, Masuda & Namby 1979, Suggs et al 1987). A honeycomb shaped matrix of cells which unfolds into a strand suitable for automatic feeding into a transplanter has been developed.

Automatic equipment has been developed to sort bare root plants on the basis of plant length (Maw & Suggs 1984a, 1984b). Eddington (1988b) developed a system which sorted plants on the basis of dual wavelength reflectance to determine which plants were of an acceptable size. An experimental sorter for detecting empty cells was also developed by Brewer and Maw (1983). Cundiff (1979) developed and evaluated a machine to dig seedlings from a row, tape them into a strand and store them on a reel. Tree seedlings have been automatically selected by means of a rotating singulator or a hook device which withdraws a single stem from a seedling mass (Graham & Rohrbach 1983).

Plants may also be grown in cells arranged in trays in a rectangular grid system. In this system the location of each cell can be sensed by the machine and used as the basis of an automatic plant feeding mechanism (Armstrong et al 1984, Edwards et al 1984, Pinkepank & Shaw 1980, Huang & Splinter 1968). Trays may be of nondegradable material from which the plant must be removed before transplanting. In contrast, if the tray is made of porous or biodegradable material it may be cut or otherwise divided into individual cells and planted with the seedling (Penley 1981).

There appears to be a trade-off between producing bare root plants in relatively inexpensive field beds versus the greenhouse production of intact root plants in strands or trays which can be automatically fed easier than the bare root plants. Investigations with tobacco (Suggs & Mohapatra 1987, Suggs &

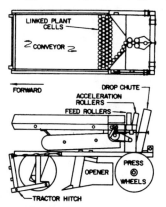

Fig. 1—Self-feeding transplanter which uses plants grown in a honeycomb of cells which unfolds into a strand or belt of cell-carrying plants.

Mohapatra 1988, Suggs et al 1988) showed that plant size, intact versus bare root, undercutting and leaf clipping did not affect crop yield or value but did have an effect on growth rate.

The objective of the work reported in this paper was to develop and evaluate an automatic feeding bare root transplanter and to modify, adapt and evaluate two models of existing automatic feeding intact root transplanters with respect to tobacco and vegetable crops.

3 MACHINE DESIGN

3.1 Intact root transplanter

Two transplanter models for feeding intact-root cell-grown plants were investigated. These machines were built in Japan for transplanting sugar beets but required extensive modification before they would dependably plant the larger plants used in tobacco and vegetable production. Both of these machines use plants grown in a honeycomb matrix of paper cells. These cells, shipped folded, are expanded and filled with soil media and seeded so that germination and growth may take place. The glue lines between the strands of cells dissolve during plant growth allowing the matrix to unfold into a strand of plant carrying cells. Unfolding is not allowed to occur until the strand is being fed into the transplanter.

For one of the machines the plant strand is passed between a pair of counter-rotating, foam covered roller which feed the plants at the desired spacing, Fig. 1. The plants then pass between a second pair of rollers having a peripheral speed several times as fast as the feed rollers. These rollers grip the plants, break the strand at between-plant perforations and accelerate it into a drop chute which guides it to the furrow just in front of the furrow-closing press wheels. Because of the relative large size of the plants being transplanted, it was necessary to design and develop belts to guide and control the plants as they approached the feed rollers. Also a pair of "spider" wheels having projecting spokes was added to comb upward through the plants to straighten, lift and untangle the plants just before they entered the feed rollers. The original foam covered feed rollers were replaced with a triple stack of spiked wheels on each side of the strand to provide a more positive feeding action and to perforate the cell for easier root emergence.

At the high ground speeds associated with high transplanting rates and wide plant spacing, the horizontal velocity of the plant caused it to overturn when it made contact with the ground. This was partially solved by angling the delivery chute forward at its discharge end. In order to accommodate varying ground speeds the lower end of the chute was pivoted and actively controlled by the centrifugal force generated by a set of rotating weights (fly-ball governor) driven by a ground wheel. Chute angle was adjusted continuously by the position of these weights communicated to the chute by a flexible push-pull cable.

On the other intact root transplanter pairs of planting fingers extending radially from a wheel (Ferris wheel fashion) grip the plant, tear it from the strand, transfer it to the furrow and release it just in front of the packing wheels, Fig. 2. The paper honeycomb plant cells are used as in the previous example with the roller feeding machine. As the plant cells are unfolded they are led down a surface and between a pair of restraining brackets which expedites plant transition from the strand to the planting fingers. During operation a pair of fingers moves into position and closes on a plant. The restraining devices then open to allow the fingers to pass but holds the strand so that only one plant will be torn from the strand. As the hands move around the circle they are opened and closed at the appropriate time by cam follower bearings riding on a circular cam.

This machine was mounted on a small two-wheel tractor and did not have sufficient traction except in very light soils. Addition of dual tires did not solve this problem.

3.2 Bare root transplanter

An automatic feeding transplanter which used bare-root transplants grown in rows in field cold frame beds was designed and built by the Transplanting Mechanization Project in the Biological and Agricultural Engineering Departmental research shop at North Carolina State University. The system

consisted of (1) a plant digger and stitching unit which sewed the plants to a strand of paper so they could be fed into (2) the automatic feeding transplanter.

Seeds were sown in rows about 15 cm apart in cold frames or field beds covered with plastic film. Plants were top clipped several times to improve

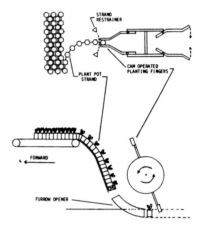

Fig. 2—Self-feeding transplanter, using plants grown in honeycomb cells, consisting of pairs of plant fingers or hands to grip the plant cells, tear them from the strand and place them into the open furrow.

uniformity and dug when about 15 cm tall. The digger had an undercutting blade to loosen the plants so they could be lifted by a pair of upwardly inclined V-belts, Fig. 3. Paper tape was fed out along each side of the plants and a sewing machine stitched through the tapes and plant leaves to form a continuous strand of plants.

This strand was fed between a pair of V-belts on the transplanter which guided them to the furrow, Fig. 4. The machine components were hydraulically driven and could be stopped and started by

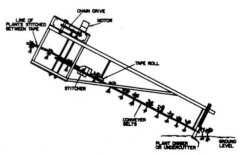

Fig. 3—Machine for digging a row of plants and stitching them between tapes into a continuous strand.

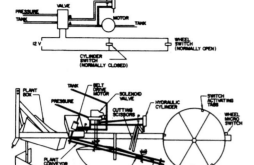

Fig. 4—Bare-root transplanter. Stitched plants passing between belts are separated from strand and planted.

they were being fed through the feed and acceleration rollers faster than the drop chute could accept them. This was alleviated by modifying the curves in the drop chute, increasing its width and modifying the upper end so that leaves would be less likely to catch on the chute edges.

Table 1. Performance of automatic roll-feeding transplanter with corn seedlings in 3 cm cells.

Speed plants/ min	Properly planted, %	Spacing index %	Spacing, cm	S.D. (n=80) cm
84	97	84.3	70.6	26.3
90	100	92.1	65.8	12.3
94	99	96.7	63.0	15.6
96	99	98.9	61.7	9.6
98	99	100.8	60.5	9.0
99	96	100.2	59.9	13.0
141	91	90.5	66.8	23.4

electrically controlled valves in the first version and by electric clutches in the second version of the transplanter. Just before a plant reaches the end of the V-belts, where it would be deposited into the furrow, the stem contacts a microswitch which stops belt motion and actuates a cutter to sever the paper tape and free the plant from the strand. When the transplanter has moved ahead one plant spacing, another microswitch is closed by a device driven from the press wheel. Closing this switch restarts the feed belts and releases the plant into the furrow. Since there is now no plant stem against the first microswitch, feed is continued until a plant moves into position and the cycle is repeated.

4 RESULTS

4.1 Intact root transplanter

Because of the continuous nature of its feeding action the roll-feeding machine functioned well at speeds up to 140 plants per minute for a one row machine, Table 1. Except at the highest planting speed, 96 to 100% of the plants were properly planted. Plants which were covered or misaligned and sites which had two plants were considered as not properly planted. A spacing index which compares the number of plants planted by the number which should have been planted ranged from 84 to just over 100%.

There was some problem with plants jamming in the delivery chute because

In an experiment with corn seedlings, Table 1, spacing, which was targeted for 61 cm, varied from 60 to 71 cm with a standard deviation of 9 to 26 cm. Plant spacing was not uniform because plant passage time down the drop tube was dependent on plant size and shape as well as how quickly all its leaves cleared the acceleration rollers. Also, there was some problem with plants being tangled in the feed and acceleration roll area. The spider wheels and feed belts mentioned earlier alleviated this problem, but the devices were completed so late in the transplanting season that it was not possible to run a replicated experiment.

On the other intact root transplanter plant location is mechanically controlled from the honeycomb cells all the way through placement in the furrow. Therefore, plant spacing is more accurate than for the roll feeding model, Table 2. Plant spacing, targeted at 60 cm, varied only a few centimeters and had a standard deviation of 1.8 to 4.8 cm. Planting speeds were varied from 40 to 100 plants per minute without serious decrease in the percent of properly planted seedlings or increase in missed, misaligned or doubles.

Table 2. Operating characteristics of automatic feeding "Ferris wheel" transplanter with tobacco seedlings in 3 cm cells.

Speed plants/ min	Properly planted, %	Spacing	
		Mean, cm	S.D. (n=30), cm
40	94	54.1	3.7
43	95	53.3	4.8
47	100	52.8	2.5
59	100	59.7	1.8
78	86	55.8	3.0
86	98	53.6	3.6
100	92	57.4	2.3

The machine functioned reasonably well at speeds approaching 100 plants per minute, Table 2. In most cases 95 to 100% of the plants were properly planted. Plants or spaces not properly planted were due to misaligned or double plants and to misses. The machine did not function well at speeds much in excess of 100 plants per minute. This was due primarily to the intermittent nature of its operation in which plant gripping fingers or hands have to close on each plant and the plant restrainers have to open to allow the fingers to pass.

4.2 Bare root transplanter

The bare root system shows potential but will require more work before it can be considered ready for commercial production. Plant variability continues to be a problem but progress in germination synchroncy and foliar clipping may soon make the growth of uniform seedlings possible.

Digging and sewing plants onto the tape have not progressed as far as the transplanter. The digger lifted and conveyed the plants but the tape dispenser and stitcher did not work well.

Whereas the microswitch successfully detected the presence of a plant, it could not differentiate between a plant or a leaf. Also if plants were loaded very close together on the tape the machine sometimes cut through the second plant or planted both at one time.

With the initial design the bare root transplanter was operated at speeds of

15 to 43 plants per minute at target plant spacing of 61 cm. The feed tapes were hand loaded with tobacco or broccoli plants. Spacing on the tape was either constant or variable.

This machine was redesigned to include computer sequencing of events to insure that an event did not start before the previous one was complete. For example, belt motion should not start until the cutter was retracted. The electric valves controlling the hydraulic flow to the various parts of the circuit were replaced with faster acting electric clutches controlling the output of constant running hydraulic motors. Also a shaft encoder was added to replace the tab on the press wheel which controlled plant spacing in the first design.

Machine performance improved appreciably from the preliminary and initial designs through the intermediate design to the final design, Table 3. In the first two trials, machine speed was limited to 43 plants per minute and only 40 to 70% of the plants were properly planted. In the intermediate design with electric clutches, electronic control and rotating saw to cut the strand, transplanting speed was increased to 50 plants/min with 64 to 85% of the plants being properly planted. For the final tests the transplanter was mounted on a tractor which had sufficient hydraulic capacity to operate at higher speeds. Speeds up to 60 plants per minute were achieved with 72 to 96% of the plants being properly planted. Plants not properly planted were separated into multiple drops, when the machine planted more than one plant in a space and misses, when the machine either did not eject a plant or the ejected plant was not properly planted in the ground.

In the final design, performance at 30 and 45 plants per minute was acceptable for most crops. However, at 60 plants per minute performance deteriorated and will need to be improved before the machine can be commercially accepted.

Random spacing of the plants on the tape caused no problem at 30 and 45 plants per minute but caused the machine to plant doubles at 60 plants per minute. At the higher speed the drive clutch did not respond fast enough to prevent a closely spaced plant from running past the microswitch before the belts stopped. This plant was either planted between the regular plant spacings or with the next plant. Plans are underway to correct the problem by rescheduling the computer and by replacing the present clutch with a faster acting one.

Table 3. Performance of bare root transplanter.

Machine	Speed plants/min	Properly planted %
Preliminary Design	15	40
	43	40
Initial Design	15	70
	43	60
Intermediate Design	20	64
	36	85
	50	73
Final Design	30	94
	45	96
	60	72

REFERENCES

Armstrong, E.C., W.A. Hahacek & T.A. Spinetti 1984. Automatic soil plug loader and feeder. U.S. Patent No. 4,443,151.

Boa, W. 1984. The design and performance of an automatic transplanter for field vegetables. J. Agri. Eng. Res. 30:123:-130.

Brewer, H.L. 1978. Automatic transplanter system for field crops. ASAE Paper No. 78-1011, ASAE, St. Joseph, MI 49085.

Brewer, H.L. & B.W. Maw 1983. Automatic seedling sorter to increase greenhouse space utilization. ASAE Paper No. 83-1090, ASAE, St. Joseph, MI 49085.

Cundiff, J.S. 1979. Mechanism to lift transplants from field beds. ASAE Paper No. 79-1074, ASAE, St. Joseph, MI 49085.

Eddington, D.L. 1988a. Automatic tobacco transplanting. Ph.D. Thesis. Biol. & Agri. Engr. Dept., N.C. State Univ., Raleigh, NC 27695-7625 USA.

Eddington, D.L. 1988b. transplant sensing using dual wavelength reflectance. Amer. Soc. of Agri. Engr. Paper 88- 3518.

Edwards, B.E., S.R. Krogman & E.J. McArdle 1984. Plant transfer mechanism. U.S. Patent No. 4,440,101.

Graham, L.F. & R.P. Rohrbach 1983. Mechanical singulation of bare root pine seedlings. TRANSACTIONS of ASAE 26(4):972-975.

Hauser, V.L. 1985. An automatic grass transplanting machine. TRANSACTIONS of the ASAE 28(6):1777-1782.

Huang, B.K. & W.E. Splinter 1968. Development of an automatic transplanter. TRANSACTIONS of the ASAE 11(2):191-194,197.

Masuda, A. & T. Nambu 1979. Apparatus for separating and transplanting paper tube seedlings from a continuous paper tube seedling assembly. U.S. Patent No. 4,167,911.

Maw, B.W. & C.W. Suggs 1984. Sorting and selections of bare root transplants. TRANSACTIONS of the ASAE 27(3):706-710 and 714.

Maw, B.W. & C.W. Suggs 1984. A seedling taping machine for bare root plants. TRANSACTIONS of the ASAE 27(3):711-714.

Moden, W.L., Jr. & V.L. Hauser 1982. An automatically-fed bandoleer transplanter. TRANSACTIONS of the ASAE 25(4):864-867.

Penley, P.A. 1981. High speed transplanter. U.S. Patent No. 4,289,080.

Pinkepank, W.R. & L.N. Shaw 1980. An automatic vegetable seedling soil block transfer system. ASAE Paper No. 80-1054, ASAE, St. Joseph, MI 49085.

Pugh, C.R. 1983. Total labor input per acre. Unpublished paper, Dept. of Economics and Business, N.C. State Univ., Raleigh, N.C. 27695.

Roths, H.C.O. 1973. Transplanting machine. U.S. Patent No. 3,719,158.

Smith, J.A., R.G. Wilson & J.G. Robb 1984. Paperpot system for mechanized sugarbeet transplanting. ASAE Paper No. 84-1516, ASAE, St. Joseph, MI 49085.

Suggs, C.W. & S.C. Mohapatra 1987. Tobacco transplants, Part 2. Effects of bare-root versus intact-root plants on yield, value, growth rate and chemistry. Tob. Sci. 31:104-108.

Suggs, C.W., T.N. Thomas, D.L. Eddington, H.B. Peel, T.R. Seaboch & J.W. Gore 1987. Self-feeding transplanter for tobacco and vegetable crops. Trans. of Amer. Soc. of Agri. Engr. 3(2):148-152.

Suggs, C.W. & S.C. Mohapatra 1988. Transplants Part 2. Effects of bare-root versus intact root on yield, value, growth and chemistry. Tob. Sci. 32:1-6.

Suggs, C.W., S.C. Mohapatra & W.H. Johnson 1988. Tobacco transplants Part 2. Effects of clippings and undercutting on yield, value, chemistry and growth.Tob. Sci. 32:24-28.

RESUME

Deux transplanteuses à alimentation automatique sont en développement. Une des deux machines utilise des plantes germées dans un quadrillage de cellules en papier qui se déploie en une chaine de cellule à plantes. La chaine passe entre deux rouleaux à alimentation qui séparent les plantes, et ensuite à travers une paire de rouleaux à acceleration qui saisissent la chaîne entre les perforations. Les plantes sont guidées le long d'une chute jusqu'à une rainure ouverte. Certaines caractéristiques de la machine sont un matériel de redressement des plantes et une chute avec un angle de sortie controlée pour compenser la vitesse de la transplanteuse. L'autre machine utilise des plantes à racines nues germées en rangées sur un chassis froid. Les plantes sont déterrées à la machine et suturées/collées en une chaine qui est introduite dans la transplanteuse, la séquence étant controlée par ordinateur. Lors d'essais pratiques avec la première machine à cellule de papier, certains succés ont été obtenus avec des vitesses de plantation aussi élevées que 140 plantes par minute. L'autre machine a une vitesse opérationnelle d'environ 40-60 plantes par minute.

Land and Water Use, Dodd & Grace (eds), © 1989 Balkema, Rotterdam. ISBN 90 6191 980 0

Untersuchungen an Reifen zum boden- und pflanzenschonenden Befahren von Grünland

K.Armbruster

Institut für Agrartechnik, Universität Hohenheim, Bundesrepublik Deutschland

KURZFASSUNG: Auf Grünland wurden an Grünlandreifen, Breitreifen mit normaler und reduzierter Stollenhöhe, Niederquerschnittsreifen und konventionellen Ackerschlepper-Treibradreifen Zugkräfte gemessen. Direkt vor dem auf das Befahren folgenden Schnitt des
Grüngutes wurde die Abnahme des Ertrags in der Fahrspur im Vergleich zum Bestand zwischen den Spuren bestimmt.

Hierbei zeigten sich sowohl zur Erzielung hoher Triebkraftbeiwerte als auch zur
Vermeidung hoher Ertragsverluste Reifen vorteilhaft, die die tragfähige Grasnarbe wenig
schädigen. Breite Reifen mit niederem Innendruck und Profile mit breiten flachen und
weit überlappenden Stollen sind besonders geeignet.

ABSTRACT: On grassland the tractive performance of special grassland tyres, wide tyres
with normal and reduced lug height, and low-profile low-pressure tyres has been investigated in comparison with the tractive performance of conventional tractor drive tyres.
Before the next cut of the grass the difference in yield of the overrun area and the
area between the tracs has been measured.

To reach high tractive forces and low yield reduction tyres that don't damage the turf
have been advantageous. Wide tyres with low inflation pressure and profiles with wide
and flat lugs and overlapping patterns are most suitable.

RESUMEÉ: En prairie la force de traction pour des pneus differnts - pneu de tracteur
ordinaire, pneu spécial prairie, pneu large avec hauteur des crampons normale et
reduite, pneu avec surface transversale reduite - était mésurée et avant la prochaine
coupe du foin le rendement dans et entre les traces était comparé.

Les résultats montrent l'avantage des pneus qui causent peu de domage au racines parce
qu'ils réalisent des forces de propulsion elevées et évitent une dépression de rendement. Les pneus larges avec préssion basse et un profil avec des crampons larges, plats
et chevaùchants sont les plus favorables.

1 EINLEITUNG

Im Gegensatz źu Produkten des Ackerbaues,
die überwiegend vom Landwirt direkt ver-
kauft werden, läßt sich der Wert des
Ertrages von Grünland nur schlecht ein-
schätzen.

Schäden zeigen sich außerdem nicht nur
wie im Extrem durch fehlenden oder stark
verkümmerten Pflanzenbewuchs, sondern
hauptsächlich durch eine Änderung der

Pflanzenarten. Ertragreiche und wertvolle
Gräser werden von verschiedenen Kräutern
verdrängt.

Bild 1 wurde etwa 6 Wochen nach der
Befahrung aufgenommen. Die Fahrspur ist
noch deutlich am veränderten Pflanzenbe-
stand zu erkennen.

Dem Ertrag von Grünland wurde deshalb

Bild 1. Veränderung des Pflanzenbestandes durch Überfahren

Bild 2. Versuchsaufbau mit Zugschlepper und Bremsfahrzeug

lange Zeit nur untergeordnete Bedeutung zugemessen.

Durch steigende Maschinengewichte und -leistungen, intensive Grünlandnutzung und Umweltschutzauflagen bei der Unkrautbekämpfung treten diese Probleme vor allem bei ungünstigen Witterungsbedingungen zunehmend stärker zu Tage.

Die Anforderungen an Ackerschlepperreifen waren zunächst von der Forderung nach gutem Traktionsverhalten auf Acker bestimmt. Neue Entwicklungen beschäftigen sich mit der Verminderung des Kontaktflächendruckes in der Aufstandsfläche zwischen Rad und Boden. Einsatzgerechte Bereifung soll die Verdichtung der oberen Bodenschichten und die Bildung tiefer Spuren vermindern.

Zahlreiche Forschungsarbeiten zum Thema der Verdichtung sind in einer KTBL-Schrift und von Soane zusammengefa t. Untersuchun-

gen zur Auswirkung auf den Ertrag von Winterweizen wurden am NIAE von Chamen u.a. durchgeführt.

Gegenüber Ackerboden zeigt sich Grünland als weniger empfindlich gegenüber Verdichtung. Zum einen wird keine Bodenbearbeitung und -lockerung durchgeführt, die durch Überfahrungen rückgängig gemacht wird, zum andern ist Grünland bei intakter und geschlossener Grasnarbe tragfähiger als Ackerboden.

Erst in jüngster Zeit werden speziell auf Grünland zugeschnittene Reifen angeboten, die sich im konstruktiven Aufbau von konventionellen Ackerschlepper-Treibradreifen unterscheiden. Das Interesse an boden- und pflanzenschonenden Bereifungen ist stark gestiegen, weil damit Befahrungen termingerecht auch bei ungünstiger Witterung möglich sind.

Im Allgäu wurden mit solchen Reifen Versuche durchgeführt.

2 VERSUCHSDURCHFÜHRUNG

Der Aufbau der Versuche ist in Bild 2 gezeigt. Als Bremsfahrzeug war ein Unimog eingesetzt. Er wurde über ein Stahlseil mit Zugkraftme dose und Ruckdämpfer von einem konventionellen Ackerschlepper mit abgeschaltetem Vorderradantrieb gezogen. Die Schlepperhinterachse war mit den jeweiligen Versuchsrädern bereift. Zur Ermöglichung einer gro en Variationsbreite an Zugkraft- und Schlupfwerten waren beide Fahrzeuge ballastiert.

Um die Spuren des Zugfahrzeuges nicht mit dem Bremsfahrzeug zu überrollen, war die Me strecke von der Länge des Stahlseiles (etwa 12 m) begrenzt. Zur Ermittlung von Zugkraft - Schlupf - Kurven wurden zahlreiche Me fahrten durchgeführt, wobei der Fahrwiderstand des Bremsfahrzeuges variiert wurde. Verschiedene Wertepaare aus gemessenem Schlupf und gemessener Zugkraft ergaben sich entsprechend.

Vor der auf die Befahrung folgenden Nutzung wurde an einigen markierten Fahrspuren der Ertrag in Grünmasse als Mittel zwischen den Spuren von linkem und rechtem Rad bestimmt. Als Referenz diente jeweils der nicht überfahrene Streifen zwischen den Radspuren.

3 VERWENDETE REIFEN

Die Versuchsreifen wurden herstellerunabhängig nur nach ihren verschiedenen konstruktiven Merkmalen ausgewählt.

Breitreifen weisen gegenüber konventionellen Ackerschlepperreifen eine bis zu

100 % größere Breite auf. Die vergrößerte Aufstandsfläche vermindert die Verdichtung der oberen Bodenschichten. Durch das erhöhte Luftvolumen haben sie eine sehr hohe maximale Tragfähigkeit und können im praktischen Einsatz mit geringerem Luftdruck gefahren werden. Extrem breite Reifen (z.B. Terra-Reifen) wurden nicht untersucht. Sie überschreiten meist die für Straßenfahrt zulässige Breite des Schleppers und eignen sich wegen ihres hohen Preises nur für sehr große Betriebe mit empfindlichen Böden.

Grünlandreifen unterscheiden sich in ihren Abmessungen im Prinzip nicht von konventionellen Ackerschlepperreifen. Ihr Unterschied liegt in der Gestaltung des Profiles. Die Höhe der Stollen ist um etwa 50 % reduziert, um ein tiefes Einschneiden in die Grasnarbe zu verhindern. Der Abstand zwischen den Stollen ist geringer. Die Anzahl der Stollen am Radumfang kann erhöht werden, wodurch mehr Stollen in Kontakt zum Boden stehen und die Stollenaufstandsfläche vergrößert wird. Dies vermindert je nach Bodenzustand ebenfalls das Einsinken in die Grasnarbe und erhöht gleichzeitig die Laufleistung bei Straßenfahrt.

Eine Reduzierung der Stollenhöhe wird von verschiedenen Herstellern vorzugsweise für Breitreifen angeboten. Mit der gleichen Zielsetzung, die bei Grünlandreifen verfolgt wird, werden die Stollen abgefräst, um vermindertes Einschneiden in die Grasnarbe mit der Reduzierung des Bodendruckes zu verbinden. Zahlreiche der bislang erwähnten positiven Eigenschaften werden von Niederquerschnittsreifen vereinigt, die von skandinavischen Herstellern angeboten werden. Bild 3 zeigt einen entsprechend bereiften Schlepper im Grünlandeinsatz.

Wegen ihrer großen Breite können sie mit niederem Luftdruck betrieben werden und verursachen nur geringen Bodendruck. Für Grünlandeinsatz angebotene Profile haben nur geringe Stollenhöhe und sehr breite, weit überlappende Stollen mit runden Übergängen zur Karkasse und zur Reifenseitenwand, um die Grasnarbe nicht zu durchschneiden. Durch ihre sehr große Stollenaufstandsfläche sinken sie am wenigsten in die Grasnarbe ein.

Nachteilig ist, daß diese Reifenausführungen auch für Forst- und teilweise für Erdbewegungsmaschinen ausgelegt sind und mit sehr steifen Karkassen gefertigt werden. Wegen der geringen Stollenzwischenräume haben die Profile bei Verwendung auf Acker nur geringe Selbstreinigung.

4 VERSUCHSERGEBNISSE

Zum Verhalten der Reifen auf anderen Böden, besonders auf Acker, wurden keine Versuche durchgeführt. Aufschluß hierzu geben Arbeiten von Steinkampf, Jahns und Dwyer.

Das charakteristische Verhalten von Reifen auf Grünland wird vom Zusammenspiel von Grasnarbe und Stollen bestimmt. Bild 4 zeigt die Spur eines ziehenden Reifens bei geringer Zugkraft. Im abgebildeten Fall entstand zwischen Rad und Boden nur wenig Schlupf, die abgescherte Grasnarbe ist nur wenig entgegen der Fahrtrichtung (zur Bildunterkante) verschoben.

Deutlich sichtbar sind die Bereiche der Stollenaufstandsfläche. Hohe Stollen haben die Grasnarbe mit der oberen Bodenschicht und dem Hauptwurzelbereich abgeschert. Daraus resultieren die später beschriebenen größeren Ertragsabnahmen. Die tragfähige Schicht aus Grasbewuchs und Wurzelgeflecht kann zur Abstützung großer Kräfte nicht genutzt werden.

Bild 3. Schlepper mit boden- und pflanzenschonender Niederquerschnittsbereifung

Bild 4. Stollenabdruck eines hochstolligen Reifens bei Fahrt mit geringem Schlupf

Bild 5. Fahrspur eines niederstolligen Reifens

Im Unterschied hierzu greifen niedrigere Stollen weniger tief in den Wurzelbereich ein, Bild 5. Die Grasnarbe wird zwar ebenfalls beschädigt, vollkommen kahle Stellen entstehen jedoch nicht.

4.2 Zugkraftverhalten

Am wenigsten von anderen Faktoren beeinflußt zeigt sich die Auswirkung unterschiedlicher Stollenhöhen im direkten Vergleich der Breitreifen, die sich lediglich in der Stollenhöhe (Normalausführung 60 mm, reduziert 35 mm) unterscheiden. Durch die große Stollenaufstandsfläche zeigen sich Unterschiede erst bei größeren Schlupfwerten, wenn die Reifen unter-

schiedlich tief in die Grasnarbe eindringen. Geringere Stollenhöhe führt hier zu höherer Zugkraft.
Größere Unterschiede im gesamten Schlupfbereich ergeben sich beim Vergleich des Grünlandreifens und des konventionellen Ackerschlepperreifens, Bild 7.
Die Stollenhöhe und -anzahl zeigt stärkeren Einfluß auf das Treibkraftverhalten als die Reifenbreite.
In Bild 8 ist auch die Triebkraft des Niederdruckreifens dargestellt. Er erreicht schon bei geringem Schlupf höhere Zugkräfte als die Vergleichsreifen.

4.3 Ertragsmessungen

Die ermittelten Erträge wurden auf die Fläche der Fahrspur bezogen.
Grundsätzlich steigen die Verluste mit zunehmendem Schlupf an. Besonders deutlich

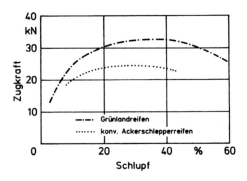

Bild 7. Zugkraft von Grünlandreifen und konventionellen Ackerschlepperreifen.

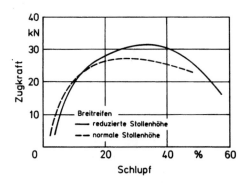

Bild 6. Zugkraft in Abhängigkeit vom Schlupf von Breitreifen mit normaler und reduzierter Stollenhöhe.

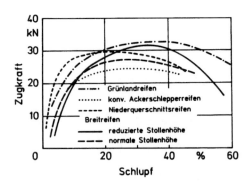

Bild 8. Zugkraftverhalten der untersuchten Reifen

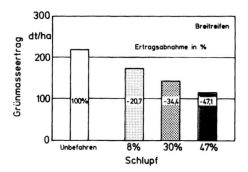

Bild 9. Ertragsabnahme bei verschiedenen Schlupfwerten des Breitreifens mit normaler Stollenhöhe

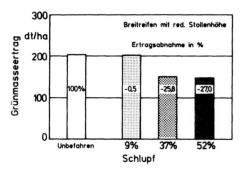

Bild 10. Ertragsabnahme in der Fahrspur des Breitreifens mit reduzierter Stollenhöhe

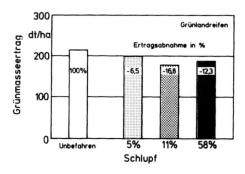

Bild 11. Ertragsabnahme des Grünlandreifens

wird dies am Beispiel des Breitreifens mit normaler Stollenhöhe. Der konventionelle Ackerschlepperreifen verhielt sich in der Tendenz ähnlich.

Eine deutliche Verbesserung wurde durch die Reduzierung der Stollenhöhe erreicht. Bei geringem Schlupf zeigt sich hier kein signifikanter Ertragsrückgang.

Nach dem Abscheren einer je nach Stollenhöhe unterschiedlich dicken Schicht von Grasnarbe und Boden bei mittleren Schlupfwerten steigen die Ertragsabnahmen auch bei noch steigendem Schlupf nicht weiter an.

Besonders verlustarm zeigte sich auch der Grünlandreifen.

Während bei allen bisher beschriebenen niederstolligen Reifen im Bereich geringen Schlupfes keine Ertragsabnahmen festgestellt werden konnten, zeigte sich beim Niederquerschnittsreifen sogar eine Erhöhung des Ertrages.

Dieser positive Effekt wird auf die günstige Auswirkung der Rückverdichtung der Grasnarbe zurückgeführt. Durch die Walzwirkung des Niederquerschnittsreifens wird ein guter Narbenschluß erzielt. Auch Jacob berichtet, daß bei häufigem Überfahren mit Reifen dieser Bauart eine homogene geschlossene Grasnarbe mit hohen Erträgen erzielt wird.

Erst bei höheren Schlupfwerten und abgescherter Grasnarbe verursacht dieser Reifen ähnlich hohe Verluste wie die Vergleichsreifen.

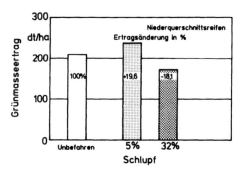

Bild 12. Ertragsveränderung bei unterschiedlichem Schlupf beim Überfahren mit Niederdruckreifen

REFERENCES

Bodenverdichtung. KTBL-Schrift 308, Vor-
träge der KTBL/FAL-Fachtagung "Bodenverdichtung beim Schlepper- und Maschinen-
einsatz und Möglichkeiten zu ihrer
Verminderung", Kuratorium für Technik
und Bauwesen in der Landwirtschaft,
1986.

Soane, B.D. Over-compaction of soils on
scottish farms: a survey. SIAE Research
Summary No. 3, 1987.

Chamen, W.C.T., Cope, R.E., Geikie, A.,
Stafford, J.V. A long-term experiment to
compare the effekt of different
tyre/soil contact pressures on soil and
crop responses when growing winter
wheat. NIAE Divisional Note 1263,
Silsoe, 1985.

Chamen, W.C.T., Chittney, E.T., Geikie,
A., Howse, K.R. The effect of different
tyre/soil contact pressures on soil and
crop responses when growing winter
wheat: year 2 - 1983 -84. NIAE Divisional Note 1296, Silsoe, 1985.

Steinkampf, H. Betriebseigenschaften von
Ackerschlepperreifen bei unterschiedlichen Einsatzbedingungen. Landbauforschung Völkenrode, Sonderheft 80,
1986.

Jahns, G. Numerische Beschreibung der
Betriebseigenschaften von Acker-
schlepperreifen. Landbauforschung
Völkenrode, Sonderheft 80, 1986.

Dwyer, M.J. Somer aspects of tyre design
and their effect on agricultural tractor
performance. NIAE Bericht 204/75.
Silsoe, 1975.

Dwyer, M.J., Evernden, D.W., McAllister,
M. Handbook of agricultural tyre performance, NIAE Report No. 18, Silsoe, 1976.

Dwyer, M.J. The tractive performance of a
wide, low-pressure tyre compared with
conventional tractor drive tyres.
Journal of Terramechanics, Vol. 24, No.
3, pp. 227-234, 1987.

Jacob, H. Grünland schonend befahren.
Württembergisches Wochenblatt für die
Landwirtschaft WWL 12/88, Seite 24-29.

Land and Water Use, Dodd & Grace (eds), © 1989 Balkema, Rotterdam. ISBN 90 6191 980 0

Pénétrométrie, choix des outils et dates de travail du sol

J.F.Billot
Cemagref, Antony, France

RESUME: Parmi les matériel de mesure relatifs au sol, on trouve maintenant disponibles dans le commerce, des pénétromètres enregistreurs simples à des prix modiques. Ces matériels, rustiques, sont adaptés aux besoins des agriculteurs et de leurs conseillers. Peu encombrants, disponibles dans la cabine du tracteur, ils peuvent être mis en oeuvre très rapidement pour tester l'état du sol dans une parcelle.

Les données recueillies sont d'ordre quantitatif aussi bien que qualitatif: valeur de la dureté du sol et forme des courbes obtenues. Les nombreuses courbes de résistance à la pénétration en fonction de la profondeur du sol informent l'agriculteur sur les variations d'état du sol à diverses profondeurs: humidité relative des diverses couches, zones tassées, discontinuités, semelles, etc... Au fil des campagnes, en confrontant ces données avec le comportement du matériel, l'agriculteur peut acquérir une expérience qui le rend rapidement apte à interpréter les courbes pénétrométriques pour choisir les outils et les périodes favorables à tous les types d'interventions dans une parcelle.

ABSTRACT: Use of penetrometers; how to choose tillage dates and implements.

Many soil measuring instruments are available on the market. Among them, simple and cheap recording penetrometers can now be found. Those penetrometers are well suited to the needs of both farmers and agricultural advisers. They can be implemented very quickly for testing the soil state in a field as they are not cumbersome and can be at hand either in the tractor cab.

Data obtained are quantitative as well as qualitative ones: i.e. soil firmness value and shape of curves. Thus, the farmer, through the numerous soil strength curves as a function of soil depth, is well informed on the soil state variations at various depths: relative moisture content of various layers, compacted areas, discontinuities, pans. When comparing the penetrometer data with the equipement behaviour, the farmer, after many uses of the penetrometer at different periods, has a sufficient background to analyse the penetrometer curves. Therefore, the farmer is in a better position to select the implements and periods best suited to the tillage operations to be carried out on a field.

ZUSAMMENFASSUNG: Penetrometer, Wahl der Geräte und der Bodenbearbeitungstermine.

Was die Messgeräte für den Boden betrifft, verfügt man heute auf dem Markt zu ermässigtem Preis über einfache Registrierpenetrometer. Ein solches widerstandsfähiges Material entspricht den Bedürfnissen der Landwirte und ihrer Berater.Man kann sie sehr schnell in Betrieb setzen, um den Bodenzustand einer Parzelle zu prüfen, weil sie raumsparend verfügbar sind und zwar in der Schlepperkabine.

Die angegebenen Ergebnisse sind sowohl quantitativ als auch qualitativ Bodenfestigkeitswert und Verlauf der erhaltenen Kurven, die zahlreichen Kurven, die den Eindringungswiderstand in Abhängigkeit von der Eindringungstiefe darstellen benachrichtigen den Landwirt über die Variationen des Bodenzustandes an verschiedenen Bodentiefen: verhältnismässige Feuchtigkeit der verschiedenen Schichten, verdichtete Zonen, Diskontinuitäten, Pflugsohlen. Indem er während der Feldarbeitsgänge diese Daten mit denen des Verhaltens seiner Geräte vergleicht, wird der Landwirt imstande sein, die penetrometrischen Kurven leicht auszuwerten, und das im Hinblick auf die Wahl der Geräte und der günstigsten Perioden für allerlei Arbeitsgänge auf der Parzelle.

1 POSITION DU PROBLEME

Une exploitation agricole est constituée d'unités économiques et morphologiques considérées comme homogènes: les parcelles.

L'agriculteur, doit tenir compte de leurs caractéristiques pédo-climatiques si il veut gérer de manière rationnelle ses interventions dans les champs. Il en est de même au moment du choix des aménagements et des équipements.

1.1 Connaissances empiriques

L'agriculteur dispose de données relatives aux parcelles de son exploitation. Tout d'abord, si il est depuis longtemps sur l'exploitation, celles qu'il a collectées au cours d'années de pratique personnelle ou familiale. Cependant, avec l'avénement de la motorisation, l'agriculteur ne suit plus à pied le travail de la charrue, observant de près le comportement du sol sous l'action de l'outil. Il y a d'autre part de plus en plus de mobilité géographique chez les agriculteurs qui doivent adapter leur pratique à de nouvelles terres. Et sur une même exploitation les systèmes culturaux sont appelés à évoluer en fonction des contraintes économiques. Aussi, ce type de connaissance empirique du sol acquise au fil des ans, et relative à des outils et à des systèmes culturaux qui évoluaient peu, a-t-il tendance à disparaître.

1.2 Données cartographiques

Parfois il dispose aussi de cartes précisant le comportement mécanique et hydrologique d'un sol donné, à partir des observations morphologiques réalisées par un expert, complétées par enquête pour tenir compte de l'expérience des agriculteurs de la région. Les cartes, à base pédologique, actuellement disponibles sont principalement utilisées pour des études à caractère régional, les agriculteurs peuvent en tirer des informations générales sur la forme des excès d'eau et l'intensité globale de l'hydromorphie. Cependant la majorité de ces cartes n'ont pas une échelle et une précision suffisantes pour fournir à l'agriculteur des données utilisables au niveau de la parcelle.

On fonde de grands espoirs sur l'utilisation des images satellites. Des tentatives ont été réalisées, avec un certain succès, pour relier les données enregistrées (luminance) aux caractéristiques pédologiques du sol. Ces travaux devraient être poursuivis, ils permettraient d'atteindre la précision recherchée dans les cartes des sols, la définition maximum des images SPOT1 étant de 10m. Mais il ne semble pas que ces images, ne serait-ce qu'en raison de leur coût, puissent être utilisées directement par les agriculteurs pour la gestion des parcelles. Par contre des photos aériennes effectuées à faible altitude à des moments bien choisis (périodes critiques de végétation) sont toujours riches d'enseignements sur la potentialité relative interparcellaire et intraparcellaire.

1.3 Praticabilité de la parcelle

Les informations dont dispose l'agriculteur, se présentent donc de manière éparse, et souvent à un stade non encore directement opérationnel à l'échelle de la parcelle. Enfin elles sont relatives à des types de sols plutôt qu'à des parcelles définies. Aussi a-t-on tenté de développer ces dernières années une logique qui permettraient de relier les données globales facilement accessibles (données cartographiques et météorologiques) à des caractéristiques propres à la "praticabilité" de la parcelle considérée. La praticabilité étant définie comme la caractéristique de l'intéraction entre le climat et le terrain qui s'évalue par "la facilité à conditionner correctement le peuplement cultivé pour un coût de mise en oeuvre donné". (Papy, Inra)

A partir de ces connaissances et vis à vis de la praticabilité, l'agriculteur se trouve confronté, dans l'exercice de sa profession, à deux types de décisions:

1. des décisions "tactiques", à prendre au jour le jour, en fonction des circonstances climatiques et de l'état du sol à un moment donné,

2. des décisions "stratégiques", qui concernent l'aménagement d'une parcelle ou le choix d'un itinéraire technique.

Nous allons aborder successivement ces deux types de démarches en analysant dans quelle mesure des tests mécaniques peuvent apporter une meilleure approche de la réalité du comportement du sol.

2 DECISIONS STRATEGIQUES

2.1 Etablissement de modèles de comportement du sol

La démarche généralement adoptée par les chercheurs consiste à élaborer des modèles de comportement du sol de la parcelle permettant de définir les "jours disponibles". C'est à dire les périodes optimales pour l'exécution des principales opérations culturales des itinéraires techniques choisis, dans le cadre du système de culture pratiqué.

2.1.1 Modèles de comportement hydrique

On commence en général par élaborer un modèle de comportement hydrique du sol en vue d'établir dans un second temps une relation entre des séquences climatiques et des séquences de données caractérisant la praticabilité du terrain. Une fois le modèle validé, on le fait fonctionner à partir des données climatiques journalières emmagasinées au cours des années, et on reconstitue ainsi année par année la séquence journalière des états de praticabilité du sol. On a ainsi une idée assez juste de la fréquence d'utilisation possible de tel passage d'outil ou de tel itinéraire technique.

Le paramètre de sortie fourni par le modèle hydrique est généralement l'humidité moyenne de l'horizon travaillé. Il s'agit alors de déterminer les humidités "seuils" qui marqueront les limites de changement de praticabilité de la parcelle. La praticabilité, quant-à-elle, dépend des matériels utilisés pour les opérations culturales et intègre les notions de portance du terrain et de "travaillabilité". C'est pourquoi, plutôt que de relier directement humidité du sol et praticabilité de la parcelle , on préfère souvent procéder en deux phases:

1. mise en évidence des relations entre l'humidité et un test standardisé destiné à évaluer les qualités mécaniques du sol,

2. mise en relation des résultats du test avec la praticabilité.

2.1.2 Emploi et choix d'un test mécanique complémentaire

L'action sur le sol, des outils et des roues, étant d'ordre mécanique, l'introduction d'un test mécanique semble indispensable. Tout d'abord, il est souvent plus facile à mettre en oeuvre aux champs un test simple et bien choisi, que de mesurer l'humidité elle même, ensuite il permet d'évaluer la praticabilité par une mesure, c'est à dire de façon relativement indépendant du jugement plus ou moins fidèle et subjectif d'un observateur.

Choix d'un test mécanique performant:

Le test choisi doit être sensible aux propriétés mécaniques des sols,et compte tenu du rôle joué par la structure du sol dans ces propriétés, on devra pouvoir le mettre en pratique dans les champs et non pas en laboratoire, sur des échantillons de sol remaniés. Parmi les tests mécaniques existants, le plus simple à mettre en oeuvre est la mesure de la résistance à la pénétration d'une pointe normalisée.

Les pénétromètres n'ont jamais bénéficiés en France d'un préjugé favorable. On leur reproche principalement de mesurer une donnée synthétique liée à la fois à l'humidité et à la porosité du sol. Ce test est pourtant de plus en plus utilisé, car la mesure pratiquée, simple et rapide, peut être effectuée un grand nombre de fois dans la parcelle pour rendre compte de sa variabilité. Divers types de matériels peuvent être utilisés et il est toujours possible d'établir des relations entre eux.

Les pénétromètre statiques enregistrent en continu ou point par point la valeur locale, en fonction de la profondeur, de la résistance à la pénétration d'une pointe conique, sur laquelle on exerce une pression continue. Les pénétromètres dynamiques s'enfoncent dans le sol sous l'effet des chocs successifs d'une masselotte tombant d'une hauteur déterminée. On considère alors que la résistance du sol à la pénétration est proportionnelle au rapport du nombre de chocs appliqués, à la profondeur d'enfoncement.

2.2 Exemple d'établissement d'un modèle de comportement à partir de tests pénétrométriques

2.2.1 Matériel et méthodologie utilisés

Ph. Collas (Cemagref Bordeaux) a utilisé un pénétromètre dynamique sur sols argileux et sur sols limoneux pour définir les valeurs de la résistance à la pénétration qui peuvent être considérées comme des seuils de comportement.

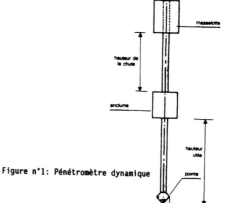

Figure n°1: Pénétromètre dynamique

Lors de ces mesures il a pris certaines précautions méthodologiques.Tout d'abord les valeurs globales de résistance à la pénétration étant difficiles à interpréter si la structure du sol est susceptible d'évoluer dans le temps, les mesures ont été effectuées uniquement sur des sols où la dernière intervention culturale datait d'au moins trois mois. Ensuite, chaque mesure pénétrométrique a été couplée avec la mesure d'humidité de l'horizon correspondant. Enfin, on a cherché à multiplier les mesures durant les périodes critiques pour les interventions dans les champs: deux fois par semaine avec cinq répétitions.

Il a ainsi pu établir la relation entre l'humidité pondérale W et la résistance à la pénétration RP exprimée en mégapascals (MPa). Elle peut être représentée par une relation de type puissance:

$$\text{-en sol argileux: } RP = e^{10,7} . W^{-3,3}$$

$$\text{-en sol limoneux: } RP = 43000 . W^{-3,8}$$

2.2.2 Etablissement d'un test pénétrométrique de praticabilité

L'humidité du sol lui confère un certain comportement mécanique, évalué par la mesure de la résistance à la pénétration, qu'il convient alors de traduire en termes de praticabilité. Un moyen simple et rapide de juger de la praticabilité d'une parcelle est de s'en remettre au diagnostic de l'agriculteur considéré comme expert sur sa propre exploitation. Le diagnostic ainsi demandé doit être exprimé en termes suffisamment simples: sol impraticable, portant, sol "travaillable". C'est aussi l'occasion de formaliser et de sauvegarder la connaissance empirique des sols encore détenue par certains agriculteurs pour la mettre à la disposition des autres agriculteurs par le biais de ces modèles de comportement.

Lors de ces déterminations il faut prendre garde que les experts-agriculteurs n'intègrent dans leur jugement de la parcelle la nécessité qu'ils avaient d'intervenir compte tenu de leurs habitudes de travail ou de leurs contraintes d'exploitation.

On constate que les résultats apparaissent semblables quelle que soit la texture du sol. La limite entre l'état impraticable et l'état portant se situe vers RP = 0,2MPa et la limite entre l'état portant et l'état travaillable vers RP = 0,3MPa.

On constate aussi que le recours systématique à des mesure pénétrométriques parallèlement à une pratique d'expert permet de rendre ce jugement plus fidèle.

2.2.3 Applications des modèles

Une fois ces relations établies entre d'une part l'humidité et la résistance à la pénétration RP, d'autre part RP et la praticabilité, on peut relier l'humidité du sol à la praticabilité. Toute séquence journalière de valeur d'humidité de la couche superficielle du sol, peut alors être traduite en une séquence d'états de praticabilité.

En définissant le comportement d'une parcelle comme celui d'un type particulier de sol pour lequel ces relations auraient été préalablement établies, on peut ainsi en déduire des informations utiles pour des prises de décision en vue d'aménagements (jugement de l'intérêt du drainage) ou de choix d'itinéraire technique.

A partir de données pénétrométriques
En ne prenant en compte que les valeurs de "RP" au champ, on peut montrer l'intérêt du drainage dans une parcelle limoneuse en comparant les jours disponibles pour un labour de printemps, sur sol limoneux en condition drainée et non drainée. On constate sur la figure n°2, en choisissant une valeur seuil de 0,3 MPa, que le sol n'a jamais été praticable pendant la période observée dans la parcelle non drainée, alors qu'il l'a été pendant 60% du temps dans la parcelle drainée.

A partir de données météorologiques
On peut aussi simuler des séquences d'état de praticabilité à partir de séquences climatiques. En effectuant pour chaque jour le décompte des observations de chacun des trois états de praticabilité (impraticable, portant, travaillable) et en reportant parallèlement les périodes d'exécution des principales opérations culturales de l'itinéraire technique choisi.

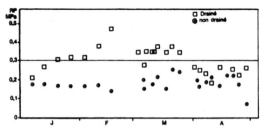

Figure n°2: Mesure de la résistance à la pénétration en sol limoneux, dans une parcelle drainée et non drainée

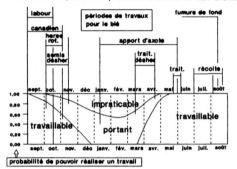

Figure n°3: calendrier de la praticabilité d'un sol très argileux.(Ph. Collas)

On y voit qu'aux alentours du 1er avril dans ces sols argileux, il y a 2 chances sur 10 d'être en situation impraticable, 5 chances sur 10 en situation portante et 4 chances sur 10 en situation travaillable.

A l'issue de cette expérimentation, qui s'est déroulée sur plusieurs campagnes agricoles, dans divers types de sols, et qui a mis en oeuvre le savoir faire de divers experts, on constate que la crédibilité d'une démarche visant à quantifier les jours disponibles pour les opérations culturales, repose principalement sur la fiabilité des diagnostics portés sur la praticabilité des sols. Il semble que cette fiabilité ne puisse être assurée que par la mise en oeuvre de mesures physiques à la parcelle, dont les résultats sont confrontés aux jugements des agriculteurs-experts, ces mesures pratiquées systématiquement permettent de fidèliser le jugement des experts en les contraignant à ne tenir compte que de critères objectifs.

3 DECISIONS TACTIQUES

Chaque intervention dans une parcelle a des effets positifs et négatifs. L'agriculteur doit pouvoir optimiser ces interventions à partir de critères de jugement sûrs, dépourvus de tout caractère subjectif.

Il convient de distinguer les opérations culturales selon les exigences à satisfaire, quant à leur date de réalisation et quant à l'état du profil de sol qui en résulte. On peut distinguer les interventions préalables à l'implantation de la culture qui tendent à créer un environnement favorable, et celles qui sont menées dans les cultures en place.

3.1 Interventions hors culture

Pour les façons allant de la récolte du précédent au semis (déchaumage, épandage, labour et travail profond, reprises de labour, semis), les bornes de dates de réalisation souhaitables sont plus ou moins étroites. Elles dépendent de la culture, de la variété, de la date de libération de la parcelle par le précédent cultural, mais aussi des contraintes liées au comportement propre du terrain. Ainsi pour les façons de travail du sol qui visent à créer un état structural du profil favorable au démarrage et au développement de la culture, le résultat des passages est très fortement conditionné par l'état hydrique du sol et la qualité du profil de départ.

3.1.1 Portance

Il convient tout d'abord que le tracteur puisse circuler de manière efficace dans la parcelle, sans laisser de profondes ornières mal comblées par l'action des outils. Il faut donc que le sol soit suffisamment portant c'est à dire qu'il résiste à la fois à la pression verticale des pneumatiques et à l'action de décohésion superficielle de leurs crampons.

3.1.2 Travaillabilité

Les conditions de travaillabilité, c'est à dire de modification positive de la structure du profil, ne peuvent être jugées qu'a posteriori sur la qualité du profil obtenu. Mais il est tout de même souhaitable de pouvoir prévoir l'effet sur le sol de tel outil ou de la circulation de tel engin. Chaque fois qu'il y a passage d'un ensemble tracteur-outil les roues du tracteur compactent le sol tandisque les outils ont des effets complexes de compactage et de fragmentation.

Compactage par les roues
Un passage de tracteur, détermine une contrainte appliquée au niveau de la surface du sol. L'épaisseur de la zone tassée augmente avec l'humidité du sol.
Le tassement affecte en priorité la porosité intermottière correspondant à l'assemblage des mottes du profil, puis la porosité interne des mottes. A même humidité du sol au moment du compactage, le résultat final dépend beaucoup de l'état structural initial du profil. Ainsi lorsque les mottes ont une structure interne très tassée, on aboutit à un état final très compact sur toute l'épaisseur de l'horizon labouré dès que la porosité intermottière a disparu.

Action des outils
Selon l'état d'humidité du sol, un même disque l'émiettera ou y découpera des copeaux, une dent fissurera les mottes ou remontera des "lards", un rouleau rappuiera un horizon trop creux ou se remplira de terre. De même, une dent passée en profondeur avec pour objectif de disloquer, ou au moins de fissurer une semelle de labour ne pratiquera qu'une saignée si le sous sol est humide.
Pour ces interventions l'état d'humidité du sol est donc très contraignant, il conditionne la réussite et, en cas d'erreur de jugement, peut pénaliser la culture. La pénalité n'est d'ailleurs pas la même dans toutes les conditions pédo-climatiques; il est des sols à fort taux de matière organique qui "pardonnent" beaucoup d'erreurs, d'autres à faible stabilité de structure ne retrouvent que lentement leur porosité initiale après un tassement trop énergique ou un malaxage trop intense de la couche travaillée.

3.2 Interventions sur des cultures en place

Les interventions en cours de culture (épandage d'azote et de raccourcisseur, traitement de desherbage, pulvérisation de pesticides), doivent être souvent exécutées dans un créneau défini par le développement de l'espèce cultivée, des plantes adventices, des maladies et des prédateurs. Dans ce cas, le critère de jugement principal sera relatif à la portance du sol au stade favorable: il faut au moins pouvoir passer. A forte humidité le passage des tracteurs provoque un tassement des horizons superficiels du profil, et même des horizons plus profonds si la charge est forte. Mais cet inconvénient est secondaire par rapport à l'intérêt d'intervenir dans un créneau de dates bien défini, soit que les roues n'affectent qu'une petite partie de la parcelle (traitements avec des matériels équipés de rampes de grande largeur), soit que le sol, une fois tassé, peut être retravaillé par la suite (cas des récoltes).

3.3 Indicateurs de l'état du terrain

L'agriculteur doit donc avoir la connaissance la plus juste possible des conditions mécaniques et hydriques de la parcelle, sur toute son étendue (présence de mouillères) et aussi selon la profondeur du sol.

3.3.1 Indicateurs visuels et tactiles

Deux types d'indicateurs sont utilisés actuellement, ils renseignent directement sur l'état hydrique du sol ou sur son comportement mécanique.

Ainsi le changement de couleur de la surface du sol constitue, moyennant certaines précautions, un indicateur performant de l'état hydrique de la couche travaillée. On doit tout d'abord pouvoir estimer (ne serait-ce qu'approximativement) l'état hydrique en profondeur; c'est le cas en sortie d'hiver où sauf dans les cas de sols que l'on sait engorgés, on peut admettre que les couches profondes sont à la capacité au champ. Il faut ensuite que l'indicateur soit référencé à une texture donnée, puisqu'elle détermine fortement les profils de dessication. Papy (INRA) a ainsi pu montrer que dans des sols limono-argileux du nord du Bassin Parisien, on peut pronostiquer de très bonnes conditions de reprise d'un labour fragmenté au moment où la totalité de la surface vient de virer au clair.

Les agriculteurs estiment aussi la portance du terrain en marchant dans le champ, la cohésion du sol en donnant des coups de pied dans les mottes, et sa plasticité d'après le comportement entre les doigts d'échantillons prélevés. Toutes ces pratiques sont autant de tests de comportement mécanique qui intègrent à la fois des effets de l'état structural et de l'état hydrique.

3.3.2 Utilisation de pénétromètres

Les performances de ces divers indicateurs dépendent a priori d'une bonne connaissance des variations intra-parcellaires. Les parcelles agricoles présentent souvent des hétérogénéités; en surface elles se traduisent par des taches plus ou moins claires ou sombres sur les sols nus. C'est particulièrement apparent sur les parcelles situées en coteaux, ou encore sur les vues aériennes. Ces taches témoignent des variations latérales des horizons pédologiques de surface ou profonds qui se traduisent par des changements de texture et entraînent des comportements physiques particuliers. L'agriculteur qui cultive son champ de manière homogène, crée alors dans la même parcelle lors des opérations de travail du sol des profils d'états structuraux et de comportement hydrique différents.

Il serait souhaitable que l'agriculteur dispose d'un moyen d'exploration fine de la parcelle; il pourrait alors, en tenant compte des hétérogénéités, utiliser son matériel agricole au mieux, par exemple en modifiant les réglages selon les zones. L'utilisation de pénétromètre semble devoir répondre à ce besoin.

4 LES PENETROMETRES, OPERATIONNELS POUR LES AGRICULTEURS

La pénétrométrie, employée, non plus uniquement pour obtenir un indice global relatif au sol, mais comme source d'informations sur la variabilité spatiale et verticale du terrain, peut contribuer à cette connaissance indispensable de la parcelle tout en apportant des informations complémentaires sur le comportement potentiel du sol en profondeur.

4.1 Description du matériel

A cette fin on utilisera de préférence des pénétromètres statiques enregistreurs, matériels plus facile à mettre en oeuvre que les modèles à percussion. On trouve maintenant disponibles dans le commerce (Agro-Systèmes, Nardeux) à des prix modiques, des matériels rustiques bien adaptés aux besoins des agriculteurs et de leurs conseillers. Peu encombrants, toujours disponibles dans la cabine du tracteur ou du véhicule utilisé lors du "tour de plaine", ils peuvent être mis en oeuvre très rapidement pour tester l'état du sol dans une parcelle.

Les variation de résistance à la pénétration en fonction de la profondeur sont tracées sur une feuille de papier au fur et à mesure que l'on enfonce la pointe dans le sol. La mesure ne nécessite qu'un seul opérateur.

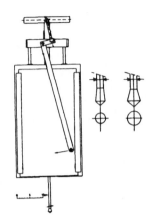

Figure n°4: Pénétromètre statique enregistreur manuel

Sur le terrain on peut utiliser la même feuille pour différentes mesures en changeant la couleur des crayons pour individualiser les courbes et faciliter le dépouillement. Selon la dureté du sol, on peut choisir entre deux pointes de 1cm^2 et 2cm^2 de section pour obtenir une courbe d'amplitude élevée et avoir ainsi le maximum de rendu des variations de dureté du sol. Le choix d'une pointe de 90° d'angle au sommet, donne plus de précision dans l'appréciation de la localisation en profondeur des discontinuités verticales du profil.

4.2 Mode opératoire:

L'agriculteur tout en circulant dans le champ (test de portance et de cohésion des mottes) effectue de nombreux relevés au pénétromètre d'abord à intervalles réguliers pour faire connaissance méthodiquement avec la parcelle, puis par coups de sonde pour confirmer (ou infirmer des hypothèses découlant de l'aspect de surface) dans des zones choisies. Il relève ainsi de nombreuses courbes pénétrométriques qu'il peut classer selon leur forme et regrouper par familles.

Figure n°6: Divers types de courbes pénétrométriques

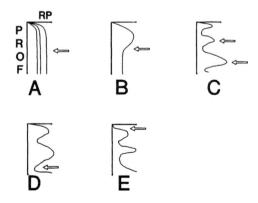

1734

A: Courbe plate et verticale. La résistance est homogène sur toute la couche arable. Selon l'intensité de la RP on a affaire à une région tassée, à un horizon labouré entièrement repris en masse, ou à une préparation du sol particulièrement homogène.

B: Courbe présentant un ventre allant en s'ammenuisant avec la profondeur. Il s'agit d'un tassement consécutif à un passage de roue dont l'influence s'exerce, avec une intensité décroissante, sur toute la couche arable.

C: La courbe présente de nombreux pics dans la partie supérieure et un brusque accroissement en fond de travail. Sur ce profil la structure de l'horizon travaillé est très motteuse et le fond de labour est très tassé (semelle).

D: La courbe reflète une nette atténuation de la RP dans la partie surmontant le fond de travail. Plusieurs interprétations sont possibles: ou bien le sol y est creux (végétation enfouie en fond de labour) ou encore, si cette particularité se retrouve sur toutes les courbes, il s'agit d'une zone humide en profondeur située juste au dessus d'une semelle imperméable.

E: La courbe présente un replat proche de la surface. Le profil accuse une nette discontinuité provoquée par un travail du sol superficiel. Ce type de profil traduit en général la création d'une semelle superficielle ou du moins d'une zone lissée proche de la surface.

4.3 Interprétation:

Ces nombreuses courbes de résistance à la pénétration en fonction de la profondeur du sol informent l'agriculteur sur les variations d'état du sol à diverses profondeurs: humidité relative des diverses couches, zones tassées, discontinuités, semelles, etc. Elles viennent compléter par des données enregistrées, auxquelles on peut se référer par la suite les informations déja apportées par les observations et tests mécaniques de surface.

A partir des diverses formes de ces courbes, de leur fréquence relative et de leur répartition sur l'ensemble de la parcelle on peut connaître le pourcentage de tel ou tel type de structure dans le champs.

Au dela de ces informations d'ordre statistique très importantes pour bien appréhender la variabilité de la parcelle, les courbes apportent des données d'ordre quantitatif aussi bien que qualitatif. La valeur de la dureté du sol informe sur son degrès de portance. Et comme nous l'avons vu, la forme des courbes obtenues renseigne sur la structure du profil. Il est alors fortement conseillé de compléter les informations partielles déduites de la forme des courbes par l'observation directe du profil cultural obtenu en faisant un trou à la bêche. C'est la seule façon d'apprécier les problèmes liés à la structure du sol et leur retentissement sur le développement de la culture.

S'il se livre souvent à ce type de mesures et d'observations l'agriculteur acquerra au fil des années une connaissance très complète du comportement du sol des différentes parcelles de l'exploitation. Pratiquant ainsi un véritable étalonnage qualitatif du pénétromètre il pourra mieux faire la relation entre l'allure des courbes enregistrées et des états particuliers du sol considéré. Il sera alors en mesure d'apprécier les risques encourrus et de décider du bien fondé d'une intervention envisagée.

4.4 Aide au diagnostic et à l'expérimentation

Les pénétromètres peuvent aussi être utilisés par les conseillers agricoles soit en cours d'expérimentation, comme appareils de mesure pour comparer l'action de divers matériels de travail du sol, soit comme aide au diagnostic sur une exploitation. Les matériels utilisés sont le plus souvent des pénétromètre enregistreurs statiques comme celui présenté sur la figure 4. Pour une analyse plus fine on pourra aussi avoir recours au pénétromètre électronique automatique, matériel de mesure mis au point par le Cemagref pour visualiser la structure du sol en place.

Figure n°7 Profil pénétromètrique: Retassement du labour et semelle épaisse et continue.

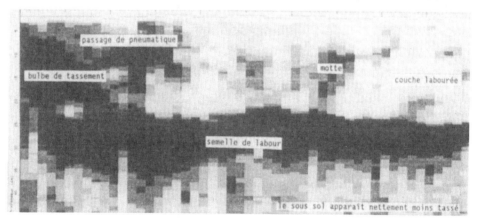

Figure n°7
Etat de la surface: Chaume de blé.
Nature du sol: Limon moyen sur limon argileux vers 40 cm de profondeur.Zone fréquemment humide, partie basse à colluvionnements limoneux. Décompactée trois ans auparavant. Nature du précédent cultural:Betterave.

On procéde dans un premier temps à une exploration rapide de la parcelle au moyen de sondages au pénétromètre manuel. Ainsi dans des parcelles où les agriculteurs rencontraient des problèmes de compactage du sol, l'examen rapide des courbes obtenues nous renseigne sur la fréquence, le type et la localisation géographique des diverses "configurations de tassement". On peut alors choisir en toute connaissance de cause les emplacements les plus caractéristiques où effectuer les observations de profils culturaux et les mesures au pénétromètre électronique, pour rendre compte au mieux du phénomène à observer.

Le conseiller s'appuiera ensuite sur ces observations et ces mesures pour en tirer des diagnotics sur les nuisances occasionnées aux cultures et formuler des conseils d'intervention tenant compte des propriétées du sol considéré:

1. dans certains cas on peut proposer le passage d'un décompacteur (sur toute la surface et dans le sens de la pente en s'assurant systématiquement des possibilités d'écoulement des eaux issues des raies de décompactage ou encore uniquement sur les ornières laissées par les passages d'appareils de traitement),

2. dans d'autres les recommandations sont d'ordre agronomique (implantation d'engrais vert, drainage de la parcelle) ou concernent les pratiques culturales (type de labour, sens de passage des outils).

CONCLUSION

Il semble donc que des tests mécaniques simples reposant sur des mesures pénétromètriques aux champs puissent venir,en complément d'autres données, renforcer et rationnaliser la connaissance du comportement des sols agricoles vis à vis de leur praticabilité. On pourrait les utiliser au cours de l'établissement de modèles de comportement des divers types de sols agricoles et notamment lors de la conversion en termes de praticabilité des modèles de comportement hydriques établis pour ces sols. Les agriculteurs pourraient aussi en tirer de précieuses informations qui les aideraient pour les prises de décisions à caractère tactique.

La démarche générale que l'agriculteur devrait suivre nous semble être la suivante:

Tout d'abord il identifie et caractérise les divers types de sols de son exploitation en s'appuyant sur les informations cartographiques et sur la connaissance empirique de ses terres complétée par des résultats d'analyses.Il affecte alors à chacune des parcelles de son exploitation les caractéristiques du type de sol qui y domine, et s'il y a lieu, quand cela est possible, il modifie le tracé des parcelles de manière à les rendre plus homogènes.

Employant ensuite les modèles de comportement relatifs à ce type de sol, établis par les chercheurs, il choisit les itinéraires techniques les plus adaptés au système de culture pratiqué. Il définit ainsi les équipements et les aménagements les mieux adaptés.

Il lui reste alors à employer au mieux son matériel. Il devra se servir des modèles précedemment établis, pour définir en fonction des données météorologiques journalières (pluviométrie, ETP) les moments les plus aptes aux interventions. Il adaptera ensuite ces données à la réalité de chaque parcelle par le biais d'indicateurs visuels ou mécaniques. Les données pénétromètriques paraissent constituer,dans cette phase de la démarche, des indicateurs d'ordre mécanique particulièrement performants

On peut aussi penser que dans un avenir assez proche, ces mesures pénétromètriques, qui rendent compte de la variabilité de la nature et de l'état du sol de la parcelle pourront être utilisées pour alimenter en données les logiciels, qui géreront d'ici peu le fonctionnement du tracteur et les modifications de réglage des outils. On disposerait ainsi d'un traitement spécifique aux différentes zones homogènes de la parcelle avant que, dans une phase ultérieure, on utilise des matériels de travail du sol "intelligents" qui modifieront automatiquement leur action sur le sol en fonction de données sur l'état du sol fournies par des capteurs.

Références:

Agro Systèmes. 1988. Le compactomètre Agro-Profil.Documentation commerciale. ZI des Gaudière BP57 37390 Mettray France.

Billot, J.F. & D.Méchineau 1988. Le pénétromètre automatique:un appareil de mesure des relations sol/climat/machine. Antony France: CEMAGREF BTMEA.32:24-32.

Collas, P. & J.R.Thiercelin 1988. Quantification des effets de l'excès d'eau en termes de jours disponibles. Paris:Perspectives Agricoles.126:31-36.

Courault, D. & M.C.Girard 1988. Analyse des sols à l'échelle de la parcelle par télédétection. Toulouse France:CNES SPOT1 Utilisation des images,bilan, résultats.

Dalleinne, E. 1974. Le pénétromètre,outil indispensable pour juger un ensemble de façons culturales à l'échelle d'une parcelle.Antony France:CNEEMA BI.202:41-44.

Nardeux Humisol. 1989. Documentation commerciale. 11 rue des Granges Galand BP 212 37552 St.Avertin Cedex France.

Papy, F. 1988. Problèmes de praticabilité dûs à l'excès d'eau: conséquences sur la conduite des cultures de blé et de maïs. Paris:Perspectives Agricoles.126:48-54.

Land and Water Use, Dodd & Grace (eds), © 1989 Balkema, Rotterdam. ISBN 90 6191 980 0

Field capacitive performance of track type tractors in plough operations under tropical conditions

F.G.Braide
Ahmadu Bello University, Zaria, Nigeria

ABSTRACT: The mechanization of land preparation using tractors is gaining popularity in Nigeria. Small scale farmers especially civil servants with less than 5 hectares who farm for pleasure and supplement the family income, prefer to hire tractors. State owned organizations and few individuals with more than 100ha own tractors. A time study of track type tractors under irrigation and rainfed conditions was conducted on two farms in the savanah ecological zone of Nigeria. The tractors used in the study were Caterpillar D4E and D4D. Under irrigation the effective field capacity was 2.1 ha/h, with a field efficiency of 92.8%. Man-machine activity was estimated at 98.2% while the man-machine productivity was 2.14 ha/h. In the case of rainfed condition the effective field capacity and field efficiency were 2.32 and 96.4% respectively. Man-machine activity was 98.6% and Man-machine productivity was 2.34 ha/h. It was found that the energy consumption under irrigation was about twice that under rainfed condition.

RESUME: La mécanisation dela préparation de terre en se servant de tracteurs devient de plus en plus populaire au Nigéria. Des fermiers a subsistance, surtout less fonctionnaires tenant moins de 5 hectares qui labourent la terre pour plaisir et l'augmentation du revenu familial, préferent louer des tracteurs. Les organisations D'etat et un peu d'individus tenant plus de 100 hectares, possédent de tracteurs. In a mene une étude de valuation de temps de tracteurs à piste dans une condition d'irrigation et dans la saison des pluies sur deux fermes situées à la zone écologique savane du Nigéria. Les tracteurs employées dans L'étude étaient les caterpillar D4E et D4D. A l'irrigation, le véritable performance au champ était 2,1 h/l'h avec un rendement au champ de 92,8%. La capacite de machine actionnée par l'homme est valuée à 98,2% tandis que le rendement homme-machine était 2,14 h/l'h. Dans le cas de la saison des pluies, le véritable performance au champ et rendement étaient 2,32 et 96,4% respectivement. La capacite de machine actionnée par l'homme était 98,6% et le rendement homme-machine était 2,34 h/l'h. On a découvert que la perte d'énergie à l'irrigation était environs deux fois celle à la saison des pluies.

KURZFASSUNG: Die Mechanisierung der Landbestellung durch Verwendung von Traktoren wird allmählich beliebt in Nigeria. Kleinbauern,vor allem Beamte mit weniger als fünf Hektar Land,die aus Vergnügen und zur Verbesserung des Familienein-Kommens Landwirtschaft betreiben, ziehen es vor, Traktoren zue mieten. Staatsbetriebe und einige wenige Einzelpersonen mit mehr als 100 ha besiten Traktoren.
Anf zwei Betrieben in der Savannah - Zone von Nigeria wurde eine Zeitstudie von "track type" Traktoren sowohl unter den Bedingungen der künstlichen traktoren Caterpillar D4E and D4D benützt.
Unter Künstlicher Bewässerung war die effektive Feld-Kapazität 2.1. ha pro Stunde und dier feld-Wireksam keitsgrad 92.8% Mensch-Maschine-Aktivität wurde af 98.2% geschatzt, während Mensch-Maschine-Produktivitat 2.14 ha pro stunde betung.
Während der Regenzeit war die effektive Feld-Kapazitat 2.32 ha pro Stunde und der Feld-Wireksamkeitsgrad 96.4% Mensch-Maschine-Aktivitat war 98.6% ind Mensch-Maschine-Producktivitat 2.34 ha pro Stunde. Es wurde festgestellt, dap der Energieverbrauch unter Künstlicher Bewässerung ungefähr doppelt so gross war als während der Rpgenzeit.

1 INTRODUCTION

Nigeria like most developing countries in Africa, is striving hard towards self-sufficiency in food production. In order to meet this goal, appropriate technologies are required. Machines are some of the technological imputs for successful agricultural production. The need to know the level of machines that will be appropriate for our agricultural production exists.

In order to increase food production in Africa, appropriate technological inputs have to be added in our production system. More land has to be cultivated and all operations completed on time. This can only be possible if mechanization is practiced. Farm machinery are becoming very expensive and complicated. The tractor which is the main source of power on the farm, is complex and many developing countries in Africa will not have the technological capability to manufacture it. They have to depend on import if ever they desire to emback on mechanization.

With the sudden reduction in oil revenue in Nigeria, agriculture is now being considered as an alternative source of foreign currency earner. This has led many individuals and organizations to go into large scale farming. For such ventures, the appropriate technology is the tractor power and the implements that go with it. This study was undertaken to solve the problems of large scale farming in Nigeria. Several studies have been conducted and reported elsewhere on large scale farming in Nigeria. Kolawole (1972) reported that the annual tractor useage in Nigeria was much lower than that of developed countries. This has not changed appreciably from recent work reported by Anazodo (1983). Table 1 shows that a greater portion of tractor power utilization is in the seed bed preparation. It is therefore necessary that as a start in machinery management studies, the capacitive performance of land preparation should be tackled first. The huge investment in tractor and machinery calls for a better understanding of the capacitive performance of some common tillage machines used in Nigeria.

Several studies on machinery performance have been carried out elsewhere. These include the work of Barnes (1960) in which machine performance was divided into four categories namely functional, mechanical, capacitive and economic. Capacitive performance describes how a machine completes a job within allowable constraints of time. The capacitive performance of an agricultural machine is measured in terms of field area covered per unit of time. This capacity measure, termed effective field capacity can be evaluated with the following equation (ASAE 1979):

$$C = SWe/10 \qquad (1)$$

where C is effective field capacity (ha/h), S field speed (km/h), W machine width (m) and e field efficiency, (decimal).

In this equation, field efficiency is the ratio of effective field capacity to theoretical field capacity, and it includes the effects of time lost in the field and failure to utilize the full width of the machine.

Table 1. Tractor power usage in Oyo and Bauchi States, modified from Anazodo (1983) Data.

Operation	Bauchi State			Oyo State		
	A*	B**	C+	D**	E++	F**
Plowing	156.2	63.0	413.6	82.5	248	41.8
Harrowing	21	8.5	29.4	5.9	112	18.9
Ridging	20.2	8.1	50.6	10.1	53	8.9
Seed-drilling	6.2	2.5	7.6	1.5	-	-
Transportation	44.4	17.9	-	-	180	30.4

*	4-year average number of farmers receiving TEHU service
**	Percent
+	4-year average land area under TEHU (ha)
++	5-year average number of farmers under TEHU
TEHU	Tractor and Equipment Hiring Unit

Time lost during field operation results from many causes. Time for turning and travel across head-lands which are related to the geometry of the field. Hunt (1973) developed equations relating field efficiency to field dimensions for several common operational patterns. Kepner et al., (1977) have shown that turning, idle travel and machine adjustment time tend to be proportional to operating time, while other delays, such as stopping to fill seed and fertilizer boxes, interruptions caused by poor field and crop conditions, and stopping to unload harvested crops tend to be proportional to area. The following equation for field efficiency was proposed:

$$e = \frac{100(TO)}{TE + TA + TH} \qquad (2)$$

where e is field efficiency, decimal, TO theoretical operating time, TE effective operating time = TO/K, K fraction of machine width utilized, decimal, TA time lost proportional to area and TH time lost proportional to TO

Renoll (1970) defined field machine index as the ratio of productive field time to productive plus turn time, as given in the equation:

$$FMI = \frac{(T - TU - TT)\ 100}{T - TU} \qquad (3)$$

where FMI is field machine index, percent, T total field time, TU unproductive field time not including turning and TT turning time.

2 METHODOLOGY

This performance study of track type tractors was conducted under irrigation and rain-fed agriculture. The irrigation field was located at Kadawa Irrigation Project which forms part of the Kano Basin Development Authority while the rain-fed study was done at Mokwa.

For the study under irrigation condition, the machine used was Catterpillar D4E equipped with a heavy duty Rome plow. Using an accurate stop watch, a time study was conducted for all the activities involved in plowing operation as shown in Table 2

Table 2.   Activity classification for machine plowing operation

Activity	Description
Plow	Machine actually plowing through the field.
Turn	Machine turn at end of field or at headland.
Miscellaneous activities	Machine not plowing due to delay activities such as minor machine adjustments or repairs, machine clearing and operator personal times

The activities in Table 2 were further classified as follows:

1 Potential operating time: The total time the machine was in the field on a given day. Idle time caused by bad weather, daily service and travel time to and from the workshop was not included so also any machine repair time of over 10 minutes.

2 Operate activity time: Time machine performing its primary function or necessary support function.
  i) Primary function activity time: Time machine is plowing
  ii) Support function activity time: Non productive job time required for primary function such as turning at field ends or headlands.
3 Delay activity time: Time due to interruptions in the operating activity due to mechanical failure, machine cleaning and management or operator stops.
Under the rain-fed agriculture the tractors used were catterpillar D4E and D4D. They were equiped with a heavy duty Rome plow. A stop watch was also used to measure the duration of activities of each observation. Soil samples were collected for moisture content determination. Fuel consumption was determined using a measuring cylinder to refil the tank after each observation.

3 RESULTS AND DISCUSSIONS

The results are given in tables 3 and 4 for the irrigated condition and rain-fed agriculture respectively. Before discussing the results, it is necessary to define some terms that were used in the analysis of the data. Theoretical field capacity Fc (ha/h) was calculated by using the formula

$$F_c = \frac{W\ V}{10} \qquad (4)$$

where W is width of plow (m), V speed (km/h)

The effective field capacity F_{ce} (ha/h) was calculated by the quotient of the measured area covered and the total time.

i.e. $$F_{ce} = \frac{A}{T_p + T_s + T_d} \qquad (5)$$

where T_p is plow time (h), T_s support time (h), T_d delay time (h) and A area covered (ha)

The field efficiency F_{eff} was calculated using the formula given below

$$F_{eff} = \frac{T_p \times 100}{T_p + T_s + T_d} \qquad (6)$$

In a similar way Man-Machine activity which actually signifies the skill of the operator was defined as the ratio of the

Table 3. Time study of track type tractor in plow operation under irrigated condition

Machine	A	B	C	D	E	F
Catter-	1	5.25	22.3	12.97	5.25	241.22
pillar	2	5.49	268	10.83	0.17	295.83
D4E	3	5.49	67.03	5.58	-	72.61
tractor	4	5.23	194.65	10.23	0.58	205.46
	5	5.23	240.25	13.75	2.00	256.0

A observation
B speed (km/h)
C plow time (mins.)
D Support time (mins)
E delay time (mins)
F Total time (mins.)

Table 4. Time study of track type tractors in plow operations under rainfed conditions

Machine	A	B	C	D	E	F
Carter-	1	6.14	194.72	3.24	1.89	99.85
pillar	2	5.43	143.86	0.50	-	144.36
D4E	3	5.53	139.11	1.50	3.32	143.93
Tractor	4	5.75	85.67	2.39	-	88.06
	5	5.45	46.68	0.73	2.34	49.75
	6	6.94	36.93	0.93	-	36.86
	7	4.14	41.28	0.10	1.00	42.38
Catter-	1	5.30	81.18	2.39	0.60	84.79
piller	2	5.56	110.98	5.65	0.18	116.81
D4D	3	7.96	32.40	1.18	1.34	34.92
Tractor	4	5.53	90.25	5.09	0.83	96.17
	5	7.90	63.97	1.77	-	65.74
	6	6.66	89.38	1.36	-	90.74
	7	5.64	158.40	1.96	0.18	160.54
	8	5.64	84.48	0.79	3.46	88.73
	9	5.40	60.79	1.40	1.00	63.19
	10	4.43	55.62	0.97	0.74	57.33

A observation
B speed   (km/h)
C plow time (mins)
D support time (mins)
E delay time (mins)
F total time (mins.)

primary and support functions to the potential operating time. This was given as

$$M_{ma} = \frac{(T_p + T_s) \times 100}{T_p + T_s + T_d} \qquad (7)$$

The Man-Machine productivity M_{mp} which may be defined as the area covered per plow and support time given as

$$M_{mp} = \frac{A}{T_p + T_s} \qquad (8)$$

With these definitions and equations, the results presented in table 5 and 6 were calculated.

A look at table 5 and 6 for rain-fed and irrigation condition respectively reveals that the effective width of work was more or less the same which shows how skilled the operators were. The effective field capacity was within the range found elsewhere in the literature (Ayres and Williams, 1981). The man-machine activity was very high both in rain-fed and irrigation agriculture. This signifies that there was minimum delay or stopage due to machine failure. Also in the case of the rain-fed conditions, the soil was of optimum moisture for plowing. The man-machine productivity as expected was consistently higher in both cases than the effective field capacity. This was expected since by definition, it is the area covered during plowing and support time.

Table 5. Performance of track type tractors in plowing operations under rainfed conditions

Machine	A	B	C	D	E	F	G
Catter-	1	3.99	2.45	2.32	94.89	98.13	2.36
piller	2	4.32	2.35	2.33	98.92	100.00	2.33
D4E	3	4.32	2.39	2.31	96.66	97.71	2.37
Tractor	4	4.32	2.50	2.36	97.30	100.00	2.36
	5	4.22	2.35	2.27	96.41	97.89	2.32
	6	4.32	2.99	2.92	97.40	100.00	2.92
	7	4.32	1.80	1.76	97.45	97.73	1.79
Catter-	1	4.17	2.27	2.13	96.44	99.29	2.15
pillar	2	4.31	2.41	2.30	95.01	99.86	2.29
D4D	3	4.10	3.26	3.03	93.10	96.55	3.15
Tractor	4	4.43	2.40	2.26	94.00	99.31	2.27
	5	4.34	2.13	2.07	97.35	100.00	2.07
	6	4.04	2.69	2.66	98.68	100.00	2.66
	7	4.12	2.32	2.21	94.96	96.15	2.29
	8	4.32	2.44	2.32	95.20	96.08	2.41
	9	4.32	2.34	2.27	97.12	99.33	2.29
	10	4.29	1.90	1.82	96.06	97.12	1.87

A observation
B effective width (m)
C theoretical field capacity (ha/h)
D effective field capacity (ha/h)
E field efficiency (%)
F man-machine activity (%)
G man-machine productivity (ha/h)

Table 6. Performance of track type tractor in plow operation under irrigation condition

Machine	A	B	C	D	E	F
Catter-	4.2	2.2	2.03	92.4	97.8	2.08
pillar	4.4	2.42	2.19	90.6	94.2	2.32
D4E	4.35	2.39	2.20	92.3	100.00	2.20
Tractor	4.0	2.09	1.98	94.7	99.7	1.99
	4.23	2.21	2.07	93.8	99.2	2.09

A effective implement width (m)
B theoretical field capacity (ha/h)
C effective field capacity (ha/h)
D field efficiency (%)
E man machine activity (%)
F man machine productivity (ha/h)

In table 7 we have the mean observations of the performances of the three machines stuided. The soil moisture which was least at 4.8% wet basis affected the effective field capacity under irrigation conditions where as at the rain-fed stage there was no significant difference.

Table 8 shows the mean energy consumption for plowing. Irrigation condition consumes about twice that of rain-fed conditions. Here again in the soil moisture was low and penetration was difficult. Since the type of soil for both cases were not determined in this study, it is not conclusive to explain the high energy consumption under irrigation condition, however it can be seen that moisture content of the soil affects the energy consumption of plowing.

The difference in energy consumption for the D4E and D4D under rain-fed condition was probably due to the age of the tractor. The D4E under rain-fed condition was a four-year old machine while the D4D was five year old.

4 CONCLUSION

Field capacitive performance of tillage tools operating under different conditions and using different prime movers have high field efficiency ranging between 92 and 97 percent. This high efficiency was due to the skill of the operator, the very long length of run which minimised the relatively low non productive time such as the supporting time and the delay time. This was mainly due to the geometry of the field.

Man-machine productivity was consistently very close to the effective field capacity in the study. This was

also affected by the geometry of the field since the non productive time which included turn time and stopages were relatively much smaller than the total productive time. Fuel consumption appeared to be affected by soil moisture and age of the tractor.

Table 7. Mean of observations and performances of machines

Machine	A	B	C	D	E	F	G	H
I	5.34	4.24	4.8	2.26	2.10	92.8	98.2	2.14
II	5.53	4.26	0.21	2.40	2.32	97.00	98.78	2.35
III	6.3	4.29	8.66	2.41	2.31	95.79	98.43	2.34

A velocity (km/h)
B implement width (m)
C soil moisture (%)
D theoretical field capacity (ha/h)
E effective field capacity (ha/h)
F field efficiency (%)
G man-machine activity (%)
H man-machine productivity

I   Catterpillar D4E Tractor
     (Irrigation)
II   Catterpillar D4E Tractor
     (Rain-fed)
III Catterpillar D4D Tractor
     (Rain-fed)

Table 8. Energy consumption

Machine	A	B	C	D	E
I	5.34	4.24	10.00	4.8	14.3
II	5.53	4.26	15.56	8.21	6.48
III	6.3	4.29	14.12	8.66	7.96

A velocity (km/h)
B implement width (m)
C plow depth (cm)
D soil moisture (%)
E fuel consumption (l/ha)

I   Catterpillar D4E Tractor
     (Irrigation)
II   Catterpillar D4E Tractor
     (Rain-fed)
III Catterpillar D4D Tractor
     (Rain-fed)

REFERENCE

Anazodo, U.G.N. 1983. Survey and
comparative analysis of farm power and
machinery management system in
Nigeria. Proc. NSAE 7(1982): 58-73.

ASAE 1979. Agricultural Machinery
Management Agricultural Engineers
Yearbook.

Ayres, G.E. and Williams, D.L. 1981.
Estimating field capacity of farm
machines. Extension bull. Pm. 696
Iowa State University.

Barnes, K.K. 1960. Proper evaluation of
machines. Implement and tractors
75(8): 48, April 16, 1960.

Hunt, D. 1973. Farm Power and machinery
management. Iowa State University
Press Ames Iowa.

Kepner, R.A., Bainer, R. and Barger. E.R.
1977. Principles of farm machinery.
The AVI Publishing Company Ins.
West Post Connecticut.

Kolawole, M.I. 1972. Economic aspects of
tractor contracting operations in
Western Nigeria J. Agric. Engg. Res.
17(4): 289-294.

Renoll, E.S. 1970. A method for predicting
field machinery efficiency and
capacity. Trans. of the ASAE 13
448-449.

Land and Water Use, Dodd & Grace (eds), © 1989 Balkema, Rotterdam. ISBN 90 6191 980 0

The economics of zero traffic systems for winter barley in Scotland

D.J.Campbell
Scottish Centre of Agricultural Engineering, Penicuik, UK

M.J.McGregor
East of Scotland College of Agriculture, Edinburgh, UK

ABSTRACT: Conventional and zero traffic systems are compared for winter barley grown for four seasons in either a ploughed or direct drilled clay loam. Results show that the zero traffic system gave more uniform crops, smaller seasonal variations in yield, better root penetration and hence water availability, reduced cultivation energy requirement and the possibility of direct drilling soils otherwise considered unsuitable. Detailed economic analysis of the results enhances the advantages of the zero traffic system.

RÉSUMÉ: Un système de préparation de semis conventionnel est comparé à système de culture en travail de sol minimum pour une culture d'orge d'hiver dans un loam argileux labouré et à semis direct. L'expérience fût poursuivie durant quatre ans. Les résultats indiquent que le système à travail minimum produisait une récolte plus uniforme, une réduction des variations inter-saisonnières du rendement, une meilleure pénétration des racines et donc une extension de l'accès aquifere, une réduction du coût energétique pour la préparation du sol et la possibilité de semis direct dans des terres qui ne seraient pas normalement considéré commes étant practicables. Une analyse détaillée des résultats renforce les avantages attribués à ce système de culture à travail minimum.

ABSTRAKT: Konventionelle und Zero Verkehrsysteme werden verglichen fuer Winter Gerste die in vier Jahreszeiten in entweder Pflug oder aber Direktsaat (Saathorn) Lehmerde waechst. Resultate haben erwiesen das durch das Zero Vekehrsystem Ernten einheitlicher waren, kleinere Jahreszeitliche Unterschiede im Ertrag der Ernte, bessere Durchdringung der Wurzeln und dadurch das Vorhandensein von Wasser, weniger Energie oder Treibstoff Einsatz sowie die Gelegenheit zur Direksaat (Saathorn) in Erde die man sonst als ungeeignet dafuer haelt. Wirtschaftliche Analyse der Einzelheiten sowie die Resultate erhoehen die Vorteile des Zero Verkehrsystems.

1 INTRODUCTION

There is increasing interest in the control of field traffic as a means of avoiding problems of soil compaction. Soil and crop responses to various traffic treatments have been tested for a range of soils, crops and climates in both Europe (for example, Soane et al. (1972), Perdok and Telle (1979), Canarache et al. (1984), Blackwell et al. (1986), Campbell et al. (1986) and Chamen et al. (1987)) and North America (for example, Raghaven et al. (1978), Voorhees et al. (1985), Taylor (1986) and Meek et al. (1988)). Generally, these experiments showed that traffic can have an adverse effect on crop growth, quality and yield although occasionally a positive yield response to traffic has been found.

More recently, suitable machinery has started to become available such that farmers could grow barley in traffic-free beds up to 12 m wide (for example, Chamen et al. (1980), Taylor (1981) and Tillett and Holt (1987)).

Campbell et al. (1986) showed that seed-bed traffic could have a substantial adverse effect on the performance of winter barley in Scotland. Since harvest traffic generally involves higher wheel loads applied to the soil when it may be much more susceptible to compaction, Dickson and Campbell (1984, 1985, 1987 and 1988) compared conventional and zero traffic systems in conjunction with either ploughing or direct drilling during 1982-86. The zero traffic system employed machines with their wheel track increased to 2.8 m. This paper examines the economic implications of their results.

2   MATERIALS AND METHODS

2.1 Location and soil

The experiment was located 10 km south of Edinburgh, Scotland where the soil is a clay loam, the altitude is 200 m and the long-term average annual rainfall is 866 mm. The soil is considered capable of average cereal production with some variation in crop performance between years since an impermeable subsoil and variation in annual rainfall may cause excessive soil wetness. Direct drilling of winter and spring cereals is generally much less successful than conventional sowing.

2.2 Machinery

John Deere 4440 (114 kW) and 4020 (85 kW) tractors with the track of their front and rear wheels modified to 2.8 m were used for soil management in conjunction with conventional implements, including a Massey Ferguson 30 drill and 130 direct drill. A Massey Ferguson 515 combine harvester was used for the first three harvests and a Claas Dominator 58 in the final harvest, both machines being modified to a 2.8 m track front and rear. Thus the crop was cut from the full width of the beds in the zero traffic system.

2.3 Treatments and experimental design

The conventional traffic system corresponded to that found on local commercial farms while, in the zero traffic system, all wheels were restricted to permanent perennial tramlines at 2.8 m centres which were established after secondary cultivation in the first season of the experiment. Each traffic system was tested in conjunction with both ploughed and, in the final three seasons, direct drilled seedbeds. There were four replications of the four treatments and each plot received the same treatment in all seasons except that all plots were ploughed in the first season.

2.4 Soil and crop management

Initially, all plots were ploughed to 250 mm depth with a two-furrow mouldboard

plough. The permanent tramlines were then established in the zero traffic plots and were thereafter used by all traffic on these plots for the duration of the experiment. This was achieved by chisel ploughing in subsequent seasons, the plough having eight low-draught tines. Secondary cultivation was by reciprocating harrow or spiked rotary cultivator according to the soil conditions prevailing in any season.

Pest, weed and disease control was satisfactory in all seasons except that, in the second season, a few plants were affected by snowrot (Typhula incarnata) in the spring. The extent of the infection was not serious and application of fertiliser eliminated the symptoms of the disease in two weeks. In all four seasons straw and stubble were burnt. Soil trace elements and pH were monitored regularly and levels were found to be satisfactory without significant variation between plots. A growth regulator was used to avoid crop lodging with the exception of the third season when, despite its use, there was some lodging in the chisel ploughed plots of both traffic systems.

2.5 Economic analysis

A gross margin approach was used to compare the differences between the treatments in each year and between years. The approach used was, however, different from that adopted widely in the United Kingdom (see, for example, SAC (1988) and Nix (1988)). The first difference is that the costs of owning and running machinery and implements - quasi-variable costs - have been included in the analysis. Each implement and item of machinery has been costed on an hourly basis to reflect depreciation, cost of financing, repairs and maintenance, fuel and labour costs. This is achieved by using the following equation developed from Witney and Oskoui (1982);

$$A = C + \frac{\sum_{n=1}^{N} (R_n P^n + F_n P^n + L_n P^n - S_N P^N)(P-1)}{P(P^N-1)}$$

where:

A = present hourly cost
C = initial capital cost
R_n = repairs and maintenance costs (including insurance and finance costs) in n^{th} hour
F_n = fuel cost in n^{th} hour
L_n = labour cost in n^{th} hour
S_N = current resale value of a tractor having done N hours

N = number of hours machine has worked
P = $(1+g_i)/(1+r)$
g_i = inflation rate
r = opportunity cost of capital

The present hourly cost is used in the calculation of the gross margins. The values used for the zero traffic system assume that the costs of operating a 12 m gantry are the same as the costs of operating the tractors and implements used in the experiments. This assumption was made necessary because gantries are not sufficiently widely used for objective data to be available at present. The result of this assumption is that the true costs of operating a 12 m gantry have probably been over-estimated.

The second deviation from the standard gross margin approach is that the gross income and variable cost values for each treatment were discounted by the opportunity cost of capital to reflect the timing of the cash flows in the crop cycle. All items of expenditure and income were discounted on a monthly basis using a post tax discount rate of 11.25% per annum. All cash flows were discounted to August 1988 and all costs and returns are those which prevailed at that time.

3. RESULTS

Full details of the soil and crop responses are given by Dickson and Campbell (1984, 1985, 1987 and 1988). The crop grain yields for the ploughed and direct drilled treatments are shown in Tables 1 and 2 respectively. The yields for the zero traffic systems are those for 12 m tramline systems. It is clear from these results that, for the period of the experiment, the zero traffic system consistently out-yielded the conventional traffic system. The mean yield for the zero traffic system (6.98 t/ha) was 0.68 t/ha (SD = ± 0.56) better than for the conventional traffic system.

Table 1: Crop grain yields for ploughed treatments (t/ha)

| Year | Traffic system | | Z-C |
	Zero (Z)	Conven. (C)	
1982-83	6.30	4.80	1.50
1983-84	8.01	7.93	0.08
1984-85	6.97	6.09	0.88
1985-86	6.64	6.37	0.27
Mean	6.98	6.30	0.68

Table 2: Crop grain yields for direct drilled treatments (t/ha)

| Year | Traffic systems | | Z-C |
	Zero (Z)	Conven. (C)	
1983-84	8.08	8.20	(0.12)
1984-85	7.00	6.98	0.02
1985-86	4.38	4.14	0.24
Mean	6.49	6.44	0.05

The yield advantage (zero over conventional) was much smaller in the direct drilled treatments. The overall mean advantage was 0.05 t/ha (SD = ± 0.15). A further significant effect shown under direct drilling is that the yield advantage improved with successive crops.

The results of the economic analysis of the experimental work are shown in Tables 3 and 4 for the ploughed and direct drilled treatments respectively. The results reflect those of the yield data. The mean advantage of the zero traffic system over the conventional traffic system is £50.48/ha (SD = ± 55.77). The loss in 1983-84 is the result of a lower yield differential in that year and higher machinery costs. The higher costs occurred because tractor power was not matched with the prevailing soil conditions and the agronomic operations being carried out.

Table 3: Gross margins for ploughed treatments (£/ha)

| Year | Traffic systems | | Z-C |
	Zero (Z)	Conven. (C)	
1982-83	152.22	25.07	127.15
1983-84	284.96	307.03	(22.07)
1984-85	185.85	112.04	73.81
1985-86	225.33	202.30	23.03
Mean	232.05[1]	272.71[1]	50.48

Note: [1]These values exclude 1982-83 results to allow comparison with direct drilled treatments

Ploughed treatments are shown to be superior to direct drilled treatments in all years in which comparisons can be made. The mean difference between the ploughed and direct drilled treatments was £27.78/ha and £17.35/ha for the zero and the conventional traffic systems respectively.

Table 4: Gross margins for direct drilled
treatments (£/ha)

| Year | Traffic systems | | Z-C |
	Zero (Z)	Conven. (C)	
1983-84	327.79	336.14	(8.35)
1984-85	226.12	215.22	10.90
1985-86	58.91	33.06	25.85
Mean	204.27	255.36	9.46

4   DISCUSSION

It is clear from the results that zero
traffic systems are superior to conventional traffic systems. The yield and financial advantage shown in the experiments
will mean that significant gross margin
increases can be expected by farmers when
gantry systems are used. The gross margin
will be enhanced when it is possible to
determine values for the following factors.

1. The costs of operating a 12 m gantry.
The use of gantries will result in higher
yields because there will be fewer tramlines per field. No information was available which would allow us to cost such a
gantry system but it is expected that when
such systems are used commercially, the
margin between zero and conventional
traffic systems will be increased even
further.

2. The actual time taken for each cultivation operation. The actual times taken
for each cultivation were not recorded in
the experiment. The economic analysis
assumed that the time taken for each cultivation operation was the same for both
the conventional and zero traffic systems.
If this assumption is varied such that
cultivation for the zero traffic systems
is assumed to be even only 10% less than
the conventional traffic systems, then the
gross margin differential would rise by
£7.50/ha for the ploughed treatment and
£2.05/ha for the direct drilled.

3. The power requirements for each cultivation. The costings relate to the tractors and implements which were actually
used in the experiment. This meant,
however, that the power used for some
operations was excessive. For instance in
1984-85, spraying for the zero traffic
systems was carried out with a 116 kW
tractor. It is estimated, from the draught
force measurements, that the difference in
power requirements between the two systems
(conventional and zero) is approximately
15 kW. If it is assumed that the cultivation, spraying and fertiliser distribution
operations for the zero traffic systems
require tractors with 15 kW less power
than the conventional traffic systems,
then the gross margin differential between
the two systems would rise by another
£7.55/ha for the ploughed treatment and
£3.45 for the direct drilled.

4. All treatments in each year, were
sown on the same date which was before the
optimum planting date for the variety sown;
as determined by the Scottish Agricultural
Colleges crop specialists. The sowing date
was between 12 and 14 days earlier than
optimum in three years and three days
earlier in one. On a commercial farm the
time at which cultivation and sowing of
winter crops occurs in Scotland is usually
very short. Failure to get a crop in the
ground at the optimum time will lead to
timeliness penalties, Oskoui (1983). It
is estimated that gantry systems would
extend the time available for cultivation.
This would then allow farmers to sow crops
nearer to the optimum sowing date for the
variety than would be possible under conventional traffic systems. The timeliness
penalties associated with delays in sowing
would also need to be assessed when comparing cultivation systems. In Scotland, the
timeliness penalty associated with a one
week delay in sowing winter barley, is
estimated to be £11.00/ha (0.13 t/ha),
Oskoui (1983).

In addition, the results show that a
soil classified as unsuitable for direct
drilling with conventional traffic may be
direct drilled satisfactorily in the
absence of traffic. The direct drilled
crop performance improved as a result of
progressively improving soil physical conditions in successive seasons.

5   CONCLUSIONS

1. The zero traffic system out-yielded
the conventional traffic system under both
ploughing and direct drilling.

2. The gross margin analysis shows that
under ploughing, the zero traffic system
was financially superior by £50.00/ha.
This assumes a 12 m gantry system is
used.

3. Although initially negative, the
advantage in gross margin for the zero
traffic systems under direct drilling
improved with sequential direct drilling
to give an advantage of £25.00/ha in the
third year.

REFERENCES

Blackwell, P.S., Graham, J.P., Armstrong, J.V., Ward, M.A., Howse, K.R., Dawson, C.J. and Butler, A.R. 1986. Compaction of a silt loam by wheeled agricultural vehicles. I. Effects upon soil conditions. Soil Tillage Res., 7: 97-116.

Campbell, D.J., Dickson, J.W., Ball, B.C. and Hunter, R. 1986. Controlled seedbed traffic after ploughing or direct drilling under winter barley in Scotland, 1980-1984. Soil Tillage Res. 8: 3-28.

Canarache, A., Colibas, I., Colibas, M. and Horobeanu, I. 1984. Effect of induced compaction by wheel traffic on soil physical properties and yield of maize in Romania. Soil Tillage Res. 4: 199-213.

Chamen, W.C.T., Collins, T.S., Hoxey, R.P. and Knight, A.C. 1980. Mechanisation opportunities likely to be provided by engineering in the 21st century. The agric. Engr 35: 63-70.

Chamen, W.C.T., Chittey, E.T. and Howse, K.R. 1987. The effect of different tyre/soil contact pressures on soil and crop responses when growing winter wheat: year 3 - 1984-85. Div. Note DN 1432, ARFC Inst. of Engng Res., Silsoe.

Dickson, J.W. and Campbell, D.J. 1984. Soil and winter barley responses to zero and conventional traffic systems after ploughing on a sandy clay loam, 1982-83. Dep. Note SIN/386, Scott. Inst. agric. Engng, Penicuik, 32 pp.

Dickson, J.W. and Campbell, D.J. 1985. Soil and winter barley responses to zero and conventional traffic systems after ploughing or direct drilling on a sandy clay loam, 1983-84. Dep. Note SIN/454, Scott. Inst. agric. Engng, Penicuik, 22 pp.

Dickson, J.W. and Campbell, D.J. 1987. Soil and winter barley responses to zero and conventional traffic systems after ploughing or direct drilling on a sandy clay loam, 1984-85. Dep. Note SIN/481, Scott. Inst. agric. Engng, Penicuik, 22 pp.

Dickson, J.W. and Campbell, D.J. 1988. Soil and winter barley responses to zero and conventional traffic systems after ploughing or direct drilling on a sandy clay loam, 1985-86. Dep. Note 6, Scott. Centre agric. Engng, Penicuik, 45 pp.

Meek, B.D., Rechel, E.A., Carter, L.M. and Detar, W.R. 1988. Soil compaction and its effect on alfalfa in zone production systems. Soil Sci. Soc. Amer. J. 52: 232-236.

Nix, J. 1988. Farm management pocketbook. 19th ed. Wye College, University of London.

Oskoui, K.E. 1983. The practical assessment of timeliness penalties for machinery selection purposes. The Agricultural Engineer, Winter: 111-115.

Perdok, U.D. and Telle, M.G. 1979. Controlled traffic in crop productions. Proc. 8th Conf. Inst. Soil Tillage Res. Organ., Stuttgart, 2: 325-331.

Raghavan, G.S.V., McKyes, E., Gendron, G., Borglum, B.K. and Le, H.H. 1978. Effects of tire contact pressure on corn yield. Can. agric. Engng 20: 34-37.

SAC, 1988. Farm management handbook 1988/89. The Scottish Agricultural Colleges, Perth.

Soane, B.D., Hutchison, P.S. and Campbell, D.J. 1972. The effect of traffic and weed control methods on soil conditions and potato yields. Dep. Note SIN/102, Scott. Inst. agric. Engng, Penicuik, 18 pp.

Taylor, J.H. 1981. A controlled-traffic agricultural system using a wide-frame carrier. Proc. 7th Conf. Int. Soc. Terrain Vehicle Systems, Calgary, 1: 385-409.

Taylor, J.H. 1986. Controlled traffic: a soil compaction management concept. Tech. Paper 860-731, Soc. Automotive Engrs, Pennsylvania, 9 pp.

Tillett, N.D. and Holt, J.B. 1987. The use of wide span gantries in agriculture. Outlook on Agric. 16: 63-67.

Voorhees, W.B., Evans, D.D. and Warnes, D.D. 1985. Effect of preplant wheel traffic on soil compaction, water use, and growth of spring wheat. Soil sci. Soc. Amer. J., 49: 215-220.

Whitney, B.D. and Oskoui, K.E. 1982. The basis of tractor power selection on arable farms. J. agric. Engng Res. 27: 513-527.

Land and Water Use, Dodd & Grace (eds), © 1989 Balkema, Rotterdam. ISBN 90 6191 980 0

Qualitative evalutation of a combined implement for minimum tillage

R.Cavalli & L.Sartori

Department of Territory and Agro-forestal systems, Padua University, Italy

ABSTRACT: A combined implement developed by Dept. Territory and Agro-forestal Systems of Padua University is proposed. Combined implement performances were established: with a 103 kW 4WD tractor, the best results are achieved at a speed of about 3 km/h and discs rotary speed of about 80 r.p.m. In these conditions, field capacity, energy requirements, burial index and clod dimensions are similar to those ones of ploughing. The better incorporation of crop residue into the soil and the possibility of breaking the soil up appear very interesting and confirm the design of the implement.

RESUME: On propose un équipement combiné developpé par le Dépt. Territoire et Systèmes Agro-forestiers de l'Université de Padoue. Les essais ont consenti de définir certains paramètres fonctionnels de l'équipement en mettant en évidence que les meilleur prestations avec une puissance optimale de 103 kW, pouvaient être reconduites à une vitesse d'avancement de 3 km/h et une vitesse de rotation des disques à pales de 80 rpm. Dans ces conditions la capacité de travail et la consommation d'énergie sont comparable au travail traditionnel de labourage, ainsi que le degré motteux et l'indice d'enfouissement des résidus des cultures. La meilleure distribution des résidus est apparue particulièrment interessante sur le profil travaillé qui, avec la possibilité d'un travail par à coups du terrain, confirme la validité de l'equipement combiné.

ZUSAMMENFASSUNG: Es wird kombinierte Ausrüstung, entwichelt von der Abteilung "Gebiet und System del Land und Forstwirtschaft" der Universität Padua angeboten. Die Tests unterstützen die Definition einiger funktioneller, Parameter der Ausstattung, hervorhebend die besten Leistungen wobei, mit einer Kraft von optimal 103 kW eine Wiederherstellung mit einer Geschwinding keit von 3 km/h und einer Rotation del palattierten Scheiben von 80 u/min erfolgen kann. Mit diesen Konditionen kann eine Konfrontation der Arbeitskapazität und der Energiekosten mit det traditionellen Arbeit der Pflügens, wie auch der Große der Schollenbildung und der Zeichen des Umgrabens der landwirtschaftlichen Überreste erfolgen. Das spezielle Interesse liegt in der bestmöglichen Verteilung der Überreste im bearbeiteten Profil, das, gemeinsam mit der Möglichkeit des Aufreißens der Erde, den Wert der kombinierten Ausstattung bestätigt.

1 INTRODUCTION

Reduced depth of tillage is a main way of achieving lower intensity of cultivation in order to reduce energy and labour requirements. At present some implements adopted for reduced tillage do not perform well in specific situations. Shallow ploughs for cultivation after crops which leave large amounts of plant residue cannot perform well in heavy and dry soils because of difficulties in maintaining constant tillage depth. Moreover, it has been demonstrated that in clay soils the action of p.t.o. driven implements, used for shallow cultivation immediately before or contemporary to drilling, causes an over pulverization of soil that may lead to crust formation.

New machines for shallow cultivation of heavy soils would carry out breaking up and shattering action using both p.t.o driven implements and shanks to obtain good stability and to improve connections between the surface water and groundwater table. They would also be equipped with chopping and turning implements in order to achieve effective incorporation of crop residue.

In order to achieve these results the Dept. Territory and Agro-forestal Systems of Padova University has proposed a coordinated activity with the Pegoraro Vittorio Company which has led to the development, design and testing of a combined implement.

2 IMPLEMENT DESIGN

The implement consists of a main frame and two gangs of tools arranged in oblique position with respect to the travel direction; the first one is a gang of 5 shanks (800 mm long and 560 mm wide) with lateral flaps and the second one is a gang of 5 bladed discs (700 mm in diameter and 560 mm wide) mounted on an axle driven by a hydraulic motor (35.8 kW maximum power) at variable speed (from 40 to 105 r.p.m.) and with same rotation wise of forward direction. The axle is connected to the frame by an articulated device that allows tillage depth adjustment of bladed discs.

A horizontal rotor is placed between the shanks and bladed discs, driven by a hydraulic motor (8.9 kW maximum power) at rotor speed of 145 r.p.m. The blades prevent buildup of soil and trash between the shanks. Two hydraulically-controlled gauge wheels are mounted on both sides of the implement to allow tillage depth of the shanks.

The hydraulic transmission has two coaxial pumps driven by tractor p.t.o.: the first one, connected to the hydraulic motor driving the bladed discs, is an inline axial piston pump with variable displacement design in order to adjust rotational speed of the bladed discs; the second one, connected to the hydraulic motor driving the rotor placed between tha shanks and bladed discs, is an external gear pump.

Frame design allows to attach a straw chopper in front of shanks, driven by tractor p.t.o., set at the same angle to the shanks and bladed discs gangs. The straw chopper has a rotor with 60 curved blades and a roller with a spiral relief. The straw chopper can be fitted on a frame mounted in front of the tractor. In this outfit shanks and bladed discs gangs can be moved forward on the frame of the implement, giving a better balancing.

Implement sizes are 3000 mm long and 3000 mm wide and overall weight is 1950 kg (1550 kg without the straw chopper).

Considering operation of the sequence of devices it should be noted that the roller with spiral relief pulls down and spreads all over the working width the crop residue which is shredded and mixed with soil by the straw chopper. The particular design of shanks allows to work the soil at a depth up to 300 mm with a chiselling action; soil built up by shanks is partially turned by lateral flaps and completely inverted by bladed discs, causing crumbling of clods and burying of crop residue; adjusting of the rotational speed of the bladed discs allows to adapt the tillage action to soil characteristics. The horizontal rotor placed between the shanks and bladed discs prevents excessive build up of soil and avoid plugging in trash.

3 MATERIAL AND METHODS

Field tests have been carried out on levelled plots with different crop residue (seed sugarbeet straw, winter wheat stubble, maize stalks); the implement was coupled to a 103 kW 4WD tractor equipped with front three point linkage and p.t.o.

The aim of the experiment was the evaluation of the implement's qualitative performance according to the following objectives:

1. analysis of the relationship between forward speed and rotational speed of bladed discs with different crop residue and soil moisture;
2. analysis of tillage quality in comparison to shallow ploughing.

In each test have been measured soil coverage by crop residue and clod sizes after tillage were measured. Soil coverage, as percentage of soil surface covered by crop residue, was determined by the photographic method proposed by Laflen et al. (1981) and by other authors (Dickey et al. 1983, Dickey et al. 1984, Dickey et al. 1985, Hartwing et al. 1978, Sloneker and Moldenhauer 1977). The burial index (Bi) was derived from soil coverage as follow:

$$Bi = \frac{Ca - Cp}{Cp} \cdot 100$$

where:
Ca = soil coverage before tillage (%)
Cp = soil coverage after tillage (%).

The method used to measure clod size and frequency is also based on photographic analysis, starting from the method proposed by Campbell (1979). The area and major dimension of each clod having a major dimension greater than 50 mm were

Table 1. Burial index with three rotational speed of bladed discs and two kinds of residue.

Crop residue	Rotational speed of bladed discs (r.p.m.)	Crop residue cover degree before tillage (%)	after tillage (%)	Burial index (%)(*)
winter wheat stubble	55	85.87	6.46	92.47
	80	85.87	9.00	89.52
	105	85.87	5.20	93.94
Seed sugarbeet straw	55	80.40	1.33	98.34
	80	80.40	2.33	97.10
	105	80.40	2.20	97.26

(*) Main effect of residue significative (P < 0.01); main effect of rotational speed of bladed discs not significative; interaction between residue and rotational speed of bladed discs not significative.

measured. Evaluation of these quantities allowed classification of clods into dimensional classes and definition of the percentage of total area covered by each class and frequency of dimensional classes. Though the method regards only a bi-dimensional analysis of clods, it was considered satisfactory for the experiment aim.

4 RESULTS

4.1 Relationship between forward speed and rotational speed of bladed discs

Results of previous field tests (Cavalli and Sartori 1988) show the opportunity to operate the combined implement at the maximum forward speed allowed by tractor power (at least 3 km/h) with rotational

Table 2. Percentage of total area covered by definite classes of clods in two types of soil and with three rotational speeds of bladed discs.

Soil moisture (%)	strength (daN/dm²)	Rotational speed of bladed discs (r.p.m.)	% of total area Clods dimensional classes (mm)				
			< 50	50-100	100-150	150-200	> 200
			(*)		(*)	(*)	(*)
24.5	12.2	55	37.50	14.78	11.75	16.53	19.44
24.5	12.2	80	40.49	14.33	14.56	11.68	18.96
24.5	12.2	105	42.75	15.30	15.81	12.97	13.27
		average	40.24	14.80	14.04	13.73	17.22
15.8	13.2	55	73.00	15.19	9.11	1.10	1.60
15.8	13.2	80	71.22	14.65	7.87	3.79	2.48
15.8	13.2	105	68.85	17.29	8.81	3.74	1.30
		average	71.02	15.71	8.60	2.88	1.70

Main effect of soil significative (P < 0.01) (*); main effect of rotational speed of bladed discs not significative; interaction between soil and rotational speed of bladed discs not significative.

speed of bladed discs of about 80 r.p.m. With these adjustments the combined implement could perform higher field capacity, with reduced slippage and lower fuel consumption. Perfect integration between the buiding up action of shanks and the inverting action of the bladed discs could also be achieved because soil can regularly flow through these devices, reducing energy requirements.

Analysis of soil cover after tillage performed in different crop residue conditions (seed sugarbeet straw and winter wheat stubble) at forward speed of 3.04 km/h and different rotational speeds of bladed discs shows (table 1) higher efficiency with loose residue such as seed sugarbeet straw (burial index 97-98%). Moreover it seems that there is neither effect of rotational speed of bladed discs nor interaction with the kind of residue. So it is really important to evenly spread crop residue on the field in order to allow its burial also at shallow depth. Consequently application of chopping and spreadind devices on combines becomes more and more important in order to operate under optimal conditions.

Cloddiness after tillage was analyzed in soils characterized by different moisture content and strength (table 2). These parameters affect clods size more than the rotational speed of bladed discs: the percentage of clods smaller than 50 mm is lower (40.24% vs. 71.02%) while the percentage of clods greater than 200 mm is greater (17.22% vs. 1.79%) in soils with higher moisture content compared to soil with lower moisture content (figure 1).

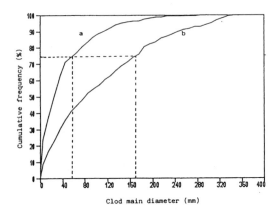

Figure 2. Cumulative frequency of area covered by definite classes of clods; soil moisture content 15.8% (a) and 24.5% (b).

Figure 2 shows the influence of soil moisture on tilth: in wet soils (24.5% moisture) 75% of total area is filled up by clods whose diameter is smaller than 173 mm; in dry soils (15.8% moisture) 75% of the total area is filled up by clods whose diameter is smaller than 58 mm.

Table 3, showing frequency of clods dimensional classes, confirms the same trend, emphasizing that in wet soils the highest frequency occurs for classes of clods with diameters ranging from 60 to 80 mm (from 50 to 70 mm in dry soils).

4.2 Comparison with shallow ploughing

The combined implement was compared with a mounted reversible 4-slatted furrow plough on maize stalks, using the same tractor and operating in the same conditions; the plough was selected in order to operate the tractor at nearly maximum power. Before shallow ploughing crop residue shredding was carried out using a straw chopper 2.2 m wide and powered by a 58 kW tractor.

Shallow ploughing gives the highest burial index (99.93%) while the incorporating action of combined implement is less efficient (burial index 92.64%), but this performance does not differ from shallow ploughing (200 mm) (Colvin et al 1986, Dickey et al. 1984, Gregory 1982) and it must be considered suitable for drilling and to avoiding crust formation.

Even if shallow ploughing produces reduced cloddiness, due to the action of slatted moulboard, observed differences in all clods dimensional classes are not statistically significant, except for the dimensional class which contains clods

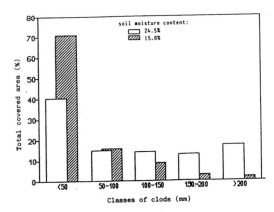

Figure 1. Percentage of total area covered by definite classes of clods in soils with different moisture contents.

Table 3. Frequency of dimensional classes of clods in two types of soils and with three rotational speed of bladed discs.

Soil moisture (%)	strength (daN/dm²)	Rotational speed of bladed discs (r.p.m.)	Frequency (%) Clods dimensional classes (mm)			
			50-100	100-150	150-200	> 200
			(*)		(*)	(*)
24.5	12.2	55	65.10	17.27	10.41	7.22
24.5	12.2	80	63.21	21.40	9.88	5.51
24.5	12.2	105	61.43	24.41	8.90	5.26
15.8	13.2	55	81.87	15.86	1.35	0.91
15.8	13.2	80	82.46	13.85	2.24	1.46
15.8	13.2	105	84.40	12.48	2.69	0.43

Main effect of soil significative (P < 0.01) (*); main effect of rotational speed of bladed discs not significative; interaction between soil and rotational speed of bladed discs not significative.

with diameter greater than 200 mm (tables 4 and 5, figure 3).

Figure 4 shows that 75% of the total area is covered by clods with diameter smaller tha 70 mm after shallow ploughing and smaller than 104 mm after working with the combined implement.

Analysis of incorporating capacity was completed by observation of soil profile after tillage: the combined implement works soil without forming hardpan and placing crop residue through all the tilled layers; shallow ploughing forms plough pan and concentrates plant residue on the furrow bottom. An appropriate vertical distribution of crop residue is an important requirement of tillage because it could affect organic matter evolution and provide good seed germination and root growth.

5. CONCLUSION

Experiments were carried out to define tillage characteristics of the combined implement that performs best at 3km/h forward speed and at rotational speed of bladed discs of 80 r.p.m. With these adjustments burial capacity and cloddiness after tillage do not differ from shallow ploughing.

The combined implement gives a better placement of crop residue through soil profile and this fact, along with absence of plough pan, confirms the effectiveness of the implement design.

Further developments are needed to study association of the combined implement with rotary cultivator for one-pass seedbed preparation; the implement should perform

Table 4. Percentage of total area covered by definite classes of clods in relation to implement adopted.

	% of total area Clods dimensional classes (mm)				
	< 50	50-100	100-150	150-200	> 200
					(*)
Shallow plough	66.17	16.34	9.03	3.51	4.68
Combined implement	59.56	13.16	10.47	6.52	10.10

(*) Significative difference (P < 0.05).

Table 5. Frequency of dimensional classes of clods in relation to implements adopted.

	Frequency (%) Clods dimensional classes (mm)			
	50-100	100-150	150-200	> 200
Shallow plough	78.09	16.20	3.57	2.13
Combined implement	72.08	18.38	4.53	4.25

primary cultivation, instead of shallow ploughing, if used for tillage delayed from drilling and should become an integral tool for primary cultivation and seedbed preparation, useful for tillage contemporary to drilling.

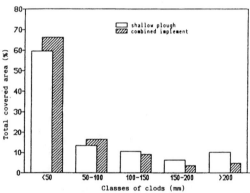

Figure 3. Percentage of total area covered by definite classes of clods after shallow ploughing and tillage with combined implement.

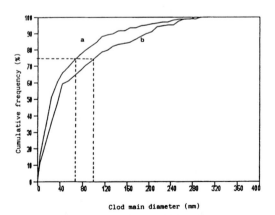

Figure 4. Cumulative frequency of area covered by different classes of clods after shallow ploughing (a) and tillage with combined implement (b).

REFERENCES

Campbell D.J. 1979. Clod size distribution measurement of field samples by image analysing computer. SIAE dept. note n. SIN/274.

Cavalli R.and L. Sartori. 1988. Nuova attrezzatura combinata per la lavorazione ridotta del terreno. L'Informatore Agrario 44 (17):47-50, 53-55

Colvin T.S., E.C. Berry, D.C. Erbach, J.F. Laflen. 1986. Tillage implement effects on corn and soybean residue. Transactions of the ASAE 29 (1): 56-59.

Dickey E.C., C.R. Fenster, J.M Laflen, R.M Mickelson. 1983. Effects of tillage on soil erosion in a wheat-fallow rotation. Transactions of the ASAE 26 (3):814-820.

Dickey E.C., C.R. Fenster, J.M Laflen, R.M Mickelson. 1984. Tillage, residue and erosion on moderately sloping soils. Transactions of the ASAE 27 (4):1093-1099.

Dickey E.C., C.R. Fenster, J.M Laflen, R.M Mickelson. 1985. Soil erosion from tillage system used in soybean and corn residues. Transactions of the ASAE 24 (4):1124-1129.

Gregory J.M. 1982. Soil cover prediction with various amount and types of corn residue. Transactions of the ASAE 25 (5):1333-1337.

Hartwig R.O.and J.M. Laflen. 1978. A meterstick method for measuring crop residue cover. Journal of soil and water conservation 33 (2):90-91.

Laflen J.M., M. Amemiya, E.A. Hintz. 1981. Measuring crop residue cover. Journal of soil and water conservation 36 (6):341-343.

Sloneker L.L and W.C. Moldenhauer. 1977. Measuring the amounts crop residue remaining after tillage. Journal of soil and water conservation 32 (5):231-236.

Land and Water Use, Dodd & Grace (eds), © 1989 Balkema, Rotterdam. ISBN 90 6191 980 0

The combined lateral and longitudinal forces on a 18.4-35/15-35 tractor tyre

H.L.M.du Plessis
Department of Agricultural Engineering, University of Pretoria, RSA

ABSTRACT: The side force characteristics of tractor tyres are needed for the dynamic modelling of tractors and combines on roads. A single axle testing trailer (20 ton) with two steerable wheels which may be free rolling, braked or hydraulically driven, was built. The side force characteristics for tyres were determined on a fairly smooth asfalt surface and test results were in close comparison to those obtained with truck tyres under simular conditions.

KURZFASSUNG: Die Eiginschaften von Seitenkräften von Schlepperreifen werden für die dynamische Modeliering von Schleppern und Mehdräschern auf Strassen benötigt. Ein Anhänger mit Einzelachse und Rädern die gesteuert werden können ist gebaut worden. Die Räder können frei rollen, gebremst oder hydraulisch angetrieben werden. Die Eigenschaften der Seitenkräfte die auf ziehmlich glatten Asfalt ermittelt wurden sind vergleichbar mit resultaten von LKW Reifen unter gleichen Umständen.

RéSUMé: Pour le modelage dynamique de tracteurs et de moissonneuses-batteuses sur la route on tient compte des caractéristiques de force latérale exercées par les pneus de tracteur.
Nous avons construit une remorque expérimentale à essieu simple (de 20 tonnes) avec deux roues dirigeables, capables aussi de rouler librement, d'être freinées ou d'être commandées hydrauliquement.
Les caractéristiques de force latérale des pneus ont été déterminées sur une route goudronnée assez peu accidentée et les résultats de l'expérience étaient étroitement comparables à ceux obtenus avec les pneus de camion dans des conditions semblables.

Acknowledgements: I wish to express my gratitude towards the Department of Agriculture and Water Supply for financial assistance on this project.

1 INTRODUCTION

The handling, steering and stability characteristics of combines and tractor/trailer combinations, both for on and off-road applications are growing in importance, especially as a result of the tendency towards higher tractor speeds on roads. In order to model these vehicles dinamically, the side force vs slip angle characteristics for heavily loaded tractor tyres are needed and a research project in this respect was undertaken.

2 THE SIDE FORCE ON A STEERED WHEEL

When a vehicle moves in an arc under on or off-road conditions, the centre lines of steered wheels diviates from the direction of movement of the wheel centre (fig. 1), causing a side force on the tire as well as a sideways displacement of the contact area with respect to the wheel centre line. On a hard surface, side slip occurs over a part of the contact area, (fig. 2) whilst on a soft surface it occurs over the whole

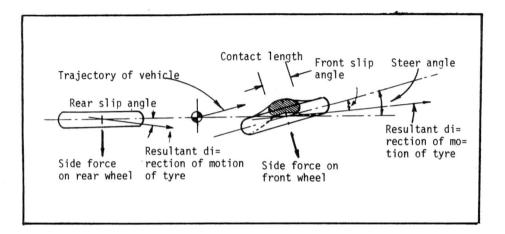

Figure 1 Slip angles assumed for a vehicle during a cornering manueuvre (after Crolla, & el Razaz, 1987.)

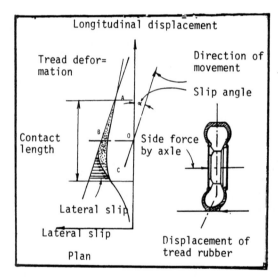

Figure 2 Tread and carcass deformation during cornering (after Crolla, et al 1987)

contact area. (Crolla, et al 1987.) This sideways movement also increases towards the rear of the contact area.

3 THE MEASUREMENT OF DYNAMIC TYRE FORCES

The dynamic tyre forces may be measured in a laboratory soil bin (Krick, 1971) or by driving the test wheel at a certain slip angle on the inside or outside of a large drum, large hori-

zontal disc or guided horizontal flat belt. Such results for force and moment measurements on large truck tyres differs considerably from practical tests, probably as a result of the larger wheel diameter when compared to drum diameter (Weber, 1987). This problem may be solved by using a special apparatus and full scale on and off-roads tests. The principles of different testing systems eg. the Cornell of Delft tyre tester (Clark, 1981) or the systems used by Schwanghart (1968) and Mc Allistar (1979) were considered, but would probably not bu applicable as a result of the large side forces on the tyres.

Weber (1987) overcame this problem by using a semi-trailer with an additional steerable axle and two opposing free rolling or braked wheels in front of the main axle.

A hydraulic cylinder was also used to change the slip angle whilst driving and the transfer function for the development of side forces may be studied.

Weber used a purpose built instrumented hub dynamometer to measure the side, vertical and longitudinal forces.

His idea of two oposing wheels would also be applicable, but the tractor wheels would not fit onto the semi-trailer.

4 TEST APPARATUS FOR MEASURING LARGE TYRE FORCES

In order to balance the expected large side forces, a 20 ton single axle trailer with two steerable wheels was built.

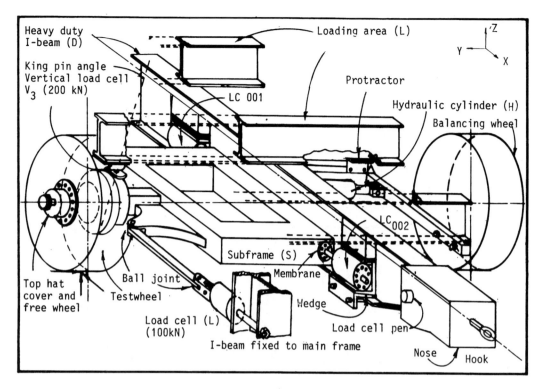

Figure 3 The test wagon

The wheels may either be used in free rolling, braked or hydraulically driven mode.

For conventional traction or brake tests, the wheels are locked in a straight ahead configuration and the traction or braking force measured as a function of slip for different wheel loads.

Traction forces, when traveling across a side slope, or forces on a driven wheel during cornering may be simulated by steering the wheels towards or away from each other, measuring the vertical, side and longitudinal forces for the driven, braking or free rolling option.

The test wagon was constructed using a heavy duty I-beam (D) fig 3 with two driven steerable wheels, mounted to two subframes. The one subframe and balancing wheel was rigidly fixed to the main frame whilst the test wheel was mounted on a free floating subframe (S), carried by two identical purpose made bi-directional force transducers LC 001 and LC 002 and a 200 kN vertical load cell just inside the test wheel. The longitudinal force was measured by a 100 kN load cell (L).

The vertical, longitudinal and side force on the wheel could be calculated by summing the forces in the three directions.

The slip angle of the wheel is controlled by two hydraulic cylinders, the one fixed to the main and the other to the floating subframe and the slip angle may therefore be changed whilst driving.

5 THE HYDRAULIC DRIVE

A 120 kW diesel engine was used to drive a double Brauninghaus pump (125 ET 22 R 21 2120) with stepless flow control. The pump was connected in two closed hydraulic circuits to two BMV 75 TF Linde hydraulic motors, each with a two stage servo control allowing a high and low ratio for the wagon. (fig 4)

The pumpflow and speed was initially controlled by a pressure sensitive system. To get more direct control, this was replaced by a mechanical system using two levers. Depending on the position of the control levers, oil flows to either

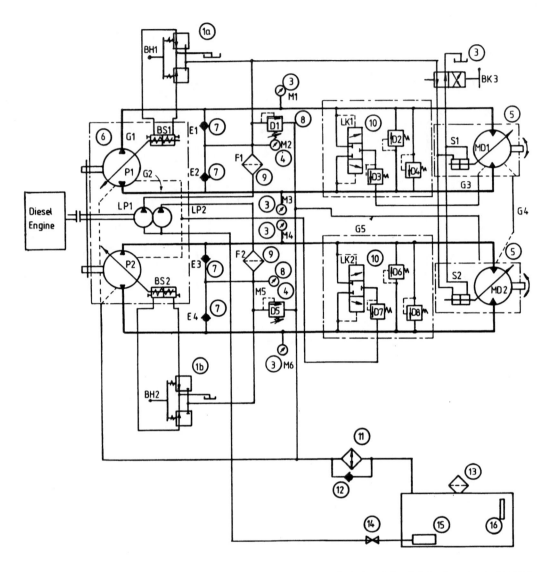

Figure 4 Hydraulic drive for tyre tester

G_1 or G_2, the one way valve E_1 or E_2^1 shuts and the motor MD_1 rotates clock or anticlockwise.

The supply pump LP_1, with savety valve D_1, supplies oil at a low pressure via filter F_1 and valve E_1 or E_2 to the low pressure line G_1^1 or G_2^1 and to the two step speed control BK_3. Low pressure oil also flows through the switch valve LK_1 and the pressure control valve D_3, to the motor frame MD_1, via G_4 to motor frame MD_2 and back through G_5 to the oil cooler 11.

IT	PART NUMBER	DESCRIPTION	QTY
	LOT 3022	Level Temp Ind	1
	OS 16	Strainer	1
	1,5"	Shut Off Valve	2
	TA4080C	Filler Breather	1
	CIT 06	NRV	
	FOG	Cooler	1
	RRV	Rel & Purge Valve	2
	274A 2N50110	Filter	2
	BGRP 087	Relief Valve	2
	CIT. 06	N.R.V.	4
	125 ET 22 R 21 2120	Pump	1
	BMV 75 TF	Motor	2
	0-6000kPa	Gauge	2
	0-40MPa	Gauge	4
		Man Motor Cont	1
		Prop Man Pump Cont	2

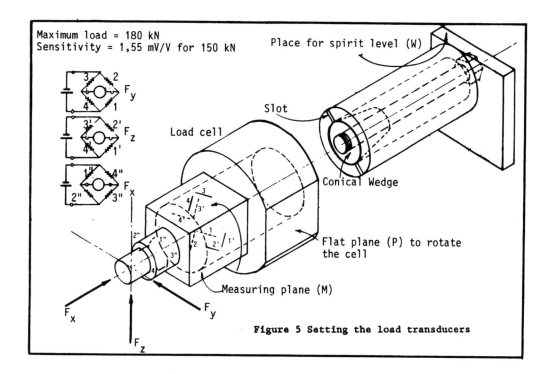

Maximum load = 180 kN
Sensitivity = 1,55 mV/V for 150 kN

Place for spirit level (W)

F_y

F_z

F_x

Load cell

Slot

Conical Wedge

Flat plane (P) to rotate the cell

Measuring plane (M)

F_x

F_y

F_z

Figure 5 Setting the load transducers

Pump LP_2 and motor MD_2 operates in exactly the same way. A part of the low pressure oil from LK_2 is used for pump lubrication.

Two Brevini double stage gear boxes (type EC 2150F) with gear ratio 31,5:1 and planetary reduction hubs (3,58:1), were used. For towing speeds above 15 km/h, the sun gear in the planetary hub is drawn sideways onto a transfer shaft, unlocking the drive system to obtain a free wheel option.

6 SPEED AND FORCE TRANSDUCERS

6.1 Speed transducers

The true ground speed was measured by using a Ono Sokki magnetic pulse pickup and steel gear, driven by a cable and drum. A simular gear and pickup was used on the shaft of the hydraulic motor.

6.2 Force transducers

The total longitudinal force was measured by mounting a load cell in a regtangular telescopic section at the front of the wagon. (fig 3)
Teflon strips were used between the telescopic parts in order to minimize the influence of friction.

The longitudinal force on the test wheel was measured by a 100 kN bi-directional industrial load cell between the main and floating subframe, attached as near as possible to the test wheel using ball joints to minimize the influence of friction on side force measurements.

The two purpose made bi-directional transducers to measure the vertical and side forces were machined from EN 30 B steel and heat treated. The vertical force F_z was measured by double strain guages 1',2' and 3',4' fixed at 45° to the neutral axis. This mode, sensitive to shear deformation, was prefered to guages used on a bending transducer as it is insensitive to the exact point of application of the shear force. (fig 5)
The side force F_y was measured by guages 1,2 and 3,4.

To prevent the transfer of side forces at the vertical 200 kN load cell, mounted above the king pin, two Amutit-S plates with two sets of octogonal needle bearings, perpendicular to each other, was used.

All purpose made load cells were laboratory checked for accuracy and cross sensitivity.

1759

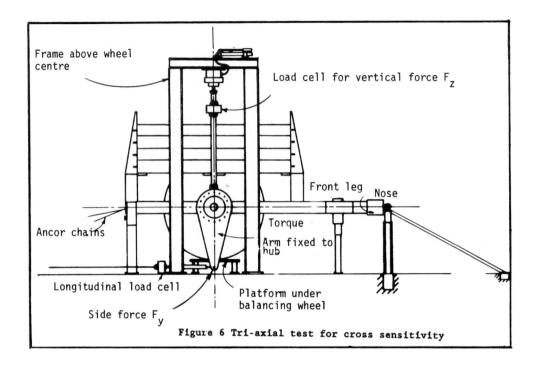

Figure 6 Tri-axial test for cross sensitivity

7 MOUNTING AND SETTING THE LOAD TRANSDUCERS

During test the floating subframe must be parallel to the test track. For this purpose the transducer mountings may be moved vertically against the I beam and wedged into position with the centre line of both transducers exactly in line and parallel to the main axis.

The measuring surfaces of both transducers must also be exactly parallel and perpendicular to the test track. For this purpose a special setting unit was clamped in the load transducer, parallel to the measuring axis. (fig 5) The transducer was then rotated by two set screws and the unit checked with a high precision spirit level.

To minimize the influence of friction, the subframe was attached to the two horizontal transducers by using two 3 mm 160 mm diameter membranes (heat treated EN 19 steel). As the membranes acted as two springs in parallel to the 100 kN load cell (L), the longitudinal calibration was done on the test wagon. (fig 3)

8 THE TRI-AXIAL TEST FOR CROSS SENSITIVITY

Under typical testing conditions, three perpendicular forces and an input torque may be applied to the test wheel. It is therefore of vital importance to measure the forces and torque without a cross influence between the transducers. For this purpose the wagon was loaded with a 12 000 kg mass, fixed to the ground and the test wheel replaced by a 910 mm torque arm. (fig 6)

Certain combinations of F_x, F_z and F_y forces, with maximum values of 20 kN, 50 kN and 20 kN were applied, measurements taken on the transducers, and the forces calculated and compared to the applied forces and wheel torque. The measured and applied forces were in close correlation, except in the case of very small side and longitudinal forces.

9 HARD SURFACE TESTS ON TRACTOR TYRES

Hard surface tests were carried out on two partly used tractor tyres on a fairly smooth asfalt surface for five static wheel loads between 24,3 kN and 53,6 kN and slip angles from 0 - 12° in steps of 2°. This was done for the free rolling, braked and driven option.

For the braked option the towing vehicle loses speed at a fast rate, causing difficulties when determining the maximum braking coefficient.

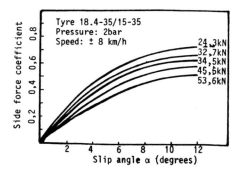

Figure 7 The influence of static load and side slip angle on side force coefficient for a free rolling tractor tyre.

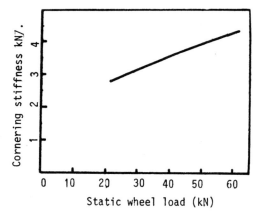

Figure 8 The influence of static load on cornering stiffnes

To prevent possible damage to the driving shafts when heavily laden, driving tests on the asfalt surface were not carried out to high slip values.

Test results for the free rolling tyres are shown in fig. 7 and 8. As a result of load transfer the dynamic wheel loads were slightly less than the static values.

10 CONCLUSION

The present tendency is towards higher road speeds for tractors and tractor trailer combinations. When modelling, the dynamic behavior of such vehicles, the side force characteristics for large heavily laden tractor tyres are of major importance.

Very little if any information is available on this aspect of tyres and a two wheel field tyre tester was built to measure the forces.

Tests were undertaken with tractor tyres on fairly smooth asfalt surfaces and the results showed a close correlation with those obtained by testing truck tyres on similar surfaces. (Weber, 1987) These tests will in the near future also be carried out on soft soil surfaces and the results will probably be available at a later stage.

LITERATURE

CLARK, S.K. 1981. Mechanics of Pneumatic Tyres. U.S. Dept. of Transportation, National Highway Traffic Safety Administration, Washington.DC.

CROLLA, D.A. & A.S.A. el-Razaz, 1987. A review of the combined lateral and longitudinal force generation of tyres on deformable surfaces. J. of Terramechanics 24 (3): 199 - 225.

KRICK, G. 1971. Behaviour of tyres driven in soft ground with side slip. Translation from ATZ 73(7): 243-246; (8): 301-306, Dunlop Ltd.

Mc ALLISTAR, M. 1979. A Rig for Measuring the forces on a towed wheel. J. Agric. Engng. Res. 24: 259 - 265.

SCHWANGHART, H. 1968. Lateral forces on steered tyres in loose soil. J. Terramechanics 5(1).

WEBER, R. 1987. Lateral forces of large truck tyres. Course on Vehicle Hardling, Laboratory for Adanced Engineering, University of Pretoria.

Influence of trampling on some physical and hydrophysiological characteristics of woodland soil

A. Ferrero

Institute for Agricultural Mechanization, Torino, Italy

ABSTRACT: A sandy loam soil in deciduous woodland was treated with three levels of mechanical compaction. Over a depth of 0-30 cm, the effects on several chemical, physical and hydrological characteristics were studied, as well as the effects on root and shoot dry matter increase of two sod-sown forage grasses, Poa pratensis and Phleum pratense. The most severe treatment increased the density and thermal conductivity of the 1-10 cm stratum, and appreciably decreased water infiltration; Poa pratensis but not Phleum pratense was adversely affected by compaction in both root and leaf production.

RESUMÉ: Un terrain limoneux sableux de taillis est soumis à trois niveaux croissants de compactage dynamique. Le long du profil 0-30 cm quelques caractéristiques chimiques et physiques, les paramètres hydrologiques et la production de substance sèche radicale et épigée ont été relevés sur deux graminacées fourragères semées sur du terrain solide. Le piétinement plus intense a augmenté la masse volumique apparente, dans la couche 1-10 cm, et la conductibilité thermique; de plus, il a réduit sensiblement la quantité d'eau infiltrée. L'effort de pénétration a augmenté avec le niveau de piétinement. Seule une graminacée, le pâturin, a réduit significativement la production de racines et de feuilles à cause du compactage; la fléole n'a pas subi d'influence.

ZUSAMMENFASSUNG: Ein schlammiger Sandboden eines Schlagholzwalds wurde drei zunehmenden Niveaus dynamischer Kompaktion unterzogen. Entlang dem Profil 0-30 cm wurden einige chemische und physikalische Eigenschaften, die hydrologischen Parameter und die Produktion von trockenem epigäischem Wurzelstoff auf zwei in festem Boden gesäten Futtergräsern aufgenommen. Das stärkste Stampfen erhöhte die Schüttdichte in der Schicht 1-10 cm sowie die Wärmeleitfähigkeit und stellte man außerdem eine deutliche Reduzierung der Wasserdurchsickerung fest. Die Durchdringungskraft nahm gemäß dem Stampfgrad zu. Nur ein Gras nämlich das Rispengras, reduzierte deutlich die Erzeugung von Wurzeln und Blättern wegen der Kompaktion. Das Wiesenlieschgras wurde dadurch nicht berührt.

1 INTRODUCTION

Broad leaved deciduous coppice is widespread in hill areas of Italy, even if it is frequently not submitted to regular forestry practices. In fact the land is often abandoned and allowed to run down as the units of woodland may be of poor quality, modest size and growing on difficult terrain.

The rational use of these woods is possible if they are regularly managed using machinery or are grazed by sheep or cattle. Grazing is more appropriate in areas that present difficultus for mechanization, and is appropriate in the context of integrated exploitation of agro-forestry resources. Experiments on these lines have been conducted on a large scale in other countries (Blackburn 1983).

Both kinds of management require periodic treatments that must take careful account of the existing relationships between air, water and solid matter in woodland soils. The soil structure may be influenced more than with farm soils by management activities such as passage of wheeled vehicles or animals, because woodland soils and their plant associations are more complex.

The effects of the passage of farm machinery over cultivated land have been widely studied. They vary with the type of machinery, soil type and wetness; repeated traffic can have cumulative effects on both soil properties and plant growth (Soane 1981).

On pastures, trampling by animals is the most important factor, and it has been shown that with medium to large sized animals, changes occur, related to the frequency of transit, resulting in changes in vegetation Animal compaction have negative effects also on hydrological parameters.

Many studies attempting to evaluate those effects have been conducted in the laboratory using artificial compaction simulating the hooves of cattle (Chancelor, Schmïdt & Shoeme 1962). Less complete investigations have been made on the long or short term effects of traffic or of animal trampling on woodland soil.

The effects of trampling are linked to the frequency and location of passage, the type of animal, and the density and spatial structure of the herd.

At present there is a need to know more about the effects of trampling in woodland, but work has been limited by the lack of analytical methods, especially for large areas (Lipiec & Tarkiewicz 1986).

A ten year programme at the Institute of Agricultural Mechanization has been set up to study the integration of semi-extensive beef raising into farming and forestry on marginal hill land. In this context, plot experiments were set up to study methods of measuring the compaction effects of traffic or cattle trampling on physical and hydrological characteristics of woodland soil, and on root development.

2 MATERIALS AND METHODS

Four plots, each of 15 m^2 were set out on a hill farm in upper Monferrato, in a coppice of oaks with prevailing floristic component belonging to Quercetalia pubescentis mixed with Robinia pseudo-acacia.

The soil, resting on marls, was sand-loamy and rich of organic matter in the upper layer. The plots were left undisturbed as controls (C_0) or were given one, two or three compaction treatments (C_1, C_2, C_3) with a pneumatic compactor of mass 76 kg, delivering an energy of about 27 kJ/m^2. The soil moisture at compaction was about 21% by weight.

At the beginning and end of each treatment, samples were taken from depths of 1-10 cm, 11-20 cm and 21-30 cm to determine particle size distribution, chemical composition, and real density. Undisturbed soil cores 7.3 cm in diameter were also taken, using a standard coring device, for measurement of apparent density of dried samples.

Resistance to penetration was measured periodically, using a hand driven penetrometer with crank advance, and a 1 cm^2 cone-base area capable of continuosly registering values from 0.5 to 9.8 MPa. At the same time soil samples were taken from the three layers for determination of soil moisture. Further gravimetric measurements of soil moisture M (%, w/w) were made throughout the season, and after rainfalls.

Undisturbed samples from the three layers were also taken, using a 100 cm^3 cylinder probe; these samples were water-satured from the bottom, to measure water content at saturation (SC), and at field capacity (FC). On other samples were measured the water retention capacity corresponding to the wilting point (WP), with a Richards porous membrane apparatus, following the method of Luppi (1973).

On bare soil of 23-28% average humidity, on the 0-30 cm profile and for each plot, the cumulative infiltration and the infiltration rate were measured, using a double cylinder infiltrometer, internal diameter 300 mm, with constant load of h = 0.03 m, adapted for use on hill sites.

Twice during the season, at the end of August and the end of September, for periods of 16 days each, the 24 h temperature cycle was measured with thermo couples at depths of 5 and 10 cm.

On one part of each plot, two grasses,

Poa pratensis and Phleum pratense were surface-sown. These species have different rates of establishment and susceptibility to trampling. Germinability, and the number of culmus and leaves 40 and 80 days after sowing, were measured. Then 24 plants of each species were extracted with a cylindrical sampler, and the dry matter production of shoots and roots was measured per unit volume, over the depth 0-25 cm.

Each cylinder of soil, with the associated plant, was immersed for 24 h in a 10% aqueous solution of sodium hexametaphosphate, and then washed with running water on a sieve, to separate the roots, using the method of Wilhelm, Mielke & Feuster (1982). The mean length of roots and leaves was also measured on these plants. The data were examined by analysis of variance and by comparing the means with the Duncan test.

3 RESULTS AND DISCUSSION

3.1 Chemical and granulometric composition

The relevant data are given in Table 1. The granulometric composition did not appear to be influenced by compaction levels, except that the proportion of coarse silt was greater in control plot and C_1. The organic carbon-total nitrogen ratio (C/N) was also not changed by the treatment.

Greater differences were found in the levels of organic matter, total nitrogen and

assimilable phosphorus but only in the two upper layers sampled.

Levels of these parameters were highest in the untreated plot or that one receiving the light compaction (C_1), statistical analysis did reveal significant differences. It seemed that repeated compaction reduces organic matter and assimilable phosphorus, especially in the upper layers.

3.2 Physical and hydrological characteristics

The apparent density of the solid portion of the soil was significantly increased (1.306 and 1.396 g/cm^3) in the uppermost layer, by treatments C_2 and C_3. The effects of repeated "trampling" on the apparent density were lessened at an intermediate depth (11-20 cm) since the data were not significantly different, for different treatments.

The soil water content (%, v/v) at saturation (SC) was higher 53.84% (v/v) though not statistically significant due to high variability, in the top layer of undisturbed plots, whereas no differences were found for other levels (Table 2).

The field water capacity (FC), the parameter most sensitive to structural conditions, did not differ in the different plots Minor differences occurred in wilting point (WP) - the level of water retention in the soil, as measured by a Richards porous membrane at 1.5 mPa (pF 4.2) - and

Table 1. Woodland soil texture and chemical characteristics measured at three different depths, on a control plot and those given three different levels of compaction (°).

SOIL CHARACTERISTICS AT THREE DIFFERENT DEPTHS		CONTROL PLOT C_0			COMPACTED PLOTS								
					C_1			C_2			C_3		
		1-10 (cm)	11-20 (cm)	21-30 (cm)	1-10 (cm)	11-20 (cm)	21-30 (cm)	1-10 (cm)	11-20 (cm)	21-30 (cm)	1-10 (cm)	11-20 (cm)	21-30 (cm)
Coarse sand	(2-0.2 mm) %	22.60	18.28	17.92	23.54	20.43	19.74	21.76	20.21	20.56	21.31	19.23	18.48
Fine sand	(200-50 μm) %	49.08	54.24	57.56	50.05	52.01	56.12	56.36	50.66	55.65	50.16	53.59	57.12
Coarse silt	(50-20 μm) %	13.24a	11.69	9.87	12.49ab	10.54	11.10	8.24	10.82b	10.23	9.36	9.35b	8.15
Fine silt	(20-2 μm) %	12.16	12.06	10.01	13.20	12.42	10.90	13.43	14.76	9.96	15.30	14.25	10.77
Clay	(<2 μm) %	3.22	3.43	4.64	3.82	3.60	4.02	3.63	4.14	4.47	3.98	4.12	4.58
pH (w/v H$_2$O)		6.88	6.93	7.14	6.94	6.88	7.06	6.95	7.01	7.18	6.98	7.16	7.22
Total nitrogen	%	0.38	0.31	0.27	0.36	0.32	0.26	0.32	0.30	0.25	0.29	0.31	0.26
Organic matter	%	7.23a	5.97a	4.81	6.33ab	5.19	4.93	6.05b	4.98b	4.78	5.78c	5.02b	4.82
C/N ratio		10.73	11.19	10.33	11.18	9.41	11.00	10.97	9.62	11.09	11.56	9.39	10.75
Available P	ppm	13.80a	11.90a	9.20	11.30a	11.43	8.23	9.66c	10.12b	8.83	9.16c	9.52c	8.24
Exchangeable K	me/100 g	0.35	0.28	0.26	0.34	0.26	0.22	0.31	0.27	0.30	0.33	0.25	0.21

(°) Mechanical compaction treatments performed once (C_1), twice (C_2) or three times (C_3). Means relating to the same soil depth, followed by different letters differ significantly by P = 0.05.

Table 2. Physical and hydrological characteristics of woodland soil three different levels of compaction (°).

SOIL CHARACTERISTICS AT THREE DIFFERENT DEPTHS	CONTROL PLOT C_0			COMPACTED PLOTS C_1			C_2			C_3		
	1-10 (cm)	11-20 (cm)	21-30 (cm)	1-10 (cm)	11-20 (cm)	21-30 (cm)	1-10 (cm)	11-20 (cm)	21-30 (cm)	1-10 (cm)	11-20 (cm)	21-30 (cm)
PHYSICAL												
Density (g/cm^3)	2.389	2.456	2.510	2.412	2.488	2.562	2.430	2.476	2.591	2.435	2.523	2.482
Dry bulk density (g/cm^3)	1.196D	1.209c	1.182	1.257C	1.218c	1.198	1.306B	1.243b	1.201	1.395A	1.297a	1.211
NYDROLOGICAL (%, v/v)												
Saturation capacity	53.84	52.75	51.34	50.99	53.15	51.16	49.62	52.70	50.72	49.20	52.52	49.81
Field capacity	40.91	39.72	37.53	40.25	39.44	37.22	39.56	41.12	38.34	38.96	39.19	36.33
Wilting point at pF 4.2	14.81	14.34	14.44	15.12	14.98	13.50	14.76	13.77	13.88	15.36	14.20	14.33
Percolatable water	12.93a	13.03	13.81	10.74b	13.71	13.94	10.06b	11.58	12.38	10.24b	13.33	13.48
Available water	26.10	25.38	23.09	25.13	24.46	23.72	24.80	27.35	24.46	23.60	24.99	22.00

(°) Mechanical compaction treatments performed once (C_1), twice (C_2) or three times (C_3). Means relating to the same soil depth, followed by different letters differ significantly by P = 0.05 (lower case) or by P = 0.01 (upper case).

this confirms that this characteristic moisture level is little influenced by human intervention, and is more dependent on soil composition (Luppi 1973).

The percolable water, defined as the difference between the content at soil saturation and at field capacity (SC-FC), can give a useful indication of the macroporosity and structural condition of the soil (Sasso 1978). It was significantly higher only at the 1-10 cm depth of plots C_0 and C_1.

Available water, calculated as the difference between content at field capacity and that at wilting point (FC-WP) - a parameter fundamental for plant development - did not show significant differences but appeared to be slightly greater in the intermediate stratum.

The characteristics of the available water and field capacity suggest that in conditions similar to those of the trial, where there are good contents of organic matter and sand, compaction treatments of the levels applied do not materially affect the distribution, of macro and micropores. Micropore distribution should be longely determined by the chemical composition and particle size distribution.

3.3 Temperature

Daily soil temperature changes at 5 and 10 cm depth (Cavazza 1967) were recorded for 16 days, from 28 August to 12 September

(Fig. 1). Mean values in all plots gave typical bell-shaped curves, less marked at deeper levels.

Highest values and greatest fluctuations were found at the 1-10 cm level in plots C_2 and C_3; these plots behaved differently from C_0 and C_1, in warming and cooling more rapidly. Differences in night minimum were smaller, and the curves more uniform.

The greatest temperature difference (2.7 °C) was found between plot C_3 and C_1, (Fig.1); the maximum temperature was generally found in plots receiving the most impaction.

At a depth of 10-20 cm, daily temperature variation was reduced, although the maximum difference (2.4 °C) was only slightly less than above. Plots C_3 and C_2 always showed the highest temperature and, in contrast to plots C_1 and C_0, tended to have temperatures permanently raised, even as regards the nightly minima. All compacted plots reached their maximum temperatures at about the same time, whereas the control plot evidenced greater thermal inertia.

The clear differences in thermal behaviour shown in the upper layers of C_3 and C_2 could be due to their water status, characterized by lower humidity; this could be responsible for the worse state of thermal inertia in these soils. In addition, the higher level of compaction could have increased thermal conductivity in the upper layers, so causing greater warming of the soil at the 10 cm level.

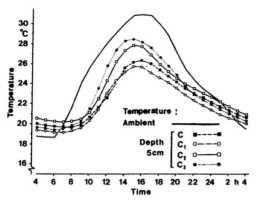

Fig.1 Mean variation of recorded temperature, over a period of 16 days, at 5 and 10 cm depth in the control plot (C_0) and in those given one compaction (C_1), two treatments (C_2) or three treatments (C_3).

3.4 Variations in humidity and water penetration

Fig. 2 shows the changes in soil moisture M (%,w/w) with time and along the depth profile, and the correlation with precipitation. Greatest differences were noted at intermediate soil moistures, but were almost abolished near field capacity and the wilting point. Differences were smallest at intermediate depth.

Plot C_3, receiving the heaviest treatment, showed consistently different behaviour, characterized by mean moisture values lower and less variable than in the other plots. Also, particularly in the top layer, there was an appreciably lower response to precipitation less than 10 mm.

The control gave moisture content values equal or slightly lower than plot C_1, with a tendency to dry out more rapidly. This was probably due to the presence of a greater number of macropores, and a smaller response to rain.

The data show that, over the period of observation, only repeated compaction had a detrimental effect on water uptake, especially at greater depths. However, light compaction appears to have retarded drying of the surface.

The amount of water percolated was low in all cases, as shown in Figure 3. This shows the cumulative infiltration I_c as a function of time t according to the following relation:

$$I_c = at^n$$

where a (water infiltrated after 1 min) and n were experimentally obtained.

C_0 and C_1 were clearly different in total accumulated infiltration after 5 h (44 and 37 mm respectively, both values higher than in plots C_2 and C_3). They also had a higher initial infiltration rate and a higher continuity in infiltration time.

The most heavily compacted plot behaved in a characteristic way. More water infiltrated in the first minutes, but the rate

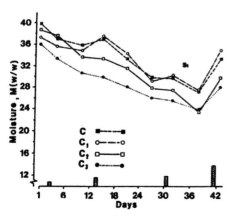

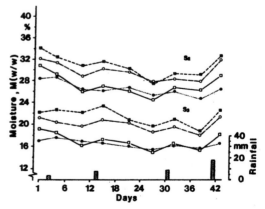

Fig.2 Changes of water content of the soil in layers S_1 (1-10 cm), S_2 (11-20 cm), S_3 (21-30 cm), for the control plot (C_0) and those given one compaction (C_1), two treatments (C_2) or three treatments (C_3).

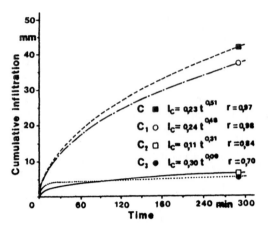

Fig.3 Water cumulative infiltration in soils from the control (C_0) and compacted plots (C_1, C_2, C_3).

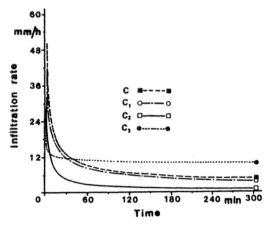

Fig.4 Water infiltration rate in soils from the control (C_0) and compacted plots (C_1, C_2, C_3).

lessened rapidly. The speed of infiltration was lower than that in C_0 and C_1, and became constant after 70 min; however in the other plots, infiltration rate continued to decrease, and reached an almost constant value of 8.78-2.44 mm/h (equivalent to the permanent infiltration rate) after 4 h (Fig. 4).

This behaviour can be explained by the infiltration behaviour observed in the field, which was rather discontinuous in plot C_3, and by inspection of the coefficient a. This value was higher (30 mm) in this plot, and suggests that the soil, at the beginning was in a more favourable condition to pass water downwards, due to lowe‾red moisture content at the surface and the presence of micro cracks in the "floor" present at a depth of 2-3 cm.

The passage of water would have closed these cracks due to soil smelling and to blockage by finer particles, so lowering permeability in a few minutes.

3.5 Resistance to penetration

Figure 5 shows the force needed to penetratrate the soil throughout the profile from 2 to 28 cm depth.

The data were obtained in four tests made at different times with a mean soil moisture M (%; w/w) of 30-33% in the first

15 cm and 24-27% at the deepest level.

Increasing compaction caused increased resistance to penetration, at comparable levels of moisture. All plots showed rather similar behaviour, with a rapid increase in force required, reaching a modest peak at about 4 cm depth for plot C and C_1, and 7 cm for plots C_2 and C_3. After this peak, re‾sistance was uniform down to 15-16 cm, and

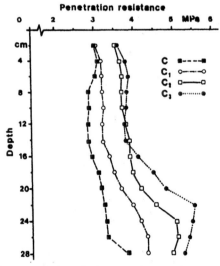

Fig.5 Trends of penetration resistance with woodland soil depth, for control (C_0) and compacted plots (C_1, C_2, C_3). Mean of 4 measurements.

in plot C_0 even tended to diminish. Below this depth, all values notably increased, with the maximum in plot C_3.

This behaviour can be explained by the differences in humidity found at different levels, and by the different relationship between penetrability and water content found in the different plots. Figure 6 shows the regression lines representing the relationship between penetration force and moisture content M (%, w/w). The force is inversely proportional to humidity, but varies according to treatment.

The graphs for the treated plots, over humidity values of 17-40% are higher up, and with a steeper slope than for the control. In particular the behaviour of

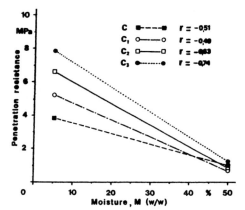

Fig.6 Regression of the relation between water content and penetration resistance in control (C_0) and compacted plots (C_1, C_2, C_3).

plots C_1 and C_2 were similar and concerned a distinctly higher range.

It was clear that all compaction treatments markedly influenced penetrability, with some variation due to moisture, within limits normally encountered in the field.

The values of the correlation coefficients (r_{c0} = -0.51; r_{c1} = -0.49; r_{c2} = -0.63; r_{c3} = -0.74) of the regression plots show that resistance to penetration was distinctly greater in treatments C_2 and C_3, and that the relation between the two variables considered was closer in these plots, probably due to greater cohesion induced by the compaction.

3.6 Growth tests

Compaction influenced seed germination and yield in different ways and to differing extents in the two grasses tested, as shown in Table 3.

The number of plants alive 40 days after seeding was significantly different only for the Poa, and was highest in the control plot. Root production and dry matter yield of the shoots was lower under compaction (C_2 and C_3). The mean root length was higher for plots C_0 and C_1, whereas leaf length did not differ.

With Phleum, differences were found only on root length and weight, which were higher in the plot C_1. Phleum reacted less to the different treatments, and in fact the lowest compaction level favoured root deve-

Table 3. Influence of compaction level of woodland soil on growth of Poa pratensis and Phleum pratense, sodseeded on the experimental plots (°).

GROWTH PROPERTIES		Poa pratensis				Phleum pratense			
		CONTROL	COMPACTED PLOTS			CONTROL	COMPACTED PLOTS		
		C_0	C_1	C_2	C_3	C_0	C_1	C_2	C_3
Emergence	%	90.8a	82.3b	86.7ab	75.3c	81.7	86,3	80,3	77,6
DM production/plant									
- roots	(mg)	50a	41a	28b	27b	27	25	23	22
- shoots	(mg)	131a	118ab	103b	99b	38b	41a	35b	33b
Mean length									
- roots	(cm)	5.8a	5.6ab	5.3b	4.5c	5.8b	6.2a	5.7b	5.5b
- leaves	(cm)	15.1	14.9	14.0	12.9	19.5	20.8	20.6	18.9

(°) Mechanical compaction treatments performed once (C_1), twice (C_2) or three times (C_3). Values for a given property and for each species are significantly different by P = 0.05 where indicated by different letters.

lopment. In contrast, Poa was more sensitive to the adverse conditions induced by repeated compaction.

The results can be interpreted in terms of the differing root patterns. In Poa these were many fine roots, clearly forming bundles. They would have been affected more strongly by compaction, tending to fail to penetrate, instead expanding laterally in plots C_1 and C_2. This behaviour was found in several test soil sections.

The Phleum plants, at the time of sampling, had fewer but larger roots. It is thus likely that under similar conditions, this species was able to develop better roots in less porous soil which was moderately compacted.

4 CONCLUSIONS

The organic matter content, total nitrogen and assimilable phosphorus were all higher in plots C_0 and C_1 than in C_2 and C_3. The apparent density was lower, in coincidence with higher organic matter levels, in C_0 and C_1, and in general these differences tended to disappear with increasing depth. However, hydrological parameters were not changed in any clear way by the treatments, although higher water percolability was correlated with lower soil density. This suggests that, relative to the hydrophysiological characteristics, woodland soil is not greatly harmed by compaction, even if repeated, in terms of plant development.

Confirmation is given by data, in time and down the soil profile, for moisture variation. The figures differed, and not very greatly, only for intermediate moisture levels, between field capacity and wilting point.

The temperature fluctuation was little influenced by compaction, with differences apparent only near midday, due to greater thermal conductivity in plots C_2 and C_3.

However, compaction, when repeated, did influence water infiltration, reducing it appreciably, even 17 times. This suggests the formation of a superficial "pan".

All treatments notably increased resistance to penetration, with different degrees, down the profile. At a depth of 10 cm, resistance was about three times greater in plot C_3.

Compaction clearly affected only one of the two grass species tested, Poa pratensis, which gave markedly lower yields on plots C_2 and C_3.

The data obtained do not permit definitive evaluation of the effects of compaction on those soil properties most important for plant growth. For this, an analysis of the efficiency and resolution of the tests used would be necessary. On the other hand, a critical study of methods available for measuring certain soil characteristics was aimed at. These should be valid indicators of the effects of trampling on the soil, and their further influences on vegetation.

The granulometric and chemical analysis, limited to the test conditions, did not result in useful answers, except for determinations of organic matter and nitrogen. Likewise, apparent soil density and hydrological measurements, using the cylinder method, did not give significant differences between the control and the highest level of treatment, due to changes brought about in the samples and the consequent high variability of the results.

Similarly, soil temperature and moisture measurements were not very sensitive.

Greater sensitivity was shown by the infiltrometer measurements, and in particular determinations on resistance to penetrability. This latter test, apart from being rapid and non-destructive, can be repeated frequently.

It gave the most clear-cut and complete information on the effects of each treatment, down the length of the profile.

The results on growth tests, which should be a synthesis of all other effects, were more complex. The data obtained were not always consistent and reliable, probably because of the short period of the test. However, with longer growth tests, there would be the risk of changes in soil conditions induced by external factors, as well as effects of the plants themselves.

It is necessary, to obtain definitive results, to repeat the tests on other soils compacted in different conditions of moisture and carried out at different season. This would also permit deeper analysis of the influence of climatic factors.

AKNOWLEDGEMENT

The author wishes to express his thanks to Prof. G. Sasso for his help in the study of soil hydrological characteristics.

REFERENCES

Blackburn, W.H. 1983. Livestook grazing impacts on watersheds. Rangelands 5: 123-125.

Cavazza, L. 1967. Fisica del terreno agrario. Torino, UTET.

Chancelor, W.J., R.H. Schmīdt & W. Shoeme 1962. Laboratory measurement of soil compaction and plastic flow. Trans. ASAE 5: 235-239.

Lipiec, J. & S. Tarkiewicz 1986. Méthode d'étude de l'influence du compactage du sol sur le développement du système racinaire. Zes. Prob. Postepow Nauk Rolniczych, 312: 271-276.

Luppi, G. 1973. Ricerche sulla determinazione delle caratteristiche idriche dei terreni. Quaderni de "La Ricerca scientifica", 79: 72-83.

Sasso, G. 1978. L'analisi idrofisiologica del terreno come ipotesi di contributo alla conoscenza agronomica dello strato coltivabile. Relazione sui primi trecento reperti sperimentali. Ann. Acc. Agric. di Torino, 120: 149-219.

Soane, B.D., P.S. Blackwell & J.W. Dickson 1981. Compaction by agricultural vehicles: a review. 2. Compaction under tyres and other running gear. Soil Tillage Res. 12: 373-400.

Wilhelm W.W., L.N. Mielke & C.R. Feuster 1982. Root development of winter wheat as related to tillage practice in Western Nebraska. Agron.J. 74: 85-88.

Land and Water Use, Dodd & Grace (eds), © 1989 Balkema, Rotterdam. ISBN 90 6191 980 0

Soil compaction in the tropics and some means of alleviating it

B. Kayombo
Sokoine University of Agriculture, Morogoro, Tanzania

ABSTRACT: The limited number of experiments on mechanical land clearing in the tropics show convincing evidence of compaction problems both with respect to increasing erosion hazard and decreasing crop productivity. If mechanical land clearing is unavoidable due to economic or social circumstances, it is likely that compaction problems could be materially reduced by the avoidance of wet season traffic, reduced topsoil movement, and a reduction of vehicular traffic. Available literature has also revealed that soil compaction originating from other causes, namely, genetically compacted soil and subsoil gravel horizons, and continuous cultivation and cropping has detrimental effects on soil structure, crop growth and yield, and it can persist for a long time if no adequate steps are taken to minimize or ameliorate it. The no-tillage system with crop residue mulch is an effective soil compaction control practice, provided it can be adopted as a package of all cultural practices that make it work.

RESUME: Un nombre limite' d'expériences portant sur le travail mecanise du sol dans les pays tropicaux mettent en évidence les problèmes avec pour conséquences: l'augmentation de l'éroson et une diminution de la productivite' des cultures. Si la mécanisation agricole est essentielle pour des raisons économiques et sociales. il paraitrait que les problèmes de compaction peuvent être reduits en évitant le passage des machines en saison de pluie. en réduisant le movement de la couche superficielle du sol et la reduction de la circulation des vehicules. Des documentations a' notre disposition nous revélent que la compacton du sol causée par d'autres phenomènes tel que la comacton génétique du sol. les horizons du gravier du sous-sol. la récolte et cultivation continues ont des conséquences tres nuisibles sur la structure du sol.la croissance des plantes et leur récolte. Ce problème peut persister pour une très longue durée au cas où les solutions immédiates ne sont pas trouvées. Le non-labour suivi de mélange des débris végétaux des cultures est une manière efficace de controller la compaction du sol. surtout quand elle peut être adoptée comme pratique commune sur toutes les cultures qui le rendent rentable.

KURZZUSAMMENFASSUNG: Die begrenzte Anzahl von Versuchen über mechanische Landrodung in den Tropen zeigt deutlich die Probleme zunehmender Bodenverdichtung, Bodenerosion und der damit einhergehenden Reduzierung der Bodenfruchtbarkeit an. Wenn auch aufgrund ökonomischer oder sozio-ökonomischer Bedingungen die mechanische Landrodung "unvermeidbar" ist, so kann doch das Ausma der Bodenverdichtung wesentlich reduziert werden indem vermieden wird, während der Regenzeit den Boden zu befahren, den Oberboden durch Bodenbearbeitung zu bewegen und ferner, indem jegliches Befahren des gerodeten Landes auf das unumgängliche Minimum reduziert oder Modifikationen an den Fuhrwerken vorgenommen weden.
Die vorhandene Literatur verweist auch darauf, da Bodenverdischtungen auch andere Gründe haben können, wie pedogenetisch verdichtete Böden, verdichtete Kies- bzw. Steinhorizonte im Unterboden. Ununterbrochene Bodenbearbeitung und kontinuierlicher Pflanzenanbau ohne Brache haben schädliche Auswirkungen auf die Boden-Struktur, auf das Pflanzenwachstum und den Ertrag. Bodenverdichtungen sind sehr persistent, wenn keine adequaten Ma nahmen zur Minimierung bzw. zur Boden-Melioration getroffen werden.

Das System der pfluglosen Bodenbearbeitung ("No-Tillage-System") mit Verbleib der Ernteteste auf dem Feld zum Mulchen, ist eine effektive Ma   nahme zur Kontrolle von Bodenverdichtungen, vorausgesetzt es kann als Ma   nahmenpaket in die Landbaupraktiken für alle entsprechenden anzubauenden Kulturpflanzen integriert werden.

1 INTRODUCTION

Numerous large-scale land development programmes with mechanized agricultural production are inevitable as attempts are made to rapidly increase food production. However, the conclusions drawn from a critical appraisal of implemented large-scale agricultural development schemes in tropical Africa, Latin America and tropical Asia are far from encouraging. Failure of such schemes has been attributed to many factors, including socio-economic conditions and infrastructure (Baldwin 1957), insufficient baseline data to enable adequate planning for resource development and management (De Wild 1967, Bauer 1978), and failure of monsoons. In addition to these exogenous factors, the ability of the soil to sustain economic yield is obviously an important factor and needs to be adequately assessed. The importance of soil factors is basic to all planning.

Soil compaction is best confined to the process whereby a mechanical load of short duration results in a reduction of soil pore volume, a modification of pore structure, a decrease in permeability and an increase of strength (Soane 1982). Compaction of agricultural soils is of world-wide concern because of increased use of heavier farm machinery and increased cultivation of new lands in developing countries, particularly in the tropics. The demand for greatly increased rates of work to provide a rapid transition from forest to crops to reduce erosion hazards has led to the use of very much heavier machinery in recent years, and the low structural stability of many tropical soils, combined with the high erosivity of rainfall increase the risks of serious soil degradation by field traffic (Lal 1984).

This paper attempts to assess the prevailing status of the soil compaction problem and identifies measures which can alleviate it in the tropics.

2 INFLUENCE OF COMPACTION ON SOIL PROPER-TIES AND CROPS

Soil compaction in tropical soils is caused by
(1) occurrence of naturally compacted soils (Martinez and Lugo-Lopez 1953, Macartney et al. 1971, Willcocks 1981),
(2) occurrence of abundant subsoil gravel horizons (Thomas 1974, Babalola & Lal 1977, Grewal et al. 1984),
(3) use of heavy machinery in land clearing (Seubert et al 1977, Hulugalle et al. 1984, Dias and Nortcliff 1985), and
(4) increased mechanization and continuous cropping that has resulted in an increased exposure of structurally fragile soils to high intensity storms and year-round high temperatures (Trouse and Humbert 1961, Ahn 1968, Kemper and Derpsch 1981, Oni and Adeoti 1986).

Soil compaction effects on tropical soils result in increases of soil bulk density (Seubert et al. 1977, Kayombo and Lal 1986a) and soil strength (Willcocks 1981, Hulugalle et al. 1984) and restrict the movement of water and air through the soil by reducing the diameter and number of large pores (Sanchez 1981, Dias and Nortcliff 1985).

The various factors causing soil compaction as outlined above, are bound to exert an enormous impact on establishment, growth and yield of crops in tropical regions. Soil compaction adversely affects seedling emergence, shoot and root growth, and grain yield of upland rice (Ogunremi 1983). Studies on paddy rice, however, indicate that soil compaction enhances better establishment and higher yield up to a soil bulk density of $1.50 - 1.60$ Mgm^{-3} but impeded above this range (Ghildyal 1978, Ogunremi 1983). Studies on maize show that plant development and grain yield are severely restricted by compaction (Seubert et al. 1977, Grewal et al. 1984, Hulugalle et al. 1984, Kayombo et al. 1986). Soil compaction adversely affects root growth of soybean by reducing the length and weight of the roots and by decreasing nodulation (Singh et al. 1971, Maurya and Lal 1979). The nutrient status and moisture regime of the soil regulate the damaging effect of soil compaction on soybean yield (Katoch et al. 1983, Noguira et al. 1983). Cowpea root growth and grain yield , however, appear to be little affected by soil compaction (Maurya and Lal 1979, Willcocks, 1981, Hulugalle et al. 1984). Root crops are no exception to the vagaries of soil compaction. Cassava tuber development is restricted by compactive force (Onwueme 1978). Studies in south western Nigeria, however, indicate that soil compaction increases tuber yield up to a soil bulk density of 1.35 – 1.50 Mg m^{-3} but impeded above this range (Kayombo and Lal 1986b). Studies conducted in Colombia show a similar observation (Cateno and Perez 1982).

3 CONTROL OF SOIL COMPACTION IN THE TROPICS

Control of soil compaction is a continual requirement in agriculture in order to maintain crop productivity. The performance of field operations necessarily involves some compaction. Hence a major task of soil management is, first, to minimize soil compaction to the extent possible, and, second, to alleviate that unavoidable measure of compaction caused by traffic and tillage once it occurs.

3.1 No-tillage farming

In no-tillage, seeds are planted in a narrow slit opened mechanically or by manually operated equipment and tools, in the killed sod or in previous crop residues without primary or secondary tillage operations (Lal 1983). The continuity of channels created by the earthworms and related biological activity, and macropores created by decaying root system of previous crops are important factors in water transmission through the profile during the high intensity of storms. It is also the stability and continuity of these channels in the untilled soil that favour deep root system development into the gravelly subsoil horizons which are otherwise difficult for roots of seasonal crops to penetrate (Babalola and Lal 1977). No-tillage benefits on soil structure maintenance and crop performance have been demonstrated in East and West Africa (Khatibu and Huxley 1979, Lal 1984) and Latin America (Kemper and Derpsch 1981, Roth et al. 1988). The data from Nigeria on an Alfisol indicate that grain yields remain high in a no-till mulch system even after 24 consecutive crops of maize (Lal 1984).

3.2 Controlled grazing

It is generally speculated that soil compaction and subsequent erosion on properly managed pastures, with controlled grazing, is not as serious as from arable lands planted to open row crops. However, there is very little quantitative data available to support this argument. Excessive grazing has resulted in denuda-

tion of the vegetative cover and in accelerated soil erosion (Greenland and Lal 1977). One substantive study in East Africa indicates that 40 percent of rainfall is lost as runoff from heavily grazed watersheds (Pereira 1973). From the point of view of soil and water conservation, grazing pasture should not exceed one cow/2.4 ha.

3.3 Avoiding compaction in land clearing

Clearly the risks of compaction in mechanical land clearing are so high that possible means of avoiding the problem must be explored. The use of improved methods of hand clearing followed by rapid establishment of cover crops seems to largely avoid the problem. Where clearing by mechanical means is the only practical method, attention to certain details in machine selection and operation controls may reduce the incidence of compaction by :
(1) timing of mechanical clearing operations so as to take place in the dry seasons,
(2) avoiding loss of surface organic matter and plant debris, and
(3) traffic and machinery operation control through restriction of traffic to predetermined tracks (Soane et al. 1982) and reduction of the distance between windrows to reduce the number of vehicle passes (Van der Weert 1974).

3.4 Root growth and biological loosening in compact soils

Many field investigations suggest that there are large differences in the ability of different plant species to overcome mechanical stress in the soil. This opens an attractive approach to the alleviation of soil compaction, namely, to look after pioneer plants with study roots and a greater ability than the average to penetrate semi-rigid soil. So far very little systematic work has been done along these lines, but several authors have pointed out that the cowpea root system is not seriously impeded by compaction below tillage depth (Maurya and Lal 1979, Willcocks 1980, Hulugalle et al. 1984). In Nigeria the roots of a leguminous shrub, Mucuna utilis, have been found to penetrate the gravel layer at 20 cm depth and grow to a depth of 30 cm (Hulugalle et al. 1984). Use of cover crops and suitable tree species followed by controlled and planned grazing has been found to restore the fertility of compacted and eroded lands in East Africa (Pereira and Thomas 1961, Pereira and Dagg 1967) and Latin

America (Kemper and Derpsch 1981, Roth et al. 1988). A legume, Stylosanthes guianensis, has been found to be particularly suitable for restoring compacted and eroded Alfisols in West Africa (Lal et al. 1979).

3.5 Mechanical loosening of compact soils

Cultivation after land clearing leads to a significant increase in structural instability of the soil. It thus becomes necessary to ameliorate soil compaction from mechanical clearing and continuous cultivation if crop production is to be sustained on these soils. Subsoiling and chisel ploughing have been found to markedly increase crop yields from severely compacted Ultisols in Peru (Alegre et al. 1982). In West Africa, periodic chiseling upto 50 cm depth in the row zone with a mole attached to the chisel plough has been recommended for compacted Alfisols resulting from mechanized no-tillage practices in order to improve infiltration and root development in the subsoil horizon (Lal 1983).

In upland tropical regions dominated by naturally compacted soils, some mechanical cultivation is necessary for good crop establishment and yield (Northwood and Macartney 1971, Willcocks 1981). In Senegal, with shallow cultivation, the bulk density of the first few centimetres decreases from 1.60 to 1.40 Mg m^{-3}. With tractor ploughing the same values reach to a depth of 10 - 30 cm (Charreau 1972). Further improvements to the amelioration of naturally compacted soils can be done through the adoption of a special controlled traffic tillage system variously known as zonal tillage (Northwood and Macartney 1971) or precision strip tillage (Willcocks 1981). In this system mechanical cultivation by a narrow subsoil tine is restricted to a strip of about 30 cm width and at 75 cm crop row intervals. This system offers the following advantages:

(1) microcatchment of water from the untilled interrow space to the tilled strip, and

(2) maintenance of a reasonable soil porosity in the crop rows since there is controlled traffic.

On the other hand, not all instances of soil compaction are necessarily harmful. In some dry climates a rather heavy compaction of highly permeable sandy soils may improve the efficiency of scarce rainfall and irrigation by reducing the infiltration rate as well as the rate of crop water use (Gupta 1981, Gulati et al. 1985). In paddy-rice production, soil puddling and compaction are essential parts of the production system (Hakansson et al. 1988).

4 CONCLUSIONS

Available literature from the tropics reveal that soil compaction originating from mechanical land clearing, mechanized continuous cultivation and cropping compounded by high intensity tropical rainfall, genetically compacted soils and subsoil gravel horizons, does indeed occur; has detrimental effects on soil structure, crop growth and yield; and it can persist for a long time if no adequate steps are taken to minimize or ameliorate it. The no-tillage system is an effective soil compaction control practice, provided it can be adopted as a package of all cultural practices that make it work. Other control measures as well as biological and mechanical means of alleviating soil compaction should be used according to local conditions.

REFERENCES

Ahn, P.M. 1968. The effects of large scale mechanized agriculture on the physical properties of West African soils. Ghana J. Agric. Sci. 1:35-40.

Alegre, J.C., D.K. Cassel, D. Bandy & P. A. Sanchez 1982. Effects of land clearing on soil properties of an Ultisol and subsequent crop production in Yurimaguas, Peru. Paper, Proc. Int. Conf. Land Clearing Development, IITA, Ibadan.

Babalola, O. & R.Lal 1977. Subsoil gravel horizon and maize root growth: I. Gravel concentration and bulk density effects. Plant Soil 46: 337 - 346.

Baldwin, K.D.S. 1957. The Nigeria agricultural project: An experiment in African development. Oxford:Blackwell.

Bauer, F.H. (ed.) 1978. Cropping in North Australia: Anatomy of success and failure. Proc. 1st NARU Seminar, 24-27 August 1977, Darwin.

Catano, A.H.O. & F.C.A. Perez 1982. Evaluacion de tres sistemas de preparacion de suelo en cuatro materiales de yuca (Manihot esculenta Crantz). Tesis Ing. Agr. Plamirg, Universidad Nacional de Colombia.

Charreau, C. 1972. Problems poses par l'utilisation agricole des sols tropicaux par des cultures annuelles. Agron. Trop. 27: 905 - 929.

Dias, A.C.C.P. & S. Nortcliff 1985. Effects of two land clearing methods on the physical properties of an Oxisol in the Brazilian Amazon. Trop. Agric. (Trin) 62: 207 - 212.

Ghildyal, B.P. 1978. Effects of compaction and puddling on soil physical properties and rice growth. In "Rice and Soil", p. 317 - 336. Manila : IRRI.

Greenland, D.J. & R. Lal (eds.) 1977. Soil conservation and management in the humid tropics. Chichester: John Wiley & Sons.

Grewal, S.S., K. Singh & S. Dyal 1984. Soil profile gravel concentration and its effecton rainfed crop yields. Plant Soil 81: 75 - 83.

Gulati, I.J., K.C. Laddha, F. Lal & R.P. Gupta 1985. Effects of compacting sandy soil on soil physical properties and yield of guar (Cyamposis tetragonoloba L.). Trans. Indian Soc. Desert Tech. & Univ. Centre for Desert Studies 10: 19 - 23.

Gupta, R.P. 1981. Management of highly permeable and low water retentive soils. Indian Agric. Res. Inst. S.P.C. Bull. 1: 1 - 14.

Hakansson, I., W.B. Voorhees & H. Riley 1988. Vehicle and wheel factors influencing soil compaction and crop response in different traffic regimes. Soil Tillage Res. 11 : 239 - 282.

Hulugalle, N.R., R. Lal & C.H.H. Ter Kuile 1984. Soil physical changes and crop root growth following different methods of land clearing in western Nigeria. Soil Sci. 138 : 172 - 179.

Katoch, K.K., G.C. Aggarwal & F.C. Gary 1983. Effect of nitrogen, soil compaction and moisture stress on nodulation and yield of soybean. J. Indian Soc. Soil Sci. 31: 215 - 219.

Kayombo, B: & R. Lal 1986a. Effects of soil compaction by rolling on soil structure and development of maize in no-till and discing systems on a tropical Alfisol. Soil Tillage Res. 7 : 117 - 134.

Kayombo, B. & R.Lal 1986b. Influence of traffic-induced compaction on growth and yield of cassava (Manihot esculenta Crantz). J. Root Crops 12: 19 - 25.

Kayombo, B., R.Lal & G.C.Mrema 1986. Traffic-induced compaction in maize, cowpea and soya bean production on a tropical Alfisol after ploughing and no-tillage: Crop growth. J. Sci. Food Agric. 37 : 1139 - 1154.

Kemper, B. & R. Derpsch 1981. Results of studies made in 1978 and 1979 to control erosion by cover crops and no-tillage techniques in Parana, Brazil. Soil Tillage Res. 1 : 253-267.

Khatibu, A.I. & P.A. Huxle 1979. Effects of zero cultivation on the growth, nodulation and yield of cowpea (Vigna unguiculata) at Morogoro, Tanzania. In R.Lal (ed.), Soil tillage and crop production, Proc. Series 2 p. 271-280. Ibadan : IITA.

Lal, R. 1983. No-till farming: soil and water conservation and management in the humid and subhumid tropics Ibadan : IITA.

Lal, R. 1984. Soil erosion from tropical arable lands and its control. Adv. Agron. 37: 183 - 248

Lal, R., G.F.Wilson & B.N. Okigbo 1979. Changes in properties of an Alfisol produced by various crop covers. Soil Sci. 27 : 377 - 382.

Macartney, J.C., P.J. Northwood, M. & R. Dawson 1971. The effect of different cultivation techniques on soil moisture conservation and the establishment and yield of maize at Kongwa, central Tanzania. Trop. Agric. (Trin.). 48: 9 - 23.

Martinez, M.B. & M.A. Lugo-Lopez 1953. Influence of subsoil shattering and fertilization on sugarcane production and soil infiltration capacity. Soil Sci. 75 : 307 - 315.

Maurya, P.R. & R. Lal 1979. Effects of bulk density and soil moisture on radicle elongation of some tropical crops. In R. Lal & D.J. Greenland (eds.), Soil physical properties and crop production in the tropics, p. 339-347. Chichester : John Wiley & Sons.

Noguira, S., S.S. Dos & S. Manfredini 1983. Influencia do compactacão do solo no desenvolvimento da soja. Pes. Agr. Bras. 18 : 973 - 976.

Northwood, P.J. & J.C. Macartney 1971. The effects of different amounts of cultivation on the growth of maize on some soil types in Tanzania. Trop. Agric. (Trin.) 48: 25 - 33.

Ogunremi, L.T. 1983. Rice response to tillage and water regimes. Ph.D. thesis, Univ. Ibadan.

Oni, K.C. & J.S. Adeoti 1986. Tillage effects on differently compacted soil and on cotton yield in Nigeria. Soil Tillage Res. 8: 89 - 100.

Onwueme, I.C. 1978. The tropical root and tuber crops. New York: John Wiley & Sons.

Pereira, H.C. 1973. Land use and water resources in a temperate and tropical climate. Cambridge: Cambridge Univ. Press.

Pereira, H.C. & D.B. Thomas 1961. Productivity of tropical semi-arid thorn-shrub country under intensive management. Emp. J. Exp. Agric. 29: 269-286.

Pereira, H.C. & M. Dagg, 1967. Effect of tied ridges, terraces and grass leys on a lateritic soil in Kenya. Exp. Agric. 3 : 89 - 98.

Roth, H.C., B.Meyer, H.G. Frede & R. Derpsch 1988. Effect of mulch rates and tillage systems on infiltrability and other soil physical properties of an Oxisol in Parana, Brazil. Soil Tillage Res. 11: 81 - 91.

Sanchez, P.A. 1981. Soils of the humid tropics. Studies in Third World Societies 4:347-410.

Seubert, C.E., P.A. Sanchez & C. Valverde 1977. Effect of land clearing methods on soil properties of an Ultisol and crop performance in the Amazon jungle of Peru. Trop. Agric. (Trin.) 54: 307 - 320.

Singh, A., J.N. Singh & S.K. Tripathi 1971. Effect of soil compaction on the growth of soybean (Glycine max Merril). Indian J. Agric. Sci. 41: 422 - 426.

Soane, B.D. 1982. Processes of soil compaction under vehicular traffic and means of alleviating it. Paper, Proc. Int. Conf. Land Clearing Development, IITA, Ibadan.

Soane, B.D., J.W. Dickson & D.J. Campbell 1982. Compaction by agricultural vehicles: A review. III. Incidence and control of compaction in crop production. Soil Tillage Res. 2 : 3 - 36.

Thomas, M.F. 1974. Tropical geomorphology: A study of weathering and land form development in warm climates. London: Macmillan.

Trouse, A.C. & R.P. Humbert 1961. Some effect of soil compaction on the development of sugarcane roots. Soil Sci. 91: 208 - 217.

Van der Weert, R. 1974. Influence of mechanical forest clearing on soil conditions and resulting effects on root growth. Trop. Agric. (Trin.) 51: 325 - 331.

Willcocks, T.J. 1980. Semi-arid tillage studies. In "Final Scientific Report of Dryland Farming Research Scheme Phase II", Vol. 1, ODA, London.

Willcocks, T.J. 1981. Tillage of clod-forming sandy loam soils in the semi-arid climate of Botswana. Soil Tillage Res. 1 : 323 - 350.

Land and Water Use, Dodd & Grace (eds), © 1989 Balkema, Rotterdam. ISBN 90 6191 980 0

Horizontal penetrometer for trench walls

G.Manor
Technion, Israel Institute of Technology, Haifa, Israel

R.L.Clark
Department of Agricultural Engineering, University of Georgia, Athens, Ga., USA

D.E.Radcliffe
Department of Agronomy, University of Georgia, Athens, Ga., USA

ABSTRACT: The hardware and software design and initial testing of a horizontal penetrometer used to measure the average cone index of each homogeneous layer of soil exposed in a trench wall is described. The objective of the horizontal penetrometer described is to develop a more complete understanding of vertical penetrometer data as related to soil layers.

1 INTRODUCTION

In recent years, the cone penetrometer (ASAE standard S313.2, 1988; Gill and Vanden Berg, 1968) has seen increased popularity as a reliable tool to quantify soil mechanical properties (Perumpral, 1987). The main reasons for this increase in reliability has been the use of hydraulic cylinders to insert the probes at a constant speed and the development of microcomputer based instrumentation systems to collect and analyze penetrometer data. These two factors have contributed significantly to the ability to collect and analyze significant amounts of penetrometer data in short time periods during which soil properties remain essentially constant.

The main uses of the penetrometer have been to predict tractive performance (Wismer and Luth, 1974; Clark, 1984, 1985; Ashmore et al. 1987) and to define the impedance of different soil layers to root growth as related to cropping and tillage systems (Bulik and Rumsey, 1987; Buscher et al. 1986; Carter, 1985; Clark and Reid, 1984; NeSmith et al. 1987; Threadgill, 1982; Radcliffe et al. 1988a, 1988b; 1985a,b, Hadas et al. 1988). For these applications, the penetrometer was inserted vertically into the soil, and a record of cone index (force divided by area) as a function of depth was maintained down to a specified depth.

To clearly understand the need for a horizontal penetrometer as compared with the vertical penetrometer, consideration must be given to normal soil structure. Soils in Georgia are usually made up of relatively homogeneous layers. Each layer normally has a consistent texture, color, and density. A trained soil scientist can clearly identify these layers on an exposed wall of a trench dug in the soil. If the soil texture, density, and moisture content are indeed homogeneous and constant in a layer, then it would be expected that the cone index in that layer would be constant.

To further understand the meaning of vertical cone index data, it is helpful to consider a theoretical cone index versus depth curve as shown in Figure 1. This

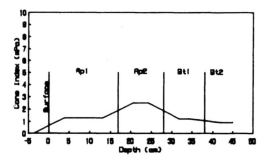

Figure 1. Theoretical graph of vertical cone index as a function of depth, assuming soil has homogeneous layers

graph was constructed assuming that there are four layers of soil, each with homogeneous properties. The layers are identified as Ap1, Ap2, Bt1, and Bt2. The ASAE Soil Cone Penetrometer Standard(S313.2, 1988) defines the soil surface as the point when the base of the cone is at the soil surface. Therefore, normally the cone index will have a positive value at the surface; the cone index would begin indicating a positive value when the tip of the cone enters the soil. As the cone enters the first layer, it is expected that the cone index will probably undergo a relatively linear change. When the cone has fully entered the layer, the cone index should level off to a given average value which represents the cone index of the homogeneous soil in that layer. When the tip of the cone begins to enter the second layer, assuming that the properties of the second layer are noticeably different from the first layer, the cone index will begin to change. In the example, the average cone index of the second layer is significantly greater than the first layer. When the cone has completely entered the second layer, it would be expected that the cone index would again level off to provide an indication of the average cone index of that layer. The graph will continue in a similar manner as the cone penetrates various layers.

Based on the above theoretical discussion of cone index as a function of depth and as related to layers, the following hypothesis was formulated:

When a penetrometer enters a homogeneous soil layer which has properties significantly different from the previous layer, the cone index will begin changing as the cone tip enters the new layer, and will then become relatively constant; this constant cone index should then represent the average cone index of that layer.

If this hypothesis is true, it may be possible to use the standard vertical penetrometer to locate soil layers without having to dig a trench and go through the normal hand process of layer identification, if the layers have significantly different physical properties. To test this hypothesis, a horizontal penetrometer was constructed to be used in measuring the cone index in the center of each layer on the vertical face of a trench wall. This paper describes the design of the hardware and software for the horizontal penetrometer. A subsequent paper will describe an experiment designed to test the hypothesis.

2 LITERATURE REVIEW

Previous research on the development of a horizontal penetrometer to measure the cone index in soil layers per se was not found in the literature. The only device approaching a horizontal penetrometer was a horizontal high speed cone penetrometer fabricated by Wismer and Luth (1972). This penetrometer was mounted rigidly on a vertical shank and measured the soil resistance as the shank was moved through a soil bin. The forces were measured between the shank and the moving structure (Luth and Wismer, 1971) including the soil resistance to the movement of the shank.

Several attempts have been made to find the relationship between the cone index and other soil properties. The main efforts have focused on correlating soil bulk density and moisture content to the cone index (Ayers and Perumpral, 1982; Gameda et al, 1988; NeSmith et al. 1987). As most of the successful correlations were found with artificial soil samples in laboratory conditions, the reliability of these results to represent the behavior of field soils is questionable (Hadas et al. 1988).

By cutting a trench and exposing the different layers, the soil profile can be identified (Nelson et al. 1975; Wolf et al. 1981; Radcliffe et al. 1988a). A horizontal penetrometer can then penetrate each homogeneous soil layers deep enough to overcome the surface effect on the soil failure plane around the cone (Ayers and Bowen, 1983 after Durgunoglu and Mitchel, 1973). Lastly, undisturbed core samples can be taken from each layer to measure the bulk density and moisture content at the same time.

3 HARDWARE AND SOFTWARE DEVELOPMENT

The overall design criteria were:

1. to follow the ASAE Standard (S313.2) in terms of cone size and speed, using the larger cone size,

2. to have a maximum penetrating depth of 15 cm.,

3. to have a short penetration drive system to fit narrow trenches,

4. to allow a working depth of up to 45 cm from the soil surface to fit the Bt2 horizon (Radcliffe et al. 1988a).

5. that the hardware was to be mounted on the carriage of an existing vertical tractor mounted penetrometer (Clark et al. 1986) for side movement and depth control,

6. to allow adjustment of the horizontal angle of the probe,

7. to have an adjustable wall surface detector for zero depth measurement,

8. to have electronic sensors for force and depth, with a microcomputer based data collection system, and

9. to be able to measure forces up to 4.5 kN.

The penetration drive system selected is an electric motor driven screw with relatively constant speed (like Tollner and Simonton, 1988) model ACPJ-I-8-12 volt made by Burr Engineering Co., Battle Creek, MI. It consists of a BPJ-Head 17-1 Gear Ratio (a), with a 0.2 m stroke, a speed of about 28 mm/s, a hall effect sensor to measure penetration depth, trunion style mount (b), and a threaded head and adjustable mounting bracket (c) (Fig. 2 and 3). Its retracted length is 0.52 m.

Figure 3. The horizontal penetrometer mounted by the trunion style mount to the vertical beam. The beam is mounted on the horizontal shaft of the vertical penetrometer and connceted to the main chassis by means of an adjustable bolt and a trolley inside a rail.

Figure 2. The horizontal penetrometer is fully penetrated into the wall of the trench. The depth of the penetrating point is measured in relation to an aluminum rigid bar.

Figure 4. The tractor mounted penetrometers are connected to a microcomputer for data collection.

The force transducer (d) is a model TC-S made by T-Hydraunics Inc., Westerville, Ohio. Its capacity is 0 to 4448 N (0-1000 lbs), and it is mounted to the actuator head by a double sided bolt, with a length of .076 m. The standard penetrometer shaft (e) was bolted to the other side of the force transducer, with a length of 0.2 m to use the full stroke.

A steel rod (f) with a flat disc at its end was mounted to the adjustable bracket. It was adjusted to define the wall surface with the cone base for zero depth measurement.

The trunion style mount was hitched to the end of the vertical beam (g) and an adjustable length threaded link (h) was used to adjust the horizontal angle.

The vertical beam was mounted to the carriage (l) of the hydraulically operated vertical penetrometer (m) (Clark et al. 1986; Clark and Reid, 1984). This system has a microcomputer based data collection system (Fig. 4). The penetrometer can beadjusted toward the wall of the trench by moving the tractor and by the long bolt (i) which connects the vertical beam (g) to a trolley (j). The trolley runs inside a horizontal rail (k) mounted to the main frame of the vertical penetrometer. The depth of the penetrometer in the trench can be adjusted by the three point hydraulic hitch of the tractor. Side movement can be achieved by the hydraulic motor which provides side movement of the vertical penetrometer.

The overall length of the retracted penetrometer is .85 m and the minimum width of the trench should be 1.0 m. The length of the vertical beam is 1.42 m to fit the lift range of the International Farmall 544 tractor that was used (Fig. 3).

The software was written in BASICA for the IBM PC computer used in the data collection system. Three main programs were written with the objectives of (1) to check and calibrate the sensors in the system, (2) to collect data and store the data on disk, and (3) to retrieve data for plotting and statistical analysis. The data collection program was written so that data is plotted on the computer monitor as the data is collected. This feature is very useful to observe potential errors in the data during data collection.

4 EXPERIMENTAL PROCEDURES

The initial tests with the horizontal penetrometer were conducted in a plot at Watkinsville, GA. The plot had a known traffic and tillage history, with fixed wheel traffic, for 8 years prior to the tests. The tillage and cropping treatment for this plot consisted of fall disking to a depth of 100 to 130 mm, followed by wheat in the winter. Following spring disking, again to a depth of 100 to 130 mm, a summer crop of grain sorghum was grown(see Langdale and Wilson(1987) for a complete description of the previous experiments).

Because of low soil moisture, a small subplot at the end of the plot was wetted with a drip irrigation system, with the objective of obtaining field capacity to a depth of approximately 50 cm. Five standard vertical penetrometer probings (see Figure 5 for location) were obtained in this wetted subplot approximately 24 hours after irrigation was discontinued. Since the soil surface was not flat, a flat aluminum bar was laid across the row ridges and was used as the reference plane for the beginning of data collection.

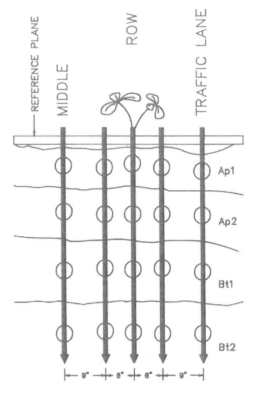

Notes:
1. Each location for vertical penetrometer sample is indicated by a vertical arrow.
2. Each location for horizontal penetrometer sample is indicated by a circle.

Figure 5. Sketch depicting location of soil layers, penetrometer sample locations, and soil sample locations.

Figure 6. The wall of a trench marked with flags to identify the horizons and the hoes made by the horizontal penetrometer at their middle.

A trench was carefully dug with a backhoe to a depth of about 60 cm., perpendicular to the normal traffic. The face of the trench towards the tillage plot was shaved about 5 cm. with a shovel to remove the effects of the backhoe and to provide a relatively smooth vertical face. The soil horizons were identified and marked by flags (Fig. 6). It was assumed that the bulk density, moisture content and texture for each horizon were homogeneous horizontally at least to a depth of .20m. The soil resistance to the penetrometer was measured approximately in the center of each layer, in the same locations in which the vertical penetrometer data was taken.

5 RESULTS AND DISCUSSION

Typical cone index readings of the horizontal penetrometer are shown in Fig. 7. Each line was measured at the Bt1 horizon (Fig. 5) at each of the 5 positions. All the readings were started with some resistance as the cone was fully pushed inside the soil when data collection began. The cone index from a depth of about 0 to 2 cm. generally shows a gradual increase, which is considered to be a function of the failure plane reaching the wall surface as explained by Ayers and Bowen (1983). The curves reach an almost straight asymptotic line. These curves differ from the well known vertical cone index curves which have no asymptotic line as the vertical penetrometer is

moving through different layers of soil while the horizontal one is penetrating a more homogeneous layer.

Figure 8 presents a graph of actual and theoretical vertical cone index as related to the discussion of theoretical cone index in the introduction of this paper. The theoretical curve was generated by assuming that the cone index would change in a linear fashion as the cone tip enters a layer, and that the cone index would attain a value approximately equal to the actual cone index in the center of the layer. The example presented shows that the actual compared with the theoretical are very close. However, this concept should be examined statistically on a broad data base.

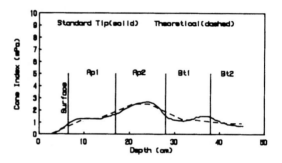

Figure 8. Actual versus theoretical vertical cone index example.

6 CONCLUSIONS

From the initial data presented in this paper, it appears that the horizontal penetrometer can be used to identify mechanical impedance of a natural homogeneous soil layer, especially in the lower horizons. This will give the ability to correlate cone index readings to soil bulk density, texture, and moisture content in field conditions.

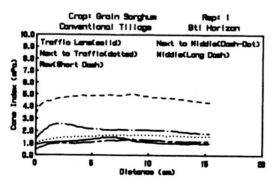

Figure 7. Typical results of horizontal cone penetrometer measurements.

7 REFERENCES

ASAE. 1988. Standard S313.2, Soil Cone Penetrometer. ASAE Standards, ASAE, St. Joseph, MI.

Ashmore, C., E.C. Burt, J.L. Turner. 1987. An empirical equation for predicting tractive performance of log-skidder tires. TRANSACTIONS OF THE ASAE Vol. 30(5):1231-1236.

Ayers, P.D., J.V. Perumpral. 1982. Moisture and density effect on cone index. TRANSACTIONS OF THE ASAE Vol. 25(5):1163-1172.

Ayers, P.D., H.D. Bowen. 1983. Predicting Soil Density Profiles Using Cone Penetration Resistance. ASAE Paper No. 83-1051.

Bulik, C. and J.W. Rumsey. 1987. Row-crop Soil Compaction Profiles-implications for Reduced Tillage Tool Design. ASAE Paper No. 87-1573.

Buscher, W.J., R.E. Sojka and C.V. Doty. 1986. Residual Effects of Tillage on Coastal Plain Soil Strength. Soil Science, Vol. 141(2) p.144-148.

Carter, L.M. 1985. Wheel Traffic is Costly. Transactions ASAE, Vol. 28(2):430-434.

Clark, R.L., D.E. Radcliffe, E.W. Tollner. 1986. Microcomputer Based Instrumentation and Software for a Tractor Mounted Penetrometer. ASAE Paper No. 86-1040.

Clark, R.L. and J.T. Reid. 1984. Topographic Mapping of Soil Profiles. ASAE Paper No. 84-1029.

Clark, R.L. 1984. Tractive modeling and field data requirements to predict traction. ASAE Paper No. 84-1055.

Clark, R.L. 1984. Tractive modeling with the modified Wismer-Luth model. ASAE Paper No. 85-1049.

Durgunoglu, H.T. AND J.K. Mitchel. 1973. Static Penetration Resistance of Soils. Research report prepared for NASA Headquarters, Washington, D.C. April 1973. Univ. of California, Berkeley.

Gameda, S., G.S. Raghavan, E. Mckyes. 1988. Soil Penetrometry for Compaction Modelling. ASAE Paper No. 88-1019.

Gill, W.R., and G.E. Vanden Berg. 1968. Soil Dynamics in tillage and traction ARS USDA. Agri. Handbook No. 316. pp 94-97.

Hadas, A., W.E. Larson, R.R. Allmaras. 1988. Advances in Modeling Machine-Soil-Plant Interactions. Soil and Tillage Research 11:349-372.

Langdale, G.W., R.L. Wilson, Jr. 1987. Intensive Cropping sequences to sustain conservation tillage for erosion control. Jour. of Soil and Water Conservation 42:352-355.

Luth, H.J. AND R.D. Wismer. 1971. Performance of Plane Soil Cutting Blades in Sand. TRANSACTIONS OF THE ASAE Vol. 14(2):255-262.

Manor, G., R.L. Clark, D.E. Radcliffe, L.T. West. 1988. Detection of hard pan depth with cone penetrometer. In Print.

Nelson, W.E., G.S. Rahi, L.Z. Reeves. 1975. Yield Potential of Soybean as Related to Soil Compaction Induced by Farm Traffic. Agronomy Jour. Vol. 67 p. 769-772.

Ne Smith, D.S., D.E. Radcliffe, W.L. Hargrove, R.L. Clark, E.W. Tollner. 1987. Soil Compaction in Double-Cropped Wheat and Soybeans on an Ultisol Soil Science Soc. of Am. J. Vol. 51(1). 1987. p.183-186.

Perumpral, J.V. 1987. Cone Penetrometer Applications - A Review. TRANSACTIONS OF THE ASAE Vol. 30(4) 939-944.

Radcliffe, D.E., G. Manor, G.W. Langdale, R.L. Clark, R.R. Bruce. 1988a. Effect of traffic and tillage on mechanical impedance in a layered soil. Proceedings, Southern Conservation Tillage Conference. Tupelo, Miss. August 11-12, 1988.

Radcliffe, D.E., E.W. Tollner, W.L. Hargrove, R.L. Clark, AND M.H. Golabi. 1988b. Effect of tillage practices on soil strength and infiltration of a Southern Piedmont soil after ten years. Soil Sci. Soc. Am. J. 52:798-804.

Radcliffe, D.E., R.L. Clark, S. NeSmith, W.L. Hargrove, AND G.W. Langdale. 1985a. Penetrometer measurements in conventional and minimum tillage. Proceedings Southern Region No-Till Conference. July 16-17. Griffin, GA. 98-102.

Radcliffe, D.E., R.L. Clark, G.W. Langdale, S. NeSmith AND W.L. Hargrove. 1985b. The effect of traffic and tillage on cone index in two Georgia soils. ASAE Paper No. 85-4041.

Threadgill, D.E., 1982. Residual Tillage Effects as Determined by Cone Index. TRANSACTIONS of the ASAE Vol. 25(4):859,863-867.

Tollner, E.W., H.R. Simonton. 1986. A Portable Cone Penetrometer for Measuring cone Index and Stress Relaxation. ASAE Paper NO. 86-1535.

Wismer, R.D. AND H.J. Luth. 1972. Performance of Plane Soil Cutting Blades in Clay. TRANSACTIONS OF THE ASAE Vol. 15(2):211-216.

Wismer, R.D. AND H.J. Luth. 1974. Off-road
Traction Prediction for Wheeled
Vehicles. TRANSACTIONS OF THE ASAE
17(1):8-10, 14.

Wolf, D., T.H. Garner, J.W. Davis. 1981.
Tillage Mechanical Energy Input
and Soil-Crop Response. Transactions ASAE,
Vol. 24(6) p. 1412-1425.

Wolf, D. and A. Hadas. 1984. Soil
Compaction Effects on Cotton
Emergence. Transactions Vol 27(3): 655-
659.

Wood, R.R. AND L.G. Wells. 1985.
Characterizing Soil Deformation by Direct
Measurement Within the Profile.
Transactions ASAE, Vol 28(6):1754-1758.

Land and Water Use, Dodd & Grace (eds), © 1989 Balkema, Rotterdam. ISBN 90 6191 980 0

Controlled traffic research: An international report

J.H.Taylor
National Soil Dynamics Laboratory, Agricultural Research Service, US Department of Agriculture, Auburn, Ala., USA

ABSTRACT: A rationale is presented for management of soil compaction based on soil-machine and soil-plant research. Agricultural tires need dry, highly compacted soil to provide flotation, mobility, and good tractive efficiency. Agricultural crops generally need just the opposite soil condition: good moisture and low compaction levels. Controlled traffic is suggested as a concept that allows optimization of soil conditions for both crops and tires. Management, not elimination, of soil compaction is the desired goal. Research results from several countries are reported.

Eine theoretische Grundlage wird hier dargestellt für die Handhabung von Erdzusammenpressung (Erddichte), die auf Forschungsergebnissen mit Erdmaschinen und mit Pflanzen basiert. Landwirtschaftliche Reifen brauchen trockene, sehr fest zusammengepresste Erde, um das Gleiten, die Beweglichkeit und gute griffige Leistungsfähigkeit zu gewährleisten. Landwirtschaftliche Ernteertrage brauchen gewöhnlich genau die gegenteiligen Erdbedingungen: gute Feuchtigkeit und lockere Erddichte. Es wird ein kontrollierter Verkehr vorgeschlagen als eine Idee, die ein Optimum der Erdbedingungen für beides, Erträge und Reifen, erlaubt. Die Handhabung der Bedingungen, nicht die Beseitigung von Erddichte ist das erwunschte Ziel.

Une analyse rationnelle est présentée pour gérer la nature du sol en fonction des recherches considérant les interactions entre le sol et les machines et matériaux. Les pneus agricoles exigent un sol sec, très compact afin de palier aux problèmes de flottement et de mobilité et pour permettre une traction plus efficace. Généralement, les cultures demandent des conditions totalement opposées: des sols bien humides et peu compacts. La circulation contrôlée serait un moyen qui permettrait d'obtenir les meilleures conditions de sol, à la fois pour les cultures et pour les pneus. Gérer et non éliminer la compacité des sols est le but recherché. Les résultats des recherches de plusieurs pays sont considérés.

1 INTRODUCTION

The problems with soil compaction have paralleled the growth in farm machinery size and weight. Soil compaction has become a problem throughout the world wherever agriculture is highly mechanized. Research has shown that subsoil compaction is related to total load on a wheel as well as unit surface pressure. Problems such as poor stands and irregular growth of plants, wet and cold soil, and ineffective tile drainage are now often recognized as symptoms of soil compaction.

Soil conditions required for good crop growth are not generally conducive to good flotation or tractive efficiency for machinery, and soil conditions suited to good flotation and tractive efficiency are not conducive to good crop growth. A crop production system employing permanent traffic lanes has great potential for improving both the crop zone and the traffic lanes rather than continuing the frustrating compromise of conventional crop production systems in which they are treated alike.

Soil compaction by tires supporting heavy machinery was recognized as a problem in the Southeastern United States in the fifties. The problem spread through the Mississippi Delta and the Southern Plains into California by the early sixties. In the seventies and eighties, the Corn Belt and Lake States became victims of soil compaction.

The initial concern was that soil compaction in the crop zone was reducing yields. In the fifties, many farmers and researchers tried to relieve the problem by subsoiling. Some years on some soils, subsoiling worked well. In other years, it was worthless in the same fields. Many farmers gave up in frustration, as did some researchers.

The weather continuously interacts with man's attempts to determine cause-and-effect relationships in soil compaction. Much of the early subsoiling research was done by people with a poor understanding of plant physiology. A plant responds as a living system, and its development is influenced by its ability to obtain sufficient and timely physiological requirements from its environment. This in turn depends upon the soil and aerial environment, the type of plant, the physiological state of development, and many other factors. Therefore, research aimed at solution of the soil compaction problem must consider the total soil-plant-machine system.

2 RATIONALE FOR CONTROLLED TRAFFIC

In most cases, soil compaction is not seriously reducing yields today because deep tillage is used annually to remove the compaction. Soil compacts easily when its structure is weakened, as by tillage or when wet. Tillage destroys the structure, and traffic following tillage frequently recompacts the soil. A cycle of tillage-traffic-tillage-traffic has developed. The deeper tillage is done to remove compaction, the deeper the next wheel compacts the weakened soil. In general, this requires slightly deeper tillage each year than the year before to remove subsoil compaction. Deep tillage used annually to correct subsoil compaction is very expensive and treats only the symptoms, not the basic problem.

Taylor (1983) defined controlled traffic as a crop production system in which the crop zone and the traffic lanes are distinctly and permanently separated. This system establishes traffic lanes that are not deep tilled and are used for wheel paths year after year. The lanes become compacted, improving tractive efficiency, flotation, and timeliness of operations, while the untrafficked crop zone, if initially well prepared, tends

to stay that way without annual deep tillage. The crop zone of a controlled traffic crop production system may be managed by conventional tillage or by conservation tillage methods. The machinery may be wide-frame (gantry) or conventional (cantilevered). Crops need lightly compacted, moist soil; tires need highly compacted, dry soil. Controlled traffic is a concept that makes it possible to optimize soil conditions for each of these directly opposed requirements in the same field.

More recent evolution of the controlled traffic concept indicates the crop zone can be further subdivided into desired levels of soil compaction, depending upon final use in the cropping scheme. The use of controlled traffic as a soil compaction management concept makes it possible to create the selective compaction levels desired for each of the various soil zones as determined by their intended use.

Discussions of an "optimum" field compaction level produces more questions than answers. Optimum for what--seedbeds, rootbeds, infiltration zones, drainage zones, or traffic lanes? Each of these may need different levels of soil compaction. Figure 1 illustrates that the controlled traffic concept combined with a gantry machine has the potential for achieving the desired "management" of soil compaction.

Soil compaction is inherently neither good nor bad; it is just one more factor that must be under management's control. Zero traffic with fluffy, loose soil may be just as undesirable as severely compacted soil. A single field has many functional uses, as illustrated in Figure 1, each of them requiring a different level of soil compaction. These zones may also change with time and crop rotations. However, a gantry operating on firm traffic lanes makes it possible to create whatever conditions are desired.

Gantry farming is not necessarily controlled traffic farming. If the harvesting or primary tillage is done by conventional machinery on different traffic lane spacings, it should not be classified as controlled traffic. Controlled traffic is a concept; the gantry is a machine. While the gantry may well be the ideal machine for instigating controlled traffic it is not the only way. Modified tractors were used for 20 years before the first gantry was built for controlled traffic.

No-till systems with continued random wheel traffic have proliferated all over

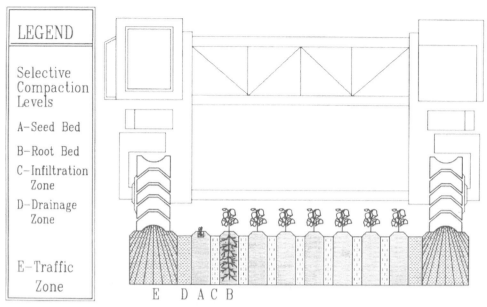

Selective
Compaction
Levels

A–Seed Bed

B–Root Bed

C–Infiltration
Zone

D–Drainage
Zone

E–Traffic
Zone

E   D A C B

Figure 1.

the USA. However, there are many complaints of poor water absorption after a few years of no-till in some soil types and geographic areas. Since most deep tillage is done to remove traffic-induced compaction, it seems reasonable that compaction problems will develop if tillage is stopped and random traffic continued.

Taylor (1981, 1982, 1983) defined and established a rationale for the entire controlled traffic concept. Controlled traffic research was initiated to increase crop yields by eliminating traffic-induced compaction from the cropping area. While some yield increases have been obtained, reduced production costs from elimination of deep tillage has proven to be of more benefit. Timeliness of operations, especially harvesting and spraying, made possible by firm, permanent traffic lanes has yet to be fully evaluated, but appears very beneficial. Burt et al. (1984) have shown that prepared traffic lanes can provide mobility up to 2 days earlier than the uncompacted crop zone following a flooded condition.

3 RESEARCH RESULTS

3.1 United States research

Controlled traffic research in the USA

was initiated three decades ago by several researchers to eliminate compaction caused by wheel traffic in the crop zone. Cooper et al. (1969) discussed the early field work at Auburn on effects of traffic compaction on cotton. They found reduced air and water infiltration, increased runoff, and reduced plant root activity, any of which can contribute to reduced yields. Deep tillage performed in 1959 gave yield increases over conventional tillage of 450 kg/ha in 1960, 205 kg/ha in 1961, and 80 kg/ha in 1962. Three years of wheel traffic had essentially eliminated the effect of the deep-tillage treatment. Experiments were then designed to determine the feasibility of controlling wheel traffic.

Carter and Colwick (1971) reported on work begun in 1962 to devise an "optimum tillage" system. They used a zone concept and limited traffic to alternate furrows. The untrafficked furrows were treated specifically for maximum water infiltration. They reported that the pattern of soil strength which developed with "optimum tillage" was quite different from that developed with "normal tillage." There was a zone of low strength in the soil under the rows which extended down to the depth of the tillage operation. The zone of high strength under the traffic furrows for "optimum tillage" was larger and closer to the surface than for "normal tillage."

Dumas et al. (1975) reported average yield increases of 15% due to controlled traffic in cotton over a 4-year period regardless of the type of tillage imposed. However, deep-tilled plots with no traffic yielded 56% more than conventionally tilled plots with tractor and sprayer traffic. While they expected controlled traffic to improve the crop zone or rootbed, improvements in mobility on the traffic lanes or roadbeds soon became evident. They reported that this improved mobility permitted timeliness of all operations irrespective of weather conditions, and that tractive efficiency was improved also.

Williford (1980) discussed the wide-bed cotton production system developed at Stoneville during the 1970's. It has proven to be the most economical system for the sandy loam soils of the Delta. The system was designed to restrict wheel traffic to specific traffic zones throughout the season. When used as a controlled traffic system, annual deep tillage was not required to maintain yield levels. Williams et al. (1985) reported estimated returns above specified expenses for the wide-bed system at $375/ha compared to $296/ha for the conventional system.

Voorhees (1979) conducted experiments on a clay loam soil in southwestern Minnesota to measure the effects of various degrees of soil compaction on tillage draft. Wheel traffic consisting of 0, 1, 3, or 5 passes over the entire plot was imposed in the spring on freshly tilled soil. Tillage draft was measured during fall moldboard plowing. The largest relative increase in draft, 25%, resulted from the first pass of the tractor wheel. Additional amounts of wheel traffic continued to increase draft, but at a much lower rate. The maximum 5 passes resulted in a 43% increase in tillage draft.

Voorhees et al. (1985) have found that wheel traffic from normal farming operations can cause significant compaction of well structured soils in the northern Corn Belt. Axle loads common to harvest and transport equipment (>9 metric tons) compacted a relatively wet soil to a depth of 60 cm which was not effectively ameliorated by freezing and thawing. Corn and soybean yields were decreased up to 27 and 19%, respectively, depending upon climatic conditions during the growing season.

Mettler and Sons Farm (Anonymous, 1983)

reported $250/ha savings in cost of production for irrigated cotton with controlled traffic in California. They used permanent traffic lanes and beds and eliminated subsoiling, chiseling, and disking. This in turn allowed them to eliminate the big 4-wheel-drive tractors and they used smaller tractors for their reduced energy tillage system. Cotton yields were unchanged.

It should be obvious that any row-crop farmer can start to modify his cultural system to manage compaction. When the effects of traffic-induced soil compaction are recognized, machinery selection and traffic patterns are more carefully scrutinized. However, there are gaps in wheel-spacing capabilities of machinery systems that only the manufacturers can eliminate.

Parsons et al. (1984) did an extensive study of wheel-spacing possibilities for ridge-till crop production systems. While most of the equipment could be made compatible, there were significant exceptions: lime spreaders, custom fertilizer spreaders and sprayers, grain buggies, and other "combine chaser" vehicles. They found that operating combine tires on the ridges could significantly affect the next year's yield even though the ridges were dry and firm at harvest time. Ridge profile was altered, affecting planter performance the following year. Corn yields were reduced 10 to 20% in years following combine traffic on ridges.

Morrison (1985) investigated equipment needs for permanent wide-bed systems. He used permanent wide beds to replace permanent ridges. He developed cropping systems for both one-track spacing and for alternating traffic lane spacings. He found that dual wheels must be eliminated or replaced by tandems and that tire width must be limited to avoid compaction of cropping areas.

Compaction management systems are not restricted to agricultural crops. Miles (1978), after studying the damage done to forest soil by logging operations, concluded that soil compaction varied considerably. He found that the major compaction occurred where primary and secondary skid trails were located, and that compaction reduction treatments must be instituted as part of the forest management system.

3.2 International research

During the past decade there has been a

dramatic worldwide increase both in the interest in deep soil compaction research and in the controlled traffic concept for soil compaction management. The United Kingdom, The Netherlands, Israel, and the United States each have at least one location working with specially designed spanning equipment to control traffic. In Japan, a completely automated, rail-supported system of equipment for crop production is used for research. Canada, West Germany, Sweden, South Africa, Australia, and the USSR also have controlled traffic research in progress plus other approaches to soil compaction research.

In The Netherlands, controlled traffic research was conducted for several years using modified conventional tractors with a wheel span of 3.0 meters (Lamers et. al, 1986). Compared to conventional random traffic systems, the controlled traffic gave consistent yield increases up to 10%. However, they concluded that the yield increases were insufficient to compensate for the loss of production on the traffic lanes. They calculated a span of 10 meters would be necessary to economically justify controlled traffic. Beginning in 1984, they concentrated their research on Low Ground Pressure (LGP) tires for management of soil compaction.

At the Federal Agricultural Research Center (FAL), Braunschweig, Federal Republic of Germany, Sommer et al. (1988) are using controlled traffic as part of their conservation tillage research. The cropping system consists of rotating corn, sugar beets, barley, and flax. The wheel spacing is 2.5 meters, and the firm traffic lanes provide mobility for field equipment such as sprayers when fields with conventional traffic patterns are much too soft and wet. The objective here is to develop conservation tillage practices that will be rapidly adopted by farmers. Their results include increased yields, reduced erosion, and improved trafficability.

In England, Chamen et al. (1988) are now using a 12-meter gantry. In earlier research using 3-meter-wide equipment they found that energy requirements for crop establishment can be reduced up to 70% by eliminating traffic from the cropped area. They also found that the loose seedbed created by zero traffic triggered a manganese deficiency causing reduced wheat yields. This was corrected by applying manganese sulphate, but could have probably been corrected by firming the soil slightly in early spring. In a recent economic analysis (Hietbrink,

1988) a 12-m gantry system showed an increased profit over conventional farming systems for a rotation of crops. In similar circumstances a 6-m tractor-based zero traffic system showed a reduced profit.

Tillet and Holt (1987) of the Horticultural Engineering group at Silsoe have developed a 9-meter-wide tracked vehicle. The vehicle was initially designed as a harvesting aid but has now been used for planting and several other tasks in vegetable production. The quality of the produce sent to market is of primary importance, and gantries straddling a wide bed can significantly improve the quality of crops such as cauliflower where repeated selective harvest is necessary.

Spoor et al. (1988) have done extensive work in sugar beets. They found that the significant yield response to timeliness of planting is greatly affected by permanent traffic lanes. What would have been a 4-day delay due to extra tillage required in a conventional tillage system became a 20-day delay because of rain on the fifth and subsequent days. Harvester and trailer rolling resistance was reduced and more efficient beet lifting was realized on controlled traffic systems. They are working on traffic lane profiles for better drainage.

In Scotland, Dickson and Campbell (1988) compared a 2.8-meter modified tractor on permanent traffic lanes with a conventional system (both ploughing and direct drill) for winter barley followed by potatoes. The zero traffic system gave increased yields for both potatoes and barley, and more uniform ripeness in the barley. Barley emergence was depressed by transient waterlogging under direct drilling. Draft forces for primary cultivation were reduced by 14% for barley and 50% for potatoes by zero traffic as compared to conventional traffic systems.

In Sweden, where much of the early research on effects of soil compaction was conducted, the research continues. They have developed the concept of an optimum state of compactness for specific soils and crops. They are concerned with the persistence of compaction and are presently involved in cooperative research with several other countries in the humid freeze-thaw area of Europe and North America. In this research they are looking at crop response to subsoil compaction for several years following a one-time application of traffic with loads of 10 tons on a single axle or 16 tons on a tandem axle. In earlier research Hakansson and Danfors (1981) reported that axle loads above 6 tons (8-10 tons on a tandem axle) may result in compaction at depths below 40 cm.

In Israel, Hadas (1987) found controlled traffic especially effective in soils of the semi-arid, subtropical region where organic matter and structural stability are low, and compaction is present under rather small traffic loads. He found that soil compaction reduced air porosity, water retention, conductance, root proliferation, and root activity. He found potential yield increases of 10-15% for controlled traffic in field crops (cotton, alfalfa, wheat), and expected similar results for high-cost vegetable crops. However, root or bulb crops were expected to have even higher marketable yield increases under Israeli conditions.

In Australia several controlled traffic and gantry farming projects are underway. Tullberg et al. (1987) at Queensland Agricultural College reported on research to assess the effects of controlled traffic on tractor energy losses and tillage requirements as well as soil and crop effects. They concluded that controlled traffic can (1) reduce the fuel cost of crop establishment by at least 40%; (2) allow similar output and capacity from a tractor of at least 30% less power; (3) maintain yields without the necessity for deep tillage operations; and (4) increase rainfall infiltration and so reduce runoff and erosion in some circumstances. This research was conducted in subtropical dryland grain production.

In Victoria, at the Institute for Irrigation and Salinity Research in Tatura, Adem and Tisdall (1984) have developed a remarkable soil management system. When removed from pasture and used for commercial tomato production, irrigated red-brown soils in northern Victoria last only about 3 years because of loss of water-stable aggregates and organic carbon. The "Tatura System" used minimum tillage on permanent, raised beds with traffic confined to the furrows. Sunflowers are double cropped with cereals. The winter cereal not only provides a crop but also roots to improve soil structure, and mulch for the summer crop to conserve water, reduce soil temperature, protect surface soil from impact of raindrops, ensure emergence, and encourage earthworm activity. The beds are wet by capillary action from shallow water in the furrow.

Japan has worked with gantries for 25

years, and at the National Agricultural Research Center in Tsukuba (Miyazawa, et al., 1987) they have the most sophisticated system known to the author. The gantry is electrically powered and operates on concrete rails on 12-m spacings. It is completely automated and under computer control from the storage area to a preselected soil plot. All operations needed for rice production are carried out by this gantry. Robotic manipulators are in various stages of development for use on this gantry. They are now developing a practical field system on rubber tires with an isolated energy system and a space location detection system.

The Union of Soviet Socialist Republics (USSR) has published numerous articles in their own literature on spanner agriculture during the last decade. They may have greater traffic-induced soil compaction problems than any other country. Large farms and heavy equipment combined with susceptible soils and weather conditions have created significant compaction problems for them. Lazovski (1984) did a historical, strategic review of spanners in agriculture and concluded that future USSR needs could not be met without them. He gives several reasons for his conclusions and states that irreparable harm to the soil may be done if present practices are continued for 25 more years.

There are investigators in many other countries conducting research to utilize controlled traffic as a soil compaction management system. Space limitations, not the quality of their research, prevents the mention of them here.

4 CONCLUSIONS

Traffic-induced soil compaction can be a problem throughout the world wherever agriculture is mechanized. High levels of soil compaction are generally best for traction and transport while low levels are more desirable for crop growth.

Controlled traffic is a concept for soil compaction management which makes it possible to optimize soil conditions for both tires and crops.

Timeliness of operations, such as spraying and harvesting, from permanent traffic lanes may be the greatest economic advantage of the controlled traffic concept.

Within the crop zone several levels of soil compaction may be desired to create proper micro-environments for seed, roots, water infiltration, and drainage. Controlled traffic is a concept; the gantry is a machine, and gantry farming is not necessarily controlled traffic.

REFERENCES

Adem, H. H., and Tisdall, J. M. 1984. Management of tillage and crop residues for double cropping in fragile soils of southeastern Australia. Soil and Tillage Res. 4: 577-587.

Anonymous. 1983. Making tracks saves plowing. Progressive Farmer Magazine, 7: 17-19.

Burt, E. C.; Taylor, J. H.; and Wells, L. G. 1984. Traction benefits of prepared traffic lanes. Agric. Engng. 65(9): 16-18.

Carter, L. M., and Colwick, R. F. 1971. Evaluation of tillage systems for cotton production. Transactions of the ASAE 4(6): 1116-1121.

Chamen, W. C. T.; Vermeulen, G. D.; Campbell, D. J.; Sommer, C.; and Perdok, U. D. 1988. Reduction of traffic-induced soil compaction by using low ground pressure vehicles, conservation tillage, and zero traffic systems. Proc. 11th Int. Conf. Soil Tillage, Edinburgh, Scotland, pp. 227-232.

Cooper, A. W.; Trouse, A. C., Jr.; and Dumas, W. T. 1969. Controlled traffic in row crop production. Proc. of the 7th Int. Cong. of Agric. Engng., Baden-Baden, W. Germany, Section III, Theme 1, pp. 1-6.

Dickson, J. W., and Campbell, D. J. 1988. Conventional zero traffic systems compared for winter barley and potatoes. Proc. 11th Int. Conf., Soil Tillage, Edinburgh, Scotland, pp. 239-244.

Dumas, W. T.; Trouse, A. C., Jr.; Smith, L. A.; Kummer, F. A.; and Gill, W. R. 1975. Traffic control as a means of increasing cotton yields by reducing soil compaction. ASAE Paper No. 75-1050, Am. Soc. of Agric. Engrs. St. Joseph, Michigan 49085, USA.

Hadas, A. 1987. Controlled traffic research in Israel. Acta Horticulturae No. 210: 43-47.

Hakansson, I., and Danfors, B. 1981. Effect of heavy traffic on soil conditions and crop growth. Proc. of the 7th Int. Conf. of the Int. Soc. for Terrain-Vehicle Systems (ISTVS), Calgary, Canada, Vol. 1: 239-253.

Hietbrink, O.; Chamen, W. C. T.; and Audsley, E. 1988. An economic comparison of gantry and tractor-based zero traffic systems used for growing a rotation of crops on a cereals farm. Div. Note DN1487, AFRC Inst. Engng. Res., Silsoe, England.

Lamers, J. G.; Perdok, U. D.; Lumkes, L. H.; and Klooster, J. J. 1986. Controlled traffic farming systems in The Netherlands. Soil and Tillage Res. 8: 65-76.

Lazovski, V. V. 1984. The use of mobile spanner systems. Mechanization and Electrification of Agriculture, Vol. 2: 3-5.

Miles, J. A. 1978. Soil compaction produced by logging and residual treatment. Transactions of the ASAE 21(1): 60-62.

Miyazawa, F.; Yoshida, T.; Sawamura, A.; and Taniwaki, K. 1987. Gantry system. Proc. Symp. on Agric. Mech. and Int. Coop. in High Tech. Era, Tokyo, pp. 109-114.

Morrison, J. E. 1985. Machinery requirements for permanent wide beds with controlled traffic. Applied Engng. in Agric. 1(2): 64-67.

Parsons, S. D.; Griffith, D. R.; and Doster, D. H. 1984. Equipment wheel spacing for ridge-planted crops. Agric. Engng. 65(8): 10-14.

Sommer, C.; Dambroth, M.; and Zach, M. 1988. The mulch-seed concept as a part of conservation tillage and integrated crop production. Proc. 11th Int. Conf. Soil Tillage, Edinburgh, Scotland, pp. 875-879.

Spoor, G.; Miller, S. M.; and Breay, H. T. 1988. Timeliness and machine performance benefits from controlled traffic systems in sugar beet. Proc. 11th Int. Conf. Soil Tillage, Edinburgh, Scotland, pp. 317-322.

Taylor, J. H. 1981. A controlled traffic system using a wide-frame carrier. Proc. of the 7th Int. Conf. of the Int. Soc. for Terrain-Vehicle Systems (ISTVS), Calgary, Canada, pp. 385-409.

Taylor, J. H. 1982. From traction research to a crop production system. Proc. of ASTM Technical Seminar, Off-the-Road Uses of Tires, Akron, Ohio.

Taylor, J. H. 1983. Benefits of permanent traffic lanes in a controlled traffic crop production system. Soil & Tillage Res. 3: 385-395.

Tillet, N. D, and Holt, J. B. 1987. The use of wide span gantries in agriculture. Outlook on Agriculture, Pergamon Press, Oxford, 16:(2): 63-67.

Tullberg, J. N., and Murray, S. T. 1987. Controlled traffic tillage and planting. Report to the National Energy Research Development and Demonstration Council, Canberra, Australia.

Voorhees, W. B. 1979. Energy aspects of controlled wheel traffic in the northern Corn Belt of the United States. Proc. of the 8th Conf. of the Int. Soil Tillage Res. Org. (ISTRO), Stuttgart-Hohenheim, F.R.G., Vol. 2: 333-338.

Voorhees, W. B.; Nelson, W. W.; and Randall, G. W. 1985. Soil and crop response to wheel traffic on highly productive soils of the northern Corn Belt. Proc. of the Int. Conf. on Soil Dynamics, Auburn, Alabama, Vol. 5: 1120-1131.

Williams, R.; Mullendore, G.; Head, B.; Thomas, J.; Musick, J.; Baker, R.; Hamill, J.; Laster, M.; Porvin, D.; Spurgeon, W.; and Williford, R. 1985. Estimated costs and returns, cotton, all areas of Mississippi. Mississippi Agricultural and Forestry Experiment Station, Mississippi State, Mississippi. 65 pp.

Williford, J. R. 1980. A controlled-traffic system for cotton production. Transactions of the ASAE 23(1): 65-70.

Land and Water Use, Dodd & Grace (eds), © 1989 Balkema, Rotterdam. ISBN 90 6191 980 0

An agricultural engineering approach to investigate crop responses to soil loading treatments

G.D. Vermeulen
Institute of Agricultural Engineering, Wageningen, Netherlands

F.R. Boone
Tillage Laboratory, Wageningen Agricultural University, Wageningen, Netherlands

W.B.M. Arts & F.G.J. Tijink
Institute of Agricultural Engineering, Wageningen, Netherlands

ABSTRACT: The use of a roller to simulate currently practiced as well as possible future loading treatments by field traffic is discussed. Soil loading with the modelled running gear means that the technical action is better defined and will result in more controlled soil conditions than in usual field studies. This improves the possibilities for understanding and future modelling of the effects of soil structure on crop performance for a variety of possible traffic treatments. The first year's result of an integrated wheel-soil-crop field experiment for a spring traffic situation is presented.

1 INTRODUCTION

The directly noticeable benefits of large-scale machinery in terms of reduced labour costs and maximum tractive performance have caused the substantial increase of wheel and axle loads and influenced the choice of wheel equipment in agricultural practice. This development increased the risk of soil compaction. High costs are currently being spent on restoring compacted fields to a condition thought better for crop production, even though the effect of such treatments on the yield has not been quantified. A more rational approach would be to quantify the effects of compaction, including their economic significance as a basis for choosing the appropriate treatment.

Although tyres were already in use in agriculture fifty years ago, methods for predicting the soil profile properties and subsequent crop growth after the passage of a tyre are still not available. The wheel-soil relationships have mostly been studied by engineers and the plant-soil relationships by agriculturalists. The methods currently used can be summarized as follows:

1.1 Wheel-soil relationships

Soane and his co-workers (1981) distinguish three types of methods: empirical, quasi-theoretical and theoretical (soil mechanical) methods. All models have in common that the initial soil conditions should be determined to some degree. The need for a very detailed description will limit the utility of a method.

Empirical methods can be reasonably accurate (Tijink 1988), but require an extensive database of field data, which is rarely available and can easily be made obsolete, e.g. by changes in tyre construction.

Quasi-theoretical methods as used by Gupta and Larson (1982) and Koolen (1976) are based on simple, reproducible relationships between soil behaviour and uni-axial stress in the laboratory. On the whole they are still approximative, because of the low accuracy of the mathematical factors introduced for influences such as shear stress and stress distributions. Bolling (1987) and Van den Akker (1988) use stress-strain relationships as measured in triaxial laboratory tests as a basis for predicting soil behaviour under wheels. Stress distribution in the wheel-soil contact area and the concentration factor in the Söhne-Boussinesq equation have to be measured or estimated. Many laboratory and field tests are necessary to develop a satisfactory equation for a particular field. Direct practical application is therefore limited, but continued research could reveal more suitable parameters for use in empirical methods.

Theoretical models, based on the description of soil as an elastic, plastic or visco-elastic medium are, in principle, unsuitable for describing the process of compaction. Critical state soil mechanics needs further development before it can be used in practice, but can provide basic knowledge for designing adequate empirical equations (Hettiarachi, 1988).

1.2 Soil-plant relationships

The relation between soil conditions and plant production is very complex, mainly because of the many parameters involved, their mutual interactions and dependence on environmental factors. This was pointed out very clearly by Trouse and co-workers in 1971 and by Boone in 1988. The development of computer models for crop growth has made it possible to handle more parameters and interactions. Therefore, continued research is necessary to quantify the effect of the most important parameters and interactions that control crop production. In his review Boone (1988) indicates that the influence of penetration resistance, aeration and their interactions with soil moisture on crop growth can now be roughly quantified (Boone et al. 1987).

1.3 Wheel-soil-plant relationships

Thus, we are still a long way away from being able to reliably predict the effect of a particular wheel action (and certainly of a traffic regime) on crop production. This severely handicaps our ability to stipulate the possibilities and limitations for future traffic systems. This was the main reason for including not only soil but also crop responses in the evaluation of promising traffic systems that began at IMAG several years ago.

Five years of Controlled Traffic research on a marine loam soil showed that "loose soil" gave, on average, a yield 6-9% higher than the soil conditions that could be found in practice (created by driving track by track with tyres loaded to 25 kN). Moderately compacted fields gave the highest yields in some cases (Lamers and Perdok 1982) (Perdok and Lamers 1985). This, and the costs involved in establishing and maintaining very loose soil encouraged us to do more research on the optimization of wheel equipment for agriculture. Perdok and Arts (1987) used some rules of the thumb for soil behaviour and crop requirements as a basis for chosing proper wheel equipment at a certain load. The prospects of a low ground pressure farming system are currently being studied (Vermeulen et al 1988).

A promising method for predicting the effect of a single wheel pass on soil and crop is the use of an empirical method that incorporates theory and similarities as much as prove to be useful for reducing the number of parameters needed. Therefore, in addition to the traffic systems research, a new approach for investigating the individual technical parameters in the relation between load, loading technique and soil and crop responses, by using a field traffic simulation technique is presented here. It should remove some of the variability in experimental results by controlling the number of technical and soil parameters in a pragmatic way. It should also prevent the derived prediction equations from being made obsolete by technological changes.

2 MODEL RESEARCH

The field traffic simulation technique should result in a smaller set of parameters with fewer sources of variation and should enable the important parameters to be varied independently between extreme values. Although scale models can be used in many cases, the presence of real-scale plants, root systems and structured soil prohibits the use of scale models in integrated wheel-soil-plant research. Consequently it was decided to use a true-scale hardware model. No effort is made to model the soil profile and soil structure, but it is attempted to reduce the variation by paying attention to uniformity during the selection, preparation and loading of the soil. As far as crops are concerned, special care is taken to ensure a uniform stand.

2.1 The hardware model

Instead of a tyre, the effects of applying a load with a hardware model is studied. The model should be a simple, well defined piece of equipment, whose effects on soil are expected to correspond with the effects of existing loading techniques.

To simulate the existing and future loading techniques, it should be possible to vary the parameters independently in the following estimated order of importance:
1. Load
2. Contact width per element
3. Configuration of elements on axle
4. Contact length
5. Number of passes
6. Slip
7. Speed
8. Vibrations
9. Others

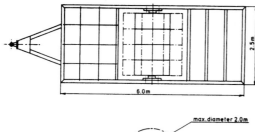

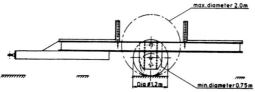

Fig. 1: **Schematic drawing of the loading frame with lined-up rollers.**

The equipment built for this purpose consists of a loading frame for applying loads (range 45 - 165 kN) on an exchangeable running gear of up to 2.0 m wide on one axle (Fig. 1). The running gear currently available consists of 4 non-driven, smooth, steel roller elements, each 0.5 m wide and 1.2 m in diameter. The elements can be mounted on the axle in various configurations.

The first three mentioned parameters will be studied first. The roller elements allow these parameters to be varied independently and are expected to give a good simulation of non-driven tyres operating at low deflection as can be expected in spring, when the soil is very soft.

A force transducer has been mounted between the tractor and the loading frame to measure rolling resistance and calculate the compactive energy, which can be used for predicting total compaction (Fattah et al. 1981).

2.2 Controlling soil parameters

Uniform soil conditions are important at the time of loading as well as during crop growth. In both cases, the first step is to select a uniform soil which, means a field as uniform as possible over the required depth on the basis of soil texture. This property is commonly used to describe soils in the absence of a more suitable classification for mechanical behaviour of soil. (Koolen & Kuipers 1983). On sandy soils, the initial soil condition can be further controlled by tillage and precompaction to a certain density. The loading frame with the lined-up rollers can be used to compact 2.0 m wide strips for this

purpose. For clayey soils, additional weathering may be necessary after tillage and before precompaction to achieve a sufficiently uniform condition.

To reveal the effect of vertical soil profile properties on crop responses, uniform soil conditions in the horizontal plane are required. The dimensions of the area should allow uniform infiltration, runoff and rooting conditions for the test crop in the horizontal plane. The loading frame with the lined-up rollers can be used for this purpose providing 2.0 m wide uniform strips. For studies of crop responses to width and configuration of elements on the axle, controlled variation of properties in the horizontal plane is required. Geometrical configuration of plants, i.e. distance to the compacted zone should be taken into account. To avoid further complications, the vertical rooting conditions should be reasonably uniform up to a certain chosen depth.

To ensure that the initial nor the soil structure after a pass with the loading frame is disturbed, the loading frame and other necessary equipment is pulled by a tractor with a track width of 3.0 m.

2.3 Field experiments

Two experiments were started in 1988. One with respect to the effects of traffic in spring on a tilled rootbed (5-25 cm depth) for an annual crop (peas). The other deals with the effects of year-round traffic on a once-tilled deep root-bed (0-50 cm depth) for permanent grassland. Some of the preliminary results of the first experiment are reported here.

3 MATERIALS AND METHODS

The spring traffic experiment was carried out on a marine loam soil (USDA classification). The experiment is layed out to apply four levels of load to the soil at three levels of soil moisture content in triplicate.

The field was tilled with a rotary digger to a depth of 25 cm in the autumn of 1987. Next spring, it was attempted to create three levels of moisture content by sprinkler irrigation. The irrigations started two weeks before the loading treatments were applied. In the lower part of the arable layer, the soil moisture level remained around field capacity both on the irrigated plots and the non-irrigated plots. Differences in moisture content were mainly restricted to the 0-5cm top layer. Moisture contents in this layer were 13.9, 15.5 and 16.7% (w/w) respectively at the time of loading. Four loading treatments (respectively 0, 45, 85 and 165 kN) were applied by the loading frame with the lined-up rollers at a speed of 1.5 m/s. Due to bulldozing effects, the load of 165 kN could only be applied after one pass with a load of 85 kN and therefore should not be considered as a single pass over soft soil for modelling purposes. After seedbed preparation with a powered rotary harrow to a depth of 5 cm, peas (cv. maxi) were sown at a depth of 5 cm and a row width of 25 cm. After establishment of the crop was completed, the arable layer was kept close to field capacity in order to simulate a wet season. Additional sprinkler irrigation was applied on all plots when natural rainfall was insufficient.

4 RESULTS

Mean soil porosity of the arable layer below the seedbed decreased curvilinear with roller load from 47 á 48% (v/v) in the unloaded soil to about 42 % (v/v) in

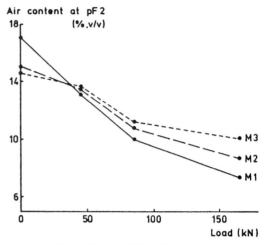

Fig. 3: The effect of loading in dependence of the moisture content of the 0-5 cm top layer (M1 - 16.7; M2 - 15.5 and M3 - 13.9 % w/w) on mean air content at pF2 of the 5-25 cm layer. LSD - 2.6 % (p - 0.05)

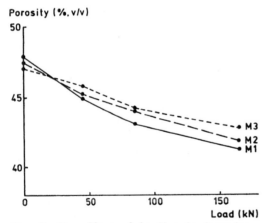

Fig. 2: The effect of loading in dependence of the moisture content of the 0-5 cm top layer (M1 - 16.7; M2 - 15.5; M3 - 13.9 % w/w) on mean soil porosity of the 5-25 cm layer. LSD - 1.7 % (p - 0.05)

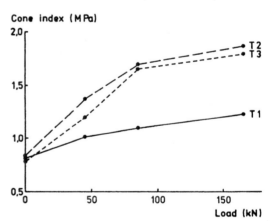

Fig. 4: The effect of loading on the mean cone index of the 5-25 cm layer as measured on three data (T1 - 1; T2 - 57; T3 - 196 days after loading). LSD - 0.24 MPa (p - 0.05)

the soil loaded with 165 kN (Fig. 2). There is a tendency that soil porosity decreased somewhat more when the surface soil had been kept slightly wetter. However, the differences due to moisture levels were not statistically significant.
Concurrently air content at field capacity (pF2) decreased from 15% (v/v) in the unloaded soil to values between 7 and 10% in the soil loaded with 165 kN (Fig. 3).

A Bush cone penetrometer with a standard ASAE type B cone was used to measure the penetration resistance at 3.5 cm depth intervals. The mean cone index was calculated as the average of the cone index values at depths of 7.0 - 24.5 cm (10 replicates). The cone index immediately after compaction was 0.8 MPa in the unloaded soil and increased asymptotically to 1.2 MPa in the most compacted soil (Fig. 4). Two months later, when the arable layer

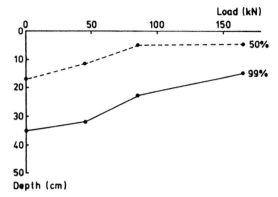

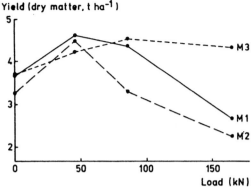

Fig. 5: The effect of loading on the depth in the soil above which 50 and 99% of all roots are present (At 16.7% (w/w) moisture in the 0-5 cm top layer at loading - M1) at day 62.

Fig. 6: The effect of loading in dependence of the moisture content of the 0-5 cm top layer (M1 - 16.7; M2 - 15.5; M3 - 13.9 % w/w) on mean dry matter yield of peas (seed, ton per ha). LSD - 0.9 t/ha (p - 0.05)

was at a very similar water content, cone index was still 0.8 MPa on the unloaded soil, but had increased considerably on the loaded soil and amounted 1.8 MPa on the heaviest loaded soil. Measurements after harvest of the pea crop, when the soil was again at field capacity, were close to these values. Obviously, cone index values increased about 30% in the first 57 days after loading by the development of water- and chemical bonds. As the crop was sown one day after loading, the increase in cone index coincided with the early development of the root system.

Plant density varied between 50 and 52 plants per m^2 and therefore was not significantly different between treatments. The first signs of retarded growth on the heaviest loaded plots were observed at 22 days after sowing when the average length of the main root was 24 cm for the unloaded plots and only 15 cm of the heaviest loaded plots. Plant growth on the heaviest loaded treatment clearly lagged behind all other treatments. Root counts two months after sowing revealed that root density in the seedbed increased but maximum root depth decreased with compaction (Fig. 5). Root density in the arable layer of the unloaded soil did not change with depth but showed a clear accumulation at the interface of the arable layer with the subsoil. Very few roots were found in the subsoil. Root density in the arable layer of the loaded soil decreased with depth and the stronger the heavier the loading had been. At that moment, shoot growth appeared to be similar on the unloaded soil and the soil loaded with 45 kN but was less on the soil loaded with 85 and considerably less on soil loaded with

165 kN. There was a tendency that negative crop responses were somewhat less on plots that were driest during loading.

High natural rainfall between 65 and 90 days after sowing occasionally increased matric water potential at 20 cm depth to values above -5 kPa which decreased oxygen supply to suboptimal levels, especially on the plots loaded with 85 and 165 kN. In the middle of this period when the pea crop was in the early phase of pod filling, shoot weight was similar on the unloaded and lightly loaded soil but considerably less on plots that received the two heavier loading treatments. Crop senescence started also earlier on both last mentioned plots.

Final yield of peas showed an optimum with loading (Fig. 6). Where the soil was sprinkler irrigated before loading was applied, highest yields were obtained on the plots loaded with 45 kN. Without sprinkler irrigation highest yields were obtained on soil loaded with 85 kN. However, in this last case, crop responses were statistically not significant. The reduction in crop yield was very drastic where sprinkler irrigated soil was loaded with 165 kN.

5 DISCUSSION

The results of one year experimentation show that the test crop peas responded sensitive to the load applied with the lined up rollers. Even small differences in moisture content in the profile of the arable layer at the time of loading could be detected in terms of yield. Heavy loading of a relatively moist top layer was much more harmful than the same action on a somewhat drier top layer. The observations available point to a too high mechanical resistance for root growth in the first part and aeration problems in the second part of the growing season.

The critical mechanical and aeration limits for crop growth found in earlier research with maize (Boone et al. 1987) seem to apply well for pea too in this first year of the experiment. There is no direct evidence that could explain the beneficial effect of a light loading on the final yield of peas. However, it has been observed that the shoot-seed ratio at harvest on the unloaded plots was smaller than on all other plots.

To quantify the effects of roller load and soil moisture content on the soil properties of this field by means of a prediction equation, more extreme moisture content profiles in the arable layer will be required. In this year's experiment the residual effects proved to be still too high as compared to the moisture effects.

Integrated wheel-soil-crop research is necessary to provide guidance for the design and application of tyres and traffic systems. The new approach to perform this type of research looks promising. The subject is very broad and it is realized that the entire system cannot be studied in a specialized manner by a single discipline. Hopefully other disci- plines will join in to investigate which technical parameters mainly control soil conditions and crop production.

Applied research should determine the similarity between currently available tyres and the model and classify them accordingly. The combined research would yield a method for quantifying the effects of traffic with currently available tyres in their various configurations and adjustments (load and inflation pressure) on crop yield.

REFERENCES

Akker, J.J.H. van den 1988. Model computation of subsoil stress distribution and compaction due to field traffic. Proc. of the 11th Int. Soil Tillage Res. Org., Edinburgh Scotland, Vol. 1:403-407.
Boone, F.R., H.M.G.van der Werf, B.Kroesbergen, B.A. ten Hag and A. Boers 1987. The effect of compaction of the arable layer in sandy soils on the growth of maize for silage. 1. Critical matric water potentials in relation to soil aeration and mechanical impedance. Neth. J. Agric. Sc. 35:113-128.
Boone, F.R. 1988. Weather and other environmental factors influencing crop responses to tillage and traffic. Soil Tillage Res. 11: 283-324.
Bolling, I. 1987. Bodenverdichtung und Triebkraftverhalten bei Reifen - Neue Mess- und Rechenmethoden. Diss. der Techn. Univ. München, BRD Forschungsbericht Agrartechnik des MEG 133.
Fattah, E.A., R.N.Yong & K.S.Ng 1981. Compactibility of soil under towed rollers. Proc. of the 7th Int. Conf. of the ISTVS, Calgary, Canada, Vol. 2:585-608.
Gupta, S.C. & W.E.Larson 1982. Modeling soil mechanical behaviour during tillage. In: P.W. Unger and D.M. Van Doren (eds.), Predicting tillage effects on soil physical properties and processes, Am. Soc. Agron., Madison, WI, p.151-178.
Hettiarachi, D.R.P. 1988. Theoretical soil mechanics and implement design. Soil Tillage Res. 11:325-347.
Koolen, A.J. 1976. De invloed van de mechanisatie op de bodemstruktuur. In: 100 jaar onderwijs voorlichting en onderzoek in de landbouw, Wageningen. (Also in Koolen & Kuipers (1983), p.41-42).
Koolen, A.J. & H.Kuipers 1983. Agricultural Soil Mechanics. Springer Verlag, Berlin.
Lamers, J.G. & U.D.Perdok 1982. Het rijbanenteeltsysteem. Bedrijfsontwikkeling 13/4:381-395.
Perdok, U.D. & J.G.Lamers 1985. Studies of controlled agricultural traffic in the Netherlands. Proceedings of the Int. Conf. on Soil Dynamics, National Tillage Machinery Laboratory, Auburn, U.S.A., Vol. 5:1070-1086.
Perdok, U.D. & W.B.M.Arts 1987. The performance of agricultural tyres in soft soil conditions. Soil Tillage Res. 10:319-330.
Soane, B.D., P.S.Blackwell, J.W.Dickson & D.J.Painter 1981. Compaction by agricultural vehicles: A review. 1. Soil and wheel characteristics. Soil Tillage Res. 1:207-237.
Tijink, F.G.J. 1988. Load-bearing processes in agricultural wheel-soil systems. Dissertation Wageningen Agricultural University, the Netherlands.
Trouse, A.C., D.H.Parish & H.M.Taylor 1971. Soil conditions as they affect plant establishment, root development and yield. In: Compaction of agricultural soils. ASAE monograph, p.225-312.
Vermeulen, G.D., W.B.M.Arts & J.J.Klooster 1988. Perspective of reducing soil compaction by using a low ground pressure farming system: selection of wheel equipment. Proc. of the 11th Int. Soil Tillage Res. Org., Edinburgh, Scotland, Vol. 1:329-334.

Land and Water Use, Dodd & Grace (eds), © 1989 Balkema, Rotterdam. ISBN 90 6191 980 0

Open-field crop experiments on soils heated with power station cooling water

L.Balemans & W.Van Der Biest
Studiecentrum voor Toegepaste Elektriciteit in Land- en Tuinbouw (IWONL-Laborelec), State University of Ghent, Ghent, Belgium

M.Devrome
Tuinbouwwerkgroep Zuid-Vlaanderen, Kruishoutem, Belgium

A.De Groote
Provinciebestuur Oost-Vlaanderen, Ghent, Belgium

ABSTRACT: Reject heat in cooling water from the Ruien (B) Thermal Power Station was used to heat the soil of two experimental fields. In a preliminary test, a 1 500 m^2 plot was used; in the consecutive large scale experiment 3/4 of a 1 ha horticultural plant was heated, the other 1/4 serving as a reference. Heating requirements of the latter field (between 15 and 30 W/m^2) were covered by a 100 m^3/h water flow through a PE-piping system buried between 0.50 and 0.60 m depth, and resulting in a 1.7 – 3.5 K temperature difference with the reference soil.

A large variety of vegetables has been tested on their yield both on the heated soil and on the non-heated reference plots. Each time the additional influence of applying covering materials was studied as well. Some trials were carried out inside a plastic greenhouse.

1 INTRODUCTION

Thermal power stations have an average maximum efficiency of about 42 %. The input energy which is not converted into electricity is commonly set free as low temperature reject heat in cooling water. Although the degree of pollution, the low temperature level and the discontinuous character of the refrigerant availability from the 980 MW Ruien (B) Thermal Power Station restrict its application potential, good use can be made of it for supplementary soil heating for open-field crops.

To study potential crop yield enhancement, preliminary tests were performed on a 1 500 m^2 test field.

More recently, the experience gained from the small plot was transported to a new 1 ha site; 3/4 of the soil are heated with a 100 m^3/h cooling water flow through a subsoil piping system; the other 1/4 of the plant serves as a reference.

2 PRELIMINARY TESTS

2.1 Project description

The surface area of the terrain is split up in a 900 m^2 heated section and a 600 m^2 unheated reference. The heating-pipe network consists of 3/4" PE pipes at 0.30 m depth and an inter-pipe distance of 0.20 m. The network is fed by an open channel distribution network.

2.2 Yield results

A large number of vegetables and varieties was tested on their reaction to soil heating. Table 1 shows the most relevant data with regard to yield increases and advancements.

The dependence of the yield characteristics on the variety used in the experiments is very well illustrated by the results registered for *celery*.

Table 1. Yield increases and advancements obtained during the preliminary tests.

Vegetable	Yield increase on heated section (%)	Yield advancement on heated section (weeks)
Turnip-rooted		
celery	57	2 – 3
Celery		
Gele barbier	23	up to 8
Cellerity	32	–
Avon Pearl	92	–
Endives	15	2 – 3
Fennel	48	4 – 5
Asparagus	42 – 65	2 – 3
Radish	17	1 – 2
Leek		
summer var.	65	–
autumn var.	19 – 48	1 – 3
winter var.	23 – 57	3 – 4
Brussels sprouts	22	6
Savoy cabbage	40*	1
Broccoli		
Green comet	37	2
SG.1	52	2
Potatoes		
early	7 – 16	–
medium early	51	–

* Relative to number of sprouts harvested

Although distinct yield advancements were recorded for *fennel*, soil heating was particularly useful for reducing the number of shooted tubers.

Harvesting of *summer leek* could start at virtually the same moment on both the heated and non-heated section; on the reference the last batch was harvested 2 weeks after the last batch on the heated section.

3 LARGE SCALE EXPERIMENT

As the yield results obtained on the 1 500 m² plot were fairly satisfactory, it was decided to repeat the experiment on a practical scale. This would allow to make an economic assessment of the benefits of soil heating.

3.1 Project description

Prior to starting the large scale experiment, the soil of the new 42 x 250 m plot was thoroughly preconditioned, and a drainage system was installed. The heating-pipe network consists of 60 pipes of 250 m length each. It was subdivided into 3 sections, each with a different lay-out, see table 2. A surface plan of the project area is shown in figure 1. Some trials were carried out inside a plastic greenhouse.

Table 2. Piping network lay-out in the 3 sections.

Sect. #	No. of pipes	Pipe diameter (mm)	Dist. between pipes (m)	Pipe depth (m)
1	20	40	0.50	0.50
2	20	40	0.50	0.60
3	20	50	0.60	0.60

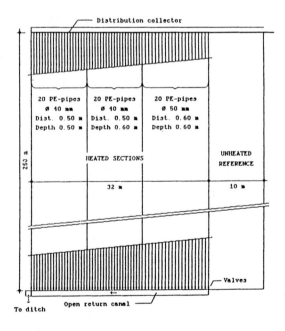

Figure 1. Surface plan of the project area.

3.2 Soil temperatures

During the summer season (March to August), the average temperature difference between the heated soil and the non-heated reference was 2 - 3.5 K at 0.00 - 0.30 m depth, 4 - 6 K between 0.50 and 1.00 m and still 4 K at 3 m below the plot's surface. Given a cooling water flow of 100 m^3/h, the heat supply to the soil was found to be 15 - 30 W/m^2, depending on the time of the year and on cooling water temperature.

From September to February, i.e. throughout the winter period, temperatures at the depths mentioned above were reduced to 1.7 - 2.5 K, 4 K and 3.4 K respectively. Wherever possible, the effect of applying a film cover was investigated. An EVA film with 250 pin holes per m^2 was observed to give a maximum temperature rise of 3.7 K.

To study the possibility of a soil heating system with a discontinuous supply of cooling water (extrapolation to other industries), the water flow through the piping system was interrupted for periods of 3 and 4 days. It was noticed that the temperature difference at 0.25 m and 0.50 m depth between the heated and non-heated sections remained virtually constant. A water supply interruption did however cause substantial accumulation of dirt inside the pipes, and consequently required rinsing of the network.

3.3 Yield results

Due to unfavourable soil texture, it has been a problem reaching the same level of production on the new field as obtained during the preliminary experiment. For some crops neither yield increases nor advancements have been recorded (early summer cauliflower, summer lettuce, radish, autumn spinage).

Moderate results were obtained for spring lettuce under cover and Brussels sprouts. With regard to sprouts, the yield advancement was reflected by a more coarse size distribution, see figure 2.

An appreciable yield enhancement was recorded for turnip-rooted cel-

ery. The yield increase of 18 % is illustrated in figure 3.

Substantial return enhancements were obtained for the crops in plastic tunnels with soil heating. It is evident that in this case the benefits obtained from soil heating are masked by the additional climate improvement by the tunnel, which also requires a supplementary investment. See figure 4.

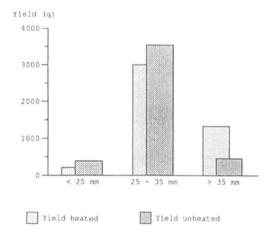

Figure 2. *Brussels sprouts*: size distribution and yields on heated and non-heated sections.

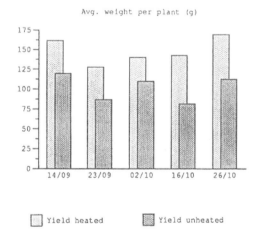

Figure 3. *Turnip-rooted celery*: average weight per plant (g) on heated and unheated sections for various harvesting dates.

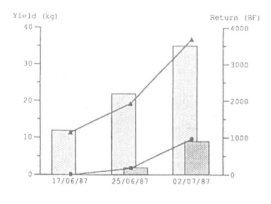

Yield (kg) ... Return (BF)

Legend:
- ☐ Yield heated
- ▲ Return heated
- ▨ Yield unheated
- ⊡ Return unheated

Figure 4. *Haricot beans*: yield and return differences on heated and unheated sections for various harvesting dates.

3.3 Costs versus financial benefits

In order to assess the viability of growing schemes, it is necessary to compare the costs for getting the soil heating system to operate as it should and the return obtained from yield increases and advancements.

Fixed costs (investment = I)
- maintenance of valves: $0.02 \times I$
- maintenance and writing-off of pumps: $0.10 \times I$
- writing-off PE piping network: $0.04 \times I$
- intrest on complete system: $0.06 \times 0.60 \times I$

Variable costs
- operation time: 9 months/year = ± 6 500 hrs
- power taken up: 24.5 kW
- correction for expansion of experimental terrain: $\times 0.75$
- correction for equivalent electricity consumption of cooling tower water pumps: $\times 0.93$

With an investment I = 2 250 000 BF VAT excl., the fixed costs amount to 22 BF/m². With an electricity price of 3.5 BF/kWh, variable costs add up to 50 BF/m².

Based on the financial returns of the individual crops, it is possible to put together viable cropping schemes. One suchlike scheme, which covers the total working costs of 72 BF/m² and offers an additional bonus, is given in Table 3.

Growing asparagus only results in a yearly average yield increase corresponding with about 100 BF/m². It is evident that a cropping scheme with vegetables in tunnels would allow even higher returns; in this case however, the additional costs for the tunnel construction have to be covered as well.

Table 3. Open air cropping scheme

Vegetable	Period	Yield increase 'heated' (BF/m²)
Cauliflower	01/02 – 31/05	10
Parsley	01/05 – 31/07	10
Strawberry	21/07 – 30/06	80
Lettuce	01/07 – 31/08	50
Spinage	01/08 – 30/10	20
	Average / year	85

4 CONCLUSION

Considering the important electricity consumption, and taking into account the substantial investment costs, the economical justification of the present project is still doubtful.

It was noticed however that extrapolation of the experiment to industries with a discontinuous cooling water supply (with e.g. weekend interruptions) would be technically possible, as this has little or no impact on the soil temperature.

Chauffage du sol pour des cultures en pleine terre par l'eau de
refroidissement d'une centrale électrique

RESUME: La chaleur rejetée dans l'eau de refroidissement de la centrale
électrique de Ruien (B) a été utilisée pour le chauffage du sol de deux
champs d'essai. Dans un essai préliminaire on a utilisé une surface de
1 500 m^2; dans les expériences suivantes une surface horticole de 3/4 ha a
été chauffée, tandis que le 1/4 restant a servi de référence. La chaleur
nécessaire durant les derniers essais (à savoir 15 - 30 W/m^2) fut obtenue
par un débit d'eau de 100 m^3/h livré par un réseau de conduits en PE placés
à 0.50 et 0.60 m de profondeur, avec comme résultat un écart de
température de 1.7 à 3.5 K par rapport à la parcelle de référence.
Une grande variété de plantes ont été testées quant à la production
obtenue sur le sol chauffé et sur le sol de référence non-chauffé. En même
temps l'influence du recouvrement du sol a été étudié. Certains essais ont
été effectués dans des serres en plastiques.

Bodenheizung von Experimentalfeldern mittels Abfallwärme ins Kühlwasser
eines thermischen Elektrizitätswerks

ZUSAMMENFASSUNG: Die Abfallwärme enthaltet ins Kühlwasser des Thermischen
Elektrizitätswerks in Ruien (B) wird benützt zur Heizung der Boden von
zwei Experimentalfelden. Ein vorangehender Versuch wurde ausgefuhrt auf
einem 1 500 m^2 Terrain. Das darauffolgende Experiment fand statt auf einen
größerem Grundstück dessen Fläche 1 ha beträgt; 3/4 davon sind gehizt, das
überige ungeheizte 1/4 gilt wie Referenz. Der Heizungsbedarf diesen
letzten Terrains (15 - 30 W/m2) wird gedeckt von einem 100 m3/h Kühl-
wasserdebit durch einen PE-rohrenkreis der entweder 0.50 oder 0.60 m tief
unter der Bodenfläche eingegraben ist. Dieses System gab einen Tem-
peraturdifferenz von 1.7 bis 3.5 K zwischen der geheizte Abteilung und der
ungeheizte Referenz.
Eine große Menge von Gemüsen wurde getestet zur Bestimmung ihren Ernte-
ertrag und Ernteverfrühung. Jedesmal wurde auch den Einfluß des Gebrauchs
von Folien aufgezeichnet. Einige Proben fanden statt in einem Plastik
Gewächshaus.

Land and Water Use, Dodd & Grace (eds), © 1989 Balkema, Rotterdam. ISBN 90 6191 980 0

Gypsum/lime slotting technique and machinery for ameliorating acid and sodic subsoils

J.Blackwell
Cassiro Pty Ltd, Wauchope, NSW, Australia

N.S.Jayawardane & R.Butler
Csiro Division of Water Resources, Griffith, NSW, Australia

J.C.Casegrain
Cassiro Pty Ltd, Wauchope, NSW, Australia

R.E.Smart
Ministry of Agriculture and Fisheries, Ruakura, New Zealand

ABSTRACT: Research has shown that massive amelioration of acidic or sodic clay subsoils in order to improve yield limiting soil physical and chemical properties is too expensive and short lived for widespread adoption.

A novel deep soil amelioration technique and machinery is described which, at relative low power requirements, offers potential for increased crop yield by increasing infiltration and air filled porosity, and protection from subsequent recompaction.

ZUSAMMENGASSUNG: Die Literatur zeigte, da massive Unterbodenamelioration von aciden oder sodischen Tonböden, um ertragslimitierende bodenphysikalische und-chemische Eigenschaften zu verbessern, unökonomisch und von kurzer Lebensdauer waren und deshalb keine weitverberbreitete Anwendung fanden.

Eine neue Unterbodenameliorationstechnik ist beschrieben, welche mit relativ geringem Energieaufwandt das Ertragspotential dieser Böden durch erhöhte Infiltration, vergröbertes Volumen der luftgefüllten Bodenporen und Verminderung der Gefahr von Wiederverdichtung, erhöht.

RESUME: Les recherches ont demontré que l'amélioration des sous-sols, des terres a forte densité de soude et des terres acides par la correction des facteurs physiques et chiemiques, qui sont des facteurs limites pour rendement coutent tres cher et sont en chomage partiel - Pour cette raison ces methodes n'ont pas été employees.

Une technique nouvelle pour améliorer ces sols est decrite. Cette technique augmente le potential de rendement avec une demande d'energie relativement petite, en augmentant l'infiltration et les volumes des poches d'air et en reduisant le dange de recompaction.

1 INTRODUCTION

Soils used for horticultural and agricultural production in Australia often feature heavy clay subsoils. Physical and chemical properties of these soils are often restrictive to plant growth and yield. In Australia 60% of the irrigated lands are duplex red-brown earths which exhibit an intrinsic sodicity problem.

Yield limiting soil properties can broadly be summarized into (i) surface crusting and hardsetting, (ii) extremely low (less than 2mm of water per day) permeability of the upper B-horizon and (iii) excessive sodicity, acidity and salinity within the root zone.

Inadequate aeration in the root zone due to poor internal drainage of the clay sub-soil is a major contributory factor to low yields of crops on these soils (Jayawardane and Meyer 1984). Field experiments (Loveday et al. 1970; Muirhead, Loveday and Saunt 1970) have shown that application of gypsum can increase the infiltration rates and crop yields on these soils. Regular gypsum application, however, is not widely practised by farmers, possibly due to the need for frequent applications, which makes it unecomomic. Loveday et al. (1970) also found that, while deep tilling can markedly increase infiltration rates and crop production on these soils, these effects tend to decrease significantly in the following seasons, apparently due to repacking of the soil under flood irrigation and tractor wheel compaction. Many of these problems can be exaggerated by agricultural practises.

Subsoil acidity and acidification can be regarded as a world wide problem counteracting the increasing demand for efficient crop production. In Australia this is particularly evident in the high rainfall areas of the east coast. Severe subsoil acidity is the major limiting factor to commercial crop production under otherwise ideal conditions (i.e. climate).

Acid soils often contain species of exchangeable aluminium which have been shown to be toxic to plants (Veitch, 1904). Much of the poor root development and consequent drought susceptibility seen in acid soils with a pH less than 5.0 in the subsoil layers is probably due primarily to aluminium toxicity, which limits both rooting depth and degree of root branching (Foy 1984). Aluminium injured roots are characteristically stubby and brittle. Root tips and lateral roots become thickened and may turn brown, with the root system as a whole showing no fine branching.

Bromfield (et al. 1987) have shown that surface incorporated lime and top-dressing lime does have an effect on subsoil acidity. This effect may be sufficiently deep and fast enough to create an environment satisfactory for shallow rooted species (i.e. pastures) growing in light textured soils where the original pH (0.01M Ca Cl) was not lower than 5 at any depth. However, for deep rooted, high value horticultural crops on soil with a pH profile of less than 4, the effects are simply too slow, and not effective to sufficient depth to be considered viable. The relatively large capital investment in surface lime application takes many years to show a return and may be one reason why more farmers do not adopt the practice.

It is evident that the adverse effects of subsoil acidity and sodicity could be, at least partially, corrected, if ameliorants could be incorporated to depth at justifiable costs. However, in the past, available methods of subsoil amelioration, such as the direct incorporation of lime into the subsoil, or thorough mixing of lime with the subsoil, were both considered impractical (Pearson, Childs and Lund 1973, Doss, Dumas and Lund 1979, Arkin and Taylor 1981). An alternative, efficacious strategy had to be found. Pearson et al. (1973) and Kauffman and Gardner (1978) demonstrated that partial (segmental) mixing of lime with the soil could produce satisfactory crop growth and yield.

Pearson et al. (1973) suggested that root growth would benefit considerably from

adequate rates of deep-placed but incompletely mixed lime in the strongly acid subsoils.

2 DESCRIPTION AND RESULTS OF TECHNIQUE

A new technique for soil amelioration using gypsum or lime enriched slots (Figure 1) has been proposed

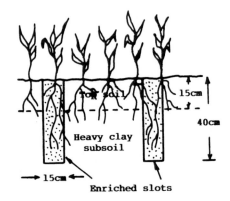

Fig. 1. Sketch of soil profile showing enriched slots

The technique was tested in a field trial comparing, control(c) no added gypsum, surface applied gypsum 8t/ha (G) and slotted gypsum 8 tonnes/hectare in 40cm deep slots 0.66m spacing (SG1), and at 1.0m spacing (SG2). Slotting showed faster and deeper wetting during irrigation (Figure 2) and improved aeration in the surface and subsurface layers (Figure 3) (Jayawardane and Blackwell 1985).

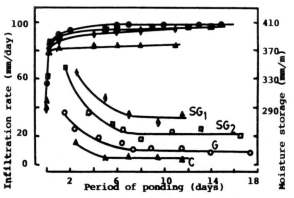

Fig. 2. Infiltration rates (open symbols) and cummulative moisture storage in 1m depth of soil (closed symbols) during ponding of plots C, G, SG$_1$, and SG$_2$.

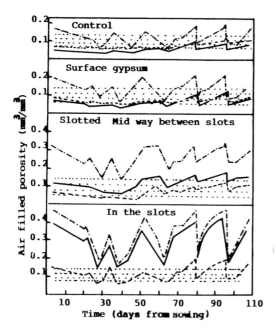

Fig. 3. Air filled porosity over a season at 3 depths for treatments C, G and SG_1. At air porosity levels greater than 0.14 (top of shaded area) oxygen flow to roots is unimpeded. At air filled porosity levels below 0.08 (bottom of shaded area) root growth is severely restricted. This caused an increase in the rate of phenological development, tillering, canopy closure and yields of a wheat crop (Figure 4 and Table 1) (Jayawardane, Blackwell and Stapper 1987)

Table 1. Grain yields and yield components in two harvest areas of plots C, G, SG_1 and SG_2

Yield components and yield	Plot C	Plot G	Plot SG_2	Plot SG_1
Harvest area A				
Yield (g/m²) 12% mc	541	587	841	791
Total dry matter (g/m²)	1055	1162	1745	1561
Harvest index (%)	45.1	44.5	42.5	44.6
Spikes /m²	367	394	580	513
Spike weight (g/shoot)	1.78	1.79	1.76	1.84
Kernels/spike	33.9	35.5	35.1	35.1
Kernel wt.(mg) oven dry	39.1	38.0	37.2	39.1
Kernels/m²	12170	13605	19890	17800
Harvest area B				
Yield (g/m²)(12% mc)	528	614	830	823

Jayawardane and Blackwell (1985) also suggested that the ameliorative effects of soil slotting could last longer than deep ripping due to the slower rate of consolidation of the slots. Blackwell et al. (1989) showed that vertical stresses under tractor wheeling were attenuated by the stronger undisturbed soil between the slots which would result in less compaction of the slots (Figure 5).

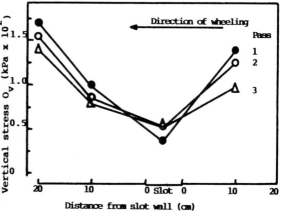

Fig. 5. Vertical soil stresses in a 70mm wide slot and either side of the slot during three consecutive transverse wheelings by an (18.4 x 38) tractor tyre loaded to 4 tonnes at a speed of 0.7 km/hr

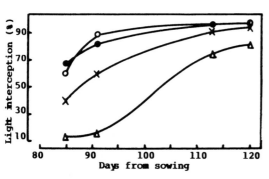

Fig, 4. Light interception in plot C, (▲), plot G (X) plot SG_1 (●) and plot SG_2 (O).

This coupled with potentially lower power requirements, since only 1/3 to 1/7 of the soil volume is tilled could be a critical factor in determining the longevity and economic viability of this ameliorative technique relative to current practice.

In irrigation areas the technique could offer an alternative to repeated land levelling

since the technique allows excess water to be removed by percolation into the lower layers. This excess water is available for plants during the following dry periods. In addition, expensive drainage structures to remove excess water are not required.

The fast internal drainage and early surface drying allows for earlier tractor access to the field, which can be a critical factor during some management practices, particularly spraying and harvesting, without the associated risk of physical damage to the subsoil when working soil that is too wet.

Many subsoils on the east coast of Australia are of such low pH as to preclude the entry of roots. Roots are often restricted to the top 15cm of soil in spite of annual applications of lime followed by deep ripping. Lime enriched slots were used in an attempt to improve rooting depth and vigour in a six year old vineyard in Port Macquarie, New South Wales which showed typical aluminium toxicity symptoms. Limed slots (150 mm wide x 400 mm deep) were created 50 cm and 80 cm away from the vine row and running parallel with it. Just before slotting approximately 4 t lime/ha was spread on the surface in a band where the slots were to be formed. The slots were imposed in late December after the initial 'flush' of vine root growth. The slots were excavated on 13 May 1988 and showed prolific healthy root growth with many fine branching roots to the full 40 cm depth of the slots. (Figure 6).

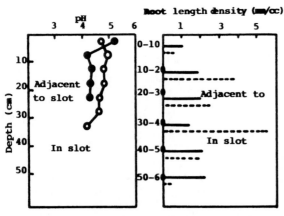

Fig. 6. Six-year-old Chardonnay vines slotted December 1987 (150mm x 400mm) 4t lime/ha. Roots sampled May 1988.

The increased number of healthy roots to depth in the slots means that deeper reserves of moisture can be tapped; initially from the slot itself thus setting up a moisture gradient from the non-slotted to the slotted zone and encouraging water flow into the slot. It is anticipated that the lime slots will confer an 'ameliorative' effect on a greater volume of the profile than that which is physically disturbed and treated with lime. In preliminary experiments gypsum slots of 100-150mm width were effective in improving the soil physical conditions within the entire root zone. (Jayawardane et al. 1985).

3 DESIGN CRITERIA OF THE SLOTTER

Slot configuration for optimum plant growth varies according to the soils, crops, climates, irrigation methods and other agronomic practics adopted. Hence a primary design requirement of the implement is sufficient versatility to provide changes in spacing, width and depth of slots with simple adjustments. It should also allow for changes in amount of ameliorant applied and the proportion applied to the slot and onto the surface of the soil.

3.1 Slot forming mechanism

The slotting mechanism should create vertical discrete slots with the adjacent soil undisturbed, to protect the slot from recompaction under trafficking (Blackwell P.S. et al. 1989). The slots should be sufficiently wide to allow rapid water entry and to provide internal drainage depending on the soil type and crop, but no wider than necessary to reduce power requirements. The maximum width of slot should not approach or exceed the minimum tyre width during longitudinal trafficking to minimise recompaction within the slots. (Blackwell J. et al. 1989).

In duplex soils with impervious layers the depth of slot should be sufficient to penetrate the "throttle" to provide water entry and aeration of the upper horizons, where the major portion of the roots occur. It is also desirable to minimize this depth to reduce power requirements. A slot depth of 400mm was sufficient to cope with prolonged winter rainfall periods in an irrigated transitional red brown earth, New South Wales, Australia (Jayawardane 1985). In high rainfall, acid country on the east coast of Australia an 80 cm depth of slot has been used. (Blackwell J. et al. 1988 b). These preliminary studies indicated that a minimum spacing of 1m was adequate. However, from the point of view

of minimising power requirements a wider
spacing is desirable.

Since the ameliorative technique is seen
as long term land development rather than
an annual tillage operation, high speed of
operation is not critical. However,
increased speed reduces costs and increases
the timeliness of operation.

3.2 Ameliorant carrying and distribution system

To minimise ferry time or the need for a
mother cart the implement should have a
large capacity hopper for carrying the
ameliorant.

The distribution system should be
capable of metering small amounts of
ameliorant accurately. For instance at the
lowest likely application rate of 4 t/ha
and slots spaced 1.2 m apart the output
would be 4 and 12 kg/min for each slot at
tractor speeds of 0.5 and 1.5 km/hr,
respectively. The metering and delivery
system must also overcome the problems
caused by the tendency for gypsum or lime
to bridge over small hopper openings and
compact, thus interrupting the smooth flow
to the metering system.

4 PROTOTYPE DESIGN

4.1 The slotting mechanism

In the initial small plot pilot trials the
slots were excavated using a conventional
"Ditch Witch" TM with a single chain and
scoop type conveyor elevator powered by a
14 KW motor. A larger implement based on a
similar principle but with multiple chains
was considered but rejected on the grounds
of low ground speed capability and high
power requirements. These machines also
have a high maintenance cost because of the
large number of moving parts. A multi-
tined design similar to that described by
Hyde (1983) was rejected, as well-defined
slots would be difficult to form. Such an
implement would have very high draft
requirements on the heavy soils likely to
need amelioration.

Rotary tillage appeared to meet all the
design requirements whilst having the
advantage of low draft requirements.
Layout of the first prototype slotter is
shown in Figure 7.

The rotor assembly consists of rotors (A)
mounted on a 3m long, 150mm diameter
hollow shaft supported at either end on

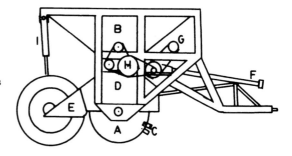

Fig. 7. Schematic side elevation of
rotary slotter.

heavy duty roller bearings. This shaft is
from a standard Howard (TM) AH120 Rotary Hoe
with flanges at 250mm intervals. One meter
diameter rotors were attached to these
standard flanges. Six pairs of 100mm long
shanked, rotary hoe blades (C) of any
required width are bolted at equal spacing
to the periphery of each rotor. This
arrangement provides a maximum working
depth of 500mm.

Each rotor is provided with a fixed
shroud (D) over the top of the rotor and an
adjustable shroud (E) attached to the fixed
shroud close to the soil surface. The
shrouds ensure that ameliorated soil is
returned to the slots formed in the soil.

The slotter was driven by the power take
off (F) (540 rpm) from the rear of the
tractor via a splined drive shaft and
multi-plate dry clutch to a right angle
drive gearbox and a heavy duty jack shaft.
These components are all parts from a
Howard(TM) AH120 rotary hoe. The jack
shaft transmits power from the gear box to
an extended chain drive (30mm pitch) which
is totally enclosed in an oil-tight chaincase forming an integral part of the main
frame. Rotational speed of the rotor is
varied by either changing the two sets of
pick off gears supplied with the Howard
Selectatilth(TM) gearbox or by altering the
ratios between the driver and driven
sprockets in the chaincase. In the initial
trials a rotor speed of 94 rpm was used.

The spacing of the slots could be
adjusted by changing the spacing of the
rotors and the slot width could be adjusted
by 'cropping' the rotor blades. For the
initial field testing 3 rotors, 100mm wide,
spaced 1.2m apart were used. The depth of
slotting could be varied by lowering or
raising the rotor assembly through

hydraulic rams (I) on either side of the machine. Using 3 rotors at maximum depth of 500mm a P.T.O horsepower of 97 kw was required in the transitional red-brown earth of the initial trial site.

4.2 Ameliorant carrying and metering system

The hopper dimensions were designed to give a payload of between 4-6 tonnes. (B) By-product gypsum is capable of bridging over a gap of 480mm. Hence a hopper base dimension of 500 x3000mm was used to prevent bridging. Traditionally gypsum hoppers have sides sloping at 70° to decrease adhesion. Auger box dimensions of 500 x3000mm would have meant a very high machine to achieve the design payload. In order to provide a smooth flow of gypsum or lime with a side wall angle of 45°, the hopper lining was fabricated from 306 grade stainless steel.

The design requirement of the metering and distribution system is to maintain constant supply to the three rotors, even at the minimum rate (4 kg/min to each rotor). Traditional belt and gate gypsum and lime spreaders may be feasible, but accurate low rates of application are difficult due to the tendency for the flat belt to slip under the gypsum and lime mass at very small gate openings. Screw conveyors are traditionally used for uni-directional transport of gypsum and lime eg. from bulk hopper to free flow end delivery. This arrangement is satisfactory for high flow rates but requires complicated speed reduction gears for low flow rates (Kernebone - personal communication).

The much simpler gypsum distribution system created for this prototype implement (Fig. 8) consisted of two adjacent 250mm contra flighted augers (J) rotating within an auger box (K) located directly beneath the hopper. Three pairs of adjustable outlets (L) were provided in the bottom of the auger box, one pair over each rotor. Auger pitch was 200mm but only the sections (N) located directly above the outlets were flighted. To overcome the problem of blockage by small lumps of gypsum a 12mm diameter 150mm long rod (M) was welded to the auger flight directly above the adjustable outlets to sweep the outlet during each rotation of the augers. A series of 12mm diameter pegs (O) with 200 mm pitch were welded to the 50mm auger tube at a spacing of 150mm. This peg arrangment moved sufficient material to ensure the flighted lengths of auger were kept full while there was gypsum in the auger box but enabled the auger to "slip" through the gypsum as the load increased.

At the ends of both augers one reverse pitch (P) was installed to aid transfer of gypsum between augers and to counteract any tendency to pack against the end walls. (Figure 8)

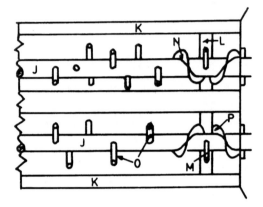

Fig. 8. Schematic plan view of ameliorant metering mechanism.

The two augers were powered hydraulically (G) with speed reduction via a system of chains and sprockets (H) to give an auger speed of 6.5 rpm at full flow. The gypsum application rate could be adjusted by changng the rotation speed of the augers and/or the size of the openings. (Figure 9)

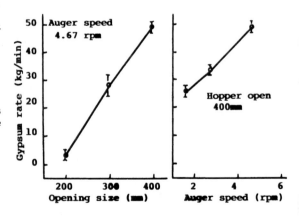

Fig. 9. Calibration of ameliorant metering at different auger speeds and hopper openings.

CONCLUSION

Slotting with the required ameliorant offers a viable alternative to current methods of improving subsoils. In vineyards in Australia the technique is an economic proposition for ameliorating soils with very acid profiles to 800 mm depth. The technique offers irrigation and dry land farmers the opportunity to increase their yields whilst decreasing run off from the property thus alleviating po9llution hazards downstream.

REFERENCES

Jayawardane, N.S. and W.S. Meyer 1984. Measurement and use of air-filled porosity profiles in an irrigated swelling soil as an index of soil aeration. In 'Root Zone Limitations to Crop Production on Clay Soils'. W.A. Muirhead and E. Humphries (eds.), p 1-8 (CSIRO Australia : Melbourne.)

Loveday, J., J.E. Saunt, P.M. Fleming and W.A. Muirhead 1970. Soil and cotton responses to tillage and ameliorant treatments in a brown clay soil. 1. Soil response and water use. Aust. J. Exp.Agric. An. Hus. 10: 313-337.

Muirhead, W.A., J. Loveday and J.E. Saunt 1970. Soil and cotton response to tillage and ameliorant treatments in a brown clay soil. 2. Growth yield and quality of cotton. Aust. J. Exp. Agric. An. Hus. 10: 325-337.

Blackwell, P.S., N.S. Jayawardane, J. Blackwell, R. White and R. Horn 1989. The protection of deep cultivated soil in gypsum enriched slots, compaction under wheels. Soil. Sce. Soc. Am. J., 53 (Accepted for publication).

Veitch, F.P. 1904. Comparison of methods for the estimation of soil acidity. J. Am. Chem. Soc. 26:637-662.

Jayawardane, N.S., J.Blackwell and M. Stapper 1987. Effects of changes in moisture profiles due to surface and slotted gypsum application to a clay soil on crop development and yield of wheat. Aust. J. Agric. Res. 38:239-251.

Foy, C.D. 1984. Physiological effects of hydrogen, aluminium and manganese toxicities in acid soil. In Soil Acidity and Liming (2nd Edition) F. Adams (ed).

Bromfield, S.M. R.W. Cumming, D.J. David and G.H. Williams 1987. Long term effects of incorporated lime and top dressed lime on the pH in the surface and subsurface of pasture soils. Aust. J. Exp. Agric. 27:533-538.

Pearson, R.W., J. Childs and Z.F. Lund 1973. Uniformity of limestone mixing in acid subsoil as a factor in cotton root penetration. Soil. Sci. Soc. Am. Proc. 37:727-732.

Doss, B.D., W.T. Dumas and Z.F. Lund 1979. Depth of lime incorporation for correction of subsoil acidity. Agronomy J. 71:541-544.

Arkin, G.F. and H.M. Taylor (eds) 1981. Modifying the root environment to reduce crop stress. St Josephs, MI: Am. Soc. Ag. Eng.

Kauffman, M.D. and E.H. Gardner 1978 Segmental liming of soil and its effect on the growth of wheat. Agronomy J. 70:331-336.

Jayawardane, N.S. and J. Blackwell 1985. The effect of gypsum enriched slots on moisture movement and aeration in an irrigated swelling clay soil. Aust. J. Soil. Res. 23:481-492.

Blackwell, J., N.S. Jayawardane, W.A. Muirhead and J.C. Cassegrain. 1988b Soil slotting for existing and future vineyards. In vineyards soil and sod culture management. Proceedings of Aust. Soc. Vit. and Oenology. Conference 24 June 1988, Mildura, Victoria. T.H. Lee, B.M. Freeman and P.F. Hayes (eds).

Hyde, L.M., J.E. George, K.E. Saxton and J.V. Simpson 1986. A slot-mulch implement design. Am. Soc. Ag. Eng. 29(1):20-25

Blackwell, J., R. Horn, N.S. Jayawardane, R. White and Blackwell P.S. 1989. Vertical stress distribution under tractor wheeling in a partially deep loosened typic paleustalf. Soil and tillage research. (Accepted May 1988).

Land and Water Use, Dodd & Grace (eds), © 1989 Balkema, Rotterdam. ISBN 90 6191 980 0

Experiments on the use of a hydrocarbon based soil conditioner

K.Bahadyr

Chaire de Bonification,Tachkent, USSR

ABSTRACT: Artificial structural formation of soil allows more economic use of water and reduces soil erosion. It gives better water, air and nutrient regimes in the soils, increases yields, improves the effectiveness of irrigation and increases the protection of the soil and water environment from pollution by fertilizers. Results obtained under natural conditions over a period of some years confirm these expectations of improved soild conservation and fertility.

1. INTRODUCTION

Intensive use of soil leads to fertility losses, - a problem more pronounced in aris zones. Irrigation requires the economic use of water, conservation of soil against erosion and protection of the environment from pollution. In the south of the country large areas are irrigated using furrow irrigation. This irrigation method is very simple and reliable when compared with complex systems of water distribution.

When irrigating fields with gradients of 0.01 to 0.22 conservation measures are needed.

We have developed a polymer (entitled series "K") which has the following advantages:

1.1. A single sprinkling of the soil with the polymer gives a soil structural formation similar to that after three years growing of alfalfa.

It means that the time required for improvement of soil fertility is reduced. Introducing the polymer dose into soil regulated water permeability by regulating the penetration of water on slopes with different degrees of steepness. It is possible to make long furrows from water distribution points at the top of the slope.

1.2. The polymer increases the air content in the soil, making it more favourable for roots of plants.

1.3. The polymer makes a protective layer of soil which cannot be seen in nature, but it keeps the soil free from crust formation.

1.4. The polymer enhances the washing out of soils with high salt contents.

1.5. When using irrigation furrows with special tractor implements the length of the furrows can be increased by up to 500 m along slopes ranging from 0,01 to 0,03. After applying the polymer in this case irrigation can be used without a network of irrigation ditches in the fields.

1.6. This structure maintains soil humidity by reducing evaporation from the upper layers of soil.

2. EXPERIMENTAL DATA AND RESULTS

Experiments on soils with hard gypsum inter layers to a depth of 0,9 m with special tractor implements and polymer dose rates of 180 kg per ha have yielded the following results: without polymer we have obtained cotton yields of 8,2 centner per ha (ent/ha), when using water irrigation rates of 10 250 m^3 per ha; with polymer - 27,5 ent/ha using irrigation of 8110 m^3 per ha (m^3/ha), while the addition of 30 tonne per ha of a manure yielded 29,6 ont/ha, using 7150 m^3/ha of water.

This experiments have been conducted under conditions of desert soils. When conducting the experiment, these soils are enhanced by silts, delivering 5 to 10 gramm per litre and containing nutrients for plants.

Experiments on the length of furrow in the three ranges 0-110 m, 110 - 370 m,

370 - 500 m, with gradients of 0,017, 0,03 and 0,013 and the use of polymer in doses of from 35 to 120 kg/ha over four years showed that irrigation may be reduced from four days to two. Irrigation is controlled automatically by means of fixed flumes mounted across the top of each field and requires no temporary distribution network or intermediate canals. After four years of polymers application, irrigation during the fifth year can be used without polymer. On the first year of irrigation by furrows, irrigation rates were 7770, 6390, 5230 m^3/ha, and for the fourth year were 6750, 5430, 5270 m^3/ha for each of the three furrow lengths used. Yields of cotton amounted to:

1st year - 29,7; 27,2; 30,9 ent/ha;
4th year - 39,7; 38,7; 36,8 ent/ha
Using normal irrigation yields were 29,2; 27,6 and 26,8; ent/ha Humus and nitrogen reserves were increased by 0,16 and 0,9 %.

An artificial structure was formed in the upper layer of stong and sandy soils with gradients 0,07-0,03 by depositing silts with nutrients. Good moisture in soil can be retained by the residues of accumulated silts which continuously improved the soil. For example, a dose of 125-167 kg/ha of polymer increases the yield of cotton crop from 9,5-12 to 17-21 ent/ha. The length of irrigation furrow was increased from 50 m to 100-250 m. The number of water retaining soil particles increased from 6 to 52 %. In each year of 3 years of experiments the following constituents of silts were kept in the soil: Humus - up to 0,2 %, nitrogen - 0,1 %, carbonate - 0,06 % (which increased soild fertility). Water consumption was decreased from 11090 m^3/ha and 14,4 tone per ha of silt was accumulated annually. In the 3rd year the yield of cotton rose to 28 ent/ga but remained at a level of 12 ent/ha without polymer.

Lands with gradients of 0,16 to 0,22 in agriculture are considered as non-profitable because of the fact that irrigation by furrows is difficult. Irrigation gives problems with soil erosion and for moistening soil to 1 m depth long irrigation times are required. Polymer allows irrigation without erosion and expedites the penetration of water along slopes. Tractor operation is difficult so we propose zigzag type furrow for treatment of steep parts of slopes with polymer. These furrows are supplied automatically with water supply from horizontal irrigation ditches.

Irrigation for a period of two years on such slopes with soils of low permeability gave yields of green mass amounting to 340 ent/ha. The recommendation to sow wheat in autumn, in spring to collect yield from winter precipitations, then to sow maize and irrigate using the above mentioned measures. Under our conditions we have obtained two yields: 17,1 ent ha wheat and 91,3 ent/ha maize in 1 year.

For land reclamation 180-500 kg/ha of polymer on loamy soils is recommended for the 1st year of irrigation. An artificial structure and loose layers can increase water permeability and assist the moving of salts into the lower soil. This measure for reclamation of desert lands reduced the time period for application of irrigation from 3 years to 1 year.

Using the polymer we can also regulate water permeability along convex and convex-concave slopes without leveling of the surface. With a slope of 0,08, which is initially convex and then ends as concave, experiments showed cotton yield of 28,5-27 ent/ha with polymer, and 22-19 ent/ha without polymer. The dose of polymer was 120-160 kg/ha. In spring when roots of plants are short, capillary side moistening of soil under the plants is accelerated. It is very important for plants which suffer from drought.

3. CONCLUSION

Artificial structural formation of soil can be achieved from wastes at hydrocarbon plants. We use such polymers to increase soil fertility and recommend that others extract these polymers from gas and use them for agricultural needs.

Land and Water Use, Dodd & Grace (eds), © 1989 Balkema, Rotterdam. ISBN 90 6191 980 0

Chaulage et modifications de la fertilité physique en sols limoneux hydromorphes

F.Kockmann
Chambre d'Agriculture de Saône et Loire, Macon, France

B.Fabre
Institut Supérieur Agricole Rhône-Alpes, Lyon, France

Résumé: Nous nous sommes intéressés aux répercussions du chaulage dans les sols limoneux hydromorphes de Saône et Loire, (France), drainés ou non. Nous montrons que l'effet du calcium au champ, évalué par l'observation du profil cultural, sur des cultures d'automne, est fonction de l'état de dégradation du profil après les travaux du sol, de la vitesse de percolation de l'eau en excès vers les couches profondes, de la quantité d'eau reçue en automne d'une part et en hiver et au printemps d'autre part. Les effets du chaulage sur la fertilité physique quand ils existent, se traduisent par une amélioration importante de la densité racinaire dans les horizons labouré et immédiatement sous-jacent.

Il apparait que drainage et chaulage sont deux techniques complémentaires, que les mesures de laboratoire doivent être doublées par des observations de terrain pour élaborer un conseil aux agriculteurs, et que les critères choisis pour évaluer l'extériorisation du chaulage sont très pertinents.

Abstract: This following paper : focusses on effects of liming in loamy soils in SAONE et Loire (France). We show that this effect, on field, evaluated by observing the agronomic soil profile, on winter crops, is linked with the state of soil degradation after tillage, the seepage speed of excessfull water in deep horizons, the rainfall during autumn, winter and spring. These effects on soil physical fertility, cause an improvment of root density.

It occurs that liming and draining are complementary, that the laboratory mesures may be doubled by observations on field and that our indicators of liming effect on soil physical properties are very pertinent.

Zusammenfassung: In diesem Text haben wir die Kalkungseindrücke in den schlammigen, entweder entwässerten oder nicht, Erden von Saône et Loire (France) geprüft.
Wir zeigen, dass die Kalkwirkung auf dem Boden (durch ein Agrarbodenprofil im Herbst beobachtet) von mehreren Faktoren abhängig ist: die nach Bodenarbeitungen Bodenprofilverschlechterungslage; die im Herbst gefallenen Wassermengen; die Ubermasswasserversickerungsgeschwindigkeit nach tiefen Bodenschichten ; die, einerseits im Herbst, andererseits im Frühling gefallenen Wassermengen.
Die etwaigen Kalkungswirkungen auf der physiken Bodenfruchtbarkeit äussern sich in einer wichtigen Wurzeldichteverbesserung im gepflügten und ganz unterliegenden Horizont.
Es ist klar, dass Entwässerung und Kalkung zwei ergänzende Techniken sind; dass die Labormassnahmen mit Ort und Stelle Beobachtungen ergänzt werden müssen, um am bestens die Landwirten zu beraten, und dass die für Kalkungsauswirkungenschätzung erwählten Kennzeichen sehr zutreffend sind.

Nous remercions pour leur collaboration D. Ortscheit et M.J. Fournet qui ont réalisé une synthèse du travail à mi-chemin de l'expérimentation (1985) puis assumé le suivi et le compte-rendu des campagnes en 1986 et 1988.

INTRODUCTION

En Saône et Loire, les sols limoneux hydromorphes posent de nombreux problèmes de
mise en valeur. Les voies les plus explo-
rées sont le drainage et la pratique des
amendements calcaires pour améliorer les
conditions de cultures.
Or, si le drainage connaît un réel succès,
le chaulage reste une pratique rare et oc-
casionnelle chez les agriculteurs ; qui
plus est, les références sont absentes.
C'est ainsi qu'une expérimentation sur les
répercussions du chaulage en limons a été
mise en place en Bresse, au printemps
1983.
Les deux techniques, drainage et chaulage,
reviennent à jouer sur la fertilité physique des terrains définie comme la plus
ou moins grande facilité de créer et de
maintenir un état physique adapté dans ses
conséquences au système de culture pratiqué, par G. Monnier, P. Stengel, et J.
Guérif (1981).

De cette définition, nous tirons dès à
présent des hypothèses de travail:
  - Les critères analytiques du laboratoire (analyses de terre, test de stabilité structurale, sensibilité au compactage) qui définissent les propriétés d'un
matériau ne sont pas suffisants pour juger
de cette fertilité et des effets des pratiques susceptibles de l'améliorer.
  - Les systèmes de cultures n'ont pas les
mêmes exigences en terme de fertilité et
on doit donc adapter les conseils en fonction des cultures présentes d'une part et
de l'état d'hydromorphie des terrains
d'autre part, outre bien évidement l'état
calcique initial, défini par le pH eau, le
pH KCl, et les teneurs en CaCO3 et CaO
échangeable.

Nous nous sommes donc placés dans plusieurs conditions de milieu, identiques
par la texture, mais différentes par
l'état calcique, (pH variant de 5,9 à 7),
la présence de drainage et le degré
d'hydromorphie du terrain, les cultures en
place (céréales d'hiver, colza d'hiver,
maïs, tournesol et prairies temporaires).
On trouvera les détails des implantations
sur 13 sites expérimentaux ainsi que
l'exhaustivité du protocole dans une pré-
cédente publication (B. Fabre, F Kockmann
1987).
Rapellons simplement que les parcelles
sont choisies chez des agriculteurs, qui
ont reçu pour seules consignes de noter
les différentes opérations culturales et
de les pratiquer sur l'ensemble de la parcelle. Sur chacune, nous délimitons deux

sous-parcelles, l'une témoin (T0) et
l'autre (T1) recevant 2 000 à 3 500 u
CaO ; outre 11 essais basés sur la comparaison "absence-présence", l'échantillon
comprend deux essais "doses", T1 2000/T2
4000. Le produit est un calcaire pulvéru-
lent (54 % CaO, 80 % passant au tamis de
0,08 mm). L'expérimentation est achevée en
septembre 1988.
Le calcium a de multiples effets : il modifie la garniture ionique et par là, le
pH ; il agit aussi sur la floculation des
argiles et par là, sur la stabilité structurale. Nous ne développerons ici que les
répercussions du chaulage sur la fertilité
physique, en privilégiant délibérément les
contrôles et les résultats "au champ".

1 MODELE GENERAL DE L'EFFET DU CA SUR LA
FERTILITE PHYSIQUE ET SON EXTERIORISATION
AU CHAMP.

1.1 Schéma d'interprétation

L'apport de calcium au sol modifie ses
propriétés chimiques et physiques. Eva-
luées en laboratoire, sur le matériau
terre, ces propriétés se traduisent au
champ par des modifications du comportement du sol, qui ont des répercussions sur
l'élaboration du rendement des cultures.
Le schéma 1 montre les différents critères
qui vont rentrer en interaction pour extérioriser un effet de l'apport de calcium.

Schéma 1: Conditions d'extériorisation du
chaulage

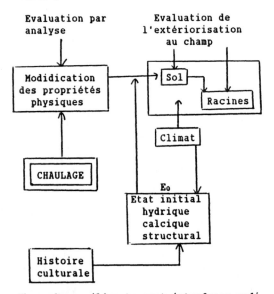

De nombreux éléments vont interferer modifiant les résultats dans un sens inattendu

si l'on n'y prend pas garde. Comme l'ont dit J. Guérif et G. Monnier (1982): "La fertilité physique d'un sol ne peut être considérée comme une donnée pédologique intrinsèque". Il faut " la replacer dans le contexte climatique et cultural dans lequel est utilisé le sol" et donc les seules évaluations de laboratoire, si elles permettent de comparer des matériaux, restent insuffisantes pour juger de l'effet d'une technique comme le chaulage. Mais avant d'aller plus loin dans l'étude, il convient de préciser quels sont les critères d'évaluation retenus pour mesurer les effets du Calcium.

1.2 Les critères d'évaluation

1.2.1 Au laboratoire

Nous avons mesuré les teneurs en calcium échangeable, en matières organiques et les pH eau et KCl à au moins trois reprises sur la durée de l'expérimentation pour surveiller l'évolution des états calcique et organique. Les résultats feront l'objet d'une communication ultérieure.
Par ailleurs, des évaluations de la stabilité structurale ont été faites en 1985 par les tests de percolation, mieux à même de mettre en évidence l'effet d'une modification de la garniture ionique,(HENIN et Al, 1969).
Fournet et Ortscheit (1985) constatent une modification significative des vitesses de percolation après chaulage. Nous n'y reviendrons pas.

1.2.2 Au champ

L'extériorisation au champ de l'effet du chaulage se voit sur deux éléments principaux, l'état du profil cultural et celui de l'enracinement.
* L'état du profil cultural
En limons battants, le chaulage améliore théoriquement la stabilité de la structure et peut modifier la sensibilité du sol à la battance en surface et la prise en masse en profondeur ainsi que la vitesse de ressuyage. L'extériorisation au champ de ces comportements est toutefois fonction de condition:
- La battance dépend de l'état E_s du lit de semence,(degré d'affinement, cohésion et humidité des mottes) et de l'intensité des pluies après le semis (Boiffin et al.,1981)
- La prise en masse dépend de l'état E_l de l'horizon labouré, (degré de fragmentation des mottes et proportion de terre fine) et de la somme des pluies entre le labour et la récolte.

- La vitesse de ressuyage dépend de la porosité des horizons labourés (H5) et sous-jacents (P1) et de la vitesse d'évacuation de l'eau que le drainage améliore sensiblement. Or l'état hydrique lors des travaux détermine fortement l'intensité des tassements, notamment sous la pression des pneumatiques. En conséquence, la vitesse de ressuyage interfère sur les risques de gâchage de la structure par les travaux réalisés après une pluie ou une séquence pluvieuse ainsi que sur la durée des états d'engorgement du profil en période humide.

Schéma 2 : Evolution des états du sol

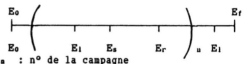

n : n° de la campagne
E_r : état après la récolte.

Pour vérifier et mesurer les différences de comportement entre les stations T0/T1 de chaque parcelle, on a centré le protocole sur la comparaison des profils culturaux en appliquant la méthode qualitative mise au point par H. Manichon (1982) Rappelons que cette méthode caractérise l'état du profil en évaluant deux critères:
- l'état interne des mottes comprenant trois états typiques de porosité croissante, δ, Γ, et reflètant une histoire et un processus d'évolution particuliers;
- le mode d'assemblage des mottes, défini par quatre états de base, M, SD, SF, F, par ordre croissant de porosité extramottière.
Le principe de base de l'observation repose sur la délimitation des unités morphologiques caractérisées par un état du même type, à l'intérieur des horizons anthropiques, lit de semence, horizon labouré, non repris et horizon sous le fond de labour.
Dans chaque station, nous faisons un profil de 4,8 m. de large pour être représentatif (Servettaz, 1978). Pour accroître la rigueur de la délimitation des unités, nous appliquons une grille de maille 5*5 cm. sur la face observée. Chaque profil est ainsi cartographié, et chaque état type est évalué en fonction de sa surface relative.
Comme nous faisons les profils culturaux à une date relativement proche de la récolte, nous limitons les comparaisons aux horizons H5 et P1.
Les différences de distribution des surfaces relatives de chaque mode

d'assemblage révèle un effet chaulage sur la prise en masse de l'horizon et nous attribuons une note de l'effet chaulage en comparant les distributions entre T0 et T1.

On considère qu'une différence supérieure à 10% entre les proportions d'état interne delta révèle un effet du chaulage sur la vitesse de ressuyage. Les mottes delta résultent uniquement des compactages sévères provoqués par les pneumatiques lors des travaux réalisés sur un sol insuffisamment ressuyé. Les limons sensibles aux tassements mémorisent fortement les accidents de gâchage de la structure, que ces accidents soient hérités, charrois de récolte du précédent, de fumier, ou récents, reprise de labour et semis....La connaissance du système de cultures, histoire et itinéraire technique, en lien avec la pluviométrie lors des travaux et la localisation des unités morphologiques dans le profil autorisent généralement à dissocier les effets hérités et récents.

En toute rigueur, la mise en évidence de l'effet du chaulage sur la battance nécessitait de caractériser le lit de semence le jour même du semis puis à intervalles réguliers. Dans les faits, nos observations furent limitées à une évaluation rapide du degré de mottosité à une date plus ou moins proche du semis, puis à un second contrôle post hivernal sur les cultures d'automne.

Nous retenons donc comme critères révélateurs de l'effet chaulage, le pourcentage de mottes delta et l'évolution des modes d'assemblages des mottes dans l'horizon labouré.

** L'enracinement

Le comportement du sol joue sur les conditions de croissance et de développement de la plante. Ses effets vont plus ou moins se marquer sur les composantes du rendement et sur les capteurs selon les dates auxquelles les conditions sont devenues limitantes.

Bien que nous ayons contrôlé les répercussions du chaulage sur les différentes étapes du fonctionnement du peuplement, nous nous limiterons ici aux seuls effets sur la profondeur et la densité de l'enracinement, mesurés à proximité de la floraison. Sur la face d'observation du profil cultural, après une préparation minutieuse, nous appliquons une grille soit de maille 5*5 cm. avec une évaluation qualitative (0, 0+, 1, 2) reflétant l'abondance des racines à l'intérieur de chaque case, soit de maille 2*2 cm. avec une notation de présence/absence.(1986, 1987, 1988). La comparaison des enracine-

ments en T0 et T1 repose sur une confrontation des cartographies. La notation simplifiée, (0,1) permet d'exprimer la surface colonisée en pourcentage pour les horizons labourés et profonds. Un écart supérieure à 10 % dans H5 (NEH5) et 20% dans P1 (NEP1) entre les deux profils racinaires est révélateur d'un effet chaulage noté + (ou ++ si les différences sont comprises entre 18 et 25 % pour H1 et supérieure à 40% pour P1). Les effets négatifs sont notés- et -- de façon symétrique.

2 RESULTATS ET EFFET DU CLIMAT

Pour illustrer notre démarche à propos du rôle des conditions climatiques sur l'impact du chaulage, nous prenons donc le cas des cultures d'automne sur les 10 sites T0/T1.

2.1 Les effets sur le sol

Deux indicateurs de pluviométrie sont pris en compte, la hauteur d'eau sur la période (P) et le nombre de décades (N) pour lesquelles cette pluviosité est supérieure à 20 mm. Trois périodes sont donc ainsi décrites: Tableau 1

- L'automne qui donne les risques de gâchage de la structure par les travaux de récolte et de mise en place des cultures; ces critères sont mis en relation avec la vitesse de ressuyage et les pourcentages de mottes Delta dans le profil.

- L'hiver et le printemps qui évaluent les risques d'engorgement du profil et de diminution globale de la porosité; c'est donc le mode d'assemblage des mottes qui doit réagir.

Tableau 1 : Pluviométrie des campagnes étudiées

	Automne		Hiver		Printemps	
	P	N	P	N	P	N
84/85	276	6	156	2	171	2
85/86	92	2	163	5	296	5
86/87	151	4	136	2	148	4
87/88	294	5	232	6	317	7

2.1.1 En automne

On constate que les conditions climatiques lors des travaux d'automne (charrois, labour et reprise de labour) sont , par rapport à l'extériorisation du chaulage, à priori favorables pour 10 situations, (années 1985 et 1988) et défavorables dans 12 cas (années 1986 et 1987).

Tableau 2 : Réponse des parcelles au chaulage

84/85

N°P	11		19		3		13		7		15		5		21		23		25	
	TO	T1	TO	T1	TO	T1	TO	T1	TO	T1	TO	T1	TO	T1	TO	T1	TO	T1	TO	T1
% δ			67	0	49	0	75	7	0	0	72	38	/	/						
Note			++		++		++		0		++		+							
%M			67	20	97	56	75	63	65	53	0	6	0	0						
%SD			30	60	0	37	22	35	32	37	72	42	75	20						
%SF			0	20	0	5	0	0	3	10	20	45	25	80						
Note			++		++		+		+		++		++							
NEH5			++		0		++		++		+		/							
NEP1			0		+		-		0		+		/							

85/86

N°P	11		19		3		13		7		15		5		21		23		25	
	TO	T1	TO	T1	TO	T1	TO	T1	TO	T1	TO	T1	TO	T1	TO	T1	TO	T1	TO	T1
% δ							34	21	0	0	0	0			0	0			83	26
Note							+		0		0				0				++	
%M							85	82	48	28	39	37			50	21			81	40
%SD							15	16	50	51	35	51			11	52			7	39
%SF							0	2	2	18	18	8			36	22			8	16
Note							0		++		0				++				++	
NEH5							+		0		0				/				+	
NEP1							+		+		0				/				+	

86/87

N°P	11		19		3		13		7		15		5		21		23		25	
	TO	T1	TO	T1	TO	T1	TO	T1	TO	T1	TO	T1	TO	T1	TO	T1	TO	T1	TO	T1
% δ	90	80			0	0	10	18	45	50	0	0			0	0			25	25
Note	+				0		0		0		0				0				0	
%M	92	88			20	11	41	31	/	/	36	14			25	20			44	54
%SD	0	10			72	55	37	35	/	/	37	70			27	31			54	46
%SF	0	2			7	33	22	33	/	/	12	2			30	37			0	0
Note	0				+		+				+				0				0	
NEH5	0				+		+				0				0				0	
NEP1	0				++		++				0				0				0	

87/88

N°P	11		19		3		13		7		15		5		21		23		25	
	TO	T1	TO	T1	TO	T1	TO	T1	TO	T1	TO	T1	TO	T1	TO	T1	TO	T1	TO	T1
% δ					71	37			80	77							69	47	52	74
Note					++				0								++		-	
%M					71	69			85	15							67	86	66	85
%SD					19	31			15	11							10	0	33	5
%SF					9	0			0	0							23	14	0	10
Note					0				0								-		0	
NEH5					-				0								--		+	
NEP1					++				--								-		+	

Les blancs correspondent à des situations en cultures de printemps. Les parcelles 15, 5, 21, 23 et 25 sont drainées. La parcelle 7 n'est pas hydromorphe, la 13 l'est peu, la 19 et la 3 sont hydromorphes et la 11 est très hydromorphe.

En 1985 et 1988, en référence au tableau 2, les différences entre les pourcentages de mottes delta apparaissent effectivement le plus souvent fortes ou très fortes (7 cas). Les parcelles 7 (85 et 88) et 21 (88) font exceptions : ces résultats sont probablemnt liés aux choix des dates d'intervention, en conditions nettement réssuyées en 85 et franchement humides en 88.

Pour 1986 et 1987, on constate que pour 8 cas, les notes sont nulles. Les situations qui font exception, c'est à dire pour lesquelles on observe un net effet ont subi des travaux en mauvaises conditions entre la date du chaulage et celle de l'observation. Ces exceptions confirment en définitive la possible présence d'effets hérités, résultant du chaulage. En conséquence, les tendances constatées démontrent que, lorsque l'automne est pluvieux, le chaulage accroît la praticabilité du milieu, en terrain hydromorphe ou drainé ; toutefois, une intervention réalisée en conditions trop humides gâche irrémédiablement le profil, chaulé ou non. En automne, à faible pluviométrie, la praticabilité du milieu est satisfaisante, drainée ou non ; le chaulage ne s'extériorise que par des effets hérités éventuels.

2.1.2 Période Hiver-Printemps

En 1985, la limitation de la dégradation de la structure dans les parcelles chaulées est systématique et indépendante du drainage ou de l'hydromorphie naturelle. Pour un même type de climat, 1987, on constate au contraire des différences de comportements :
- En sites drainés, l'effet apparaît nul voire faible ; or, les profils culturaux ne sont pas dégradés (faible proportion de mottes delta, contrairement à 85) ; le drainage suffit à lui seul pour évacuer l'eau en excès.
- En sites hydromorphes, l'effet est modeste mais réel (sauf si le profil en T0 et T1 est gâché comme en 11) ; le chaulage, améliorant la percolation, atténue alors la dégradation de la structure.
En 1986, la forte pluviosité entraîne un effet chaulage dans les sites drainés ou peu hydromorphes à une exception près (15). L'hydromorphie de la parcelle 13 suggère que l'engorgement fréquent du profil annule l'effet du chaulage et l'évolution vers un état massif est inévitable. On retrouve cette conclusion pour l'année 1988, très humide aux deux périodes : les profils T0, T1 déjà dégradés à l'entrée de l'hiver, n'ont pu avoir

qu'une évolution défavorable en toutes situations drainées ou non. En effet, la percolation de l'eau n'est jamais assez rapide pour éviter les excès momentanés. Dans le tableau suivant, rassemblant les résultats, on mesure que l'extériorisation du chaulage sur le ralentissement de la dégradation de la structure, dépend de l'état du profil cultural créé au semis - dégradé (D) ou non dégradé (ND)- de la pluviosité en hivr et au printemps -forte (F) ou faible (f)- ainsi que du fonctionnement hydrique du sol.

Pluv.	Terrains drainés			Terrains hydromorphes		
	T0T1 D	T0D T1ND	T0T1 ND	T0T1 D	T0D T1ND	T0T1 ND
F f	0 0	++ ++	++ +à0	0 0	0 ++à+	0 +

2.2 Les effets sur l'enracinement

Les principaux effets se font sentir sur les enracinements, dans l'horizon labouré et dans les premiers horizons pédologiques.

Généralement, il y a une bonne répercussion des modifications de comportement du sol sur l'enracinement (tableau 2). Lorsqu'on a une extériorisation favorable du chaulage sur le sol, on observe un effet positif ou très positif sur l'enracinement des premiers horizons (8 cas) ; la seule exception (23 en 88) semble s'expliquer par la notation (présence/absence), inadaptée à un enracinement pivotant.

Quand il n'y a pas d'extériorisation sur l'état physique du sol, il n'y a pas d'effet sur les racines (6 cas), sauf quand on a une prolifération de vers de terre en T1, liée à l'amélioration de l'état chimique du milieu par le chaulage (4 cas).

On constate que plus la dégradation du profil est précoce, plus la dynamique de l'enracinement est perturbée et plus il y a un effet du chaulage.

2.3 Remarque

Pour les cultures de printemps, nous avons limité nos observations aux seuls profils culturaux (15 cas). Concernant l'amélioration de la vitesse de réssuyage, nous retrouvons les mêmes résultats que pré-cités en 2.1.1. Par contre, l'effet du chaulage sur le ralentissement de la prise en masse du profil est assez rare, dans la mesure où la somme des pluies post-semis est sensiblement plus faible que pour les cultures d'automne.

CONCLUSION

Globalement, la vérification de nos hypothèses montre que les critères choisis pour qualifier l'état structural sont pertinents. Un nouveau protocole d'observation du profil cultural (Manichon, 1986 et Manichon, Gautronneau, 1987) prend en compte les différentes strates latérales créées par les passages d'engins. Il permet de mieux différencier les effets hérités de l'histoire culturale des effets des passages récents. On pourra donc mieux comprendre les conséquences du chaulage sur la fertilité physique.
On peut, d'une manière générale, prévoir la tendance de la réponse au chaulage. Plus l'état calcique initial est faible, plus les critères au champ comme au laboratoire réagissent. Mais la mise en place d'une expérimentation de chaulage ne peut en aucun cas se limiter à la mesure au laboratoire de certaines caractéristiques du sol. Il est indispensable d'évaluer les conditions d'extériorisation au champ de cette technique.
En effet, le fonctionnement hydrique d'un sol interfère sensiblement sur les possibilités d'expression des effets du chaulage. Schématiquement, on peut affirmer que drainage et chaulage constituent deux techniques complémentaires : le drainage assure l'évacuation de l'eau en excès, le chaulage maintient la stabilité d'un réseau de porosité permettant la percolation de l'eau. Nous avons bien vérifié que l'effet du calcium sur la fertilité physique correspond à ce phénomène.
La prévision des conséquences du chaulage doit donc intégrer l'état d'hydromorphie du terrain et la pluviosité de l'automne d'une part, de l'hiver et du printemps d'autre part.
De plus, nous montrons que plus il y a des cultures d'automne dans le système de culture, meilleure est la réponse au chaulage. On doit donc, dans ces systèmes là, davantage prendre soin de l'état calcique. Nous constatons encore que, si les itinéraires techniques conduisent à un profil dégradé (charroi et récoltes en période humide), on a de grandes chances de ne pas extérioriser l'effet du calcium. Il faudra donc, dans un conseil, prendre en compte les systèmes de culture et leur maîtrise technique par l'agriculteur.

Pour conclure par rapport à l'effet chaulage, on peut dire qu'il peut être suffisant dans les sols peu ou non hydromorphes mais que le drainage reste prioritaire dans les sols hydromorphes.

BIBLIOGRAPHIE

J. Boiffin, B. Fabre, Y. Gautronneau, M. Sebillotte, 1981. Les risques de mauvaise levée du colza d'hiver en terre battante. Analyse d'une expérimentation de longue durée en vue de leur prévision. Informations techniques. CETIOM n° 73, p. 12-28.

M. Coppenet et al. 1986. Etat calcique et fertilité: le chaulage. Acta, Paris, 170 p.

B. Fabre, F. Kockmann, 1987. Relance du chaulage en Bresse Chalonnaise. BTI n°416 p.3-19

J. Fauche, M. Cristin, 1981 .Effets des fertilisants minéraux sur l'acidité du sol. TUA N°8:5-8

M.J. Fournet, D. Ortscheit, 1985. Répercussions des amendements calcaires sur les limons battants en Bresse Chalonnaise. MFE ISARA Lyon, 95 pages.

J. Guérif, G. Monnier, 1982. Evolution de la fertilité physique des sols dans les systèmes de cultures fourragères de l'ouest de la France, in Forum des fourrages de l'Ouest ITCF p.113-129.

S. Henin, R. Gras, G. Monnier, 1969. Le profil cultural Masson, Paris 332 p..

F. Kockmann, B. Fabre, 1988. Chaulage sur limons en Bresse in 3° Forum de la fertilisation raisonnée, COMIFER p.131-134

H. Manichon, 1982. Influence des systèmes de cultures sur le PC : élaboration d'une méthode de diagnostic basée sur l'observation morphologique. Thèse INAPG, 214 pages.

H. Manichon, 1986. Observation morphologique de l'état structural et mise en évidence d'effet de compactage des horizons travaillés. Colloque CEE. Soil degradation. Septembre 85, Avignon.

H. Manichon, Y. Gautronneau, 1987. Guide méthodique du profil cultural. ISARA Lyon, 70 pages.

G. Monnier, P. Stengel, J. Guérif, 1981. Recherches de critères de la fertilité physique du sol et de son évolution en fonction du système de culture. CR du Coll. Agrimed CEE. Evolution du niveau de fertilité des sols. Bari, 28-29/9/81.

3.2 Harvesting and handling of agricultural products

Land and Water Use, Dodd & Grace (eds), © 1989 Balkema, Rotterdam. ISBN 90 6191 980 0

Keynote paper:

Separation, sorting and grading during harvesting and handling of agricultural products at farm level

E.Manfredi
Farm Machinery and Mechanization Institute, University of Bologna, Italy

ABSTRACT : Technological progress in harvesting and primary handling of agricultural products is brought about by economic and operative needs. The speaker briefly discusses applications which lead to the automation of selection and sorting operations for many products, from seeds to fruits and vege__tables, both during harvesting and after. Objectives achieved and further development in the main agricultural sectors are also discussed.

1 INTRODUCTION

The evolution of agricultural methods is marked in all sectors by machine, instrument development for: the reduction and improvement of human labor required; product quality improvement in terms of both the elimination of impurities and waste and the selection of products according to precise standards; the efficient utilization of by-products both in the soil and in farming methods and economy in terms of the labor and energy involved in product storage, transport for processing and fresh market consumption and so on.

However, technical progress has been motivated fundamentally by economic interests (production cost reduction), social realities (substitution for lack of manual labor availability) and human needs (improvement ofworking conditions, consequential reduction in physical work load, improvement in health and safety conditions at the work place, improved precision in repetitive and st_ressful operations).

These motivations find and have found ever more precise application in operations which, at harvest time, regard the separa - tion of the product (seeds, fruit, etc.) from the plant,product cleaning, grading and transport to storage facilities for either food (cereals, legumes and seeds, tubers, roots, vegetables, fruits, etc.) or industrial (tobacco, cotton,etc.) use.

2 AUTOMATION AND THE FUTUR OF HARVEST AND POST-HARVEST TECHNIQUES

Electronic and computer applications in the sector of agricultural machinery affects mostly the objective regarding work quality. Real time controls allow for the automatic correction of functional conditions so that the operation is carried out within the limits set by agronomic needs. In combines, for instance, sensors control grain loss or or the feed flow on the anterior platform, allowng for automatic forward speed regulation. Cutting bar height can also be controlled as can be the directional guides with sensors. Thus, the new development of these applications is born of the need to automate product operations during harvest, for example, fruit selection, elimination of stones and soil from tubers, roots, etc.

We are going, then, in the direction of selective harvest automation and the substitution of repetitive and støessful labor when sorting products in the field at harvest and on the farm after harvest.

The use of automatic selection and sorting systems is strictly tied to the development of research on the physical, mechanical, electrical and optical characteristics of the various agricultural products.

It is from the knowledge of these characteristics, which have required the development of methodologies and specialized equipment, given the nature of the material, that systems based on electrical and optical sensors have found utilization in harvest machinery and selection and sorting systems so that such operations can be carried out with less manual labor, safely and thus with higher productivity levels and final product quality.

3 SEPARATION,SORTING AND CLEANING IN THE HARVESTING OF CEREALS AND SEEDS

The need to improve combine-harvester performance while increasing the thresher and sieve surface as well as the need for increased polyvalency with the different grain and seed products have brought about remarkable modifications in the threshing and successive separation and cleaning systems.

A case in point is the threshing cylinder of the conventional combine and the axial – flow rotary separation.

Other solutions place the threshing cylinder on the cutting platform with the axial flow of the straw in a co-axial separating grate. This allows for a more complete separation and grain cleaning cycle and more efficient recovery of and re-threshing of the wheat ears.

Self-leveling and transversal and longitudinal systems are important for work in hilly areas. Self-leveling systems for the threshing cylinder, straw walkers and sieves are also important for work efficiency on sloping terrain.

The recent innovations in combines, including those for increasing polyvalency, are related to increased cutting width and thus to their potentially very high productive capacity. The work capacity of the combine is, then, affected by the efficiency of the separation and cleaning apparatus and thus by product flow.

The development of automatic product loss control systems which allow the machine to proceed at optimal forward speed during the various stages of harvest and elaboration is a response to these needs.

4 PEA AND GREEN BEAN HARVESTING
4.1

The modern machines, which have taken the place of the dual-phase harvesting process which includes the use of the windrower followed by picker-sheller, are the automatic pod-picker-sheller combines.

With these modern combines it is possible to time operations according to ripening time and thus to achieve product quality improvement with less losses and lower costs.

The speed of this process plays a decisive role in the development of production for freezing which demands a higher degree of tenderness than for canning.

Moreover, the most recent pod-picker pea combines, thanks to opportune adaptations of accessories and to adjustments in the harvesters and shellers, can be used for the harvesting of fresh bean.

In the latest generation combines the product is separated from the plant and unloaded on lateral conveyors (continuous belts or header augers) which brings it to the central elevator(an inclined continuous belt or two conveyor belts with a jump in between them).

The shelling chamber with axial product flow is equipped with a beating drum and a rotary separator cylinder (rotating in the direction of the thresher but more slowly) which allows the peas and impurities to pass through.

In some cases the inclination of the shelling chamber can also be regulated to adapt the product passing time to both the shelling characteristics of the particular cultivar and climatic conditions. The self-leveling system also responds to this need.

The sorted product then drops from rotary separator onto one or two inclined continuous belts (inclination is adjustable according to the product) which allow the peas to roll toward the longitud nal transporter and heavier and flatter impurities to be unloadedtoward the exterior. During the passage on the longitudinaltransporter, the product undergoes an initial ventilation cleaning to remove light fragment impurities. The product is then cleaned with a secondary cleaning fan.

The automatic sorting and cleaning takes place after the elevator in another rotary separator. The peas fall through the mesh

while any pods that remain are brought back
to the shelling chamber feeding belt for
product recovery.

4.2The combines for green beans can operate
on one or two rows with longitudinal picking rotors or more rows with a frontal
head with a picking reel.

Product separation fom impurities which
begins with product unloading onto the elevator belt, is distinguished by the ventilation modalities and the succession of mechanical separation(from the vine) before
the green beans arrive at the loading tank.

In some, the first separation of leaves
harvested with the green beans takes place
in combination with a picking reel on the
head through aspiration before they are unloaded onto a conveyor belt where a second
cleaning takes place with blown air.

Unsatisfactory adjustment of the pneumatic apparatus and very irregular calibration
can cause a rather high product loss.

Thus, automation of such regulation is
necessary for factory sorting and cleaning
work reduction.

5 TOMATO HARVESTING AND SORTING

The sector tomato harvesting, especially
with complete machines, has entered a phase which is dense with initiative and innovation. The technical and economic results
are very significant. Continuous improvement,
however, depends on both adequate varietal
choices on the basis of the pedoclimatic
environment and on rational cultural techniques.

A reduction in industrial tomato production costs relies on mechanical harvesting
and the organization of factory delivery
according to methods in use for other industrial vegetables which include cultivation
planning and investments in the field.

The harvesters, towed or self-propelled,
carry out all tomato harvesting operations
in one passage. They can be divided according to product sorting modalities:

a) manual selection on board

b) selection at the farm or factory after
washing

c) automatic selection on board.

Themost widely used machines are with manual selection on board. They are equipped

with one or two lateral sorting belts where
from 4 to 12 workers remove undesiderable
material (damaged and rotten or green fruit,
soil, etc.).

For fruit selection the tomato harvester
can be equipped with photo—electric sorting
apparatus which identifies the color red and
with electro-pneumatic elements for the e-
limination of green fruit which are activated automatically by the photo—elettric sensors.

The electronic optical system consists of
modules with optical circuits connected to
electronic circuits which are responsible
for the selection operation. There are four
photo—sensor in each module:two identify impurities, the other two ascertain the color
of inspected tomato. These last two read the
relative reflection co—efficient given off
in the green and red spectrum for each tomato that falls from every channel of alligning
belt. They also provide eletrical signals
proportional to the intensity of the radiation reflected in the green and red wavelen-
ths. These signals are sent to an electronic
circuit where they are coded and compared to
a standard co—efficient which can be pre-se-
lected and regulated and is interpreted as
the machine limit value. If the color of the
tomato is differet than the limit value the
computer sends a signal to the system for
activation of elimination of the rejected
fruits.

The expulsion system is composed of high
speed pneumatic cylinders which intercept
both rejected tomatoes and other foreign objects in their fall trajectories. The cylinders hit the objects with extreme precision
causing them to fall to the ground.

6 TUBER AND ROOT HARVESTING AND SORTING

6.1 Complete potato harvest machines must:
lift a great amount of soil together with
the tubers; separate the tubers from soil
and stems; clean and sot the product with
separation and elimination of impurities and
undersized tubers and convey the tubers
to containers or hoppers on the machine.

After having been harvested alog with clods,
stem and weeds, the tubers are conveyed on
the sieve-conveyor where they are separated
from most of the soil and any stems and clods
are broken up.

Different solutions for the vibrational separating apparatus can be employed on the basis of physical propertys of the soil; the difficulty of separation of the tubers from the soil, and the type of cultivar. Possible tuber damage is considered in this phase.

Product flow continues on the cleaning apparatus which separates the impurities and then on the sizing apparatus which sorts the tubers according to size.

In addition to the automatic operations carried out beginning with harvesting, diverse clod and pebble separator variations exist: semi-automatic with either manual or mechanical sorting. There are also systems which separate automatically with a command circuit activated by X-rays which penetrate the product flow.

On less complex machines, it is possible to carry out initial mechanical sizing so that the undersized products can be separated and only products of an acceptable size are conveyed to the manual sorting point.

A comparison of machines of differing complexity has indicated, then, that the automatic harvesters (with no workers aboard) can assure higher productivity without compromising product quality as long as the automatic sorting systems are regulated correctly. The anterior conveyor speed is very important and should not exceed the forward speed so as to avoid tuber damage.

In fact, the work quality of a potato harvester is evalueted according to damaged product and soil quantity.

6.2 Sugarbeet harvesting is carried out, according to the environment, with different machines: towed and self-propelled.

The industry accepts only roots which are free of leaves, tops and soil. This is a big problem both during harvesting and at the sugar factory. During harvesting the work is burdensome because of the time loss involved. At the sugar factory it renders washing and soil and water removal difficult. Thus a compromise among harvesting time, maximum product quantity obtainable and soil removal has been established.

The soil quantity is satisfactorily eliminated by ever more efficient cleaning apparatus on both self-propelled harvesters and on loaders in the last phase.

On large self-propelled machines which tubers and roots, automatic control is attempted through the use of hydrostatic trasmissions during operation (load sensing and automotive systems) in that the operation itself depends in large part on soil type and condition.

7 FRUIT HARVESTING AND SORTING

Fruit harvesting can be either semi-mechanical (manual separation) or mechanical and is partially selective.

In many cases manual harvesting is aided with equipment on the field (apples,pears, etc.)and sorting is carried out at the farm or storage facility. In some cases manual harvesting is aided with equipment which allows for product preparation in ready-for-marketing containers after sorting.

Some robot prototypes for fruit harvesting (apples, oranges) have recently been presented.

When the harvesting is mechanical (olives, prunes, cherries,peaches,almonds,etc.) with shaker machines,the separation takes place through the vibrations induced to the tree.

In any case the cleaning, selection and grading with specialized equipment depending on the type of fruits and its destination (fresh-market or industry) is then necessary.

In fact at the farm or the storage facilities can be imployed machines for cleaning, sorting, calibration and final preparation for the market.
Particulary the calibration of the fruits can be carried out according to size, likewise the systems for potatoes, or weight. The last system is more suitable for fruits with big size, like apples, oranges, etc., but requests particular attention on the setting. The final sorting of fruits, at storage facilities for preparation to market, is possible to use electronic systems, which operate with the same principles indicate for sorting red tomatoes or potatoes from the soil. For this operation there are also more complex systems utilized for internal damage and deterioration identification after storage.

8 NON FOOD PRODUCT HARVESTING(COTTON AND TOBACCO)

8.1 Cotton, a dual aptitude crop (fiber and

oil), can be selectively harvested in that modern pickers which pick the single capsules with ripe fiber and exclude the others. These can be harvested later without damaging the plant.

In this way the fiber is harvested at the right stage of ripeness and with a remarkable degree of cleanliness which reduces ginning costs.

8.2 The tobacco harvesting systems are particulary differentiated according to the variety. For example, Virginia Bright tobacco can be harvested in separate leaves by machines which operate selectively according to maturation or which operate on all of the leaves after the plant pruning. In the first case, the machine harvests the leaves that are unloaded for the preparation of frames for curing operation, while in the secon case the leaves are onto machine deposited in to a container for curing.

Harvesting is manual for most of the other varieties With more or less complex machines which prepare the leaves on frames for bulk-curing. A case apart is when the plant is harvested and leaf separation and sorting take place after curing. The tobacco leaves must be after curing selected for quality and color. This operation, usually carried out manually, can be perfected and less laborious with the help of photoelectric cells and a computerized system for the selection of leaves according to color (from blonde to dark brown).

9 HARVESTING AND SELECTION OF SOME TROPICAL CROPS

9.1 Cereal harvesting, particularly rice, corn, sorghum, etc., can be carried out with a "stripper" machine which removes the grains with separation in a group of cylinders and successive cleaning before sacking.

9.2 Peanut shelling and cleaning can be done at a fixed location or with picker-shellers which operate after the peanuts have been separated from the soil and windrowed for initial drying in the field.

9.3 Post-harvest coffee operations include pulp removal with machines/separate the cof-
which
fee beans with successive sorting.

10 CONCLUSIONS

This keynote paper has outlined on the some of principal problems of theme 2.

The different subjects of this theme are developed in the papers presented by the many speakers.

Land and Water Use, Dodd & Grace (eds), © 1989 Balkema, Rotterdam. ISBN 90 6191 980 0

Harvesting and cleaning machine for olives dropped on the ground

A. Arrivo
Istituto di Genio Rurale e Forestale, Potenza, Italy

F. Bellomo
Istituto di Meccanica Agraria, Bari, Italy

ABSTRACT: In order to obtain an integral mechanization with an appreciable labour reduction in harvesting olives from the ground, a particulary machine, able to execute at the same time harvesting and cleaning of olives, was designed. This machine can be proposed both in a self-propelled model and on tractor-mounted one. A self-propelled model mounted on a motor vehicle on purpose equipped is described. At present this model is in advanced stage of building. The main characteristic of designed machine is the harvesting unit adaptability to all ground irregularities.

1 INTRODUCTION ON THE CURRENT STATE OF OLIVE HARVESTING MECHANIZATION

A research on olive harvesting practices in Apulia (Italy)(1) and hence an economic and technical analysis of the results obtained (2) have both highlighted the current state of harvesting mechanization as well as the factors mostly affecting its development. Both in the province of Bari and Foggia where there is prevalence of varieties more sensitive to detachment by shaking and where picking from the tree is performed by one single operation, the use of a shaker with help of nets has been shown to entail far lower labour requirements, with appreciable results as to the work quantity and quality. The limit to the spread of shakers in such areas is to be attributed exclusively to economic appraisals.

In the Brindisi, Lecce and Taranto provinces where gradually ripening varietis unsuited to shaking-related abscission are prevailing and harvesting is done from the ground on more times, the use of shakers involvs as yet no saving of labour so that its use even in large-sized farm enterprices should come justifiable. This being the case, the use of the shaker may be felt fit only because it makes it possible to shorten harvest time, compared with that of olive natural drop, which may last quite a long time. In the areas of south Italy machines allowing a much easier harvesting from the ground as well as cleaners have in fact turned out to be successful.

Furthermore, our work study has enabled us to detect the boundaries not only of shaker application, but also of the commonest equipment employed for harvesting from the ground.

The utilization of this last machinery has been recognized to require stll a lot of labour, though alongside an increase in working capacity. Data thereof have further revealed that there is not a harvesting organization method among those analyzed clearly prevailing over the rest in terms of utilization since all operations strictly connected with harvesting are not fully mechanized, however organized currently available machine labour may turn out to be.

Brush-windrowers have great working capacity; but the work is done only in part, owing to the way they operate, which -obviously - is to be combined to quite onerous operations, like, for example, heaping, harvesting and setting olives in boxes.

Brush-harvesters have low working capacity. They are in fact slower than brush-windrowers since they are mostly drawn manually by a walking operator, in addition to being expected to gather crop. Moreover, they have to brush the whole harvest area, rolling below the tree.

Cleaners are currently distinct devices; hence they need labour in order to be fed - both in the field and oil-mill. It should be noted that the harvested crop is almost esclusively loaded manually into trailers to be then conveyed to the processing unit.

2 A PROPOSAL FOR HARVESTING-CLEANING MACHINE

Once ascertained the lake of utility of shakers in southern areas and the technical inadequasy of machines for harvesting from the ground more suited to operate in these areas, we felt it necessary to probe our work study into the last type of equipment so as to enhance the chance of relevant operations and performances being successful.

Besides, once pointed out the inefficiencies of existing machines for olive harvesting from the ground through the analysis of the different work organization methods, which are possible with their use, we finally planned a new machine able to eliminate all manual operations such as heaping, handling, transfer to a cleaner and, lastly, crop loading into a trailer if such a machine were to be equipped with a dumping container or a conveyor for unloading the harvested crop directly into a trailer.

In short, the machine should allow full mechanization of olive harvesting from thr ground. It goes without saying that the mechanical implement has to enter the properly rolled harvest area below the tree, without causing any damage. Hence the need to draw particular attention to technical characteristics like mass, size, manoeuvrability of the machine, which could operate in the following ways:

1. Harvesting of olives arranged in straight windrows by means of a brush-windrower as illustated in fig.1

2. Harvesting of olives arranged in circular windrows around the tree trunk by means of a brush-windrower

3. Harvesting of olives brushing directly the whole harvest area below the tree

In the first as well as second case, a big advantage would lie in a higher working capacity, in the third case, however, no use of brush-windrowering machine would be made.

A Harvesting method using a harvest-cleaning machine generally expects a number of men ranging from 3 to 5 to be working together - in relationship to the system assigned for the crop transport - to attain the same working capacity of the method using the only brush-windrowering machine, which domand a supply of about 15 men.

The proposal to employ such type of machine as well as to modify partially the current harvesting practices stems from the fact that, actually, olive-growing, whose characteristcs are unsuited for the use of shakers, is rather frequent in Apulia. Therefore it would be totally inconceivable to alter it in the short run so as to adapt it to harvesting by shaking; hence the need for more rational solutions to the problem of harvesting from the ground with a view to reducing production costs as much as possible. And this in view of a regional plan envisaging both development and conversion of olive-growing, also in relation to the full mechanization of tree-picking.

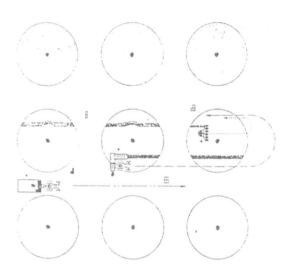

Figure 1 Working organization using an harvesting and cleaning machine.

The problem lies in the design of a machine suitable for harvesting from the ground, which is up to requirements in that it should eliminate all major deficiencies of the machines marketed today. A second step should provide for the carrying out and testing in order to check whether what stated beforehand as to saving of labour proves true or not. Not only does the carrying out of the above machine come into play, but also a different laying of the crop on the ground, just prior to the passage of the harvester in order to make a more sensible use of it.

This could entail at most changed harvest areas below the tree so that straight or round windrows could easily be shaped, which would further allow for a rapid and easy use of harvester. It stand to reason that modification to harvest areas is far simpler in straight-rowed olive-groves where it is easier to shape long continuous windrows for endless working. Special models of harvesters have already been developed and tested in olive-groves "reconverted" in this way.

3 TECHNICAL CHARACTERISTICS OF THE PROPOSED MACHINE

An harvesting-cleaning machine is composed basically of:
- a harvesting unit
- a conveying unit
- a cleaning unit

For a higher automation whereby the number of workers would be kept to a minimum, the machine will be further equipped with a container catching harvested olives or a conveyor fit to transfer olives directly into a trailer operating side by side with a harvesting machine throughout working.

Otherwise olives can be conveyed to the outlet of the cleaner into a small box which, once filled, is replaceable. Full boxes are left on the ground and subsequently elevated into a trailer, as shown in the harvesting method reported in fig.1.

Obviously, the choise of the most suitable solution will depend largely on technical evaluation referring to tha machine handling and lightness specifically. The solution, which envisages the unloading into boxes, is certainly the most time-saving, although it demands a larger supply of workers for boxes to be moved and loaded into a trailer.

The machine has to be basically adapted to uneven as well as noy completely dry land. The harvesting apparatus must be always touching the ground, the whole lenght of it more specifically, as a result, it cannot be rigidly connected to the supporting frame of the machine. The harvesting unit is made up of a roller equipped with brush-like flexible elements, rotating around a horizontal axis perpendicular to the direction of motion. To be precisely adapted to the soil structure, the roller needs swinging around two horizontal axes set perpendicular to one another; it must therefore be necessarily linked to the supporting frame of the machine through a double hinge. Ultimately, the machine can be modelled on two frames linked to each other by a double hinge: one of which being stiff with the vehicle carrying

structure, the other to be conversely positioned deliberately with respect, to the former.

One of the peculiarities of the harvesting unit lies in that it will be touching the ground at the harvest stage and be lifted at the transfer stage. And this by means of an hydraulic jack between the two frames as above.

The conveying apparatus must be stiff-fastened to the harvesting unit. It can be made up of a mesh conveyor or a cochlea. Conversely, the cleaning unit will have to be mounted on the supporting frame of the machine.

The above double hinge can be made by either a ball hinge with blocking of rotation around the vertical axis or two distinct plane hinges.

Once stressed the basic features of the operating machine, one may guess a great number of constructive solutions with respect to the type of traction and drive.

First of all, we have to discriminate between a self-propelled and a tractor-linked machine. Clearly, the former enables us to attain a product more suited to the requirements for the harvesting of olives from the ground since all proportions of the machine can be adjusted. In addition to operating members, power and propelling units can be designed and proportionately adjusted, the result of which being a machine unquestionably more suitable for lightness, handling and power purposes, in view of the work to be done.

The solution of a tractor-linked machine causes some problems in handling as well as in approaching to the harvest area throughout working, with respect to the performances of the tractor employed and the quality of land to work on. Both trailed and tractor-mounted machine can be taken into account; at any rate, the harvesting unit must always be linked to the main frame of the machine by a double hinge.

Harvesting unit should be located in the front part of the tractor to operate best. This is virtually difficult to do since tractors, in general, are not supplied with front power-lift. Consequently, we must opt for solutions based on a side or rear connection, which are surely cumbersome and unconfortable for the tractor-driver. However, the most suitable tractors should be small-sized and low-powered. Motion drive to the harvesting unit cannot be performed by means of stiff members; hence hydraulic drive should be desiderable.

4 DESIGNED MODEL OF A SELF-PROPELLED MACHINE

After analysing the problem of olive harvesting from the ground and envisaging all possible solutions as to carrying out an integral mechanical harvesting, we switched to the design and construction of a model of harvesting machine.

We felt it fit to start with the model of a self- propelled machine which surely gives a technical advantage over the tractor-linked one. Anyway, other models, if any, should be easily designed on the basis of the data obtained from testing the self-propelled model.

The machine was designed in cooperation with the Meridional Meccanica firm in Monopoli (BA) also involved with the construction. The scheme of the machine from the above and side view has been reported in Fig. 2 and 3. The scheme points out very limited over-all dimensions and a compacted model machine.

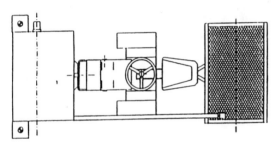

Figure 2 View from above of the self-propelled machine

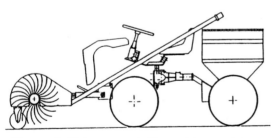

Figure 3 Side-view of the self-propelled machine

In addition to harvesting, conveying and cleaning units, as already reported in the general description of a clean-harvester, the model is made up of a two-wheel drive propelling unit of 10 kW power. A rear axle linked to the propelling unit by a double hinge completes the carrying

structure of the motor vehicle: a vertical axis hinge allows the vehicle to steer in small spaces; a longituninal axis hinge makes it possible for rear wheels to adapt to the ground roughness, thus ensuring the vehicle's stability.

The harvesting apparatus consists of a roller supplied with flexible members, enclosed into a casing. The roller is mounted on a horizzontal axis perpendicular to the advancing motion of the machine. Throughout working the roller and the vehicle weels rotate in opposite direction; olives are therefore pushed against the casing which - in turn - directs them to the conveyor. The support of the harvesting unit is connected with a frame by means of a plane hinge which causes the unit to swing araund an axis running in parallel to the motion direction, passing through the roll center, for the brush to be well adapted to ground cross irregularity.

The above frame is -in turn- connected with the propulsor structure through a plane hinge which causes the harvesting unit swivel around a transversal horizontal axis so that the brush can adapt to ground longitudinal irregularity. A hydraulic jack makes it possible for the harvesting unit to be always touching the ground at the work- stage and to lift it over the transfer period of the machine.

Olives are transferred to cleaning unit through a cochlea located on one side of vehicle. The cleaning apparatus is composed of two vibrating grids which serve to separete impurities from olives according to the different sizes.

Motion trasmission to the machine's various operating members occurs via a hydraulik system which foresees the use of pumps mounted on the driving shaft and of hydraulic motors mounted directly on the shafts of operating members. Obviously, the system is supplied with all elements serving to govern and adjust the speed of the different revolving members throughout working, so as to adapt performances of the machine to a great number of conditions under which it would operate.

5 PERFORMANCE OF A HARVESTING-CLEANING MACHINE

The construction of a model machine suitable for harvesting olives from the ground should be completed by the current harvest-time. So it may, in the short run, be tested in the field. Anyway, it is possible for us to measure its performance on the basis of the performances of machine available on today's market, which -though separately- perform the same operations. If the machine harvests in already rolled straight windrows, its advancing speed may stand at some 11.5 km/h, which is nearly half the speed of a mechanical brush-windrower. The same speed will be a bit lowered if windrows are round-shaped. Harvesting capacity should prove to be ranging 0.5 to 1.0 ha/h - namely from 5 to 10 q/h if we consider the mean quantity of olives present on the ground. Obviously, the cleaning unit must be carefully proportioned to this value of olive yield per hour.

Labour employment may be confined to one single tractor driver if the machine is supplied with a container or conveyor for continuous unloading into a trailer. The workforce may number 23 men in case of unloading into boxes.

Thus, if a harvesting method is organized using the following machines:
  - self propelled brush-windrower
  - self- propelled harvesting- cleaning machine

- tractor-drawn trailer
with an employment of labour totalling to 35 workers, the same performances are obtained as in harvesting method using a self-propelled brush-windrowers and a cleaner, which requires a labour employment of 1520 workers.

The use of an harvesting-cleaning machine may therefore be claimed to do work of at least 15 men under favourable operating conditions. If the machine is to operate over the whole surface of the harvest area below the tree, then the working capacity will be surely lower.

REFERENCES

Arrivo, A. & F.Bellomo 1982. Cantieri per la raccolta delle olive. Macchine & motori agricoli. N.1 (1).

Arrivo, A., G.Grittani & L.Limongelli 1984. La raccolta meccanica delle olive. Studio tecnico-economico. Notiziario Agricolo Regionale. Regione Puglia. N.1-2 (2).

Land and Water Use, Dodd & Grace (eds), © 1989 Balkema, Rotterdam. ISBN 90 6191 980 0

Citrus automatic grading device

F.Barresi
AID SpA, Robotic and A.I. Division, Catania, Italy

G.Blandini
Istituto di Meccanica Agraria, Università di Catania, Italy

ABSTRACT: A.I.D. has recently developed an automatic machine ("Videograding") which carries out, in real time, grading of oranges according to quality and appearance. Laboratory and field trials results, never previously obtained with automatic devices, allow a reduction up to 70% of grading workers in citrus packinghouses.

RESUME: DISPOSITIFS DE SELECTION QUALITATIVE AUTOMATIQUE DES AGRUMES. Récemment l'A.I.D. a réalisé la selectionneuse automatique "Videograding" la quelle, par un ordinateur qui analyse, grace á un camera, l'image des agrumes, permet la sélection des fruits d'aprés leur qualité. Les épreuves ont donné des résultats considerables et, bien qu'il faut compléter manuellement la sélection, la productivité s'accroît de 2 ÷ 3 fois.

ZUSAMMENFASSUNG: VORRICHTUNGEN FÜR DIE QUALITATIVE UND AUTOMATISCHE SORTIERUNG DER ZITRUSFRÜCHTE. A.I.D. hat neulich die automatische Sortiermaschine "Videograding" geschaffen, die, dank einer besonderen Datenverarbeitungsmaschine die die Fersehaufnahme der auf spezielle Förderer rollender Zitrusfrüchte analysiert, den selben Forderer mit Steuerungen ausstattet, die die Früchte in mehrere Qualitätsstufen teilt. Die Versuche haben hervorgehoben, dass die Maschine die Sortierung mit Ergebnissen vollzieht, die gewiss erheblich, auch wenn nicht optimal, sind, und die nie zuvor erreicht worden waren. Obwohl es noch zur Zeit nötig ist, die automatische Sortierung mit der Handsortierung zu ergänzen, hat die Produktivität des dazu bestimmten Personals zwei oder dreimal gesteigert.

1 FOREWORD

Fruit selection holds a relevant role in the product industrialization and marketing process to ensure the economic success of the activity. In fact it has the aim of giving homogeneous industrial peculiarities to a product that is naturally non homogeneous, in respect of the end user request.

In such a way the peculiarities that could be interesting are: size, shape, color, defects, mould, etc. Often it is taken into consideration only a subset of these grading parameters, sometimes only one, depending on the end users the products are addressed to.

Some of these grading activities are carried out manually (on external features based), some mechanically (sizing machines); others are not carried out at all, because the status of art does not give an affordable way to perform a non-intrusive analysis on the whole fruit (juice contents, acid/sugar ratio biological analysis, etc.).

Nevertheless the actual situation shows a positive trend: there are a lot of R&D teams that are aimed to realize automatic machines to perform these kinds of functions, although the results are, at this moment, not satisfactory.

The A.I.D. recently developed a machine that performs the automatic grading functions for citrus fruits, based on the external features: this kind of operation actually is carried out manually.

In this paper, after a technical description of the two tested machines we explain the obtained experimental results.

These results seem to be very interesting and encouraging for the economic and social evaluation of the product: in fact, these machines aim to guarantee high standard fruit quality and grading performan-ces, obtaining also a cost reduction in grading activities.

Finally it must be said that the operators, the machine will take the place of, carry out a seasonal, repetitive and unsatisfactory job.

2 AUTOMATIC GRADING BY IMAGE PROCESSING

2.1 Notice about the technique

The image processing is the technique used to make decisions and it is a very complex process.

The process begins by grabbing an image from a camera and sampling the incoming signal for the next digital computation. The image is seen by the image processor as a 2-dimension array of picture elements called "pixel", whose values represent the brightness of the corresponding areas.

The first analysis phase is called "segmentation" and it consists of a classification of all the pixels in separated categories, often using criteria of the probability theory, depending on some particular features: brightness, colour, differences from the neighbor elements, etc.

Afterward higher level features could be extracted from the categories obtained: homogeneous class connection, related population, shape, spatial frequencies.

The last analysis is called "recognition": it consists of marking the recognized regions on the a priori knowledge basis. For example the process will decide that a pixel region corresponds to a strawberry stem or to a zucchini mould or to a mechanical support.

At the end the analysis of the recognized parts of the scene gives information to a decisional process that decides the operative actions to perform a specific function (i.e. to eject a not mature fruit).

2.2 Structure

A general problem in image processing application is to find the best compromise between processing rate and cost effectiveness. Single scene analysis requires a large number of elementary instructions, almost some millions, number that can increase depending on the complexity of the recognition algorithms adopted.

Apart some particular cases, it is not possible to obtain acceptable performances using commercial computers, also of supermini class. There are two possible solutions: the first is to adopt dedicated image processors; the other is to share the process in multiple layers, the lower level functions are executed by dedicated hardware boards, meanwhile the higher level functions are executed by commercial fast microcomputers.

The first solution now is entering in the market, until few time ago it was reserved only for research team; it is implemented to solve very complex problems for its prohibitive cost. The latter, less expensive, uses fast computing devices and standard microcomputers for extensive calculations.

2.3 Automatic grading

This application intends to carry out the function traditionally carried out by human eyes: to locate defects or peculiarities related to the external aspects of the examined object.

Substituting the human eye, the machine has some advantage but also some limitations: it will not be possible to equipe an analysis system with the power and the criteria (not always known) of a human brain; from an other point of view, it is very difficult for a man to maintain a constant efficiency also for few hours, wherease the machine is faster, perfectly repetitive, work-time independent, and it can provide other services, like sizing, counting, job management, etc.

An automatic grading system analizes a repetitive scene for the position of the object, the constant light and the brightness, but must satisfy high speed and complex analysis requirements.

A common problem in the analysis of objects similar to rotation solid is to examine the whole surface of the object minimizing overlapping problems and perspective distortions, obtaining a uniform lighting of the object. For this reason a so called "vision chamber" must be carefully designed: it is the place in which the object are correctly positioned and illuminated to capture the image.

The output of such a grading machine is a multiple channel shunting system: the channels correspond to the different quality classes, in which the products must be sorted.

3 THE "VIDEOGRADING" FOR ORANGES

3.1 Structure

The "Videograding" is an automatic grading machine for oranges, designed and developed by A.I.D. to be used in the packinghouses to make a complete analysis of the fruit surface, despite the wide range of colour, pigmentation, dimension, shapes and varieties. It is designed even to grade internal characteristics such us the juice content.

The "Videograding" is made of 3 functional subsystems:
- electromechanical;
- vision;
- computer.

3.2 Mechanical subsystem

The function of the mechanical subsystem are:
- machine to environment interface;
- fruit singulation and input to the system;
- presentation of the oranges in the camera inspection area;
- routing of the fruits on appropriate channels as from computer decision.

The main problem is the fruit translation and positioning in the vision area; oranges, while translating, must rotate 180 degrees between one camera shot and an other.

In order to meet this requirement, a special roller conveyor has been developed with a special shape which allows the correct fruit rotation for whatever diameter.

An other hard problem is related to fruit output system due to the high conveyor speed (up to 8.33 fruits/s).

3.3 Vision subsystem

The vision subsystem acquires an image of the object under inspection and sends it to the computer system. It is made of a metallic structure called the "vision chamber", inside which are located all the components of this subsystem:
- lighting group made of high efficiency fluorescent lights;
- optical group made of diffusers, filters and mirrors; the filters discriminate defects from good areas and work fine for all orange varieties, with background from yellow to red;
- acquisition group made of a CCD camera which interfaces the image processing system. The camera has a high resolution and an electronic shutter with exposure time of 1 ms; its syncro signal is sent to a servo controller wich controls the conveyor speed in order to assure the correct timing of shooting.

3.4 Processing subsystem

The processing subsystem supports both data processing and image processing and it is made of dedicated processors networked to a PC compatible computer which acts as system supervisor and controller.

The architecture of this subsystem is hi-

ghly flexible and modular: it is possible to define a base configuration with 1 dedicated processor up to 8.

The PC compatible controller uses a MS-DOS operating system. Its main functions are the process control, the user interface, the statistical functions and the diagnostic system.

The dedicate computers are the kernel of the system: they receive the image of the object and apply the typical image processing functions (transformations, digital filtering, convolution, edge detection) in order to determine the object features (defects, shape, colour) for the classification needs.

After the grading process, the dedicated processors send the appropriate commands to the mechanical subsystem.

3.5 Prototype general specifications

The "Videograding" has a modular architecture: the PC compatible controller can manage 2 up to 8 dedicated computer, with a step of 2.

The control software of the machine is structured in order to let the user modify the parameters of the operation: then it is possible to have in output the desired grading.

The man-machine interface is organized with pull down menu and its use is easy even for not specialized operators.

The machine can analyze surface areas of about 1 mm² and the minimum defect area detected is 2 mm². For each channel the maximum speed is 8.33 fruits/s; the basic 2 channel machine has a nominal capacity of 8 to 9 t/h, related to the average fruit mass.

The weight function (optional) is available through the use of electronic load cells which can be mounted on the machine. It gives information which, used in combination with others from the image processor, can lead to an evaluation of the fruit juice contents; this function is useful also for administrative purposes.

The "Videograding" has a powerfull auto diagnostic system and it can print sophisticated reports on its operation.

The prototypes developed had 2 channels with one common vision chamber, and neither the weigth nor the juice analysis functions were activated.

4 TRIAL METHODOLOGY

Trials have been performed with the same methodology in laboratory and then in two packinghouses; in the laboratory we used a mechanical add on in order to let the same fruits recirculate in the machine whilst in the packinghouses the "Videograding" was on line.

The results from laboratory and field were quite similar, so we report just only one average set of results.

The trials have been performed separately on the 3 most common orange varieties in Sicily (Tarocco, Moro and Sanguinello); all of them have pigmented pulp and yellow peel with sometimes red stripes.

The quality standard is derived from the actual law of the Foreign Trading Minister, 13 may 1971, which define the quality levels for oranges to be exported.

There are 4 categories of decreasing quality: Extra (E), First (I), Second (II), Third (III). Only Extra and First categories go to the fresh fruit market, the other two being processed or sold on the local market.

In order to semplify the trials, II and III categories were grouped togheter in a common II class.

For each trial, the oranges were accurately selected by specialized people in order to have 1000 fruits with a distribution considered as the average in the real conditions: 60% of Extra, 30% of First and the remaining 10% of Second. These oranges, without any pre-calibration, were well miscelated before the trials.

The conveyor speed was set at 6.25 fruits/s; in order to verify any sensitivity to speed variation we made some trials at 4.16 and 8.33 fruits/s.

Other parameters, in addition to the ones indicated by the A.I.D., have been used for a better understanding of the grading errors.

The fruits graded by the machine have been carefully checked by specialized personnel in order to classify the errors in relation to fruit diameter and colour and conveyor speed.

The real capacity of the grading machine has been calculated evaluating the filling coefficient of the conveyor and then, for each trial, the average fruit mass.

The system capability of lodging the grading process data have been very useful to identify the error causes and it has been heavily used to understand how to improve the machine.

A parallel test of the usual manual grading process has not been performed as this process has a large performance spread depending on packinghouse and actual market demand.

The above referred law tells that in the Extra cannot be present more than a 5% of First and in the First cannot be present more than 10% of the Second. Generally, this performance is manually achieved with

a worker productivity of 0.4 t/h$_{man}$ up to 1.2 t/h$_{man}$, depending on the degree of general mechanization of the packinghouse.

5 RESULTS AND DISCUSSION

First of all it was highlighted that the "Videograding" has shown real working capacity near enough to the nominal ones. In fact, at the speed of conveyor equal to 6.25 fruits/s for channels, the coefficient of filling gave a result quite independent from the variety of the oranges and, as an average number, equal to 84%.

Since the average mass of each fruit has ranged from 0.14 to 0.16 kg, the real working capacity of the machine, for two channels, resulted of $5.3 \div 6$ t/h, a value which seems adequate to the potential of the packinghouses and a furter growth of 30% of the capacity is possible with the increment of the conveyor speed up to 8.33 fruits/s.

Concerning to the quality of sorting (table 1), it seems enough satisfying for the category Extra, which contain about the 12% of fruits of I and almost none of II; equally acceptable are the results obtained for the I, in which was found 14% of orange Extra and 10% of II, and for II in which was found the 23.1% of fruits of I. In fact, although the machine still is not able to assure the quality standard expected in reference to the law above referred, his presence is positive.

It seems important to analyze the errors in relation to some variables (table 2). As an example, it is not important the variety of oranges, whilst it is important the ripeness that, progressing, determinates especially for some variety the intensification of the red stripes on the peel; unfortunately, this phenomenon determines an increasing of fruits of Extra downgraded as First (10%).

The grading performance is not good for small diameter fruits: this is not a big problem, as they are not a major share of

Table 1. Average values of output according to actual fruit quality (%)

Category	E	I	II
E	87.4	13.6	0.2
I	12.1	76.1	23.1
II	0.5	10.3	76.7
Total	100,0	100,0	100,0

Table 2. Grading errors according to fruit varieties, colour, size and to conveyor speed (%)

Variable	E -> I	I -> E
VARIETY:		
-- "Tarocco"	5	27
- "Moro"	4	26
"Sanguinello"	5	26
PEEL COLOUR:		
-- Yellow-orange	5	26
-- Bright orange	4	24
- Delicate red stripes	5	27
- Intense red stripes	10	28
DIAMETER (mm):		
- < 60	9	31
- 60 ÷ 70	5	26
- 70 ÷ 80	4	25
- 80 ÷ 90	4	25
- > 90	6	27
CONVEYOR SPEED (fruits/s)		
- 4,16	5	26
- 6,25	5	26
- 8,33	6	26

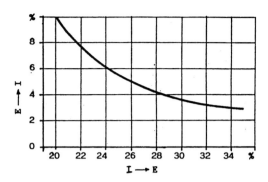

Fig.1 – Grading parameters mutual variation

tion of each pixel;
- B: errors due to low contrast between good and defective areas;
- C: errors caused by a not perfect image composition due to missing or overlapping parts;
- D: errors due to a not perfect stem analysis which led to a defect misclassification.

As evident from table 3, often the detected errors are due to a sum of causes. The prevalent errors are of A and B types, but some improvement to reduce their impact has been already defined, as described in the next paragraph.

It is important to report that, after an initial scepticism, the commercial operators seemed to be quite satisfied with the "Videograding" performance.

In any case, in order to meet the manual process performance, it has been necessary to integrate the machine with some manual labor; this manual labor means in any case a reduction of about 70% respect to old full manual process.

While the final goal of the "Videograding" is to fully automate the grading process, even with these first prototypes an increasing of 2 up 3 times of the workmen productivity has been obtained.

the incoming fruits. In any case, the best performance of the "Videograding" is obtained with the most common sizes, from 70 up to 90 mm of diameter.

The machine performance does not decrease when varying conveyor speed: at the maximum 8.33 fruits/s speed, only a small 1% increment of Extra in First downgrading is detected. This means that the "Videograding" can be used also usually at the maximum speed with a great working capacity.

With reference to the trials methodology, the machine performance has been modified varying the grading parameters (fig. 1): the "Videograding" demonstrated a good flexibility, coping with different trading needs.

The analysis of the error causes, made with the help of the system computers, led to the following error classification:
- A: error in image acquisition due to boundary effect, not complete fruit rotation, bad camera focusing, non linear image acquisition depending on the central or peripheral posi-

Table 3. Error classification (%)

Error cause	Up-grading	Down-grading
A - Image acquisition	46	40
B - Contrast	28	39
C - Image composition	26	19
D - Stem analysis	18	13

6 CONCLUSIONS AND PROSPECTS

The "Videograding" machine performs the grading of the oranges with results, although not excellent, certainly good and never obtained before.

The activity at this moment, moreover, is direct either to the enhancement of the machine, with the implementation of the other evolutions, and to the realization of versions for other products.

For the enhancement of the performances, the results of the experiments have allowed to set off some modifications for the next realization, which it is foreseen to be able to reduce the errors of about 50% and, consequently, to eliminate any manual operation in the grading.

These modifications concern:
- methods for image acquisition based on 4 images instead of 2: it would decrement the errors on the image acquisition (type A) of more than half;
- analysis of the single defect, to reduce the errors of type D;
- correlation between several images of the fruits that, associated with the previous point, would decrease the errors type A, C and D;
- multiple filters, selectable by operator in function of the variety and ripeness of the fruits, to limit errors of type B.

About the application for other vegetable products, versions for apples and tomatoes are under research; the system architecture is good enough, while other subsystems will be modified, in particular the conveyor and the application software.

Another machine is under research too for the automatic grading of a wide range of products based on several features (quality, dimension, weight, shape, colour, quantity of juice, etc.); the analisys performed at this moment is encouraging for the possibility of his feasibility.

REFERENCES

Castelli E. & Rovetta A. 1986. Riconoscimento automatico di forme di oggetti in movimento. Pixel n. 8

Kaplan H.J. & Scopatz S.D. & Miller W.M. 1984. Analysis of manual and machine grading operations in California lemon packing houses. Proc. 1984 Intern. Citrus Congress, S.Paolo, Brasil

Parrella C. 1986. Automaticamente: dalla raccolta al supermercato. Progettare, n.84, may

Vecchi V. 1986. Verso la computerizzazione delle centrali ortofrutticole. Terra e Vita, n. 16

Land and Water Use, Dodd & Grace (eds), © 1989 Balkema, Rotterdam. ISBN 90 6191 980 0

Performance evaluation of stationary threshers

L.L.Bashford & O.El Gharras
Institut National de la Recherche Agronomique, Settat, Morocco

K.VonBargen & D.V.Byerly
University of Nebraska, Lincoln, Nebr., USA

ABSTRACT: A procedure to evaluate the performance of stationary threshers is discussed. This procedure was then used to evaluate two stationary threshers. Differences in performance between the two machines were observed and are reported.

INTRODUCTION

In many developing countries, much use could be made of small stationary threshers for cereals and legumes. These machines are simple in function and operation and are used in areas that are not accessable to a combine or in areas where the economics do not support the use of expensive combines. Small stationary threshers would be particularly appropriate where crops have been harvested and collected at some central location for threshing by animals.

Procedures and methodology to evaluate the performance of stationary threshers are nonexistent. However, procedures that have been developed by the American Society of Agricultural Engineers (ASAE) can be modified to fit the requirements and operation of a stationary thresher.

The Prairie Agricultural Machinery Institute (PAMI) has been one of the more active institutions testing combine performance and making reports of the test results available to the public. PAMI uses a combine designated as a reference machine as the comparison to the test combine. This reference machine, when used in varying crop conditions, serves as the standard or normalization for comparing combines tested in different years and crop conditions.

OBJECTIVE

The objective of this paper is to present the procedure used to evaluate two stationary threshers and report the results.

PROCEDURE

Two stationary threshers were obtained for testing in cereal crops in Morocco. One of the threshers was manufactured by a small manufacturer in Morocco and the second thresher was imported. The two machines were different in operational characteristics but both accomplished the same function.

The threshers were evaluated using Saada wheat. The procedures used to evaluate the two threshers were developed from ASAE STANDARD: ASAE S396 Combine Capacity Test Procedure. The standard for testing combine capacity calls for the comparative testing of one machine relative to another in a particular crop condition. The thresher manufactured in Morocco was designated the test machine and the other thresher was designated the comparison machine.

Performances of the threshers were evaluated by varying total feed rates starting at some minimum total feed rate and increasing until some upper limit was reached. On-farm demonstrations of the test machine during the previous year, indicated a wide range of threshing capacities but with an average grain feed rate of 200 kg/h. Therefore, both threshers were adjusted during pretest runs for optimum performance at that rate.

The total feed rate of a combine can be easily controlled by the forward speed. However, for a stationary thresher where the input is accomplished manually, the total feed rate is more difficult to control. As a result of preliminary tests with manual input, preweighed material of approximately 100, 150 and 200 kg was used for the test runs. The intention was to manually input these preweighed stacks in 15 minutes.

Samples of material other than grain (MOG) and grain were collected at their respective discharge sites. Sample duration time was 30 seconds for both MOG and grain with 3 or 4 samples being obtained during each test run. A box was fabricated to catch all MOG discharged from the machine during the sampling period. Grain samples were obtained by catching at the grain discharge. A grain catch was obtained after each MOG catch.

All MOG catches were evaluated for processing losses. These catches were weighed, sieved and the sieved material passed through a winnowing fan to remove any small debris. The grain recovered from each MOG catch was then weighed and recorded.

All grain catches were subsampled and examined to assess the damage due to cracked and crushed grain. The remainder of the grain samples were winnowed to determine the amount of MOG in the clean grain.

Preliminary test results from the test machine indicated that good threshing with minimal grain damage resulted at a threshing rotor speed of 775 rpm. Grain damage was excessive at the recommended rotor speed of 850 rpm. However, a series of performance tests were later run at the recommended rotor speed.

Preliminary test results from the comparison machine indicated that a rotor speed of 750 rpm was adequate for the threshing process. Rotor speeds in excess of 800 rpm were initially tried but lowered with no loss in threshing efficiency. Grain damage was not the criteria for rotor speed as no grain damage occurred during the preliminary tests.

DESCRIPTION OF THE MACHINES

The terminology of ASAE Standard: ASAE S343.1 Terminology for Combines and Grain Harvesters was used. Table I lists the specifications for the two machines. A review of this table illustrates that the machines are approximately the same size. The threshing rotor diameters and lengths are relatively the same. The comparison machine has more threshing area because the separating grate is functional for 360 degrees but the cleaning screen is considerably smaller. The specifications for the engine on the test machine were not known.

The test machine is basically a modified hammer mill. Threshing is accomplished by impact with rotor spikes. All input material exits the threshing area through a screen. Thus the straw is reduced into lengths usually less than 5 cm. The thresher has no tailings return. All grain not removed from the cleaning screen is lost in the straw discharge. The thresher has three cleaning fans, two providing wind for the cleaning screen, and one providing a final cleaning for the grain prior to the bagging auger. The wind volume is adjustable by dampers on all fans but the direction is fixed.

The comparison thresher has more components than the test machine, Table I. Operation of the machine was smoother. Threshing is also accomplished by impact with a set of spikes. The straw is, in effect, augered through the threshing area by the rotor and discharged with a paddle fan. The straw length is usually greater than 5 cm. The comparison machine has a tailings return. The thresher uses two fans to provide wind to the cleaning screen. The wind volume is also adjustable by dampers, but the direction is fixed.

Table I. Specifications of the test and comparison threshers.

Machine ID	test	comparison
Rotor diameter, mm:	530	520
Rotor length, mm:	1155	1060
Separating grate:		
width, mm	780	1060
arc length x		
diameter	655 x 540	1948 x 620
area, m^2	0.51	2.06
hole diameter x		
spacing	20 x 26	-----
rod diameter x		
spacing	-----	8 x 18
Cleaning screen:		
width, mm	880	480
length, mm	1080	300
area, m^2	0.95	0.144
hole diameter x		
spacing	10 x 14	13 x 18
Tailings return:	No	Yes
Rotor spikes:		
rows x no / row	8 x 22	8 x 14
length, mm	75	80
Engine:		
manufacturer	Lister	Kubota
Bhp x rpm	est. 10	11 x 2400
displacement, cc	=====	598
Thresher length, m:	3.14	3.10
Thresher height, m:	2.43	2.66

RESULTS AND DISCUSSION

The weights of the unthreshed input material and of the grain obtained from the test run are given in Table II. The MOG was then determined from the weight differences. The feed rates calculated from the MOG and grain catches obtained during the 30-second sampling periods are given in Table III. Each value is the average of all samples obtained during a test run.

Comparing total feed rate in Table III to the weight of unthreshed material in Table II suggests some inconsistencies. For instance, tests H1131 and H1231 were both run with the same amount of input material. Yet comparing the average total feed rates between the two tests would suggest otherwise and illustrates the difficulty of maintaining a steady input of material.

Another measure of the varying inputs during the testing is the MOG/G ratio. In Table II, except for test H0003 and H0528, the MOG/G ratios are fairly consistent when based on total weight. Yet when MOG/G ratios are based on feed rates, as in Table III, more variation is apparent and illustrates the wide variation of total feed rates when material is input by hand.

It was evident from the very beginning of the technical evaluation of the comparison machine that there were problems adjusting the wind volume. The wheat used for the performance tests had a solid stemmed straw. Therefore the straw was heavier than normal. When the wind volume was increased to remove the straw from the cleaning screen, excessive grain was carried off with the straw discharge. When the wind volume was decreased so that grain losses were acceptable, the volume of straw in the tailings return was excessive and plugged the return auger. Therefore, the critical adjustment for this machine and with this particular variety of wheat was the wind. Because of the plugging at acceptable grain loss levels, the comparison machine was evaluated at only two feedrates.

The primary measure of threshing capacity is the maximum rate at which a threshing machine can operate while maintaining an acceptable loss level. PAMI uses a processing loss of 3% as an acceptable loss level for rating combines. Therefore, the results obtained from the two threshers were arbitrarily compared to the PAMI loss reference.

Figure 1 illustrates the processing losses of the two machines as a function of the total feed rate. There appears to be no difference in processing losses between the two rotor speeds of the test machine.

Table II. Throughput determined by weighing the unthreshed material and the threshed grain for each test run.

ID	Grain+ MOG kg	Grain kg	MOG kg	MOG/G ratio	Rotor speed RPM
H0002	100	36	64	1.78	785
H0003	148	63	85	1.35	785
H0528	148	46	102	2.22	785
H0628	95	33	62	1.88	785
H0728	185	67	118	1.76	785
H0830	144	52	92	1.77	843
H0930	184	62	122	1.97	843
H1030	97	33	64	1.94	843
H1131	191	67	124	1.85	843
H1231	190	74	116	1.57	785
T0001	92	34	58	1.71	750
T0004	128	44	84	1.91	750

ID beginning with H is test machine
ID beginning with T is comparison machine

No difference would be expected unless the faster rotor speed further reduced the straw causing a different relationship among the MOG, grain and wind on the cleaning screen. A simple linear analysis regressing processing losses on total feed rate was run to sense changes occurring at different feed rates. Even though the measure of fit of the processing loss versus total feed rate was very low, the trend for the processing loss to decrease as total feed rate increased was evident. This result is opposite to that expected. Usually in combines, an increase in feed rate results in an increase in the processing losses. It is conceivable that, with larger feed rates, a thicker mat of material forming on the cleaning screen might have a tendency to retard the wind volume. Therefore, less grain would be discharged over the cleaning screen. This same reasoning would suggest that at larger

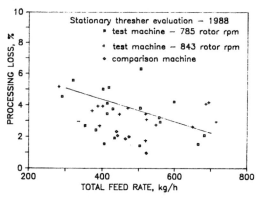

Figure 1. Total feed rate versus processing loss.

Table III. Average of respective samples of designated variables.

ID	MOG feed rate kg/h	Grain feed rate kg/h	Total feed rate kg/h	Processing losses %	Damaged grain %	MOG in grain %	MOG/G ratio
H0002	270	235	505	2.4	6.9	0.7	1.15
H0003	344	166	510	3.8	6.7	1.9	2.25
H0528	229	120	349	3.8	5.6	1.2	2.60
H0628	222	167	389	2.3	4.4	1.3	1.35
H0728	231	261	492	3.5	2.7	0.6	1.10
H0830	220	144	364	3.9	15.8	2.1	1.56
H0930	318	188	506	3.0	11.3	3.2	1.70
H1030	251	174	425	3.2	16.7	3.6	1.43
H1131	419	265	684	3.5	5.1	1.7	1.61
H1231	296	208	504	4.4	4.9	0.7	1.52
T0001	---	276	---	1.6	0.6	1.2	----
T0004	---	225	---	2.0	1.7	0.3	----

Comparison machine MOG rate, as measured by sampling, is not known because of the excessive straw in the tailings return.

feed rates, the MOG in the clean grain would be greater and this was not the case. Using 3% as an acceptable processing loss, approximately half of the samples were in excess of 3% and half less than 3%.

With the comparison machine, processing losses were very acceptable. However, as previously mentioned, with the wind volume properly regulated to minimize processing losses, the return auger continued to plug with the excessive straw in the tailings. For this thresher, the 5 samples obtained from the comparison machine illustrate that the processing losses were all less than 3%.

The importance of keeping processing losses at a level less than 3% may not be critical. In many countries, straw is used as an animal feed, therefore, some grain in the straw would not be a loss.

Another measure of thresher capability is the cleanliness of the grain after threshing. Figure 2 illustrates the MOG in the clean grain as a percent of the weight of the total grain sample versus the total feed rate. A simple regression analysis indicated no trends in response of MOG in the clean grain to total feed rate for the test machine. Very small amounts of MOG were observed in the clean grain for the comparison machine, with all samples being less than 2% of MOG in the grain. Irrespective of the feed rate, rotor speed or type of thresher, the MOG percentage in the clean grain was usually less than 3%.

A measure of visible grain damage occurring during the threshing operations is the amount of cracked and crushed grain

in the clean grain. At the higher threshing rotor speed of the test machine, 843 rpm, there is a definite relationship between damaged grain and total feed rate, Figure 3. The results of a linear regression analysis indicated that the percent damaged grain is reduced by 3.5% for every 100 kg/h increase in total feed rate. The results of the analysis at the slower threshing rotor speed illustrates that the percent damaged grain is reduced by 0.4% for every 100 kg/h increase in feed rate. Therefore, the speed of the threshing rotor should be no faster than required to efficiently thresh the grain. One hypothesis might be that at low feed rates, the grain in the threshing area loses the cushioning effect due to the presence of the straw. As feed rate increases, the presence of larger amounts

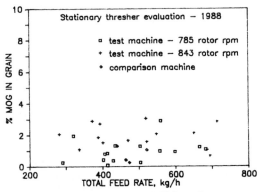

Figure 2. Total feed rate versus MOG in grain.

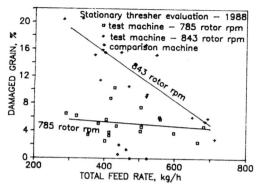

Figure 3. Total feed rate versus damaged grain.

of straw cushions the impact of the rotor spikes on the grain thereby reducing grain damage. At the faster rotor speed, the necessity to keep the machine operating at full capacity is apparent. At the slower rotor speed, 784 rpm, there is also an advantage to keeping the machine operating at full capacity. Irrespective of rotor speed, these threshers, like combines, should be operated near capacity for best results.

The comparison machine caused very little damage to the grain, with all observations of damage being less than for the test machine. The reason for this is that the comparison machine functions more like a conventional combine, and the test machine functions more like a hammer mill.

CONCLUSIONS

The following conclusions can be made:
1. Both threshers can do an adequate job of threshing.
2. Total feed rates, estimated at 500 kg/h, were obtained with the comparison machine. Actual total feed rates, based on sampling, were not known because of excessive straw in the tailings return. Processing losses were less than 2% and damaged grain was nil.
3. The comparison machine had a major problem with excessive straw in the tailings return at acceptable processing loss levels. This problem could probably be solved by a design change.
4. Total feed rates up to 700 kg/h were observed with the test machine with processing losses averaging less than 4%. Damaged grain averaged less than 7% at moderate threshing rotor speeds.
5. The test machine can cause excessive grain damage at fast threshing rotor speeds and small total feed rates. This machine functions best when operated near capacity.

6. A big advantage of the test machine is that straw is reduced during the threshing process. This makes the straw very desirable for animal feed.
7. As new varieties of cereals are developed, attention needs to be focused on the effect of the straw on pneumatic separation.

OBSERVATIONS

1. The test machine is underpowered. The machine has more capacity than available power.
2. When testing these machines, it was very important for those persons involved with the actual test to have sufficient practice in material input and sampling. This results in a much smoother input of material and eliminates slugging the machine.
3. The collection of the MOG catches has to be accomplished without interfering with the wind volume and direction. If the catch devices are too restraining, alteration of machine performance occurs easily.

ACKNOWLEDGEMENT

Support for this research was received from USAID Project No. 608-0136.

Land and Water Use, Dodd & Grace (eds), © 1989 Balkema, Rotterdam. ISBN 90 6191 980 0

Performance changes of the cleaning unit of a combine harvester during field tests

T.Beck

Institute for Agricultural Engineering, University of Hohenheim, FR Germany

KURZFASSUNG: Die Leistungsfähikeit der Trennprozesse im Mähdrescher wird erheblich von den Stoffeigenschaften des Erntegutes und den Umgebungsbedingungen beeinflußt. Leistungsänderungen während eines Tages werden vor allem von der Oberflächenfeuchte des Erntegutes bestimmt. Diese Oberflächenfeuchte kann bis jetzt nicht direkt gemessen werden und wird deshalb aus den mittleren Gutfeuchten und den Klimadaten über der Erntefläche abgeleitet. Diese Werte und die Durchsatzleistung einer Mähdrescher Reinigungsanlage wurden im Feldversuch gemessen. Die Methode zur Aufnahme der Durchsatz/Verlust-Kennlinien und eine elektronische Wetterstation werden vorgestellt. Die während der Feldversuche 1988 gewonnenen Erkenntnisse werden dargestellt und diskutiert.

ABSTRACT: Performance of the separating processes in a combine harvester depends heavily on crop properties and on environmental conditions. Performance changes within one day are mainly caused by varying surface moisture contents of the crop. Surface moisture content cannot be measured directly, but can be derivated from the mean moisture content of the crop particles and from climate data over the crop in the field. These properties were measured together with the performance of the cleaning unit of a combine during field testruns. The method for measument of feedrate vs. loss curves and an electronic weather station for field tests are described. The results of the 1988 test runs are discussed.

RESUMEE: L'efficacité des systèmes de séparation dans une moissoneuse-batteuse depend pour grand part de la qualité du materiel récolté et aux conditions environnementalles pendant la récolte. Changements de l'éfficacité pendant une journée sont causé avant tout par l' humidité au surface du materiel recolté. Cette humidité au surface ne peut pas être mesuré directement, mais il est possible de la correlé à l'état hygrométrique du materiel récolté et des conditions climatiques sur le champ. Cettes quantités et la capacité d'un nettoyage dans une moissoneuse-batteuse étaient mesuré pendant des essaies en champ. La méthode pour gagner les courbes de perte et la construction d'une station météorologique électronique sont présentés. Les résultats trouvés pendant les essaies en champ 1988 sont présentés et discutés.

1 INTRODUCTION

A modern combine harvester is today one of the most expensive agricultural machines. High costs for purchase and operation call for optimum performance to match the needs for efficient grain production. The characteristic quality of a combine that is most important for evaluation of performance is the capacity of the seperating processes at a certain level of grain losses, generally described by a feedrate vs. loss curve. Unfortunately feedrate vs. loss curves for a single combine obtained

under different conditions change significantly due to certain variables (fig.1).To evaluate the performance of a combine it is important to know the crop properties and harvesting conditions under which capacity was measured. Changes in these variables often result in significant shifts of capacity. Crop properties and harvesting conditions both consist of a set of interdependent quantities of which the influence on capacity is widely unknown.

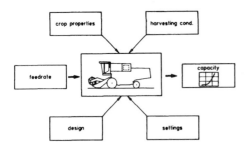

fig.1 Influences on the seperation
processes in a combine harvester

Measurement of feedrate vs. loss curves
reveals two phenomena that are caused by
changes in crop properties and harvesting
conditions. The capacity of a combine
varies in different years and fields and
also changes within one field during the
same day. The latter problem makes it
difficult to evaluate feedrate vs. loss
curves measured at certain times of a day,
because it is unknown how much the capacity was influenced by changes in crop
properties and harvesting conditions.
Results of combine tests are therefore
often influenced by uncertainties that
make it difficult for test institutes to
give reliable figures for capacities. The
problem becomes even worse when the test
institutes are forced to abandon the
operation of a reference combine for
economic reasons. Research and development
departments of combine manufacturers face
similar problems. In addition the influence of crop properties must be known more
exactly when control devices for combines
are developed to compensate for disturbances that occur during operation.

2 APPROACH

Changes in the combine's capacity in
different years and fields depend mainly
upon crop properties that are determined
through the development of the crop during
the growing period.However changes of
capacity within one day in the same field
must be caused by faster changing properties. These can only be properties that
are influenced by the microclimate over
the harvested field.
  The microclimate determines the moisture
conditions in the crop and therefore the
friction coefficients between particles in
the seperation processes of the combine.
The friction coefficient is of crucial
importance for the quality of the separa-

tion processes in the combine, because all
separation processes are based on the
penetration of kernels through a straw or
chaff mat.
  Mean moisture content of the crop particles is thus generally recorded when the
capacity of a combine is measured, because
it is assumed that mean moisture content
affects the friction coefficient. Huisman
(1971) however found only poor correlations
between friction coefficients and mean
moisture contents of freshly harvested
straw. Relative humidity of the air is also
considered to have a major influence on
combine capacity and is therefore often
used to interpret capacity shifts.These
correlations are generally better than
those with mean moisture content.
  Hall and Husman (1981) observed a significant effect of moisture conditions on
combine performance. They correlated both
relative humidity and mean moisture contents as single variables with capacities
of the seperation devices of a combine.
They found correlations for both properties
but no reliable trends. Similar results
were obtained by Srivastava, Mahoney & West
(1986). They also found that the sensitivity of the combine's cleaning unit to
changes in the crop properties is higher
than that of the straw walkers.
  Friction coefficients and thus the
capacity of the separation processes are
obviously influenced by both quantities.
Because there are no other variables that
fluctuate within one day, moisture quantities must be further examined to find the
reason for capacity changes within the
same day. A closer look on the frictional
events in the seperation processes reveals
that only the surfaces of the crop
particles are involved in the processes.
Friction coefficients of the particles are
thus not determined by the mean moisture
content but by the moisture content of the
particle surface or near surface areas.
This surface moisture content may significantly differ from the mean moisture
content that is normally measured. Peak
values of 100% are reached when water is
accumulated on the surface of the crop
(dew point). The adhesive effects may
then cause much higher contact forces
between single particles than is possible
through frictional forces.
  The surface moisture content is determined by the drying and remoistening behaviour of the crop during the day. The drying
process of the crop in the field depends
on convective effects and on direct heat
accumulation through radiation energy.
Hofacker (1986) examined the drying process
of wheat kernels with regard to hot air
drying in post harvest dryers. He developed

a model that can also be used to better understand the drying process in a grain field. This drying process operates below temperatures of 40 deg C, so that the balance between surrounding air and particles is the main influence. Depending on condition of the air and particle moisture contents, moisture changes in both directions is possible, enabling both decrease and increase of particle moisture content and influencing surface moisture content heavily. Crucial properties are relative humidity of the air, temperature of the air, air velocity and mean moisture content of the crop particles plus energy from solar radiation.

The complex thermodynamics in a grain field can not be described by a mathematical model. Experimental studies as conducted by Hofacker (1986) in the laboratory are no longer useful when transfered to a grain field. It was therefore tried to describe the surface moisture content through the more easily measured crucial quantities

- mean moisture contents
- relative humidity of the air
- temperature of the air
- air velocity
- radiation energy

to evaluate capacity changes of the cleaning unit of a combine harvester. To find changes in capacity the cleaning unit was operated at different times of day while continuously recording climatic data with an electronic weather station.

3 METHODS AND EQUIPMENT

3.1 Measurement of climate data

To match the specific needs for measurement of climatic conditions in a grain field an electronic weather station was built (fig.2). While mechanical devices are only able to continuously record relative humidity and teperature of the air, an electronic station allows for the recording of all necessary climatic data. Further advantages are higher accuracy and lower time constants of the electronic sensors and easy storage of data.

The sensors were attached to a steel mast which was set into the earth an held by four ropes. To avoid border effects the mast was placed in the middle of a 15 x 15m area in the field.

Relative humidity and temperature of the air were measured with polymercapacitor and thermistor assemblies, one located 2m above the ground the other inside the crop.

fig.2 electronic weather station during field tests

The data gathered from the sensor in 2m height represent an international meteorological standard, while the sensor inside the crop was supposed to more effectively describe the surface moisture effects. In all test runs significantly different readings were obtained from both locations. Most important however was the timelag between the two curves in the ascending and descending branch (fig.3). Solar radiation was measured with a highly sensitive pyranometer; air velocity was measured with a cup anemometer. The recording of the data was done by a battery powered data logger which also supplied the sensors with electric energy. After the test runs all data were transfered to a personal computer.

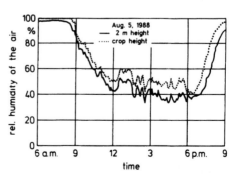

fig. 3 course of relative humidity of the air above and in the crop

3.2 Measurement of capacity of the cleaning unit

To record the feedrate vs. loss curve of the cleaning unit a standard middle class combine harvester was used. The combine was supplied with an automatic system to measure grain feedrate and grain losses during a single testrun with the combine maintaining a constant ground speed. The measurement system was controlled by a computer to guarantee constant test run conditions (fig.4).

To determine grain losses the entire chaff of a testrun was collected in a bag and later processed to determine the amount of grain remaining in the chaff sample. Grain feedrate was measured by collecting the entire grain of a test run in a seperate container inside the grain tank of the combine. At least seven single test runs were carried out to calculate a feedrate vs. loss curve.

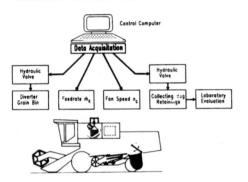

fig. 4 measurement system for grain loss and grain feedrate

To obtain capacity changes three curves were taken at different times (fig. 5). The first curve was measured before noon when relative humidity of the air an mean moisture contents of the crop decreased. The next curve was taken in the afternoon under generally optimum harvesting conditions, while the last curve was taken at sunset when remoistening of the crop was assumed.

4 RESULTS

The measured capacities for the cleaning unit (=feedrate at 0.5% losses) before noon were usually the lowest of the day. A maximum was obtained by the afternoon test runs, while capacities measured at sunset were already below the maximum. This corresponds to the empirical knowledge for the climatic conditions of a sunny day without clouds. Mean moisture contents, relative humidity of the air and energy of radiation (fig.5&7) for an ideal sunny day show good correlations with the measured capacity changes when used as single variables.

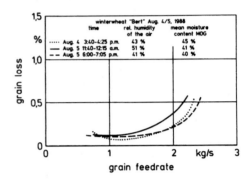

fig.5 feedrate vs. loss curves of the cleaning unit for one test series

The observation of the course of climate data becomes only then important when climatic conditions differ from those of the ideal sunny day.

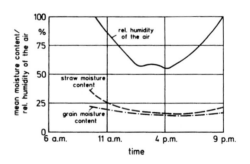

fig.6 course of moisture quantities for an ideal sunny day

For illustration the data in table 1 shows the measured mean moisture contents for a day that deviated significantly from the ideal sunny day. The courses of relative humidity of the air and energy of radiation for the same day are shown in figures 3 and 7.

time	mmc S	mmc C	mmc G
9:10 a.m	53.4	-	-
11:50 a.m	41.4	-	-
0:35 p.m	43.8	10.2	17.5
4:30 p.m.	43.7	10.6	16.2
6:30 p.m	40.5	9.8	15.4
7:50 p.m	43.3	15.4	14.9

mmc = mean moisture content, S = straw, C = chaff, G = grain

table 1 mean moisture contents for a cloudy day

Remarkable for that day are the increasing mean moisture contents in the afternoon. Relative humidity of the air remained on a relatively high level and did not reach the minima measured for other days. The reason for these observations was the appearance of heavy clouds shortly after noon that remained for nearly two hours. The course of radiation energy correlated exactly with the appearance of the clouds (fig.7).

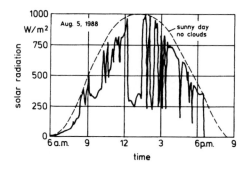

fig.7 radiation energy over one day

Instead of the usual strong drying during behaviour this time of day a slight remoistening was recorded. Because transport of water in the crop requires a certain amount of time, the remoistening could only spread to the near surface areas of the crop.

Surface moisture content however cannot be measured directly, but its quantity and influence can be derived from the measurement of mean moisture contents and climatic conditions. A statistical model to calculate capacities of the surface moisture content was employed on the basis of a regression routine. For all test runs good correlations were found when relative humidity of the air and mean moisture contents were used as independent variables in the model.

The dependence of the capacity on both variables is illustrated through the feedrate vs. loss curve for Aug,4,1988 in figure 4. Although mean moisture content of material other than grain (MOG) is considerably higher, the capacity measured under optimum drying conditions and respectively low surface moisture content, was nearly as high as the measured maximum capacity.

With regard to quantitative statements the already discovered correlations must be proofed in aditional field tests, because the data basis is still too small for statistical statements of high significance.

REFERENCES

Hall, J.W. & Husman, J.F. 1981. Correlating physical properties with combine performance. ASAE Paper 81-3538.

Hofacker, W. 1986. Trocknungsverhalten und Qualitätsveränderungen von Weizen. Dissertation Universität Stuttgart.

Huisman, W. 1977. Moisture content, coefficient of friction and modulus of elasticity of straw in relation to walker losses of a combine harvester. Proceedings of the ASAE/CIGR international grain and forage harvesting conference, sept 25-29,1977 Ames/Iowa

Srivastava,A.K, Mahoney, W.T. & West, N.L 1986. Effect of crop properties on combine performance. ASAE Paper 86-1583

Land and Water Use, Dodd & Grace (eds), © 1989 Balkema, Rotterdam. ISBN 90 6191 980 0

Durchsatzbestimmung für eine Regelung der Mähdrescher-Reinigungsanlage

S.Böttinger
Institut für Agrartechnik, Universität Hohenheim, Stuttgart, Bundesrepublik Deutschland

KURZFASSUNG: Kornverlust und Durchsatz sind elementare Eingangsgrößen für eine Teil-prozeßregelung der Mähdrescher-Reinigungsanlage. Eine neu entwickelte Meßeinrichtung für die Korndichte kann durch die Kompensation der im Betrieb des Mähdreschers auftretenden Schwingungen und Neigungen die Dichte auf ca. ± 1 % genau bestimmen. Kombiniert mit einem volumetrischen Durchsatzmeßgerät wird der Kornmassendurchsatz kontinuierlich ermittelt. In einem zweiten Meßverfahren wird die Korrelation des statischen Druckab-falls unter dem Obersieb der Reinigungsanlage mit dem Durchsatz verwendet. In Abhängig-keit von der Gutbeladung und der Gebläsedrehzahl tritt ein mittlerer statischer Druck-abfall von bis zu 40 Pa auf. Ohne nennenswerten Zeitversatz wird aus dem Druckwert der Gesamtdurchsatz der Reinigungsanlage und nach einer Kalibrierung der Korndurchsatz des Mähdreschers berechnet.

ABSTRACT: Grain loss and feedrate of combine harvesters are essential parameters for a back loop control of the subprocess 'cleaning unit'. With compensating vibrations and declinations of the combine in the field, a newly developed grain densimeter can estima-te the density with an accuracy of ± 1 %. Together with a volumetric feedrate meter, grain feedrate can be determined continously. For a second measuring method the cor-relation of the static pressure drop under the chaffer of the cleaning unit to feedrate is used. An average static pressure drop up to 40 Pa is related to sieve loading and fan speed. Total feedrate of the cleaning unit and, after calibration, grain feedrate can be calculated without an appreciable time lag.

RESUMEE: La perte de grain et le débit sont les paramètres principaux pour un réglage de la moissoneuse-batteuse et specialement le contrôle du processus du nettoyage. Un nouveau système pour la determination de la densité des grains est capable de la mésurer avec une précision de ± 1 %. Cette précision est achevée à cause de la compensation des vibrations et des inclinaison de la moissoneuse-batteuse pendant les travaux champétres. En combinaison avec un débitmètre volumétrique, le débit en grain peut être calculer continuellement. Un second système utilise la correlation de la chute de pression statique dessous la grille du nettoyage avec le débit. En fonction du chargement du nettoyage et les tournements du ventilateur cette chutte de pression monte jusqu'au 40 Pa. Connaissant cette correlation on peut calculer- presque sans délai - le débit total du nettoyage et avec une calibration le débit en grain.

1 EINLEITUNG

Für die mechanisierte Ernte von Getreide hat sich der Mähdrescher in den meisten Ländern der Welt durchgesetzt. Kontinuier-lich steigende Flächenerträge und die zunehmend angespannte wirtschaftliche Situation der landwirtschaftlichen Betrie-be verlangen eine ständige Leistungsstei-gerung dieser Schlüsselmaschine. Durch die bei den Groß-Mähdreschern erreichten Außenabmessungen ist eine Leistungserhö-hung durch Größenwachstum der Dresch- und Trennorgane nahezu ausgeschöpft. Verstärkt wird an der weiteren Optimierung der bestehenden und an der Entwicklung und Einführung alternativer Dresch- und Trenn-verfahren gearbeitet.

Der Mähdrescherfahrer hat eine der teuersten Maschinen in der Landwirtschaft zu bedienen. Deshalb muß er ihre Ernteka-pazität maximieren ohne jedoch einen bestimmten Grenzwert für die Kornverluste zu überschreiten. Der Fahrer hat häufig nicht genügend genaue Informationen über die aktuellen Durchsatz- und Verlustwerte, um auf diesen eine Entscheidung aufzubauen.

Durch die Entwicklung von Regel- und Informationssystemen soll dem Fahrer die Bedienung des Mähdreschers vereinfacht bzw. teilweise abgenommen werden. Für derartige Systeme sind Gutdurchsatz und Kornverlust ebenfalls elementare Größen. Bisher entwickelte Durchsatzmeßeinrichtungen haben sich zum größten Teil nicht durchgesetzt. Hauptursache ist häufig die starke Abhängigkeit des Meßprinzips von den Guteigenschaften wie Stroh- und Kornfeuchte (Kutzbach 1988, McGechan, Glasbey 1982).

2 REGELUNG DER MÄHDRESCHER-REINIGUNGS-ANLAGE

Ein Regelsystem für den Mähdrescher läßt sich untergliedern in Teilprozeßregelungen für die einzelnen Förder-, Dresch- und Trennorgane und in das übergeordnete Gesamtregelsystem. Durch die Teilprozeßregelung der Reinigungsanlage wird deren Arbeitsverhalten bei gegebenen Bedingungen (Durchsatz, Gutzusammensetzung) durch Variation der Gebläsedrehzahl optimiert. Kann durch eine Verstellung des Gebläses keine befriedigende Arbeitsqualität erreicht werden, erfolgt eine Meldung an das übergeordnete Regelsystem des Gesamt-Mähdreschers. In Abhängigkeit des Arbeitsverhaltens der anderen Teilprozesse kann das Gesamt-Regelsystem eine Durchsatzänderung durch eine Änderung der Fahrgeschwindigkeit veranlassen, Bild 1.

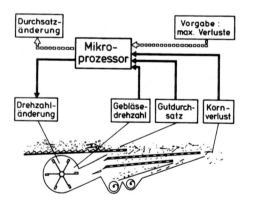

Bild 1   Teilprozeßregelung der Reinigungsanlage

Bisher vorgeschlagene und entwickelte Regelsysteme haben meist das Ziel, den Durchsatz des Mähdreschers durch eine

Änderung der Fahrgeschwindigkeit auf konstant hohem Niveau zu halten (Friesen, Zoerb, Bigsby 1965). Teilweise wird auf Durchsatzschwankungen auch durch eine Verstellung der Dreschtrommeldrehzahl reagiert (Herbsthofer 1971). Regelsysteme für die Reinigungsanlage wurden entwickelt, um die Hangempfindlichkeit der Siebe zu kompensieren (Kutzbach 1985, Stahl, Freye, Kutzbach 1981). Im Labor wurde die Gebläsedrehzahl einer Reinigung geregelt um bei einem vorgegebenen Durchsatz die Verluste zu minimieren (Berner, Grobler 1986).

Das Arbeitsverhalten der Reinigungsanlage wird neben der Reinheit vor allem durch den auf den Kornduchsatz bezogenen Kornverlust beschrieben. Die Probleme bei der Messung des Kornverlustes sind bekannt (Berner, Grobler, Kutzbach 1985), an Verbesserungen wird gearbeitet (Lunty, Leonhard 1986, Wang, Zoerb, Hill 1988, Böttinger 1988).

Der Korndurchsatz kann direkt als Massendurchsatz und indirekt aus einer mit ihr korrelierenden Größe ermittelt werden, (Kutzbach 1988, Schueller, Mailander, Krutz 1985).

3 ERMITTLUNG DES MASSENDURCHSATZES

Auf dem Markt sind volumetrische Durchsatzmeßgeräte erhältlich, die den Volumendurchsatz an Korn z.B. zwischen Kornelevator und Korntankbefüllschnecke messen. Ihr Arbeitsverhalten kann als gut bezeichnet werden (Rust 1987). Bae et. al. (1987) verwendeten dieses Meßgerät zur Erstellung von Ertragskarten. Für eine genaue Ermittlung des Kornmassendurchsatzes wird vom Mähdrescherfahrer eine häufige Überprüfung und Kalibrierung der Korndichte verlangt, um Änderungen durch den Einfluß der Kornfeuchte und -größe zu erkennen. Bei Hafer z.B. treten Dichteänderungen von ca. 25 % innerhalb eines Tages auf (Sörlin 1985). Mit einer automatischen Dichte-Meßeinrichtung kann dieses System in der Handhabung und in seiner Eignung für Regelsysteme deutlich verbessert werden.

Eine kontinuierliche Wägung im fahrenden Mähdrescher wird durch die auftretenden Schwingungen und Neigungen erschwert. Es wurde deshalb ein System mit einer Kompensationseinrichtung entwickelt. Ein gabelförmiges Element besteht aus zwei mit Dehnmeßstreifen beklebten Biegebalken, Bild 2. Während sich an einem Biegebalken eine definierte Masse m_0 befindet, wird in einen am zweiten Biegebalken befestigten Behälter ein definiertes Volumen V an Korn gefüllt. Auf beide Systeme wirken die

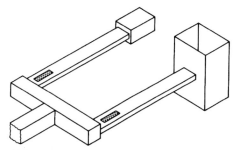

Bild 2   Dichte-Kalibriereinrichtung,
Skizze

gleichen, durch die Fahrt des Mähdreschers
über das Feld hervorgerufenen Beschleuni-
gungen und Neigungen.

Die Signalaufbereitung erfolgt für jeden
Biegebalken getrennt, Bild 3. Die Aus-
gangsspannung der DMS-Brücke wird verstärkt und gefiltert. In Vorversuchen
wurde ein Tiefpaßfilter 2. Ordnung mit
einer Grenzfrequenz von 0,07 Hz ausgewählt. Ein Mikroprozessor fragt die
Signale über einen A/D-Wandler ab und
berechnet aus dem Verhältnis beider Werte
die Korndichte im Behälter. Dieser Wert
kann dem volumetrischen Durchsatzmeßgerät
und anderen Regeleinrichtungen zur Verfügung gestellt werden. Der Prozessor des
volumetrischen Durchsatzmeßgerätes bzw.
der für Regelaufgaben eingesetzte Prozessor kann diese Aufgaben mit übernehmen.

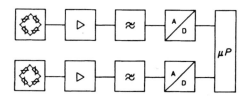

Bild 3   Signalaufbereitung für die Dichte-
kalibrierung

Beide Arme des Dichte-Kalibriergerätes
müssen, mechanisch und elektrisch betrachtet, die gleiche Empfindlichkeit aufweisen. Die Dehnmeßstreifen müssen an der
selben Stelle des Biegebalkens parallel
zur Längsachse aufgeklebt sein. Die
Schwerpunktlage der Masse m_0 und des
gefüllten Behälters müssen bezüglich des
jeweiligen Biegebalkens identisch sein.
Insbesondere dürfen die elektrischen
Bauteile zum Aufbau der Filter nur sehr
geringe Abweichungen voneinander zeigen.

Zur Erprobung im Labor wurden mehrere
Dichtemessungen durchgeführt. Für jeden
gewählten Dichtewert erfolgten statische
und dynamische Messungen. Die statische
Messung wurde bei exakt horizontal ausge-

richteter Meßeinrichtung durchgeführt, für
die dynamische Messung wurde die mit einem
Handgriff versehene Einrichtung stark
bewegt. Es wurde darauf geachtet, daß
neben harmonischen und ruckartigen
Schwingbewegungen auch Neigungen in allen
Richtungen auftraten.

Die Ergebnisse für die statischen Messungen sind sehr genau. Die Abweichungen
von der vorgegebenen Dichte betragen
maximal ± 0,2 %. Durch die dynamische
Anregung erhöht sich die Abweichung von
der vorgegebenen Dichte, sie beträgt
maximal ± 1 %. Eine günstigere Anpassung
der auftretenden Spannungssignale an den
Meßbereich des A/D-Wandlers und eine
höhere Auflösung des Wandlers kann diese
Ergebnisse etwas verändern und präzisie-
ren.

Für die Anwendung im Mähdrescher muß der
zyklische Meßablauf (Probennahme, Meß-
phase, Entleerung, Nullmessung) automati-
siert werden. Die Linearbewegung für die
Probennahme soll durch einen Hydraulik-
zylinder erfolgen, die zur Entleerung
notwendige Öffnung des Behälterbodens kann
am günstigsten durch eine federbelastete
Klappe, ausgelöst durch die Linearbewegung, durchgeführt werden. Zur Kompensation des Nullpunktdriftes bei Gleichspan-
nungsverstärkern wird zwischen jeder
Messung eine Nullmessung mit leerem
Kornbehälter eingeschoben.

4 INDIREKTE ERMITTLUNG DES KORNDURCHSATZES

Bekannte indirekte Durchsatzmeßverfahren
sind die Messung der Antriebsdrehmomente
von Schneidwerk, Einzugsschnecke, Ketten-
förderer und Dreschtrommel sowie der Aus-
lenkung des Kettenförderers. Durch sie
kann nach einer Kalibrierung auf die Guteigenschaften der Strohdurchsatz bestimmt
werden. Durch die Messung des statischen
Druckes unter dem Obersieb der Reinigungsanlage kann dagegen auf den Gesamtdurchsatz der Reinigung und, unter Annahme
konstanter Bedingungen, auf den Korndurchsatz geschlossen werden.

Segler, Freye (1978) maßen den statischen Druck am Gebläsediffussor der
Reinigung und verwendeten dieses Signal
für eine Auslastungsanzeige der gesamten
Anlage. Berner, Grobler (1986) maßen den
statischen Druck unter dem Obersieb um für
ihre Gebläsedrehzahlregelung aus diesem
Signal den Korndurchsatz zu bestimmen.

4.1 EINFLÜSSE AUF DEN STATISCHEN DRUCK

Die vom Reinigungsgebläse erzeugte Luft-
strömung wird durch Einbauten im Siebka-

sten (Leitbleche, Siebe etc.) und durch die Gutschicht auf den Sieben gedrosselt. Der Zusammenhang zwischen dem Druckabfall Δp_E durch feste Einbauten und der Luftgeschwindigkeit w_L bzw. der Gebläsedrehzahl n_G wird durch eine Potenzfunktion beschrieben (Leva 1959):

$$\Delta p_E = B_1 * w_L{}^{C_1} \qquad C_1 = 2 \qquad (1)$$

$$\Delta p_E = B_1 * n_G{}^{C_2} \qquad C_2 < 2 \qquad (2)$$

Für geringe Luftgeschwindigkeiten gilt dieser Zusammenhang auch für den Druckabfall Δp_G durch das Reinigungsgut. Ab einer für das jeweilige Gut charakteristischen Luftgeschwindigkeit w_{FI} befindet es sich aber im fluidisierten Zustand. Der Druckabfall Δp_G eines fluidisierten Gutes ist abhängig von der Gutbeladung m und unabhängig von der Luftgeschwindigkeit w_L, Bild 4.

$$\Delta p_G = B_1 * w_L{}^{C_1} \qquad 0 \leq w_L \leq w_{FI} \qquad (3)$$

$$\Delta p_G = D * m \qquad w_L \geq w_{FI} \qquad (4)$$

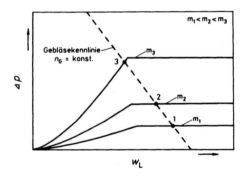

Bild 4 Einfluß der Gutbeladung auf den Druckabfall, Gebläsekennlinie

Zur Erreichung optimaler Trennergebnisse wird ein solcher fluidisierter Zustand des Reinigungsgutes angestrebt.

Bei konstanter Gebläsedrehzahl n_G und zunehmender Gutbeladung m verschiebt sich der Arbeitspunkt des Gebläses durch die stärkere Drosselung zu geringeren Luftgeschwindigkeiten w_L, Bild 4: Punkt 1->2->3. Solange die sich einstellende Luftgeschwindigkeit die charakteristische Fluidisierungsgeschwindigkeit des Gutes nicht unterschreitet ist der Druckabfall von der Gutbeladung proportional abhängig.

4.2 MESSERGEBNISSE

Der mittlere statische Druck unter dem

Obersieb ist sehr gering, in Abhängigkeit von der Gutbeladung und der Meßstelle über der Sieblänge beträgt er 5 - 40 Pa. Durch die Schwingbewegung des Siebkastens verändert sich zyklisch dessen Volumen sowie die Luftgeschwindigkeit relativ zum Sieb und zu den mit dem Sieb mitbewegten Druckmeßstellen. Die hierdurch auftretenden Schwankungen von ca. ± 50 - 100 % des Druckwertes sind zusätzlich von geringen Druckspitzen überlagert, die von den einzelnen Schaufeln des Gebläses herrühren. Sehr hohe Messraten von ca. 500 Messungen/sec. sind erforderlich um diese Druckverläufe darstellen zu können. Durch eine mechanische Dämpfung der Druckspitzen in den Meßschläuchen und eine Tiefpaßfilterung des Analogsignals der Drucksensoren kann die Meßrate je Sensor auf ca. 50 Messungen/sec. reduziert werden, Bild 5.

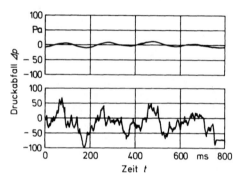

Bild 5 Statischer Druckverlauf unter dem Obersieb, oben mit, unten ohne Filterung/Dämpfung

Die am Mähdrescher auftretenden Schwingungen und Neigungen machen den Einsatz von piezoresistiven Sensoren erforderlich. Die auf dem Markt erhältlichen Sensoren befriedigen allerdings nicht bezüglich des Meßbereiches; bei den eingesetzten Honeywell-Sensoren Serie 160PC mit integrierter Signalverstärkung beträgt er ± 1250 Pa mit einer Auflösung von 2 mV/Pa. Die Drücke werden über 1 m lange Plastikschläuche mit 4 mm Durchmesser von den senkrecht zur Sieblänge unter dem Obersieb angebrachten Druckmeßstellen den neben der Reinigung angebrachten Sensoren zugeführt. Die Messwerterfassung erfolgt mit einem PCI-1680 Datenlogger und einem HP-Rechner Serie 9000/310, die Mittelung der Druckwerte erfolgt über minimal 50 Messungen.

Neben Messungen in einem Mähdrescher bei Feldversuchen wurden im Labor Reinigungsversuche mit einem definierten Gutgemisch aus Korn und Kurzstrohhäcksel durchgeführt und das Druckprofil über der Sieblänge aufgezeichnet. Der Korndurchsatz wurde zwischen $\dot{m}_K = 0,5$ und $3,5$ kg/m*s, der Kornanteil zwischen 70 und 80% und die Gebläsedrehzahl zwischen $n_G = 500$ und 700 1/min variiert.

In Bild 6 ist der Druck von 5 Meßstellen über der Sieblänge l_S in Abhängigkeit vom Korndurchsatz bei konstanter Gebläsedrehzahl aufgetragen. Mit zunehmender Sieblänge nimmt die Gutbeladung durch die Absiebung und somit der sich einstellende Druckwert ab. Mit zunehmendem Durchsatz erhöhen sich die Druckwerte nahezu linear. Aus dem Druckverlauf über der Sieblänge kann der Absiebungserfolg beurteilt werden.

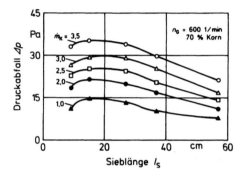

Bild 6 Einfluß des Korndurchsatzes $\dot{m}_K$ auf den statischen Druckverlauf über der Sieblänge l_S

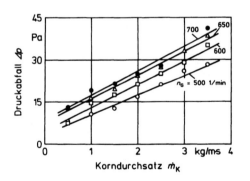

Bild 7 Einfluß des Korndurchsatzes $\dot{m}_K$ auf den statischen Druck unter dem Obersieb in Abhängigkeit von der Gebläsedrehzahl n_G, Kornanteil 70 %

In Bild 7 ist der Druck der Meßstelle 2 (Sieblänge $l_S = 0,15$ m) über dem Korndurchsatz $\dot{m}_K$ aufgetragen. Mit ansteigenden Gebläsedrehzahlen ergeben sich durch den Einfluß des Strömungswiderstandes des Siebkastens Ausgleichsgeraden mit größeren Steigungen.

Mit den Ergebnissen aus insgesamt 60

Versuchen wurden die funktionalen Zusammenhänge

$$\Delta p = f(n_G, \dot{m}_K) \qquad (5)$$

$$\Delta p = f(n_G, \dot{m}_{ges}) \qquad (6)$$

untersucht. Um den linearen Einfluß des Massendurchsatzes m und den potenzierten Einfluß der Gebläsedrehzahl n_G zu berücksichtigen wurde als Regressionsfunktion die Gleichung (7) bzw. Gleichung (8) vorgegeben:

$$\Delta p = A + B * n_G{}^C + D * \dot{m}_K \qquad (7)$$

$$\Delta p = A + B * n_G{}^C + D * \dot{m}_{ges} \qquad (8)$$

Mit den aus Gleichung (4) berechneten Parametern ergibt sich:

$$\Delta p_{ber} = -1,017 + 9,21*10^{-5}*n_G{}^{1,547} + 0,447*\dot{m}_{ges}$$

$$r^2 = 0,92$$

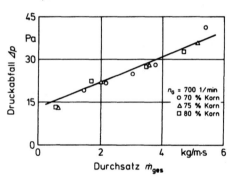

Bild 8 Vergleich der gemessenen Druckwerte für verschiedene Kornanteile mit ermittelter Regressionsfunktion, n_G=konst

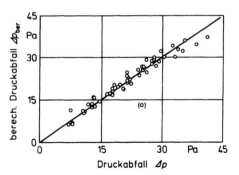

Bild 9 Vergleich der berechneten und gemessenen Druckwerte

In Bild 8 ist die Regressionsgleichung für

eine konstante Gebläsedrehzahl und die entsprechenden Meßpunkte aus Versuchen mit verschiedenen Kornanteilen am Reinigungsgut über dem Gesamtdurchsatz aufgetragen. In Bild 9 sind die aus der Regression berechneten Druckwerte Δp_{ber} über dem tatsächlich gemessenen Druck Δp aufgetragen.

Wird die Regression mit Gleichung (7) durchgeführt, ergibt sich ein geringfügig besseres Bestimmtheitsmaß von $r^2 = 0,937$. Aus der Residualanalyse ist aber zu erkennen, daß der Einfluß des Kornanteils durch diese Regressionsfunktion nicht vollständig erklärt wird.

5 ZUSAMMENFASSUNG

Die Eignung der volumetrischen Durchsatzmessung in Kombination mit einer automatischen Dichtekalibriereinrichtung konnte im Labor nachgewiesen werden. Die Meßeinrichtung ist relativ aufwendig und der Zeitversatz zwischen Festlegung des Durchsatzes (Eintritt des Gutes in den Mähdrescher) bzw. den zu regelnden Dreschund Trennorganen und der Durchsatzmessung ist sehr groß.

Die Messung des statischen Druckes unter dem Obersieb der Reinigungsanlage zur Durchsatzbestimmung wurde im Labor und bei Feldversuchen erfolgreich getestet. Es ist eine Kalibrierung auf die Guteigenschaften und auf die Einstellung der Reinigungsanlage (Sieböffnungen, Stellung der Luftleitbleche) notwendig. Die Zusammensetzung des Reinigungsgutes ist unter anderem vom Mähdrescherdurchsatz abhängig, wodurch die Bestimmung des Korndurchsatzes aus dem Gesamtdurchsatz der Reinigungsanlage etwas fehlerbehaftet wird.

Die Druckmessung unter dem Obersieb ist ideal für eine Teilprozeßregelung der Reinigungsanlage, da nahezu kein Zeitversatz zwischen dem Beladen des Siebes und der Meßwerterfassung besteht. Anhand der Druckwerte kann das Befüllen und Leerlaufen der Reinigung am Feldanfang bzw. Feldende erkannt und die Gebläsedrehzahl zur Reduzierung von Ausblasverlusten geregelt werden.

LITERATUR

Bae, Y.H., G.C. Borgelt, S.W. Searcy, J.K. Schueller, B.A. Stout 1987. Determination of spatially variable yield maps. ASAE-Paper No. 87-1533.

Berner, D., W.H. Grobler, H.D. Kutzbach 1985. Sensoren zur Messung der Kornverluste von Mähdreschern. Grundl. Landtechnik 35:127-132.

Berner, D., W.H. Grobler 1986. Gesteuerte

adaptive Regelung einer Mähdrescher-Reinigungsanlage. Grundl. Landtechnik 36: 73-78.

Böttinger, S. 1988. Regelkonzepte für die Mähdrescher-Reinigungsanlage. VDI/MEG Kolloquium Landtechnik, Heft 6, Mähdrescher: 157-169.

Friesen, O.H., G.C. Zoerb, F.W. Bigsby 1965. Automatic feedrate control of combines. ASAE-paper No. 65-167.

Herbsthofer, F.J. 1971. Combine ground speed and drum speed automatic control. U.S. Patent 3,609,947.

Kutzbach H.D. 1985. Verringerung der Körnerverluste bei Mähdreschern im Hangeinsatz. Landtechnik 40: 278-280.

Kutzbach, H.D. 1988. Entwicklungstendenzen bei der Regel- und Informationstechnik an Mähdreschern. VDI/MEG Kolloquium Landtechnik, Heft 6, Mähdrescher: 121-135.

Leva, M. 1959. Fluidization. New York, Toronto, London: McGraw-Hill.

Lunty, J.W., J.J. Leonhard 1986. Instrumentation to measure straw walker grain loss. CSAE/SCGR-Paper No. 86-204.

McGechan, M.B., C.A. Glasbey 1982. The benefits of different speed control systems for combine harvesters. J.agric.Engng.Res. 27: 537-552.

Rust, M. 1987. Untersuchungen an einem volumetrischen Durchsatzmeßgerät. Unveröffentlichte Studienarbeit Universität Stuttgart.

Schueller, J.K., M.P. Mailander, G.W. Krutz 1985. Combine feedrate sensors. Transactions of the ASAE 28:2-5.

Segler G., T. Freye 1978. Entwicklung einer Meß- und Kontrolleinrichtung für den Gutdurchsatz der Reinigungsanlage im Mähdrescher. Grundl. Landtechnik 28:148-151.

Sörlin, S. 1985. Teknik för mängdbestämning. Swedish Institut of Agricultural Engineering, Meddelande nr. 407.

Stahl, T., T. Freye, H.D. Kutzbach 1981. Automatic control of cleaning fan speed to reduce sidehill losses. ASAE-Paper No. 81-1617.

Wang, G., G.C. Zoerb, L.G. Hill 1988. A combine separation loss monitor. Transactions of the ASAE 30:692-694.

Land and Water Use, Dodd & Grace (eds), © 1989 Balkema, Rotterdam. ISBN 90 6191 980 0

Arbeitsqualität von Axial- und Tangentialdreschwerken

P. Wacker

Institut für Agrartechnik, Universität Hohenheim, Stuttgart, Bundesrepublik Deutschland

KURZFASSUNG: Für die Ernte von Körnerfrüchten werden Mähdrescher mit konventionellem Tangentialdreschwerk und Hordenschüttler und seit einigen Jahren solche mit Axialdreschwerk eingesetzt, bei denen die Entkörnung und die Kornabscheidung in einem Arbeitsorgan zusammengefaßt sind. Während über das Betriebsverhalten von Tangentialdreschwerken und Hordenschüttlern zahlreiche Informationen vorliegen, wird ein ähnlicher Informationsstand für Axialdreschwerke erst langsam erreicht. Es wird über Versuche berichtet, bei denen diese beiden Dresch- und Abscheidesysteme unter vergleichbaren Bedingungen im Labor mit erntefrischem und eingelagertem Gut untersucht wurden. Dabei wurde unter anderem der Einfluß des Gutdurchsatzes auf Kornabscheidung, Körnerverluste, NKB-Abscheidung und Leistungsbedarf ermittelt.

ABSTRACT: Combines with conventional tangential threshing cylinders and straw walkers and for several years combines with axial rotary threshers are used for grain harvesting. In the latter, threshing and final grain-straw separation are performed in one aggregate. While there is numerous information on the operational characteristics of tangential threshing cylinders and straw walkers, a similar level of information on rotary threshers is only slowly attained. Tests, in which both threshing systems were investigated under comparable laboratory conditions with fresh and stored harvest material, are reported about here. Through these investigations the influence of the feedrate level on grain separation, grain losses, MOG-separation and power requirements was ascertained.

RESUMEE: Aujourd'hui la moissoneuse-batteuse est la machine clef pour la récolte des grains. Depuis quelques années on ne trouve pas soulement des moissoneuses-batteuses conventionelles avec batteur transversel et secoueurs, mais aussi des machines axiales, qui reunient le battage et la séparation dans une seule roteur. Il y a informations sur la performance des moissoneuses-batteuses conventionelles, mais on ne les trouve guère sur les machines axiales. L'article fait un rapport sur les essaies comparatifs avec une machine conventionelle et une machine axiale qui étaient examinées dans le laboratoire avec materiel frais et stocké. On a éxaminé la séparation des grains, les pertes des grains, la séparation de la paille et la puissance necessaire du batteur.

1 EINLEITUNG

Die heute angebotenen Mähdrescher mit Tangentialdreschwerk und Hordenschüttler sind komplexe Erntemaschinen, die einen hohen Entwicklungsstand erreicht haben. Sie führen die Getreideernte mit einer vor 30 Jahren nicht erwarteten hohen Arbeitsqualität und hohen Durchsatzleistungen durch. Leistungssteigerungen wurden in erster Linie durch Vergrößerung und Optimierung der einzelnen Bauelemente erzielt. Da inzwischen vor allem durch den Hordenschüttler die zulässigen Grenzen der äußeren Abmessungen erreicht sind, werden von verschiedenen Her-

stellern Mähdrescher mit Axialdreschwerken angeboten (Wacker 1985, Kutzbach 1983). Bei diesen Mähdreschern wird die Funktion von Dreschwerk und Schüttler von einem Arbeitsorgan übernommen. Das Gut wird in axialer Richtung zwischen einem Rotor und einem feststehenden Mantel geführt und durchläuft mehrmals schraubenförmig einen Dresch- und einen Trennbereich.

Über die Vor- und Nachteile der Axialdreschwerke gegenüber den Tangentialdreschwerken mit Hordenschüttlern wird viel diskutiert. Während aber über das Betriebsverhal-

ten von Tangentialdreschwerken und Hordenschüttlern zahlreiche Untersuchungen vorliegen, muß ein ähnlicher Informationsstand für
Axialdreschwerke erst aufgebaut werden.

Untersuchungen, vor allem im Feldeinsatz
zeigten, daß Axialmähdrescher im Vergleich
zum Tangentialmähdrescher ein günstigeres
Verlustverhalten bei geringerem Körnerbruch
aufweisen. Als Nachteile wurden stärkere
Strohzerkleinerung und als deren Folge eine
höhere Belastung der Reinigung sowie ein er-
höhter Leistungsbedarf festgestellt, (Spiess
1981, Rumpler 1984).

Um die beiden Dresch- und Abscheidesysteme
umfassend vergleichen zu können, werden beide Systeme in Hohenheim unter gleichen Bedingungen mit unterschiedlichen Gutarten untersucht. Die Versuche werden im Labor mit
frischem und eingelagertem Gut durchgeführt.
Bei frischem Gut werden beide Systeme am
gleichen Tag untersucht. Dabei wird unter
anderem der Einfluß des Gutdurchsatzes auf
die folgenden Bewertungskriterien ermittelt:
Kornabscheidung, Ausdrusch- und Trennverluste, Kornbeschädigung, Abscheidung an Nichtkornbestandteilen (NKB), Strohzerkleinerung
und Leistungsbedarf. In diesem Beitrag wird
an einem Beispiel über den Einfluß des NKB-
Durchsatzes auf das Betriebsverhalten der
beiden Dresch- und Abscheidesysteme bei erntefrischem und eingelagertem Weizen berichtet.

2 VERSUCHSAUFBAU UND -DURCHFÜHRUNG

Für die Untersuchungen wurden zwei Versuchsstände aufgebaut, (Wacker 1988). Den Aufbau
des untersuchten Axialdreschwerkes zeigt
Bild 1a. Der Rotor hat eine Länge von 2,82 m
und einen Durchmesser von 0,61 m. Das Gut
wird dem Rotor über einen Schrägförderer
axial zugeführt. Zur Ermittlung der Abscheidung des Dresch- und Trennbereiches wurde
die Reinigungsanlage ausgebaut. Um die Abscheidung beobachten zu können, sind die Sei-
tenteile teilweise durch Plexiglas ersetzt.
Ein konventioneller Mähdrescher mit Tangentialdreschwerk und Hordenschüttler wurde in
ähnlicher Weise für die Untersuchungen umge-
baut (Bild 1b). Er hat eine Dreschtrommel-
breite von 1,03 m und einen Trommeldurchmes-
ser von 0,56 m. Der vierteilige Hordenschüttler hat eine Abscheidefläche von 3,5
m² und wird bei den Untersuchungen von einem Elektromotor angetrieben. Das Leistungs-
vermögen des untersuchten Tangentialdreschwerkes mit Hordenschüttler ist durch diese
Abmessungen ungefähr 30 bis 50% niedriger
als das des untersuchten Axialdreschwerkes
und kann dadurch nicht direkt mit ihm verglichen werden. Deshalb werden die Diagramme
einzeln gezeichnet, die Tendenzen der Kur-
venverläufe ermöglichen aber einen Vergleich

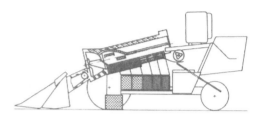

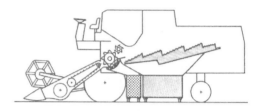

Bild 1   Aufbau der Versuchsstände
         a) Axialdreschwerk
         b) Tangentialdreschwerk
            mit Hordenschüttler

der Arbeitsqualität der beiden Dresch- und
Abscheidesysteme.

Bei den Versuchen wird das Versuchsgut
mit Ähren voran gleichmäßig auf einem 25 m
langen und 2 m breiten Förderband verteilt.
Unterschiedliche Gutdurchsätze werden durch
entsprechende Massenbelegung vorgegeben. Bei
einer Versuchsdauer von 20 Sekunden ist die
Ein- und Auslaufphase im Vergleich zum stationären Betrieb gering. Das durch den
Dresch- und Trennbereich bzw. durch den
Dreschkorb und Hordenschüttler abgeschiede-
ne Gut wird getrennt aufgefangen. Außerdem
werden die Ausdrusch- und Trennverluste, der
Leistungsbedarf der Dreschtrommel oder des
Rotors und die Stroh- und Kornbeschädigung
bestimmt.

3 VERSUCHSERGEBNISSE

Da im wesentlichen die Nichtkornbestandteile
(NKB) das Betriebsverhalten beeinflussen,
wird der NKB-Durchsatz zur Beurteilung verwendet, der bei den dargestellten Versuchen
beim Axialdreschwerk zwischen 1,2 und 18,5
t/h und beim Tangentialdreschwerk mit Hordenschüttler zwischen 1,2 und 12 t/h variiert wurde. Höhere Durchsätze waren beim
Tangentialdreschwerk nicht sinnvoll, da die
Körnerverluste über 5% anstiegen. Beim
Axialdreschwerk begrenzte die übertragbare
Antriebsleistung eine weitere Erhöhung der
Durchsätze.

Die Umfangsgeschwindigkeit des Rotors und
der Dreschtrommel betrug bei den beschriebenen Versuchsreihen jeweils 30,1 m/s. Der

Dreschspalt war beim Axialdreschwerk auf 50 mm im Einlauf und 8 mm im Auslauf des Dreschkorbes, beim Tangentialdreschwerk auf 12 mm im Einlauf und 6 mm im Auslauf eingestellt.

Das erntefrische Versuchsgut (Winterweizen, Sorte Basalt) hatte eine Kornfeuchte von 11 und 15% und eine Feuchte der NKB von 19 und 25%. Das vom selben Feld eingelagerte Versuchsgut hatte bei den Versuchen einen Korn- und NKB-Feuchtegehalt von 11%. Das Korn/NKB-Verhältnis betrug jeweils 1:1,2 bei einer Halmlänge von 0,71 m.

3.1 Kornabscheidung und -verluste

Die Kornabscheidung des Dreschkorbes nimmt beim Tangentialdreschwerk von fast 90% annähernd linear auf 65% ab (Bild 2). Zwischen dem eingelagerten Gut (U_{NKB} = 11%) und dem erntefrischen Gut (U_{NKB} = 19%) zeigen sich nur geringe Unterschiede; bei dem etwas feuchteren Versuchsgut ist die Kornabscheidung durch den Dreschkorb geringfügig höher.

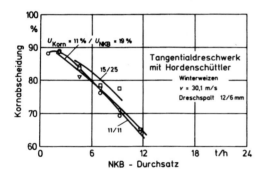

Bild 2   Einfluß des NKB-Durchsatzes auf die Kornabscheidung durch den Dreschkorb bei einem Tangentialdreschwerk

Bei niedrigen NKB-Durchsätzen können die mit dem ersten Schlag der Schlagleisten ausgedroschenen Körner die relativ dünne Strohschicht sehr früh durchdringen und durch den Dreschkorb abgeschieden werden. Mit zunehmendem NKB-Durchsatz wird die Strohbelegung im engen Dreschspalt höher, außerdem nimmt, wie unsere Messungen gezeigt haben, die mittlere Gutgeschwindigkeit im Dreschspalt von ungefähr 6 m/s auf 12 m/s zu, (Schneider 1987). Dadurch wird die durch die Schlagleisten übertragene Schlagenergie gedämpft und sowohl die Verweildauer als auch durch die kleinere Differenzgeschwindigkeit zwischen Gut und Schlagleiste die Schlagfrequenz verringert.

Der Ausdrusch erfolgt deshalb später und die begrenzte Dreschkorblänge ermöglicht nur noch einem geringeren Teil der Körner den Durchtritt durch den Dreschkorb. Außerdem werden die Ähren nicht mehr vollständig ausgedroschen, die Ausdruschverluste nehmen zu (Bild 3)

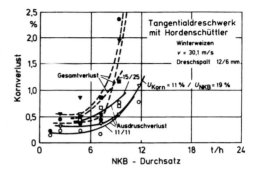

Bild 3   Einfluß des NKB-Durchsatzes auf die Körnerverluste bei einem Tangentialdreschwerk mit Hordenschüttler

Die Abscheidung der nicht durch den Dreschkorb abgeschiedenen 10 bis 35% Körner erfolgt durch den Hordenschüttler, der das Stroh auflockert und fördert (Wacker 1985). Die Fördergeschwindigkeit ist jedoch gering und die Strohbelegung dadurch hoch. Bei Überschreiten der Kapazität des Schüttlers kann die Strohmatte nicht mehr ausreichend aufgelockert werden, die Körnerverluste nehmen ab einem NKB-Durchsatz von 7 t/h überproportional zu (Bild 3). Bei niedrigeren NKB-Durchsätzen wirkt sich bei dem eingelagerten Gut die hohe Strohzerstörung (Kap. 3.3) günstig auf die Verluste aus; durch die dünne Strohmatte können die Körner leicht abgeschieden werden und die Verluste sind niedrig. Beim feuchteren Versuchsgut sind die Ausdruschverluste zwar etwas höher, für den Hordenschüttler ergaben sich aber bei dem untersuchten Gut vor allem bei hohen Durchsätzen bessere Abscheidebedingungen, so daß wegen der geringeren Trennverluste die gesamten Körnerverluste niedrig sind.

Einen anderen Verlauf der Kornabscheidung hat das untersuchte Axialdreschwerk. Es zeigt im Bereich zwischen 4 und 9 t NKB/h ein Maximum mit annähernd 85 und 88% Kornabscheidung im Dreschbereich, die Kornabscheidung im Dreschbereich ist aber auch bei anderen NKB-Durchsätzen größer als 77%. Dieses Maximum verringert sich mit zunehmendem NKB-Feuchtegehalt und verschiebt sich zu höheren Durchsätzen. Beim eingelagerten Versuchsgut wird die Kornabscheidung stärker durch den NKB-Durchsatz beeinflußt (Bild 4).

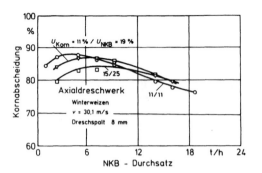

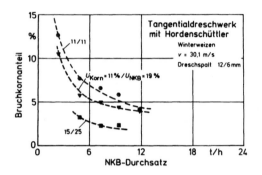

Bild 4 Einfluß des NKB-Durchsatzes auf die Kornabscheidung des Dreschbereiches bei einem Axialdreschwerk

3.2 Kornbeschädigung

Auf den gegensätzlichen Verlauf von Kornverlusten und den visuell bestimmten Kornbeschädigungen beim Tangentialdreschwerk mit Hordenschüttler weist Bild 6 hin. Die Kornbeschädigungen nehmen mit zunehmendem NKB-Durchsatz degressiv ab, da durch die höhere Strohbelegung die Schlagenergie der Schlagleisten vermindert wird und außerdem das Strohpolster die Körner vor Beschädigungen schützt.

Bild 6 Einfluß des NKB-Durchsatzes auf den Gesamtbruchkornanteil bei einem Tangentialdreschwerk mit Hordenschüttler

Bei kleinen Durchsätzen ist der Ringspalt zwischen Rotor und Mantel nicht gefüllt, so daß das Gut den Schlägen der Schlagleiste ausweichen kann und dadurch nicht intensiv genug bearbeitet wird; bei hohen Durchsätzen erschwert die dickere Gutschicht die Kornabscheidung. Außerdem macht das Gut bei kleinen Durchsätzen nur vier Umläufe im Axialdreschwerk, bei höheren Durchsätzen dagegen steigt die Zahl der Umläufe auf fünf bis sechs an (Wacker 1984).

Im Axialdreschwerk wird auch im Trennbereich die Kornabscheidung durch hohe Fliehkräfte unterstützt und durch die mechanische Bearbeitung des Gutes durch die Trennleisten ein weiterer Ausdrusch bewirkt. Dadurch bleibt auch bei sehr hohen Durchsätzen die Funktion der Trenneinrichtung erhalten. Wie Bild 5 zeigt, nehmen die Körnerverluste nur gering mit dem NKB-Durchsatz zu. Der Anstieg der Verluste bei geringen Durchsätzen wird durch die zu geringe Füllung des Ringspaltes und damit zu geringer Bearbeitung des Gutes durch die Schlag- und Trennleistung verursacht. Die etwas höheren Gesamtverluste bei dem feuchteren Versuchsgut sind vor allem auf den Anstieg der Ausdruschverluste zurückzuführen.

Der Gesamtbruchkornteil ist mit 6-7% bei eingelagertem bzw. 2 - 3% bei feuchtem Versuchsgut im mittleren Durchsatzbereich hoch. Ursache hierfür ist vor allem die vorher besprochene geringe Kornabscheidung des Dreschkorbes bei zunehmendem NKB-Durchsatz. Dadurch ist der Anteil der Körner, der den engen Dreschspalt vollständig passieren muß und erst vom Hordenschüttler abgeschieden wird, relativ groß. Durch diese intensive Bearbeitung im Dreschspalt werden vor allem vom Hordenschüttler relativ viele beschädigte Körner abgeschieden. Mit abnehmendem Feuchtegehalt der Nichtkornbestandteile und damit höherer NKB-Abscheidung durch den Dreschkorb und größerer Strohzerstörung ist die Wirkung des Strohpolsters geringer und die Kornbeschädigungen nehmen zu.

Beim Axialdreschwerk nimmt der Bruchkornteil mit steigendem NKB-Durchsatz wesentlich stärker ab. Der Feuchtegehalt der Nichtkornbestandteile wirkt sich geringer aus als beim Tangentialdreschwerk mit Hordenschüttler (Bild 7).

Im mittleren Durchsatzbereich ist der Bruchkornanteil bei den untersuchten NKB-Feuchten um über 50% niedriger als beim Tangentialdreschwerk mit Hordenschüttler. Ein Grund hierfür ist die hohe Kornabscheidung im Dreschbereich. Dadurch werden diese Körner einer weiteren Schlagbeanspruchung entzogen.

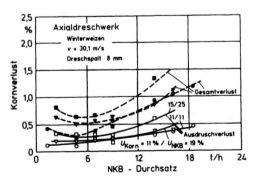

Bild 5 Einfluß des NKB-Durchsatzes auf die Kornverluste bei einem Axialdreschwerk

Außerdem wirkt sich die gegenüber dem Horden-schüttler intensivere Bearbeitung des Gutes im Trennbereich nicht so stark aus, da im Trennbereich der Spalt zwischen Rotor und Trennkorb 5 cm beträgt.

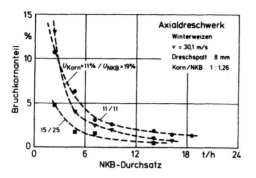

Bild 7   Einfluß des NKB-Durchsatzes auf den Gesamtbruchkornanteil bei einem Axialdresch-werk

3.3 NKB-Abscheidung und Strohzerkleinerung

Die NKB-Abscheidung, d.h. die Belastung der nachgeschalteten Reinigungsanlage durch die abgeschiedenen Nichtkornbestandteile wird ebenfalls durch den NKB-Durchsatz beeinflußt. Sie ist auch ein Kennzeichen für die gesamte Strohzerstörung.

Beim Tangentialdreschwerk mit Hordenschütt-ler nahm der auf die gesamte Masse der Nicht-kornbestandteile bezogene Anteil der durch den Dreschkorb und vom Schüttler abgeschiede-nen Masse von Spreu und Stroh mit zunehmen-dem NKB-Durchsatz annähernd linear ab (Bild 8). Es wurde jeweils ungefähr der gleiche Anteil an Nichtkornbestandteilen vom Dreschkorb und vom Schüttler abgeschieden. Der Einfluß des Feuchtegehalts der Nichtkorn·bestandteile ist deutlich erkennbar.

Durch den langen Weg des Gutes zwischen Rotor und Korb wird das Stroh beim Axialdreschwerk stärker zerkleinert und vor allem beim trok-keneren Versuchsgut wesentlich mehr Nicht-kornbestandteile (5-10%) als vom Tangential-dreschwerk mit Hordenschüttler abgeschieden (Bild 9). Die NKB-Abscheidung im Trennbe-reich ist etwas höher als im Dreschbereich. Der Einfluß des Durchsatzes auf die NKB-Ab-scheidung ist beim Axialdreschwerk mit Aus-nahme beim eingelagerten Versuchsgut aller-dings verhältnismäßig gering.

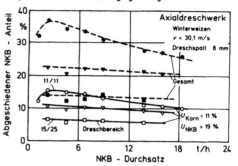

Bild 9   Einfluß des NKB-Durchsatzes auf die NKB-Abscheidung bei einem Axialdreschwerk

3.4 Leistungsbedarf

In Bild 10 ist der auf den Durchsatz der Nichtkornbestandteile in t/h bezogene spezi-fische Leistungsbedarf des Tangentialdresch-werks aufgetragen. In diesen Angaben ist nur der Leistungsbedarf der Dreschtrommel, nicht aber der Leistungsbedarf der Strohleittrom-mel, des Hordenschüttlers und des Leerlaufs enthalten. Wie dieses Bild zeigt, steigt der spezifische Leistungsbedarf mit zunehmendem NKB-Durchsatz annähernd linear von 1,2 kWh pro t_{NKB} auf 1,8 kWh pro t_{NKB} an. Der Ener-giebedarf beim feuchteren Versuchsgut liegt um ungefähr 20% niedriger.

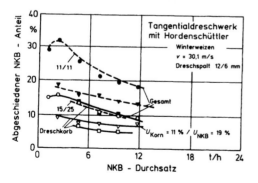

Bild 8   Einfluß des NKB-Durchsatzes auf die NKB-Abscheidung bei einem Tangentialdresch-werk mit Hordenschüttler

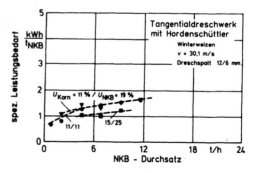

Bild 10 Einfluß des NKB-Durchsatzes auf den Leistungsbedarf bei einem Tangentialdresch-werk mit Hordenschüttler

Fast doppelt so hoch ist der spezifische Leistungsbedarf (ohne Leistungsbedarf Leerlauf) des Axialdreschwerkes (Bild 11). Er nimmt linear von 1,8 kWh pro t_{NKB} mit zunehmendem NKB-Durchsatz auf 3,5 kWh pro t_{NKB} zu. Durch die stärkere Strohzerkleinerung und -abscheidung ist beim eingelagerten Versuchsgut der Einfluß des NKB-Durchsatzes geringer.

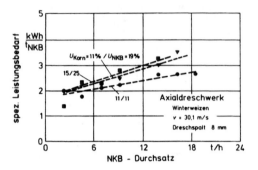

Bild 11 Einfluß des NKB-Durchsatzes auf den Leistungsbedarf bei einem Axialdreschwerk

4 ZUSAMMENFASSUNG

Diese Vergleichsuntersuchungen unter identischen Bedingungen zeigen Vor- und Nachteile beider Dresch- und Trennsysteme auf.
- Hinsichtlich des Durchsatzes hatte das untersuchte Axialdreschwerk ein Optimum für die Kornabscheidung im Bereich zwischen 4 und 9 t/h NKB-Durchsatz, während die Kornabscheidung des Tangentialdreschwerkes stark abnahm.
- Im Gegensatz zum Tangentialdreschwerk zeigte das Axialdreschwerk mit zunehmendem Durchsatz einen wesentlich flacheren Anstieg der Körnerverluste. Bei geringen Durchsätzen mit kleinen Strohmengen wird das Gut beim Axialdreschwerk nicht intensiv genug bearbeitet und die Verluste bleiben relativ hoch bzw. steigen wieder an.
- Der Anteil der beschädigten Körner ist beim Axialdreschwerk nur ungefähr halb so groß wie beim Tangentialdreschwerk.
- Durch die intensivere Beanspruchung des Strohs beim Axialdreschwerk ist die Strohbeschädigung höher und die Belastung der Reinigungsanlage durch die abgeschiedenen Nichtkornbestandteile größer.
- Der spezifische Leistungsbedarf ist beim Tangentialdreschwerk deutlich niedriger als beim Axialdreschwerk.
- Das Axialdreschwerk reagiert -zumindest im untersuchten Feuchtigkeitsbereich- empfindlicher auf die Gutfeuchte als das Tangentialdreschwerk mit Hordenschüttler.

5 LITERATUR

Kutzbach, H.D. 1983. Dresch- und Trennsysteme neuer Mähdrescher. Landtechnik 38, H. 6, S. 226-230.

Rumpler, J. und P. Rabe. 1984. Axialflußmähdrescher - Betrachtungen zum Leistungsstand. Agrartechnik, Berlin 34, H. 6, S. 274-277.

Schneider, G. 1987. Vergleichsuntersuchungen an einem Tangential- und einem Axialdreschwerk. Unveröffentlichte Diplomarbeit, Universität Hohenheim.

Spiess, E. 1981. Axialfluß-Mähdrescher IH 1460 im Vergleich. dlz 32, S. 846-851.

Wacker, P. 1984. Laboruntersuchungen an einem Axialdreschstand. Unveröffentlichter Institutsbericht des Instituts für Agrartechnik, Universität Hohenheim.

Wacker, P. 1985. Untersuchungen zum Dresch- und Trennvorgang von Getreide in einem Axialdreschwerk. Dissertation Universität Hohenheim.

Wacker, P. 1985. Einflüsse auf die Dreschleistung von Mähdreschern. Landtechnik 40, H. 6, S. 273-277.

Wacker, P. 1988. Vergleich von Axial- und Tangentialsystemen in Getreide. Landtechnik 43, H. 6, S. 264-266.

Land and Water Use, Dodd & Grace (eds), © 1989 Balkema, Rotterdam. ISBN 90 6191 980 0

Development of power and pedal operated peanut shellers

K.Gore
Jain Irrigation Systems Ltd, Jalgaon, India

C.P.Gupta & G.Singh
Asian Institute of Technology, Bangkok, Thailand

ABSTRACT: Newly developed power and pedal shellers have three units (sheller, blower and grader) combined together unlike others in present use. Breakage of kernels has been reduced considerably by using shoe rubber sole pads with cushioning material on rasp bars. President Jimmy Carter, USA, and H.M. The King of Thailand have visited and deemed the sheller performance as useful for marginal farmers in developing countries. Pedal sheller is especially useful in remote rural areas where electricity is not available and diesel engines or electric motors are cost prohibitive.

1 INTRODUCTION

Peanut (Arachis Hypogaea L.), known by many names worldwide like; groundnut, earthnut, monkeynut is a soil enriching nitrogen fixing legume adapted to diversity of soils and temperature zones. It is one of the most useful plants in the world; and grown commercially in both hemispheres for its fruits or pods. Countries like India, Nigeria, Senegal, Sudan, China and USA are the main producers, however, India is the largest producer. More than half of this production and over 60% of the export value in global trade, comes from the developing countries. On the worldwide average 60% of the production is marketed for direct consumption and oil production; 32% used as food locally, and 8% is retained as seed.

Peanuts are aptly described as nature's masterpiece of food values; as possibly no other crop in the world has as many combined advantages as a food ingredient as does peanut. Chief among these are pleasing aroma and flavours, crunchy texture; high energy value; high in proteins, minerals and nacin; capable of being stored or shipped to any point of the world and suitable for being made into hundreds of products for serving any time of the day at all occasions.

1.1 Need for peanut sheller

Peanuts have traditionally been an important food legume crop which provide and have the potential of providing more cash income and food sources for small scale farmers and urban poors in the developing countries. They are probably the most important oil-bearing seed in the world ; and have rapidly become valuable source of plant protein. So for all these processing operations and also for direct consumption; shelling is an important basic operation to be carried out.

The price of kernel is approximately two to three times higher than the dry pods. If the farmer shells the peanuts and sells the kernels, he gets higher price leading to a higher income. This higher income indirectly leads to intensive cropping. Besides this the cost of transportation of peanuts to city areas and finished products back to villages also reduces, followed by lower retail price

The traditional methods of shelling i.e. seed separation by hand are uneconomical, time consuming and labourious. Farmers are now a days aware of increasing cost and seasonal shortage of labour and timeliness in operations in the double cropping. Though many kinds of manual as well power operated shellers are in the use; very little attention is being paid to the efficiency of the machine and the overall performance which is due to lack of technical knowledge, individual preferences, prevailing socio-economic conditions, land resources etc. which play an important role in the purchase of peanut sheller.

Again in presently available shellers there are only shellers or winnowers; while graders are not being used at all. Good kernels, broken kernels and immature kernels are separated out by hands which is time consuming and labourious. On the other hand, considering the high unemployment or underemployment in many developing countries; pedal power (operated by human legs) has a tremendous potential to operate the small agricultural machines in remote rural areas where electricity is not available and diesel engines are cost prohibitive.

2 TYPES OF SHELLERS

The groundnut shellers can be classified as (1) manually operated and (2) power operated depending on their power source. Based on shelling action, the shellers can be divided into (1) reciprocating type and (2) continuous or rotary type.

2.1 Manually operated groundnut shellers

A semirotary type groundnut sheller was originally developed at the Tropical Products Institute and is now produced in many countries. It mainly consists of a hoper, a wire mesh, shelling bar and reciprocating arm. The hopper is a semi-cylindrical trough. The wire mesh is attached to the bottom of the hopper. The shelliing bar is attached to the lower end of the reciprocating arm. It can be activated from the centre within the hopper.

A hand operated groundnut decorticator was originally developed by Dandekar Brothers, Maharashtra (India). It consists of a shelling cylinder which is rotated by hand. It has been modified by the Dept. of Agril. Engineering, Khon Kaen University, Thailand.

UPLB peanut sheller was developed at the University of Philippines. The shelling unit consists of a stationary hopper with built in spring loaded shelling bar and underneath it is a reciprocating slotted screen. It is fitted with a bi-cycle chain drive and a blower to separate the shells from the kernels. Its capacity is about 40-80 kg/hr.

A foot operated groundnut sheller was developed by Hindsons Private Ltd., Punjab (India). It is fitted with a fly wheel for easier operation and a blower to separate shells from kernels. It is operated by one person and has a capacity of 25 kg/hr.

2.2 Power operated groundnut shellers

TNAU groundnut decorticator was developed by Tamilnadu Agricultural University, Coimbatore, India. The machine consists of a hopper, double crank lever mechanism, oscillating unit and a blower assembly, all fitted to a frame. In the oscillating unit a number of cast iron pegs are fitted. The groundnut pods are shelled between the oscillating unit and a perforated concave sieve. The husk is blown away by a blower and clean kernels are collected through a spout at the bottom. The clearance between the oscillating unit and the concave is adjustable. It has capacity of 400 kg/h of pods or 260 kg/h of kernels. The percentage of breakage, threshing efficiency and cleaning efficiency were 4.5%, 95% and 98% respectively.

An automatic groundnut decorticting machine was manufactured by Harrap, Williamson Ltd., Salford, U.K. It consists of a hopper, beating chamber and cleaning fan. A ribbed feed roller feeds the pods into the beater chamber below. From the feed roller, the pods fall into the beater chamber where these are struck by rotating flexible beaters. The broken shells and kernels are forced out through the perforated cylindrical steel shelling screen. After leaving the beater chamber, the kernels and broken shells fall into a duct which has a wire mesh delivery chute at its lower end.

A fan mounted below the mesh blows air up to the duct. The draught is sufficient to blow the shells upward and out of the shell outlet spout, but will not lift the kernels which continue to fall into the wire mesh delivery chute.

BPI groundnut sheller with cleaner was developed at the Bureau of Plant Industry, Metro Manila, Philippines. It is powered by a 2.2 kw electric motor and has capacity of 30 kg/h. It has an assembly of three oscillating screens. The groundnuts are fed through a hopper and pass between two rollers with just enough clearance to crack the pods. The shelled and unshelled groundnuts move down the oscillating inclined trough, passing below a suction duct. The kernels radially drop into a secondary trough and then into the container. The shells of the groundnut are sucked upwards by the suction fan and blown into the duct for discharge. This machine could perform shelling without causing damage to the nut.

Kittichai (1984) developed a power operated groundnut sheller at AIT, Bangkok. The shelling cylinder consists of 12 sets of 10x20 cm rubber tyre shoes which are 30 degree apart. The diameter and width of shelling bar are 54 cm and 22 cm respectively. For the optimum performance of shelling, the blower settings were 1000 rpm blower shaft speed and 30 degree blower chute angle. The best performance of the sheller was achieved at 20 mm clearance and shelling bar speed of 180 rpm. At these settings, shelling cpacity, shelling efficiency and percentage breakage were 210.5 kg kernels/hour, 98.03 and 5.3% respectively. The power consumption of the sheller was about 1.0 to 1.1 kw.

3 OBJECTIVES

The research project was taken up with the following objectives :

1. To modify the components in the peanut sheller developed by kittichai (1984) for better performance.

2. To design and develop a power and manually operated peanut sheller with mechanism to separate good, broken and immature kernels and unshelled pods.

3. To make comparative cost analysis of the power operated sheller with a manual sheller.

4 MATERIALS AND METHODS

The average size of peanut pods used in the shelling test was 27 mm in length and 12 mm in dia. The distribution of large, medium and small size of the peanuts were 51, 37, and 22 percent of the total pods respectively.

4.1 Modification of concave

The wire mesh concave used in the peanut sheller developed by Kittichai (1984) was rough and bolted at 32 places along the frame to keep it in proper alignment. Therefore a new sturdy concave was made with radial bar spacing of 9 mm.

4.2 Development of sieve assembly

Screens with slots along the direction of oscillations were tried. Slot size were 85x9 mm and 85x5 mm. The G.I.sheet used was of 1.5 mm thickness. Blockage was found to be reduced as the oscillating motion pushed the seeds forward.

But still the seed separation or grading was not efficient due to much load on the screens. To reduce the load, a set of three trays of size 62x30x3.7 cm were made. The screens were then fitted into the trays. The screens are replaceable to suit various varieties of groundnut. The first screen scalps the unshelled pods and and other impurities more than 9 mm to the left outlet. The second screen grades good kernels from 9 to 5 mm in dia.from rest of the mass to the front outlet while third screen screens broken and immature kernels having less than 5 mm dia. from the remaining mass. Trials were taken for optimum range of speed of oscillation and accordingly the belt size and pulley sizes were determined.

The width of blower chute was more in the original design, which resulted in loss of air volume and need for higher rpm of the blower speed, requiring more power for the same cleaning. The width of the blower chute was reduced and a wire-mesh 10x10 mm was fitted inside to control the blower loss.

4.3 Shelling cylinder rasp bars and hopper

Air cushioning pad was placed between

shoe rubber sole and rasp bar plate. The curvature to the rasp bars was made uniform to facilitate better shelling action. This type of rasp bar worked satisfactorily as shoe rubber sole has adequate elasticity and rough surface for breaking the pods without slipping.

A new hopper with 5 kg capacity and 57 degree inclination with the horizontal was made.

4.4 Screw mechanism clearance adjustment

For proper and easy adjustment of clearance, a screw mechanism was developed. By screwing or unscrewing, one can vary the clearance between concave and shelling cylinder rasp bars.

4.5 Phase II - Pedal operated sheller

Pedal power, a power generated by human legs is of sufficient higher magnitude and can be sustained over a longer period of time than that of by human hands. Bicycle pedal or chain sprocket mechanism is the best and efficient way to do this.

The pedal-chain sprocket mechanism run on anti friction bearings causes very low frictional losses. Pedal mechanism would be a very suitable standby system where electrical energy is frequently unobtainable in remote rural areas and electrical motors or diesel engines are most expensive. While pedalling the hands of operator are free to perform other tasks at the same time.

Fig. 1 Shelling cylinder (rasp bars) of the pedal sheller

Shelling cylinder (Fig. 1) is the heart of the sheller. To be easily operated by two operators (as the three units ; shelling, winnowing and grading are combined) the shelling cylinder diameter was kept 300 mm and length more than 500 mm; following the general rule of 1:1.5 ratio of diameter to length.

Concave is semicircular grate type design. Experience gained in power operated sheller was used in developing the manual sheller. Two persons pedalling seperately the ordinary bicycle pedal chain sprocket assembly at 45 rpm, drive the shelling cylinder shaft at about 135-140 rpm; blower at 550 rpm and grading assembly at 290-300 rpm.

5 EXPERIMENTAL TECHNIQUE

Performance trials for power sheller were taken after incorporating the above mentioned modifications. In the shelling mechanism test 4 kg of groundnuts were filled into the hopper in each test in three continuous runs. The feed rate was varied by the flap arrangement inside. The shelling bar speed was adjusted by a variable speed motor and clearance by a screw mechanism. Power and time for shelling were recorded. For these tests, the product was analysed for shelling efficiency, percentage of breakage and shelling capacity. The moisture content of the pods was varied by natural drying in the sun and shade.

In cleaning mechanism test, 4 kg of groundnuts were fed in three continuous runs and sample was taken after finishing each run. Blower speed and blower chute inclination were set and power consumed was recorded in every test.

For pedal operated sheller, 6 kg of peanuts were filled into the hopper for each test in three continuous runs. The feed-rate was varied by the flap arrangement inside the hopper and uniform feeding was ensured by feed role mechanism. Shelling cylinder speed was varied approximately by trial and error method as while pedalling there might be little variation. Approximately three pedalling speeds of 40, 45, 50 rpm were selected for evaluating operator's performance of pedalling. For these tests the product was analysed for shelling efficiency, percent breakage and shelling capacity with power consumption. While pedalling the operator themselves can feed 10 kg capacity hopper, provided; gunny bagged peanuts are within the reach of their hands. The handles, hopper controller, concave flapper and other arrangements are also within their reach, while they are pedalling.

The shelling cylinder speed (s), cylinder concave clearance (C), kernel moisture content (M) and feed rate (F) were taken as independent parameters for the study (Table 1 and 2).

Table 1 Levels of independent parameters : power operated sheller

Sr.No.	Shelling cylinder speed (rpm)	Cylinder concave clearance (mm)	Kernel moisture content (%)	Feed rate (kg/h)
1	S1 = 160	C1 = 14	M1 = 7	F1 = 340
2	S2 = 180	C2 = 16	M2 = 13	F2 = 400
3	S3 = 200	C3 = 18	M3 = 20	F3 = 460

Table 2 Levels of independent parameters : pedal operated sheller.

Sr.No.	Shelling cylinder speed (rpm)	Cylinder concave clearance (mm)	Kernel moisture content (%)	Feed rate (kg/h)
1	S1 = 120	C1 = 14	-	F1 = 160
2	S2 = 135	C2 = 16	-	F2 = 180
3	S3 = 150	C3 = 18	-	F3 = 200
4	-	C4 = 20	-	-

Percent shelling efficiency, percent breakage of the kernels, capacity and power consumption were taken as dependent parameters for the study to find out the optimum values for better performance. For pedal operated sheller; investigations for various moisture contents were not carried out.

5.1 Shelling performance index (SPI)

This measure is used to compare the relative shelling performance of various shellers under identical conditions or to determine the overall performance of a particular sheller under various operating conditions.

$$SPI = \frac{Shl.efi. \times cap. \times \left(1 - \dfrac{Per.breakage}{100} \right)}{Power \ (KW)}$$

6 RESULTS AND DISCUSSIONS

6.1 Power operated sheller

Typical peanut shelling process is well explained by the material flow chart (Fig. 2) for the peanut sheller operation.

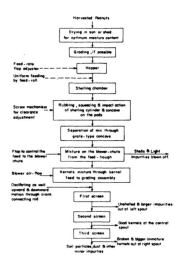

Fig.2 Flow chart for peanut sheller operation.

After modifying the concave and hopper, the shelling efficiency was increased considerably, and the power consumption was also reduced but the breakage was not much affected. Moreover when the rear end clearance was reduced to 1/2 of the front end, the shelling was improved. With radial bar concave, as the shells were not getting crushed again and again, the amount of shelled pods in unit time was more for unit mass of pods. The power consumption and time of shelling was less.

It was observed that for the 180 rpm of the shelling cylinder, 280 rpm of sieve assembly shaft was satisfactory for separation and grading. The proper angle of inclination for top, middle and bottom screens were found to be 8 degree, 5 degree, 13 degrees respectively. The third screen has round holes of 5 mm dia through which small broken hulls, soil particles and very immature kernels could pass easily.

The impact given to the pods on coming in contact with new rasp bars, made up of shoe rubber sole with cushioning material, was less. The cushioning material takes care of uneven layering of pods while crushing; which enabled to shell at lower clearance, resulting in higher capacity and higher feed rate. In general breakage was less as compared to rasp bars made of rubber tyre material used in previous model. Maximum shelling efficiency was found at 200 rpm at 14 mm clearance.

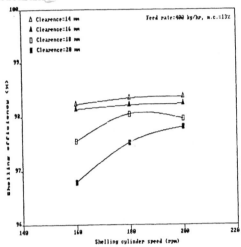

Fig. 3 Shelling efficiency variations at different concave clearances

Fig. 3 Shows that as the clearance increases, shelling efficiency decreases. Similarly shelling efficiency is significantly affected by feed rate and moisture content. Shelling efficiency decreased in general with the increase in moisture content. With increasing moisture content, shelling becomes difficult because much of the energy received is used for elastic deformation of the pods. So there is tendency for less breakage of pods with increased moisture content.

Fig. 4 shows that as shelling speed increases, breakage increases for the same clearance, feed rate and moisture content. The increase is due to higher impact force imparted to the pods during shelling process at higher speed.

With increased clearance, percent breakage is reduced at all speeds. In general higher feed rate resulted in more breakage, at all clearances and shelling bar speeds and moisture contents. The higher breakage may be due to clogging of the material in the cylinder concave assembly at higher feed rates and more bruising of kernels by rubbing between the layers of unshelled and shelled pods in the concave.

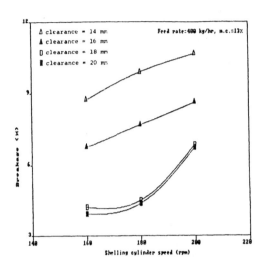

Fig. 4 Effect of shelling speeed on kernel breakage

Shelling capacity is, in general, increased with increasing feed rate at the same clearance and shelling speed. It revealed that power consumption has increasing trend for increase in feed rate at all combination of other parameters. At 180 rpm speed and feed rate of 400 kg/hr for concave clearance 18 mm, the power consumption was found to be 0.75 kwh.

6.2 Pedal operated sheller

The various results for the shelling performance were found in accordance with the results obtained in power sheller. The shelling efficiency was found to be increased with the increased speed of pedalling for all combinations of feed rates and clearances.

Overall efficiency was around 98% and with graded pods it was found to be better than this. The proper concave clearance (2:1 for front to rear) and the position of shelling cylinder with its shafts placed below the main frame angle played an important role; as about 3/4 of its effective circumference was supposed to be used for actual shelling.

Breakage was found to be less than 3%. This was probably due to more or less proper impact of the rasp bars in the shelling chamber and proper concave design. Only the kernels/pods shelled by the direct impact with rasp bar usually get damaged but those by indirect impact were slightly or not at all damaged.

In the work performance test; it revealed that from the operators' comfort point of view, 130-140 rpm (40-46 rpm of pedalling) of shelling cylinder with 180 kg (pod)/hr feed rate was desirable for the operatiion. The capacity was found to be 66 kg; 80 kg (kernel)/hr and 60 kg(kernels)/hr; for the selected three ranges of speeds.

Power consumption measurements for comfortable operation obtained through variable speed motor and gear reduction unit, were about 320 watts. With pedalling speed of 40-45 rpm at no load conditiion, grading assembly power consumption was about 30 watts, sheller 30 watts and blower to 80 watts.

7 COMPARATIVE COST ANALYSIS

The cost of shelling using power operated sheller was compared with that of manual sheller and winnower, pedal sheller and power sheller with power tiller (with electric motor). The shelling capacity of both shellers, 280 kg (kernel)/hr for power operated sheller and 40 kg (kernel)/hr for manual sheller and winnower, and 80 kg (seed)/hr for the pedal sheller were used in the calculations of operation cost. The manual sheller was assumed to be operated in conjunctiion with a winnower which had cleaning capacity of about 280kg (kernel)/ hr.

Fig. 5 General view of AIT developed power and pedal shellers

General view of the power and pedal shellers is shown in Fig. 5. For the manual sheller and winnower, the labour requirement were one and two men respectively. The labour requirement for the power operated sheller was two men, one for feeding peanuts into hopper and another for the kernel output. The pedal operated sheller was operated and fed by two persons.

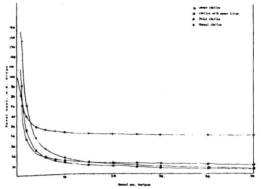

Fig.6 Relationship between total costs and annual use of the shellers

Fig. 6 shows the relationship between total cost of shelling per ton and annual use in tons per year. It is evident that the manual sheller can be replaced by the power sheller when the annual use is more than 3 tons. The operation cost, of shelling using power sheller was approximately half of the manual sheller at the annual use of 13 tons and this cost would be reduced if the annual use is more.

The survey of the farmers in Thailand indicated that average planted area per household was 0.48-0.64 ha, at the average yield of 1200 kg/ha. The minimum annual planted area for the ownership of power and pedal sheller was found to be 3.68 ha and 1.0 ha respectively. Farmer can earn extra income by custom-hiring. In other sense co-operative ownership is to be encouraged.

8 CONCLUSIONS

1. The performance of the sheller was improved by using the grate type round bar concave instead of wire mesh concave.

2. The concave front and rear end clearances kept in the ratio of 2:1 had considerable effect for better shelling.

3. The slotted screen assembly was found to be working well at 280 rpm, when shelling cylinder shaft speed was 180 rpm. Kernels were graded acording to whole, broken and immature kernels.

4. The rubber shoe sole pad with cushioning material pasted on the rasp bar gave the best result for reducing breakage.

5. From the test it was observed that as the moisture content decreased below 10% or increased more than 15%, the breakage was more. For lower moisture, splitting of kernels was very common and for higher moisture, bruising and hull damage was observed. Hence shelling is recommended to be done at 10-15% moisture content range.

6. The newly designed and developed pedal sheller operated and fed by two persons was found to be working well, with capacity of 80 kg (kernels)/ hr and power consumption of 318 watts at 180 kg (pods)/hr feed rate and pedalling speed of 40-45 rpm.

7. The overall performance of the (three unit sheller, winnower and grader) combined shellers thus presented practical models to be adapted by the small scale farmers. The pedal sheller is very much suited to remote rural areas where electricity is many times not available and electric motor or diesel engines are cost prohibitive.

9 REFERENCES

Chauhan, A.M., Singh, S. and Sandhar, N.S.1977. Mechanical decortication of groundnut. Journal of Agric. Engg. Indian Society of Agricultural Engineers 14(4) : 135-140.

Final Report 1987. Groundnut shellers/ strippers project. Dept.of Agril. Engg., Faculty of Engg., Khon Kaen University, Khon Kaen, Thailand.

Gore, K.L. 1987. Development of power and manually operated peanut sheller. M.Engg. Thesis No. AE-87-48, Asian Institute of Technology Bangkok, Thailand.

Kittichai, T. 1984. Development and testing of a power operated groundnut sheller. M.Engg. Thesis No. AE-84-11, Asian Institute of Technology, Bangkok. Thailand.

Madan, A.K., Singh, S. and Sharma, V.K. 1985. Studies on axial flow threshing system for groundnut. Proc. ISAE, SJC II : II-180.

Singh, B., Varshney, B.P., Ahmed, N. and Tiwari, G. 1985. Development and testing of groundnut decortiicator. J. Agri. Engg. 22(3):15-21.

Singh, G. and Thongsawatwong, P. 1983. Evaluation and modification of two peanut shellers. AMA, 9(4): 77-81.

Singh, H. and Verma, S.R. 1972. Development and performance of an experimental groundnut thresher. J. Agri. Engg. 9(1) :50-58.

Singh, M., Chauhan, A.M. and Sharma, V.K. 1985. Studies on seperation and cleaning system of groundnut thresher. Proc. ISAE, SJC 1:II-173-179.

Singh, R., Shrivastava, R. and Singh, B. 1978. Design and testing of groundnut decorticator. AMA, 9(4): 77-81.

Turner, W.K., Suggs, C.W. and Dickns, J.W. 1985. Impact damage to peanuts and its effect on germination, seedling development and milling quality. Trans. of ASAE, 10(2): 248-249.

Land and Water Use, Dodd & Grace (eds), © 1989 Balkema, Rotterdam. ISBN 90 6191 980 0

Structural and operating features of devices for mechanical almond shelling

P.Guarella & A.Pellerano
Institute of Agricultural Engineering, Bari, Italy

ABSTRACT: Due to the great variation in the size of almonds and to the structure and consistency of the almond shell itself, almond shelling is a rather complicated process. This text discusses several of the functional structural features of shelling machines, with particular reference to machines available on the Italian market.
If the wide range of cultivars at present in use were reduced to include only the most efficient from the agronomic and commercial point of view, this would not only simplify the machines themselves but would reduce both shelling costs and later processing requirements.

RÉSUMÉ: L'opération de décorticage des amandes est plutôt compliquée du fait de la taille très diverse des fruits ainsi que de caractéristiques de structure et de dureté de la coque.
On illustre et aborde ici certains aspects fonctionnels et constructifs des différents dispositifs de décorticage, se référant en particulier aux dispositifs adoptés dans les machines disponibles sur le marché italien.
La réorganisation et la réduction des cultivars meilleurs du point de vue des caractéristiques agronomiques et commerciales, permettraient non seulement de simplifier les machines mais de réduire aussi les coûts de décorticage et la plupart des transformations industrielles successives.

ZUSAMMENFASSUNG: Das Mandelschälen ist ziemlich kompliziert auf Grund sowohl verschiedener Fruchtgrössen als auch der Struktur- und Festigkeitseigenschaften der Schale.
Im Zusammenhang werden einige Bau- und Betriebsseiten der verschiedenen Schälvorrichtungen in bezug auf die mit den heutigen Binnenmarktsmaschinen einteiligen Vorrichtungen erläutert und erörtert.
Eine Neuordnung sowie Herabsetzung der zur Verfügung stehenden Cultivars zugunsten der besten Cultivars laut agronomischer kommerzieller Eigenschaften würde zu einer Vereinfachung der Maschinen, einer beträchtlichen Reduktion der folgenden industriellen Bearbeitungen beitragen.

1.FOREWORD

Almond shelling and seed parting thereof – if carried out manually – do call for a large-scale employment of labour generally greater than a man-hour per kilogram; the manual execution of either operation commonly adopted not so long ago would now not be economically feasible on account of its cost which is definitely less competitive than the market price of the finished product.

The problem of mechanical shelling turns out too complex on the basis of too domestic heterogeneous varietal an output, namely the Apulian one, and this concerning not only the size (lenght and thickness) of fruits, but also both the shape and consistency of the shell.

And, what's more, there is no certain steady connection between the sizes of the various almonds, since the latter greatly vary even within the same cultivar, strictly depending on their real provenance, cultivation schemes and, conditions being equal, each year's crop.

2. WORKING UNITS AND OPERATING GUIDELINES.

Machines suitable for mechanical almond hulling (shell breaking) being or today's market, to be regarded mostly as the fruits of farmers' and smallscale handicraft businesses' creativity, resort to different working units made up mainly of:
   a) rollers;
   b) cutting devices;
   c) jaws;
   d) half-chambers.
Each one of the above solutions - referring to different operating guidelines - is presented with pros and cons as regards the seed integrity, the sequence of shell breaking and the sizes of the fragments; the latter - though seemingly negligible - getting noteworthy at the subsequent seed parting and grading stage in particular, the working capacity of ad hoc units as well as the demand for labour (1).

In the countries of the Mediterranean area only the last two ones have been put into practice, which has finally resulted in the setting - up of industrial - type equipment.

Counter - revolving straight conical roller working units complying with US operating systems, have proved to be unsuccessfull to almond shelling nationwide given a kind of shell much thicker and smoother than Californian cultivars.

Machines, themeselves, for shell breaking, equipped with cutting devices - whether with saws or blades, have found no industrial use, owing, above all, to their too low working capacity and ability with respect to their complex structural arrangement necessary to guide is advance and then support the almond during breaking.

Other limiting factors peculiar to machines as above lie in that the distributor element tends to get flooded, and the cutting device is endowed with low durability, which causes frequent work stops to occur.

In principle, one type of shelling machine, which has long been made and patented by a private individual, proves to be interesting, because of the undeniable advantages we would enjoy during the following stage of seed shell parting. Such machine - once the almond is correctly positioned - allows for in half-broken shells, due to a combined action

between a cross cutting and a longitudinal traction exerted by the blades themselves serving as wedges.

The working unit (Fig. n. 1), which is composed of two adjustable opposing blades firmly mounted on a special support, bounds a junction, namely a bottleneck through which each shell fruit is forced to pass upon thrust of a cam-driven special expeller.

During shelling the almond must necessarily rest vertically, thereby handling the blades by keeping its greater cross diameter orthogonal to the cutting edge; a spoon-shaped special mobile positioner will keep the almond blocked against the expeller and will drop it once the blades have been crossed.

As already tried out, shelling may be thought of as accurate when the free area between the blades is by one-third less than the one of the cross size of an almond in the section where cutting occurs. And this is the most delicate feature, or, if we like, the "weak point" of said working unit since all this should be indicative of a preliminary gauging allowing not only for the equatorial size of every single almond, but also the longitudinal one: maximum cross size being equal, a rether elongated, irregular shape cannot allow the positioner to be well

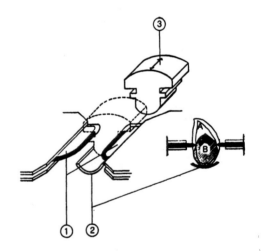

Fig. n. 1 - The fixed-blade shelling device suitable for any long-shaped shell fruit (A) may certainly prove unsucessfull to any round-shaped one (B).
   1 - Fixed blades;
   2 - saddle on which vertically arranged shell fruits are resting;
   3 - mobile expeller pushing the almonds against the blades.

kept and, what's more, it may make the cutting unit unfit, this causing contact with the blades to occur on a plane different from the one laid out during adjustement.

3. - SHELLING MACHINES

Shelling machines on today's market are composed mainly of a sizing sieve and shelling units. Generally, they are further supplied with conveyor belts for the automatic charging of entire almonds and for the disposal of the shelled fruit.

The only discrimination between them lies in the type and number of sheller elements (six to sixteen): the latter must be equal to that of the sizes commonly referred to as "gauges" whereby engineers have felt it necessary to separate the product.

Some acceptance processing quality unlikely to be achievied by means of one only sheller element, i.e. at the undifferentiated stage on account of a high size variability of the almonds.

The features these machines share in common with one another are the presence of parallel - arranged hulling devices varying in number, which are fed by a gravity - related distributor consisting - if necessary - of so many sectors.

The machines which are mostly used nationwide are those of Delle Foglie's - types DF 6 (Fig. n. 2) - DF 12 and DF 16

A1, equipped with rocking jaw shelling units, as well as those of Stramaglia's - type STR 12 - supplied, conversely, with opposed crushing half-chambers. The different modes of shelling available entail specific machines capable of performing shell seed parting. The latter - once coupled in series with those suitable for hulling - are generally provided by the same firm since they have been specially developed, strictly depending upon prevailing shell breaking features of each machine.

3.1 - Rocking jaw hulling machine.

The almonds to be shelled, once deposited on a hopper which, in turn, may also be flush-mounted as to the bearing surface of the unit so as to make charging operations easier, finally get to a rotating sizing grid located on top of the machine (see Fig.n 3 and 4) through the agency of a bucket elevator belt.

This distributor element - formed by a forced sheet metal, the shape of which resembling a horizontal axis drum - consists of as many sectors as sheller elements.

Hole sizes get bigger and bigger from the opening to the end section; hence, only smaller - sized almonds pass through the initial section and, progressively, almonds with a size consistent with the

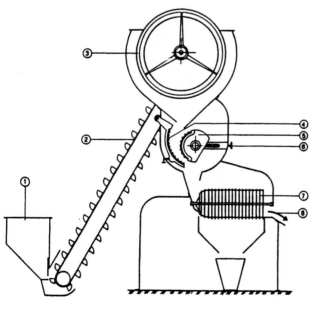

Fig. n. 2 - A schematic cross-section of Delle Foglie's shelling machine.
1 - Shell fruit hopper;
2 - bucket elevator-conveyor;
3 - rotating sizing grid;
4 - fixed jaw;
5 - mobile jaw;
6 - adjustment knob;
7 - grid for dust and small-fragmented shell disposal;
8 - outlet of rhe shelled fruits to be sent to the shell seed parting unit.

grid holes, which special internal projecting ridges push forward.

Thus, there follows a subdivision into "gauges" related to the average of the peak cross sizes of almonds, which leads to a more uniform size by groups, each reaching its working unit respectively.

The shelling divice is made up of two jaws; the first one being integral to the loom of the machine while the second – which is cam - driven - rocking at a 160 – cycle per minute frequency.

The bending radii of the two jaws are by no means similar; consequently, the space circumscribed by the two facing surfaces is not constant but decreasing and adjustable from upwards to downwards.

Almonds fitting by gravity into the shelling device (Fig. n. 3) move toward the narrow section where they are compressed and trimmed, owing both to the relative movement and the geometry of the mobile v. the fixed unit.

The two jaws are further equipped with longitudinal - toothed interchangeable wearing shields in order to cause the almond to move, increase the action of breaking and, finally, facilitate tha shell seed parting.

By adjusting the position of the mobile element via an external knob, we may achieve the highest standard of shelling, so that the number of partly broken almonds, to be than recycled would be kept to the minimum, as would chipping and seed crushing which nearly always involve loss and/or the product depreciation in the subsequent stage parting and grading.

It stands to reason that both the increase in almond uniformity and smoothness and the decrease in the breadth of each size range can be adjusted more accurately: in that case costs of the machine are on the rise since it is necessary to have it equipped with a greater number of sheller elements; generally speaking, twelve sizes account for a fair compromise solution between these two sharply contrasting requirements.

3.2 – Half-chamber shelling machines.

The model adopted by Stramaglia consists of a ring fixed to the machine framework and of an internal disc able to rock around a coaxial shaft.

Peripheral to the disc and inside the ring are variable-sectioned truncated cone-shaped half-cavities which, when facing, turn into correspondingly shaped passing chambers (Fig. n. 4).

The shell of the almonds falling in the above chambers gets broken thanks to the

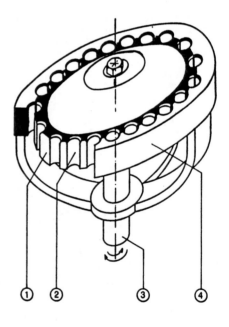

Fig. n. 4 - Opposed half-chamber shelling device.
    1 - Rocking disc;
    2 - mobile half chamber;
    3 - rocking disc splined to the shaft;
    4 - integral machine loom ring.

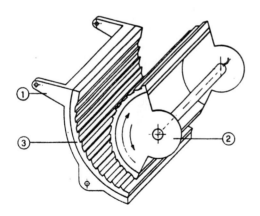

Fig. n. 3 - Structural arrangement of the jaw shelling device.
    1 - Loom fixed support;
    2 - rocking jaw;
    3 - toothed interchangeable wear shield.

mobile disc rocking versus the fixed ring. The amplitude of deviation is limited so as to avoid shell breaking as well; therefore, in the various sheller elements, it can be compared with the mean value of the standard size in order to define the proportions of the relative shelling chamber.

Obviously, also in this case the number of shelling devices the machine is fitted with, must be equal to that of the sizes considered.

The mobile disc is caused to move by means of two connecting rods, one of which being governed by the so-called crank-mechanism with an adjustment knob radius: so it is possible to change the oscillation amplitude to get the best out of shelling (Fig. n. 5).

The opening cross section of every chamber is greater than the lower discharge section, which makes it possible for shell fruits to better enter, and not to be discharged, just prior to crushing (Fig. n. 6).

One structural featuere shared by both units herewith examined lies in their capacity for separating sizes singled out into two groups. And this with a view to achieving a higher standard of shell seed parting.

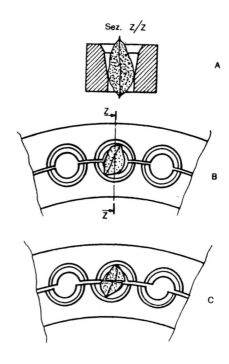

Sez. Z/Z

Fig. n. 6 - Detail showing the shape of shelling chambers (A) and a sequence of operations as to charging (B) and breaking (C) position.

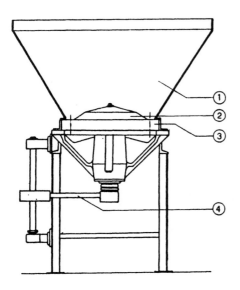

Fig. n. 5 - A schematic cross-section of an opposed half chamber hulling device.
1 - Control hopper;
2 - deflector allowing shell fruit pouring;
3 - shelling device;
4 - connecting rod for rocking disc operation.

The shelled fruit is therefore separated into two fractions coupling "small-sized" fruits with "big-sized" ones; said fractions being then conveyed to two distinct parallel - operating parting units.

Apart from a higher structural complexity specific to the working line and, especially, to the working unit which requires as many differently sized shelling chambers as sizes, in Stramaglia's machine distributor-gager sieves are separated into two opposite and parallel-arranged sections so as to reduce the unit longitudinal size; the first gager drum is further preceded by a supplementary fine-drilled grid serving to dispose of impurities, if any, as well as fragments and hull waste in particular, which might cause the shelling unit, especially elements destined for smaller-sized shell fruits, to get flooded.

Shelling tests have been performed by means of a Spanish-made machine working on the same principle as Stramaglia's, though structurally different. For the working unit consists of shelling half-chambers made out of prismatics bars - fixed or, alternatively, mobile - rocking back and forth special supports; the above-standing separator distributor sieve is then composed of two vibrating slanting grids whereby increasing diameter - drillings (Fig. n. 7) are performed, corresponding to the four hulling sections.

Cultivars cultivated in other countries of the Mediterranean area do not always exhibit similar shell structural mechanical dimensional features; hence, units developed with great difficulty by the above firms nearly always come to prove oversized or guarantee too low a quality and capacity for processing.

The reduction of the number of cultivars, as wished elsewhere, should allow a higher diffusion of cultivars noteworthy on the grounds of both agronomic and commercial features; hence structurally simpler-designed machines would be on the market, as well as a reduction in the employment of labour coupled with a drop in the costs of shell seed parting, the latter to be regarded as rather a complex, but only partly mechanized operation.

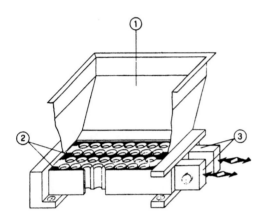

Fig. n. 7 - Structural arrangement of a shelling device of the spanisch-made Jumbus machine.
1 - Control hopper of sized almonds;
2 - fixed bars;
3 - mobile bars.

4. - CONCLUSIONS

The problem of mechanical almond shelling may be felt to have been satisfactorily coped with since the marketed units have achieved some good reliability and significant structural improvement these last few years.

This is quite an exceptional achievement because the number of machines being marketed is very low.

One of the aspects slowing down productive activity lies as yet in the impossibility of meeting foreign markets: as a matter of fact, units have been developed on the basis of the most widespread cultivars at a regional level, or, at most, nationwide, generally being characterizer by a type of shell ranging from "stone-hard" to "medium hard", according to the US scale of hardness; in some cases, such type of shell resists over 100-daN crushing forces.

1) P. Guarella, A. Pellerano - "Industrial equipment for almond shelling and grading in Apulia". - Proceedings of the 4th Domestic Congress of rural engineers, Porto Conte - Alghero - 1988.

Land and Water Use, Dodd & Grace (eds), © 1989 Balkema, Rotterdam. ISBN 90 6191 980 0

Machine vision grading of tree seedlings

G.A.Kranzler & M.P.Rigney
Oklahoma State University, Stillwater, USA

ABSTRACT: A computer vision technique was developed for measuring morphological characteristics of tree seedlings. Pine seedlings were inspected on a conveyor belt at nursery production rates. Classification as acceptable or cull was based on minimum criteria for stem diameter, shoot height, and projected root area. In laboratory tests, performance exceeded that of typical manual graders in terms of both accuracy and speed.

RESUME: Une technique de vision par ordinateur a été développée pour mesurer les caractéristiques morphologiques des jeunes brins. De jeunes brins de pins ont été classés sur une courroie de transport à des taux de production de pépinière. La classification comme acceptable ou à éliminer était basée sur des critères minimums portant sur le diamètre de la tige, la hauteur du sauvageon et la largeur prévue des racines. Dans des expériences en laboratoire, les résultats ont été meilleurs que ceux d'un trieur à main typique en ce qui concerne et la précision et la rapidité du triage.

ZUSAMMENFASSUNG: Es wurde ein computergesteuertes Leseverfahren zur Messung der morphologischen Charakteristika von Baumsetzlingen entwickelt. Auf einem Förderband wurden Tannensetzlinge bei baumschulüblicher Produktionsgeschwindigkeit klassifiziert. Die Einstufung richtete sich nach Mindestwerten bezüglich Stammesdurchmesser, Schösslingshöhe und Wurzelausprägung. Bei Labortests wurde das manuelle Klassifizierungsverfahren in Genauigkeit wie Schnelligkeit übertroffen.

1 INTRODUCTION

In nations around the world, well over a billion tree seedlings are produced annually in commercial, federal, and state nurseries. Most of these seedlings are graded at harvest to remove inferior stock and select seedlings with superior production potential after transplanting.

Grading is typically performed by inspectors at grading belts who visually examine individual seedlings for dimension and appearance. Manual grading is labor-intensive and costly. The work

Paper No. 2915 of the Oklahoma Agricultural Experiment Station.

tends to be tedious and tiring. Classification is subjective and susceptible to human error. Valuable production data such as classification statistics and seedling count are difficult to obtain. Limitations of manual grading have spurred increasing interest in automated alternatives.

Automated systems have been designed for measurement of tree seedling characteristics (Buckley et al. 1978) and seedling grading (Ardalan and Hassan 1981, Lawyer 1981). Mechanical and opto-electronic techniques were used to successfully measure morphological features. Throughput, however, was considerably slower than commercial nursery production rates.

Digital image processing has been successfully implemented in many industrial and agricultural inspection processes. It has demonstrated high accuracy and throughput, permitting 100% inspection where product sampling was previously the only feasible method of quality control. Machine vision inspection offers significant potential for addressing the challenge of automating tree seedling grading.

This study was initiated to investigate the ability of machine vision to grade pine tree seedlings under nursery production conditions. Our objective was to develop and implement a machine vision algorithm to measure key morphological characteristics and grade seedlings at typical production-line rates of approximately one seedling per second. We wanted to evaluate performance in terms of measurement speed, precision, and accuracy of classification.

2 METHODS AND MATERIALS

Several assumptions were adopted concerning the environment in which commercial grading would be performed. First, seedlings would be singulated, permitting only one seedling to appear within the camera field-of-view at a given time. Second, variations in shoot orientation and lateral position would be loosely constrained to +/- 30 deg and +/- 6 cm, respectively, from optimum. Finally, it was assumed that a black conveyor belt would be used to transport seedlings beneath the cameras.

2.1 Equipment

Equipment included a conveyor belt, machine vision computer, cameras, lenses, and lights. To simulate production grading operations, a variable-speed belt conveyor was constructed to transport seedlings for inspection.

A machine vision computer work station was used for project development. Images were digitized into an array of 256 X 240 picture elements (pixels) with 256 grey levels. A high-speed hardware coprocessor performed computationally intensive operations such as image filtering and edge detection, runlength encoding, and moments calculations.

Two solid-state black-and-white video cameras were employed for image acquisition. Camera 1 was used to obtain a close-up image of the seedling root collar zone. A field-of-view (FOV) approximately 12.8 cm square provided a 0.5 mm pixel resolution. Camera 2, with a FOV approximately 51 cm square and resolution of 2.2 mm, acquired an image of the entire seedling.

Illumination was provided by fluorescent room lighting and strobed xenon flash. Relatively low-level room lighting was adequate for detection of the moving seedlings in the FOV of camera 2. When a seedling was detected, synchronized strobe lamps were triggered to obtain a "frozen" image with each camera.

2.2 Grading scheme

Morphological characteristics are used for grading most nursery stock. These characteristics include stem diameter at the root collar, shoot height and weight, root weight or volume, root fibrosity, foliage color, presence of terminal buds, shoot/root volume ratio, and ratio of top height to stem diameter (sturdiness ratio) (Forward 1982, May et al. 1982). Stem diameter, shoot height, and root volume are generally given priority and were adopted as the grading criteria for this project. Of these three, stem diameter is typically considered most important.

To meet image processing time constraints, we emphasized stem diameter measurement accuracy and obtained a close approximation of shoot height and an index of root volume as indicated by projected root area (Fig. 1). Recent comparison of projected root area with measured root volume resulted in a correlation coefficient of 0.80 (Suh and Miles 1988).

Seedlings were graded into two classes; acceptable and cull. Seedlings were classified as acceptable if measured parametes met or exceeded the following minimum criteria:

Stem diameter	-	3.0 mm
Shoot height	-	16 cm
Projected root area	-	200 pixels

Fig 1 Seedling classification was based on stem diameter, shoot height, and projected root area.

3 ALGORITHM

The grading algorithm is composed of several separate tasks. These operations are: calibration, seedling detection, shoot orientation measurement, root collar location, diameter measurement, projected root area measurement, shoot height measurement, grade classification, and recording of seedling statistics. A detailed description of the algorithm is presented by Rigney (1986).

Selection of FOV for camera 1 required a compromise between diameter measurement precision and the probability of the root collar appearing within the FOV. Because the lateral position of the root collar was not tightly constrained, a relatively wide FOV was necessary. We decided to make the FOV as large as possible, while maintaining a measurement precision of at least 0.5 mm.

3.1 Seedling detection

Under ambient lighting conditions, a sequence of images acquired with camera 2 was processed. Each image was masked by a template defining a window in which seedling presence was tested (Detect Window, Fig. 2). Seedling detection initiated image acquisition from each camera, using strobe illumination.

3.2 Seedling orientation

Image 2 was first processed to determine seedling orientation on the conveyor belt. Area moment calculations provided the angle between the seedling major axis and the vertical axis of Figure 2. Subsequent measurements of stem diameter and shoot height were corrected for orientation angle. Because measurement error becomes excessive at large angles, seedlings were not graded if the orientation angle was greater than thirty degrees from vertical.

3.3 Root collar

Accurate location of the root collar was crucial for subsequent measurement of stem diameter, shoot height, and projected root area index. Image 1 was processed by an iterative algorithm employing runlength encoding to find a collar location satisfying heuristic criteria. Conceptually, each horizontal line in the image was considered to be composed of pixel strings belonging either to the foreground (seedling) or background (conveyor belt). The frequency of a line was defined as the number of transitions between the foreground and background. The root collar could be expected to reside in a region of low frequency lines bounded by areas of higher frequency. Needles, branches, and roots would contribute many transitions to horizontal lines in Figure 3 (high frequency). Lines in the root collar zone contain significantly fewer transitions. The width of individual pixel strings was also exploited, because we had *a priori* knowledge of the expected stem width.

Line frequencies were processed to build a list of root collar candidates. The resulting list was processed to locate the root collar both vertically and horizontally (Fig. 4). The algorithm located the root collar of most seedlings in one iteration. Seedlings with branches, needles, or roots in the root collar zone sometimes required two iterations.

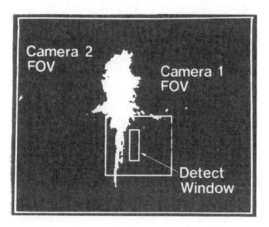

Fig. 2 Seedling moves through camera field-of-view. Image acquisition is triggered by detect window.

Fig. 3 Close-up image from camera 1 details root collar region.

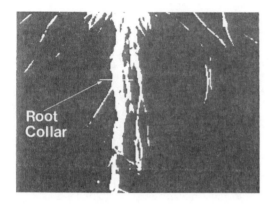

Fig. 4 Algorithm locates root collar.

The scale and mapping relationship between images 1 and 2 was previously determined during system calibration. After locating the root collar in image 1, the collar location was mapped into image 2.

3.4 Stem diameter measurement

Diameter measurement processing was performed inside a region around the root collar location in image 1. Region size was defined by the set of candidate collar lines found in the collar location subroutine. The region was convolved with an edge detector favoring vertical edges. Average distance between the strongest edges bracketing the root collar was calculated as the collar diameter (Fig. 5).

3.5 Root area measurement

Root area was measured from that portion of image 2 below the root collar (as mapped from image 1). An edge detector sensitive to all edge orientations was applied. The weakest edges detected were the result of noise in the image background (conveyor belt). Above the noise level, however, roots much smaller than the pixel resolution (2.2 mm) could be detected because of their contribution to the brightness of corresponding pixels. Pixels with edge intensities greater than the noise level were summed to yield projected root area (Fig. 6).

3.6 Shoot height measurement

Image 2 was processed to determine shoot height. Starting at the top of the image, each line was tested to detect transitions indicating the presence of the seedling top. The top was assumed to be located when four consecutive lines met this criterion. Shoot height was defined as the distance between the seedling top and root collar location (previously mapped into image 2).

3.7 Main program

Inside the main program loop, values returned by subroutines were tested to control program flow. If all grading subroutines were successful in their respective tasks, a grade classification was assigned to the seedling.

Whenever a subroutine failed its task, the seedling was recorded as not gradable. Finally, measured seedling parameters, grade, and count were written to a statistics file.

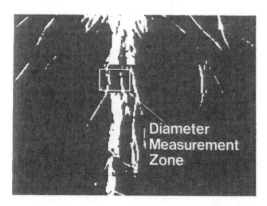

Fig. 5 Image is processed to identify stem edges in root collar zone.

Fig. 6 Image is processed to enhance seedling roots.

Table 1. Percent misclassification of 100 seedlings, 20 replications

Manual grade	Acceptable		Cull		Total		
	#	mis.	#	mis.	#	mis.	n.g.
Borderline	6	31.7%	11	18.6%	17	23.2%	2.6%
Easily Classified	63	2.2%	20	2.0%	83	2.2%	2.3%
All	69	4.7%	31	7.9%	100	5.7%	2.3%

mis. = misclassified   n.g. = not gradable

3.8 Calibration

Proper calibration of threshold values and scale factors was essential for optimum algorithm performance. The calibration subroutine initialized sixteen parameters with default values. The user is then provided an opportunity to interactively alter the default values. A wooden dowel of known diameter and length was used to calibrate scale factors. Grey level thresholds were set using a representative seedling.

4 EVALUATION

A reference set of 100 loblolly pine (Pinus tacda L.) seedlings was manually measured and graded. Stem diameters ranged from 2.3 to 6.0 mm. Performance of the machine vision system was then evaluated by grading each of the seedlings twenty times.

Time required for the algorithm to grade a seedling averaged approximately 0.5 seconds. To facilitate manual placement of the seedlings on the conveyor, tests were conducted at a belt velocity of 0.5 m/s.

Classification error rate averaged 5.7 percent for the set of 100 seedlings (Table 1). This performance significantly bettered manual grading operations, which have an average misclassification rate of seven to ten percent (Boeckman 1986). As expected, a large part of the classification error was attributable to seedlings which straddled the borderline between acceptable and cull, with respect to diameter and root area. Such seedlings comprised 17 percent of the grading test set and had an average misclassification rate of 23.2 percent. The remaining 83 seedlings had an average misclassification rate of 2.2 percent. Since there is no significant penalty for misclassification of borderline seedlings, 2.2 percent misclassification

may be a better indicator of algorithm performance.

The coefficient of variation for 20 measurement repetitions averaged 7.6, 12.2, and 4.1 percent for stem diameter, root area, and shoot height, respectively. This result indicates a standard deviation of 0.23 mm for a 3.0 mm stem diameter.

The few seedlings which showed the largest deviations in measured parameters were characterized either by needles extending down past the root collar, or by roots bent upward past the root collar, or both. The subroutine which located the root collar performed inconsistently on such seedlings. A few of these seedlings could not be graded.

In subsequent work, projected shoot area was measured, allowing calculation of shoot/root area ratio. Though not tested, we anticipate a close correspondence between projected area and mass, allowing fast and non-destructive estimation of conventional shoot/root mass ratio. Suh and Miles (1988) reported high correlation (0.95) between seedling projected area and measured weight. Calculation of the sturdiness ratio (diameter/height) has also been implemented and could provide another useful parameter for classification.

The measurement precision demonstrated by the algorithm suggests use for classification of seedlings into several acceptable grades. Additional grade definitions could be optimized for specific planting sites. Further, we expect that the comprehensive statistics collected in a commercial implementation would make machine vision grading a valuable nursery management and research tool.

While this study was limited to bare-root pine seedlings, the inspection techniques developed are directly applicable to other tree species and plant types, including container grown seedlings.

5 SUMMARY AND CONCLUSIONS

This study has demonstrated that machine vision can provide accurate production-rate grading of harvested pine tree seedlings. Singulated seedlings were transported on a conveyor belt, with shoot orientation and root collar position loosely constrained. Seedlings were classified as acceptable or cull on the basis of stem diameter, shoot height, and projected root area.

Tests with loblolly pine seedlings revealed excellent system performance. Seedlings were graded in approximately 0.5 seconds, with an average classification error rate of 5.7 percent. These results exceed manual grading performance, which typically requires one second per seedling with an error rate of seven to ten percent. Misclassification was largely due to seedlings with borderline diameter and/or root area, and the occurrence of branches or roots in the root collar zone. Measurement precision was adequate for seedling classification into several grades, suitable for specific planting sites. The technology promises increases in both measurement accuracy and speed.

REFERENCES

Ardalan, S.H. & A.E. Hassan 1981. Automatic feeding and sorting of bare root seedlings. ASAE Paper No. 81-1082. ASAE, St. Joseph, MI.

Boeckman, W. 1986. Personal communication. Weyerhaeuser Forest Regeneration Center, Fort Towson, OK.

Buckley, D.J., W.S. Reid & K.A. Armson 1978. A digital recording system for measuring root area and dimensions of tree seedlings. Transactions of the ASAE. 21(2):222-226.

Forward, P.W. 1982. Stock production specifications - bare root stock. Artificial Regeneration of Conifers in the Upper Great Lakes Region. Michigan Technological University, Houghton, MI. pp. 260-268.

Lawyer, J.N. 1981. Mechanization of nursery production of bare root deciduous planting stock. Forest Regeneration. ASAE Publication 10-81. ASAE, St. Joseph, MI. pp. 30-37.

May, J.T., E.W. Belcher, Jr., C.E. Cordell, T.H. Filer, Jr. & D. South 1982. Southern Pine Nursery Handbook. Forest Service, USDA.

Rigney, M.P. 1986. Machine vision for the grading of pine seedlings. M.S. Thesis. Agricultural Engineering Department, Oklahoma State University, Stillwater.

Sang-Ryong S. & G.E. Miles 1988. Measurement of morphological properties of tree seedlings using machine vision and image processing. ASAE Paper No. 88-1542. ASAE, St. Joseph, MI.

Land and Water Use, Dodd & Grace (eds), © 1989 Balkema, Rotterdam. ISBN 90 6191 980 0

Grain cleaning and grading

J.J.Lenehan
Agricultural Engineering Department, Teagasc, Oak Park Research Centre, Carlow, Ireland

ABSTRACT: The use of grain cleaning and grading equipment is important for the cereal producer as an aid to marketing his crops. The operation of a range of equipment is described. Tests were performed on barley samples in the laboratory using a two-screen aspirated cleaner. Increases in hectolitre weight ranged from 0.8 to 8.4 kg/hl when samples were cleaned over a 2.2 mm slotted screen. Increases in hectolitre weight ranged from 2.0 to 13.6 kg/hl when samples were cleaned over a 2.5 mm slotted screen. The market for malting barley requires grain at a protein content of 11.5% or less for processing. Experiments showed the potential of grading samples to meet this standard.

RESUME: L'utilisation de matériel de nettoyage et de calibrage de grain est importante pour le producteur de céréales en tant qu'aide à la commercialisation des ses récoltes. Ce document décrit l'exploitation d'une gamme de matériels. Des tests ont été effectués en laboratoire sur des échantillons d'orge à l'aide d'un appareil de nettoyage par aspiration á deux tamis. Les augmentations de poids par hectolitre variaient de 0.8 à 8.4 kg/hl lorsque les échantillons étaient nettoyés sur un crible à rainures de 2.2 mm. Les augmentations en poids par hectolitre variaient de 2.0 à 13.6 kg/hl lorsque les échantillons étaient nettoyés sur un crible à rainures de 2.5 mm. Le marché d'orge de brasserie exige du grain d'un taux protéinique de 11.5% pour la transformation. Les expériences montrent le potentiel de calibrage des échantillons susceptible de se conformer à cette norme.

AUSZUG: Der Einsatz von Korn-Reinigungs- und Sortiergeräten ist für den Getreidebauer ein wichtiges Hilfsmittel für den Verkauf seiner Ernte. Hier werden einige Geräte beschrieben. In einem Labor wurden Versuche an Gerstenproben mit einem Doppelsieb-Ansaugreiniger vorgenommen. Der Anstieg des Hektolitergewichts reichte von 0,8 bis 8,4 kg/hl, wenn die Proben über einem 2,2 mm Spaltsieb gereinigt wurden. Bei der Reinigung der Proben über einem 2,5 mm Spaltsieb betrug der Hektolitergewichtsanstieg zwischen 2,0 und 13,6 kg/hl. Der Markt für Braugerste erfordert fur die Verarbeitung Korn mit einem Proteingehalt von 11,5%. Die Versuche haben gezeigt, daß die klassifizierten Proben das Potential hatten, dieser Norm gerecht zu werden.

1 INTRODUCTION

In many production processes it is necessary to separate loose material, including granular mixtures, into different fractions. This is also the case in agriculture, e.g. large quantities of seeds are separated, fruit are size-graded and potatoes are separated from lumps of soil and stones (Grochowiez 1980).

In grain production all the seed is cleaned at least once, i.e. in the combine harvester. In many cases, the grain also undergoes further cleaning and grading, as the initial processes carried out on the harvesting equipment do not produce a sample which meets particular market requirements. This is not a result of ineffective mechanisms on the harvester. Its aim is to harvest the total amount of seed produced in the field. In doing so, some chaff, weed seeds and other waste material, along with the proportion of the crop which is below the required quality, is mixed with potentially higher grade material. It is possible to produce higher grade samples on the combine; however, this would inevitably lead to an increase in harvesting losses and thus is not an economic proposition.

2 LEVELS OF CLEANING

Certain markets must be more selective on quality because of the intended use for the grain. Millers and maltsters require higher quality grain than that which is for use in animal feedstuffs. The quality requirements of the various markets are well documented, and include specifications on the level of screenings, hectolitre weight, protein content, Hagberg Falling Number (HFN) of wheat, etc., as detailed in Tables 1 and 2.

Table 1. E.C. Intervention quality standards for cereal grains

	Milling wheat	Feed wheat	Barley
Max. moisture content (%)	15.5	15.5	15.5
Min. hectolitre weight (kg/hl)	72	72	63
Max. impurities (%)	10	12	12
Hagberg Falling Number	220	–	–
Standard protein (%)	11.5	–	–

Table 2. Quality standards for malting barley (export requirements)

Quality characteristic	
Maximum moisture content	13.5%
Maximum protein content	11.5%
Maximum admixture	2.0%
Minimum over 2.5 mm screen	90.0%
Maximum through 2.2 mm screen	3.0%

Grain, as harvested, is always accompanied by various contaminants, e.g. chaff, straw, weed seeds, dust and sand. These contaminants reduce the value of the product. By cleaning and/or grading grain it is possible to reduce the level of

contamination. In general, cleaning operations fall into 3 major categories:
1. Pre-cleaning
This rudimentary process is aimed at removing straw, chaff, plant fragments, small seeds, dust and sand. It is the minimum treatment recommended before drying.
2. Commercial cleaning
This is a more intensive operation than pre-cleaning, resulting in the production of samples for sale capable of meeting commercial standards.
3. Cleaning of seed samples
The aim here is to clean to a degree which ensures that a potential seed sample is not rejected or lowered in value because of impurity or non-uniformity.

3 DIFFERENTIATING CHARACTERISTICS OF SEEDS EMPLOYED IN SEED CLEANING EQUIPMENT

The large number of differentiating characteristics of seeds, as outlined by Wallace and Naylor (1981), suggests a free choice in selecting a seed cleaning process. However, if a satisfactory sorting process is to be established, it is necessary to choose a differentiating characteristic which enables the required sample to be produced at a commercially acceptable throughput. The differentiating characteristics used are:
1. Aerodynamic properties
2. Geometrical characteristics
3. Specific gravity
4. Surface characteristics
Characteristics 1 and 2 are used in the majority of grain cleaning equipment. Characteristics 3 and 4 are used in more specialised equipment.

3.1 Aerodynamic properties

In order to determine the movement of a seed in an air stream, it is convenient to simplify matters by assuming that the seed is falling through a vertically ascending air stream, as described by Henderson and Perry (1955). The behaviour of seeds and contaminants in the air stream are a result of differences in their aerodynamic properties. The forces acting on a particle in a vertical air stream are indicated in Fig. 1.
The particle in the air stream is also influenced by the force of gravity, G, which is exerted vertically downward. In the case of a vertically ascending air stream, the forces, R and G, are parallel but act in opposite directions, leading to

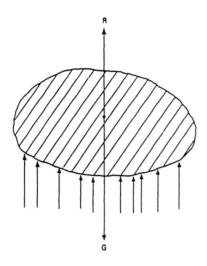

Fig. 1 Forces acting on a particle in a
vertical air stream

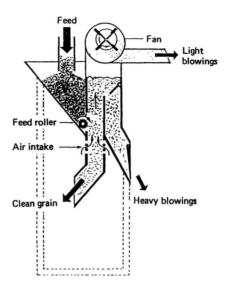

Fig. 2 Simple aspirator

three possibilities:
1. R < G, the particle will accelerate
downward.
2. R > G, the particle will move upward
with the air stream.
3. R = G, the particle will remain
suspended and the absolute velocity of the
seed will be zero.
Under the conditions of a turbulent air
stream, the drag force, R, can be
expressed as:

$$R = kAPc^2$$

k - dimensionless coefficient of
   aerodynamic drag
A - cross-sectional area of seed
   perpendicular to the direction
   of the air stream (m^2)
P - specific gravity of air (kg/m^3)
c - relative velocity of seed and air
   stream (m/s)

For separation purposes, the light seeds
and impurities must be carried away and
the good seed must fall through the air
stream for collection. Thus, the velocity
of the air stream must be selected so that
for the impurities R is greater than G,
and that for the good grain R is less than
G. The cut-off point, where R is equal to
G, is given when the absolute velocity of
the seed is zero. The velocity of the air
stream in this situation is termed the
critical velocity. Fig. 2 illustrates a
simple aspirator. Grain enters the
machine through an intake hopper; its flow
being controlled by a feed roller. Air
flow is regulated by a control flap. Good

grain falls through the air stream for
collection. Heavy trash is separated in
the plenum and discharged. Light trash is
discharged through the fan.

3.2 Geometrical characteristics

The geometrical characteristics of
seeds, as indicated by Lenehan (1985), can
be defined by describing three dimensions:
thickness, width and length, as shown in
Fig. 3. Width and thickness are
determined from the largest cross-section
of the seed. The thickness of seed is
used as the differentiating
characteristic in the case of screens with
slotted openings, and the width of seed is
used in the case of screens with round
openings, as indicated in Fig. 4. The
difference in length of particular seeds
is used as the differentiating
characteristic when the separation is made
in an indented cylinder.
Fig. 5 illustrates a two-screen cleaner
with inlet aspiration. After passing down
the aspirating leg, the grain with heavy
contaminants, and undersized and imperfect
seeds, is fed onto the upper end of the
first of two inclined reciprocating
screens. This screen has apertures large
enough to allow the biggest seeds to fall
through, but carries all larger material
over its lower end. The grain, together
with the smaller contaminants, falls
through the upper screen onto the lower
screen. The size and shape of the

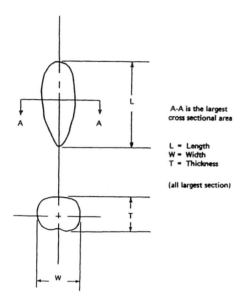

Fig. 3 Dimensions of a seed

A-A is the largest
cross sectional area

L = Length
W = Width
T = Thickness

(all largest section)

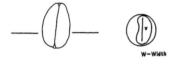

W—Width

SEED SIZING ON ROUND SCREEN

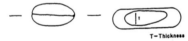

T—Thickness

SEED SIZING ON SLOTTED SCREEN

Fig. 4 Sizing seeds on screens

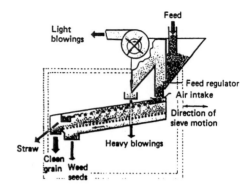

Fig. 5 Two-screen cleaner with head aspirator

apertures in this screen determine the proportion of the grain passing over the lower end for collection as clean grain. The type of process being carried out, e.g. pre-cleaning or grading, will determine the screens being used. Some of these machines use a three-screen layout to allow further sub-grading of the sample.

During the screening operation, the apertures in the screens will tend to get blocked. Slowly moving strip brushes, which transverse under the bed of the screen, can be used to clear lodged seed. Another method of cleaning the openings consists of employing the forces of inertia of rubber balls placed in cages under the screen being cleaned. During screening, these balls bounce and strike

against the screen, thus dislodging any particles from screen openings.

A second group of screen cleaners are the rotating screen machines, which are also referred to as rotary grain cleaners. These machines include types with vertical and horizontal axes.

The action of the rotating screen agitates and presents the material being processed to the screen. The machines can be used both for pre-cleaning and/or grading, depending on the screen configuration. A typical horizontal axis machine is shown in Fig. 6. Compared to flat screens, the cylindrical screens are distinguished by the following advantages:
- simple drive
- dynamic balance
- ease of cleaning of the openings by roller brushes

The use of the length of seeds is the third geometrical characteristic employed in the separation of seeds. It is used in machines termed indented cylinders. The indented cylinder separating mechanism consists of an open-ended metal cylinder with regular indentations, either drilled or pressed in its inner surface. The cylinder is inclined at a small angle to the horizontal and is rotated slowly on its axis. The indented cylinder selects fractions of the sample which are short enough to be carried up in the indentations, whereas grains, etc., which are too long cannot be so lifted from the bulk, as shown in Fig. 7. The contaminated seed is fed into the cylinder at its upper end. The shorter particles in the sample are picked up in the indentations and are carried around with the cylinder until they fall by gravity into a collecting tray. They are transferred out of the cylinder to a

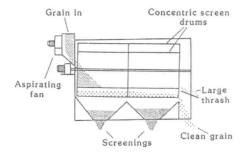

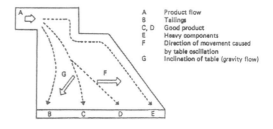

A	Product flow
B	Tailings
C, D	Good product
E	Heavy components
F	Direction of movement caused by table oscillation
G	Inclination of table (gravity flow)

Fig. 6 Rotary grain cleaner

Fig. 8 Plan of specific gravity separator illustrating flow of material in relation to pitch, influence of gravity flow and direction of stroke

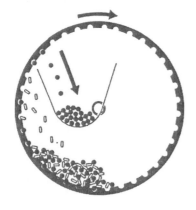

Fig. 7 Principle of separation with indented cylinder

collecting point. Particles which are too long to be retained by the indentations are collected as they leave the lower end of the cylinder.

3.3 Specific gravity

Specific gravity can be used as a differentiating characteristic in many cleaning operations. Gravity separating devices use two principles:
1. The ability of grain to flow down an inclined plane.
2. The fluidising effect produced on a bed of grain by the upward motion of air.

The prime unit of this separator is a trapezium-shaped perforated table, as shown in plan view in Fig. 8. The table is baffled underneath so that air being fed up through it is evenly distributed. The volume of air is controllable, depending on the species of

crop being handled. The operation of the specific gravity separator is described by Bailey and Graham (1987).

The table has a reciprocating motion that moves any material upon it in the direction of conveyance. The table has vertical adjustments allowing it to be tipped towards the front and towards the left, the net pitch being such that a sphere placed on the table would roll in the direction indicated by the gravity flow arrow, G.

In operating, material is introduced into the feed box. Air is blown up through the perforated table at such a rate that the material is partially lifted from contact with the table. Lighter, smaller particles are lifted somewhat higher and 'float' down the table towards the discharge edge. The large and heavy particles are not lifted by the air. The oscillating motion of the table moves them in the direction of conveyance, and they are discharged at the right edge of the table. Other material that is only partially lifted touches the table at frequent or infrequent intervals, and is discharged at an appropriate intermediate point.

3.4 Surface characteristics

Inclined belt separators have been developed specifically for the removal of wild oats from cereal crops (McLean 1980). Such mechanisms consist of a short conveyor belt; the fabric from which the belt is constructed has a surface texture to which wild oats seeds will adhere. Contaminated seed is fed onto the middle of the belt and the smooth round grains of cereals and similar crops roll down the belt against the direction of rotation.

By contrast, the awns of wild oats seeds prevent their movement on the belt, the outcome of which is that they are carried up the machine in the opposite direction, for subsequent discharge. Throughput and separating efficiency can be adjusted by changing both the angle and speed of rotation of the conveyor belt.

4 POTENTIAL OF UPGRADING GRAIN SAMPLES BY CLEANING

4.1 Materials and methods

In this study, samples of barley were selected from the 1988 harvest. The following operations were performed on the samples:

1. Moisture content test: The moisture content was assessed in accordance with I.S.O. standards (130°C for 2 hours) using a Brabender oven.

2. Hectolitre weight test: The hectolitre weight was assessed using the EASI-WAY hectolitre weight tester.

3. Cleaning/grading operation: These operations were performed on a Rober-Mini Petkus laboratory cleaner. This machine is a 2-screen aspirated cleaner.

4. Protein content test: This analysis was performed using an NIR (Near Infra-Red Diffuse Reflectance Spectroscopy) instrument. Light at a predetermined number of wavelengths is sequentially shone onto a ground sample of the material being tested. Some of this light is reflected directly back at the light source, while the remainder undergoes multiple reflection and adsorption in the ground material before being reflected back. This diffuse reflectance component contains information on the protein content of the sample (Dwyer 1986).

4.2 Results and discussion

Hectolitre weight is a measure of the bulk density of grain, i.e. the weight of a given volume of grain, and is expressed as kilograms per hectolitre. Work by Browne (1962) indicates that the hectolitre weight is dependent on the moisture content of the sample. It is an important index of grain quality as it reflects the 'plumpness' of grain, and the amount of immature grains and contaminants in a sample. Immature grains, or screenings, and other contaminants effect the packing efficiency of grains in a sample and thus lower the hectolitre weight (Bayles 1977).

The hectolitre weight of the raw samples of barley was measured. These samples were then subjected to cleaning operations on the laboratory cleaner and the hectolitre weight of the resultant samples was measured. The cleaner was fitted with a 3.5 mm slotted top screen and a 2.2 mm slotted bottom screen for the initial test. Subsequently, the samples were re-mixed and cleaned over a 3.5 mm slotted top screen and a 2.5 mm slotted bottom screen. The top screen removed any large contaminants from the sample. The 2.2 mm and 2.5 mm screens are standard screens over which malting barley is graded in practice. In both cases, the grain passed through the aspiration section on the cleaner. The results are presented in Table 3.

Table 3. Hectolitre weight of barley cleaned on a 2.2 mm screen and a 2.5 mm screen. (Figures in brackets are percentage of sample in category)

			Hectolitre weight	
Sample no.	M.C. (%)	Bulk	Over 2.2 mm screen	Over 2.5 mm screen
1	18.8	65.2	66.0(97.6)	67.2(91.5)
2	21.7	57.2	63.6(93.9)	66.0(84.8)
3	21.2	64.4	67.2(95.9)	67.6(88.3)
4	19.7	62.4	66.0(92.9)	66.8(79.4)
5	22.8	60.0	65.2(87.0)	66.0(69.0)
6	19.5	62.0	67.2(94.6)	68.0(85.0)
7	20.8	62.8	66.4(92.6)	67.2(79.2)
8	20.7	64.8	67.6(92.5)	67.6(81.8)
9	18.8	59.2	64.4(87.5)	65.6(64.6)
10	18.8	60.8	66.8(89.1)	67.6(74.7)
11	17.1	62.8	67.2(95.6)	68.0(81.8)
12	17.7	62.8	66.4(92.6)	67.2(77.2)
13	17.4	63.2	67.2(92.5)	68.0(82.0)
14	17.1	63.2	67.2(94.2)	68.0(83.6)
15	19.5	62.4	66.0(91.7)	66.8(73.2)
16	16.1	54.8	63.2(86.9)	68.4(58.9)
17	16.1	55.2	62.4(63.5)	65.2(18.3)
18	16.0	62.4	66.8(94.5)	66.8(80.2)
19	16.2	57.6	62.4(75.6)	65.2(36.1)
20	17.1	58.0	63.6(86.8)	66.0(69.8)

The increase in hectolitre weight ranged from 0.8 to 8.4 kg/hl when samples were cleaned over a 2.2 mm screen with corresponding recoveries of top grade grain of 97.6% and 86.9%. When cleaned over a 2.5 mm screen, the increase in hectolitre weight ranged from 2.0 to 13.6 kg/hl with corresponding recoveries of top grade grain of 91.5% and 58.9%. Of the 20 samples tested, only sample 1 satisfied, in its raw state, the screenings test requirement as specified by malting barley export specifications. This indicates the need for growers aiming at such markets to consider installing cleaning equipment to increase the value of their produce.

Protein content is another important quality factor of malting barley. The maltster requires a low protein content for satisfactory processing. This is reflected in the purchasing standards for malting barley where 11.5% would be the normal maximum protein content allowed. Price penalties or rejection would be enforced on samples not attaining this standard.

As a result of their physiological make-up, the larger barley grains in a sample have the lowest protein content. To investigate this, samples were prepared for protein analysis by grading over a 2.5 mm slotted screen and a 2.2 mm slotted screen. Four protein tests were performed on each sample - bulk material, grain over 2.5 mm screen, grain over 2.2 mm screen and grain through a 2.2 mm screen. The results are presented in Table 4.

In all cases, the barley graded over the 2.5 mm screen had the lowest protein content.

5 CONCLUSIONS

There is a comprehensive range of grain cleaning machinery available to grain producers. This cleaning machinery can be used effectively to upgrade the quality of grain. However, the potential of cleaning grain to enhance its quality is dependent on the quality of the unprocessed sample. Laboratory tests on representative samples would indicate the economics of undertaking any particular cleaning operation. The results show that, in the majority of cases, samples as taken off the combine harvester will require cleaning if they are to achieve top price on the high quality markets, e.g. malting barley.

REFERENCES

Bailey, J. and K. Graham 1987. A pilot study of the use of a Kamas gravity separator to improve the Hagberg of a range of wheat samples. Report by ADAS Mechanisation Officers, Chelmsford.
Bayles, R.A. 1977. Poorly filled grains in the cereal crop. J.Natn.Inst.Agric.Bot. 14:241-247.
Browne, D.A. 1962. Variation of the bulk density of cereals with moisture content. J.Agr.Eng.Res. 7:288-290.
Dwyer, E. (Department of Food Science and Technology, Teagasc) 1986. Testing for grain quality at intake points.
Grochowiez, J. 1980. Machines for cleaning and sorting of seeds. U.S. Department of Commerce, National Technical Information Service.
Henderson, S.M. and R.L. Perry 1955. Agricultural Process Engineering. John Wiley and Sons Inc., New York, Chapman and Hall Ltd., London.
Lenehan, J.J. 1985. Grain cleaning. ADAS report.
McLean, K.A. 1980. Drying and storing combinable crops. Farming Press Ltd., Ipswich.
Wallace, W. and J. Naylor 1981. Introduction to seed and grain cleaning. ASAE paper no. 81:355 5.

Table 4. Protein contents of different fractions of barley (measurements on d.m. basis)

	% Protein			
Sample no.	Bulk	Barley over 2.5 mm screen	Barley over 2.2 mm screen	Screenings through 2.2 mm screen
1	9.50	9.25	9.50	12.35
2	12.30	11.70	11.95	15.70
3	11.00	10.50	11.00	14.40
4	12.95	11.65	11.95	15.20
5	12.70	12.15	12.40	14.80
6	10.75	10.50	10.60	14.15
7	11.90	11.60	11.75	14.65
8	10.20	9.80	9.90	12.65
9	11.13	10.05	10.85	13.10
10	10.60	9.45	9.90	12.60
11	11.25	10.85	11.05	13.60
12	10.70	10.30	10.40	12.90
13	10.15	9.70	9.75	12.10
14	10.60	10.40	10.50	13.20
15	10.70	10.45	10.70	12.40
16	13.60	12.00	12.95	16.50
17	12.15	11.00	11.50	14.10
18	10.50	10.15	10.40	12.45
19	13.70	12.10	13.34	14.80
20	11.95	10.20	10.70	13.35

Land and Water Use, Dodd & Grace (eds), © 1989 Balkema, Rotterdam. ISBN 90 6191 980 0

Mise au point d'un décortiqueur à mil-maïs-sorgho au Senegal

H.M.Mbengue
ISRA – CNRA, Bambey, Senegal

ABSTRACT : In the past, despite efforts to mechanize decortication operation of local cereals, no machinery had given total stisfaction. Thanks to an IDRC grant, ISRA and SISMAR initiated investigations on a huller model mini PRL to adapt it to the processing of traditionnel cereal crops. These investigations led to the design of a dehuller capable of adequately processing small amounts of non calibrated grains. The testing of nine pilot units under real conditions gave satisfactory results and led to the improvement of the equipment in terms of robustness, simplicity and purchasing cost.

RESUME : Malgré les efforts entrepris dans le passé pour mécaniser le décorticage des céréales locales autres que le riz au Sénégal, aucun modèle n'avait donné pleine satisfaction. L'ISRA et la SISMAR, sur financement CRDI, ont donc initié des études sur le décortiqueur mini PRL afin de l'adapter aux céréales traditionnelles. Ces recherches ont débouché sur un appareil capable de décortiquer correctement de petites quantités de grains non calibrés. Le suivi de neuf unités pilotes placées en milieu réel a donné des résultats satisfaisants qui ont permis, en outre, d'améliorer ultérieurement le matériel sur les plans robustesse, simplicité et coût d'acquisition.

ZUSAMMENFASSUNG : Trotz der in der Vergangenheit unternommen Versuche einheimische Getreide (ausser Reis) mechanisch zu schälen, hat kein Gerätemodel zufriedenstellende Resultage geliefert. Die ISRA und die SISMAR welche durch den CRDI finanziert wurden haben Studien unternommen um die Schälmaschine "mini PRL" an die lokalen Getreide anzupassen. Das Forschungsresultat ergab ein Gerät das kleine Mengen unkalibrierte Körner korrekt schälen kann. Die neun Piloteinheiten (Einsatz unter Realbedingungen) haben zufriedenstellende Resultate geliefert. Gleichzeitig wurde zudem das Material in Sachen Robustheit, Einfachheit und Ankaufspreis verbessert.

I INTRODUCTION : UTILISATION DES CEREALES LOCALES, CONTRAINTES DE TRANSFORMATION PRIMAIRE ET EXPERIENCES D'INTRODUCTION DU DECORTICAGE MECANIQUE AU SENEGAL

Les mils, sorghos et maïs demeurent l'essentiel de l'alimentation des populations rurales sénégalaises. Les divers plats constituant cette alimentation de base sont le couscous, le lakh et le gnéleng faits à partir de farine et sankhal déri-, vés de ces céréales (8). Mème si ces plats ont des méthodes de préparation différen-tes, ils ont une caractéristique commune qui est le décorticage préalable du grain. Traditionnellement, il est fait à l'aide d'un mortier et d'un pilon, avec addition d'eau afin que le péricarpe cède facile-ment sous les coups du pilon. Le grain est ensuite vanné, lavé à grande eau, puis ressuyé avant d'être moulu. Les produits ainsi obtenus ont une forte teneur en eau (28 p. 100 base sèche) et de ce fait, ne se conservent pas, doivent être consommés rapidement. Il en découle une pratique quasi quotidienne du décorticage et de la mouture, opérations auxquelles la femme sénégalaise consacre en moyenne 3 à 4 heures par jour, soit 1/3 de son temps de travail. Le suivi des opérations de décor-ticage manuel fait ressortir des débits horaires moyens de 8 kg pour le mil et le sorgho, et de 7,5 kg pour le maïs, ceci à des taux de décorticage respectifs de 22,8 p. 100 et 19,7 p. 100 (1, 3, 4).

Ainsi, le décorticage manuel est carac-térisé par :
- son caractère quotidien rendu nécessaire par le manque de stabilité des produits obtenus ;

- sa pénibilité ;
- et son inadaptation au contexte urbain où les populations, dans leur grande majorité, sont encore attachées aux habitudes alimentaires traditionnelles.

Devant ces importantes contraintes, la Recherche s'est très tôt intéressée à la mise au point de principes de décorticage et de mouture à sec, ceci afin d'obtenir un produit stable, soulager la femme et lui permettre de s'adonner à d'autres activités moins contraignantes et plus lucratives.

Malgré les efforts consentis, les modèles de décortiqueurs connus au Sénégal n'ont pas donné satisfaction au niveau de la transformation artisanale. Il s'agissait principalement des modèles COMIA-FAO et PRL-HILL THRESHER SUPPLY. En plus des problèmes techniques rencontrés (usure rapde des organes abrasifs et nécessaire calibrage des grains pour le modèle COMIA-FAO, batch minimal de 10 kg pour le modèle PRL alors que les apports individuels sont en moyenne de 4 kg), les tests d'introduction en milieu réel ont montré que :
- les débits et les coûts de revient des prestations sont trop élevés pour justifier une utilisation au niveau de la plupart des villages du Sénégal : 35 à 55 F CFA/kg pour des quantités journalières variant de 30 à 90 kg ;
- les cibles préférentielles ne peuvent être que les zones urbaines, péri-urbaines et semi-urbaines, ainsi que les gros villages pour du travail à façons sur de grandes quantités et/ou pour une commercialisation des produits semi-finis (3, 4, 5, 6, 7, 8).

C'est à partir de ce constat de semi-échec qu'a été élaboré le projet de "Création d'un décortiqueur adapté aux besoins de transformation des céréales locales au Sénégal". Financé par le Centre de Recherches pour le Développement International (CRDI), ce projet est mené conjointement par l'Institut Sénégalais de Recherches Agricoles (ISRA) et la Société Industrielle Sahélienne de Mécanique, de Matériels et de Représentations (SISMAR).

II OBJECTIFS DE L'ETUDE

Il s'agit d'étudier en détail les aspects techniques, économiques et sociaux liés à l'introduction d'un système de décorticage à sec des céréales locales dans les villes et les villages pour du travail à façons.

Les objectifs spécifiques sont :
- mettre au point un mini-décortiqueur adapté aux conditions locales de décorticage des mils, maïs et sorghos, ceci à partir du modèle mini-PRL ;
- introduire le modèle retenu dans le milieu

et en étudier les différents aspects ;
- assurer une large diffusion des résultats définitifs en relation avec les divers services de vulgarisation et de développement du Sénégal.

III METHODOLOGIES

3.1 Mise au point du matériel

La démarche adoptée est caractérisée par des échanges "triangulaires" continus et directs entre la Recherche, le constructeur et les utilisateurs-consommateurs. Les résultats des essais en laboratoire et des tests de comportement en milieu réel accompagnés des propositions de modifications servent de base à une meilleure conception du modèle définitif, ce dernier devant être techniquement fiable et adapté aux réalités socio-économiques du pays.

3.2 Essais de décorticage mécanique

Pour chaque modèle de décortiqueur et pour chaque type de grain, on détermine l'influence de la nature du disque abrasif, de **son régime** de rotation, du temps de travail et de la quantité de grains sur la qualité du décorticage et sur la consommation spécifique de carburant. Les grains décortiqués sont également soumis à l'appréciation d'un échantillon de ménagères afin qu'elles les comparent à ceux décortiqués traditionnellement.

3.3 Suivi des unités pilotes

Les unités tests sont placées auprès de groupements féminins choisis en fonction de leur degré d'organisation et de leur esprit d'initiative. Le suivi des unités se fait à travers des fiches où sont notées les quantités traitées pour chaque type de céréales, les consommations de carburant et de lubrifiants, les pannes et leurs causes, les anomalies de fonctionnement, les difficultés rencontrées dans la gestion quotidienne de l'unité, les dépenses et les recettes, ainsi que les contraintes sociales et économiques rencontrées par les utilisateurs.

IV RESULTATS ET DISCUSSIONS

4.1 Mise au point du décortiqueur

Les essais menés sur le modèle PRL ont abouti à la mise au point du mini-décortiqueur ISRA-SISMAR. La principale innovation par rapport au mini-PRL est la présence d'un système de nettoyage du grain décortiqué composé :
- d'un séparateur-nettoyeur constitué d'un

cylindre muni d'un tamis et de deux brosses ;
- et d'un ventilateur relié d'une part au séparateur-nettoyeur, et d'autre part à un cyclone par deux conduites flexibles.

La chambre de décorticage est constituée d'une boîte métallique avec couvercle dans laquelle une série de 10 disques sont montées sur un axe horizontal à des intervalles de 3 cm. Le décorticage se fait par abrasion, à sec. Les grains décortiqués sont recueillis dans une trémie par basculement de la chambre, puis acheminés vers le séparateur-nettoyeur par gravité et vis sans fin. Les brosses, dans leur mouvement de rotation, nettoient les grains et forcent le son au travers du tamis. Le son tombe au fond du cylindre où il est aspiré puis refoulé vers le cyclone par le ventilateur, tandis que le grain est récupéré par une goulotte fixée sur le couvercle du cylindre.

Les résultats technico-économiques du suivi des unités tests ont permis d'apporter les modifications suivantes sur ce modèle :
- suppression du système de basculement de la chambre ;
- élimination du séparateur-nettoyeur ;
- division de la chambre en deux parties afin de permettre un meilleur contact surface abrasive/grain, et réduction des disques de 10 à 8 unités ;
- aspiration directe du son à partir de la chambre ;
- et aménagement d'un dispositif de vidange à volets renforcés au fond de la chambre de décorticage.

Ces modifications ont débouché sur un appareil d'une capacité utile maximale de 10 kg, pouvant être mû par un moteur thermique ou électrique, et s'adaptant parfaitement aux conditions technico-socio-économiques de l'environnement aussi bien rural qu'urbain du Sénégal. Par rapport au modèle ISRA-SISMAR précédent, le coût d'acquisition a été réduit de 25 p. 100 au moins.

4.2 Essais sur mini-décortiqueur PRL

Pour des charges de grains allant de 0,5 à 8 kg, des régimes de rotation de 1500 à 2000 tours/mn et des temps de travail de 3 à 5 minutes, les meules en carborundum ont donné des taux de décorticage variant de 5,65 à 20,52 p. 100 pour le mil et le sorgho, et de 6,25 à 17,84 p. 100 pour le maïs, et des consommations spécifiques de gas-oil de 18 à 4,13 ml/kg pour le mil et le sorgho, et 17,5 à 4,36 ml/kg pour le maïs. Le taux de décorticage est proportionnel à la charge de grains, au temps de travail, et à la vitesse de rotation des

disques. La consommation spécifique de carburant est inversement proportionnelle à la charge de grains, et directement proportionnelle au temps de travail et au régime des meules.

Avec les disque en résinoïde et pour des régimes allant de 2000 à 3000 tours/mn, les autres facteurs restant les mêmes, le taux de décorticage varie de 6,4 à 35 p. 100 pour le mil et le sorgho, et de 13 à 31,5 p. 100 pour le maïs, tandis que la consommation spécifique va de 13,3 à 2,4 ml/kg pour le mil et le sorgho, et de 18,5 à 4 ml/kg pour le maïs. Le taux de décorticage est proportionnel à la vitesse des disques et au temps de travail ; il est inversement proportionnel à la charge de 0,5 à 4 kg, et directement proportionnel de 4 kg à plus ; ainsi, les plus faibles taux de décorticage sont obtenus avec 4 kg. Les variations de la consommation spécifique de carburant sont identiques à celles obtenues avec les meules en carborundum, c'est-à-dire directement proportionnelles au temps de séjour et à la vitesse des disques, et inversement proportionnelles à la charge de grains.

En comparant les résultats des deux types de disques, on constate que les disques en résinoïdes réalisent des taux de décorticage plus élevés et des consommations spécifiques d'énergie plus basses que les meules en carborundum ; d'autre part, les meules s'usent beaucoup moins vite que les disques en résinoïdes en raison de leur épaisseur et de leur résistance à l'usure plus grandes.

4.3 Essais sur mini-décortiqueur ISRA-SISMAR

Ces résultats concernent le dernier prototype équipé de 8 disques en résinoïde. Les disques en résinoïdes ont été choisis à cause de leur plus grande disponibilité au Sénégal, mais également pour leur faible prix d'achat par rapport aux disques en carborundum.

Le taux de décorticage varie de 15 à 27 p. 100 pour le mil et le sorgho, et de 12 p. 100 pour le maïs avec des charges allant de 0,5 à 10 kg, des temps de travail de 1 mn 30 s à 2 mn 30 s et des régimes de rotation de 1800 à 2200 tours/mn. Comme avec le mini PRL, les taux maxima sont obtenus avec 0,5 kg et les minima avec 4 kg, toutes conditions étant égales ailleurs. Les consommations spécifiques de gas-oil dans les mêmes conditions varient de 13 à 2,5 ml/kg pour le mil et le sorgho, et de 17,5 à 4,5 ml/kg pour le maïs. Elles sont inversement proportionnelles au temps de travail et au régime des risques.

4.4 Acceptabilité des produits

Tous les lots décortiqués ont été moulus puis distribués à un échantillon de 300 ménagères aussi bien en zone rurale qu'en milieu urbain. Dans leur grande majorité, les ménagères ont apprécié très favorablement les produits dont le taux de décorticage était égal ou supérieur à 18 p. 100. Les échantillons dont le taux de décorticage, était inférieur à 18 p. 100 étaient moins appréciés ou franchement rejetés lorsque le taux de décorticage était inférieur à 10 p. 100, surtout lorsqu'il s'agissait de sorgho ou de mil. Les échantillons les plus appréciés étaient les plus décortiqués, ce qui confirme les enquêtes antérieures effectuées dans ce domaine (3, 6, 8).

4.5 Suivi des unités tests

Les unités ont été implantées à partir de Juin 1987 dans 8 villages et 1 ville. Les sites ont été choisis de façon à représenter les principales situations technico-économico-sociales du Sénégal : milieu urbain, village-carrefour, village enclavé, village à revenus élevés, village relativement pauvre, etc.

La principale panne a concerné le système de basculement de la chambre de décorticage, en particulier le levier de l'embroyage qui permet d'arrêter la rotation des disques au moment du remplissage et de la vidange de la chambre. Les principales pièces d'usure se sont révélées être les disques, le tamis et les brosses. C'est sur la base de ces observations qu'il a été décidé de supprimer le système de basculement et d'aspirer de la chambre. Ces modifications ont entraîné une réduction globale du coût de l'appareil de 25 p. 100, et une plus grande maniabilité due à la simplification des principales manoeuvres.

Les quantités journalières traitées varient de 25 à 200 kg avec une moyenne de 20 à 25 jours ouvrables par mois, 240 à 300 jours par an. Une analyse sommaire des prestations fait ressortir 2 types d'unités suivant la qualité des utilisateurs et la destination du produit :
- les unités villageoises où la clientèle est presque composée exclusivement de ménagères et le produit destiné à l'autoconsommation ;
- et les unités commerciales qui satisfont également les besoins des commerçants qui vendent le produit ailleurs : on en compte trois sur les neufs implantées. A l'exception des commerçants et de quelques particuliers, les quantités apportées pour le travail à façons sont de l'ordre de 3-4 kg par personne. Ceci a une double conséquence : l'élévation de la consommation de carburant et l'usure rapide des disques, ces derniers ne travaillant qu'avec les extrémités au lieu d'utiliser toute la surface abrasive disponible.

Les calculs faits à partir des données du suivi donnent des prix de revient variant de 10 à 35 F CFA/kg. Ce sont les unités villageoises qui atteignent les prix les plus élevés, principalement à cause de coûts fixes unitaires et de consommations de carburant plus grands que dans les unités commerciales.

Si l'on ne considère que les résultats économiques obtenus après 18 mois, on concluerait que ce matériel n'est pas rentable dans la plupart des villages-test. Faudrait-il pour cela le limiter aux sites urbains et semi-urbains ? Nous pensons que non !! Le décorticage mécanique est une nouveauté au Sénégal et, comme toute innovation technologique, il y a des barrières techniques, économiques et psychologiques à surmonter.

Vers 1944, des industriels proposèrent d'installer des moulins à vent à Saint-Louis pour soulager les femmes. Les marabouts s'y opposèrent parce que les femmes tomberaient dans l'oisiveté et les vices y découlant. L'objection fut jugée respectable par l'administration coloniale et les moulins à vent furent prohibés jusqu'en 1857 (2). Les premières enquêtes socio-économiques que nous avons réalisées montrent que les femmes sont très satisfaites des services du décortiqueur et que s'il arrive qu'elles ne l'utilisent pas, c'est uniquement par insuffisance de moyens financiers, la priorité étant donnée à la mouture mécanique. Elles s'organisent d'ailleurs pour une appropriation vraie et propre de cette innovation : collecte des surplus de production pour la vente du grain décortiqué, champ collectif, embouche bovine et ovine, etc. Afin de diminuer le prix de revient du décorticage, il convient d'étudier le système décorticage/mouture afin de voir les possibilités d'utilisation d'une seule cellule motrice pour les deux machines. Il faudra surtout étudier les possibilités de revenus supplémentaires qui autoriseront une utilisation régulière de l'ensemble décortiqueur-moulin par les ménages ruraux et urbains.

V CONCLUSION

Les travaux menés par l'ISRA et la SISMAR ont permis de mettre au point un décortiqueur adapté aux besoins de la transformation artisanale des céréales locales autres que le riz au Sénégal. Les résul-

tats techniques et sociologiques obtenus
sur une période de 18 mois permettent
d'envisager une plus large diffusion de
cet appareil. La rentabilité économique en
milieu rural pourra être assurée par une
bonne organisation des groupements utilisateurs, et la mise en place de circuits
d'approvisionnement en pièces détachées
et de maintenance appropriés, ainsi que
l'utilisation d'une seule cellule motrice
pour l'ensemble décortiqueur/moulin.

REFERENCES BIBLIOGRAPHIQUES

1. DIOP, A., 1980 - Essais d'ajustement
   du moulin JACOBSON et paramètres de
   base pour le décorticage. ISRA-CNRA/
   BAMBEY - Sénégal.
2. DUVAL, J. 1859 - Politique coloniale de
   la France. Revue des deux mondes, tome
   XIII.
3. MBENGUE, H.M., 1985 - Projet de technologie post-récolte EP-79-0066 phase II.
   Rapport final. Document de travail
   D/Systèmes N° 85-10, 48 p. ISRA, D/Sys-
   tèmes, Dakar, Sénégal.
4. MBENGUE, H.M., 1986a - Les équipements
   et matériels de traitement post-récolte
   des céréales au Sénégal. D.T. D/Systè-
   mes N° 86-5, 39 p. ISRA, D/Systèmes,
   Dakar, Sénégal.
5. MBENGUE, H.M., 1986b - La mécanisation
   de la transformation des céréales au
   Sénégal : aspects techniques et nutri-
   tionnels. D.T. D/Systèmes N° 86-7,
   23 p. ISRA, D/Systèmes, Dakar, Sénégal.
6. MBENGUE, H.M., HAVARD, M., 1986a - La
   technologie post-récolte du mil au
   Sénégal. Etudes des différents niveaux
   de mécanisation. D.T. D/Systèmes N°
   86-2, 46 p. ISRA, D/Systèmes, Dakar,
   Sénégal.
7. MBENGUE, H.M., HAVARD, M., 1986b - Résultats d'enquêtes sur la technologie
   post-récolte des céréales au Sénégal.
   D.T. D/Systèmes N° 86-6, 40 p. ISRA,
   D/Systèmes, Dakar, Sénégal.
8. YACIUK, G., 1977 - Résultats de l'enquête sur la technologie post-récolte
   en milieu rural au Sénégal. ISRA, CNRA,
   Bambey, Sénégal.

Land and Water Use, Dodd & Grace (eds), © 1989 Balkema, Rotterdam. ISBN 90 6191 980 0

First approaches to robot utilisation for automatic citrus harvesting

G.Blandini
Istituto di Meccanica Agraria, Università di Catania, Italy

P.Levi
AID SpA, Robotic and A.I. Division, Catania, Italy

ABSTRACT: This paper describes an application of sensors and artificial vision systems to a first prototype of electric robot, designed and built by A.I.D. for automatic citrus fruit harvesting. The robot, made up with standard components currently available on the market , has a single arm and it has successfully operated both in laboratory and in an orange grove. The trials have given very useful information for the design and manufacturing of a second more efficient and multiarm prototype.

RESUME: Les premiers essaies pour l'utilisation des robots dans la récolte des agrumes. Ce travail décrit l'application de senseurs et d'un systeme de vision artificielle,le premier de cette sorte réalisé par l'A.I.D, qui est utilisé pour la retolte àutomatique des agrumes. Ce robot a seul bras a ete essaye au laboratoire et dans les plantations des agrumes.les epreuves nous laissent des informations utiles dans l'objectif de la réalisation d'un robot prototype multibras.

ZUSAMMENFASSUNG: ANFÄGLICHE VERWENDUNG DES ROBOTERS FÜR AUTOMATISCHE ERNTE DER ZITRUS-FRÜCHTE. Hier wird über der Einsetzung von Sensoren und künstlichen Seheenssystemen auf einen ersten Prototyp von elektrischen, mit Zylinderkoordinaten Roboter berichtet, der von A.I.D. für die automatische Ernte der Zitrusfrüchte geschaffen worden ist. Dieser Roboter, mit Standardkomponenten geschaffen, die leicht aufindbar auf den Markt sind, hat einen einzigen Arm und ist schon vielmals im Laboratorium und in einer Agrumenpflanzung geprüft worden. Die Versuche haben uns eine Reihe von interessanten Aufklärungen gezeigt, die besonders nützlich für die zukünftliche Planung und Leistung eines zweiten, vielarmigen, wirksameren Prototyp sein können.

1 FOREWORD

In Italy there are about 165,000 ha of citrus orchards, 100,000 located in Sicily; the overall yearly production is about 3 million of tons and it is mainly for the fresh fruit market.

The harvesting cost rose up to 35% of the total production cost and it is on charge of the trading companies.

The use of special machines, such as hidroelevators, grove trucks or pallets, is limited due the low economic saving related to the high investment and the small grove sizes.

The mechanical harvesting with air blowers or mechanical shakers and the complementary chemical treatment, never passed the experimental stage even in countries where the production is mainly for processing as it is difficult to detach the citrus fruit and to avoid damages to fruits and trees.

As an alternative research, institutes and industries are carrying out feasibility studies and developing prototypes of robotic harvesting machines, with a major look to the citrus area.

The problem is very hard: industrial robots driven by vision operate in controlled environments, with known and modifiable parameters; on the other hand, grove conditions spread out over a large range, such as illumination, weather, scene complexity; the intrinsic arm task of positioning, stem identification and cutting is very complex.

Some experimental studies have been already carried out. Parrish and Goksel developed a system for apple harvesting, based on optical filtered images acquired by a computer and then binarized; Harrel et al. developed a hydraulic driven robot arm for orange harvesting, driven by a color based vision system with statistical pixel classification; Grand D'Esnon et al. designed and built a sophisticate hydraulic robot for apple harvesting, with vision system based on optical filtered images digitally tresholded to obtain a binarized image.

The main problem is to select a reliable, cheap and fast vision system, able to manage the harvesting task with efficiency and to minimize investment and operative costs.

The A.I.D. S.p.A., after a long feasibility study, developed a first prototype with cylindrical coordinates for oranges harvesting; its vision system is based on color processing, edge detection and scene interpretation through models.

This paper describes the prototype and its operation and gives a report of the first trials carried out both in laboratory and in grove.

2 THE EXPERIMENTAL PROTOTYPE

2.1 Specifications

The robot is made of the following subsystems (fig. 1):
 - control;
 - vision;
 - actuator;
 - arm;
 - end effector;
 - system software;
 - application software.

The control subsystem is built around a 32 bit microprocessor and it hosts the software development environmment and the application programs; other than the CPU, there are the central memory, the comunication interface and the motion control board, which feeds the control signal to the servoamplifiers and gets motion information from the encoders and limit switches.

The vision subsystem is made of a dedicate hardware in order to give a separate

computing engine to the vision tasks: the architecture is modular and special purpose processors can be added in order to implement new functions. Data flows to the system bus, the video bus and the in/out video channels are managed by a control board. The input video channel is made of a chroma key module, a video buffer, an A/D converter and a syncro generator. The video output channel uses a D/A converter; this subsystem hosts the frame grabber, a custom edge processor and a model matching board. The latter one identifies the position in the scene of a prestored model, where the difference between the model pixel values is minimum.

The image sensor is a CCD RGB camera, with an 8 mm autoiris lens; the autoiris is driven from the video signal. The RGB signals are sent to a chroma key module which generates a monochromatic, video composite, signal whose level is roughly proportional to the difference of the hue detected and a reference hue; in other words, this signal peaks when the scene hue is orange.

The actuators subsystem gets the power from a bulk unregulated DC source driven from an insulating tree-phase transformer. A command signal is generated from the digital motion controller and it is transformed into a power pwm signal which drives the DC motors of the arm.

The arm is very simple and it is assembled on a small trailer, toghether the electronic cabinets and the power source: the trailer is driven by a tractor (fig. 2). The arm has 3 degrees of freedom and it is a cylindrical coordinates system (fig. 3): it can rotate by mean of a rotating table driven with a gearmotor (axis θ), elevates (axis z) and extends (axis r); axis r and z are made of a couple of linear ball screw units linked by a cross connection slide. The arm hosts the end effector, the motors, sensors and the camera, this one being fixed to axis z.

The end effector, located at the end of axis r, is made of a mobile elix and a fixed cylinder (fig. 4); the end effector is moved in the proximity of the orange by the arm, under control of the vision system; an infrared optical sensor signals that it is time to activate the elix: the stem is captured between the mobile elix and the fixed cylinder and then cut.

The software system manages all the pick cycle and it is organized as a main module supervising the global process of scanning the whole canopy, a vision module wich gives the centroid of detected orange and a harvest module which drives the arm to detach the target fruit.

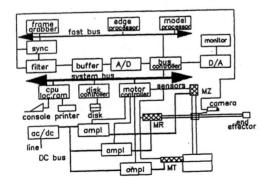

Fig.1 - System block diagram

Fig.2 - A.I.D.robot for citrus automatic harvesting

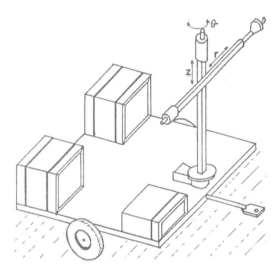

Fig.3 - Prototype plan of A.I.D. robot mounted on trailer

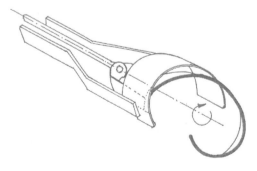

Fig.4 - Plan of A.I.D. patented manipulator

2.2 Working conditions

The working area over the canopy is divided in 20 smaller areas, called Logical Vision Fields (LVF): the arm makes a complete harvest cycle over LVF before moving to the center of an other one. The LVFs are fixed to the tree, so their projection move inside the CCD and the frame grabber as the z and θ axes move; to keep always a LVF inside image memory of the frame graber, the LVF has been dimensioned to give a 180x192 pixel projection.

When working over a LVF, the vision system analises the scene looking for an orange centroid; when the centroid has been detected, axes z and θ are activated and the vision process acts only a small area of interest surrounding the position of the orange, until the centroid is aligned with the camera.

During the pointing phase, the system activates a dinamic tracking of the orange coping with small deviation of the orange from its position due to wind or other causes; a real time speed adjustment is continously made in order to have a fast but smooth motion profile. Once the orange is aligned to the camera, axis r is aligned and extended, the end effector reaches the operating area and the fruit is gently detached.

3 TRIAL RESULTS

3.1 Generality

The design of a trial methodology has been very hard due to the difficulty of reproducing in the laboratory the grove conditions and the constraints derived from the biological cycle and the severe operating outdoor conditions.

A preliminary activity of laboratory and grove data acquisition has been made in order to make a rough evaluation of the performance of the vision system.

A special care has been put in the laboratory evaluation of the sensitivity of the vision system to the lights, the background, the color and shape of the oranges, the presence of obscuring leaves or branches and the cluster of fruits.

3.2 Laboratory trials

The laboratory trials have been characterized by the following parameters:
- orange shape: circular and squared
- orange diameter: 66, 83, 115 (mm)
- max error in model matching: 90
- incident light degree: 0, 20, 40, 60,

90, diffused
- reference filter hue: orange
- background: black, blue
- neon distance: 1.5 m
- direct light distance: 1 m
- light types: 0 -> Neon;
            1 -> 2000W, 2500K filament
            2 -> 500W, 2000K halogen
            3 -> 1000W, 5600K halogen

The following tests have been carried out using a plastic tree:
  A) centroid determination while varying orange dimension and shape;
  B) centroid determination for a fixed dimension orange, against a black background, while it is progressively obscured by leaves;
  C) scene analysis with 6 oranges and lights 1, 2 and 3.
  D) centroids determination ot two oranges of fixed dimension, with increasing overlapping, against a black background and with light 2.

The A test showed a not satisfying response in shape discrimination whist, for a selected shape, is quite not sensitive to different dimension.

The results of the B tests showed that the error in centroid identification increase when the visible part of the range becomes smaller; this is obvious because the centroid position is calculated as the center of mass of the projected area.

Tests C gave good results regard system tolerance to light condition variations, this being due to the sophisticated edge processing. In any case, a strong direct light gives some problem both for the effect of the objects and the dynamic of the image sensor: this leads to a reduction of the capability of well contrasting the orange hue.

Results from D tests showed that up to a 5% overlapping the two centroids were detected fine, but when the overlapping reached 25% the system identified alternatively two centroids, leading to an arm instability. For an overlapping greater than 50%, only one centroid was detected as the system was no more able to distinguish two oranges.

3.3 Field trials

Grove tests have been carried out in a Valencia orange grove (fig.5):
  - during the day, at different times;
  - during the night: with artificial lights 1, 2 and 3.

Due to the different reflectance between the artificial tree and the real one, optimum parameter values were different between laboratory and grove.

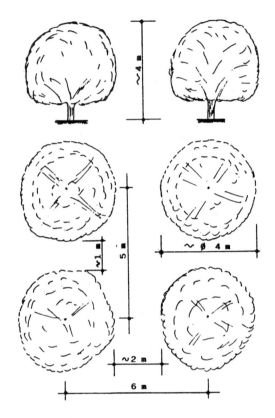

Fig.5-Citrus grove lay-out

Two new parameters were introduced:
-G (camera gain):   A (auto), 0, 6, 12(db)
-WB (white balance): A (auto), K (3200 K),
                     I (indoor), O(outdoor)

Day-time tests have given satisfactory results, particularly during the hours when the tree was uniformly illuminated without sunshine through the leaves. A particular problem has been due to the re-greening phenomenon of Valencia oranges because grove tests have been carried out on July, at the end of the ripening cycle; this phenomenon caused little errors in centroid determination with subsequent bad positioning of the end-effector and failure in picking the fruit. Further improvement, both in vision process and end-effector error recovery capability, will solve the problem.

Best results have been obtained with G=A and WB=O, as the best contrast between oranges and other objects was achieved. With G=A and WB=K, the performance was accettable only with an auxiliary artificial light with 3200 K color temperature. If G=A and WB=I, the results became unacceptable as the leaves were not filtered out of the scene. Other parameter combinations gave unsatisfactory results.

Although these tests were very rough, obviously the night operations have given better results, because it was possible to determine the best suited artificial light.

Besides the vision process itself, during the grove tests the overall automatic harvest cycle has been tested: the average pick cycle required 10 s and 70% of the oranges recognized from the operator was identified by the machine; 65% over this 70% has been harvested.

This performance seems to be unacceptable, but it is enough because the prototype was a first tentative to apply such innovative technologies to a very hard application area.

4. CONCLUSIONS

The results obtained by A.I.D. developing this first electrical robot prototype demonstrate the feasibility of the automation of the fruit harvesting by mean of a machine controlled by a microcomputer which uses an artificial vision system.

Major limits are evident in the vision system precision and in end effector performance; the prototype will be upgraded and a test methodology will be defined.

A future tree year project will lead to a multiarm prototype of a robotic machine for citrus harvesting, which will operate in groves with a regular layout. The new machine will adopt a real time, goal oriented, vision system with 3D capability: the control computer will be able to drive the arms to the fruit, position an intelligent end effector, detach the orange, cutting the stem and conveying it to the bin.

The new project spcificates a machine with autonomous navigation and a throughput of 2,000 oranges per hour at a 65% efficacy. In order to facilitate the automatic harvesting, new growing techniques will be studied, such as special mechanical pruning, in order to have the fruits mainly on the outside canopy.

REFERENCES

Grand D'Esnon A. 1985. Robotic harvesting of apples. Procuration of Agrimation, ASAE

Grand D'Esnon A. & Pellenc R. & Rabatel G. & Journeau A. & Aldon M. 1987. Magali: a self propelled robot to pick apples. ASAE paper n.87-037

Harrell R.C. & Adsit P.D. & Slaughter D.C. 1985. Real time vision-servoing of a robotic tree fruit harvester. Winter Meeting ASAE, Chicago

Harrell R.C. & Adsit P.D. & Slaughter D.C. & Pool T. 1986. Image enanchement in robotic fruit harvesting. ASAE paper n.86-3041

Kawamura M. & Mamikawa K. & Fujiura T. & Ura M. 1983. Microcomputer controlled manipulator system for fruit harvesting. Research report on Agricoltural Machinery, n.13

Levi P. & Falla A. & Pappalardo R. 1988. Image controlled robotics applied to citrus fruit harvesting. Proc. RoviSeC-7, Zurich

Parrish E.A. & Goksel A.K. 1977. Pictorial pattern recognition applied to fruit harvesting. Trans.of the ASAE, n.5

Pejsa Y.H. & Orrock Y.E. 1983. Intelligent robot systems: potential agricultural applications. ASAE First Internat. Conf.

Tutle E.G. 1983. Image controlled robotics in agricultural environments. ASAE First Internat. Conf.

Land and Water Use, Dodd & Grace (eds), © 1989 Balkema, Rotterdam. ISBN 90 6191 980 0

The impact of transportation on forage harvesting and processing procedures

H.D.Bruhn & R.J.Straub
University of Wisconsin, USA

R.G.Koegel
USDA/ARS Forage Research Center, Madison, Wisc., USA

K.J.Shinners
University of Wisconsin, USA

ABSTRACT: While forage quality is of extreme importance, quality is no longer the only factor affecting forage marketability. Density and ease of handling in shipping and feeding are now factors in forage marketing. The standard rectangular bale of field cured hay with a density of only 240 kg/m^3 may be unsatisfactory both on basis of quality, cost of transportation, and labor required for loading, unloading, and feeding. while the large cylindrical rolled bale and the large rectangular bales of higher density may be well suited for specialized situations, they are by no means universally acceptable. Present research on stem maceration and mat formation at cutting time has potential to increase the field curing rate of forage (alfalfa) and to prevent leaf shattering and loss in gathering. The secondary compressing of bales as well as such procedures as pelleting, wafering, cubing, and briquetting have the potential of converting cured forage of a density from 80-240 kg/m^3 to bulk densities in the 640-720 kg/m^3 range. In extreme conditions it may be feasible to compress forage to considerably greater density and re-shred for feeding.

RÉSUMÉ: Quoique la qualité du fourrage à une très grande importance, ce n'est plus le seul facteur qui détermine sa valeur commerçiale. La densité et l'aisance de manutention au transport et à l'affouragement sont aussi des facteurs. C'est possible que la balle rectangulaire traditionelle de foin séché au soleil avec une densité de 240 kg/m^3 ne convienne pas non seulement au point de vue qualité et frais de transport, mais aussi coût de main d'oeuvre pour la manutention et l'affouragement. Les grandes balles cylindriques ou rectangulaires de haute densité conviennent bien au applications spécialisées cependant, elles ne sont pas acceptées partout. Les recherches courantes en macération des fourrages et formation de "nattes" nous donnent la possibilité d'un séchage beaucoup plus rapide au champ et permet d'éviter l'effeuilage et la perte des feuilles au remassement. La compression secondaire des balles par les procédes de formation de pellets, gaufrettes, cubes, ou de briquettes, donne la possibilité d'augmenter la densité de la luzerne de 80-240 kg/m^3 à 640-720 kg/m^3 environ. Sous conditions extrêmes, il est possible de comprimer des fourrages à des densités encore plus élevées et de les broyer avant l'affouragement.

ZUSAMMENFASSUNG: Die Grünfutterqualität ist zwar sehr wichtig, sie ist aber nicht der einzige Faktor für den Verkauf. Dichte und einfache Handhabung sind jetzt auch wichtig. Die normalen Pressballen von feldgetrocknetem Heu mit einer Dichte von nur 240 kg/m^3 mögen hinsichtlich Qualität, Transportkosten, und Arbeit beim Laden, Entladen, und bei der Fütterung nicht genügen. Die grossen Rund- und Kubischenballen von höhere Dichte mögen für Sonderfällen gut geeignet sein, sie sind aber keineswegs überall verwendbar. Gegenwärtige Versuche mit intensiver Aufbereitung beim Mähen und anschliessendem Mattenpressen bieten die Möglichkeit der Schnellfeldtrocknung und der Vermeidung von Bröckelverlusten beim Ernten. Das Nachpressen von Ballen und andere Verfahren wie Pelletieren, Brickettieren, uns. geben die Möglichkeit die Dichte von feldgetrocknetem Heu von 80-240 kg/m^3 bis auf 640-720 kg/m^3 zu steigern. Notfalls ist es möglich noch höhere Dichten zu erreichen, mit Nachzerkleinerung fürdie Fütterung.

INTRODUCTION

Since the time of primitive agriculture the primary objective of making changes in forage harvesting procedures has been the

reduction of labor required, with a minimum
loss in the quality of the final product.
From an initial herd size of one goat or
one cow agriculture has gradually changed
to much larger herds, in many cases to two
or three hundred cows per herd, and in
some cases over 1000.

While crop production per hectar has
improved with better varieties and new
cultural practices, this has by no means
kept pace with the increase in herd size,
so we are now gathering forage from a
continuing greater distance.

Also there has been the factor of speci-
alization where livestock is concentrated
in certain areas because of climatic conditions, available labor, or a host of
other reasons. The results are that
forage as well as grain production may
become more and more a "cash" crop.

TRANSPORTATION

While this need for greater transportation
facilities has developed, our transportation system has grown adequately to meet
these demands. The short hauls pose no
real problems. Transportation within the
operating range of the field harvesting
machinery is quite adequate. The problem
develops where transportation extends to
major highway travel for moving forage
through commercial hay markets or overseas
shipment. Here the fact that forage (hay)
is both a bulky and a perishable commodity
becomes a serious problem.

FORAGE DENSITY

Hay as it was initially harvested and
stored in a mow has a density of 64–80
kg/m3 (4–5 lb/ft^3). However, in this
condition the amount of harvesting labor
is high, and the hay has very limited
commercial value. Field baling or other
methord of compaction could have reduced
the harvesting labor, reduced the storage
space required, and, in addition, made it
a product suitable for sale at market as
it can be more readily transported and
inspected.

A common on-farm field baling operation
increases the density to the range of
128–160 kg/m^3 (8–10 lb/ft^3). If this hay
had been baled in anticipation of sale
and transport, the density could have
been increased to 192–240 kg/m^3 (12–15
lb/ft^3). If the bale shape and dimensions
were such that they stacked with no voids,
this would be sufficient for truck loading.
It would be only about 1/3 of the density
desirable for overseas transport.

EXCESS MOISTURE AND DETERIORATION

While the gains achieved by increasing the
density to the above range have provided
some gain in transportation, storage space,
labor reduction, and ease of inspection,
there was a negative factor in that the
hay had to be reduced from the acceptable
25–27% moisture (wet basis) content for
long loose hay storage, to 20–24% for
small rectangular bales, and to the 15–19%
range for the large dense bales. Exceeding
these moisture ranges will result in mold
growth or discoloration within the bale.

Both the large round bale and the large
rectangular bale weighing in the 400–800 kg
range require more extensive field drying
either by longer exposure, by a higher
field drying rate, or by additional drying
immediately following the baling operation.

LOW MOISTURE AND HARVEST LOSS

In the areas of low rainfall (irrigated
areas) longer exposure time can be tolerated, although the lower moisture level
does result in greater leaf shattering
loss during the baling operation. When
mean bale moisture was decreased from
32.9 to 13.6%, mean total losses increased
from 16.4 to 19.3% (Koegel et al 1985).

In more humid regions where forage crops
grow very well because of frequent rainfall, longer exposure to achieve sufficiently dry hay for large bales is not a desirable solution because of the probability
of rain damage during the field drying
period.

BALE DRYING

Batch drying of the standard small rectangular bale following baling has been practiced in a small number of instances. Some
special crop drying buildings have been
used for this purpose (Bruhn and Matthias
1959a). Bales of 96–128 kg/m3 (6–8 lb/ft^3)
at 35% moisture content were successfully
dried to 18% using both solar or a combination of solar and off-peak electric heat
(Parker and Collins 1988). At least one
continuous bale drier was designed. In
common practice small bales were usually
stacked loose enough in storage to allow
considerable ventilation.

An investigation was conducted at the
University of Wisconsin of all of the
spontaneous forage fires in the state in a
two-year period (Bruhn 1970). This study
indicated that the probability of spontaneous ignition in small rectangular bales
in storage was only 1/3 of that of long
loose hay in a mow, while the probability

of a barn fire in chopped hay was four times that of long loose hay.

Because of greater tendency of the large dense bales to retain heat, especially when covered or closely stacked, they do present problems leading to spontaneous ignition. However, mold growth or heat discoloration appear under much less severe conditions, and as a result there are numerous studies of larger round bale drying in attempts to develop a system to prevent deterioration in quality during both transportation and storage. Brandemuehl et al (1988) reported more uniform drying and a 20-hour time reduction with radial drying compared with axial flow drying and justified drying because of higher quality.

ACCELERATING FIELD DRYING

In the midwest dairy belt of the USA the corrugated roll forage crusher has accelerated the field drying rate of alfalfa to the point where hay can be sufficiently dry for low-moisture silage in a few hours and dry enough for small rectangular bales on the second day (Bruhn 1955a).

A newer development, that of a much more intensive shredding of the forage stem and the formation of a cohesive mat laid back on the stubble to dry, promises to bring about a much higher field drying rate, while at the same time reducing the loss of fine leaf particles (Shinners et al 1985). An in-depth report of this new process is being presented by Dr. Kevin Shinners at this meeting in his paper entitled "Rapid field drying of forages utilizing shredded mat technology."

This process promises to provide not only a faster rate of drying, but, in addition, less losses in gathering the forage for baling, increase in digestibility, and the extension of field drying to the lower moisture level desirable for the large dense bales. This may make it feasible to carry compaction in the field to an even greater density.

ULTRA DENSIFICATION OF CONVENTIONALLY CURED FORAGE

A delayed secondary compaction of field baled forage offers some advantages in preparing forage for marketing and transportation. The delay allows for some natural additional moisture reduction and the possibility of being performed when labor and facilities are more readily available.

Buchele and Marley (1978) reported the experimental compacting of "small round alfalfa bales" from a length of 0.66 m (26 inches) to a length of 0.38 m

(15 in) with a force of 907N (2040 lb) and compacting large alfalfa bales to approximately 30% of the initial length with a force of 3.5×10^6N (787,000 lb).

A recent study at the University of Wisconsin (Keleny et al 1985) reports that previously baled forage at 79ºC, when pressed with a 30-second hold time in a 15.9 cm x 15.9 cm (6.25 x 6.25 in) chamber at 6.9 MPa (1000 psi) was compacted to a density of 740 kg/m^3 46.2 lb/ft^3 but when compression was released, expanded to a relaxed density of 239 kg/m^3 (14.90 lb/ft^3).

At a compacting pressure of 20.7 MPa (3000 psi) the compressed density reached 1081 kg/m^3 (67.5 lb/ft^3) and expanded to a relaxed density of 460 kg/m^3 (28.7 lb/ft^3). While this degree of relaxed density does not reach the state of compaction desirable for overseas shipment, it would be sufficient under most highway regulations to load a truck to its maximum weight limit without exceeding the maximum height, width, or length limit.

If a standard farm-type bale of 35.6 cm x 45.7 cm x 91.4 cm (14 in x 18 in x 36 in) with a density of 160 kg/m^3 (10 lb/ft^3) were compressed to a density of 460 kg/m^3, its compressed length would be 31.9 cm (12.5 in). Modification in truck bed construction for bale loading would be necessary for these short bales.

Random dumping into an enclosed box-type container would result in a lowering of the bulk density to about 75% of the individual relaxed compressed bale density, but this could be an advantage because of labor saving on loading and unloading.

PELLETS, WAFERS, AND CUBES

In cases where ultra high density is desired, such as for overseas shipment or for a bagged product for sales in smaller quantity, a solution is the formation of pellets, wafers, or cubes.

The most dense type of forage is the hard pellet normally made from dehydrated ground alfalfa and formed with an extrusion type die. These pellets usually range in diameter from 2.4 m to 25 m (3/32 – 1 in) and have a length of one to three times the diameter. Instantaneous pressure during extrusion may reach 448 MPa (65,000 psi) and may attain a bulk density of 640–720 kg/m^3 (40–45 lb/ft^3) (Dobie 1959). Because of the high production cost and the fact that these pellets are not suitable to supply the entire forage ration of a dairy cow, they are not generally adapted to dairy production (Bruhn et al 1959b).

A larger type of pellet made of chopped

but not ground forage (Bruhn 1957) is much more applicable to both domestic and overseas shipment of forage for dairy cattle. This chopped unground forage is frequently compressed into forms referred to as pellets, wafers, or cubes. Fundamentally the characteristics of all are quite similar. These pellets may be 50-65 mm (2-2½ in) in diameter and of a length approximately equal to the diameter. The wafers range from 100-130 mm (4-5 in) in diameter and the thickness is approximately 0.2-0.25 times the diameter. Cubes are approximately 35 mm x 35 mm x 45 mm (1 3/8 x 1 3/8 x 1 3/4 in).

CATTLE ACCEPTANCE OF HIGH DENSITY FORAGE

While the cattle appeared to have some difficulty eating pellets compressed at pressures ranging from 55 MPa to 69 MPa (8000-10,000 psi) to a density of 800-900 kg/m³ (50-60 lb/ft³), individual pellet density, the difficulty was not sufficient to cause any reduction in dry matter intake (Bruhn 1955b).

Wafers and cubes of the same density seemed to cause no problem to the cattle. The shape factor of wafers and cubes seems to be advantageous.

Pellets, wafers, and cubes made from unground hay must be in the 10-14% moisture (WB) range for long term storage. However, it is feasible to produce these hay forms from material of higher moisture content and then continue the drying by circulating heated air through the storage area.

MOISTURE LIMITS COMPRESSIBILITY

Water is incompressible and during compaction tends to be expressed from high moisture plant cells, causing alfalfa to lose its cohesive characteristics. This accounts for the very low density of compaction and the extensive loss in density of the 36% moisture pellets shown in Fig. 1. At a given pressure dry forage compacts more readily than damp forage and also expands less for a shorter period of time after being removed from the compression chamber. Some of these compacting machines fail to operate on alfalfa above 20% moisture content (WB).

Alfalfa at 10-12% moisture (WB) can be compacted into pellets, wafers, or cubes at a pressure of 35-55 MPa (5000-8000 psi) to an individual pellet density of 800-880 kg/m³ (50-55 lb/ft³), that is, a bulk density of 600-660 kg/m³ (37-41 lb/ft³). This provides a satisfactory feed for dairy cattle, loads, and unloads with grain handling conveyors, and falls

in the minimum overseas transport cost category of 640-700 kg/m³ (40-44 lb/ft³).

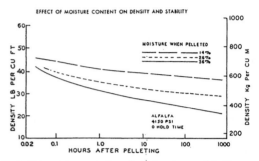

Fig. 1 High moisture content results in less dense pellets (dry matter basis) and also in greater expansion of the pellets which causes further divergence in density over a period of time.

Molitorisz and McColly (1972) report an ingenious roll formed and sliced type of wafer compacted to a density of 640 kg/m³ (40 lb/ft³) with a radial compacting force of 2240N (500 lbs) and a density of 800 kg/m³ (50 lb/ft³) with a radial force of 5376N (1200 lb). At a moisture content of 20% the energy required was reported at 8.2 kW/tonne (10 hp hr/ton) when compacting to a density of 480 kg/m³ (30 lb/ft³) and increased very rapidly above 640 kg/m³ (40 lb/ft³).

TRANSPORTATION LIMITATIONS

Major long distance transportation of forage material is either by highway truck or by sea-going vessels. Truck weight and size limitations in the USA are under state regulation, but because of interstate travel, these regulations have become quite uniform. Except in special cases the allowable truck size limits are 2.79 m (8 ft 6 in) wide, 4.43 m (13 ft 6 in) high and 13.12 m (40 ft) long for single trucks and 19.69 m (60 ft) long for truck tractor and trailer (Wisconsin Department of Transportation July 1987). For practical purposes this is a load approximately 2.62 m (8 ft) wide by 2.62 m (8 ft) high by 15.75 m (48 ft) long. Gross weight limitations are based on number of axles, spacing, etc. This includes the weight of the vehicle which varies. An interstate trucking firm indicated that the loading of their 15.75 m (48 ft) semi-trailers was limited to 20,408 kg (45,000 lb). This is equivalent to a density of 234.7 kg/m³ (14.65 lb/ft³).

This density could be attained by precise stacking of the ultra compressed conventional bales and also by the random dmuping of these bales into an open box-type truck trailer body.

Balers are available that make bales appoximately 1.31 m x 1.31 m x 2.62 m (4 ft x 4 ft x 8 ft) at a density of 192-240 kg/m^3 (12-15 lb/ft^3). There is little prospect of any easy method of modifying 1.6 m or 1.97 m (5-6 ft) diameter or length cylindrical bales for void free truck loading. Complete reprocessing into wafers, cubes, or pellets seems to be the only solution.

OVERSEAS TRANSPORT

For overseas shipment minimum costs dictate bulk densities approaching 640-700 kg/m^3 (40-44 lb/ft^3). The small hard pellets made from ground dehydrated alfalfa will meet this criteria. However, because of dehydrating costs and limited applicability to dairy production, there may be limited demand for this product.

The alfalfa cube, the wafer, and the large unground alfalfa pellet are all well adapted for both dairy cow feeding and horse feeding. The lower density of the desired shipping range is approximately equal to the upper limit density that dairy cows readily consume and also approaches the upper limit of manufacture.

The alfalfa cube is well suited to both bulk handling similar to grain and particularly well suited to containerized shipping. An additional feature of pellets, wafers, and cubes is that concentrates can be blended with the forage as a part of the compaction process. Thus, if desired, a complete ration can be provided with little opportunity for sorting by the animal.

QUALITY DETERIORATION IN TRANSPORT

Usually baled hay is transported uncovered on flat bed trucks. In many cases the bales were stacked outside uncovered prior to shipment. It is possible that enough moisture could be absorbed by a bale during a heavy rain so that, if that bale were to be restacked deep in a pile of dense bales, spontaneous ignition could result.

Many people seem willing to accept deterioration if it is the result of a labor saving activity and the forage is used near the point of production. However, if forage is to be transported considerable distances, inferior products cannot be tolerated.

PRESERVATIVES

Maceration and densification seem to provide conditions that accelerate microorganism activity. With more dense baling, cubing, etc. more discoloration and heat demage can be expected. Numerous preservatives are available to be applied to baled or cubed alfalfa prior to or in the compaction process.

The effectiveness of a new product to be applied to cubed alfalfa prior to shipment was recently reported by Huhnke and Brusewitz (1988). The purpose of this product is primarily the reduction of the moisture absorption rate in humid climates and the retarding of mold growth.

SUMMARY

The need of the dairy farmer for high quality hay, the rejection of poor quality by the long distance customer, and the price differential between good and poor quality forage in the commercial hay markets may cause a general over-all desire by the hay producers to provide higher quality forage. In all probability this will stimulate development in quality hay harvesting processes.

REFERENCES

Brandemuehl, S.L., Straub, R.J., Koegel, R.G., Shinners, K.J. and Fronczak, F.J. 1988. Radial drying of high moisture large round bales. American Society of Agricultural Engineers paper 88-1066, St. Joseph, Michigan 49085-9659, USA.

Bruhn, H.D. 1955a. Status of hay crusher development. Agricultural Engineering 36: 165-170.

Bruhn, H.D. 1955b. Pelleting grain and hay mixtures. Agricultural Engineering 36: 330-331.

Bruhn, H.D. 1957. Engineering problems in pelletized feeds. Agricultural Engineering 38: 522-525.

Bruhn, H.D. 1970. Chopped hay biggest fire hazard. News and Features, University of Wisconsin, June 16.

Bruhn, H.D. and Matthias, V.W. 1959a. An all-purpose crop conditioning structure. Transactions of the ASAE 2: 58-60.

Bruhn, H.D., Zimmerman, A. and Niedermeier, R.P. 1959b. Developments in pelleting forage crops. Agricultural Engineering 40: 204-207.

Buchele, W.F. and Marley, S.J. 1978.
Analysis of a bale compactor. Proceed-
ings International Forage Harvesting
Conference, Ames, Iowa, USA, 308,
International Congress on Agricultural
Engineering and American Society of
Agricultural Engineers.

Dobie, J. 1959. Engineering appraisal of
hay pelleting. Agricultural Engineering
40: 76-93.

Huhnke, P.E. and Brusewitz, G.H. 1988.
Performance of Nutri-shield in protecting
alfalfa hay cubes. American Society of
Agricultural Engineers, paper 88-1552.

Keleny, C., Straub, R.J., Schuler, R.T.
and Koegel, R.G. 1988. Compression of
baled hay. American Society of Agricultural Engineers paper 88-1551.

Koegel, R.G., Straub, R.J. and Walgenbach,
R.P. 1985. Quantification of mechanical
losses in forage harvesting. Transactions
of the ASAE 28: 1047-51.

Molitorisz, J. and McColly, H.F. 1972.
Analyses of channel systems for rolling-
compressing hay wafers. Transactions of
the ASAE 15: 24-27.

Parker, B.F., White, G.M., Gates, R.S. and
Collins, M. 1988. American Society of
Agricultural Engineers paper 88-4539.

Shinners, K.J., Barrington, G.P., Straub,
R.J. and Koegel, R.G. 1985. Forming
mats from macerated alfalfa to increase
drying rates. Transactions of the ASAE
28: 374-377 & 381.

Wisconsin Department of Transportation 1987.
4801 Sheboygan Ave., Madison, Wisconsin
USA. Trucking Wisconsin Style. Regulations
and requirements.

Land and Water Use, Dodd & Grace (eds), © 1989 Balkema, Rotterdam. ISBN 90 6191 980 0

KSU date palm service machine

S.A.Al-Suhaibani & A.S.Babeir
Agricultural Engineering Department, King Saud University, Saudi Arabia

J.Kilgour
Silsoe College, Cranfield Institute of Technology, UK

ABSTRACT: Dates are one of the major crops in Saudi Arabia. Dates production system needs several operations which needs the man to climb to the top of the tree in spite of problems of this method. These operations require mechanization to replace the traditional method. This paper presents the priliminary field test of the designed and constructed date palm service machine.

1 INTRODUCTION

Dates are one of the major crops in Saudi Arabia where Saudi Arabia is ranked fourth among date producing countries. Al-Suhaibani et al (1). The number of palm trees in the Kingdom is now 11.4 millions of which 7.98 millions are productive with an estimated crop yield of 400,000 tonne of dates per year.

Several operations usually required for productive date palm trees such as dethorning, pollination, thining, bagging, pruning, and harvesting. These operations require mechanization to replace the traditional method which requires the man to climb the tree and work at a considerable height above the ground, which is a dangerous, slow and compounded by a severe labour shortage.

Mechanization of a date palm trees become one of the concerns of Agricultural Engineers in Saudi Arabia. In response to this, the Agricultural Engineering department of King Saud University (KSU) took an initiative to solve this problem and cooperated with Silsoe College (Cranfield Institute of Technology) to design, construct and test a date palm service machine to elevate the labour to the proper height for performing the operations safely and satisfactorily.

2 THE SURVEY

A survey was carried out in Saudi Arabia on 19 farms scattered in three regions in the Kingdom. Dates of tree spacing,

height, trunk diameter, and bunch spacing of a total of 1118 trees were collected in addition to the ground profiles. The maximum height of the trees for the 19 farms were ranged from 7.73 to 17.43, while the average maximum 12.7. Figure 1 shows the cumulative frequency distribution of height of the trees measured.

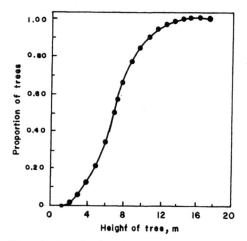

Fig. 1. Cumulative frequency distribution of height of 1118 palm trees from 19 farms

The trees in old farms are usually randomly scattered so the analysis included the degree of dispersion (or regularity) by calculating the nearest neighbour index R and corresponding statistic C. The nearest neighbour analysis is a way of measuring pattern in terms of the arrangement of a set of points in two

dimensions. The nearest neighbour index R can vary between 0 and 2.15 indicating a completely random pattern and a complete dispersed pattern respectively. There were nine fields with R ranging from 1.4 to 1.7 indicated fields with reasonably regular tree growing pattern. Value of C greater than 1.96 indicate a very regular dispersion in planting patterns. The higher the value of C the more regular the pattern. Correspondingly, large values of C indicated the consistent regular dispersion of trees with in the farms where only minor tree removal would give good access to the machine. Figure 2 shows the tree distribution in one of the fields which was typical of many of the farms.

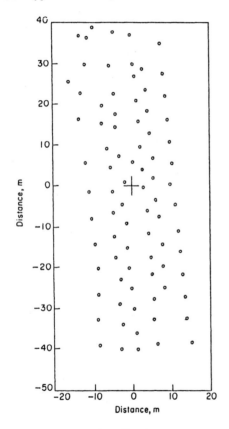

Fig. 2. Tree distribution,
planting density .
406 trees/ha. +: position
theodolite

3 DATA FOR GROUND PROFILES

For each site a number of clear sight lines were chosen in the general direc-
tion of the main slope of the land and a series of spot heights at each change of level were recorded. About 80% of the sites had small bunds and shallow channels.

Almost without exception the lines of the channels and bunds and lines of palms coincided so that it would be easily to either drive down rows going parallel to obstruction or to cross them at right angles. Diagonal crossings could lead to sideways stability problems which would be best avoided.

The average size of the obstructions in the 80% category was as follows:

Bunds:	Height	Width
	20 cm	50 cm

Channels:	Depth	Width
	30 cm	50 cm

4 DESIGN

From the survey the overall dimensions of the machine were chosen so that the machine would be suitable for most farm situations.

The principal dimensions are:- overall length 5.50 m, overall width 2.10 m, overall height 2.70 m, wheel base 3.50 m, track 1.75 m, underneath clearance 500 mm, Al-Suhaibani et al (2).

Figure 3 shows the machine in the field position with the stabilisers in the extended position. With the boom fully extended, the basket can reach trees in the following positions:- tree height 10m, row spacing 7m, tree spacing in row 5 m. For trees of this height and spacing the machine can reach 6 trees from one position as the top chassis has 360 degrees movement and can reach in all directions. For a single tree the maximum height of reach is 14 m. The machine can reach 98% of all trees found in the survey.

The basket is U-shaped in the plan view and can be angled sideways through 180 degrees so that it can fit round a tree just below crown of leaves. From this position the operator should be able to reach all the bunches of dates easily.

The dimensions are:- width 2.10 m, height 1.70 m. Tree space-width 0.75 m, length 1.10 m.

In the floor are two smaller baskets to contain the dates, of dimensions:- width 52 cm, length 93 cm, depth 50 cm, which can be lowered to the ground by hydraulic winch.

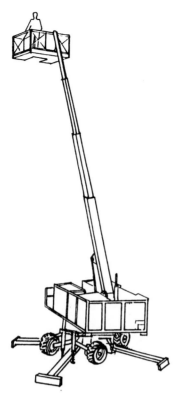

Fig. 3. Date palm service machine in the field position

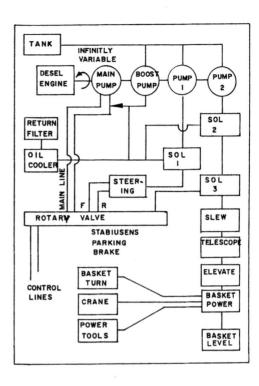

Fig. 4. Hydraulic circuit - main chassis

5 THE HYDRAULIC SYSTEM

All the functions of the machine are powered by a hydraulic system driven by a 52 KW Perkins diesel engine mounted transversely across the back of the machine. The circuit is divided into three main parts. The pumps, tanks, oil coller, and controls for the elevation of the basket are on the top chassis, see Figure 4. The wheel motor and steering circuit is shown in figure 5. The controls from the drivers' position may select four wheel drive with the differential lock, low ratio or high ratio or two wheel drive, low or high ratio to suit the ground and slope conditions.

A poclain H23 variable displacement pump with a remote hydraulic joystick controller is used together with Poclain series 2 dual displacement wheel motors. The rear motors are fitted with a fail safe hydraulic disc brake.

Both front and rear wheels may be steered independently of each other so that the machine may be steered front only at high road speed and crabbed or opposite for

tight turns in the field. Pump number 1 provides power for the steering or the oil cooler via solenoid valve 1 when the machine is stationary.

The stabilisers are operated by Pump number 2 when the machine is stationary. This pump also drives the oil cooler when the steering is being used. When the stabilisers out valve is operated all the stabilisers telescope out to their maximum position. When they are all out the pilot valve is actuated and then each stabiliser may be pushed down to take the load. To fold up the stabilisers the valve levers are operated in the opposite directions, see Figure 6.

When the pressure switches in the stabiliser down rams register the correct pressure solenoid number 3 allows oil to the basket controls. This valve is a Danfoss PV460 electrically controlled load sensing valve so that the various functions can be controlled very accurately and at variable speed and give easy positioning of the basket next to the trees.

The rams are fitted with lock valves to prevent collapse in the event of a pipe

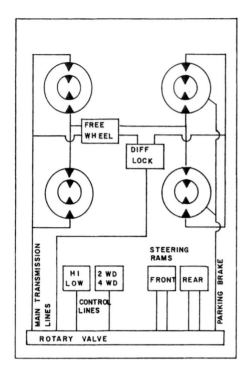

Fig.5. Hydraulic circuit - bottom chassis
transmission and steering

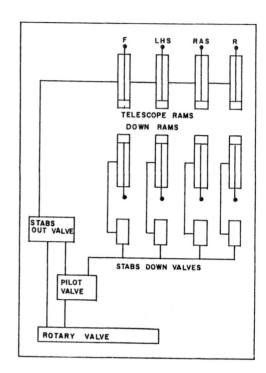

Fig.6. Hydraulic circuit bottom chassis
stabilizers

failure. If there is an electrical failure the basket controls can be operated
manually from the main valve bank situated
near the engine. If the engine also fails
the boom telescope, elevate and slew may
be operated by a hand pump or if in a
suitable position they can be let down by
using the bypass knobs on the lock valves.
An emergency ladder is provided in the
basket as a last resort.

By operating other valves the basket can
be turned 90 degrees left or right to
facilitate easy positioning at the tree.
They can also operate the crane to let the
boxes of dates down to the ground and they
can also be used to control as set of
Stanley hydraulic powered hand tools
supplied with the machine.

6 SAFETY FEATURES

The control unit only allows the machine

to be operated in certain ways with the
objective of making it safer for the
operator and bystanders.

Main features of the system are:
-Transport only possible with stabilisers
 fully retracted.
-Hi wheel motor ratio only possible with
 rear wheel straight.
-Operation of stabilisers only possible
 with harvesting machinery closed.
-Limited harvest actions if machine base
 becomes unsafe during harvest.
-No motion is possible when emergency
 stop switch activated.

There is a visual display on a panel by
the driver's seat which shows the status
of the various activities by a series of
red and green lights on a diagramatic
planview of the machine. So it is
possible for the operator to determine for
instance which stabiliser is not carrying
its minimum load so that he will know
which one to adjust.

7 WEIGHTS

With the basket in the fully extended position 200 Kg of dates and one operator can be carried. The floor baskets can hold 50 Kg of dates to be lowered to the ground.

On either side of the boom on the top chassis are two payload containers with a total capacity of approximately 1 tonne of dates.

8 PRELIMINARY FIELD TEST

A field test was conducted in March 1989 at a date palm farm near King Saud University Campus at Riyadh after a short training of three persons.

8.1 TERRAIN MOBILITY

On this particular site, the machine could easily drive from the roadways down with the fields and was able to climb over the minor irrigation channels without touching the chassis. Care was necessary on the drier parts of the field to drive over the channels at right angles to avoid side ways instability problems. The typical height of the top of the bund to the bottom of the channels was 30 cm.

8.2 TREE SPACING

The machine could fit between the trees and except on the first position could easily turn 360 degrees to any position required. At the first position one of the trees was out of line from the rest and the machine could be turned only enough to reach 4 of the possible 5 in that position.

8.3 SOIL STRENGTH

On the areas which the machine entered, the cone penetrometer reading was 42.5 Kpa between the wheel tracks and 49.7 Kpa along the wheel tracks (average of 48.9Kpa reading). Some slight increase in compaction of 5% in the softer areas of the field but not in the harder areas. At the lowest part of the field where it was wettest the rut depth was approximately 15 cm and had the machine been driven further in this area it might have got stuck. The soil on this site had a relatively high clay content and it would not have been possible to drive on a flooded soil of this type.

8.4 POSITIONING TIMES

Data were collected for three sets of trees, where the times for each activity were recorded for each tree. The height of the tree to the crown was also recorded. table 1.

Table 1. Service time (T, sec.) for 3 sets of trees and tree heights (H, m)

Set	1		2		3	
Tree No.	H	T	H	T	H	T
1	7	72	12.6	116	11	151
2	9.6	102	12	128	11	232
3	10.1	115	11.4	450	9	135
4	12	154	7.6	167	11.3	136
5	8.4	166	12	258	12.8	177
6			11.7	242	12	133
7					9.6	187
Average	9.4	122	12.9	227	11	164
T/H	12.9		17.6		15.0	

The time per tree is widely variable because of the operators had different background on driving farm equipment. In the first position the machine was operated with the best operator among the three while the second one was the least profficient. Figure 7 shows set No. 2 (position # 2).

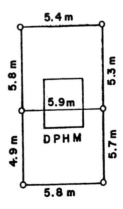

Fig. 7. Set No. 2 tree distribution.

The preparation time required in each position are given in table 2.

Table 2. Machine preparation time for
different positions

Main function	Position		
	1	2	3
Stabilizers			
Extend	64	64	54
Down	38	28	27
Level	119	13	13
Movement			
Foldup	216	130	162
Transfer		244	106
Moving distance (m)		17.2	17.2

8.5 COMMENTS ON THE TIMES

The stabiliser time figure was rather long
due to the stabiliser feet the load was
partly taken by the tyres again so that
the warning system showed an unsafe situation necessitating bringing the stabilisers further down. With more experience,
the operator should have lifted the machine slightly higher on the stabiliser in
the first instance.

8.6 HYDRAULIC HAND TOOLS

The tools were tried out on an area of old
dilapidated trees so as not to cause
damage to the productive trees until the
operators had developed a reasonable
skill.
  1) 60 cm chain saw. This could cut
easily through a tree but it was necessary
to 'Vee' out the cut progressively as the
dead tree still had a high moisture content and the saw would have jammed easily.
  2) Shears. This could cut through the
base of the leaf fronds easily and the
jaws were large enough for any in this
area.
  3) Circular saw. This was used to cut
off the leaf bases from the main trunk of
the tree. It worked well and was easily
manageable.
  4) Strimmers with nylon line. This was
used to cut off the spikes near the bases
of the leaves. It was totally useless as
the nylon dissappeared in a few seconds as
the work was too hard for it. Odd leaf
fronds became tangled in the shaft which
would have caused overheating and damage
to the seals in a very short time. Further
development is necessary for removing the
spikes.

CONCLUSION

The following were concluded from the
priliminary test:
1. The mobility and manoeuvrability of the
machine was satisfactory. However,
special attention is necessary when the
channels are dried by driving the machine
straight.
2. The compaction increase was 5%.
3. The recorded activities time is considerable, reasonable taking into consideration the short training of the operators.
4. Chain saw, shears and circular saw
worked well, while the strimmers which was
used to cut the spikes near the bases of
the leaves was totally useless.
  The major field test of the machine is
planned in July and August of 1989 during
date harvesting season.

REFERENCES

1. Al-Suhaibani, S.A.; A.S. Babeir;
  J. Kilgour; J.C. Flynn. 1988. The Design
of a Date Palm Service Machine. J. Agric.
Engng. Res. 40, 143-157.
2. Al-Suhaibani, S.A.; A.S. Babeir;
  J. Kilgour. Design specification of a
Date Palm Service Machine. AMA. (in
press).

Land and Water Use, Dodd & Grace (eds), © 1989 Balkema, Rotterdam. ISBN 90 6191 980 0

Fully mechanized winter pruning of grapevines

E.Cini, A.Cioni, D.Vannucci & M.Vieri
Institute of Agricultural Mechanization, University of Florence, Italy

ABSTRACT: The prototype here shown was developed at the Institute of Agricultural Mechanics of Florence and is characterized by a new system of blade contrast based on cut diameter selection, which permits pruning close to the stakes, and avoids fraying canes. The initial device was designed in the United States together with Prof. H. Studer, of the Agricultural Engineering Dept. of the University of California at Davis (1985). The cutting technique is based on the contrasting motion of the "daisy", which is free to rotate only with the advancement of the tractor, and the blades rotate in the opposite direction. On the basis of the first set of results from the field tests, the authors propose the coupling of this machine with other commercial pruning devices to improve their efficiency.

ZUSAMMENFASSUNG: Vollmechanisce winterbeschneidung von weinstoecken
Der hier vorgestallte Prototyp wurde im Institut fuer Landmaschinenbau in Florenz entwickelt und zeichnet sich durch ein neuaretiges Kontrastblattsystem aus, das auf dem variablen Snittdurchmesser beruht. Das System ermoeglicht das Bescneiden rund um die Pfaeble, so dass durch Fransenschnitt herforgerufene Schaeden vermieden verden. Das erste modelle wurde 1985 in den Vereinigten Staaten in Zusammenarbeit mit Prof. H.Studer vom Departement fuer Landmaschinenbau der Universitaet Davis in Kalifornien entworfen.
Die Scneidetechnik beruht auf der gegenlaufigen Bewegung eines Rades, das sich nur in der richtung des fahrenden Traktors frei dreht, waerend die Blaetter im Gegensinn rotieren. Auf der grundlage der ersten praktiscen Te"stergebnisse schlagen die Autoren vor, diese Maschine mit anderen handelsueblichen Beschneidegeraeten zu koppeln, um zu noch hoeheren Leistungen zu gelangen.

RESUME: Mecanisation integrale de la taille hivernal des vignes
Le prototype qui est ici présenté a eté réalisé par l'Institut de Mecanisation Agricole de Florence; il est caractérisé par un nouveau système de lames opposées, basé sur la sélection du diamètre à couper: il permet d'élaguer autour des échales et évite les dommages entrainés par un coupe effrangée. Le modèle initial a été projeté en 1985, en collaboration avec le professeur H.Studer du Département d'Ingénierie Agricole de l'Université de Davis (Californie).
La technique de coupe est basée sur les mouvements opposés d'une "marguerite", qui ne peut tourner librement que si le tracteur avance, et des lames, qui tournent en sens contraire. Les premiers résultats d'essais sur le terrain conseillent de coupler cette machine avec d'autres élagueuses offerte par le marché afin d'en augmenter l'efficacité.

1 INTRODUCTION

The machines usually employed for pruning vines are, generally speaking, closely analogous in their operational principles and in the results obtained. They differ primarily in their cutting mechanisms, whose action gives the row a hedge-like form which is usually rectangular in transverse section.
Most of the pruning machines currently on the market require pre-positioning on the row, and as a result, work at a fixed height and with a fixed space between two

cutting mechanisms set opposite each
other, one to either side of the row.
The circumstances described above impose
upon the vegetation a uniform, straight
disposition so as to avoid that the
cutting mechanisms come in contact with
the trunk of the plant or with the
stakes.
Some machines feature systems which
automatically center the cutting
apparatus to give the pruning action a
measure of symmetry even when the
disposition of the plants does not ensure
optimal regularity.
On the basis of the observations made by
the authors, the operational limitations
of the machinery currently available for
the winter pruning of vineyards may be
summarized as follows: a)inadequate
capacity to select the parts of the
plants which are to be pruned; b)
difficulty in making a cut in the
immediate vicinity of obstacles (trunks
and vine-stakes); c) the risk of a badly
executed cut which leaves the spur frayed
and broken.
Our research on the mechanical pruning of
vines trained to an espalier system was
already underway in 1978 at the Institute
of Agricultural Mechanics of the
University of Florence and has led to the
construction of a number of experimental
machines, the most recent of which is the
object of the present comunication. The
aim is to overcome the problems listed
above without applying technologies whose
complexity would inevitably lead to high
costs for the machinery incorporating
them.
From the very beginning, the major goals
of the research were to achieve a high-
quality cut and to determine the ideal
structure to support a cutting system to
be used on a variety of training systems.

2 MACHINE COMPONENTS

The prototype consists of a chassis, a
hydraulic system and two oppose pruning
units (Fig. 1).
The type of chassis used in building the
prototype here described is avalaible
commercially and is routinely used to
support vineyard-equipment such as tying
machines.
The hydraulic system is responsible for
all of the component movements (Fig.2).
The pump of the tractor supplies pressure
to the flexible tubing of the hydraulic
circuit, feeding the two motors arranged
in series to power the two pruning units,
and driving a dual-action jack to adjust

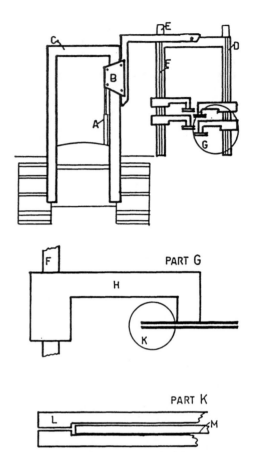

Fig.1 A- Lifting jack mounted on the
chassis. B- Lifting jack used to regulate
the cutting height of the pruning
machine. C- Chassis carried by the
tractor. D- Drive-shaft for the pruning
units. E- Hydraulic motor. F- Frame
mounted on the chassis (c). G- Pruning
head (detail). H- Supporting arm for
the stars. K- Pruning element (detail).
L- opposed plates. M- Blades.

working height within a range of 0.60 m
above ground level.
The driver uses two controls mounted on
the tractor which operate the two motors
and the jack; there is also a valve which
regulates the flow of oil to the motors,
and thus their rotating speed, up to 250
r.p.m.
Since the machine is designed "over the
row", it is equipped with two identical
pruning units (to cut from either side of
the vine-row) each one composed of two
heads. Each head consists of a
supporting arm and a daisy equipped with

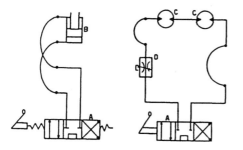

Fig.2 Hydraulic system. A- Slide valve.
B- Jack regulating the cutting height of
the pruning machine. C- Hydraulic motors.
D- Valves.

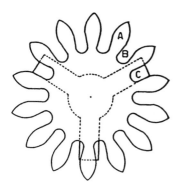

Fig.3 Daisy (cutting element). A-
Petals. B- Cutting indentations. C-
Blades.

blades, the elements which actually cut
the canes (but even saws can be used).
Each pruning unit is driven by one of the
hydraulic motors noted above, which have
vertical axes of rotation and propel a
0.150 m shaft.
The motors are set atop of the chassis
supported by a specially-designed bearing
inserted into a horizontal support plate
which is part of the chassis.
The supporting arms are locked onto the
shaft, at the appropriate pruning height.
Each arm is made up of three tubes of
varying diameter and thickness welded
together at right angles to form a "U".
The arm houses two shafts which transmit
the movement by means of two pairs of
gear drives to a third shaft which drives
the blades. The daisy is connected by
bearings to the U-shaped tube; its
diameter is 290 mm, its thickness 30 mm,
its weight 1650 g.

Each daisy is composed of a pair of iron
plates superimposed to create a space for
the blades.
The contours of the two plates are
matched to prevent the wires of the
vineyard from entering the space occupied
by the blades.
The indentations in which the canes are
cut are 50 mm deep and 24 mm wide
(Fig.3).
The petals have rounded ends to help
avoid the obstacles without damaging
them.
The blades for each daisy are two in
number, 7 mm thick, rectangular (70 x
35mm).

3 OPTIMIZING THE DESIGN

To arrive at the optimal shape and
dimensions of the petals and of the
spaces between the petals above
described, and thus to obtain the optimal
efficiency and selectivity,it was
necessary to design a series of field
surveys that would furnish valid
information on: 1) the average diameter
of canes; 2) the average distances
between the canes, the disposition of
canes with respect to a system of spatial
coordinates XYZ, with plane XY parallel
to the ground and the plane XZ coinciding
with the vertical extension of the row;
3) the dynamic behavior of the mass of
canes in response to the action of the
cutting mechanism.
As regards point (1) and (2), field
measurements were taken concerning the
various cultivars characteristic of
Tuscan vineyards producing Chianti wines.
Moreover, the authors have developed a
particular technique whereby a simple
analysis of a series of photographs can
determine an operational scheme which
responds to the overall configuration of
the vegetation.
The photographs were taken with a 50 mm
lens at a predetermined distance of 2 m,
from three orthogonal positions for each
row sample examined and with a white
screen as a background. (Fig.4)
The projection of the resulting slides
onto a screen which features a system of
trigonometric coordinates and the scale
of distances made it simple to read the
dimensions, orientation and spacing of
the canes.
Because of the different cultivars in the
Tuscan vineyards, it was necessary first
of all, to verify if it was possible to
have statistically homogeneous average
vegetation parameters, taking into

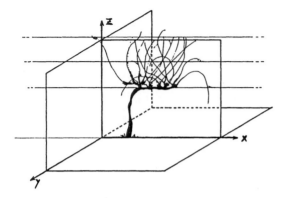

Fig.4 Schematic view of the row with the three planes at which the photographs were taken.

Table 1. Average statistical values (100 readings for each vegetation parameter) for the geometric parameters of the Chianti vineyard - tested for pruning - (*) this parameter is about the same for every cultivar. For the meaning of S and G look at eq.{1}.

Cultivar	Plants %	Mean vine branch diameter [mm]	σ	Mean distance between wine branches [mm]	σ	Vine branches angle from horizzontal [rad] (*)
sangiovese	80	14,8	5.0	128	3,2	Π/2 ÷ Π/15
canaiolo	10	17,6	9.6	133	3,5	
trebbiano	8	16,8	3,5	117	10,4	
malvasia	2	13,0	3,5	122	14,4	

account the different cultivars. To obtain this result, the mean value of each cultivar parameter Gi(i=1,2,.m; i = identification number of cultivar) was calculated.

We then calculated the weighted mean G as if all the values were statistically homogeneous (here the weight is the percentage presence of a single cultivar in the vineyard producing Chianti wines, and the reading number is the same for every cultivar). After the computation of the variance (σi^2) for each kind of parameter, we adopted the criterion expressed in equation {1} to be certain that the parameters were statistically homogeneous.

$$S = \sum_{i=1}^{m} (\mu i / \sigma i^2) * (G - Gi) \approx 0 \qquad \{1\}$$

where: S is the evaluation coefficient, and μi is the number of measurements adopted to calculate each Gi.

This procedure, easily implemented on a P.C., permits one to obtain values for each kind of statistically homogeneous parameter.

The results are summarized in table 1. Concerning point (3), it was necessary to perform a series of experimental field tests with a physical model of the prototype so that it was possible to determine empirically the dimensions of the compression spring which controls the daisy position on the row.

These tests also revealed that slight variations in the diameter of the daisy, can optimize the interception of the canes to be cut, if the foreward progress

of the tractor falls within a certain range of speeds (1,5÷2 km/h) which varies whit working conditions.

4 OPERATION

If the machine is to operate in different sorts of vineyard, its various heads must first be set at the correct intervals in order to act upon the vines in a way appropriate to the training system in question.

Then, as work progresses the driver can use the controls to change the position of the pruning units and tailor the action of the machine to the irregularities of the vegetation.

The forward motion of the machine pushes the daisies against the row by means of a spring which acts upon their support arms. The blades, which are concentric with the daisies, thus find the opposition necessary to make the cut and are blocked if a cane gets in the daisies' way.

As the machine moves ahead the tips of the petals penetrate the vegetation and every cane that enters the cutting field is severed by the blades.

When the cutting mechanism encounters a stake or the trunk of a vine, the star avoids it and cuts only the canes, even where they are much tangled (Fig.5).

The distinctive shape of the daisy allows only those canes with a diameter inferior to a given figure to enter the notches between the petals (the cutting field). Consequently, in designing the petals one can pre-determine the maximum diameter of the canes to be cut.

The distances between the petals can be chosen in response to the growth pattern of the vegetation in the vineyard, the kind of stake used, etc. It then becomes necessary to adopt these daisies whose

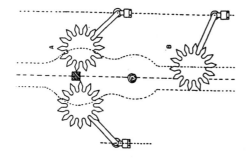

Fig.5 Operational Diagram. A- Position of the daisy by-passing a stake. B- Daisy in working position.

Table 2. Dimensions of the daisy derived from field parameters.

Canes diameter [mm]	Petals number n°	Petals width [mm]	Width indentation [mm]	External diameter [mm]
16	14	20	16	280.5
18	14	20	18	289.5
20	14	20	20	298.3
22	15	20	22	320.6
30	16	20	30	374.7

specifications are most appropriate to work in the particular vineyard in question.

To this end, we must add that as one varies the specifications of the daisy one can also vary the number of petals, their length, their thickness, etc.

To illustrate these factors the following Table 2 gives the specifications of some of the more widely applicable daisies.

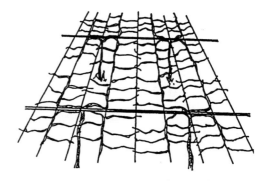

Fig.6 Arbor as modified for mechanical pruning.

5 APPLICATION TO DIFFERENT TRAINING SYSTEMS

The most important features of the prototype are: a) the independence of the cutting elements as modular units, permitting the adaptation of the action of the machine to the particular vegetation to be cut; and b) the distinctive construction of the cutting element, permitting the automatic avoidance of the structures supporting the rows and the trunks of the vines.

By taking advantage of these features one will be able to remove a great quantity of canes with a substantial reduction in the manpower necessary for completing the work.

The prototype here presented can be used for pruning espalier training systems such as the Casarsa system, various types of spurred cordon systems, palmettes, etc.

With the correct chassis, however, the pruning heads can be set for pruning vines with other training systems without costly modifications.

It could thus be used with the Gobelet and Double Espalier systems and with the new "Lyre" (form which have been used on a trial basis in France for a number of years).

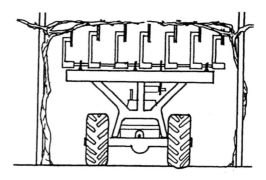

Fig.7 Diagram: the machine at work on an arbor.

In addition, with certain modifications in the training system or in the supporting structure, it is also possible to prune the traditional Arbors and Trentine Pergolas.

Let us now examine in some detail the modifications necessary to achieve this goal.

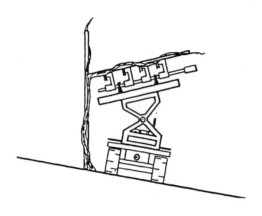

Fig.8 Diagram: machine for pruning a modified pergola.

5.1 The Arbor

A number of modifications in the traditional arbor have already been suggested to allow mechanical harvesting. The additional changes required for the introduction of mechanical pruning involve only the training wires (Fig. 6). Perpendicular rows of wires are carefully spaced to form a framework of which it would be necessary to eliminate only those wires which are transverse with respect to the main horizontal branches of the arbor.

These wires must be replaced by wires appropriately thicker in diameter but also more widely spaced (one new wire for every three or four rows in the old framework). In addition, these new wires must be sheathed in tubes of synthetic material whose diameter is sufficiently wide to avoid penetrating the cutting system of the daisies.

Once the vineyard is made structurally adequate, the new wires can also be made to correspond with the main distribution lines of the irrigation system.

The pruning heads will thus be turned by a drive shaft set horizontally below the vegetation and mounted upon it so as to cut the canes at the point desired (Fig.7).

Since the daisies would turn on a vertical plane and reach a certain distance up into the mass of vegetation, they would be able to cut not only the protruding canes but also some of those growing horizontally.

5.2 The Pergola

As with the arbor, a number of modifications have been proposed in the traditional pergola structure to permit mechanized harvesting. Of these, one in particular is immediately useful for winter pruning using the proposed prototype (Fig. 8).

This is the "Espalier Pergola," with its fruit-bearing zone lying upon a single surface inclined more or less sharply from the horizontal while the new shoots grow upward, guided by an espalier with one or two wires.

6 FINAL CONSIDERATIONS AND CONCLUSIONS

The testing performed on the prototype here presented has fully confirmed our expectations: at a speed of about 1.8-2km/h and at 150 rpm the cutting efficiency of the machine reaches 99% in Tuscan spur cordons, which permits us to place it among those pruning machines which require only minimal supplementary manual labor.

In particular, the vine-obstacle avoiding system and the system for selecting the diameter of the cut, together with the shape of the notch and the size of the blades, make for very precise cuts even in the immediate vicinity of the obstacles, above all the stakes.

Our opinion is that the experience and the results gained through the experimental development of the prototype here presented can lead to further progress, which can be summarized as follows:

1) combining a number of such machines at different levels to construct pruning machines suitable for pergola systems and requiring only minor modifications to avoid cutting the training wires;

2) the construction of automatic pruning machines for various types of orchards where a high-quality cut and diameter selection are essential;

3) the varied applications of combined obstacle-clearing and drive (spring, belt) systems to a series of tasks which require low power-levels and greatly reduced oscillation after clearing obstacles.

REFERENCES

Becker ,E. 1984. Geeignetes Unterstuetzungs material fuer den fortschrittlichen Weinbau. Die deutche Weinbau, n°6.

Clingeleffer,P.R. 1984. Production and growth of minimal pruned Sultana vines. Vitis, n°1.

Girardeau,D. 1984. Fonction décision-commande d'un robot manipulateur destiné à tailler automatiquement la vigne. Académie Agriculture de France, n°5.

Gould,I.V.;Whiting,J.R. 1987. Mechanisation of Raisin Production with the Irymple Trellis System; Transection of the ASAE, n°1 1987.

Lavezzi,A.;Vannucci,D. 1986. Montpellier 1985 e le nuove tendenze della viticoltura francese. Vignevini,n°3.

Sevilla,F. 1984. Etude et réalisation d'une machine a tailler la vigne a partir d'un modèle tri-dimensionale. Académie Agricolture de France, n°5.

Studer,H.E. et al. 1982. The alternating duplex- a training system for raisin production. Grape and Vine Centennial Symposium Proocedings 1880-1980. University of California, Davis.

Vagny,P. 1984. Demonstration de matériels de travaux à bois morts. Vignes et Vins, Feb.-March.

Vannucci,D. 1984. Proposta di una macchina rotativa per la potatura di vigneti a controspalliera. Rivista di Ingegneria Agraria, n° 1.

Vannucci,D.; Cioni,A 1988. A winter pruning device for various vine training systems.2nd International seminar on mechanical pruning of vineyards-February 1988;Motta di Livenza - Italy.

Vannucci,D.;Cini,E.;Vieri,M 1988. Fully mechanized winter pruning of grapevines. AgEng - Paris.

Land and Water Use, Dodd & Grace (eds), © 1989 Balkema, Rotterdam. ISBN 90 6191 980 0

Research on edge-curve angles of rotary blades

Ding Weimin & Peng Songzhi
Agricultural Engineering College, Nanjing Agricultural University, People's Republic of China

ABSTRACT: This paper gives a detailed description of the edge-curve angles of the rotary blades and the relationships between these angles, derives a dynamic equation for the sidelong portion of the blades. This equation is universally applicable for rotary blades. It incorporates all factors in one equation. By altering the parameters of the equation, other relevant equations can be obtained. This provides a convenient and flexible method for rotary blade design.

RÉSUMÉ: Ce travail donne une déscription détaillée des angles coupe-courbe des lames rotatoires et des relations entre ces angles, dérive une équation dynamique pour la partie de côté des lames. Cette équation est universellement applicable pour les lames rotatoires. Elle incorpore tous les facteurs dans l'une équation. En changeant les variables de l'équation on peut obtenir des autres équations applicables. Ce changement fournit une méthode confortable et flexible pour le plan des lames rotatoires.

ZUSAMMENFASSUNG: Dieser Beitrag beinhaltet eine detaillierte Beschreibung des Krümmungswinkels von rotierenden Schneiden und des Verhältnis zwischen diesen Winkeln. Es wird eine dynamische Gleichung für die Bestimmung des seitlichen Teils der Schneiden hergeleitet. Diese Gleichung ist für Rotationsschneiden allgemein anwendbar. Alle Faktoren werden in einer Gleichung berücksichtigt. Werden die Parameter der Gleichung geändert, so können andere relevante Gleichungen erhalten werden. Dies liefert eine brauchbare und flexible Methode für die Konstruktion von Rotationsschneiden.

1 INTRODRUCTION

Rotary tillers are widely used for soil cultivation. Rotary blades are the working parts of the tiller. When the rotary tiller is used in grassy paddy fields, in which soils are soft and grasses are difficult to cut, the edge-curve angle, or slide-cutting angle, is an important design parameter for the rotary blade (Sakai 1977). In order to prevent grass and straw from coiling around the blades and to keep the tiller working successfully, the edge-curve angle of the rotary blade should be greater than the sliding angle of the grass and straw so that grass and straw can slip out along the edge curve of the rotary blade (Wu 1981).

2 EDGE-CURVE ANGLES OF ROTARY BLADES

The edge-curve angle, or slide-cutting angle, of the sidelong portion of the rotary blade is defined as the acute angle between the velocity vector and the normal plane of one point on the edge curve of the rotary blade (Peng & Wu 1982), expressed by τ. This angle is on the plane constructed by the velocity vector and the tangent line of the edge curve. If the blade itself is also located on this plane, like the lengthwise portion of the rotary blade, the edge-curve angle is defined as the angle between the velocity vector and the normal line of the edge curve (Wu 1981).

The complement of the edge-curve angle, i.e., the angle γ between the velocity vector and the tangent line of the edge curve, is called the cutting angle.

When the rotary tiller is working, the rotary blades rotate around the rotary axle with the angular velocity ω and they also move forward simultaneously with the velocity u together with the tiller. The

absolute velocity v of any point on the
edge curve is the resultant of the linear
velocity rω and the forward velocity u,
i.e., v = rω + u, where r is the turning
radius of any point on the edge curve.
The value of rω is greater than the
value of u. Sometimes for simplicity it is
assumed that u = 0 and v = rω. In this
case the edge-curve angle of the rotary
blade is the angle between the linear
velocity and the normal plane (or normal
line), and it is named as the static edge-
curve angle, expressed by τ_s. The static
edge-curve angle is only an assumed,
approximate edge-curve angle. The actual
edge-curve angle is distinguished as the
dynamic edge-curve angle, expressed by τ_d.
    Corresponding to the static and the
dynamic edge-curve angles, are the static
and the dynamic cutting angles, expressed
by γ_s and γ_d respectively.
    The edge-curve angles and the cutting
angles of the lengthwise portion are shown
in Fig.1. t-t is the tangent line of the
edge curve and n-n, the normal line.
Because OL ⊥ rω and t-t ⊥ n-n, θ = τ_s.
Thus the static edge-curve angle of the
lengthwise blade can also be defined as the
angle between the turning radius and the
tangent line of one point on the edge curve
(Sakai 1977).

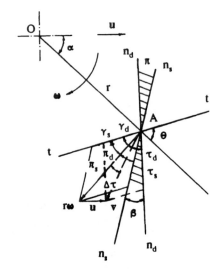

Fig.2 Edge-curve angles and cutting angles
of the sidelong portion

edge curve is not parallel to the plane of
the picture. The vertical plane of t-t,
i.e., the normal plane π (the shaded part
in Fig.2) of the edge curve, is not
perpendicular to the plane of the picture
either. The plane π_d constructed by t-t and
the velocity vector v, and the plane π_s
constructed by t-t and the linear velocity
rω are not the same plane. Neither are
they parallel to each other. On the normal
plane π of the edge curve, there is a
angle β between the intersecting line n_d-n_d
of π and π_d and the intersecting line n_s-n_s
of π and π_s. The angle between v and the
normal plane π, i.e., the angle between v
and n_d-n_d, is the dynamic edge-curve angle
τ_d. It is on the plane π_d. The angle
between rω and the plane π, i.e., the
angle between rω and n_s-n_s, is the static
edge-curve angle τ_s. It is on the plane π_s.
Because τ_d and τ_s are located on the
different planes, they do not have the
relationship of $\tau_d = \tau_s - \Delta\tau$.

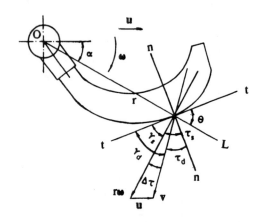

Fig.1 Edge-curve angles and cutting angles
of the lengthwise portion

    It can be seen in Fig.1 that because of
the forward velocity u of the rotary
tiller, the dynamic edge-curve angle of the
lengthwise blade is smaller than the static
edge-curve angle and the difference is $\Delta\tau$,
i.e., $\tau_d = \tau_s - \Delta\tau$.
    Fig.2 shows the edge-curve angles and the
cutting angles of the sidelong portion. In
Fig.2, the tangent line t-t of the sidelong

3 A BASIC EQUATION

For distinguishing the angles of the
sidelong blade from the angles of the
lengthwise blade, apostrophes are used as
superscripts to express the angles of the
sidelong blade. For example, τ_d represents
the dynamic edge-curve angle of the
lengthwise blade, while τ_d' represents the
dynamic edge-curve angle of the sidelong
blade.
    The edge curve of the sidelong portion of

the rotary blade is a three-dimensional curve (Fig.3), it can be expressed as:

$$\begin{cases} x = r(\phi)\cos\phi \\ y = r(\phi)\sin\phi \\ z = z(\phi) \end{cases}$$

where ϕ is the polar angle and r is the polar radius of the edge curve, $r = r(\phi)$.

The radius vector from the origin to one point on the edge curve is:

$$\begin{aligned} \rho &= x\mathbf{i} + y\mathbf{j} + z\mathbf{k} \\ &= r(\phi)\cos\phi\,\mathbf{i} + r(\phi)\sin\phi\,\mathbf{j} + z(\phi)\mathbf{k} \end{aligned}$$

The tangent line of any point on the edge curve is:

$$\begin{aligned} \mathbf{T} = d\rho/d\phi &= [\dot{r}(\phi)\cos\phi - r(\phi)\sin\phi]\mathbf{i} \\ &+ [\dot{r}(\phi)\sin\phi + r(\phi)\cos\phi]\mathbf{j} + \dot{z}(\phi)\mathbf{k} \end{aligned}$$

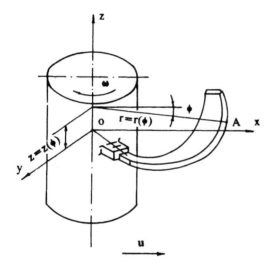

where $\dot{r}(\phi) = dr(\phi)/d\phi$, $\dot{z}(\phi) = dz(\phi)/d\phi$.

When the rotary tiller is working, the rotary blades rotate around the axle with the angular velocity ω and they also move forward with the velocity u. In this case the transformation of the coordinates can be used:

$$\begin{cases} x^* = x\cos\phi - y\sin\phi \\ y^* = x\sin\phi + y\cos\phi \\ z^* = z \end{cases} \text{and} \begin{cases} x^* = x + a \\ y^* = y + b \\ z^* = z + c \end{cases}$$

where x, y and z are the old coordinates before transformation, x^*, y^* and z^* are the new coordinates after transformation; ϕ is the rotation angle, $\phi = \omega t$; a, b and c are moving distances along x, y and z axes, $a = ut$, $b = c = 0$, and t is the time.

The radius vector of any point on the edge curve is:

$$\begin{aligned} \rho^* &= (x\cos\omega t - y\sin\omega t + ut)\mathbf{i} \\ &+ (x\sin\omega t + y\cos\omega t)\mathbf{j} + z\mathbf{k} \\ &= [r(\phi)\cos(\phi+\omega t) + ut]\mathbf{i} \\ &+ r(\phi)\sin(\phi+\omega t)\mathbf{j} + z(\phi)\mathbf{k} \end{aligned}$$

The tangent line of the edge curve is:

$$\begin{aligned} \mathbf{T}^* &= d\rho^*/d\phi \\ &= [\dot{r}(\phi)\cos(\phi+\omega t) - r(\phi)\sin(\phi+\omega t)]\mathbf{i} \\ &+ [\dot{r}(\phi)\sin(\phi+\omega t) + r(\phi)\cos(\phi+\omega t)]\mathbf{j} + \dot{z}(\phi)\mathbf{k} \end{aligned}$$

The velocity vector of any point on the edge curve is:

$$\begin{aligned} \mathbf{v} &= d\rho^*/dt \\ &= [u - \omega r(\phi)\sin(\phi+\omega t)]\mathbf{i} + \omega r(\phi)\cos(\phi+\omega t)\mathbf{j} \end{aligned}$$

According to the definition, the angle between the velocity vector **v** and the tangent line is the dynamic cutting angle, therefore:

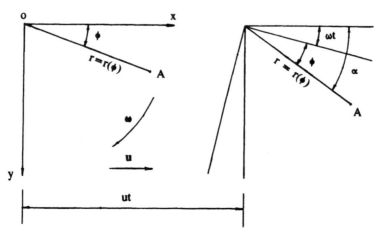

Fig.3 Edge curve and movement of the blade

$$\cos \gamma'_d = \cos(v, T^*) = v \cdot T^* / vT^*$$

where

$$v = |v| = \sqrt{u^2 + \omega^2 r^2(\phi) - 2u\omega r(\phi)\sin(\phi+\omega t)}$$

$$T^* = |T^*| = \sqrt{r^2(\phi) + \dot{r}^2(\phi) + \dot{z}^2(\phi)}$$

and

$$v \cdot T^* = \omega r^2(\phi) - ur(\phi)\sin(\phi+\omega t) + u\dot{r}(\phi)\cos(\phi+\omega t)$$

thus

$$\cos \gamma'_d = \frac{\omega r^2 - ur\sin\alpha + u\dot{r}\cos\alpha}{\sqrt{u^2+\omega^2 r^2-2u\omega r\sin\alpha}\ \sqrt{r^2+\dot{r}^2+\dot{z}^2}}$$

$$= \frac{\lambda r^2 - Rr\sin\alpha + R\dot{r}\cos\alpha}{\sqrt{R^2+\lambda^2 r^2-2\lambda Rr\sin\alpha}\ \sqrt{r^2+\dot{r}^2+\dot{z}^2}} \tag{1}$$

This is the equation of the dynamic cutting angle of the sidelong blade, where $\alpha = \phi + \omega t$, $r = r(\phi)$, $\dot{r} = \dot{r}(\phi)$ and $\dot{z} = \dot{z}(\phi)$; R is the maximum radius of the rotary blade and λ is the ratio of peripheral velocity to forward velocity, $\lambda = R\omega/u$ and $\lambda > 1$ (Hendrick & Gill 1974 & 1978).

Because the edge-curve angle is the complement of the cutting angle, then:

$$\tau'_d = \pi/2 - \gamma'_d, \quad \tan \tau'_d = \cot \gamma'_d = 1/\tan \gamma'_d$$

where

$$\tan \gamma'_d = 1/\cos^2\gamma'_d - 1$$

therefore

$$\tan \tau'_d =$$

$$\frac{\omega r^2 - ur\sin\alpha + u\dot{r}\cos\alpha}{\sqrt{(\omega r\dot{r}-ur\cos\alpha-u\dot{r}\sin\alpha)^2+\dot{z}^2(u^2+\omega^2 r^2-2u\omega r\sin\alpha)}}$$

$$= \frac{\lambda r^2 - Rr\sin\alpha + R\dot{r}\cos\alpha}{\sqrt{(\lambda r\dot{r}-Rr\cos\alpha-R\dot{r}\sin\alpha)^2+\dot{z}^2(R^2+\lambda^2 r^2-2\lambda Rr\sin\alpha)}} \tag{2}$$

Equation (2) is the equation of the dynamic edge-curve angle for the sidelong

portion of the rotary blade, it can be used to calculate the edge-curve angles of the rotary blades.

4 DISCUSSION

If the edge curve is on, or parallel to the turning plane of the rotary blade, the sidelong blade becomes the lengthwise blade. If the forward velocity of the rotary tiller is zero, the dynamic edge-curve angle equals the static edge-curve angle. Therefore the equation of the dynamic edge-curve angle of the sidelong blade is a universally applicable equation for rotary blades. By altering the parameters of this equation, other equations can be obtained very easily.

In equation (2), if z = constant, i.e., $\dot{z} = 0$, it becomes the equation of the dynamic edge-curve angle of the lengthwise blade:

$$\tan \tau_d = -\frac{\omega r^2 - ur\sin\alpha + u\dot{r}\cos\alpha}{\omega r\dot{r} - ur\cos\alpha - u\dot{r}\sin\alpha}$$

$$= -\frac{\lambda r^2 - Rr\sin\alpha + R\dot{r}\cos\alpha}{\lambda r\dot{r} - Rr\cos\alpha - R\dot{r}\sin\alpha} \tag{3}$$

In the coordinate system shown in Fig.3, $\dot{r} < 0$ for the lengthwise edge curve, therefore the denominator of equation (3) is less than zero. The equation (3) is achieved by taking the negative square root in equation (2).

In equation (2), if u = 0, it becomes the equation of the static edge-curve angle of the sidelong blade:

$$\tan \tau'_s = r / \sqrt{\dot{r}^2 + \dot{z}^2} \tag{4}$$

If in equation (2) $\dot{z} = 0$ and u = 0, or in equation (3) u = 0, or in equation (4) $\dot{z} = 0$, the equation of the static edge-curve angle of the lengthwise blade can be obtained:

$$\tan \tau_s = -r / \dot{r} \tag{5}$$

The equation (5) is also called the differential equation of the edge curve. By substituting τ_s as a function of ϕ, i.e., $\tau_s = f(\phi)$, the properties of different edge curves can be obtained (Sakai 1977, Peng & Wu 1982).

The equations (1) to (5) are derived in the coordinate system shown in Fig.3, in which $dr(\phi)/d\phi < 0$. Different coordinate systems can be selected and sometimes $dr(\phi)/d\phi > 0$ occurs (Fig.4). In this case

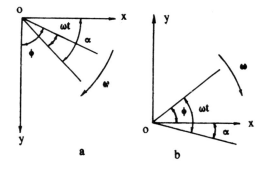

Fig.4 Different coordinate systems

all the terms containing $\dot{r}$ in the equations should have the opposite signs (changing + into – or vice versa), e.g., in Fig.4 equations (1) to (5) should be (equation (4) does not need to change):

$$\cos \gamma_d' = \frac{\omega r^2 - ur\sin\alpha - u\dot{r}\cos\alpha}{\sqrt{u^2 + \omega^2 r^2 - 2u\omega r\sin\alpha}\ \sqrt{r^2 + \dot{r}^2 + \dot{z}^2}}$$

$$= \frac{\lambda r^2 - Rr\sin\alpha - R\dot{r}\cos\alpha}{\sqrt{R^2 + \lambda^2 r^2 - 2\lambda Rr\sin\alpha}\ \sqrt{r^2 + \dot{r}^2 + \dot{z}^2}}$$

(1′)

$$\tan \tau_d' =$$

$$= \frac{\omega r^2 - ur\sin\alpha - u\dot{r}\cos\alpha}{\sqrt{(\omega r\dot{r} + ur\cos\alpha - u\dot{r}\sin\alpha)^2 + \dot{z}^2(u^2 + \omega^2 r^2 - 2u\omega r\sin\alpha)}}$$

$$= \frac{\lambda r^2 - Rr\sin\alpha - R\dot{r}\cos\alpha}{\sqrt{(\lambda r\dot{r} + Rr\cos\alpha - R\dot{r}\sin\alpha)^2 + \dot{z}^2(R^2 + \lambda^2 r^2 - 2\lambda Rr\sin\alpha)}}$$

(2′)

$$\tan \tau_d = \frac{\omega r^2 - ur\sin\alpha - u\dot{r}\cos\alpha}{\omega r\dot{r} + ur\cos\alpha - u\dot{r}\sin\alpha}$$

$$= \frac{\lambda r^2 - Rr\sin\alpha - R\dot{r}\cos\alpha}{\lambda r\dot{r} + Rr\cos\alpha - R\dot{r}\sin\alpha}$$

(3′)

$$\tan \tau_s' = r / \sqrt{\dot{r}^2 + \dot{z}^2}$$

(4)

$$\tan \tau_s = r / \dot{r}$$

(5′)

5 CONCLUSIONS

The edge-curve angle, or slide-cutting angle, is an important design parameter of the rotary blade. Meeting the slide-cutting requirement of the edge curve, i.e., keeping the edge-curve angle of the rotary blade greater than the sliding angle of the grass and straw, is a necessary design consideration when the rotary blade is used in grassy paddy fields. The equations derived in this paper can be used to calculate the edge-curve angles of every point on the edge curve of the rotary blade to see whether the slide-cutting requirement is satisfied.

The equation of the dynamic edge-curve angle for the sidelong portion of the rotary blade is a basic equation and it is universally applicable for rotary blades. It incorporates all the parameters of the edge-curve angles. By altering these parameters, other relevant equations can be obtained.

The edge-curve angles of rotary blades are divided into two types: the dynamic and the static edge-curve angles. When the rotary tiller is working, the dynamic edge-curve angle is the actual edge-curve angle of the rotary blade. For the lengthwise portion of the rotary blade, $\tau_d = \tau_s - \Delta\tau$, therefore the dynamic edge-curve angle is smaller than the static edge-curve angle. This should be considered for designing rotary blades. For the sidelong portion of the rotary blade, the relationship of $\tau_d' = \tau_s' - \Delta\tau$ does not exist.

REFERENCES

Hendrick, J.G. & W.R. Gill 1974. Rotary tiller design parameters, Part 4--Blades clearance angles. Trans. of the ASAE 17: 4-7.

Hendrick, J.G. & W.R. Gill 1978. Rotary tiller design parameters, 5: Kinematics. Trans. of the ASAE 21: 658-660.

Peng, S. & D. Wu 1982. Research on design methods for working parts of rotary tillers. J. Zhenjiang Inst. Agric. Mach. No.2 (in Chinese).

Sakai, J. 1977. Some design know-hows of edge-curve angles of rotary blades for paddy rice cultivation. Agricultural Mechanization in Asia 8(2): 49-57.

Wu, D. 1981. Slide-cutting analysis and eccentric circle design method for edge curves of rotary blades. Postgraduate thesis, Zhenjiang Inst. Agric. Mach., P.R. China (in Chinese).

Land and Water Use, Dodd & Grace (eds), © 1989 Balkema, Rotterdam. ISBN 90 6191 980 0

High speed movie observation of an axial flow combine harvester cylinder

E.Gasparetto, P.Febo, D.Pessina & E.Rizzato
Institute of Agricultural Engineering, University of Milan, Italy

ABSTRACTS: The behaviour of the axial flow cylinder of the Laverda MX combine harvester was studied using high speed cinematography, at about 2800 photograms/s.

The shootings were carried out on a stationary combine harvester, examining five different zones of the rotor. The tests were carried out on wheat, varying the cylinder rotation speed (600 and 700 rpm), with different positioning of the shooting axis (radial and tangential to the rotating drum).

By analysing the shootings with a special moviola, the kinetic characteristics of the trajectories of grains and ears/straw were drawn.

The results were compared with those achieved during the high speed movie observation of a conventional tangential threshing mechanism.

The kinetic and dynamic quantities (in the two different threshing methods) related to the detachment of the grains from the ear, the separation of the grain through the concave grid and the breaking of the kernels are discussed.

ETUDE DU COMPORTEMENT DU CYLINDRE AXIAL D'UNE MOISSONNEUSE-BATTEUSE A L'AIDE DE LA CINEMATOGRAPHIE ULTRA-RAPIDE

Le comportement du cylindre axial de la moissonneuse-batteuse Laverda MX a été étudié au moyen d'un filmage ultra-rapide d'environ 2800 photogrammes/s.

Les prises de vue ont toutes été exécutées sur cinq différents champs du cylindre, avec la machine fonctionnant sur place. Les essais ont été effectués avec du blé et on a pris en examen deux vitesses de rotation du cylindre (600 et 700 t/min), avec deux positions différentes de l'axe de prise de vue (radial et tangentiel par rapport au cylindre).

Visionnés avec un appareil spécial, les films ont permis de relever les caractéristiques cinématiques des trajectoires suivies par les grains, les épis et la paille.

On a comparé les résultats avec ceux qu'avait donnés une autre étude, toujours effectuée avec l'auxiliaire de la cinématographie rapide, sur un groupe de battage du type traditionnel (tangentiel).

On discute les caractéristiques cinématiques et dynamiques des deux systèmes de battage pour ce qui concerne: le détachement des grains des épis; la séparation des grains à travers la grille du contre-batteur; la cassure des grains.

UNTERSUCHUNG MITTELS HOCHGESCHWINDIGKEITSFILMS EINER AXIALEN MÄHDRESCHTROMMEL

Das Verhalten der axialen Mähdreschtrommel Laverda MX ist an Hand von Hochgeschwindigkeitsaufnahmen von etwa 2800 Bildern pro Sekunde untersucht worden.

Die Aufnahmen sind auf stehend funktionierender Maschine durchgeführt worden, und man hat dabei fünf verschiedene Trommelzonen untersucht. Die Versuche sind auf Weizen vorgenommen worden bei zwei verschiedenen Drehungsgeschwindigkeiten der Dreschtrommel (600 und 700 U/min) und zwei verschiedenen Positionen der Aufnahmeachse (radial und tangential zur Dreschtrommel).

Durch Untersuchung der Filmaufnahmen mittels eines besonderen Geräts ergaben sich die kinematischen Eigenschaften der Flugbahnen der Karyopsen, der Ähren und des Strohs. Die Ergebnisse werden mit denen einer anderen Studie verglichen, die auch mit Hochgeschwindigkeitsaufnahmen an einer Gruppe Trommel-Dreschkorb gewöhnlicher Art (tangential) durchgeführt wurde.

Beide Dreschsysteme werden auf ihre kinematischen und dynamischen Eigenschaften hin untersucht: die Abtrennung der Körner von der Ähre; die Trennung der Körner durch das Sieb des Dreschkorbes; der Bruch der Karyopsen.

1 FOREWORD

Studies and research on cereal behaviour during threshing are usually operations limited in their approach. On the one hand, they try to establish a link between the efficiency and mechanical and agronomic characteristics; on the other hand, theoretical studies of the force, energy and speed necessary for removing the grains from the ear, for the grain passage through the grid and for the breaking of the grains exist.

The parameters concerning work optimization of the cylinder were studied from a kinetic and dynamic point of view.

1.1 Removal of the grains from the ear

In a cereal, the existing link between the peripherical speed of the cylinder and the removal process of the grains from the ear, according to Kolganov (1956), depends on the square root of the work necessary for the removal itself. The average mass of 1000 fully grown grains reaches 40 g, whereas it is only 20 g for 1000 unripe grains. The corresponding removal work was calculated as being 6 and 12 mJ, and the removal speeds are (in the same order of reference) 17 and 34 m/s. In order to prevent breaking, these speeds must be lower than the grains critical breaking speed. Moreover, research by Hamdy et al. (1967) with the aid of centrifugal threshing, verified that 0.9 N of applied force are, in any case, sufficient to remove 98% of wheat grains. According to this research methodology, the acceleration is identical for each individual grain, whereas the force modulus depends on the mass. This means that the 0.9 N needed to remove 98% of wheat grains corresponds to an acceleration of more than 23,000 m/s^2. On the contrary, with an acceleration of 4000 m/s^2, equal to 0.16 N of applied force, only 20% of grains should be removed. These acceleration values appear to be much too high, when compared to those of other authors, and they possibly depend on the particular type of survey adopted by the authors; Finzenkeller, quoted by Königer et al. (1955) in fact, found out that 0.36 N are sufficient to remove almost all the grains.

1.2 Passage of the grains through the concave grid

According to Vasilenko (1956) the maximum relative speed, below which a single grain considered as a rotating ellipsoid, passes through the mesh of a sieve and beyond which it slides under the action of its inertia, depends on the opening diameter of the sieve mesh. Taking into account the size of the grains and putting the distance between two consecutive bars of the conventional concave at 40-50 mm, we have v_{max} = 2 ÷ 2.5 m/s. With a concave grid of 20-25 mm (as in the Laverda MX), v_{max} = 1 ÷ 1.25 m/s.

1.3 Breaking of the grains

According to Kolganov (1956) 75% of the fully grown grains break if they undergo an impact at a speed of over 36 m/s. However, during the threshing, this critical speed is increased by the protective action of the ears and straw, which on a large scale makes a direct impact improbable. Nevertheless, in spite of low drum speed on combines, a certain percentage of grains suffer damage. According to Mitchell et al. (1964) the broken grains are fewer when the moisture rate is higher and, vice versa, they increase proportionally to the impulse received and, consequently, to the impact speed.

On the other hand, there is a lack of research capable of putting together the results achieved so far. In other words, there is scarce knowledge of the absolute values of the kinetic and dynamic quantities which interfere with the single components (grains, ears, straw) when the latter undergo normal processes.

The estimate of such parameters is not an easy task because of the difficulties of using proper instruments during threshing operations, when the motion

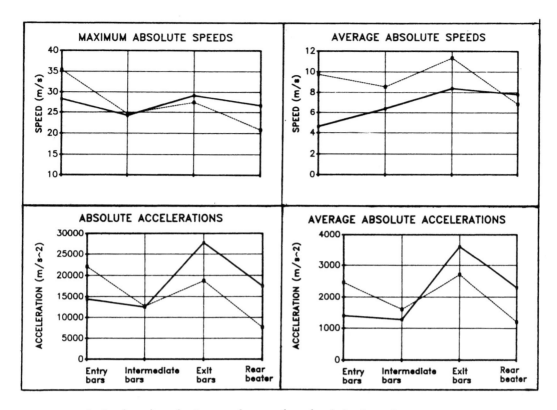

Fig. 1. Grain (———) and straw and ears (----) behaviour in a conventional combine (Gasparetto, 1977).

speed is high and cereal is present in large quantities. Some researchers, in order to solve this problem, have made use of high-speed cinematography. Scheigmann (1939), Königer (1955) and Schulze (1956 & 1964) did so by carrying out some quantitative estimates during threshing. The measurements were mainly taken on individual grains and led to confirm (as other authors had partially pointed out) that the speed of the ears is not, on the whole, over 10 m/s and that grain accelerations reach 8000 m/s^2.

Later Gasparetto (1967 and 1977) measured the speeds and accelerations of grains, ears and straw (fig. 1) within a conventional threshing mechanism (cylinder, concave and rear beater) working on wheat. The research was carried out by means of a high speed movie and Berbenni et al. (1972) discussed the specific technique.

Kutzbach (1983) compared the kinetic and dynamic characteristics (fig. 2) of the grain in both the conventional and the axial cylinders.

2 TESTS CARRIED OUT AND RESULTS

The research was carried out on the axial combine Laverda MX in 1985, to compare the results with those obtained by Gasparetto (1967 and 1977) and with the data shown by Kutzbach (1983). The shootings were carried out on a stationary combine harvester working on wheat; the bonnet, covering the upper part of the cylinder, was removed to permit observation. The particulars and features of the work carried out are described by Gasparetto et al. (1988).

The cylinder is 2.31 m long, with 0.65 m diameter. The shootings were carried out at 600 and 700 rpm, with a drum peripherical speed of 20.4 and 23.8 m/s. Along its axis the cylinder is subdivided into two parts: the first one is the threshing unit (1.04 m long); the second (1.27 m) one is the separation unit. The concave surrounds the drum, and it is made of a grid (25 mm) with square holes. On its interior, there is a series of bars, mounted at a 56° angle with the axis, which permits axial movement.

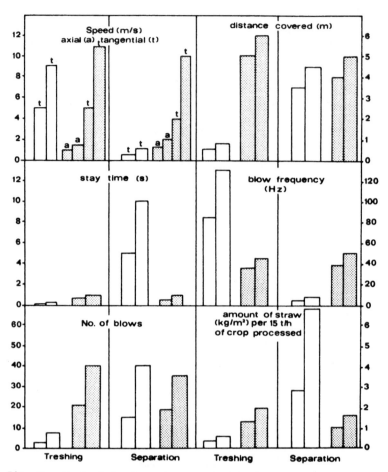

Fig. 2. Grain behaviour in conventional and axial combines (Kutzbach, 1983). (☐ = conventional combine; ▨ = axial combine).

The shootings were carried out at about 2,800 photograms/s, with different positioning of the shooting axis (radial and tangential to the rotating drum). The radial shootings (grain feeding, threshing, beginning, half-way and end of separation) make it possible to observe the grain and ear motion between cylinder and concave. The tangential shootings (threshing; beginning and end of separation) allowed the product coming out of the concave to be visible.

High speed cinematography applied to a threshing process is characterized by two disadvantages:

– the shooting field must be limited in order to allow a sufficient enlargement and to distinguish the various parts of the cereal. The shooting field, in relation to the cylinder axis, must, therefore, be decided beforehand. If the zone is next to the concave, the phenomenon may be influenced by boundary conditions and thus not represent the actual reality; on the other hand, if the shooting takes place far from the wall, stalks and ears can interfere with the focus zone and the camera lens, confusing the film. Therefore in this research the first of the two possible solutions was preferred;

– it is not possible to estimate movements parallel to the shooting axis. Therefore, all trajectories, speeds and accelerations are lacking in the components parallel to the shooting axis. To overcome this disadvantage, radial and tangential shooting were carried out.

The shootings were analysed with a special moviola and the kinetic characteristics of the trajectories of grains and ears/straw were drawn.

Grains and ears/straw were followed with a minimum of 30 to a maximum of 150 photograms, measuring their coordinates (according to the motion characteristics) at regular intervals between 2 and 20 photograms. By means of data processing, the following parameters were obtained:
- ears/straw and grain trajectories (fig. 3);
- average angle of the trajectories with the vertical axle, perpendicular to the cylinder axle (fig. 4);
- ears/straw speed in different shooting zones (fig. 5 left);
- grain speed in different shooting zones (fig. 5 right);
- ears/straw absolute acceleration in different shooting zones (fig. 6 left);
- grain absolute acceleration in different shooting zones (fig. 6 right).

From a qualitative point of view, threshing appears to be a process in which, at first, while the grains are removed from the ear, a series of factors, such as collision of the grains with the cylinder and concave bars, shaking caused by indirect kinetic action and rubbing between the grains, are combined.

The graphic description of some trajectories is shown in fig. 3. We can

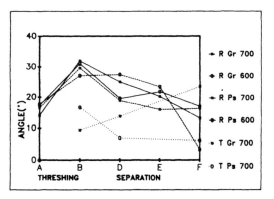

Fig. 4. Average angle of the trajectories to the vertical axle. (R = radial shootings, T = tangential shootings; Gr = grains, Ps = straw and ears; 600-700 = rotor rotation speed, rpm).

immediately notice the generally uneven motion; it is extremely irregular at every instant, due to collision with mechanical parts, internal rubbing and air streams which, in spite of their low energy, determine the movement of the different parts.

Therefore, the grain and the ears/straw are seldom found inclined toward the theoretical direction. In fact the trajectory can be inclined towards the right or left to the same extent; often after a collision with a metal part or when at the base of the concave bars (where a vortex of limited size can occur) the direction happens to be contrary to the theoretical, or else the grain remains in the air rotating vortically around its own centre of gravity.

The average angle of the trajectories (fig. 4) is seldom inclined like the bars mounted on the concave (34°). Advancing along the drum from the beginning of treshing (A) to the end of threshing (B) and from the beginning of separation (D) to the central (E) and final (F) separation, it may be observed (radial shootings):
- during threshing, the angle increases from 14-18° at the beginning to its maximum values (27-32°) at the end;
- during separation the angles decrease constantly from 21-27° at the beginning to 3-17° at the end;
- the motion is generally organized and uniform during threshing, and heterogeneous during separation;
- there is no significant difference between 600 and 700 rpm of rotating speed.

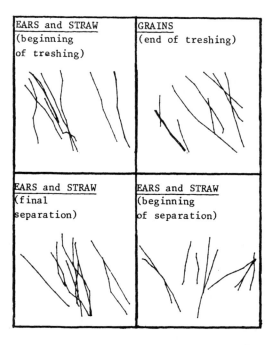

| EARS and STRAW (beginning of treshing) | GRAINS (end of treshing) |
| EARS and STRAW (final separation) | EARS and STRAW (beginning of separation) |

Fig.3. Some examples of the trajectories examined.

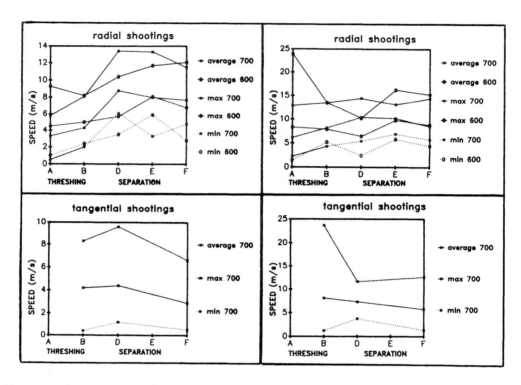

Fig. 5. Straw and ear (left) and grain (right) speed versus various shooting zones. (600-700 = rotor rotation speed, rpm).

Fig. 5 shows average, minimum and maximum speeds of grain and ears/straw. Minimum and maximum speeds are absolute values; average speeds represent the average value of the average of ten trajectories. In radial shootings the results are as follows:
- max. values of 13.5 m/s for straw and ears;
- averages of 3-5 m/s during threshing and of 5-9 m/s during separation (straw and ears);
- max. values of 22 m/s for grains;
- average grain values of 6-8 m/s during threshing and of 6.5-10.5 m/s during separation.

In tangential shootings, outside the concave, speeds are generally smaller.

Fig. 6 shows absolute average and maximum accelerations. Average values are usually within the range of a few hundred m/s². Grains are characterized by a larger spectrum, in comparison with ears and straw. Maximum values, with some exception, are around, or less 2000 m/s².

3 DISCUSSION

Cereal threshing is strictly related (as noted in the foreword) to the optimization of three factors; the removal of the grains from the ear, the separation of the grain through the grid and the breaking of the grains.

Regarding the first factor, during the tests, the rotating speed was always kept constant (20.4 and 23.8 m/s), at an intermediate value between the values mentioned by Kolganov (1956) as necessary for the separation of unripe and ripe grains (17 and 34 m/s).

Generally speaking results are in accordance with those obtained on a conventional cylinder by Gasparetto (1967 and 1977) and with the data exposed by Kutzbach (1983) in his comparison between the conventional and the axial drums. The exception is represented by the acceleration values (fig. 1 and fig. 6), whereas the data obtained with a conventional cylinder are much higher than those measured on the axial cylinder.

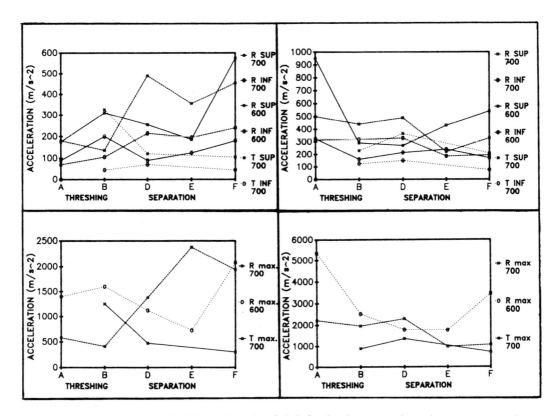

Fig. 6. Straw and ear (left) and grain (right) absolute acceleration versus various shooting zones. (R = radial shootings, T = tangential shootings; 600-700 = rotor rotation speed, rpm). Top: range of average values (SUP = upper limit; INF = lower limit). Bottom: maximum values.

The reasons may derive from:
- the different operative systems of the two cylinders (the distance cylinder-concave is higher on the axial drum);
- the variation in the shooting methodology (side shooting for the conventional cylinder and peripherical for the axial one);
- the difference in the data analysis and processing methods (direct processing in 1967-1977; successive approximations to determine the most likely position of every point in the present study).

Following Hamdy (1967) 23,000 m/s^2 are necessary to remove 98% of grains from the ear and 4,000 m/s^2 to remove 20%; Finzenkeller, quoted by Königer (1955) estimates that the two values are 9,200 and 1,600 m/s^2. Accelerations given in fig. 6 were measured along a 0.1-0.2 m trajectory. If the total distance covered by the product during the axial threshing is considered (5-6 m: Kutzbach 1983), the **probability of reaching the values of**

acceleration necessary for the removal of the grain from the ear is very high.

The second factor concerning the optimization of the cylinder is the separation of the grain through the grid of the concave. The theoretical values studied by Vasilenko (1956) explain the reason why slowing the speed and/or a wider grid help the separation. All the measured tangential speeds are higher than the critical limit speed of 1-1.25 m/s (see foreword). On the other hand, the percentage of grains separated compared with the grains present (in every part of the cylinder and of the concave) does not depend only on the critical crossing speed of the grid and on the average tangential speed of the grain, but also on the grains present in relation to the total product, **which is indeed a factor of great importance.**

The breakage of the grains is related to their removal from the ears and to their separation through the grids. More precisely, grain damage is facilitated by direct impact, the high rotation speed and the narrowness of the threshing chamber. Therefore the higher the speeds (and accelerations) for detachment from the ear, the higher the possibility of damage, which decreases with the deceleration of the tangential speed, when separation through the grid is optimized.

According to Mitchell (1964) as mentioned above, with 15% of humidity the damage-free grains account for 72% if the collision speed is 29 m/s and for 91% if the speed is 14 m/s. Therefore grain damage is in theory very serious; although the latter value ascribed to the speed (14 m/s) is rare, as seen when considering the absolute average speed data (fig.5 right).

4 CONCLUSIONS

The tests carried out permitted the calculation of the parameters related to wheat threshing, and more precisely the kinetic and dynamic quantities connected to the removal of the grains from the ears, to the separation of the grain through the grid of the concave and to the breaking of the grains.

Unfortunately, the three parameters mentioned cannot be optimized at the same time, due to their antithetical characteristics; indeed at each stage of the operation, the adjustment should depend on the instantaneous operative factors, such as the flow of the organic material, the straw-grain ratio, the moisture content, the presence of weeds, the degree of ripening, etc.

REFERENCES

Berbenni, A., Gasparetto, E. 1972. Tecniche cinematografiche ultraveloci nello studio delle macchine agricole. XVII Giornata della Meccanica Agraria, Bari.

Gasparetto, E. 1967. Ricerche sul comportamento di un gruppo battitore-spagliatore in lavoro su frumento con l'ausilio della cinematografia ultraveloce. Quaderno ISMA n. 29, Milano.

Gasparetto, E., Zen, M., Guadagnin, A. 1977. Ultra-high speed movie observation of a conventional threshing mechanism (cylinder-concave-rear beater) working on wheat. ASAE Grain and Forage Harvesting Conference, Ames, Iowa.

Gasparetto, E., Febo, P., Pessina, D., Rizzato, E., Coraluppi, A. 1988. Studio sul comportamento del battitore assiale della mietitrebbiatrice Laverda MX in lavoro su frumento con l'ausilio della cinematografia veloce. IV Convegno Nazionale A.I.G.R., Alghero.

Hamdy, M.Y., Stewart, R.E., Johnson, W.H. 1967. Theoretical analysis of centrifugal threshing and separation. Transactions of the ASAE 10(1), 87-90.

Kolganov, K.G. 1956. Mechanical damage to grain during threshing. Sbornik Trudov Po Zemled. Mekhanike, Vol. 3, Moskow.

Königer, R., Schulze, K.H. 1955. Entkörnerung durch Schlagleistentrommel. Institut für den Wissenschaftlichen Film, Göttingen.

Kutzbach, H.D. 1983. Dresch- und Trennsysteme neuer Mähdrescher. Landtechnik, No. 6, 226-230.

Mitchell, F.S., Rounthwaite, T.E. 1964. Resistance of two varieties of wheat to mechanical damage by impact. Journal of Agricultural Engineering Research, No. 4.

Scheigmann, P. 1939. Arbeitvorgänge in der Dreschmaschine. Institut für den Wissenschaftlichen Film (No. 292), Göttingen.

Schulze, K.H. 1956. Kinematographische Untersuchung des Dreschvorganges in einer Schlagleistentrommel. Grundlagen der Landtechnik, Heft 7.

Schulze, K.H. 1964. Untersuchungen über den Dreschvorgang an verschiedenen gestalteten Schlagleistentrommeln. Grundlagen der Landtechnik, Heft 21.

Vasilenko, I.I. 1956. Investigation on the passage of grain through a sieve. Sbornik Trudov Po Zemled. Mekhanike, Vol. 3, Moskow.

Land and Water Use, Dodd & Grace (eds), © 1989 Balkema, Rotterdam. ISBN 90 6191 980 0

Electronic sorting on industrial tomato harvester

G.Giametta
Institute of Agricultural Engineering, University of Reggio Calabria, Italy

ABSTRACT: Tomato mechanical harvesting requires too much manpower, although it's more convenient than the manual one, because of hand sorting (up to 24 people are engaged if sorting is done on the machine). So the interest in automatic sorting systems, and particularly in electronic sorting, has increased. The electronic sorting machine compares the red/green characteristics of each tomato by photo-optical sensors and decides whether to accept or reject it. Green tomatoes or non-tomato objects such as clods and wines,are rejected by high-speed pneumatic cylinders. The harvester used carried out trials with a 24 sorting channel photoelectronic machine. Hand presorting at belts guaranteed full working order with an average field production of 91 t/ha.

RESUME: Triage photo-électronique sur ramasseuse de tomates industrielle. Tout en ayant beaucoup plus d'avantages que la récolte manuelle, la récolte mécanique demande un emploi de main-d'oeuvre très élevé à cause de la nécessité d'effectuer un triage manuel (jusqu'à 24 ouvriers si le triage a lieu à bord de la machine). C'est pour cela que l'on observe avec attention croissante les systèmes automatiques de sélection du produit. En particulier, on montre un vif intérêt pour la méthode de triage colori- métrique qui permet d'identifier, à l'aide de capteurs photo-électriques, la couleur des baies et d'éliminer les baies vertes à travers des éléments électropneumatiques dont la mise en marche est automatique. Pendant les essais avec la ramasseuse employée, le triage a été fait par le système photo-électronique à 24 canaux de sélection. En outre, la prè- sélection manuelle aux bords des bandes a garanti l'efficacité maximum du système auto- matique en présence d'une production unitaire sur le champ très élevée (91,5 t/ha).

ZUSAMMENFASSUNG: Photoelektronische Sortierung auf Tomatenlesenanlage für Industrie. Die meckanische Tomaternte, auch wenn vorteilhafter als die manuelle Ernte, braucht doch zu viele Arbeiter, da sie die manuelle Sortierung benötigt (bis zu 24 Arbeiter, wenn die Sortierung in der Machine durchgeführt wird). Deshalb finden die automatischen Erntesortierungssysteme heute immer größeres Interesse. Hauptsächlich das größte Interesse ist für die grün/rot Sortierungsmethode: sie erlaubt die Früchtenfarbe, dank photoelektronischen Senneren zu erkennen, und die grünen Tomaten durch elektro- pneumatische Teile, die in Gang äutomatisch treten, zu verwerfen. In den Proben mit der benutzten Zusammenlesenanlage wurde die Sortierung durch das photoelektronische System mit 24 Sortierungskanälen durchgeführt. Außerdem versicherte die manuelle Vorsortierung zu den Bäden die höchste Leistung fähigkeit automatischen Systems bei enier hohen Feldproduktion (91,5 t/ha).

1. FOREWORD

Tomato crop is of great importance in

Italy, indeed our country,with 120 thousand hectars of farming land planted with tomatoes with a yield of about 5,4 million tons, ranks first in EEC (accounting for more than 60% of the total production) and second in the world, after the United States of America.

In the last few years, manpower shortage and its lacking availability have made the need for mechanization, particularly mechanical harvesting, more and more urgent. Actually manual harvesting alone represents 50% of the production cost. Therefore mechanical harvesting machines, in a large number of different models suitable for the different areas in which tomato is grown in Italy have been introduced in the market for some ten years.

Presently in Italy there are machines working by single-stage harvesting chains, where the produce is sorted in the harvester and unloaded onto a trailer placed side by side, and machines working by two-stage harvesting chains, where sorting is carried out, at fixed site, in tanks placed in the farm or in industries. Both systems, however, require a high number of workers.

Therefore there is an increasing interest in tomato automatic sorting which allows to reduce considerably the number of workers. Particularly two sorting methods have been explored: the former based on the different tomato density (absolute gravity of fruits) at the different ripening phases, according to which tomatoes, placed in water (sorting and cleaning tank at the same time), follow different trajectories; the latter (colorimetric sorting) allows to recognize the colour of the fruits by means of photoelectric sensors, and to reject green fruits by means of electropneumatic modules entering into operation automatically.

The greatest interest has been arisen particularly by this latter sorting method; actually most American harvesting machines have already been equipped with such electronic systems allowing a higher working capacity and, a smaller number of workers. In spite of the high cost, this sorting method, especially installed on relatively small machines,

could be convenient in our country, too, and could have spread in the near future.

In the light of these facts, the Institute of Agricultural Engineering at the University of Reggio Calabria, has carried out industrial tomato harvesting tests using a harvester, recently imported from the United States, equipped with a photoelectronic sorting system, in order to evaluate its working quality.

2. FEATURES OF THE TESTED MACHINE

The model 8600 Johnson self-propelled Electronic Tomato sorter was used in the tests. It is made up of the following units: cutting and conveyor system; pick up and separation unit; dirt chain conveyor belts; electronic sorting; unloading elevator. (Fig. 1)

Fig. 1 - The tomato harvester Johnson mod. 8600.

- The system for product cutting and conveying is equipped with a sickle bar, whose width of operation amounts to 1700 mm, and cutting plants at 25-35 mm below the ground; then tomatoes are elevated to the shaker bed by a conveyor chain which allows to minimize the amount of soil and dirt entering the harvester;
- the unit of pick up and separation

of the fruits from the plants, a new concept in tomato harvester design, is made up of a rotating cylindral drum, equipped with vibrating flexible tines;

- dirt chain conveyor belts separate the product from dirt and deliver fruits to two side sorting conveyor belts, from which tomatoes, on either side of the machine, pass on an elevator at the top of which a suction fan removes leaves and vines;
- the electronic sorter takes care of the automatic rejection of greens, clods and vine material;
- the elevator system unloads the product onto a trailer placed side by side to the harvester. A lowerator (decelerator) is attached to the end of the elevator; it is meant to deliver tomatoes gently into a bin or bulk truck.

The machine, weighing about 11 tons, has a turbo-diesel engine, of 84 kW at 1850 r.p.m.

3. ELECTRONIC TOMATO SORTER

The Model 2024 Johnson/ICORE Electronic tomato sorter (Fig. 2) is a 12 channel

inspection machine designed for efficient automatic sorting and rejection of green tomatoes and foreign objects directly on an in-field harvester. Two units are used to profide the harvester with 24 channels of capacity, positioned at the rear of the machine and corresponding to the side sorting belts. As the flow of tomatoes leave the elevator transfer belts, they pass through an upward flow of air created by the suction fans situated at their ends, and fall onto two singulator belts corresponding to the other two, but a bit higher level than the two side sorting belts. Tomatoes are aligned and divided into 12 rows on each belt, equipped with 12 longitudinal grooves. The fruits of each row, falling off the end of the belt - running toward the rear of the machine at a speed of 168 f.p.m. - pass in front of an optical sensor which compares the red/green characteristics of each tomato with those preset by the machine as reference, and decides whether to accept or reject it (Fig. 3).

Non-tomato objects, such as dirt clods, dried stalks or vines, are detected and, along with green tomatoes, are rejected in mid-air by an air cylinder

Fig. 2 - Overall view of electronic tomato sorter: one unit with 12 channel of capacity.

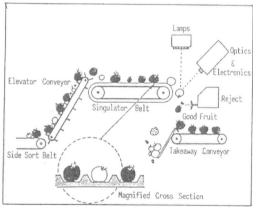

Fig. 3 - Electronic tomato sorter operation.

bopper and fall onto the ground. On the contrary acceptable tomatoes fall onto the rear sort belt for final hand sorting if necessary. As they leave the rear sorting belt , they are picked up by the unloading elevator and elevated up to be placed into the transport trailers.

In particular, the electronic tomato sorter consists of three main systems: lighting, electronics/optics, and rejection system.

The unit's design maximizes service-ability by using a modular concept. Each system is replaceable in part or as a whole in minutes, if necessary.

The lighting system consists of six sealed beam lamps which provide the illumination for the optics to scan the incoming stream of tomatoes as they fall from the singulator belt.

The electronics/optics system. There are 12 signal processing modules in each system with optics and associated electronics circuitry to provide the sorting function. In every module there are four solid-state photosensors, two of which detect the presence of foreign objects and dirt; the other two read the colour of the inspected tomatoes. In particular , these two read – for every tomato falling from each channel of the singulator belt – the relative green/red reflectance of the tomatoes. The signal processing electronic circuitry compares the reading with an adjustable standard reflectance and, if required, forwards a signal to the reject system. The adjustable standard is controlled by a red/green adjustment on the outside of the unit.

The heart of the reject system is 12 high-speed pneumatic cylinders which reject tomatoes and dirt with rifle-shot accuracy. These cylinders and their respective valves are mounted on removable plates for rapid replacement in case of damage or wear.

The pneumatic system includes high performance filtering and oiling units to deliver clean, lubricated air to the cylinders.

It is possible to change the sensitivity of the sorter toward red and green tomatoes by means of a dial which can be adjusted so as to accept only red tomatoes rejecting completely green and ripening tomatoes, or to accept fruits including besides red, a certain percentage of green and ripening tomatoes, too.

A dial enables the sorting system to detect and to reject foreign objects; also in this case the sensitivity of the system toward this material can be adjusted. There may be certain very dark soils which cannot be reliably sensed with the normal setting of this control. This can also happen in wet or muddy conditions.

In such cases it is necessary to increase the sorter sensitivity in order to have a dirt-free product.

It is possible to make the sorter so sensitive that it will reject a mud-coated ripe tomato.

Another dial controls the ability of the unit to ignore small bits of vines or leaves, which can be a problem under dry conditions. In fact, while under normal conditions it is unnecessary to use this dial, sometimes the vines will have large quantities of dried leaves on them which look to the sorter like dirt or green tomatoes. Some of these, along with small pieces of vine, may get to the sorter. If this happens too often, the reject boppers will be making many false strokes. The result will be a less efficient sort. In this case it is necessary an adequate dial setting until the sorter is ignoring the small debris. But if the dial is set too high it is possible to cause the sorter to start ignoring small green tomatoes and letting them into the load.

A further problem influencing the sort efficiency is represented by the most suitable time to strike dirts in order to rejects them. A delay dial allows to control properly the timing of the rejection bopper action. If the bopper is activated too early, it will strike the objects while they are high in their fall trajectory; if on the contrary the bopper is put into operation too late it will tend to strike them nearly at the end of their fall.

Since the rejected objects have different sizes and shapes, they don't always follow exactly the same trajectory. When you look in from the side across, all 12 channels, you will see a fan-shaped pattern (Fig. 4). If the boppers strike earlier than the right time, the fan will tilt upward and some of the rejects will travel upward (Fig. 4a), and bounce back off the singulator belt. In the opposite situation, when the reject strokes occur later, some of rejects will travel downward bouncing back off the flow separator; in this case the fan will tilt downward (Fig. 4c). When the delay dial is set correctly, the fan will appear roughly centered (Fig. 4b).

In order for the electronic sorter to be successful it is important for the singulator belt to be properly aligned: sideways drift or misalignment with respect to the optics (which should be always pointing at the centre of the singulator belt groove) will cause the reject objects, struck by the reject boppers to travel upward or downward, following wide trajectories.

The sort efficiency is basically affected also by the spacing of the reject boppers with respect to the product stream as it falls through the inspection zone. If the boppers are too far away, they cannot reach out far enough to hit smaller reject objects (Fig. 5a), so that there will be too much dirt and green in the load. If the reject system is too close to the product stream, large ripe fruits will bounce off the housing of the sorting system and go down the reject chute (Fig. 5c). The result: to much

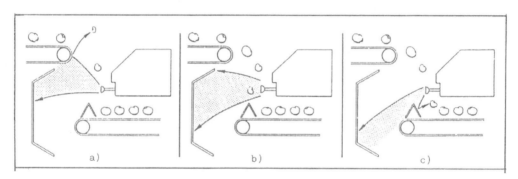

Fig. 4 - Effect of delay dial on the sort efficiency.

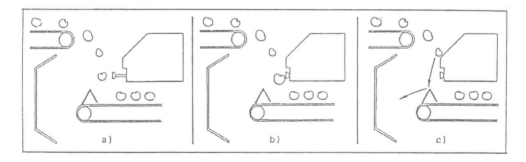

Fig. 5 - The sort efficiency can be affected by the spacing of the reject boppers.

red on the ground.

4. TEST METHODOLOGY

The tests with the said machine were carried out in the second decade of August, in the countryside around Foggia, and were performed on level, rectangular plots, of some 450 m in length for an area of about 4 hectars.

The medium soil had been accurately prepared for mechanical harvesting.

As far as the planting method is concerned, the plants where arranged in paired rows at a distance of 40 cm between the rows and 1.7 m between the centres of two contiguous paired rows. The plants, of the American cultivar "UC 82", were in good condition and ripe fruits accounted for 92%. The harvesting chain consisted in a harvester together with a trailer for unloading produce.

The working team was made up of 8 persons: 2 drivers, 4 workers for the facultative hand presorting at the side conveyor belts (when performed, this operation, even with a low number of workers, increases the automatic sorting system efficiency) and 2 workers devoted to hand sorting, it too facultative, at the rear sort belt, placed just behind the electronic sorting system.

During harvesting, the electronic sorting system was set at 5/10 in order to get the ripening fruits, too, which however had reached a sufficient ripening degree.

The data gathered, concerned field yield at harvesting, machine operational features, product losses at harvesting according to its different parts, harvested product features, with respect to the different parts of good and bad fruits and to foreign material and finally the fruit damaging caused by the harvester.

5. RESULTS AND DISCUSSION

The results of the harvesting tests are show in table 1 - 5. The yield capacity with more than 91 t/ha of fruits (Table 1) was rather high. Taking into account that the harvested produce was meant to industry, for processing in concentrates, chopped tomatoes, and pureed tomatoes, the processable product constited of ripe as well as of ripening fruits, at a sufficient ripening degree.

The quality of the field product was satisfactorily good with more than 92% ripe fruits and 4.5% ripening tomatoes while the green (2.3%) and the over-ripe fruits (0.90%) amounted to about 3%.

The real average speed of the harvester was 1.9 km/h. The almost total lack of weeds and the not very thick vegetation allowed the machine to work without plugging; moreover the soil, on account of lack of recent rainfalls and of a proper use of irrigation was not wet; this, besides influencing the machine working regularity, affected successuflly the produce quality, as it helped not to soil the fruits. All these factors, together with the considerable length of the plots contributed

Table 1. - Tomato ripening degree at the harvesting.

FIELD YIELD		GREEN TOMATOES		RIPENING TOMATOES		RIPE TOMATOES		OVER-RIPE TOMATOES	
t/ha	%	t/ha	%	t/ha	%	t/ha	%	t/ha	%
91.50	100.00	2.10	2.30	4.10	4.48	84.52	92.32	0.78	0.90

to limit the accessory times and to achieve a rather high working capacity: actually it amounted to 0.31 ha/h, equal to 26.8 t/h of the harvested produce (Table 2). The machine working productivity, taking into account that 7 workers were employed, was 3.8 t/h·man.

The amount of fruits left on the ground after mechanical harvesting, (Table 3) is equal to 5.2% of the total

A large part (30-35%) of the ripe lost fruits was made up of loose tomatoes or of fruits strippen from the plant after being cut and slipped sideways from the elevator belt of the harvesting system and fell onto the ground. Another part of ripe fruits,(35-40%) expecially the smaller ones, went lost in the vibrating shaker because, although being stripped from the plant, they rolled down with

Table 2. - Results of the tests by "Johnson mod. 8600" harvester.

FIELD YIELD	REAL FORWARD SPEED	WORKING CAPACITY	TOMATOES HARVESTED BY THE MACHINE		HOURLY YIELD	MACHINE WORKING PRODUCTIVITY
(t/ha)	(km/h)	(ha/h)	(t/ha)	(%)	(t/h)	(t/h·man)
91.50	1.94	0.3094	86.72	94.77	26.83	3.83

Table 3. - Tomatoes left on the ground after mechanical harvesting.

QUANTITIES AND PERCENTAGES	GREEN TOMATOES	RIPENING TOMATOES	RIPE TOMATOES			OVER-RIPE TOMATOES	TOTAL
			DAMAGED	INTACT	TOTAL		
t/ha	1.83	1.37	0.55	1.01	1.56	0.02	4.78
PERCENTAGE OF TOTAL FIELD YIELD AT HARVESTING %	2.00	1.50	0.60	1.10	1.70	0.02	5.22

produce; in particular, the lost ripe and ripening fruits amounted to 1.56 and 1.37 t/ha respectively, equal to 1.7 and 1.5% of the total amount of tomatoes.

vine material. Other ripe tomatoes went lost by the dirt chain conveyor. Only a negligible percentage went lost in the electronic sorting, particularly when the product stream was greater.

The harvested produce accounted for 94.8% of the total yield and of this 96% was represented by ripe tomatoes, 3% by ripening tomatoes, and slightly more than 1% by over-ripe and green fruits together (Table 5). Furthermore, among the harvested ripe tomatoes 1% was represented by damaged fruits whereas the ripening ones were virtually damage free .

Damaged ripe fruits – but only in a mild form mainly because of their remarkable resistance to the machine mechanical stressing – can be, however, used by the processing industry. The percentage of foreign material in the harvested produce remained around 1.5%, 0.9% being clods etc. and 0.6% vine material (leaves, stalks, etc.) (Table 4).

Table 4. – Working quality of the harvested and selected produce by "Johnson mod. 8600"(1)

FOREIGN MATERIAL			TOMATOES							
					RIPE FRUIT					
CLODS, STONES, ETC.	VINE MATERIAL	TOTAL	GREEN TOMATOES	RIPENING TOMATOES				OVER-RIPE TOMATOES	TOTAL
					DAMAGED	INTACT	TOTAL		
%	%	%	%	%	%	%	%	%	%
0.90	0.60	1.50	0.30	3.10	0.90	93.33	94.23	0.87	100.00

(1) % of the machine harvested produce, including foreign objects.

Table 5. – Composition of the different parts of harvested and selected tomatoes by "Johnson 8600".

HARVESTED TOMATOES														
GREEN			RIPENING			RIPE			OVER-RIPE			TOTALS		
	ON GREEN TOMATOES TOTALS	ON HARVESTED TOMATOES		ON RIPENING TOMATOES TOTALS	ON HARVESTED TOMATOES		ON RIPE TOMATOES TOTALS	ON HARVESTED TOMATOES		ON OVER-RIPE TOMATOES TOTALS	ON HARVESTED TOMATOES		ON TOTAL FIELD YIELD	ON HARVESTED TOMATOES
(t/ha)	(%)	(%)	(t/ha)	(%)	(%)	(t/ha)	(%)	(%)	(t/ha)	(%)	(%)	(t/ha)	(%)	(%)
0.27	12.86	0.31	2.73	66.58	3.15	82.96	98.15	95.66	0.76	97.44	0.88	86.72	94.78	100.00

6. CONCLUSIONS

Though being more convenient than manual harvesting, tomato mechanical harvesting requires too much manpower, because of the need to perform hand sorting (up to 24 workers, if sorting is carried out on the machine). Therefore automatic tomato sorting systems are increasingly spreading.

In the harvesting tests by the machine employed, sorting was performed by the 24 channel photoelectronic system. Furthermore hand presorting at the belts ensured the highest efficiency of the automatic system with a considerable field yield (91.5 t/ha) thus allowing a remarkable machine forward speed (1.94 km/h). In this way and under favourable crop conditions (dry soil, right product ripening degree, etc.) a high working capacity (0.31 ha/h, equal to about 27 t/h of harvested produce) was achieved thanks also to the harvester rotating shaker. Moreover 98% of ripe tomatoes compared with the total field ripe produce was harvested. Also the harvested fruit quality was satisfactory with 1.5% of foreign material and a small percentage of fruits that cannot be utilized.

Besides, even reducing or eliminating the additional hand sorting, under harvesting favourable conditions (like those under which the tests were performed) satisfactory results are expected to be achieved, if possible reducing the forward speed with a high yield.

REFERENCES

Consiglio Nazionale delle Ricerche. Progetto finalizzato per la Meccanizzazione agricola. Subprogetto n. 2. Meccanizzazione della raccolta del pomodoro da industria, Quaderno n. 11 1979.

Consiglio Nazionale delle Ricerche. Progetto Finalizzato per la Meccanizzazione agricola. Subprogetto n. 2. Meccanizzazione della raccolta del pomodoro da industria. Quaderno n. 11, 1981.

Di Ciolo S., Graifenberg A.. Raccolta meccanica del pomodoro da industria, Macchine e Motori Agricoli, 1981, n. 8, pp. 21-66.

Galli G., Spugnoli P., Zoli M.. Risultati acquisiti nella cernita del pomodoro con una macchina a flottazione. Rivista di Ingegneria Agraria, 1981, n. 2, pp. 107-117.

Gasparetto E., Patuzzo L.. Prove d'orientamento su una macchina per la raccolta integrale del pomodoro, Atti della XVI Giornata della Meccanica Agraria, Bari, 1971, pp. 245-253.

Manfredi E., Bentini M., Guarnieri A., Stato della ricerca sulla meccanizzazione del pomodoro da industria, L'Informatore Agrario, 1986, n. 8, pp. 73-82.

Moini S., M. O'Brien, P. Chen. Spectral properties of mold and defects of processing tomatoes. Transactions of the ASAE 23(4), 1980, pp. 1062-1064.

Ruiz M. Chen P., Use of the first derivate of spectral reflectance to detect mold on tomatoes. Transactions of the ASAE, 1982, vol. 25, N. 3.

Land and Water Use, Dodd & Grace (eds), © 1989 Balkema, Rotterdam. ISBN 90 6191 980 0

The theory of free cutting and rotary cutters

E.Güzel & Y.Zeren
Çukurova University, Adana, Turkey

ABSTRACT : Rotary cutters are produced by some small agricultural machinery factories in Çukurova règion of Turkey; they have free cutting blades.   Although, rotary cutters are similar, there are some differencies among them. The aim of this study was to determine   some properties of rotary cutters. These are peripheral velocity, energy of impact, curves of free cutting, cutting points, and force of cutting. This study was done in the laboratory and on the field conditions.

RESUME : Les broyeurs de tige du coton de coupetournante sont fabriquées par certains petits ateliers dans la région de Çukurova en Turquie. Ces machines s'exercent d'après le systeme de libre-coupe. Autant dire que les machines de coupetournante sont similaires entre elles mais il y a certaines différences. L'objectif de ces ètudes est déterminer certaines propriétées des machines de coupe-tournante. Ce sont : la vélocitée périphérique, l'énergie d'impacte, les courbes de libre-coupe, le point de coupe et la force de coupe. Cet étude s'est executé à la fois au laboratoire et que pour le champ.

ZUSAMMENFASSUNG : Heutzutage werden baumwollstengelhaecksler von einigen kleineren werkstaetten in der Çukurova Region von der Turkei gebaut. Solche maschinen fonktionieren nach den senkrechter drehachse der schlagelement. Es gibt doch einige aehnlichkeiten und unterschiede zwischen verschiedenen baumwollstengelhaecksler. Ziel dieser arbeit ist feststellung der eigenschaftlichen darstellung von solcher maschinen. Diese eigenschaften beziehen sich auf geschwindigkeitsclinie, schlagenergie, kurven des freien schneidens, schneidenspunkt und schneidenskraft.
    Diese arbeit wurde sowohl unter praktischen bedingungen auf dem feld als auch im werkstatt durchgeführt.

1 INTRODUCTION

Rotary cutter is an implement in the mechanization chain of cotton production. After harvesting the cotton, these machines cut the cotton stalk and distribute them on the field surface.

The cotton which has pivotal root holds the soil well. Moreover moisture is reduced in the cotton stalks and some of the stalks can be dried wholly. Elastic structure of the stalk becomes brittle. When the cutting height is 10 or 15 cm, the diameter of stalks range between 10 to 25 mm.

Rotary cutter is a blade system which pivots horizontally on the vertical shaft and moves drawning on the field, Figure1. Running taken from p.t.o. after passing over the conical gear box is transferred to the blade shaft. Cutting system is settled by four blades and clearance angle is 90^0. Peripheral velocity of the blade is changed between 60–80 ms^{-1}, and running speed is about 800–1000 r.p.m.

2. FREE CUTTING and RELATIONS of PERIPHERAL and FORWORD VELOCITY

As it is shown in Figure 2, the blade's axis which is capable of rotating about point "O" in horizontal plane, at the same time it translates along on X axis line. Forward and peripheral motion of blade's edge aa' sweeps the area between number "1" cycloids Güzel and Zeren (1981)

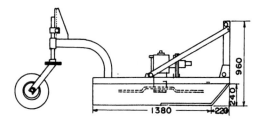

Figure 1. Side view of rotary cutter

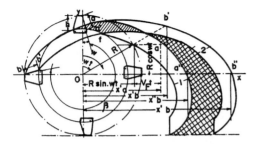

Figure 2. Running of the rotating blade's edge

At the same time the following edge bb'sweeps the area between number "2" cycloids. When the peripheral velocity is constant the width of the hatched area depends on the forward velocity of tractor. When the forward velocity is increased, some of the stalks can not be cut. For this reason, a suitable ratio is found between the peripheral velocity of the blade and the forward velocity of the tractor. Equations of the path of point "a" are as follows by Dincer (1971)

$$x'_a = V_f t + R \sin {}^w t \tag{1}$$

$$y'_a = R \cos {}^w t \tag{2}$$

Hence,

V_f = Average velocity of tractor

R = Radius of tracing blade end.

w = Angular velcity of blade end

The situation of point b',

$$x'_b = V_f t + R \sin (\beta + wt) \tag{3}$$

$$y'_b = R \cos (\beta + wt) \tag{4}$$

Because of the phase differences between the blades, as $\frac{\pi}{2}$, $ww_t = \frac{\pi}{2}$ and $t = \frac{\pi}{2w}$

Substituting this value in the expressions (1) and (2)

$$x''_a = V_f \frac{\pi}{2w} + R \sin w \frac{\pi}{2w}$$

$$x''_a = V_f \cdot \frac{\pi}{2w} + R \tag{5}$$

The second follower blade's edge bb' will cut x,

$$w t' = \frac{\pi}{2} + 2 \quad -\beta = \frac{5}{2}\pi - \beta$$

Calculating the time t' and substituting the expressions (3) and (4),

$$x''_b = \frac{V_f}{w} (\frac{5\pi}{2} - \beta) + R \tag{6}$$

For eliminating the unmoved area,

$$x''_b - x''_a = h = L. \cos \tau \tag{7}$$

Where;
  L = length of the blade edge, mm
  h = length of the active blade edge, mm
  τ = the angle between the blade edge and the travel direction

Substituting the values (5) and (6) in the expression (7), for $\beta = \frac{3}{2}$ and $V_m = wR$

$$\frac{V_f}{w} (\frac{5\pi}{2} - \beta) + R - V_f \frac{\pi}{2w} - R = h = L. \cos \tau$$

$$\underset{(1)}{\frac{V_f \pi}{w}} - \underset{(2)}{\frac{V_f \pi}{2w}} = h = L. \cos \tau$$

$$\frac{2 \, V_f \pi - V_f \pi}{2w} = h \qquad \frac{V_f \pi}{2w} = h$$

Substituting this value to the over expression $w = \frac{V_m}{R}$

$$\frac{V_f \pi}{2\frac{V_m}{R}} = \frac{V_f \pi}{1} \cdot \frac{R}{2V_m} = h$$

$$\frac{2}{\pi R} \cdot \frac{V_f \cdot \pi \cdot R}{2 V_m} = h \, \frac{2}{\pi R}$$

$$\frac{V_m}{V_f} \geqslant \frac{\pi R}{2h} \geqslant \frac{\pi R}{2 L \cos} \tag{8}$$

Duration of the movement of the follower blade edges can be expressed as,

$$t = \frac{2 \pi R}{4 V_{nz}}$$

$$V_f t = L. \cos \tau \; ; \quad t = \frac{L \cos \tau}{V_f}$$

If the resistance of cotton stalk is assumed as "R", according to the Wieneke (1972), Kanofojski (1976), and Mutaf (1976),

$$R \leqslant m \frac{V}{\Delta t} + K \leqslant m.a + K \tag{9}$$

Where
  m : mass of the cotton stalk, N s^2/m
  Δt: impact duration of blade
  K : bending resistance of stalks at impact height,N,
  In this study, bending and cutting resistance were found by a dynamometer and an impact testing machine As a result, cutting procedure depends on bending resistance. If the bending resistance is higher than cutting resistance, cotton stalks may be cut by rotary cutters. Power consumption which is taken from tractor is measured by a torquemeter. This torquemeter. This torquemeter is established between the rotary cutter and the tractor. Chinging in torque of the p.t.o. is shown in Figure 3.

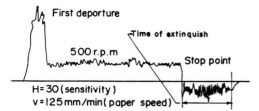

Figure 3. Changing of torque on the p.t.o

Other basic engineering data for these cutters is the cutting velocity. These values are found according to the position of free cutting curves in Figure 4. Peripheral velocity of the blade is culculated according to the first derivative of the horizontal and vertical distances with respect to time.

$$\frac{dx'}{dt} = R \ w \ Cos \ wt + V_f$$

$$\frac{dy'}{dt} = - R \ w \ Sin \ wt$$

$$V = \sqrt{V_{x'}^2 + V_{y'}^2} \tag{10}$$

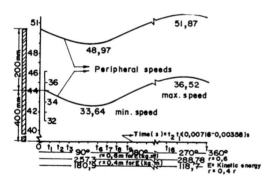

Figure 4. Changing of V-T diagrams of blade.

REFERENCES

1. DİNÇER, H., 1971. Cutting systems for grass and crops. Published by the agricultural faculty. University of Atatürk. Erzurum. Turkey. Published no.118 (in Turkish).
2. KANOFOJSKI, T.y et all. 1976. Agricultural machines theory and construction. Vol.2. Warsaw. Poland.
3. MUTAF, E., 1976. Agricultural implements and machinery. Lecture notes. Published by the Agricultural faculty. University of Ege. İzmir, Turkey (in Turkish).
4. GÜZEL, E., ZEREN, Y. 1981. A research on the effective work of rotary cutters at produced in Çukurova region. Presented at the G. National Symposium of agricultural mechanization. İzmir. Turkey. (in Turkish).
5. WIENEKE, L., 1972. Verfahrenstechnik der Halmfutter production. Göttingen. Germany.

SUMMARY

As a result of this study, some basic engineering data were determined for rotary cutters. Energy consumption of blades is ranged between 240 – 289 kgm. Power consumptions were found as 4.2 HP for the departure, 1.74 HP at 540 r.p.m. and 0.86 HP at extinguishing Also maximum and minumum velcities for blade were found between 46.97-51.87 and 33.64-36.52 ms⁻¹.

Land and Water Use, Dodd & Grace (eds), © 1989 Balkema, Rotterdam. ISBN 90 6191 980 0

Modified environment windrow harvesting of rice

B.M.Jenkins

Agricultural Engineering Department, University of California, Davis, Calif., USA

ABSTRACT: A new technique for harvesting rice from the windrow has been developed and tested over three years of research in California. Swath harvesting has been adopted by some growers in the state as a means to reduce the cost of drying grain harvested at moistures high enough to satisfy the requirements for high grain quality (as measured by whole kernel or head rice yield). The use of conventional open windrows, while allowing the grain to dry in the field, unfortunately, leads to unacceptable loss rates in whole kernel yield with more than one day exposure in the field. The principal reason for the deterioration in quality with time is the diurnal rewetting of the grain by dew forming on the upper surface of the windrow, a phenomenon well known to cause cracking in rice. The new technique modifies the local environment of the grain in the windrow by placing a thin but uniform layer of residual stem material over the surface of the windrow, thereby intercepting condensation at the surface and reducing the exposure of the grain to rewetting by this mechanism. This method does not, however, reduce the drying rates of the grain relative to that in the open windrows. Medium grain rice swathed at 28% moisture and harvested at 14% moisture after 3 days exposure yielded 59% head rice compared to 48% head rice for grain harvested from open windrows. Standing crop harvested at 16.6% moisture yielded 42% head rice. An economic model developed to compare the modified swath harvesting technique with direct harvesting shows an 8-10% advantage to swath harvesting relative to direct harvesting at 24.5% moisture, considered to be near optimal for direct harvest of medium grain rice. The modified swath harvesting technique also permits earlier harvesting of the grain, and may improve equipment utilization, thereby reducing capital investment required.

1 INTRODUCTION

In an effort to reduce the costs of drying rice, some growers in California have started to windrow harvest at least a portion of their crop. Swathing the rice into a windrow permits the grain to dry in the field, thereby reducing the amount of commercial energy at the dryer required to achieve final moisture for storage. Because the price received by the grower for the grain depends heavily on the amount of unbroken grain in the lot after milling, growers also desire to maintain high head rice, or whole kernel yields. With direct combine harvest of rice, and particularly with short and medium grain varieties, high head rice yields are generally achieved at high moisture content. Often, however, a dryer may have insufficient capacity to dry all the rice delivered by the growers when the grain is at high moisture. A moisture limitation may be imposed on the growers by the dryer so that grain must be delivered with a moisture content at or below that prescribed by the dryer. To meet this restriction, a grower may need to delay harvest and thus may sacrifice grain quality. As the mature grain in the standing crop continues through successive diurnal wetting and drying events, fissuring of an increasing fraction of grain kernels occurs, which leads to lower head rice yields at milling. Even without the restriction, harvesting at high moisture implies higher drying costs because of greater energy expenditure. For direct harvesting, the goals of harvesting at reduced moisture to decrease drying costs or increase dryer capacity are counter to the goals of harvesting at high moisture to maintain high head kernel yield.

Windrow harvesting is viewed as a means of increasing the rate of drying in the field so that harvesting is not unduly delayed. The moisture loss is intended to be sufficient to meet dryer imposed limitations or to reduce the cost of drying. With good weather typical of the early portion of the harvesting season, one day of exposure may be sufficient to drop the grain moisture from 26-28% wet basis to 20-22%, which is generally acceptable at the dryer even during peak season.

Unfortunately, commercial swathing differs little from that practiced fifty years ago for rice, at least in the structure of the windrow and the distribution of grain in the windrow. This significantly impacts grain quality when the windrow is exposed for longer than one day, either because of insufficient equipment capacity, equipment malfunction, or a desire to achieve lower moistures. Both auger and draper header type swathers place most of the grain at the surface of the windrow. This leads to higher drying rates during the day because of higher temperatures achieved under direct insolation in the windrow compared to grain in the standing crop. More importantly, however, it also exposes the grain to direct rewetting by condensation at night when dew forms, in addition to rewetting by vapor phase adsorption of water from the air. As the grain continues to dry in the windrow, rewetting by both processes, but especially by condensation, leads to greater damage of grain kernels. Rewetting is well known from field and laboratory experiments to cause fissuring in rough rice kernels if below a critical moisture content believed to lie in the range of 13-16% wet basis (Stahel 1935; Rhind 1962; Craufurd 1963; Kunze and Choudhury 1970; Bhattacharya 1980; Siebenmorgen and Jindal 1986), although not all varieties of rice are equally susceptible to fissuring (Bhattacharya 1980; Geng et al. 1984; Kunze 1985; Chattopadhyay and Kunze 1986). Results of these studies have been applied to standing grain crops, where the competing objectives of low drying costs and high head rice yields, coupled with the nonuniformity of grain moisture on the panicle, have led to the development of optimal moisture content recommendations for harvesting the grain (Bhattacharya 1980; Chau and Kunze 1982; Geng et al. 1984; Kunze and Calderwood 1985).

To reduce the damage by rewetting, a number of limited attempts have been made to protect the grain from dew formation. Ranganath et al. (1970), for example, experimentally evaluated spreading sheets over the standing crop at night to reduce condensation on the crop. When windrow harvesting of rice was still common in California, Bainer (1932) observed a greater level of fissuring in the upper layers of the windrow. Bainer described an attempt (unsuccessful) by one grower to develop a machine to invert the windrow and place the grain on the lower surface. The motivation for this was apparently to reduce the daytime temperatures and drying rates of the grain in the windrow, the consequences of rewetting by dew evidently unrecognized. While this method would be useful in preventing dew forming on the grain, it is now seen as being less desirable than other techniques described here because of the potential for increased shatter loss during the inversion process and at harvest.

There exists another motivation for swath harvesting rice. In California, as in many other areas of the world, much of the straw following harvest is burned in the field so as to reduce the difficulties of seed bed preparation for the subsequent crop. Open burning is a source of atmospheric pollution. Several alternatives have been investigated, such as straw collection and utilization for power generation. Successful implementation of some of these alternatives requires the straw to be stored and used at low moisture. Field sanitation also dictates that straw recovery should be high to reduce recurrent infestation by pathogens, such as rice stem rot. Collection of straw following direct harvest of the grain is made difficult by the high moisture of the straw and standing stubble. Recovery of straw from the tracks of harvesters and grain transport vehicles (bank-out wagons) is also very difficult. Swathing and windrowing the straw with the grain avoids the loss to the tracks of equipment, and permits drying the straw in the windrow along with the grain. This latter aspect would conceivably permit total, or simultaneous, harvest of the straw and grain. The need to protect grain quality is paramount in any attempts to utilize the straw.

Described here is the development of a process for modifying the environment in the immediate vicinity of the grain in the windrow. The technique is relatively simple and can be done mechanically. It consists of covering the upper surface of the windrow as formed by the swather with a layer of straw or stubble cut from beside the windrow, which can be done either on the swather or subsequently with a separate apparatus. This cover serves as an insulating layer above the grain to maintain grain temperature above the dewpoint even as the surface is cooled below the dewpoint by nocturnal radiation. The method does not appear to increase shatter or harvesting losses, does not reduce drying rates for the grain relative to open or unmodified windrows, yet maintains higher head rice yields over extended exposures relative to open windrows and even to standing rice at low moistures. The results of three years of field trials with two variations of the technique are discussed, as is a preliminary economic feasibility analysis intended to test the potential of the method under various conditions.

2 MODIFYING THE WINDROW ENVIRONMENT

The observation that kernel fissuring is reduced in the lower regions of the windrow implies that damage may be controlled by virtue of the position of the grain in the windrow. Laboratory experiments clearly indicate that kernel cracking is due primarily to moisture induced stresses developed as a result of nonuniform water concentration throughout the kernel.

Thermal stresses contribute substantially less to the failure mechanism. In the field, vapor phase absorption of atmospheric water by the grain is difficult to control, but adsorption of surface condensation may be controlled by limiting the condensation on the grain kernels. Grain in the lower regions of the windrow is apparently sheltered to some extent from direct condensation which occurs readily at the surface at night because of the large temperature difference between the surface and the sky. Without solar radiation to warm the surface, at night the surface will cool below the dew point temperature of the air. Rice farming in California is done using flooded culture, and fields are drained about two weeks prior to harvest. Soil moisture remains very high throughout the harvesting period, and results in high relative humidities near the surface at night. Dew point temperatures are fairly high, and the surface of the windrow need cool only a few degrees below air temperature to induce condensation. Beneath the surface, the windrow is insulated from above and warmed by the soil below which cools more slowly than either the air or the surface. For this reason, condensation is not typically observed at depth within the windrow. Condensation rewetting of the grain would therefore be avoided if the grain were placed at a level in the windrow other than the surface.

Because of the difficulties experienced by others in inverting the windrow, two other methods have been developed and tested for isolating the grain from the surface. The first technique, tested in 1985, involved folding the windrow over the center so as to position the grain between an upper and lower layer of straw. This technique proved difficult to perform mechanically, and the narrow, dense windrows produced were difficult to thresh. The head rice yields were elevated relative to open windrows, but the grain moisture was also somewhat elevated because of reduced drying rate in these dense windrows. A simpler technique was devised in 1986 and tested in 1986 and 1987. This method involves cutting the residual stubble or stem material left standing beside the windrow after swathing and spreading it in a uniform layer over the original surface of the windrow. The latter technique has been performed mechanically, does not reduce the rate of drying relative to open windrows, and gives elevated head rice yields compared to open windrows with continuing exposure in the field. Full descriptions of the experimental trials and results using these techniques appear in Jenkins (1987), and are summarized below.

2.1 Folded Windrow Technique, 1985

2.1.1 Field trials using the folded windrow technique were performed in mid-September in a medium grain variety (M-201). Swathing was started at 30% grain moisture (wet basis). A total of 0.8 ha were swathed. Three windrow treatments were tested: 1) open or unmodified windrows, 2) windrows folded immediately following swathing, and 3) windrows folded one full day following swathing. The latter treatment was tested to evaluate how equipment delays might affect head rice yields if folding were done with separate equipment after swathing. The first two treatments were harvested at intervals of 1, 2, 3 and 4 days after swathing. Windrows treated by the third method were harvested at intervals of 2, 3, and 4 days after swathing. Harvesting was accomplished with a Deere Model 105 combine harvester equipped with a rasp bar cylinder. Cylinder speed was set at 500 rpm. Grain samples were collected from the grain auger at the grain tank, taken to the laboratory, air dried without heat to 14% moisture if necessary, and milled according to standard (USDA, 1977). Moisture content at harvest, head rice yield, total milled rice yield, and dockage were determined for each sample. During the field trials, an automatic weather station monitored local weather data.

2.1.2 Results of the 1985 field trials are given in Table 1. Drying rates were lower in the folded windrows than in the open windrows, although head rice yields were still higher after 4 days. At equivalent moistures, however, there were no substantial differences in head rice yields. Because the method was difficult to accomplish mechanically, no further attempts were made to develop this technique. Of interest, however, were the results of the windrows folded one day following swathing. During the first day, the decrease in head rice yield was similar to that of the open windrows. After folding the window, there was no significant decline in head rice with continued exposure, even to moistures lower than in the windrows folded immediately. This result confirmed the observation that sheltering the grain could lead to improved head rice yields over time.

Table 1. Head rice yields and grain moistures for open windrows, folded windrows, and windrows folded after one day during the 1985 M-201 field trials.

Days after Swathing	Open		Folded		Folded after 1 day	
	Head	Moisture	Head	Moisture	Head	Moisture
0		30.0		30.0		30.0
1	63.5	19.7	64.2	23.6	- - -	- - -
2	58.3	17.4	64.2	20.6	59.0	17.8
3	50.0	15.2	59.5	19.8	57.9	17.5
4	46.3	13.6	57.7	·18.3	58.5	16.9

2.2. Covered Windrow Technique, 1986-1987

2.2.1 A simpler method was developed following the folded windrow tests of 1985. In order to increase the drying rate, the surface area of the windrow as originally formed was maintained, and only a thin, uniform layer of straw was spread over the surface of the windrow to intercept dew forming on the surface. Covering was accomplished mechanically by cutting straw beside the windrow with a flail chopper, pneumatically conveying the chopped straw into a special spreader mechanism which distributed the straw over the windrow with a high speed rotating beater similar to the straw spreader on a combine harvester. Depending on the stand density and the height of cut when swathed, the spreader mechanism left a covering layer some 25 to 50 mm in depth over the original surface of the windrow. Windrows were 1.5 m wide, and the lineal density of the cover was on the order of 1120 g/m for a cover bulk density of 14.5 kg/m^3. Swathing was started at 28% moisture following maturation of the grain.

During October, 1986, a nine day trial was conducted in a medium grain rice, M-9. Alternating open and covered windrows were harvested at days 1, 3, 5, 7, and 9 after swathing. Harvesting was performed with a Deere 7700 combine harvester equipped with a spike tooth cylinder at speeds ranging from 800 to 1050 rpm. The cylinder speed was decreased as the grain moisture declined so as to reduce damage to the grain by the harvester. Grain samples were taken from the grain auger at the grain tank of the harvester and analyzed in the same manner as the samples of 1985. Total grain recovery from each windrow was measured gravimetrically. Loss measurements were made and weather data continuously recorded on site.

A similar test was made during September, 1987 in another medium grain rice, M-202. Exposures of up to 17 days were made for both open and covered windrows. These windrows were harvested in the same manner as the 1986 trial on days 1, 3, 6, 8, 12, and 17 following swathing. Standing grain was also harvested on days 5 and 18 for comparison. A cylinder

speed of 775 rpm was used up to day 6, but was dropped to 600 rpm on day 8 to reduce potential damage due to the low grain moisture achieved by that date. The depth of windrow cover obtained in this crop was about half that obtained in 1986, and averaged about 22 mm with 480 g/m. Grain samples were collected and analyzed as before. Yield and loss measurements were made as before. During this trial, a dew sensor recorded by the automatic weather station was left on the surface of the windrow as in previous trials. From day 12 to day 17, however, the dew sensor was moved to beneath the straw cover to determine if condensation could be detected at this level. Visual observations were also made to detect the presence of dew beneath the straw cover.

2.2.2 Results of the 1986 trial are included in Table 2. Head rice yields in the covered windrow were not observed to change over the course of the 9 day trial. Head rice yields in the open windrows began to decline after 5 days of exposure. Drying rates for the grain in both types of windrows were equivalent. This was also true for the straw portion of the windrow. The period of this trial was characterized by relatively cool weather, typical of the latter part of the harvesting season. The amount of dew forming was observed to be rather light. Moistures acceptable for final storage (14%) were not achieved until day 7 after swathing. A light rain on the evening of day 8 may have influenced the head rice yields of the open windrows, but did not appear to affect the head rice yields of the covered windrows.

Results for the 1987 trial shown in Table 3 were substantially different that those of the previous year. This period was characterized by extremely warm, dry days and cold nights with heavy dew formation. The dew sensor recorded dew on the surface every night of the period it was exposed on the surface. When under the straw cover, no incident of dew formation was recorded by this sensor. Temperature measurements made throughout open and covered windrows during the course of the trial confirm the results of the dew sensor. While surface temperatures dropped below the dew point temperature on every night of the trial, the temperature of the grain below the straw cover remained consistently above the dew point temperature. This is illustrated in Figure 1 for the night of day 10 during the trial of 1987. Visual inspection of the windrow also confirmed a lack of dew at the level under the straw cover.

Table 2. Head rice yields and grain moistures for open windrows and covered windrows during the 1986 M-9 field trials.

Days after Swathing	Open		Covered	
	Head	Moisture	Head	Moisture
0		28.0		28.0
1	55.1	24.7	56.6	25.2
3	53.7	20.3	55.2	20.3
5	54.0	16.0	56.8	16.0
7	52.2	14.3	55.2	14.3
9	50.5	15.1	56.0	14.9

Table 3. Head rice yields, grain moisture, and total milling yield for standing crop, open windrows, and covered windrows of 1987 M-202 field trials.

Days after Swathing	Standing			Open			Covered		
	Head	Moisture	Total	Head	Moisture	Total	Head	Moisture	Total
0		27.9			27.9			27.9	
1	- - -	- - -	- - -	64.7	22.7	68.9	64.4	23.5	67.7
3	- - -	- - -	- - -	48.8	13.7	69.7	58.9	14.4	69.2
5	65.0	25.6	68.6	- - -	- - -	- - -	- - -	- - -	- - -
6	- - -	- - -	- - -	36.5	14.0	67.8	46.6	13.9	68.9
8*	- - -	- - -	- - -	38.8	13.3	68.2	51.4	13.0	69.6
12	- - -	- - -	- - -	34.5	10.8	69.1	45.0	10.3	70.6
17	- - -	- - -	- - -	24.1	9.8	68.6	42.9	10.2	70.6
18	42.0	16.5	69.7	- - -	- - -	- - -	- - -	- - -	- - -

*Cylinder speed reduced from 775 to 600 rpm.

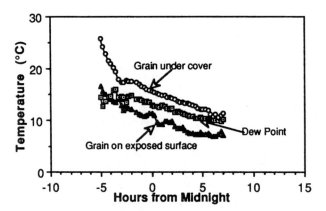

Figure 1. Temperatures of grain on the surface of the windrow and under the straw cover relative to dew point temperature during the night of day 10, 1987 trial.

The rate of head rice decline during the first three days of the trial in the open windrows was on the order of 0.05 per day (from 64.7% head rice to 48.8% head rice). The decline in the covered windrows was 0.02 per day (64.4% to 58.9%), or less than half the rate of the open windrows. The drying rates for the grain were equivalent, as were the rates for the straw. The grain had achieved 14% moisture by day 3, remained in equilibrium at this level until after day 8, when an increase in daytime temperature and decrease in daytime relative humidity drove the equilibrium moisture to 10%. With continuing exposure past day 8, the rate of head rice decline shows a deceleration for the covered windrows, but an acceleration for the open windrows. This is believed to be due to the difference in the rewetting mechanisms for grain in the windrows. In the case of the covered windrow, the principal rewetting mechanism is by vapor phase adsorption of water from the atmosphere. As the relative humidity in the field declines as the field dries, the moisture stresses in the kernels are less severe, and the rate of kernel fissuring therefore declines. In the open windrows, kernels at extremely low moisture are subjected to repeated wetting by condensation. The rate of kernel fissuring would therefore be expected to increase as observed in these windrows. The increase in head rice yields at day 8 for both types of windrows is due to the drop in harvester cylinder speed on day 8. The low moisture of the grain at day 3 indicates that an improvement in head rice yields may have been possible if the change in cylinder speed had been made earlier. It is significant that head rice yields remained comparably high in the covered windrows even when harvested at 10% moisture content.

Head rice yields in the standing grain were relatively constant at least through the first 5 days following swathing of the windrowed rice. At day 18, when the grain moisture had reached 16% in the standing crop, the head rice yields were on the same order (42.0%) as those for the covered windrows (42.9%) at 10% moisture.

In both the 1986 and 1987 trials, the loss data showed no significant differences between the open and covered windrows. There is perhaps some trend to higher processing loss with the covered windrows due to the higher straw to grain ratio. This difference might be offset by a small increase in cylinder speed for covered windrows without increasing kernel damage. Nor did there appear any significant differences in harvested yield between the windrowed rice and the standing rice. However, these results require large scale confirmation. The tests of 1986 indicated there might be increased dockage when harvesting covered windrows because of inadequate separation of the chopped straw. This was not observed in the trial of 1987 after adjustment of the harvester, and adequate separation was evidently obtained.

3 PRELIMINARY ECONOMIC ANALYSIS

To evaluate the economic feasibility of the modified swath harvesting technique, a comparison was made between it and direct harvesting. The economic analysis includes three cost categories and the revenue from the sale of the grain. All production costs up to harvesting are considered to be equivalent, and only the harvesting, transportation, and drying costs are assumed to vary. The grain revenue depends on the head rice yield, and on a per unit grain basis, is of the form (see section on nomenclature for definitions of symbols):

$$z_v = y_h p_h + (y_t - y_h)p_b \qquad [1]$$

All costs and revenues in this analysis have been normalized to zero moisture, although computed on the basis of the moisture at the time each activity occurs. The integrated net revenue for the season is:

$$Z = Z_v - Z_h - Z_t - Z_d \qquad [2]$$

The subscripts refer to the grain revenue or value, v, and the costs of harvesting, h, transportation, t, and drying, d. The costs and revenues may be computed in the manner shown below. The derivation of the equations is given by Jenkins (1987).

3.1. Direct Harvesting

3.1.1 Grain Revenue

The total grain revenue over the season, Z_v, is expressed as the sum of the daily revenues:

$$Z_v = \sum_{t_s}^{t_f} G(y_h p_h + (y_t - y_h)p_b) \qquad [3]$$

where the daily grain production function, G, is

$$G = 3600 \, h \, Y \, (1-y_l) \, n_h v_h w_h \eta_h \qquad [4]$$

and the summation is taken from the time the harvest starts, t_s, to the time the harvest finishes, t_f:

$$t_f - t_s = \frac{A \, Y \, (1-y_l)}{G} \qquad [5]$$

The moisture content of the grain at the time of harvest influences the head rice yield as noted above. The moisture content here is assumed to be found as an exponential decay with time such that:

$$m = m_f + (m_o - m_f) \, e^{-\frac{t}{\tau_d}} \qquad [6]$$

The head rice is assumed to decline linearly with moisture after achieving an average moisture $m_{h,o}$, above which the head rice is constant, as observed in the field trials and reported by Luh (1980):

$$y_h = y_{h,o} - r \, (m_{h,o} - m) \qquad [7]$$

This relationship for head rice yield is only valid during the period harvesting occurs above the equilibrium moisture content, m_f. Beyond this point, head rice will continue to decline with time, but average daily moisture will not. The result of integrating equation [3] is:

$$Z_v = Z_1 + Z_2 \qquad [8]$$

where

$$Z_1 = G(y_{h,o}p_h + (y_t - y_{h,o})p_b)(t_{h,o} - t_s) \qquad [9]$$

$$t_s = -\tau_d \ln\left[\frac{m_s - m_f}{m_o - m_f}\right] \qquad [10]$$

$$t_{h,o} = -\tau_d \ln\left[\frac{m_{h,o} - m_f}{m_o - m_f}\right] \qquad [11]$$

$$Z_2 = G\left[c_1(t_f - t_{h,o}) + c_2\tau_d(e^{-\frac{t_{h,o}}{\tau_d}} - e^{-\frac{t_f}{\tau_d}})\right] \qquad [12]$$

$$c_1 = (p_h - p_b)(\alpha + r \, m_f) + p_b y_t \qquad [13]$$

$$c_2 = r(m_o - m_f)(p_h - p_b) \qquad [14]$$

$$\alpha = y_{h,o} - r \, m_{h,o} \qquad [15]$$

3.1.2 Cost of direct harvesting

The harvesting cost computed is the sum of the fixed and variable costs of operating the the the harvester, as this is the only component of the direct harvesting system relevant to the difference between the two systems. The fixed costs are computed on the basis of an allocation factor, which expresses how much of the cost of the equipment is actually allocated against the rice harvesting operation This factor is included to enable variations in this quantity because the same machine might be used for another crop or the total amount of acreage harvested with a single machine might vary. The total cost of harvesting is:

$$Z_h = n_h\left[\frac{\Phi C}{\xi} + c_h \, h \, (t_f - t_s)\right] \qquad [16]$$

$$\Phi = \frac{i \, (1 + i)^n}{(1 + i)^n - 1} \qquad [17]$$

3.1.3 Transportation cost

The transportation cost is a function of the moisture content of the grain. Trucks are normally weight limited, and as the moisture content of the grain declines, the total truck payload expressed in terms of dry rice increases. The transportation cost is:

$$Z_t = \frac{G \, p_t \, d \, \tau_d}{1 - m_f} \left[\ln \frac{(m_s - m_f)(1 - m_{t_f})}{(m_{t_f} - m_f)(1 - m_s)} \right] \qquad [18]$$

where the moisture at $t=t_f$ is given by equation [6].

3.1.4 Drying cost

The drying cost includes both handling and energy costs. For this reason, rice delivered to a dryer will still incur a handling charge even though the moisture may be at or below 14%. Using a published rate schedule and regressing the cost in the form of the energy consumption above 14% moisture, the cost of drying is:

$$Z_d = G \left[\beta \tau_d \, \ln \left(\frac{1 - m_{t_f}}{1 - m_s} \right) + \gamma(t_f - t_s) \right] \qquad [19]$$

and is constant below 14% moisture.

3.2. Windrow Harvesting

3.2.1 Grain revenue

The grain value for the windrow harvesting system is computed in a similar manner as for the direct harvesting with the exception that the rate of head rice decline is taken in proportion to the time spent in the windrow after swathing. The residence time of the grain in the windrow varies depending on the capacities of the swather and the harvesters, and the number of harvesters allocated to each swather. The moisture content of the grain in the windrow is also taken as an exponential decay in time, with the characteristic time constant τ_w substituted for τ_d in equation [6]. The seasonal grain revenue is then:

$$z_v \, G \left[(p_h - p_b) \, (y_{h,o} - r_w \, t_w) + p_b \, y_t \right] \left(t_{h,o} - t_o \right)$$

$$+ G \left[c_1' \, (t_{\delta 1} - t_{h,o}) + c_2 \, \tau_d \left(e^{-\frac{t_{h,o}}{\tau_d}} - e^{-\frac{t_{\delta 1}}{\tau_d}} \right) \right] +$$

$$+ G \left[c_1' \, (t_{f,w} - t_{\delta n}) + c_2 \, \tau_d \left(e^{-\frac{t_{\delta n}}{\tau_d}} - e^{-\frac{t_{f,w}}{\tau_d}} \right) \right]$$

$$\cdots\cdots\cdots\cdots\cdots\cdots\cdots\cdots\cdots\cdots\cdots\cdots [20]$$

$$c_1' = (p_h - p_b)(\alpha + r \, m_f - r_w \, t_w) + p_b \, y_t \qquad [21]$$

$$G = 3600 \, h_w \, Y \, (1 - y_1) \, n_{h,w} \, v_{h,w} \, w_s \, \eta_{h,w} \qquad [22]$$

where the individual terms of equation [20] account for delays in the swathing operation as harvest progresses, the standing grain moisture declines, and the number of days required in the windrow to achieve final moisture also declines. The value of t_w is computed as the time to achieve final moisture in the windrow. The operating time of the swathers relative to the harvesters is:

$$h_s = h \left[\frac{n_{h,w}}{n_s} \, \frac{v_{h,w}}{v_s} \, \frac{\eta_{h,w}}{\eta_s} \right] \qquad [23]$$

and the total number of days required to harvest the crop is:

$$t_n = \frac{A}{n_s \, h_s \, v_s \, w_s \, \eta_s} \qquad [24]$$

3.2.2 Harvesting cost

The cost of harvesting is similar to that for direct harvest, but incremented by the cost of the swathing operation:

$$Z_h = n_{h,w} \left[c_{h,w} h_w t_n + \frac{\Phi_{h,w} \, C_{h,w}}{\xi} \right] +$$

$$n_s \left[c_s h_s t_n + \frac{\Phi_s C_s}{\xi_s} \right] \qquad [25]$$

3.2.3 Transportation cost

With grain moisture approximately constant, the transportation cost is:

$$Z_t = A \, Y \, (1 - y_1) \, \frac{p_t \, d}{1 - m_{w,h}} \qquad [26]$$

3.2.4 Drying cost

With harvesting from the windrows done at final moisture, the drying cost is essentially the handling charge:

$$Z_d = A \, Y \, (1 - y_1) \, \gamma_d \qquad [27]$$

If the moisture at harvest is above the final moisture content, then a formulation similar to equation [19] is used.

3.2.5 Results of the Economic Model

To compute the overall cost differential between the modified windrow harvesting system and direct harvest, the economic model was tested with the assumptions in Table 4. Swathing is assumed to take place with the grain at 28% moisture. Direct harvest is assumed to occur at 24.5% moisture which is typical of moisture limitations imposed by dryers and reportedly near optimal (for cost) for medium grain. After reaching maturity, rice apparently may go through a period of time at relatively constant moisture (about 28%) before beginning to dry on the standing crop. This period may last up to 10 days. If swathing is started at the beginning of this period, there is an improvement in head rice yield because more rice is swathed at high moisture, and swathing can be initiated well ahead of direct harvest. The results of the economic analysis for this case show that drying costs are reduced about 64% relative to direct harvest, harvesting costs are increased about 50% because of the added swathing operation, transportation costs are reduced 7.5% and total costs are reduced 34%. Gross revenue improves by about 2.4%, and net revenue increases by 8.8% (Table 5). If swathing is started at the end of the constant moisture period, so that moisture declines throughout the harvesting period, the net revenue increases by 5%. This is lower because of the somewhat reduced head rice occurring as a result of swathing later. As the grain moisture for initiating direct harvest declines, the net revenue increase for swathing relative to direct harvest increases due to a greater differential in head rice (Figure 2).

Table 4. Assumptions for the economic analysis.

Parameter	Value
A	10^6 m^2
Y	0.80 kg-m^{-2}
y_t	0.70
$y_{h,o}$	0.65
y_l	0.03
m_o	0.28
$m_{h,o}$	0.23
m_f	0.14
$m_{d,f}$	0.14
$m_{w,o}$	0.28
$m_{w,f}$	0.10
τ_d	10 d
τ_w	1.86 d
r	3.50
r_w	0.02 d^{-1}
h	6 d^{-1}
v_h	1.12 m-s^{-1}
$v_{h,w}$	1.34 m-s^{-1}
v_s	2.7 m-s^{-1}
w_h	5.49 m
w_s	4.88 m
η_h	0.7
$\eta_{h,w}$	0.6
η_s	0.8
d	10 km
β	0.0572 \$-kg^{-1}
γ	0.0132 \$-kg^{-1}
γ_d	0.0064 \$-kg^{-1}
p_h	0.2654 \$-kg^{-1}
p_b	0.14615 \$-kg^{-1}
p_t	0.0001 \$-kg^{-1}-km^{-1} (\$0.18/ton-mi)
$c_h, c_{h,w}$	22.75 \$-h^{-1}
c_s	16.50 \$-h^{-1}
$C, C_{h,w}$	100 000 \$
C_s	55 000 \$
i	0.10 y^{-1}
n	10 y

As the equipment utilization varies (varying allocation factor) the differential in net revenue varies in the manner shown in Figure 3. Under the assumptions used, there appears a break even point at about 120 ha. Below this point, the added cost of the swather cannot be justified by improvements in quality and reduced drying costs alone. Contract harvesting might be an alternative at this scale.

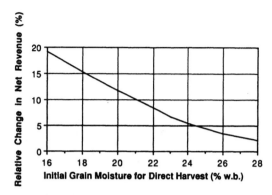

Figure 2. Relative increase in net revenue for windrow harvest relative to direct harvest as the moisture content at the start of direct harvest declines.

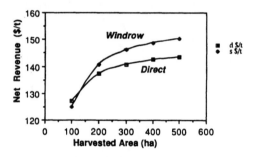

Figure 3. Net revenue for direct and windrow harvesting in relation to the scale of the farming operation.

Table 5. Costs and revenues for swath and direct harvest methods.

Category	Direct Harvest (\$/t)*	Swath Harvest (\$/t)*	Relative Difference (%) (Swathing relative to Direct)
Costs:			
Drying	15.39	5.51	-64.2
Harvesting	5.14	7.72	+50.1
Transportation	1.08	0.99	- 7.5
Total	21.61	14.22	-34.2
Gross Revenue:	145.09	148.51	+ 2.4
Net Revenue:	123.48	134.29	+ 8.8

*Grain at 14% moisture content wet basis.

4 ENERGY CONSUMPTION

The large savings in drying cost accruing from the windrow harvesting technique are predominantly due to savings in commercial energy. At the conditions tested, the reduction in commercial energy amounts to 500 to 700 kJ/kg of rice. The energy for transportation is also reduced because of higher dry weight payloads in the case of windrow harvesting. The energy savings are of much lower magnitude however, on the order of 10 kJ/kg for a 40 km haul. Energy for harvesting is increased because of the additional swathing operation. The incremental energy consumption is 36 kJ/kg. Overall, the energy savings are still dominated by the drying energy reduction.

5 DISCUSSION

A number of factors influence the results of the economic model. When direct harvesting is carried out at reasonably high moistures, the incremental benefits due to windrow harvesting are relatively small, and the costs of both systems may be seen to be roughly equivalent. At lower moistures, the potential for improved head rice with windrow harvesting appears to amplify the economic gains of the system. Additional factors would also affect the economic analysis. The implications of enhancing flexibility in harvest timing have not been fully investigated. Windrow harvesting permits earlier entry into the field, which could extend the effective length of the season, and thereby reduce the total equipment allocated to a harvest. It could also be used as a means of stabilizing head rice if insufficient harvesting equipment exists and the standing crop begins to dry rapidly. Heated air drying of rice commercially reduces head rice by up to 2 points. This has not been included in the economic analysis here. If dried to final moisture in the field, subsequent heated air drying would not be needed. On the other hand, final drying of direct harvested rice could be performed using ambient air drying, and might reduce the loss in head rice and perhaps reduce some of the cost of drying for growers who have their own drying systems.

The effect of rain on the windrows has also not been fully tested. If heavy rains occurred after the crop had been windrowed, kernel damage could be increased. To deal with this uncertainty, flexibility to revert to direct harvest in inclement weather should be retained, although further investigations are required to understand the implications for windrow harvesting. Lodging also presents a problem for swathing and is another reason to maintain flexibility to carry out direct harvesting.

The depth of cover required to shelter the grain in the windrow appears to be surprising low. The test of 1987 was carried out with an average depth of only 20-25 mm. The benefits of such cover are readily apparent however. Condensation on the windrow appears to be largely a surface phenomenon, and adequate straw material is available in the field after swathing to provide acceptable cover. The differences in the head rice yields between the 1986 and 1987 trials are attributable perhaps to both weather and the varieties tested. Varietal differences have been noted in fissuring potential (Bhattacharya 1980; Geng et al. 1984; Kunze 1985; Chattopadhyay and Kunze 1985). There may be some varieties better suited to windrow harvesting than others.

Mixing wet and dry rice has also been shown to lead to kernel fissuring (Kunze and Prasad 1978). This could perhaps explain some of the loss of head rice seen in the covered windrows after the first day and up until the windrow has dried significantly. During the first day, the grain is wet enough that adsorption of moisture from the windrow due to wet straw would not subject the kernels to excessive stress. As the windrow dries, moist conditions in the windrow over the next day or two could lead to high moisture stress. However, the uniformity of moisture in the windrow after drying compared to that in the standing crop might permit improved harvester operation and control, with less machine damage to the kernels.

6 SUMMARY AND CONCLUSIONS

The modified environment in the windrow provided by the straw cover does appear to reduce the damage of the grain with continued exposure in the field. Apparently only light covers are sufficient to yield substantial gains in head rice at low moistures compared to open or conventional windrows. Drying rates are not affected by covering to a depth of 50 mm or less. In good weather, moistures acceptable for final storage should be obtained in about 3 days, and up to 5 days in cooler weather. In the later part of the season, the risk associated with rainfall would suggest that direct harvest capabilities should be maintained. In general, the modified windrow technique provides a tool for managing grain quality and harvesting events, and perhaps reducing overall cost of production. These early field trials have shown that reductions in drying energy and costs are possible without sacrificing grain quality. Additional work is required to evaluate the system on a commercial scale.

7 REFERENCES

Bainer, R. 1932. Harvesting and drying rough rice in California. Bulletin 541, University of California Printing Office, Berkeley, CA.

Bhattacharya, K.R. 1980. Breakage of rice during milling: a review. Trop. Sci. 22(3):255 -276.

Chattopadhyay, P.K. and O.R. Kunze. 1986. Fissuring characteristics of parboiled and raw milled rice. TRANS. of the ASAE 29(6):1760-1766.

Chau, N.N. and O.R. Kunze. 1982. Moisture content variation among harvested rice grains. TRANS. of the ASAE 25(4):1037-1040.

Craufurd, R.Q. 1963. Sorption and desorption in raw and parboiled paddy. J. Sci. Food Agric. 14:744-750.

Geng, S., J.F. Williams and J.E. Hill. 1984. Harvest moisture effects on rice milling quality. California Agriculture 6:11-12.

Jenkins, B.M. 1987. Feasibility of modified swath harvesting techniques for rice. American Society of Agricultural Engineers Paper No. 87-6510, ASAE, St. Joseph, MI 49805-9659.

Kunze, O.R. and M.S.U. Choudhury. 1970. Moisture adsorption related to the tensile strength of rice. American Association of Cereal Chemists, St. Paul, MN 55121.

Kunze, O.R. and S. Prasad. 1978. Grain fissuring potentials in harvesting and drying rice. TRANS. of the ASAE 21 (2):361-366.

Kunze, O.R. 1985. Effect of variety and environmental factors on milling qualities of rice. International Rice Research Conference Paper, International Rice Research Institute, Philippines.

Kunze, O.R. and D.L. Calderwood. 1985. Rough Rice Drying. Chapter 6 in Rice: chemistry and technology, B.O. Juliano, editor, American Association of Cereal Chemists, St. Paul, MN.

Luh, B.S. 1980. Rice: production and utilization. AVI, Westport, Conn.

Ranganath, K.A., M.K. Bhashyam, Y. Bhaskar Rao and H.S.R. Desidachar. 1970. J. Food Sci. and Tech. 7:144-147.

Rhind, D. The breakage of rice in milling. a review. Trop. Agric. 39:19-28.

Siebenmorgen, T.J. and V.K. Jindal. 1986. Effects of moisture adsorption on the head rice yields of long-grain rough rice. TRANS. of the ASAE 29(6):1767-1771.

Stahel, G. 1935. Breaking of rice in milling in relation to the condition of the paddy. Trop. Agric., Trin. 12:255-260.

USDA. 1977. United States standards for rough rice, brown rice for processing, and milled rice. U.S. Department of Agriculture, Federal Grain Inspection Service, Washington, D.C.

8 NOMENCLATURE

		Units
A	Area	m^2
C	Capital cost	$
G	Daily grain production	$kg\text{-}d^{-1}$
Y	Harvestable grain yield	$kg\text{-}m^{-2}$
Z	Seasonal cost or revenue on area A	$\$\text{-}y^{-1}$
c	hourly operating cost	$\$\text{-}h^{-1}$
d	transportation distance	km
h	operating hours per day	$h\text{-}d^{-1}$
i	interest rate on invested capital	y^{-1}
m	moisture content (wet basis)	—
n	number of equipment items, also	—
	number of years of life	y
p	price of grain, also	$\$\text{-}kg^{-1}$
	unit cost of transportation	$\$\text{-}kg^{-1}\text{-}km^{-1}$
r	rate of head rice decline, standing grain	—
	and rate of head rice decline, windrow	d^{-1}
t	time	d
v	velocity	$m\text{-}s^{-1}$
w	width	m
y	yield mass fraction	—
z	unit cost or revenue	$\$\text{-}kg^{-1}$
α	constant defined by equation [15]	—
β	coefficient of the drying cost approximation	$\$\text{-}kg^{-1}$
γ	coefficient of the drying cost approximation	$\$\text{-}kg^{-1}$
δ	time of delay event	d
η	field efficiency	—
ξ	allocation factor	—
τ	time constant	d
Φ	capital recovery factor	—

Subscripts

b	broken kernels
d	direct harvest, drying
f	final
h	head rice, harvest, harvester
l	grain loss
n	generalized time index
o	initial
s	start, swather
t	total, transport, time
v	grain value (revenue)
w	windrow

RÉSUMÉ: Sur une période de trois années, on a développé et testé en Californie une nouvelle technique pour la récolte duriz en javelles. Dans cet état, plusieurs cultivateurs ont adopté la récolte en andains comme un moyen de réduire le coût de séchage des grains récoltés avec une teneur en eau suffisamment élevée pour satisfaire les exigences d'un grain de haute qualité (celle-ci étant mesurée au grain entier ou au rendement dépis de riz). Tandis qu'elle permet au grain de sécher dans les champs, l'utilisation des "open windrows" (javelles ouvertes), conventionnelles, conduit malheureusement à un taux de pertes inacceptable dans la production des grains entiers, soumis à une exposition de plus d'une journée dans les champs. La raison essentielle pour cette détérioration qualitative liée au facteur temps est la réhumidfication diurne du grain par la rosée se formant sur la partie supérieure de la javelle - un phénomène bien connu pour provoquer "cracking" (le craquellement)* duriz. La nouvelle technique modifie le milieu local du grain dans la javelle par le dépôt d'une couche fine mais uniforme de "residual stem material" (substance résiduelle de tige)* sur la surface de l'andain. On intercepte ainsi la condensation à la surface et on réduit, par ce mécanisme, l'exposition du grain à la réhumidification. Cette méthode ne diminue pas cependant le taux de séchage du grain relatif à celui obtenu en javelles ouvertes. Un riz á grains moyens coupé en javelles à 28% d'humidité et récolté à 14% apres trois jours d'exposition a produit 59% dépis de riz comparés à 48% d'épis pour un grain récolté à partir de javelles ouvertes. La récolté sur pied faite a 16,6% d'humidité a produit 42% d'épis de riz. Un modèle économique développé dans le but de comparer la technique de récolte en javelles modifiées avec la récolte directe indique un avantage de 8 à 10% pour la récolte en javelles par rapport à la récolté directe à 24,5% d'humidité, considérée presqu'optimale pour la recolte directe d'un riz a grains moyens. La technique de javelage modifié permet egalement de recolter le grain plus tot, et ouvre la possibilite d'ameliorer l'utilisation de l'équipement, reduisant par ce moyen l'investissement de capital requis.

* Note du traducteur. La parenthese contient une traduction litterale et possiblement imprecise de l'expression americaine entre guillemets.

ABRb: Eine neue Technik zum Ernten von Reis von der Windreihe ist in einem dreijährigen Forschungsprojekt in Kalifornien entwickelt und getestet worden. Grasnarbenernten wurde von einigen Züchtern im Staat übernommen, als ein Mittel zur Reduzierung der Kosten von getrockneten Körnern, die in ausreichender Feuchtigkeit geerntet werden müben, um die Voraussetztungen für Qualitätskorn zu erfüllen (wie bei Vollkorn- oder Kopfreisernten ermessen). Der Gebrauch von herkömmlichen offenen Windreihen, während die Körner in den Feldern trocknen, fuhrt leider zu untragbaren Verlustquoten in der Vollkornernte mit einer Aussetztung im Feld länger als einen Tag. Der Hauptgrund für die Qualitäts-verschlechterung mit der Zeit ist die tägliche Wiederwässerung von Getreide auf Grund von Taubildung auf der Oberfläche der Windreihe, ein Phanomen, das als Ursache für das Bersten von Reis gut bekannt ist. Die neue Technik modifiziert diei örtliche Umgebung des Getreides in der Windreihe mittels der Plazierung einer dünnen, jedoch gleichmäbigen Schicht übriggebliebenem Stammmaterials über die Oberflache der Windreihe, womit die Kondensation an der Oberfläche aufgefangen und die Absterbequote des Getreides, im Vergleich zu dem in offenen Windreihen, reduziert wird. Mittelkorn Reis, das bei 28% Feuchtigkeit abgearbt und bei 14% Feuchtigkeit nach dreitägiger Aussetzung geerntet wurde, erzeugte 59% Kopfreis im Vergleich zu 48% Kopfreis von geerntetem Getreide aus offenen Windreihen. Stehendes Getreide, das bei 16,6% Feuchtigkeit geerntet wurde, erzeugte 42% Kopfreis. Ein ökomomisches Model, entwickelt zum Vergleich der modifizierten Grasnarbenerntetechnik gegenüber direktem Ernten, zeigt einen 8-10% Vorteil für Grasnarbenernten uber direktem Ernten bei 24,5% Feuchtigkeit, was nahezu als optimal fur direktes Ernten von Mittelkörnigem Reis angesehen wird. Die modifizierte Grasnarbenerntetechnik gestattet auberdem früheres Ernten von Getreide und kann die Ausnutzung von Maschinen verbessern und somit die erforderliche Kapitalanlage verringern.

Land and Water Use, Dodd & Grace (eds), © 1989 Balkema, Rotterdam. ISBN 90 6191 980 0

Development of new equipment for coffee harvesting

D.A.Ometto
ESA 'Luiz de Queiroz', University of Sao Paulo Pirachicaba, SP, Brazil

ABSTRACT: Brazil is the first country in the world to employ a mechanical coffee harvester with full success.

Its technology started in the Agricultural Research Institute of Sao Paulo and was developed by one farm machinery.

This machine is a combine, in a portico shape that operates on top of row plants. Vibratic stems acts on the branches on both sides of the coffee tree. They are mounted on rotors, engage in the plants and the fruit are drawn by them, being collected by a set of rectractile blades. A fan system removes the impurities and final product is sacked.

The machine efficiency is about 90% and replaces about 60 to 80 men.

Stripper machines are suitable for small areas; the grains are dropped on the ground. The specifications and basic components for both types are shown in this paper.

1 INTRODUCTION

Mechanical coffee harvesting is being used in Brazil with success. There are large capacity machines termed combines and a stripper type for use in smaller areas.

These machines are now operating over 30 million coffee plants mainly in States of Sao Paulo and Minas Gerais.

They are the first machines specialised for coffee harvesting that are operating at this date in the world.

The initial experiment began at Colorado Farm, in Sao Paulo State with good results and was approved by coffee farmers.

This combine, is a type of harvester in a portico shape, that operates on top of the row plants. Vibratic stems acts on the branches in both side of coffee plants. These stems, mounted on rotors, engage in the plants, and the fruits are drawn by them, resulting a low rotation movement, synchronized with the displacement machine. A horizontal vibration is superposed on this movement in order of 800 to 1000 oscillations per minute, causing the fruits to drop. These are collected by a set of retractile blades that close the spaces under lowest branches. A fan system removes the impurities and the final product is sacked.

The other machine, a stripper type is mounted on a tractor specifically used for a coffee crop and suitable for smaller areas with a normal level of mechanisation. The action of the machine causes fruits to drop to the ground only on one side of the tree. Only one rotor vibrator is mounted in a lateral position with respect to the coffee plant and in a continuous drops one side of the fruit from the coffee plants to the ground in each step as it travels along the row.

The machine is mounted on the tractor or rear supported by one free wheel in system named "semi-wagon" and moved by the tractor P.T.O. Additional speed reducers are mounted on the tractor in order to get low speeds need to drop the fruits. Recently, there are three solutions:

a) planetary reducer placed between the final axle and the tractor wheels;

b) additional reduce box placed between gear box and the differential of tractor;

c) transmission with reduced speeds.

The more recent type of stripper is also mounted in a portico shape on the plants and drops all the fruits at same time on the ground (double stripper) with two sets of rotor vibrators, acting on both side of the plant.

This machine uses one man more, placed

on the top of the stripper for vibrator controls, height and direction.

The vibrations exciter is done by one pair of masses that turn through 180°. The mass forces resulting from centripetal acceleration are composed, so the tangential components form a binary Mt = T.1. and the normal forces N are cancelled.

The rotor is free to move in bearings, resulting in a free vibration around its central axle and transmitted by the stems to the coffee plants.

The stems are geared into the plants and are drawn by them, resulting in a movement composed of a slow rotation syncronized with the harvester displacement and a vibration caused by the exciter.

The tendency of turn due the attribution action in the exciter is avoided by a band brake.

PERFORMANCE PARAMETERS

The good performance of the harvesters depend on the perfect interaction in the three systems: machine - crop - plant. Some factors must be considered:

a) access roads must have 4 meters wide (minimum);

b) the ideal plant height for stripping fruits, is about 2.0 to 3.0m, but harvesting efficiency will be decreased with higher plants. Smaller coffee varieties as Catui and Catimor have eliminated more side branches than the higher varieties as Mundo Novo and Icatu, which need a top pruning to make harvester operation easire;

c) the presence of branches render the collection of grains difficult; the minimum height of branches over ground is 0.4 m.

d) smaller row spacing of plants will result in an increased harvesting rate per hour. Rank plantations are recommended.

e) the operation must be directed following the ground, in order to improve the stripping off and gathering of the grains.

Crops grown on level ground facilitate the operation and increase machine performance. The combine can compensate for sloping ground up to a maximum gradient of 10%. Harvester output depends on the plants spacing in the row and speed of operation.

Starting harvesting, with coffee berries in the form of "cherry", the forward speed is 600 to 700 m/h. Finishing it, these values rise to 1500 to 2000 m/h and

normally registered a daily production of 500 to 1500 plants for the combine and 250 to 800 plants for the strippers.

The stripper efficiency is about 80 to 85% at beginning harvesting and 95 to 100% at the end, depending on the state of the fruit. Damage to plants are similar to hand harvesting and are not significant.

SPECIFICATIONS

Stripper (one side)
machine height: 3.20 m
width: 2.70 m
length: 2.80 m
maximum height operation: 2.75 m
minimum height operation: 0.27 m
weight: 1.920 kg
speed operation: 500 to 2000 m/h
vibration frequency: 700 - 1100 rpm
tractors: M. Ferguson - Coffee type
          Valmet - Coffee type
          Agrale - 4300

Stripper (both sides)
machine height: 4.00 m
width: 3.44 m
length: 4.57 m
height of harvesting: 3.50 m
weight: 3.350 kg
speed operation: 500 to 1500 m/h
vibration frequency: 700 - 1100 rpm
Coupling - 3 point-hitch

Combine
engine - Perkins - 4 cyl. - diesel
consumption - 7-10 1/h
transmission: hydrostatics
total weight: 7000 kg
length: 5.80 m
total width: 3.10 m
width operation: 2.70
speed operation: 0.5 - 3.0 km/h
speed transportation: 15 km/h
output performance: 0.2 - 1,2 ha/h
row spacing plants: 3.0 m (minimum)
vibration frequency: 700 - 1100 rpm

RESULTS

The combine operational costs, was considered about US$ 0.10 per bag (60 liters), while the manual harvesting was estimated US$ 0.60 a bag from last crop. It has distinct advantages for areas greater than 250 ha (400.000 plants) and productivity above 30 bags cleaned coffee/1000 plants.

The stripper types show advantages in crops greater than 60 ha in area and are more economical than the combine.

It is estimated that the machine investment could be amortized in 3 crops.

CONCLUSIONS

The coffee harvesters offer new options to the Brazilian farmers in order to minimize the operational costs and improve the worked area.

A training programme is being developed for the farmers to ensure that the coffee plant spacings match the machine perform-ance in the field.

There are many crops in good conditions for mechanization, with respect to productivity and topography.

Despite a reluctance by the farmers to the uptake of the machine we hope that gradually, farmers will realise and appreciate the advantages that can result from its adoption:

- good quality of final product, similar to hand harvesting, without significant injuries to the plants,
- reduction of harvesting time and operational costs,
- sooner crop liberation for a new agricultural year,
- others facilities for harvesting administration.

REFERENCES

1. Fridley, R.B. Evaluating the feasibility of mechanizing crop harvest. Trans. ASAE, 11(3): 350-352. 1968.
2. Fujiwara, M. Consideracoes sobre a colheita mecanica do cafe. Revista de Mecanizacao Rural - Ano I - no 3, 19-22. 1980.
3. Sartori, S. Boletim da "Jacto". 1983.

Coffee harvester - front view

Diesel motor

vibrators

Diesel tank

controller

operations controls

oil - hydraulics system

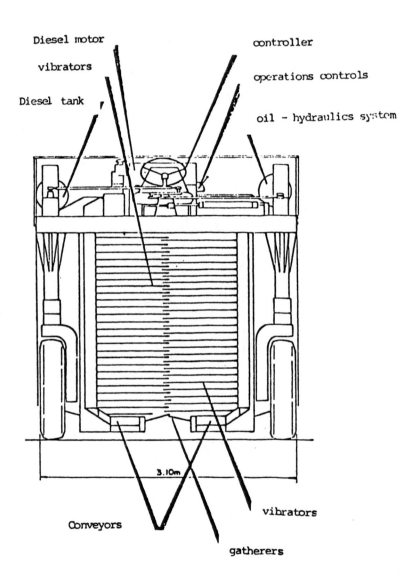

3.10m

vibrators

Conveyors

gatherers

1970

Coffee Stripper (Kokinha)

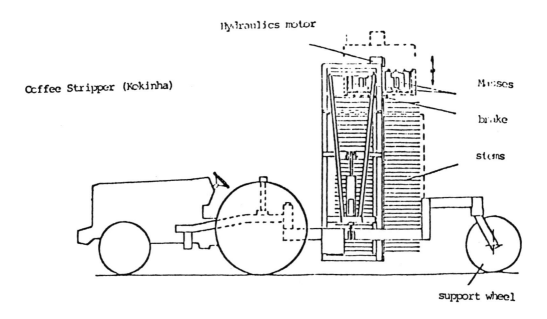

Hydraulics motor

Masses

brake

stems

support wheel

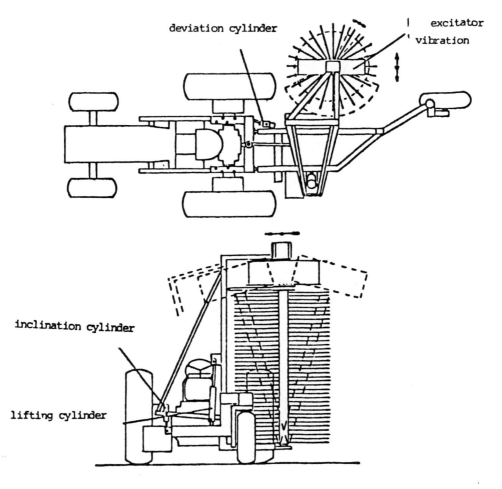

deviation cylinder

excitator
vibration

inclination cylinder

lifting cylinder

1971

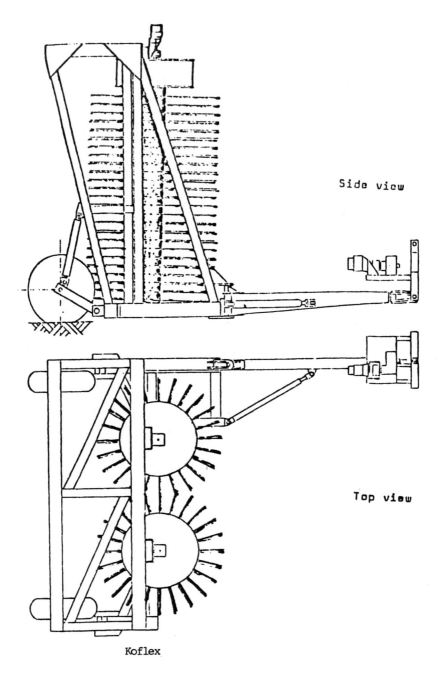

Side view

Top view

Koflex

DOUBLE STRIPPER COFFEE

Land and Water Use, Dodd & Grace (eds), © 1989 Balkema, Rotterdam. ISBN 90 6191 980 0

Einfluß der Stoffeigenschaften auf die Leistungsfähigkeit landwirtschaftlicher Maschinen

H.D.Kutzbach
Institut für Agrartechnik, Universität Hohenheim, Stuttgart, Bundesrepublik Deutschland

KURZFASSUNG: Stoffeigenschaften beeinflussen den Wirkungsgrad landwirtschaftlicher Maschinen und Geräte erheblich. Je nach Maschinenart kann dieser Einfluß relativ einfach durch Einstellvorrichtungen ausgeglichen werden, wie beispielsweise bei der Ballenpresse. Bei komplexen Maschinen jedoch, wie etwa dem Mähdrescher, führen die Änderungen der Stoffeigenschaften, auch bei optimalen Maschineneinstellungen, zu starken Durchsatzänderungen. Da die Wirkprinzipien vor allem noch nicht quantitativ belegt werden können, sind weitere Arbeiten zur Ermittlung der Zusammenhänge zwischen der Größe der Stoffeigenschaften und der Leistungsfähigkeit landwirtschaftlicher Maschinen und Geräte notwendig.

ABSTRACT: Crop properties strongly influence the efficiency of agricultural machines and equipment. In some machines this influence can be compensated for by use of simple setting devices, as seen with the baler. For more complex machines, like the combine harvester, crop property changes have a great influence on feedrate and efficiency even if the machine settings are optimal. Because the knowledge of these influences does not yet cover the quantitative aspects further work to explain the interrelationship between crop properties and efficiency of agricultural machines and equipment is necessary

RESUMEE: Les characteristiques des matières recoltés (CM) ont une influence importante sur l'éfficacité des machines agricoles. Selon la construction de la machine cette influence peut-être ajusté facilement (exemple: ramasseuses-presse), mais pour les machines plus complexes comme la moissoneuse-batteuse un changement du CM provoque des variations fortes dans la quantité travaillé même avec un reglage bien adapté. Parce que les relations et interactions de CM avec les processus dans les machines ne sont pas encore connu au fond, plus des efforts dans la recherche sont necessaires dans ce domaine pour ameliorer le fonctionement des machines agricoles.

1 EINLEITUNG

Die physikalischen Eigenschaften landwirtschaftlicher Stoffe beeinflussen die Leistungsfähigkeit landwirtschaftlicher Maschinen und Geräte, d.h. deren Arbeitsleistung bei vorgegebenem Bauraum oder vorgegebener Antriebsleistung, sehr stark. Der Konstrukteur versucht jedoch, die Maschinen durch gute Konstruktion, durch Zusatzausrüstungen, durch Einstellmöglichkeiten oder durch Regeleinrichtungen möglichst unabhängig von Änderungen der Stoffeigenschaften zu machen. Nach dem augenblicklichen Stand der Technik lassen sich landwirtschaftliche Maschinen und Geräte auf das zu bearbeitende Gut einstellen, beispielsweise die Sämaschinen auf die Korngröße des Saatgutes, die Ballenpressen auf Gutart und Gutfeuchte des zu pressenden Gutes oder der Mähdre-

scher auf die zu dreschende Frucht und deren Gutfeuchte. Eine einmal ermittelte gute Einstellung bewirkt zu anderen Zeitpunkten nicht immer die gleichen, guten Ergebnisse. Es treten Änderungen in den individuellen Guteigenschaften auf, die u.a. auch von der jeweiligen relativen Luftfeuchte, dem Wachstumsverlauf und weiteren Einflußgrößen abhängen und auf deren momentanen Wert die jeweilige Maschine individuell einzustellen ist. Die Zusammenhänge sind bis jetzt im wesentlichen allerdings nur qualitativ bekannt. Der Maschinenbenutzer ist deswegen bei der Einstellung auf seine Erfahrung angewiesen. In verschiedenen Forschungsinstituten wird versucht, auch die quantitativen Zusammenhänge zu ermitteln, um die Maschineneinstellung zu erleichtern und so die Voraussetzungen für die Anwendung von Regeleinrichtungen zu schaffen (Kutzbach 1988).

2 SYSTEMATISIERUNG DER STOFFEIGENSCHAFTEN

Beck(1988), schlägt folgende Gruppen von Stoffeigenschaften vor:

- einfache physikalische Eigenschaften
- komplexe physikalische Eigenschaften
- technologische Eigenschaften
- Referenzeigenschaften

Zu den einfachen physikalischen Eigenschaften zählen diejenigen, die an einzelnen Partikeln des Erntegutes bestimmt werden können, wie beispielsweise Reibbeiwerte, Dichten und geometrische Abmessungen.

Die komplexen physikalischen Eigenschaften umfassen solche, die sich aus den Wechselwirkungen zwischen den einzelnen Teilchen des betrachteten Gutes ergeben und oft auch mehrere einfache Eigenschaften zusammenfassen. Zu dieser Gruppe gehören beispielsweise die innere Reibung und die Schwebegeschwindigkeit.

Als technologische Eigenschaften sollen solche bezeichnet werden, die mit einer Meßmethode bestimmt werden, die einen Teilprozeß der jeweiligen Maschine vereinfacht nachbilden. Dazu zählen beispielsweise der spezifische Pflugwiderstand, die Verdichtungsfunktion und die Fluidisierungsgeschwindigkeit.

Schließlich wird im Referenzprozeß an einer über mehrere Jahre benutzten Referenzmaschine die Referenzeigenschaft als Kombination verschiedener Eigenschaften direkt bestimmt.

Der Aufwand zur Berechnung der Leistungsfähigkeit aus den Stoffeigenschaften nimmt mit der Komplexität der jeweiligen Stoffeigenschaft ab, jedoch nimmt der Aufwand zur Messung der Stoffeigenschaft mit deren Komplexität zu (Bild 1).

Bild 1 Aufwand zur Bestimmung von Stoffeigenschaften und Leistungsfähigkeit von landwirtschaftlichen Maschinen (Beck 1988)

So kann beispielsweise aus dem Referenzprozeß (Referenzmähdrescher bei Mähdrescherfeldversuchen) ohne weitere Berechnung auf die aktuellen Dresch- und Trenneigenschaften geschlossen werden; der Versuchsaufwand wird durch den Referenzprozeß jedoch verdoppelt.

Im folgenden sollen beispielhaft für Ballenpressen und Mähdrescher der Einfluß einzelner Stoffeigenschaften auf die jeweilige Leistungsfähigkeit dargestellt werden.

3 WEITERE EINFLUSSGRÖSSEN AUF DIE LEISTUNGSFÄHIGKEIT

Die Leistungsfähigkeit landwirtschaftlicher Maschinen und Geräte wird außer durch die Stoffeigenschaften durch weitere Einflußgrößen bestimmt (Bild 2).

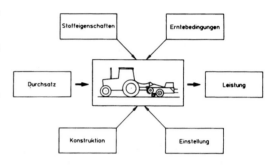

Bild 2   Einflüsse auf den Arbeitsprozeß (Beck 1988)

Die Erntebedingungen wirken über Bestandsdichte, Unkrautbesatz, relative Luftfeuchte, Strahlung und Luftgeschwindigkeit indirekt auf die aktuellen Stoffeigenschaften, durch Hangneigung und Bodenunebenheiten aber auch auf die Funktion der einzelnen Arbeitselemente der landwirtschaftlichen Maschinen und Geräte.

Die Konstruktion hat Einfluß auf den Wirkungsgrad durch die Wahl eines geeigneten Wirkprinzips und die funktionsgerechte Gestaltung der Arbeitselemente sowie die gegenseitige Abstimmung der einzelnen Arbeitselemente aufeinander.

Mit der Einstellung hat der Benutzer die Möglichkeit, die Maschine an die aktuellen Stoffeigenschaften und Erntebedingungen anzupassen und damit eine hohe Leistungsfähigkeit zu erreichen. Allerdings ist der Benutzer weitgehend auf sein Wissen und seine Erfahrung angewiesen, da es bis jetzt kaum Anzeigegeräte zur Unterstützung bei der Einstellung gibt. Erst vereinzelt werden Landmaschinen mit Regeleinrichtungen ausgerüstet, die den Benutzer von einzelnen Einstellroutinen entlasten.

Bei Halmgutballenpressen wird das Gut gegen den Widerstand des vorher gepreßten Halmgutstranges verdichtet (Bild 3).

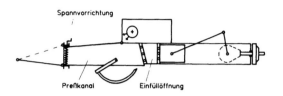

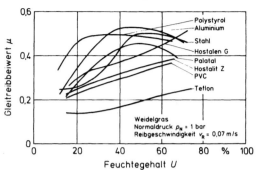

Bild 3 Aufbau einer Hochdruck-Ballenpresse

Bild 4 Gleitreibbeiwerte von Weidelgras (Kutzbach 1972)

Der Widerstand des Halmgutstranges ergibt sich aus den Reibkräften zwischen Strang und Preßkanalwänden. Durch eine einstellbare Kanalverjüngung können die Reibkräfte und damit auch die gewünschten Ballendichten vorgewählt werden. Nach Erreichen dieser Dichte verschiebt der Preßkolben den gesamten Halmgutstrang. Die Funktion der Ballenpresse wird, wie in Kap. 3 allgemein beschrieben, durch die Stoffeigenschaften, die Erntebedingungen sowie Konstruktion und Einstellung bestimmt. Hier soll insbesondere auf die Stoffeigenschaften eingegangen werden. Dabei kann unterschieden werden zwischen solchen, die den Widerstand des Halmgutstranges festlegen, und solchen, die den Zusammenhang zwischen Verdichtungsdruck und Preßdichte bestimmen.

4.1 Widerstand des Halmgutstranges

Für einen sich nicht verjüngenden Preßkanal berechnet sich der zum Verschieben des Halmgutstranges notwendige Druck nach (Kutzbach 1972) zu

$$p = p_b \cdot e^{\mu \cdot \lambda \cdot l \cdot U/A}$$

wobei p_b der Bodendruck ist, der beispielsweise durch die Ballen auf der Ladeschurre hervorgerufen wird. Umfang U, Fläche A und Länge l des Preßkanals sind durch die Konstruktion der Ballenpresse vorgegeben.
Die Reibbeiwerte μ und die Druckverhältnisfunktion λ hängen stark vom Feuchtegehalt des Gutes ab, aber auch von weiteren Größen wie Gutart, Reibpartner, Normaldruck und Gleitgeschwindigkeit. In den Bildern 4 und 5 sind beispielhaft Reibbeiwerte und Druckverhältnisfunktionen angegeben.

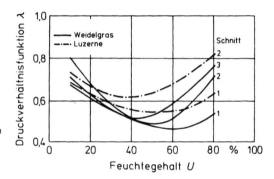

Bild 5 Druckverhältnisfunktion (Kutzbach 1972)

Durch die Kanalverjüngung von Ballenpressen wird i.a. der Seitendruck zwischen Halmgut und Preßkanal über äußere Federkräfte vorgegeben, so daß sich die Einflüsse unterschiedlicher Druckverhältnisfunktionen nicht so stark auswirken. Änderungen in den Reibbeiwerten haben, da die Reibbeiwerte als Exponent in der e-Funktion auftreten, erheblichen Einfluß auf die Längskräfte und damit auf die erreichten Preßdichten.

4.2 Zusammenhang Verdichtungsdruck/Preßdichte

Der Zusammenhang zwischen dem Verdichtungsdruck und der Preßdichte wird durch die Verdichtungsfunktion beschrieben. Im Verdichtungsdruckbereich von Ballenpressen gilt das Skalweitsche Potenzgesetz (Skalweit 1938)

$$p = C \cdot \varrho^m,$$

wobei C und m stoffspezifische Daten sind, die auch stark vom Feuchtegehalt des Gutes abhängen.

Da beim Verdichten einzelne Halme brechen, Zellverbände aufreißen und Zellflüssigkeit austritt, sind die Verdichtungsvorgänge stark zeitabhängig. Bereits die beim Verdichten erreichbare Preßdichte hängt von der Verdichtungsgeschwindigkeit ab. Der notwendige Verdichtungsdruck nimmt etwa linear mit der Verdichtungsgeschwindigkeit nach

$$p(v) = p_0 \quad (1 + 0,5 \ v)$$

zu, wobei v etwa der Kolbengeschwindigkeit entspricht und in m/s einzusetzen ist. Die Verdichtungsfunktion für Weidelgras ist in Bild 6 angegeben.

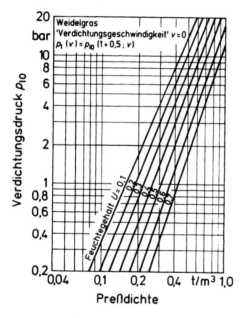

Bild 6 Verdichtungsfunktion (Kutzbach 1972)

Im Preßkanal steht das Halmgut jedoch längere Zeit unter Druck. Außerdem wird es bei jedem Kolbenhub erneut zusammengeschoben und verdichtet. Damit haben die Relaxation, d.h. die Längsdruckabnahme bei konstanter Dichte, und die Nachverdichtung, d.h. Dichtesteigerung bei konstantem Längsdruck, als weitere Stoffeigenschaften Bedeutung für die erreichte Preßdichte und damit die Leistungsfähigkeit der Ballenpresse.

Bei konstanter Dichte nimmt unabhängig von der erreichten Dichte, nach Wolf-Regett (1988), der Verdichtungsdruck für Weizen-

stroh durch Relaxation entsprechend Bild 7 ab, während bei konstantem Verdichtungsdruck durch Kriechvorgänge im Material die Preßdichte entsprechend Bild 7 durch die Nachverdichtung zunimmt. Die Verdichtungsfunktion verschiebt sich dadurch zu geringeren Drücken bzw. zu höheren Dichten.

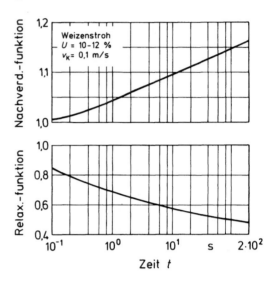

Bild 7 Relaxations- und Nachverdichtungsfunktion von Weizenstroh (Wolf-Regett 1988)

Solche isochronen Verdichtungsfunktionen sind in Bild 8 angegeben. Natürlich haben Gutart, Gutfeuchte und weitere Gutparameter Einfluß auf das Zeitverhalten nach der Verdichtung.

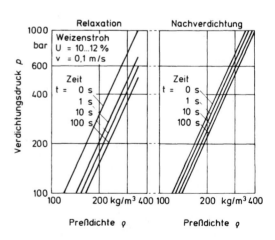

Bild 8 Isochrone Verdichtungsfunktionen für Relaxation und Nachverdichtung (Wolf-Regett 1988)

Die Stoffeigenschaften haben also bereits
bei einfachen Landmaschinen, wie beispielsweise der Ballenpresse, erheblichen Einfluß
auf die Leistungsfähigkeit. Bei Ballenpressen können diese Einflüsse durch Verstellung der Vorspannung der Preßkanalseitenwand
jedoch relativ einfach aufgefangen werden.

5 MÄHDRESCHER

5.1 Verlustverhalten

Die Arbeitsgüte des Trennprozesses von Mähdreschern ist durch den Zusammenhang zwischen Durchsatz und Kornverlusten charakte-
risiert. Die Leistungsfähigkeit eines Mähdreschers ist als maximaler Durchsatz bei
einem akzeptablen Kornverlust definiert.
Die Verlustkurven beschreiben diesen Zusammenhang. Allerdings können Verlustkurven
verschiedener Mähdrescher nur miteinander
verglichen werden, wenn sie zur selben Zeit
im gleichen Bestand ermittelt wurden, weil
einzelne z.Zt. nicht genau definierbare
Stoffeigenschaften die Verlustkurven stark
beeinflussen. Die in Bild 9 angegebenen
Verlustkurven desselben Mähdreschers zeigen
den Einfluß verschiedener Erntejahre mit
Differenzen im maximalen Durchsatz bis zu
50%.

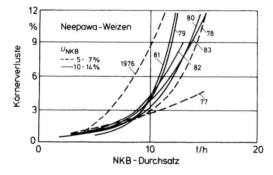

Bild 9 Verlustverhalten eines Referenz-
Mähdreschers (Wacker 1985).

Diese Verlustkurven werden während der
Ernte ermittelt, in dem bei verschiedenen
Durchsätzen die Masse der Verlustkörner aus
dem Strohschwad und dem Reinigungsübergang
bestimmt werden. Zur Analyse des Verlustverhaltens sind die Übergänge der Trenn-
elemente zur Korn-Stroh-Trennung und zur
Korn-Spreu-Trennung getrennt aufzufangen,
was die Versuchsarbeit erschwert.
    Bei den verschiedenen Instituten werden
deswegen Nachdrescher eingesetzt, durch die
die Verlustbestimmung zum Teil automatisch
erfolgen kann.

Zur Bestimmung der Leistungsfähigkeit von
Mähdreschern haben sich auch Laborversuche
mit erntefrischem Gut bewährt, weil dann
die einzelnen Stoffströme im Mähdrescher im
Labor genauer analysiert werden können. Für
die Ermittlung des Einflusses der Stoffeigenschaften auf die Trennprozesse müssen
bei den Feld- und den Laborversuchen,
gleichzeitig zu den Untersuchungen der Leistungsfähigkeit der Dresch- und Trennele-
mente, die Stoffeigenschaften mit einfachen, reproduzierbaren Meßverfahren bestimmt werden. Dabei hat sich ein mobiles
Meßlabor für Feldversuche bewährt

5.2 Einflüsse auf den Dreschprozeß

Zur Beurteilung des Dreschprozesses werden
Ausdruschverluste (in den Ähren verbliebene
Körner), Bruchkornverluste und die Abscheidung der Körner durch den Dreschkorb herangezogen. Nach Wieneke (1964) haben die Kornfeuchte, die Strohfeuchte, Art und Sorte
des Getreides, der Durchsatz und Grüngutbe-
satz wesentlichen Einfluß auf die Arbeitsqualität des Dreschwerks. Neuere Untersuchungen in Hohenheim haben gezeigt, daß
diese Einflüsse auch für das Axialdreschwerk gelten. Beispielsweise ist in Bild 10
der Einfluß des Grüngutanteils auf den Kornverlust für ein Axialdreschwerke angegeben.
Durch den Grüngutanteil (in den Versuchen
wurde statt Unkraut Rotklee verwendet)
verschlechtert sich die Arbeitsqualität der
Dreschwerke aufgrund der örtlich austreten-
den Zellsäfte und dadurch erhöhten Reibbeiwerte und aufgrund des Verstopfens einzelner Sieböffnungen wesentlich. Dieses Verhalten ist besonders zu beachten, wenn bei
Rücknahme des Herbizideinsatzes die Bestän-
de zukünftig wieder stärker verunkrautet
sind.

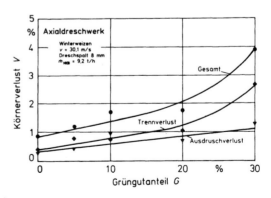

Bild 10 Kornverluste eines Axialmähdre-
schers in Abhängigkeit vom Grüngutanteil
(Wacker 1988)

5.3 Einflüsse auf die Korn-Stroh-Trennung

Die Arbeit des Schüttlers wird besonders beeinflußt durch die geometrischen Daten des zu trennenden Gutes (Halmlängen) und durch die Reibbeiwerte zwischen Korn und Stroh. Weiterhin beeinflussen aber auch die Elastizität der Strohhalme und die Strohmasse die Funktion, da eine gute Auflockerung und damit eine gute Entmischung von Stroh und Körnern nur erreicht wird, wenn zwischen den Eigenschwingungen der Gutmatte (angenommen als Mehrmassenmodell, Baader e.a. 1969, Sonnenberg 1970) und den Schüttlerbewegungen Resonanz auftritt. Dies ist auch der Grund für den starken Anstieg der Verluste bei Überschreiten des Durchsatzoptimums, Bild 9.

Großen Einfluß auf die Arbeitsqualität des Hordenschüttlers hat auch die relative Luftfeuchte. Bild 11 zeigt Verlustkurven, die im gleichen Bestand mit demselben Mähdrescher zu unterschiedlichen Tageszeiten bei unterschiedlicher relativer Luftfeuchte aufgenommen worden sind (Caspers 1987).

geführt. Der Einfluß von Änderungen der Strohfeuchte und des Winkels der inneren Reibung der Körner auf die Leistungsfähigkeit des Hordenschüttlers ist nach Messungen von Srivastava in Bild 12 angegeben.

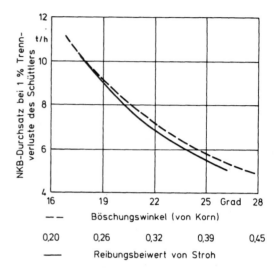

Bild 12   Durchsatz eines Hordenschüttlers in Abhängigkeit von Böschungswinkel und Reibbeiwert (Srivastava e.a. 1986)

Einflüsse auf die Korn-Spreu-Trennung

Die Korn-Spreu-Trennung erfolgt bei den heutigen Mähdreschern auf luftdurchströmten Schwingsieben. Damit spielen für die Trennung nicht nur die geometrischen Daten und die Reibbeiwerte eine Rolle, sondern zusätzlich strömungstechnische Daten wie Schwebegeschwindigkeit und Auflockerungsgeschwindigkeit sowie Hohlraumvolumen und damit auch der Anteil von Nichtkornbestandteilen, der u.a. auch von der Arbeitsgüte von Dreschwerk und Schüttler bestimmt wird. Reinigungsanlagen reagieren deswegen besonders empfindlich auf Änderungen der Guteigenschaften, die nur zum Teil durch Einstellen der Gebläsedrehzahl und der Sieböffnungen ausgeglichen werden können.

Bild 13 zeigt Verlustkennfelder einer Reinigungsanlage für zwei verschiedene Winterweizensorten bei nahezu identischen Feuchtegehalten. Die Verluste werden zwar erheblich von Durchsatz und Gebläsedrehzahl beeinflußt, jedoch überwiegt in diesem Fall der Sorteneinfluß beträchtlich.

Leistungsänderungen können jedoch auch aufgrund von Änderung des Mikroklimas im Bestand während eines Tages auftreten, wie Bild 14 zeigt. Unter gleichen Bedingungen

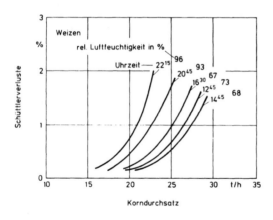

Bild 11   Einfluß der relativen Luftfeuchte auf das Verlustverhalten eines Mähdreschers (nach Caspers)

Der genaue Wirkmechanismus ist z.Zt. noch nicht bekannt, es ist jedoch zu vermuten, daß sich mit der relativen Luftfeuchte die Oberflächenfeuchte von Korn und Stroh relativ schnell ändern und damit auch die Reibeigenschaften zwischen Korn und Stroh. Weiterhin könnte die relative Luftfeuchte durch Änderungen der Halmelastizitäten das Verlustverhalten beeinflussen. Eingehende Untersuchungen zur Quantifizierung dieser Einflüsse wurden von (Huisman 1977, Hall e.a. 1981 und Srivastava e.a. 1986) durch-

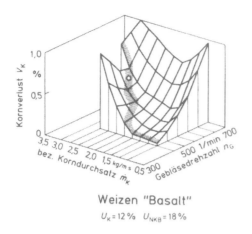

Weizen "Basalt"

$U_K = 12\%$ $U_{NKB} = 18\%$

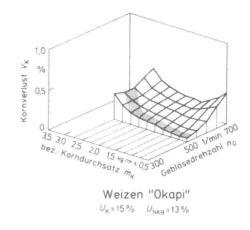

Weizen "Okapi"

$U_K = 15\%$ $U_{NKB} = 13\%$

Bild 13 Verlustkennfelder einer Mähdrescherreinigungsanlage (Böttinger 1987)

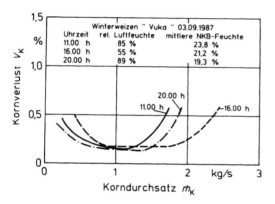

Bild 14 Einfluß von relativer Luftfeuchte
und Gutfeuchte auf das Verlustverhalten
einer Mähdrescherreinigungsanlage
(Beck 1988)

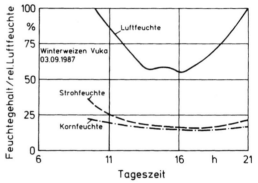

Bild 15 Stoffeigenschaften von Winterweizen und Verlauf der relativen Luftfeuchte (Beck 1988)

wurden zu verschiedenen Zeiten Verlust-
messungen an einer Reinigiungsanlage in
Feldversuchen durchgeführt, und bei Ver-
lusten von 0,5% dabei Unterschiede im
Durchsatz bis zu 40% gemessen, die ebenfalls auf Änderungen der Reibeigenschaften
der Oberflächen und weiterer feuchteabhän-
giger Eigenschaften zurückgeführt werden
müssen. Doch reichen die in Bild 14 angegebenen Daten für relative Luftfeuchte und
Gutfeuchte allein nicht zur Erklärung dieses Verlustverhaltens aus. Der geringste
Durchsatz wurde morgens um 11 Uhr gemessen, jedoch ist der Durchsatz abends bei
höherer relativer Luftfeuchte aber geringerer NKB-Feuchte noch wesentlich höher.

Der höchste Durchsatz wurde bei geringerer
relativer Luftfeuchte erreicht. Bei diesen
Feldversuchen wurde gleichzeitig der Winkel
der inneren Reibung der geernteten Getrei-
dekörner gemessen, Bild 15. Dieser zeigt
eine gute Korrelation mit den gemessenen
Durchsatzänderungen. So konnte bei diesen
Versuchen der Durchsatz aus dem Winkel der
inneren Reibung mit einem Fehler von nur 7%
vorhergesagt werden. Auch Srivastava e.a.
haben unter amerikanischen Bedingungen
einen starken Einfluß des Winkels der inneren Reibung auf die Leistungsfähigkeit der
Reinigungsanlage festgestellt, der in etwa
identisch ist mit dem Einfluß der Reibung
der Strohpartikel, Bild 16.

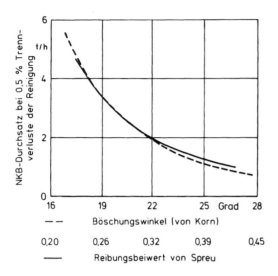

Bild 16 Durchsatz einer Reinigungsanlage in Abhängigkeit von Böschungswinkel und Reibbeiwert (Srivastava e.a. 1986)

Bei einer komplexen Erntemaschine, wie beispielsweise dem Mähdrescher, beeinflussen die Stoffeigenschaften die Leistungsfähigkeit erheblich. Diese Einflüsse lassen sich auch durch jeweils optimale Einstellung der Arbeitsorgane nicht immer ausgleichen, so daß die Leistungsfähigkeit derselben Maschine bei ungünstigen Stoffeigenschaften und Erntebedingungen bis zu über 50% niedriger sein kann.

LITERATUR

Baader, W., Sonnenberg, H. u. Peters H. 1969. Die Entmischung eines Korngut-Fasergut-Haufwerkes auf einer vertikal schwingenden, horizontalen Unterlage, Grundl. Landtechnik Bd. 19, Nr. 5, S. 149-157.

Beck, T. 1988. Einfluß der Stoffeigenschaften auf die Korn-Stroh- und Korn-Spreu-Trennung von Mähdreschern, VDI/MEG-Kolloquium Landtechnik, Heft 6, S. 107-119.

Böttinger, S. and Kutzbach, H.D. 1987. Performance characteristics of a cleaning unit under various crop conditions. ASAE-Paper No. 87-1512

Caspers, L. 1987. Neue alternative Dreschsysteme mit besonderer Beschreibung des Mehrtrommel-Abscheidesystems und seiner Leistungscharakteristik. Grundl. Landtechnik Bd. 37, Nr. 4, S. 117-156.

Hall, J.W. and Husman, J.F. 1981. Correlating physical properties with combine performance. ASAE-Paper 81-3538

Huisman, W. 1977. Moisture Content, coefficient of friction and modulus of elasticity of straw in relation to walker losses in a combine harvester. ASAE/CIGR International Grain and Forage Harvesting Conference, Sept. 25-29, Ames, Iowa, USA.

Kutzbach, H.D. 1988. Entwicklungstendenzen bei der Regel- und Informationstechnik an Mähdreschern. VDI/MEG-Kolloquium Landtechnik Heft 6, S. 121-135.

Kutzbach, H.D. u. Poppy, W. 1988. Automatisierungstendenzen bei mobilen Arbeitsmaschinen. GME-Fachbericht Nr. 3, Mikroelektronik-Schlüssel zur Automatisierung, S. 85-94.

Kutzbach, H.D. 1972. Die Grundlagen der Halmgutverdichtung. Fortschritt-Bericht VDI, Reihe 14, Nr. 16.

Wacker, P. 1985. Einflüsse auf die Dreschleistung von Mähdreschern. Landtechnik 40 H. 6, S. 273-277.

Wacker, P. 1988. Einfluß eines erhöhten Grüngutanteils auf die Arbeitsqualität von Dreschwerken. VDI/MEG-Kolloquium Landtechnik H. 6, S. 59-72.

Wieneke, F. 1964. Das Arbeitskennfeld des Schlagleistendreschers. Grundl. Landtechnik, Heft 21, S. 33-34.

Wolf-Regett, K.-P. 1988. Verdichtung und Expansion von Halmgut. Grundl. Landtechnik Bd. 38, Nr. 2, S. 58-65.

Sonnenberg, H. 1970. Korn-Stroh-Trennung mit Doppelkurbel-Hordenschüttler, Grundl. Landtechnik Bd. 20, Nr. 6, S. 161-166.

Skalweit, H. 1938. Kräfte und Beanspruchungen in Strohpressen. 4. Konstrukteur-Kursus (RKTL-Schriften, Heft 88) Berlin, S. 30-35.

Srivastava, A., W.T. Mahoney and N. West. 1986. Effect of crop properties on combine performance. ASAE-Paper 86-1583.

Land and Water Use, Dodd & Grace (eds), © 1989 Balkema, Rotterdam. ISBN 90 6191 980 0

Maize straw chopping with cylinder-type cutting head

M.L.Martinov & M.L.Tešić
Faculty of Technical Sciences, Novi Sad, Yugoslavia

ABSTRACT: The most convinient solution for maize straw chopping is the cylinder-type cutting head with a recutter screen . The experiment was realized in order to obtain data on: effect   of   chopping and feeding specific energies when chopping maize. Maize straw of two hybrids (FAO 400 and FAO 600) was taken at three stages of maturity (late doughy ripeness, rath-ripeness and wilted maize straw). Chopping were performed on six different recutters and without screen. It appeared to be unjustifiable to use recutter screens with holes smaller than 50 mm. The most suitable are the round holes. Chopping specific energy increases remarkably when recutter screens are used (1,2-4,2 times more than without recutter screens).   The higest increase is achieved when wilted maize straw is chopped.

RÉSUMÉE: La meilleure solution de broyage des tiges de maïs est le broyage avec un tambour hacheur muni de grille de recoupe. L´essai a été réalisé pour déterminer le degré et l´énergie spécifique, de broyage et d´alimentation. Les tiges de mais provenainent de deux variétés hybrides (FAO groupe 400 et 600) avec trois phases de maturité pasteuse tardive précoce et tiges fanées). Les essais etaient effectués avec 6 grilles de recoupe et sans grille . Les résultats ont montré que les grilles a ouvertures inférieures a 50 mm n´ont pas d´interet et que les ouvertures circulaires étainet les meilleures. L´énergie spécifique de broyage augmente sensiblement avec l´utilisation des grilles de recoupe (1,2 a 4,2 fois par rapport aux broyage sans grille) en particulier pour les tiges de maïs fanées.

ABSTRACT: Die günstigste Lösung für das Maisstrohhäckseln ist der Einsatz der Trommel-häcksler mit Nachschneidesieben, weil sie die Gleichmässigkeit von Partikellängen beziehungsweise den begrentzen Anteil an langen Partikeln in der gehäckselten Masse ermöglichen. Die Versuchsanlagen ermöglichten die Durchführung folgender Verfahren: Das Messen der spezifischen Energie des Häckselns und des Einzugs des Maistrohs während des Betriebes ohne Nachschneidevorrichtung und mit sechs unterschiedlichen Nachschneidesieben und Bestimmung und Darstellung des erzielten Häckselgrades in der Abhängigkeit von der Reifestufe beziehungsweise der Feuchtigkeit des Maisstrohs (angefangen von der späten Wachsreife bis zum angewelkten Maisstroh). Es zeigte sich, dass der Einsatz der Nachschneidesiebe mit Löcherdurchmesser kleiner als 50 mm nicht sinwoll ist und dass Rundlöcher geeigneter als quadratische und nutformige sind. Im Falle wenn die Hauptaufgabe steht, den Anteil an zu langen Partikeln in der gehäckselten Masse zu begrenzen, lassen sich auch Siebe mit grösseren Löcherdurchmessern. Die spezifische Energie des Häckselns wächst bedeutend beim Einsatz des Nachschneidesiebes (1,18 bis 4,14-mal im Bezug auf den Betrieb ohne Sieb) und sie erreicht den grössten Anstieg beim Häckseln des angewelkten Maisstrohs.

1 INTRODUCTION

Maize production in Yugoslavia is 9-12 million tons of corn per year. The rest of plant is straw, 75-115% comparing to amount of corn (Tešić 1983).   Chopping is a phase in the technological chain of straw processing. Until now, research has not brought out enough data on this issue. Various orientation of corn stalk axis and other straw parts at the moment of entering the cutting area is the reason why additional chopping is needed in order to obtain particles of uniform length. The research has up to shown (Belov 1986, Dju In Ju & Teherov 1979, O´Doherty 1984, Saqib & Finer 1982), that the angle between stalks and direction of feeding, influences on the chopping length the most of all.

Devices used for additional chopping are intended, in the first place, to provide the expected corn cracking which allows required fermentation at sillage preparing. Only the use of recutter screens can reduce the share of overlength particles and the mean chop length as well. Recutter screens can be used only on cylinder-type cutting heads, and that is why such cutting head type is chosen as a device for maize straw chopping. The objectiv of this reseach was to get valid data on maize straw chopping, i.e. its quality and energy required.

2 INDICATORS AND VARIABLES

Basic indicators of chopping results are: chopping effect (riched chopped-up condition) and chopping and feeding specific energies.

The most important variables are (Kromer 1983): Stage of maturity, i.e. moisture content of material, chopping device (in this case the type of used recutter screen), and the maize hybrid to be chopped.

2.1 Chopping effect (riched chopped-up condition)

According to experiences (Ghulman & Marschall 1982, O Dogherty 1982, O Dogherty 1984), distribution of the length of chopped material (from forage harvesters) coincides with a log-normal distribution, in first approximation. Parameters which describe this distribution are: median, standard deviation and additionally introduced evidence on share of particles longer than 40 and 100 mm (conditional share obtained on the basis of accepting the log-normal distribution).

2.2 Chopping and feeding specific energies

Chopping specific energy e_s is chopping energy assesed by substracting the energy for cutting head idle motion from the total energy required for cutting head performing, reduced to mass unit of the chopped straw. If specific energy is assesed on the dry matter basis, the e_{ssM} is obtained.

Feeding specific energy e_u, i.e. e_{uSM} is assesed by using the same method and is the energy require for moving the feeding mechanism (by maize straw feeding).

2.3 Chopping device

Dimensions of recutter screen holes are 23-75 mm and for dimension of 50 mm three shapes of holes were chosen: round, groove and square, in order to compare their efficiencies at maize straw chopping. For the purpose of comparison, it was also practised without the recutter screen.

2.4 Stage of maturity- moisture content

Tree stages of straw maturity were chosen: in late doughy corn ripeness (further in text: doughy ripeness), in rath-rippeness of corn (further in text: ripe maize straw) and wilted maize straw.

The doughy ripeness was chosen for the purpose of comparing the process with chopping the whole maize plant for sillage. In the rath-ripeness phase harvesting begins for both corn and straw, and that is the time when straw chopping begins too. After corn being harvested, maize straw remains in the fields and wilts under favourable climate conditions, as they usually are in Yugoslavia.

2.5. Hybrid

Two hybrides charakteristical for Yugoslavia were chosen, one being from the FAO 400 group and other from the FAO 600 group. The chopped maize straw had no cobs (corn picker-husker harvesting residues).

3 EXPERIMENTAL DEVICES

3.1 Cutting head with recutter screens

The experimental cutting head is shown in fig.1a). By form and characteristical dimensions it is similar to cylinder-type cutting neads usual on forage harvesters. The position of recutter screens provide the 0,5-1,5 mm distance to the blade tips. The cutting head had no impeller blower but the chopped material was gathered in a basket out of which samples were taken for further analysis. Blade peripheral velocity was about 24 m/s and the machine setting chop length was 12,2 mm.

Shape and dimension of holes on the recutter screens are shown in fig.1b).

Torque transducers with strain gauges and inductive RPM transducers on the chopper shafts provided the assesment of chopping and feeding specific energies. Data obtained in this way were evidenced by means of visicorder and tape-recorder (which made

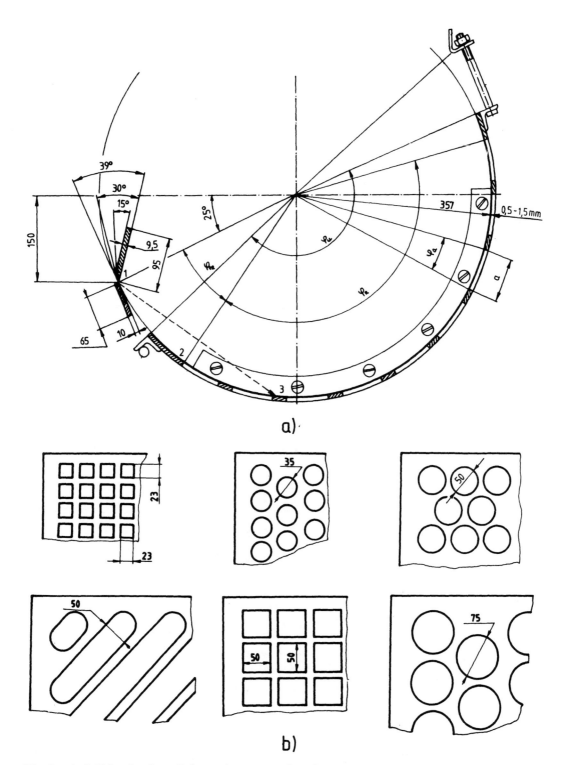

Fig.1. a) Cutting head used for maize straw chopping
b) Recutter screens used in the experiment

1983

the computer data processing possible).

Sample mass for particular measuring was approximately 100 kg of maize straw. The straw was thrown manualy on to the feed conveyor, situated in front of the feeding mechanism.

3.2 Chopped particle length classifiers

The chopped material was classified, according to the length, by using machine classifiers, because of the great number of samples and because of the fact that one kilogram of chopped material consisted of approximately 500.000 particles (O´Dogherty 1982). Following devices were prepared to be used in classification per length:

1. Conventional sieve shaker (nominal size openings: 63, 40, 25, 16 and 10 mm (described by Kromer 1983).

2. Cascade classifier made follow the model - wheat straw and grass separator, NI-AE-Silsoe (Gale & O´Dogherty 1982), with particle length classes being >90, 54-90, 33-54, 20-33, 8-20 and <8 mm.

These two classifiers cover the same particle length area and were both used in the experiment so as to conclude which one is more convenient, for neither of them was previously used for maize straw particles classification.

3. Conventional sieve shaker (usual when spheroid and elipsoid particles (Batel 1971) are classified), with sieve openings: 6,3; 5,0; 4,0; 2,5; 1,6; 1,0 and 0,63 mm.

4 RESULTS

4.1 Straw chopping effects (reached chopped-up condition)

The performance of conventional sieve shaker with shieve openings of 10-63 mm was not satisfactory. Straw particles were not devided into appropriate length classes. Hand sorting showed very significant deviation not submitted to any rule so that error could not be avoided by atributing different limits to a class, as it was done in the case of wheat straw (Kromer 1983).

After having checked the results obtained with a cascade classifier, the error was found not to be over 2.5%, so it was used for classifying the particles longer that 8 mm.

The particles shorter than 8 mm are elipsoid-shaped, so for the purpose of calssifying them the conventional sieve shaker was succesfully used (sieve openings being 0.63-6.3 mm). It was noticed that error appears for the length class under 2.5 mm,

because of longer pinlike particles found in material.

The obtained data on mass share of particles in particular length classes were used for calculating the sum mass frequency. Value pairs for: the upper length group boundary and cumulative undersized by mass were introduced on the log-normal distribution graf paper. The points marked can be connected with a straight line which means that it is justifiable to describe the particles distribution as a log-normal one. By means of analytical approximation representative straight lines were defined (as in fig.2), the coeficient of correlation being in all the cases over 0.9.

At maize straw chopping in doughy and ripe phases, approximately same chopping is attained, while for wilted maize straw the results were the worst (when the recutter was used Ø75 mm mediana is 8.8, 9.2, and 12.6 mm respectively).

No significant difference regarding chopping effect for two chosen hybrides was observed.

Comparison of chopping effects with round, groove and square openings proved recutters with round openings to provide the best chopping effect, and recutters with groove openings - the worst one (mediane for the wilted maize straw are: 8.1, 18.5, and 12.3 mm respectively)

Agregatic data on the chopping effect for work with particular recutters and without recutter screen, assesed as mean values for all the stages of maturity and for both hybrides are given in tab.1.

Recutter screens ø 23, Ø35 and Ø50 mm provide approximately same chopping effect (median is 5.7, 6.3, and 6.6 mm respectively). Although median is slightly higher when the Ø50 mm recutter is used, the share of particles longer than 40 and 100 mm is smaller because of lower standard deviation, so that the best chopping effect is obtained.

For work without recutter screen median is 19.7 mm and the share of particles longer than 40 mm is 27%.

In the last column of tab.1 the rations of specific particle cover surface are given. The speicfic particle cover surface is the ratio of particle cover surface to the particle volume cm^2/cm^3. It is calculated taking into account the median, standard deviation and particle shape factor. Since the particle shape factor is unknown, only the ratios of specific particle cover surface are calculated. The unit value was taken for the work without recutter, supposing the shape factor remains unchanged no matter what the chopping effect was.

In tab.1 it is shown that the specific particle cover surface when the Ø75 mm re-

Tab.1. Data on chopping effects (reached chopped-up condition)

	Median /mm/	Standard deviation	Share of particle longer than		Ratio of specific particle cover surface
			40 mm /%/	100 mm/%/	
Without recutter	19,2	1,20	27	8,0	1,00
Recutter ⌀ 23 mm	6,1	1,10	4,5	0,5	2,72
Recutter Ø35 mm	6,6	0,98	3,3	0,3	2,29
Recutter Ø50 mm	6,9	0,93	2,3	0,2	2,06
Recutter groove 50 mm	11,8	1,06	11,5	1,9	1,39
Recutter Ø75 mm	11,6	1,02	10,5	1,7	1,36

(Tab.1. does not include data on work with 50 mm square openings recutter, since it was used only for wilted maize straw)

cutter was used, was 36% higher than with the particles chopped without recutter. When the Ø50 mm recutter was used this increase in specific particle cover surface was more than doubled.

The bigger specific surface provides more rapid chemical and biological processes and material drying (in a drum drier, for instance).

Data from tab.1. are given in fig.2 on the log-normal distribution graf paper for the recutter screens Ø50 mm, Ø75 mm and for work without recutter.

4.2 Chopping and feeding specific energies

Data on change of chopping and feeding specific energies, in dependence on maize straw stage of maturity are given in fig.3.

The lowest values for chopping and feeding specific energies were found for the ripe straw, and the highest - for wilted maize straw. When data are reduced to dry matter, the specific energy lowest values were found for the ripe straw (the exception was work without recutter, hybrid FAO 400) and the highest - for the wilted maize straw (the exception was in mean values of chopping specific energy, for recutters, hybrid FAO 600).

If values of chopping specific energy for the work without recutter are substracted from mean values of chopping specific energy for work with recutters (dashed line), the result is specific energy required for additional chopping, which for both hybri-

des has its highest points in the case of wilted maize straw.

In fig.4 multiplication factors of chopping specific energy for the work with recutter screens and for the stages of maturity are given related to the mean value for work without recutter (shown as the unit value). With dashed lines the mean values of multiplications coeficient for all three stages of maturity are shown.

The mean value of multiplications factor for a recutter ⌀ 23 mm is 3.4, for recutter Ø50 mm is 2.4, for a grooveholes 50 mm iz 1.7 and for a sieve Ø75 mm is 1.2. Difference between multiplication factors at doughy ripeness phase and wilted maize straw is a significant one. Obviously, during drying the properties of maize straw change, so as to affect unfavourably the conditions for additional chopping, while the conditions for primary chopping remain unchanged, or are even improved (fig.3).

Comparison of the three recutters with 50 mm openings can be performed for wilted maize straw, because the recutter with square openings was used only for this material. Multiplication factor for a recutter with round openings is 3.1, for a recutter with the square ones is 2.5 and for the recutter with groove openings is 2.0.

5 RESULTS DISCUSSION

To use machines in sorting chopped maize straw particles longer than 8-12 mm requires a special device which is adjusted for

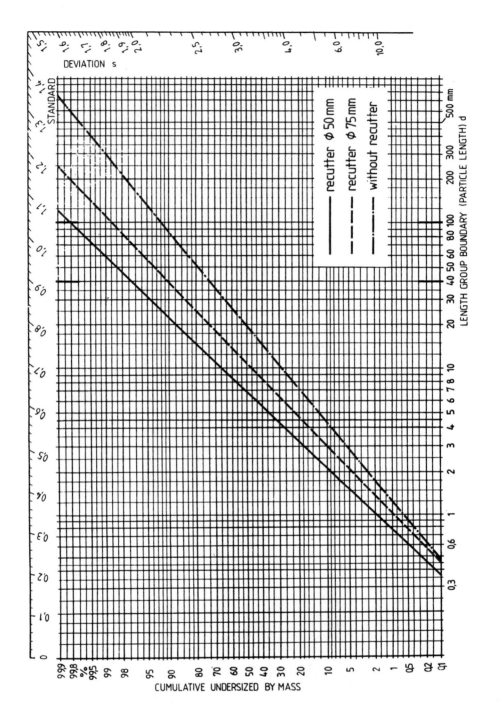

Fig.2. Chopped particle lengths distribution on the log-normal distribution graf paper

straw-like materials (one dimension is sig- to same experiences with wheat straw (Kro-
nificantly over the other two). In contrast mer 1983), conventional sieve shakers are

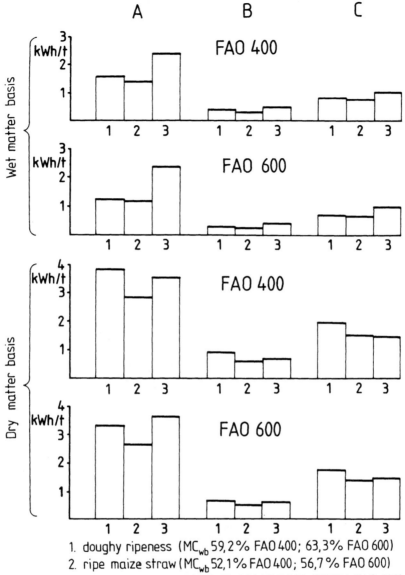

1. doughy ripeness (MC_{wb} 59,2% FAO 400; 63,3% FAO 600)
2. ripe maize straw (MC_{wb} 52,1% FAO 400; 56,7% FAO 600)
3. wilted maize straw (MC_{wb} 32,2% FAO 400; 35,3% FAO 600)

Fig.3. Changes in chopping and feeding specific energies in dependence on stage of maturity, i.e. on maize straw moisture content
    A) chopping specific energy, mean values for all recutter used
    B) feeding specific energy, mean value for all measurements
    C) chopping specific energy for work without recutter

not convenient for this purpose. A cascade classifier (Gale & O Dogherty 1982), if previously calibrated, and with repeated classifications (partly interventions with hand sorting which prolongs the time for classification to nearly 30 minutes) showed good results with the share of incorrectly classified particles up to 2,5%.

Devices and methods usual for spherical and elipsoid particles (Batel 1971, DIN

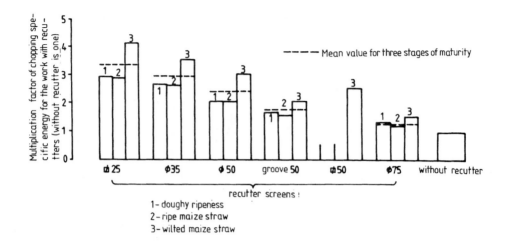

1 - doughy ripeness
2 - ripe maize straw
3 - wilted maize straw

Fig.4. Increase of the chopping specific energy at work with recutter screens, in comparison to work without recutter

66141, ISO 565, ASAE S319, ASAE S424), can be used for sorting the maize straw chopped particles shorter than 8-12 mm.

It has been proved that length distribution of chopped maize straw particles corresponds to log-normal distribution, which coincides with the results obtained for other materials (O´Dogherty 1982), but is opposite to Belov's (1986) opinion.

By using recutter screens chop length uniformity, median reduction and reduction of the share of particles longer than 40 and 100 mm are obtained.

The results show that round openings on

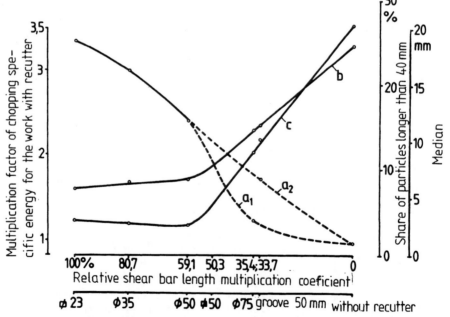

Fig.5. Chopping specific energy multiplikation factor, median and share of particle longer than 40 mm dependance on relative multiplication of shear bar length (wilted maize straw)

recutter screens are most convenient for maize straw chopping, which does not coincide with the opinions of Efimova (1977), and Reznik (1980). The Kromer´s (1979) conclusion, that the most important characteristic of the recutter screen is the shear bar length multiplication coefficient (shear bar length multiplication coefficient is: ratio between sum of all recutter holes projection on cutting head cylinder line and shear bar length), has been confirmed. In fig.5 (similar presentation is given by Kromer (1979) for wheat straw chopped with recutter screens), are given the dependence of chopping specific energy multiplication factor (in relation to specific chopping energy for work without recutter), median and share of particles longer that 40 mm on the relative shear bar length multiplication coefficient (for the recutter ⌀23 mm multiplication coefficient being 100%). The chopping specific energy multiplications factor decreases continually (curve a_1 for the recutter ⌀75 mm, and a_2 for the recutter groove 50 mm). Indicators of the chopping effect (median and share of particles longer than 40 mm) remain almost unchanged in the area which is valid for the recutters ⌀23, ⌀35 and ⌀50 mm. This indicates that for maize straw chopping the use of recutter screens having openings smaller than 50 mm is not justifieble; no better chopping effect is obteined and the chopping specific energy increases.

When the reduction of median is not require but only limitation of overlength particles share, recutter ⌀75 mm should be used. Efficiency of recutters with even bigger openings should be tested for instance ⌀90 or ⌀100 mm).

The chopping specific energy values obtained on the dry matter basis for the work without recutter are 1,4-1,9 kWh/t (5-7 MJ/t). The upper limit overcomes previously measured values of chopping specific energy for different plant matter (O´Dogherty 1982), and is significantly higher than the chopping specific energy for the realised chop length of 19 mm being 2,2 MJ/T (Voss 1970).

When maize straw is chopped in the doughy phase with the recutter screen ⌀35 mm, chopping specific energy was 2,2 kWh/t. Dernedde and Peters (1976) had measured 6 kWh/t for chopping the whole maize plant (similar conditions). However, direct comparison of such data is not correct, since in addition to differences related to chopped matter and different experimental operation, papers often ommit to state precisely what does chopping specific energy include.

Although the experiences have up to date

(Dernedde & Peters 1976) shown that in work with recutters dimension of openings excert the most important influence on chopping specific energy, it should be tested whether chopping specific energy might be reduced by increasing the machine seting chop length (which would be convinient under the condition that maize straw chopping effect does not grow worse).

Chopping specific energy on a dry matter basis changes during ripening - it has its highest values in the stage of late doughy ripeness, and its lowest - in the rath-ripe stage. This does not coincide with the opinions according to which chopping specific energy if reduced to dry matter, remains constant (Kromer 1967).

6 CONCLUSIONS

The use of recutter screens at maize straw chopping provides the decrease of mean chop length, uniformity of chopped particle lengths and decrease in share of particles longer than 40 mm and 100 mm. The best chopping effect (the lowest median and share of long particles) is achieved in rath-ripeness phase, when chopping specific energy has its lowest values. The worst chopping effect is obtained at work with wilted maize straw, when specific energy of recutting reaches its highest values.

No significant difference in the chopping effect is observed between the chosen hybrides (FAO group 400 and 600). The best chopping effect is provided by recutters with round openings.

Recutter screens with holes smaller than 50 mm should not be used at maize straw chopping, for the chopping effect is not better and the chopping specific energy increases. When the recutter ⌀50 mm is used, median near 7 mm is obtained, share of particles longer than 40 mm is 2.3%. With recutter ⌀75 mm median is 11.6 mm and share of particles longer than 40 mm is 10.5% (mean values for three stages of maturity). The increase of chopping specific energy in comparison to work without recutter screens is, for a recutter ⌀50 mm 72%, and for a recutter ⌀75 mm 23% (mean values for three stages of maturity). In the case when only the share of overlength particles is to be limited, a recutter ⌀75 mm should be used; the possibilities for usage of sieves with bigger openings - 90 and 100 mm - should be tested.

REFERENCES

Batel,W. 1971. Ein Fürung in die Korngrös-
senmesstechnik. Berlin-Heidelberg-New
York: Springer Verlag

Belov,M.I. 1986. Metodika raschota cilindri-
cheskogo izmelchajushchego apparata.
Traktory i selhozmashiny. 55: 27-28

Dernedde,W. & H.Peters 1976. Wirkung und
Leistungsbedarf von Nachschneid-syste-
men für Exakthäcksler. Grundlagen der
Landtechnik. 26: 23-30

Dju In Ju & P.Z.Teherov 1979. Opredelenie
zavisimosti dlini rezki ot parametrov
izmelchajushchego apparata. Traktory i
selhozmashiny. 48: 22

Efimova,M.G. 1977. Izmelchenie materijala
na kormouborochnyh mashyn. Mehanizacija
i elektrifikacija soc.sel.hoz. 47: 58-59

Gale,G.E. & M.J.O'Dogherty 1982. An appa-
ratur for the length distribution of
chopped forage. J.agric.Engng.Res. 27:
35-43

Ghulman,S.S. & F.F.Marschall 1982. Simulated ideal length of forage harvesters.
Transactions of the ASAE. 25: 1237-1238

Kromer,K.H. 1967. Untersuchungen der Mate-
rialförderung in und nach Schneid-Wurf-
-Trommeln. Frankfurt am Main: Kuratorium
für Technik in der Landwirtschaft

Kromer,K.H. 1979. Möglichkeiten der Nachzerkleinerung bei Exaktfeldhäcksler.
Grundlagen der Landtechnik. 29: 166-175

Kromer,K.H. 1983. Zerkleinerung von Mais
in Trommelschneidwerken. Freising. Landtechnik Weihenstephan

O'Dogherty,M.J. 1982. A rewiew of research
on forage chopping. J.agric.Engng.Res.
27: 267-289

O'Dogherty,M.J. 1984. Chop length distribution from forage harvesters and a simulation model of chopping. J.agric.
Engng.Res. 29: 165-173

Reznik,N.E. 1980. Kormo-uborochnye kombaj-
ny. Moskva: Mashinostroenie

Saqib,G.S. & M.F.Finer 1982. Simulated ideal length of cut for forage harvesters.
Transactions of the ASAE. 25: 1237-1238

Tešić,M. 1983. Postupci i mašine za ubira-
nje, transport i manipulaciju sporednih
proizvoda ratarstva, Novi Sad, Fakulty
of technical sciences

Voss,H. 1970: Ermitlung von Stoffgesetzen
für Halmgut. Braunschweig: Technische
Hochschule Braunschweig

Land and Water Use, Dodd & Grace (eds), © 1989 Balkema, Rotterdam. ISBN 90 6191 980 0

Harvesting potatoes – Harvesting damages and labour utilization on farms

P.Mattila
Work Efficiency Institute, Finland

ABSTRACT: This study looks into the mechanical damage done to potatoes and the harvest losses in practice. Twenty-eight harvest situations were studied in the autumn of 1985 and nineteen in 1986. Bruising of potatoes was determined on the basis of peeling and flesh wounds attributable to the harvesting stage. The strength of potato surface was measured by subjecting each harvested batch of potatoes to a mechanical test.

On the average, 8,8 % of the skin of tubers had became peeled by the time they reached the manual sorting conveyor of harvesters and 1,2 % of the harvested potatoes had bruises or cuts. Harvest loss averaged 2600 kg/ha (i.e. 7,6 % of the entire crop).

The degree of maturity of the crop had a considerable influence on the amount of peeling which took place. Peeling and flesh wounds were independent of one another. Little peeling was evident when working with wet soil. Flesh wounds were abundant when harvesting from low beds. Peeling became profuse if obstacles were used on the first elevator to intensify screening. In the case of "one level"-harvesters, most of the peeling took place on the screen elevator and with "double level"-harvesters at the section where stems and clods were separated.

Gegenstand dieser Untersuchung waren bei der Kartoffellese entstehende mechanische Beschädigungen der Kartoffel und Ernteverluste. Im Herbst 1985 wurden 28 Lesevorgänge, im Herbst 1986 19 Lesevorgänge untersucht. Als Kriterium zur Feststellung der Beschädigungen an den Kartoffeln diente die Bestimmung von Abschälungen an der Knolle und von Schäden am Inneren bzw. Herz der Kartoffel. Die Handhabungsfestigkeit der Kartoffeln wurde mit einer Stoßprobe bei jedem Lesevorgang gemessen.

Von der Oberfläche der Knollen, die die Kartoffelerntemaschine durchlaufe hatten, wiesen am Ende der Handsortiereinrichtungim Durschschnitt 8,8 % Abschälungen auf, und 1,2 % Prozent der geernteten Kartoffeln hatten Quetschungen oder Schrammen. Der Ernteverlust betrug durchschnittlich 2600 kg/ha bzw., 7,6 % von der Ernte.

Der Reifegrad der geernteten Kartoffeln war von wesentlicher Bedeutung für das Ausmaß der von der Kartoffelerntemaschine verursachten Abschälungsschäden. Andererseits besteht keine Interrelation zwischen dem Ausmaß von Abschälungen und Herzbeschädigungen. Unter feuchten Bedingungen kam es nur zu geringer Abschälung. Bei der Lese aus niedrigen Reihen kam es zu reihlichen Schäden am Kartoffelherz. Abschälungen waren häufig, wenn die Sortierungseffektivität durch die Verwendung von Sperren am Siebelevator gesteigert wurde. Was die Unterschiede bei den Kartoffelerntemaschinen angeht, so traten Abschälungschäden bei Verwendung von einschichtigen Maschinen am Meisten im Siebelevator und bei Verwendung von dopelschichtigen Maschinen beim Trennen von Stielen und Erdkluten auf.

La présente étude a trait aux défaults mécaniques et aux pertes á l'arrachage et au ramassage liés á la récolte pratique des pommes de terre. L'enquête a porté sur 28 récoltes de l'automne 1985 et 19 récoltes de l'automne 1986, l'endommagement étant déterminé sous la forme de

l'écaillage des tubercules et des dommages subis par la chair, La
résistance à la manutention a été mesurée par mise à l'épreuve au choc d'un
échantillonnage de chaque récolte.

Une moyenne de 9,8 % de la surface des tubercules récoltés par
arracheuse-ramasseuse a été constatée écaillée avant de parvenir enbout de
chaîne de traitement manuel et 1,2 % des pommes de terre récoltées
présentaient des meurtrissures ou des entailles. La perte à l'arrachage
s'est ainsi chiffrée àune moyenne de 2600 kg/ha soit 7,6 % de la récolte.

L'importance de l'écaillage dû à la machine est apparue comme étant
considérablement favorisée par la maturation de la pomme de terre. Aucune
interrelation ne s'est révélée entre l'écaillage et l'endommagement de la
chair. Les dommages subis par la chair se sont montrés nombreux pour
l'arrachage des bancs bas. L'écaillage s'est avéré minime dans les
conditions de forte humidité et abondant dans les cas où le triage était
intensifié par l'usage d'obstacles sur l'élévateur trieur. Dans
lesarracheuses-ramasseuses à un étage, l'écaillage a été le plus important
au niveau de élevateut trieur et dans les machines à deux étages au niveau
de la separation des tiges et des tubercules.

1. BACKGROUND

This study is a part of a joint
study financed by the Ministry of
Agriculture and Forestry in Finland
which aims at improving the quality
of table potatoes. The study was
conducted by the Work Efficiency
Institute.

The study dealt with the amount
of damage caused to potatoes by
harvesting machines and the harvest
losses resulting from it, how the
damage was caused and the factors
involved. Field work for the study
was carried out on 24 farms
producing table potatoes in August
and September of 1985 and 1986. All
in all, 46 visits were made to the
farms in connection with the study.

2. METHODS

Mechanical damage to potatoes was
examined by determining, on the
basis of sampling, the average
peeled percentage and the proportion
with flesh wounds. Samples were
taken from four points on the
harvesting machine and also from
boxes or trailer following the
emptying of the harvester's bin. The
latter two samples were not taken in
1986. Discharging of potatoes into
storage was not examined.

The strength of the skin of the
potatoes with respect to handling
was measured by subjecting each
harvested batch of potatoes to an
identical skin-dislodging bruising
test using a cement mixer. The
different harvest occasions and

machines could not, due to variable
conditions, be directly compared to
one another. The peeling which took
place in the harvesting machine was,
on the other hand, compared to the
outcome of the bruising test to
which the same batch of potatoes was
subjected to.

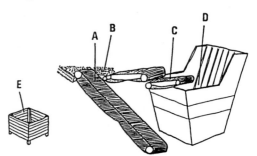

Fig. 1. Places where samples were taken.

3. RESULTS

After having passed through the
harvester, 1 - 31 % (average 8,8 %)
of the skin of the tubers had peeled
off by the time they reached the
manual sorting conveyor and 1,2 % of
the harvested potatoes had received
injuries or cuts into the flesh. On
the average, 40 % of the harvested
potato crop was without any defect.

The amount of peeling attributable
to the harvesting machine was
clearly influenced by the degree of
maturity of the crop. Only the early
varieties had matured normally prior

to being harvested. Indeed, tubers of late varieties peeled profusely in the harvester, unless the crop had been force-matured by cutting down the stems.

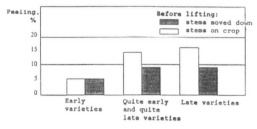

Fig. 2. Peeling when stems has been moved down. Many varieties (autumn 1985).

3.1. Harvesters

The sieving web on one-level-harvesters were noted to cause most peeling. The major cause of peeling in double level machines was the section for separating stems and clods.

The various harvester types differed from one another in the amount of peeling they caused. A sieving web, rising at a smaller angle at the rear, caused less peeling than an elevator working at a steady angle. Short sieving elevators were also noted to cause only little peeling. The stem and clod screening section of harvesters tendency to peel potatoes increased as its structural efficiency increased. The separator was at its gentlest in small machines where clods had to be separated entirely by hand.

Only minor peeling occurred on manual-sorting elevators and on separator rollers for small potatoes. Peeling was also only slight when potatoes dropped to the bottom of the elevator-bottom bin. Tubers had a long way to drop on the case of tipper bins. In addition, since the padding in the tipper bins was worn out, peeling was more extensive than in the case of dropping into the elevator bin.

On most farms the harvester's bin was emptied into 1 m^3 storage boxes. Regardless of the type of bin used, potatoes fell 110 cm to the bottom of the box. Discharging the bin caused approx. 20 % of the peeling attributable to the harvesting stage. Internal damage caused to the

potatoes by being dropped a long way was not examined.

3.2. Adjustments

Neither harvester speed nor sieving elevator speed were noted to have any obvious effect on the amount of peeling which took place. The sieving elevator's speed was, on the average, 1,16 m/s and it averaged 2,8 times the driving speed of the harvester. Peeling became profuse when screening was intensified by using obstacles above the sieving elevator.

3.3. Conditions

Harvesting conditions were also noted to have an effect on the amount of peeling. When harvesting took place with the soil wet, there were less peeling than when operating in dry easily screened soil. Soil temperature had no significant effect on peeling.

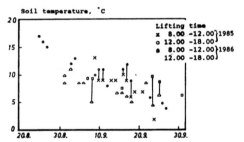

Fig 3. The temperatures of the potato beds when lifting.

Peeling of potatoes was not dependent on the number of tubers with bruises or cuts. Tubers with flesh injuries were more numerous in both harvest losses and harvested crop when using a machine with a stem roller for separating stems in comparison to a stem elevator machine. Flesh injuries were common when harvesting potatoes from low beds.

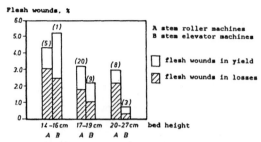

Fig. 4. Flesh wounds (%) when lifting from beds with different height.

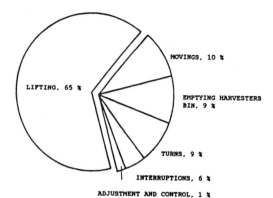

Fig. 5. Work consumption and how it devides.

3.4. Harvesting losses

In the harvest situations studied, harvest loss was, on the average, approx. 2600 kg/ha. Only one quarter of the harvest loss consisted of small (< 30 mm) tubers. On the average, 6 % of all tubers > 30 mm remained unharvested. Bruising and losses were independent of one another. Somewhat more potatoes remained unharvested when using stem roller harvesters in comparison to stem elevator harvesters. Stem roller harvesters were used, however, in immature crop as well. In cases where stems had been gotten rid of at an early enough stage, harvest losses were generally small.

3.5. Work consumption

According to measurements, made in conditions a little more difficult than normally, the work consumption of medium sized harvesting machines was, on average, 13,7 h/ha, and the total work consumption of human labour 41,5 h/ha. The work consumption varied greatly depending on the circumstances. The work consumption did not correlate with the lifting losses or with the injuries under the lifting. The actual lifting time was 65 of the total work consumption,%. Turns and movings took 19 %, the emptyings of the reservoir 9 %, interruptions 6 %, and adjustments and control 1% of the working time.

References

Mattila, P., Mattila, T., Hennola, P. & Peltola, A. 1988. Perunankorjuu - nostovioitukset ja työnkäyttö tiloilla. Summary: Harvesting potatoes -harvesting damage and labour utilization on farms. Work Efficiency Institute Publications 298. Pp.116. Helsinki.

Land and Water Use, Dodd & Grace (eds), © 1989 Balkema, Rotterdam. ISBN 90 6191 980 0

Proposal for a model to assist in the choice of forage harvesting and conservation techniques: Methodological and applicative aspects

F.Mazzetto
Institute of Agricultural Engineering, University of Milan, Italy

ABSTRACT: The choice of technique to be used for the production of forage should be based on the need to increase the quality of the product on the one hand and minimize the cost of production on the other. On the basis of this assumption, which in practice mirrors the problem of input-output choices, the present study proposes a model for technical and economic evaluation according to which the choice of a haymaking technique is analyzed in terms of organizational, energetical and productive aspects. The model, whose initial formulation is still purely deterministic, has been applied to a comparative analysis that concerned only those haymaking techniques that resulted in the production of hay, such as: traditional haymaking, two-phase haymaking and dehydration. Results from the initial application demonstrated the need to improve the hay yield produced on the afrm by selecting haymaking techniques that were alternatives to traditional haymaking. Of these alternatives, two-phase haymaking (especially the method utilizing solar collectors) best satisfied this requirement, both from an energy/operational and an economic standpoint. Considering the importance of this assumption in terms of a dairy farm's operative choices (especially with regard to the climatic conditions prevalent in Northern Italy), the model should be refined further in order to overcome its current analytical limitations (developing the model in probability terms).

RESUME': Le choix d'une technique de conservation des fourrages doit se faire en fonction de la nênessité d'augmenter la qualité des produits d'une part et, de l'autre, de minimiser le cout de production. En se basant sur cette affirmation qui, en pratique, ramène au problème des choix facteur-produit, on propose un modèle d'évaluation technico-économique selon lequel le choix d'une donnée technique est analysé sous les aspects de l'organisation, de l'énergie et de la production. Le modèle dont la première formulation est encore de type purement déterministe, a été appliqué dans une analyse comparée qui n'a pris en considération que les techniques de conservation permettant la production de fourrage sec, à savoir: la fenaison traditionnelle, la fenaison en deux temps, la déshydratation. Les résultats de cette premiére application ont fait ressortir la nécessité d'améliorer le rendement des foins produits par l'exploitation en recherchant des techniques de conservation autres que la fenaison traditionnelle. Parmi elles, la fenaison en deux temps - et notamment la solution à capteurs solaires - est celle qui répond le mieux à ces nécessités du point de vue de la consommation d'énergie, de l'organisation et de l'économie. Etant donnée l'importance que revet une telle affirmation pour le choix opératoire de l'exploitation laitière (surtout dans les conditions climatiques de l'Italie du nord), on a l'intention de perfectionner ultérieurement le modèle en vue de dépasser les limites actuelles de l'analyse (dèveloppement d'un modèle stochastique).

INHALT: Die Wahl einer bestimmten Technik der Futtererhaltung muß auf der Grundlage der Notwendigkeit, díe Produktqualität zu steigern einerseits und die Produktionskosten aufs äußerste herabzusetzen andererseits beurteilt werden. Ausgehend von dieser Voraussetzung, die in der Praxis dem Problem der Wahl Faktor-Produkt gleichkommt, wird ein technisch-wirtschaftliches Beurteilungsmodell vorgeschlagen, nach dem die Wahl einer bestimmten Technik an Hand folgender Aspekte untersucht wird: Organisationsaspekte, Energieaspekte, Produktionsaspekte. Das Modell, das in dieser ersten Fassung noch rein deterministisch ist, wurde in einer vergleichenden Analyse angewandt, die nur die Erhaltungtechniken für die Produktion von Trockenfutter betroffen hat, d.h.: traditionelle Heuernte, Heuernte in zwei Zyklen, teilweise oder völlige Entwässerung. Die Ergebnisse dieser ersten Anwendung haben die Notwendigkeit angezeigt, die Heuergiebigkeit durch alternative Erhaltungstechniken zu verbessen. Die Heuernte in zwei Zyklen, insbesonders die Solar-Lösung, ist die Technik, die vom Standpunkt der Energie, der Organisation und der Wirtschaftlichkeit am besten dieser Anforderung entspricht. Unter Berücksichtigung der Bedeutung einer solchen Voraussetzung für die Arbeitsentscheidungen eines Milchviehzuchtbetriebs (vor allem unter den norditalienischen Klimabedingungen) wird das Modell weiter verfeinert werden, um seine gegenwärtigen Analysengrenzen zu überwinden (Entwicklung eines Zufallmodells).

1 FOREWORD

Methodologies currently in use for the conservation of forage are only concerned with moisture control (haymaking and drying) or microorganisms (ensilage). Temperature control (refrigeration) is considered to be too costly in relation to the value of the forage.

For several years now, this particular aspect of agricultural production has kept scientists and industrial engineers busy looking for solutions which on the one hand would minimize production costs and the amount of labor required and on the other would improve qualitative and quantitative yields. As we know, the distribution of forage to animals requires considerable use of vehicles and personnel for the production of a commodity, forage, whose market value is not easy to quantify in relation to its qualitative aspects.

This situation is due to several factors: in the majority of cases, an Italian farm produces enough forage to satisfy its own consumption requirements, and the quantity of product available on the market is rarely evaluated on the basis of its intrinsic qualities.

Things become more complicated when we consider the fact that every conservation technique requires the use of specific machinery. In addition, the various possibilities offered by each technique are difficult to categorize because of the wide range of machinery available on the market (this is especially true in the case of traditional haymaking). Consequently, there are numerous possible combinations of machines, and these are influenced by countless factors (the size of the farm and its level of mechanization, the size and shape of the fields, the distance between fields, the degree of product contamination and mistreatment, operation on a flat surface, hill or mountain, the farmer's personal preferences, etc.).

Therefore, the selection of a conservation technique turns into an extremely complex problem based on the weight the farmer attaches to the following objectives:
- optimization of quantitative and qualitative yields, with limited losses;
- containment of production costs connected with harvesting, storage and subsequent treatment and distribution in the manger; consequently, guaranteed quality becomes a secondary consideration.

The current trend tends to favor mechanization chains that are able to reduce production costs through increased productivity. On the other hand, selection of the optimal solution is rarely considered in overall terms; instead, the farmer focuses on a specific machine, or a limited number of machines, rather than on the entire productive chain.

Naturally, it would be more logical to view the problem from an overall standpoint, that is, based on a model with an **integral approach** that simultaneously evaluates the three main aspects of the various processes (production, organization and energy).

2 THE PROPOSED MODEL

2.1 The various aspects of the problem

The aim is propose a model for evaluation in which the above-mentioned "integral approach" is based on analysis of the following aspects:
- **production**, that is, the quantitative and qualitative level of product obtainable from the cultivated surface area;
- **organization**, that is, the use of machines, facilities and labor that are proportionate to the specific farm situation;
- **energy**, that is, the costs of energy from various sources (mechanical, thermal and electric) used in every mechanization chain.

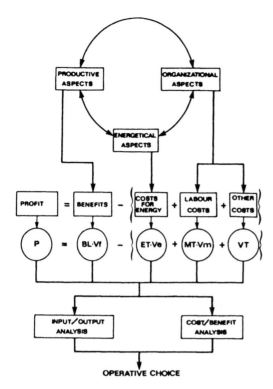

Fig.1 - Logic diagram of the defined model.

Since these three aspects influence each other, their quantification must always be considered in relation to a specific case (Fig. 1). For example, the obtainment of a higher qualitative level generally requires that the harvested product be damper, which in turn necessitates the use of different machines from those normally used in traditional haymaking. In addition, this means that the harvesting chain and product transport and storage must be organized differently, and energy must be consumed for artificial drying (the amount of energy depends on the dampness of the product). In other words, once the characteristics or requirements of one aspect have been determined, those relating to the other two aspects have to be modified as well to take into account all the factors that will influence the economic result in terms of the farm's final balance.
Returning to Fig. 1, the equation of the economic performance (identifiable with the concept of profit, P(k) in lire/ha) of

the kth operative solution may be seen as the difference between the gross profit obtainable - BL(k) - and total costs - CT(k):

$$P(k) = BL(k) - CT(k) \qquad (1)$$

CT(k) represents the production cost of the forage (from harvesting to storage of the completely processed product) and is composed of three expressions:

$$CT(k) = ET(k)*Ve + MT(k)*Vm + VT(k) \qquad (2)$$

where:
ET(k) = total energy consumption (kWh/ha);
MT(k) = total labor requirement (ulh/ha);
Ve = unit energy value (lire/kWh);
Vm = unit cost of labor (lire/kWh);
VT(k) = other costs connected with mechanization of the technique under consideration, including various expenses (wear and tear, maintenance and insurance) and the interest on all machinery and facilities used, as well as other miscellaneous expenses (lire/ha).
Given the interactive process in Fig. 1, the energy-related aspects are quantified by calculating the sum of the following values:
- Em(k) : the consumption of **mechanical energy** by mobile users (field users), which is related to the consumption of fuel for operation of the tractor's endothermal motor and/or a specific self-propelled machine (kWh/ha);
- Et(k) : the consumption of **thermal energy** in conservation techniques that include hot air drying (kWh/ha);
- Ee(k) : the consumption of **electrical energy**, which refers to practically all fixed operations carried out at the farm using electric motors (kWh/ha). Total annual energy consumption, ET(k), can only be expressed in terms of **primary energy** utilized, that is, the quantity of fossil fuel that must be consumed to obtain 1 kWh of mechanical, thermal or electrical energy available on the farm.
Production and distribution efficiency must be taken into consideration when calculating electrical energy. If an average value of 35% is assumed for distribution efficiency, the following equation is obtained:

$$ET(k) = Em(k) + Et(k) + 2.86*Ee(k) \qquad (3)$$

The **organizational aspects**, in their turn, lead to quantification of parameters concerning labor requirements, MT(k), and the use of various kinds of equipment (mechanical and construction) , VT(k), required by the kth operative solution. In general, this evaluation is carried out by considering that the entire series of operations can be divided into various phases, no matter which conservation technique is chosen. The phases are: a) cutting, b) haymaking in the field, c) gathering, d) transport, e) processing at the farm, and f) storage and conservation. Techniques differ in their varied use of resources during each phase. For example, traditional haymaking involves limited use of resources at the farm, and technological developments in this area have mainly contributed to solutions aimed at increasing work productivity during the gathering and transport phases. With dehydration, on the other hand, the majority of operations are carried out at the farm. Two-phase haymaking, finally, has some of the features of each of the other two techniques, with a sharp decrease in field drying times as compared with traditional haymaking and longer farm processing times than those required by dehydration.

In summary, concomitant analysis of organizational and energy-related aspects leads to a classic determination of the cost of forage production, that is, as the sum of the variable (ET*Ve + MT*Vm) and fixed costs (VT). While calculation of the production cost does not generally present any problems, the same cannot be said about evaluation of the economic profits, BL(k), obtainable at the level of production offered by the kth technique.

In fact, analysis of the various productive aspects is based on a commodity, forage, whose market value is not always easy to quantify in relation to its qualitative aspects. The gross profit obtainable from the product would turn out to be:

$$BL(k) = UFr(k)*Vf \qquad (4)$$

where:
UFr(k) = the total quantity of feed energy gathered annually per ha (UFL/ha);
Vf = the unit value of the milk feed unit (lire/UFL).
The difficulty involved in correct calculation of BL(k) lies in choosing the right value for Vf. Indeed, since the forage produced is an internal reinvestment for the farm, it does not directly appear in the figures that make up the GSP on the farm's balance-sheet. In this case, the qualitative and quantitative levels of the forage consumed on the farm are indirectly evaluated through the economic values of animal production (milk or meat). These values

also largely depend on other factors (the breeds in question, selection, the composition of the feed, its distribution procedure), which go beyond the productive capacity of the conservation techniques adopted.

Nonetheless, the productive aspect cannot be ignored when making estimates. The problem can be skirted (although only partial information is obtained in this way) by not considering the final zootechnical transformation and focusing attention on the qualitative and quantitative characteristics of the forage obtainable. While this method is more immediate, it again proposes the problem of evaluation of the market value of forage in relation to its qualitative characteristics. As we know, this value varies from one area to another and is also a function of the type of forage (alfa-alfa hay, hay from permanent meadows, etc.) and cut which is being referred to (hay produced from first cuts usually has a higher value). By way of indication, values on the order of 340-400 lire/UFL can be assumed at present.

An alternative to market value might be consideration of the UFL's production cost. This would offer the possibility of fixing Vf so that the unit value of the UFL did not affect the sale price of the milk (or meat) beyond a certain percentage. This approach would certainly prove to be more useful if it was necessary to compare forage with extremely different physical and chemical characteristics (e.g., hay and ensilage).

Two concomitant factors have to be taken into account for an evaluation of productivity, UFr(k): qualitative losses (decrease in content in UFL per kg of d.m. ??) and quantitative losses (harvest of a

Table 1. Quantity and quality variations in an alfa-alfa hay produced by traditional haymaking.

| | Fresh for. dryed hay | | | |
	A	B	B/A	K
(kg/ha)	2200	1390	0,63	Ks=0,37
(UFL/kg)	0,84	0,60	0,71	Kv=0,29
(UFL/ha)	1850	830	0,45	KT=0,55
gross protids:				
(g/ha)	480	250	0,52	Kp=0,48

lower quantity of d.m. than the productive potential of the green forage). These losses normally occur at the level of the vegetable tissue and the nutritive elements that compose it. They are caused by the following phenomena: a) respiration, b) mistreatment of mechanical origin, and c) fermentation.

In analytical terms, where:

Kv = losses (expressed in % or as a decimal) in nutritive value per unit of d.m.;

Ks = losses in d.m. (expressed in % or as a decimal) referring to hectare of cultivated surface area;

So = production of dry matter per hectare (kg or t/ha) verifiable at the level of green forage;

Vo = nutritive value of green forage with reference to d.m. (UFL/kg d.m. or UFL/t d.m.),

the total quantity of feed energy, UFr in UFL/ha, which is harvestable per hectare is provided by the following equation:

$$UFr = (1 - Ks)*(1 - Kv)*So*Vo \qquad (5)$$

The total losses in feed energy, KT, in % or decimal form, are equal to:
KT = 1 - UFr/(So*Vo)
from which:

$$KT = Ks + Kv - Ks*Kv \qquad (6)$$

Thus, for example (Table 1), it may be observed that the amount of energy available for potential use by the animal is equal to 1850 UFL/ha for the initial green forage (So*Vo = 2200 kg d.m./ha*0.84

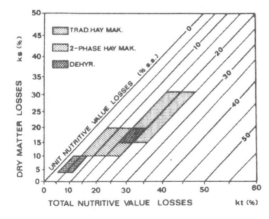

Fig.2 - Range of variation in losses of dry matter (Ks) and nutritive value (Kv) and in total losses (KT) for the traditional haymaking, the two-phase haymaking and the dehidration.

UFL/kg d.m.). In reality, the available nutritive value per ha drops by 55% (830 UFL/ha) if the qualitative (Kv = 29%) and quantitative (Ks = 37%) losses are also taken into consideration. Indeed:
KT = 0.37 + 0.29 - 0.37*0.29 = 0.55.
In addition, this loss is further aggravated by the simultaneous loss of protein (KTp = ·48%).

In conclusion, KT is the parameter which best summarizes the productive aspects of a given technique. A summary of the losses Ks, Kv and KT for the three techniques under consideration here is provided in Fig. 2.

The following aspects should be emphasized:
- when all the work sites used for traditional haymaking are managed with identical operative ability, they supply products with the same characteristics. However, the results obtainable vary widely in relation to numerous factors (dampness at the time of baling, contamination of the forage, cleanliness during turning and swathing operations, weather conditions, etc.);
- work sites that handle product drying (total or partial) provide a higher quality product which is much less susceptible to variation.

2.2 Input-Output Analysis

Once the productive aspects of the various techniques have been analyzed on the basis of the approach described above, the first step towards formulation of the final operative choice must consider the extent to which each productive factor used in the kth technique can influence the final yield. More specifically, it is necessary to determine the technical and economic consequences that variations in the use of a single variable input factor will have on the productivity of the forage crop. In addition, the optimal quantity of this factor (i.e., corresponding to maximum profit) must be calculated.

In the case under examination, special attention should be paid to two input factors: labor or MT(K), and energy consumed or ET(K). Examination of the influence of the overall costs of mechanization, VT(K), will be left to the cost-benefit analysis.

The production function that connects yield to variations in the input factor used may have two different forms, depending on whether or not there is a correlation between input and output. If

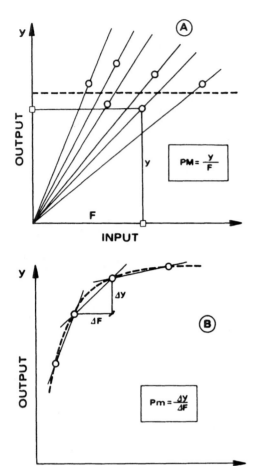

Fig.3 - Input-output analysis:
A) No correlation between input and output: the best solution is identified by the highest average production (PM) corresponding to the straight line with the maximum slope;
B) The output depends on the input according to a decreasing marginal productivity law.

there is no correlation between the two, an increase in yield could be caused by both an increase and a decrease in input (Fig. 3A). This means that although the input factor in question deserves to be analyzed, it does not have a direct influence on the quantity of output, since that value is influenced by other factors. This situation is typical of the study of the production function:

$$Ur(k) = f[MT(k)] \qquad (7)$$

which relates the productivity of the forage crop to the total labor requirement.

In other cases, when the input factor has a direct influence on the quantity of output obtainable, the slope of the production function follows the decreasing productivity law (Fig. 3B). Here, the increase in yield tends asymptotically towards the limit value represented by the UFL/ha contained in the original green forage (So*Vo). This is the result that is generally obtained from the equation:

$$UFr(k) = f[ET(k)] \qquad (8)$$

which connects the yield in UFL/ha to the total quantity of energy consumed by the entire mechanization chain.

In the two situations described above, a different criterion has to be adopted to determine the optimal working condition. In the first case, reference should be made to the concept of <u>average product</u> (PM) of the variable input factor, which is provided by the ratio between total output (y) and the quantity of input utilized (F):

$$PM = y/F \qquad (9)$$

Graphically, this equation corresponds to the slope of the segment that joins the origin of the curve in Fig. 3A to the specific point under consideration. The optimal situation is identified by the solution with the highest PM value (maximum tgα).
In the second case, however, reference should be made to the concept of <u>marginal product</u> (Pm), which corresponds to the ratio between increases in output (Δy) and variable input (ΔF):

$$Pm = \Delta y/ \Delta F \qquad (10)$$

When these variations are defined, Pm graphically corresponds to the trigonometric tangent of the line joining two contiguous points on the input-output plane (Fig. 3B). With infinitesimal variations, the result naturally is:

$$Pm = dy/dF \qquad (11)$$

Since this situation is typical of the equation connecting the use of increasing energy levels to the total forage yield, once equation (2) has been substituted into equation (1), the optimal quantity of ET to be used to maximize the profit P is obtained by assigning a value of zero to the partial derivative of P with respect to ET:

$$\delta P/\delta ET = 0 \qquad (12)$$

that is:
$$\delta P/\delta ET = [\ \delta UFr/\delta ET\ ]*Vf\ -\ Ve\ =\ 0$$
where the expression in brackets is the marginal product of the yield with respect to energy consumption. Hence:

$$Vf*Pm\ =\ Ve \qquad (13)$$

In other words, the optimal quantity of energy to be consumed is represented by the point at which the economic value of the marginal product of UFL/ha with respect to the energy consumed is equal to the unit cost of that energy. This condition corresponds to the tangent to the curve UFr/ET of a line with a slope equal to (Fig. 4):

$$tg\alpha\ =\ Ve/Vf \qquad (14)$$

In conclusion, the input-output analysis makes it possible to express an initial judgment about the validity of employment of a certain quantity of productive input. Naturally, this judgment should then be integrated with economic considerations obtained from a cost-benefit analysis.

2.3 Cost-Benefit analysis

This analysis is used to determine the quantity of product for which the highest profit is obtained from the forage crop.

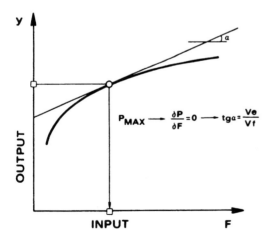

Fig.4 - If the output depends on the input on the basis of a decreasing productivity law, the best solution is identified by the condition according to which the marginal product (Pm) equals the ratio between the unitary cost of the input and the unitary value of the output.

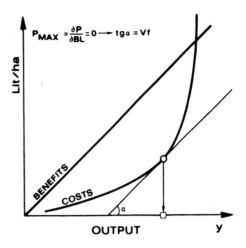

Fig.5 - Cost-Benefit analysis: the maximum profit is determined by the point at which the tangent to the cost curve is parallel to the benefit's line.

This value is provided by a comparison of revenue and costs, in overall terms, on the basis of equations (2) and (3).

The curve depicting gross profit, BL(k), as a function of the quantity of product obtainable, UFr(k), is obviously straight, since it is provided by equation (4). The total cost, on the other hand, is generally represented by a hyperbolic curve (Fig. 5), which reflects the hypothesis about the production function with decreasing marginal products.
The production cost (calculated on the basis of the operating costs of each operation from the cutting phase through conservation) is the economic indicator that represents all the organizational and economic aspects examined above.
Maximum profit is the point at which the tangent to the total cost curve is parallel to the line representing profits (Fig. 5). Indeed, with reference to equations (1) and (2):

$$\delta P/\delta UFr\ =\ Vf\ -\ \ \delta CT/\delta UFr \qquad (15)$$

Since the maximum value of P is reached when equation (15) equals zero, and since CT/ UFr represents the slope of the generic line tangent to the cost curve, the optimal volume of production is determined by the equation: $tg\alpha\ =\ Vf$.
This information, in addition to the values obtained from the input-output analysis, represents all the elements necessary to make a final operative choice as a function of the farm objectives mentioned above.

3 INITIAL APPLICATION OF THE MODEL

The methodology proposed by the model has been applied to harvesting and conservation techniques that produce <u>dried forage</u>. These are:
- traditional haymaking;
- two-phase haymaking;
- dehydration.

These three techniques are similar in that they are based on processes aimed at control of product dampness through the employment of increasing energy levels. As the amount of energy increases, a greater quantity of nutritive elements is preserved in the original green product (and the possibility of recovering more dry matter is increased). Therefore, since the product obtainable by using the three techniques is homogeneous from a physical standpoint (dry product), these represent the optimal conditions under which to carry out an **input-output** analysis. In any case, the analysis that follows may be extended to the production of ensilage; in that situation, the various methods for use of the product as feed must be taken into consideration.

Our example refers to a 30-ha plot of alfa-alfa hay which is cut four times a year. The plot has a productive potential of 13 t/ha/year of initial d.m., with a nutritive value of 0.84 UFL/kg of d.m. (So*Vo = 10,900 UFL/ha).

Two alternative solutions referring to the work conditions typical of a medium-sized or large dairy farm in Northern Italy were evaluated for each technique. Each solution was identified by an abbreviation:

FT1: traditional haymaking in the field with: a conditioning-mower, haymaking with a wheel tedder (working width: 3.5 m) and comb side-delivery rake (3 m), a gatherer-baler (prism-shaped bales), gathering with an automatic-loading cart, unloading in individual bales with manual stacking in the hayloft. Ks = 25%, Kv = 20%, UFr(FT1) = 6530 IFL/ha.

FT2: traditional haymaking in the field with: cutting and haymaking as in FT1, baling with rotor baler, harvesting with front fork loader and transport by trailer (load capacity: six bales), stacking with a gripper mounted on the front of the tractor. The yield is assumed to be the same as that produced by FT1.

FAC: two-phase haymaking with gathering of 50% damp, loose product by an automatic-loading trailer; completely automatic loading of the damp product into the drier (measuring box, feed belt,

pneumatic blower, collapsible piping), conventional Diesel fuel heat generator (130 kWt) sufficient to guarantee an average increase in temperature of approximately 6°C. Heating is only employed during the night. Ks = 12%, Kv = 10%, UFr(FAC) = 8580 UFL/ha.

FAS: two-phase haymaking: identical to FAC but with solar air collector drier for daytime heating of ventilation air. The conventional generator functions only at night during the first few days after loading and during prolonged periods without sunshine. The yield is assumed to be the same as that obtainable with FAC.

DSP: partial dehydration with gathering of 70%-damp product after use of conditioning-mower and side-delivery rake; low-temperature dehydrator (run on Diesel fuel) with double belt and attached grinding and [pellettatura ??] machine for the dried product. Ks = 8%, Kv = 6%, UFr(DSP) = 945 UFL/ha.

DST: total dehydration with cutting and gathering of 80%-damp product using a mower-cutter-loader. The equipment is identical to that used in DSP. Ks = 5%, Kv = 4%, UFr(DST) = 9950 UFL/ha.

4 RESULTS OBTAINED

An analysis of operative aspects based on the typical sequence of operations (from cutting to storage) and on the work capacity of the individual machines has led to quantification of work times and total labor requirements for each solution examined. The result is a significant

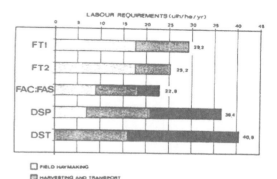

Fig.6 - Labour requirements (ulh/ha/yr) for each forage harvesting and conservation technique considered.

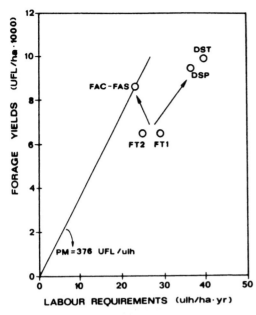

Fig.7 - Forage yields (UFL/ha) plotted against overall labor requirements (ulh/ha/yr): maximum average production (PM = 376 UFL/ulh) corresponds to the FAC and FAS solutions.

disparity between the labor requirements of the various solutions (Fig. 6).
When these requirements are considered (as part of the input-output analysis) as a function of total production with nutritive value (Fig. 7), it is observed that substantial increases in yield can be obtained with two-phase haymaking and the same amount of labor as that used in traditional haymaking. In fact, the labor requirement tends to decrease noticeably, despite the fact that additional operations have to carried out at the farm.
With dehydration, on the other hand, increased yield is always connected with a concomitant increase in labor times, which have to be divided into daily shifts because of their continuity.
The soundness of solutions FAC and FAS is also demonstrated by the fact that they are associated with the highest average product (Fig. 7), which is equal to 376 UFL/ulh. This may be compared with values between 220 and 260 UFL/ulh produced by the other solutions. In terms of energy, increases in consumption are connected with the level of mechanization, since the complexity of the operative sequences is determined by the timeliness

with which the drying process is to be finished. Timeliness requires both the presence of sufficient motive power (electric motors) able to handle large quantities of forage in short periods of time and the consumption of fuel to heat drying air. In the latter case, consumption is higher the faster the water is removed from the vegetable mass (Fig. 8).
The ratio of the increase in yield to the increase in consumption produces a slope in which UFr asymptotically tends towards the limit value (So*Vo = 10,900 UFL/ha), which corresponds to the total productive potential of the forage crop (Fig. 9). It may be observed that the increase in UFr decreases significantly when shifting from solutions FAC and FAS to those involving dehydration.
Since the optimal level of energy use is identified by the equation which states that marginal product equals the ratio of unit energy cost to the value of a unit of forage, it is clear that the solution providing this result would be characterized by energy consumption similar to that found in two-phase haymaking.
Indeed, assuming that:
- the economic value of a UFL of alfa-alfa hay is on the order of 350 lire/UFL;
- the cost of a kWh of the primary energy used is on the order of 60 lire/kWh,
the following equation is obtained:

$$\delta UFr/\delta ET = Ve/Vf = 60/350 = 0,17 \ UFL/kWh$$

If we now consider the curve plotted by interpolating the points relating to all

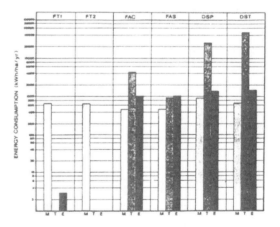

Fig.8 - Mechanical (M), thermal (T) and electric (E) energy consumption for each solution analyzed.

the solutions under consideration on the UFr/ET plane, the tangent to this curve with an angular coefficient of 0.17 UFL/kWh identifies a point (ET = 5000kWh; UFr = 8150 UFL/ha) that describes the working conditions typical of two-phase haymaking and corresponds to the optimal quantity of energy to be used (Fig. 9).

In conclusion, on the basis of the two input-output analyses carried out, the following points may be made:

a) two-phase haymaking is the one conservation technique that makes it possible to optimize the balance between total product quantity with nutritive value obtainable from 1 ha of crop, the overall labor required and energy consumption;

b) the advantage offered by two-phase haymaking is further increased by the use of solar air collectors, which make it possible to save significant amounts of primary energy. In Fig. 9, the points relating to FAS are located to the left of the UFr = f(ET) curve [of a value equal to the thermal contribution of the solar collector]. This situation means that the working conditions of the FAS solution are even more favorable than the optimal conditions determined by using the

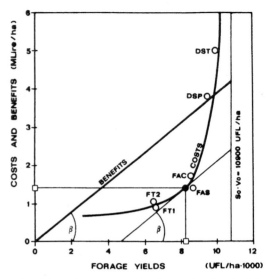

Fig.10 - Assessment of optimal yield (UFL/ha) from a cost-benefit analysis: the two-phase haymaking is once again identified as the best solution.

method described above, since UFr(FAS) = 8500 UFL/ha and ET(FAS) = 4600 kWh/ha. In other words, in this case yields are 5% greater than what has been calculated as the optimal value, with the consumption of 8% less energy than that level determined as optimal consumption;

c) solutions DSP and DST (dehydration) are unusable, especially from an energy standpoint. In fact, even when a unit value of 400 lire/UFL is considered for the forage (i.e., a higher value than the previous one), the conditions required for the economic success of partial dehydration would only be present with a unit energy cost between 14 and 20 lire/kWh.

These comments are confirmed by the cost-benefit analysis (Fig. 10). On the basis of the method described above, the optimal volume of production corresponds to the productive levels typical of two-phase haymaking. Specifically, the optimal situation is represented by a production cost of 1.45 Mlire/ha, as long as a yield of 8300 UFL/ha is guaranteed. Once again, the FAS solution results in working conditions that are better than optimal, since with a cost of 1.4 Mlire/ha it ensures higher yields.

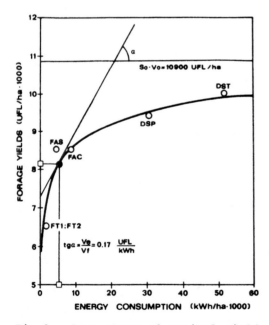

Fig.9 - Assessment of optimal yield (UFL/ha) plotted against overall energy consumption (kWh/ha). The method identifies the two-phases haymaking as the best solution.

5 FINAL REMARKS

These aspects were brought to light by an

examination of the model's applicative results (despite their limited validity as a result of the specificity of the example considered):

a) the need to improve hay yields produced on farms has led to a search for harvesting and cutting techniques which may be adopted as alternatives to traditional haymaking:

b) two-phase haymaking (especially with the use of solar air collectors) is the technique that best meets this need from an organizational, energy and economic standpoint;

c) partial dehydration (and to an even greater degree, total dehydration) is unusable because of the high investment and management costs, which are out of proportion when it comes to a product which is to be re-used on farms.

Therefore, in the short- and medium-term, it appears likely that the conditions that have favored two-phase haymaking (improved by the use of solar collectors) will continue to exist. The permanence of these conditions is compatible with the fact that the economic value of feed-related energy (especially for livestock) is bound to increase at a faster rate than the unit costs of energy. Considering the importance that statements of this sort have in terms of the operative choices to be made on dairy farms (especially with the climatic conditions encountered in Northern Italy), the proposed model should be further refined in order to overcome its present analytical limitations. These limitations mainly concern the highly deterministic nature of the model itself, which means that interrelationships between productive, organizational and energy aspects (the interactive process in Fig. 1) can only be developed on the basis of input parameters with average values. It is impossible to consider any type of risk variable (effects of climate, machine breakdowns, operator ability, management skill, etc.).

This obstacle may be overcome by first developing the model in probability terms, as well as providing it with the capacity to automatically acquire input parameters from data banks set up for this purpose. In this way, the model could be transformed from a simple calculation tool into an actual simulation model.

Finally, the analysis could be extended through the obtainment of other types of output data from the model, such as optimization of the tractor/operative machine system and consideration of the energy incorporated into the various technologies in relation to their useful life.

REFERENCES

Guidobono Cavalchini, A. 1977. Valutazione delle varie tecniche di raccolta e conservazione dei foraggi prativi in funzione del rendimento energetico. Riv. Ingegneria Agraria 3-4, p.179-192.

Rees, D.V.G. 1982. A discussion of sources of dry matter loss during the process of haymaking. Journ. of Agr. Eng. Res., p.469-481.

Riva, G. & F.Mazzetto 1985. Essiccazione artificiale dei foraggi: e' opportuno riscaldare l'aria di ventilazione? L'Informatore Agrario 25, p.29-34.

Riva, G. & F.Mazzetto 1985. Construction aspects of four solar plants for forage drying installed in Northern Italy. Proc. of UNESCO/FAO Work. Gr. Meeting Solar Drying, 27-30 June, Perpignan, France, p.26-31.

Land and Water Use, Dodd & Grace (eds), © 1989 Balkema, Rotterdam. ISBN 90 6191 980 0

A novel grain crops harvester for Bangladesh

M.M.R.Mollah
Windsor, Victoria, Australia

C.D.Watt
Silsoe College, Silsoe, Bedford, UK

ABSTRACT: A significant quantity of grain in Bangladesh is lost because of late harvesting, caused by general labour shortages and uncertain weather conditions at the time
of harvest. In separate studies the need for machine harvesting (quick harvesting) in
Bangladesh was identified. A prototype grain crop harvesting machine was designed and
developed at Silsoe College, UK. The machine was tested on Bangladesh farms for harvesting rice and wheat. The cutting mechanism of the hand powered machine consisted of two
counter-rotating cutting discs. The machine cut the crops which were planted in rows as
it was pushed forward. The cut crops were windrowed or collected into a box to the side of
the machine. An extensive number of tests were carried out on rice and wheat harvesting
to investigate the variables affecting the machine design and its performance. The optimum combinations of the machine parameters for harvesting rice and wheat were identified in this research. The prototype machine successfully cut and windrowed both wheat and
rice crops. The total power required to push the machine while cutting crops was high for a
single person. Very little power was needed to sever the crop stems. A significantly high
percentage of power, with respect to the total power applied, was used up in the rolling
resistance and the transmission system of the machine.

RESUME: Une quantité importante de céréales est perdue au Bangladesh due à leur récolte
tardive causées par une manque générale de main d'oeurve et des conditions climatiques in-
certaines qui règnent à l'époque de la récolte. Dans des études séparées on a identifié la
nécessité d'avoir une moissonneuse (qui peut faire la travail très vite) au Bangladesh.
Une moissonneuse prototype a été réalisée et développée au Colège Silsoe, Grande Bretagne. La machine a été mise en essai au Bangladesh sur des fermes pour récolter du riz et
du blé. Le méchanisme coupant de la machine actionnée à la main est composée de deux disques tranchantes tournant en sens inverse. La machine fait la récolte qui était alignée
pendant qu'on poussait la machine. Les céréales ainsi récoltées ont été mises en position
et placées dans une boite à côté de la machine. Il y eut des tests intensifs comprenant le
riz et le blé pour apprécier le comportement da la machine et des avaries qui pourraient
affecter le dessin de la machine. Lors des recherches les résistances optimales de prame-
tres de machine pour récolter le riz et le blé ont été identifiées. La machine prototype
a coupé et mis en place avec succès les récoltes de riz et de blé. La force totale requise
pour pousser la machine pendant la récolte était élevée pour une seule personne. Une quantité appréciable de pourcentage de force, en fonction de la totalité de force appliquée, a
été utilisée par la force mouvante et le système de transmission de la machine.

AUSZUG: Ein bemerkenswerter Anteil de Getreides in Bangladesch geht verloren aufgrund zu
später Ernte, hervorgerufen durch allgemeinen Arbeitskräftemangel und unbestimmte
Wetterlage wärend der Erntezeit. In verschiedenen Studien wurde die Notwendigkeit der
maschinellen Ernte (schnellen Ernte) in Bangladesch festgestellt. Das Modell einer
Getreide-Ertemaschine wurde am Silsoe College, U.K., entworfen und entwickelt. Die Maschine wurde in landwirtschaftlichen Betrieben in Bangladesch bei der Reis- und Weizenernte
erprobt. Der Schneidemechanismus der handbetriebenen Maschine bestand aus zwei in entge-
gengesetzter Richtung rotierenden Schneidescheiben. Die Maschine schnitt das in Reichen gepflantze Getreide im Vorwärtsschieben. Das geshnittene Getreide wurde in einen
Behälter an der Seite der Maschine geblasen oder gesammelt. Eine große Anzahl von Test
wurden durchgeführt, um den Einfluß verschiedener Variablen auf die Konstruktion und Leistung der Maschine in der Reis- und Weizenernte zu untersuchen. In dieser Studie wurde die optimale Kombination von Maschinenparametern für die Reis- und Weizenernte festgestellt. Die
Modellmaschine schnitt und blies sowohl Reis- als such Weizengetreide erfolgreich aus. Die
Gesamternergie,die für das Schieben der Maschine während des Erntevorgangs nötig ist,ist von
einer Person leistbar. Zum Schneiden der Getreidestiele wurde sehr wenig Energie benötigt.
Ein bemerkenswert hoher Energieanteil im Vergleich zum gestamten Energieaufwand wurde durch
den Rotationswiderstand und den Getriebemechanismus der Maschine verbraucht.

1 INTRODUCTION

Bangladesh is an agriculturally based country with 10 million hectares of rice land and three different rice seasons called 'Amon' (Oct.-Jan.) 'Boro' (April-June) and 'Aus' (June-Sept.) depending on the time of harvest (BRRI, 1986).

Although Bangladesh's labour force seems to be very high, labour shortages are always experienced at the peak period of harvesting seasons. The shortages of labour at harvesting time is the main cause of delays in harvesting rice. The probable reason is that everybody wants them at the same time which also accounts for the high wage rates. As more high yielding varieties (HYV) of rice are being introduced to farmers, intensive rice cultivations are being practiced which demand more labour at the time of harvest. There is also a trend going on among the urban labourers to move to the town for better work that is less labourious and more prestigious.

A Bangladesh farm is typically small and highly fragmented with average farm size and plot size of 2.5 ha and 0.18 ha respectively (Mollah, 1988). Bangladeshi agriculture is characterised by a low level of technology. The country suffers a chronic food deficiency of 1.5 million tons (BRRI, 1985). Bangladesh needs to import rice and wheat to meet this food deficiency. To cope with this crisis the efforts made so far have concentrated on raising agricultural production by inputting HYV, fertilizer, pesticides and irrigation water. But very little planned attention has been given to the post harvest loss of the crop. There is a way to increase the availability of food in the country by minimising the post harvest loss. Haque et al. (1985) show that loss of rice quantity is 2.5% in 'Amon', 1.14% in 'Boro' and 2.01% in the 'Aus' season due to the harvest operation. Very often farmers fail to harvest in time then these loss figures are high because the loss figures increase due to overmaturity and uncertain climate, such as hail storms, heavy rain, gusty winds, flash floods etc. Overmaturity makes the straw lodge which brings the panicles close to the soil or standing water of the field and eventually the grain sprouts or becomes mouldy and discoloured.

To minimise the loss during harvest appropriate technology should be introduced for harvesting the rice crops suitable for Bangladesh farmers. Western and developed Eastern technologies have failed in Bangladesh because of the socio-economic conditions of farmers, topography of land, soil conditions, land fragmentation and availability of resources in Bangladesh.

It is anticipated that the farmers of Bangladesh desire a low-cost, low-powered device to harvest crops like rice, wheat etc. with a higher rate of work. Considering these factors the hypothesis of the study was taken as "A machine can be designed that is powered by one man and can harvest rice at a significantly higher rate of work than traditional hand harvesting."

2 MAIN OBJECTIVE OF THE STUDY

The main objective of this study is to investigate the design parameters of a low-cost, low-powered machine able to cut and bunch/windrow grain crops at a significantly higher rate of work and be suitable for manufacturing in Bangladesh and elsewhere.

2.1 The specific objectives

1. To identify and quantify the parameters affecting the cutting efficiency and operation of a counter-rotating disc grain crops harvesting machine.
2. To assess the field operating performance of the machine in Bangladesh.

3 THE MACHINE

The machine consists of two serrated/smooth edged discs. One disc was mounted at the lower end of a vertical shaft which was mounted on low friction bearings. The shaft was powered from the ground wheels by means of driving pulleys, guide pulleys and polyurethane belts. The rotation in the direction of the wheel axle was converted into rotation about a vertical axis by means of cross belting. The other disc was mounted at the top of a freely rotating short vertical shaft. The discs were overlapped to each other. The unpowered disc was driven by friction from the powered disc.

The machine cut only crops in rows and cut one row in a single run/pass. The machine will only windrow crops to its right and therefore the machine must be operated around the perimeter of the fields.

3.1 Power input for the machine

The available power input for the machine was chosen to be the power of a single man as the work done by the author on the field simulator test rig for harvesting wheat showed that the maximum average power requirement of the machine is 63.6 watts (Mollah, 1984). The power was less than the power (74.6 watts) produced by a man during continuous work (Bhatnagar & Hoda, 1985).

3.2 Width of the machine

The width of the machine was determined by the row spacing of the crops as it cuts a single row. The maximum width of the machine was fixed at 350 mm which would give 75 mm clearance between the row and the

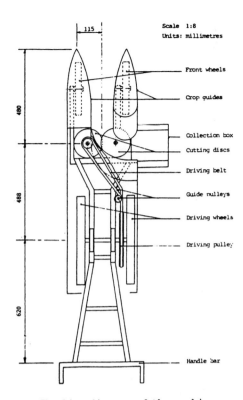

Fig.1 The line diagram of the machine.

machine for the 250 mm row spacing recommended by BRRI (BRRI, 1986). Square planting, i.e. row spacing = plant spacing is recommended for machine harvesting.

3.3 Estimation of the cutting disc parameters

The freebody diagram in fig. 2 leads to the following relationships.

$$\cos\theta_c = (R - l)/(R + r),$$
$$i.e. \quad R = (r\cos\theta_c + l)/(1 - \cos\theta_c), \quad (1)$$
$$\mu = \tan\theta_c. \quad (2)$$

Where θ_c = the angle at the point of contact; c-c = points of contact between the disc and the material to be cut; $2l$ = disc overlappings; R = radius of the cutting disc; r = effective radius of the material to be cut; μ = dynamic coefficient of friction at the point of slip between the cutting disc and the material to be cut; F_T = tangential cutting force; F_R = radial cutting force and ω = angular speed of the disc.

The relationship between the disc overlapping and the diameter of the cutting disc for different coefficients of friction from equations (1) and (2) is shown in fig. 3.

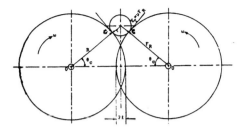

Fig.2 Freebody diagram of the cutting discs with the material to be cut at the point of slip.

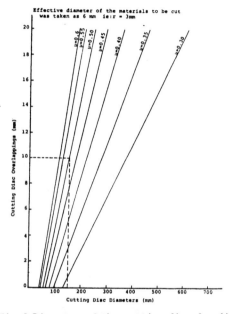

Fig.3 Diameter of the cutting disc for different disc overlappings at different coefficients of friction.

The diameter of the disc was taken as 150 mm leaving adequate clearance for the framework of the machine for the maximum width of 350 mm. From fig. 3 it can be seen that 150 mm dia. discs below 10 mm disc overlapping can work for the coefficient of friction (c.f) 0.5 or below. In practice the c.f is expected to be more than 0.5.

3.4 Forward speed of the machine

The forward speed of the machine was taken as the walking speed of its operator at 3 km/h (measured at normal walking pace while pushing a jab planter which was thought to be an equivalent working load of the machine).

3.5 Size of the ground wheel

It was decided to test the machine at various speeds of the cutting discs by using pulleys of different sizes. Therefore, two ground wheels of 500 mm diameter and 50 mm width were selected on pure judgement which could give 8% to 80% higher distance coverage by the cutting disc with respect to the forward distance moved by the machine with pulleys of different sizes.

3.6 Disc overlappings

An experiment on the field simulator test rig in 1985 (Mollah, 1988) showed that the overlapping could be extended up to 10 mm for clean cuts with one serrated edge disc and one smooth edged disc. Therefore, it was decided to operate the machine with a disc overlapping of less than 10 mm.

4 INSTRUMENTATION OF THE MACHINE

In order to measure the test parameters of the pushing force, torque to the disc and disc separating force, strain gauged beams were designed and mounted onto the machine. The gauge bridges were powered from a specially designed portable power box. The voltage outputs from the bridges were measured and recorded by a digital data logger/meter called Squirrel SQ-8-4V (Grant Instruments, 1985). The data logger was made specially to record a sample rate of 10 readings per second. It could record the voltage from 0 to 10 mV. The data recorded into the data logger was then transferred into a portable computer, an Epson HX-20 (Epson, 1985) for analysis and storage.

5 THE FIELD EXPERIMENTS

The machine tests were performed on wheat and rice in Bangladesh. A wheat field with uniform crop density was selected to test the machine's wheat harvesting ability. The wheat was grown in rows with 200 mm row spacing. The machine was tested on three varieties of rice grown at three different plantation spacings (20cm × 20cm, 22cm × 20cm and 25cm × 15cm).

A split-split-split plot design was used for the wheat harvesting experiments. This design was chosen to complete the tests within the shortest possible time to avoid the risks of hail storms. A four factor factorial design was considered for rice harvesting experiments.

It was observed that during wheat harvesting smooth edged or serrated cutting discs could be used. Subjectively serrated discs gave a better cutting performance than smooth edged discs because of the combined sawing and scissor action. The smooth driving wheel could also be used with little wheel slip. But only lugged wheel and serrated cutting discs could be used in order to make the machine work satisfactorily for rice harvesting.

All the strain gauged beams were calibrated against known dead loads before being used in fields. The voltage outputs from the bridge circuits were zeroed by a preset potentiometer before putting the driving belt on. For each run the data logger was used to record the voltage outputs after the machine started to harvest and was stopped before the machine stopped harvesting. This technique was used to avoid transient readings caused by the acceleration and retardation of the machine.

The effective capacity and field efficiency of the machine was tested for rice harvesting on a typical plot for a day. The same operator who operated the machine during field capacity tests was selected for the hand harvesting tests.

6 DATA ANALYSIS AND RESULTS

The data recorded for each treatment in the wheat and rice experiment was analysed by the Epson HX-20 portable computer to get a mean value. The mean was taken as the reading for the treatment. The analysis of variance was conducted on both the experimental design for wheat and rice harvesting. The analyses were performed using the Minitab statistical package (Minitab Inc, 1985) and the Genstat statistical package (NAg, 1983). Both packages were available on the VAX 11/750 minicomputer at Silsoe College.

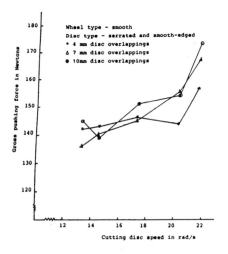

Fig.4 The relationship between the cutting disc speed and the gross pushing force for different disc overlappings during wheat harvesting with smooth wheel.

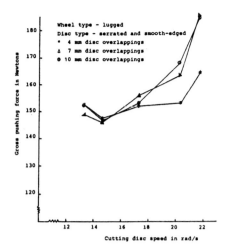

Fig.5 The relationship between the cutting disc speed and the gross pushing force for different disc overlappings during wheat harvesting with lugged wheel.

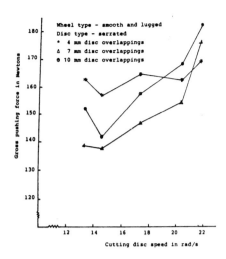

Fig.7 The relationship between the serrated cutting disc speed and the gross pushing force at different disc overlappings during wheat harvesting.

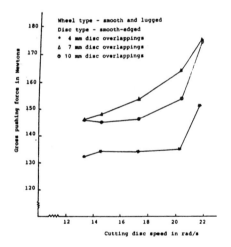

Fig.6 The relationship between the smooth edged cutting disc speeds and the gross pushing force at different disc overlappings during wheat harvesting.

The figures 4 and 5 show that the mean pushing force during wheat harvest is less in the case of the smooth driving wheels. The reason is that the rolling resistance is higher with the lugged wheels. From figures 6,7,8 and 9 it can be seen that wheat harvesting with the smooth edged cutting discs needs less pushing force and less power than the serrated cutting

discs. This variation is because of the different cutting process. The figures show that in some cases the lowest cutting disc speed (13.03 rad/s) requires more pushing force. The fact is this is the lowest cutting disc speed. At this speed the cut crop was not conveyed satisfactorily. The cut crops tried to lean forward causing blockage at the cutting point which required higher pushing force at this speed.

The figures 6 & 7 show that smooth edged discs with 4 mm overlap need the least pushing force whereas serrated discs with 4 mm overlap need the greatest pushing force. This discrepancy is probably due to different cutting actions; with serrated discs more crops are pulled.

Though the power required to cut the crop does not change significantly with disc speed, the total loss in the driving system is higher with higher disc speeds. At higher belt speeds the polyurethane belt stretches and its power transmission efficiency decreases (fig. 10).

The gross pushing force required for rice harvesting varied significantly with the rice varieties and the row spacing of the plantation. Smaller row spacing needed a higher pushing force because of the thicker crop density which causes higher friction between the guide of the machine and the crops of the adjacent row. As cutting disc overlap and speed has no effect on the gross pushing force for rice

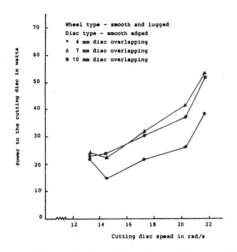

Fig.8 The relationship between the smooth egded cutting disc speed and the power to the cutting disc at different disc overlappings during wheat harvesting.

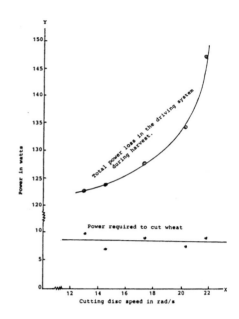

Fig.10 The power loss in the driving system and power required to cut wheat at different cutting disc speeds.

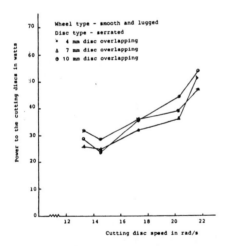

Fig.9 The relationship between the serrated cutting disc speed and the power to the cutting disc at different disc overlappings during wheat harvesting.

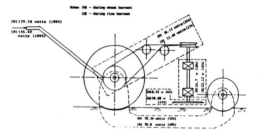

Fig.11 Power used up in different parts of the machine during harvesting experiments

harvesting then any disc overlapping and speed can be chosen.

The optimum combinations for harvesting:
Wheat:
Smooth wheel; serrated cutting discs; 7 to 10 mm cutting disc overlap and 20% higher disc peripheral distance coverage with respect to the forward distance covered by the machine, i.e.: 14.47 rad/s disc speed and 0.9 m/s forward speed of the machine.

Rice:
Lugged wheel; serrated cutting disc; 4 mm cutting disc overlap and 20% higher dis-tance coverage with respect to the forward distance covered by the machine, i.e.: 13.4 rad/s disc speed and 0.84 m/s forward speed of the machine.

Fig. 11 shows the details of the power used up in different parts of the machine during the harvesting experiments.

The measured field capacity and field efficiency of the machine are 0.03 ha/h and 60% respectively for rice harvesting. The measured hand harvesting rate was 0.01 ha/h.

7 CONCLUSIONS

From the investigations carried out on wheat and rice harvesting during the field experiments, the following general conclusions can be made.

1. The machine is able to harvest three times faster than harvesting by hand.

2. The power required to push the machine in a rice field with soft soil conditions was nearly double the power a man can produce for 8 hours continuous work. With the existing prototype machine a man would be able to do continuous work for 45 minutes, calculated on the basis of $HP = 0.35 - 0.092 \log$ "(time in minutes)" (Bhatnagar & Hoda 1985).

3. Very little percentage of power, 6% for wheat and 14% for rice is required to cut the crops by the machine with respect to the total power used in the harvesting mechanism. The power used up to overcome the rolling resistance of the machine is 48% to 50% of the total power required. From 23% to 26% power is used up in the transmission system (pulleys and belt). Finally 16% to 18% power is used up in the disc bearings.

8 FURTHER RECOMMENDATIONS

Studies should be carried out to reduce the power loss in the driving system by introducing lighter wheels with good traction aids for low rolling resistance, fewer pulleys in the transmission system and a belt with a good transmission efficiency, i.e.: does not stretch under high loads and can be used for cross-belting.

REFERENCES

BRRI. 1985. Proceedings of the workshop on experiences with modern rice cultivation in Bangladesh. 2--4 April 1985. Bangladesh Rice Research Institute.

BRRI. 1986. Rice research in Bangladesh. Bangladesh Rice Research Institute.

Bhatnagar, S.C. and Hoda, M.M. 1985. Tools and manually worked equipment in agriculture. Appropriate Technology Development Association, India.

Epson. 1985. About your Epson executive. Epson (UK) limited, Dorland House, 388 High Road, Wembly, Middlesex, HA9 6UH, UK.

Grant Instruments. 1985. Squirrel meter/logger user manual. Grant Instruments (Cambridge) Ltd., Barrington, Cambridge, CB2 5QZ, UK.

Minitab, Inc. 1985. Minitab Reference Manual, second edition Minitab Inc. 3081 Enterprise Drive. State College. PA 16801 U.S.A.

Mollah, M.M.R. 1984. An improved cutting and bunching machine for harvesting grain crops in Bangladesh. M.Sc. thesis. Cranfield Institute of Technology, Silsoe college, Silsoe, UK.

Mollah, M.M.R. 1988. A novel grain crops harvester for Bangladesh. Ph.D. thesis. Cranfield Institute of Technology, Silsoe College, Silsoe, UK.

The Numerical Algorithms Group Limited (NAg). 1983. A general statistical program. 1983 Lawes Agricultural Trust (Rothamstead Experimental Station).

Land and Water Use, Dodd & Grace (eds), © 1989 Balkema, Rotterdam. ISBN 90 6191 980 0

Performance evaluation of a single-axle tractor powered cowpea harvester

K.C.Oni

Department of Agricultural Engineering, University of Ilorin, Ilorin, Nigeria

ABSTRACT: A prototype cowpea harvester with a reciprocating cutterbar was designed to be powered by a 6kW walking type, single-axle tractor. The harvester was designed to reap two plant rows of cowpea per pass at a plant row spacing of 50cm. Three crop varieties, three machine forward speeds and two plant heights of cut were investigated at an average pod moisture content of 14% (wet basis). Field losses of cowpea grains were determined in terms of shattering, stubble and lodging losses. Statistical analyses of variance for the crop variety, machine speed and plant height of cut were carried out at 95 percent confidence limit. Results show increasing trends, either linearly or curvilinearly, of shatter, stubble and lodging losses as forward speed was increased, for each plant height of cut. Results further show that varietal differences, machine speeds plant heights of cut and their interactions significantly influenced field harvesting losses of cowpea, under the conditions investigated, using the prototype development.

RESUME: Un prototype de moissonneuse de haricots avec Faucheuse à mouvement alternatif a été conçu pour être actionné par un tracteur de 6kW, à essieu unique. La moissonneuse a été conçue pour moissonner deux rangées de haricots à la fois, espacées de 50cm. On a étudié trois variétés de semence, trois vitesses de machine différentes et deux hauteurs de coupe différentes à un degré moyen d'humidité de la gousse de 14%. On a déterminé les pertes en graines   sur le terrain du fait de l'écrasement, des graines perdues et du chaume. On a fait des analyses statistiques de variation du genre de semence, de la vitesse de la machine et de la hauteur de coupe de la plante à 95% de fiabilité. Les résultats montrent un acroissement, linéaire ou curvilinéaire des pertes par écrasement. graines perdues et chaume avec l'augmentation de la vitesse, pour chaque hauteur de plante coupée. De plus les résultats montrent que des différences de variétés, des différences de vitesse de la machine et de hauteur de coupe et leurs interactions ont influencé de façon importante les pertes de haricots sur le terrain, dans les conditions d'investigation et en utilisant le développement prototype.

ZUSAMMENFASSUNG: Es wurde der Prototyp eiues Mähbinders für cowpea mit einem gegenläufigen Sensenbalken gebaut, angetrieben vou einem 6kW starken, einachsigen Tractor. Der Mähbinder wurde für die Ernte von 2 Reihen cowpea in einen Abstand von 50cm konstruiert. Drei verschiedene Erntemöglichkeiten, drei verschiedene Geschwindigkeiten und zwei verschiedene Schitthöhen wurden untersucht, bei einem durchschnittlichem wassergehalt der Hülsen von 14% (nasse Basis). Ernteverluste wurden festgestellt anhand der Erschütterungs-, stoppel- und Ertragsverluste. Statistische Analysen der Varianz für die Ernteart, Maschinengeschwindigkeit und Schnitthöhe wurden durchgeführt mit einer 95% igen Vertrauensgrenze. Die Ergebnisse zeigen austeigende Tendez der Erschütterungs-, stoppel- und Enteverluste mit Erhöhung der Geschwindigkeit bei jedr Schnitthöhe. Die Erjebnisse teijen weiterhin, dass unter dem Versuchsbedingungen bei der Entwicklung des Prototyps die verschiedenen Varianten der Geschwindigkeit, Schnitthöhe und deren Wechselwirkung signifikant die Ernteverluste der cowpea beeinflusst.

1 INTRODUCTION

Cowpea (Vigna unguiculata) is a predominantly
hot weather crop that is well adapted to
the semi-arid region, prefering temperatures
of between 68°F and 98°F (20°C and 35°C).
Most production is under rainfed condition.
The short duration, determinate varieties
may be grown under rainfall conditions of
600 to 2000mm. Cowpea production represents
about 2% of the total world output, with an
average of 1.1 million tonnes in 1970-1974
(Daisy, 1979). In Nigeria, cowpea is a
highly valued legume crop because of its
rich source of protein. The yield of cowpea
in Nigeria is influenced by cultural practices and the method of planting but may vary
between 900 and 2500 kg/ha. Navasero and
Ajirenike (1986) reported a yield of 2200
kg/ha under experimental condition. It was
reported that square planting with plant
density greater than 18 plants/cm^2 yielded
more than rectangular planting (Anonymous,
1973).

Research results have shown that determinate upright cowpea can be cut mechanically
at about 5cm above ground with field harvesting loss of not greater than 13% (Anonymous,
1981). Field harvesting loss however
increased rapidly when the height of cut was
greater than 5 cm and was as high as 35%.
Field harvesting has been a subject of study
for many years (Lamp et al., 1961; Dunn
et al., 1973; Nave and Yoerger, 1975; Nave
and Hoag, 1975). Development effort on
cowpea harvesters in Nigeria has been
limited to modifying small machines for the
purpose. Navasero and Ajirenike (1986)
reported work done on modified 2.6 kW lawn
mower for cutting and windrowing cowpea
plant. Development efforts in other parts
of the world have similarly been devoted to
modifying conventional small grain combines
for such upright legume crop as soybean.

The cutterbar is the major offender of
field harvesting loss, accounting for about
80% of combine header harvesting loss
(Quick, 1973; Dunn et al., 1973). Research
results by Dunn et al. (1973) on soybean
asserted that harvesting loss was caused by
the cutterbar and about 64% of the field
loss was due to shatter. Nave and Hoag
(1975) investigated different cutterbar and
knife configurations and knife frequencies
in a laboratory test stand. Their results
showed that knife section and guard spacing
of less than the conventional 76.2 mm would
reduce plant stem acceleration and result
in shatter loss.

Different cutting mechanisms have either
been designed or modified from existing
machines for harvesting legume crops.
Ichikawa (1981) reported work on rotary and
reciprocating cutting mechanisms while
studies on the use of rotary slashers for
simultaneous rotary slashing and windrowing of legume plants were reported by
Navasero and Ajirenike (1986). Results of
these investigations have highlighted
further need for the development of mechanical harvesters for the harvesting of upright
legume crops in Nigeria.

2 DEVELOPMENT OF THE COWPEA HARVESTER

The cowpea harvester was designed to
harvest two plant rows of cowpea at each
pass. The objective of the mechanical
harvesting of cowpea was to remove as much
of the cut pod-bearing plant materials as
possible as the machine advances through
the field. The three major aspects of the
harvester that facilitate its field performance are the cutting mechanism, the conveying unit and the gathering unit, respectively.

2.1 The cutting mechanism

Power to drive the harvester is derived from
the 6 kW petrol engine of a single-axle
tractor. Power is transmitted through belt
drives, from the engine's 70 mm diameter
V-belt pulley to an intermediate pulley of
the same diameter and thence to a 300 mm
diameter V-belt pulley on the harvester
unit. This 300 mm diameter V-belt pulley
carries the rear cylindrical roller shaft
of the conveyor belt and also transmits
power to the pitman which is fitted to a
cam. The pitman, operating through the cam,
converts rotational motion of the rear
cylindrical roller shaft to reciprocating
motion of the cutterbar. The reciprocating
knives (sickles) are made from 3 mm mild
steel sheet with the upper lip of each knife
bevelled for sharpening. Each knife is
triangular in shape, with a pitch of 50 mm
and reciprocating inside knife guards that
are casted out of bronze. The knife guard
ledger is hardened for enhanced durability.
The isometric view of the harvester is shown
in Figure 1 while the schematic of the
harvesting unit is illustrated by Figure 2.
The cutterknives, through the action of the
reel, cut (snap) the plant stems during the
cutting process and pass cut materials to
the conveyor belt. Adjustable front
mounted wheels on each side of the cutterbar enables the equipment to be varied in
height for different heights of cut.

1 HANDLE
2 REVERSE DRIVE CABLE
3 FORWARD DRIVE CABLE
4 UPPER ROLL PULLEY
5 DRIVE PULLEY
6 TRACTION WHEEL
7 LOWER ROLLER PULLEY
8 GUIDE WHEEL
9 KNIFE GUARD
10 REEL
11 SIDE GUARD
12 FLIGHT
13 CONVEYOR BELT

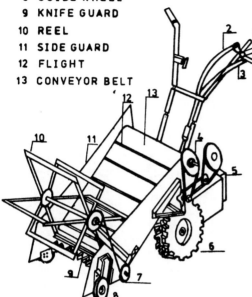

Fig. 1 Isometric view of cowpea harvester

A DRIVE PULLEY
B CONVEYOR BELT
C KNIFE SECTION
D KNIFE GUARD
E REEL

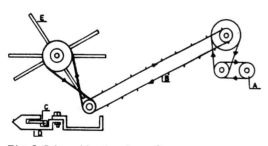

Fig.2 Schematic drawing of cowpea harvester

2.2 The conveying unit

The conveying unit consists of a rubber belt with an overall belt length of 3600 mm and an overall belt width of 590 mm. It has rubber flights along its length (Figure 1).

2.3 The gathering unit

The gathering unit consists of a wire mesh container which is attached to the side of the harvester unit. Materials from the conveyor belt, at the end of their travel, are diverted into the container which is emptied at some point, when full, into a central collection unit. The deflector that allows plant materials into the container is adjusted manually.

3 FIELD PERFORMANCE EVALUATION

Field trial with the prototype machine was conducted during the 1987/88 harvesting season. The field is located in the University town of Ilorin 04° 29'E, 08° 26'N in the derived Savannah vegetation zone of Nigeria with an annual rainfall of 1250-2000 mm. There are two growing seasons due to the bimodal nature of rainfall distribution. The first is from late March to late July while the second is from early August to early November. The soil type is sandy loam and the slope is gentle. The experimental lay out was a split plot, randomized complete block design with 9 blocks and 5 plots per block. Each plot size was 10 m x 3 m. There were four rows of plant per plot with a between-the-row spacing of 0.5 m. The three varieties of cowpea grown on the plots were, respectively, Ife Brown, TVX 3236 and IT 84E-124, all of which are of the brown type. During harvesting operation, three machine forward speeds and two heights of cut of the three cowpea varieties were investigated with three replications. There were a total of 54 treatment combinations.

The throttle settings for the engine r.p.m. corresponding to the desired forward speed were predetermined before field harvesting, using a tachometer. The corresponding settings were marked as appropriate on the throttle lever. Crop materials were harvested and collected during each pass. At the end of each pass, losses were determined in terms of shattering, stubble and lodging losses, respectively. Shattering loss was that due to the impact of the reel and the cutterbar on the plant during harvesting process. Any whole pod dislodged from the plant was collected and accounted for as shattering loss.

Stubble loss was due to pods still left on the stubbles, a function of the lowest poding height while lodging loss was due to grains in pods still carried by lodged plants which could not be reached by the harvesting unit during harvesting process. A tarpoulin with rubber flights was used as conveyor belt due to non-availability of rubber belt of right specification. The pod moisture content at harvest time averaged 14% (wet basis). The three forward speeds under investigation were 0.5, 0.7 and 0.9 m/sec, respectively, while the two heights of cut were 4 cm and 10 cm above the ground. Shattering loss was determined at premarked locations on each plot while stubble and lodging losses were determined at the end of each pass.

4 RESULTS AND DISCUSSIONS

4.1 Shattering loss

The shattering loss showed an increasing trend with increase in forward speed of the harvester as shown in Figure 3. This trend is attributable to increased momentum of the machine and that of the reel. Ife Brown variety had greater propensity to shatter than the other two varieties. The analysis of variance for shattering loss show a significant effect of variety and variety-speed interaction when tested at 95% confidence limit as shown in Table 1. The regression equations of the best fit for each curve in Figure 3 are as shown in Table 2.

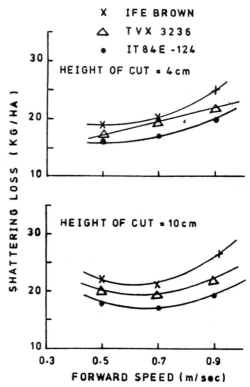

Fig. 3 Effect of machine speed on shattering loss of three cowpea varieties

Table 1. Summary of Analyses of Variance for shattering, stubble and lodging losses.

Source	df	F-values[a]		
		shatter-ring	stubble	lodging
Total	53			
Variety (V)	2	27.99*	11.46*	3.76*
Plant-height (H)	1	0.12	2.17	1.40
Machine speed (S)	2	2.85	2.42	4.98*
V x H	2	2.04	6.51*	1.50
V x S	4	2.75*	16.33*	5.72*
H x S	2	0.27	1.40	0.52
V x H x S	4	1.04	6.65*	0.86
Residual	36			

a * significant effect at 95% Confidence limit

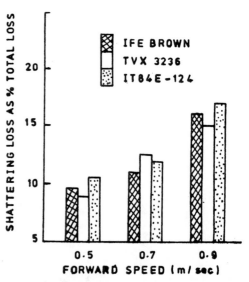

Fig. 4 Histogram of percent shattering loss versus speed for three cowpea varieties (Height of cut = 4 cm)

Table 2. Regression models for cowpea harvesting losses (L) at the three forward speeds (S) of harvester shown in Figure 3.

Variety	Regression model	R^2
(a) Height of cut = 4 cm:		
Ife Brown,	L=13.11 EXP(0.69S)	0.89
TVX 3236,	L=10.95 + 12.5S	0.98
IT84E-124,	L=11.86 EXP(0.56S)	0.98
(b) Height of cut = 10 cm:		
Ife Brown,	L=17.10 EXP(0.42S)	0.60
TVX 3236,	L=17.18 EXP(0.24S)	0.64
IT84E-124,	L=16.26 EXP(0.14S)	0.54

Table 3 presents a summary of the Analyses of Variance for the shattering and lodging losses as percent of the total loss. The effects of crop variety, machine speed and the interaction of plant height with speed were quite significant at 95% confidence limit. The histograms of shattering loss at each speed level for the three varieties are shown in Figure 4. These bar charts show higher percent loss as speed was increased for each variety, further emphasising the significance of the crop variables and machine parameters under study.

Table 3. Summary of Analyses of Variance for shattering and lodging losses as percent of total loss.

Source	df	F-values[c]	
		shattering	lodging
Total	53		
Variety (V)	2	10.23*	9.13*
Plant-height (H)	1	0.35	4.62*
Machine-speed (S)	2	5.68*	1.68
V x H	2	0.41	7.18*
V x S	4	0.61	1.20
H x S	2	5.14*	0.38
V x H x S	4	1.99	2.77*
Residual	36		

c * significant effects at 95% confidence limit

4.2 Stubble loss

The effect of height of cut was particularly pronounced for Ife Brown variety at low speed of 0.5 m/sec as shown in Figure 5. There was an increasing trend of stubble loss as speed was increased, up to 0.9 m/sec. The summary of statistical Analyses of Variance for stubble loss presented in Table 1 shows that crop variety and the interactions between variety and plant height, variety and forward speed, and variety speed-height of cut were all significant at 95% confidence limit. However, the Analysis of Variance for stubble loss as percent of total loss showed no significant effects at the confidence level investigated. The regression equations of best fit for the curves in Figure 5 are as presented in Table 4.

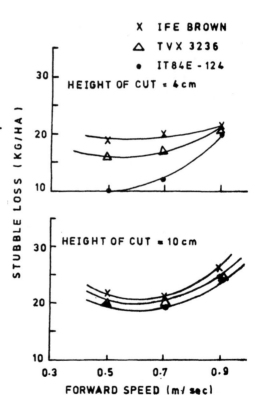

Fig. 5 Effect of machine speed on stubble loss of three cowpea varieties

Table 4. Regress models for cowpea harvesting losses (L) at the three forward speeds (S) of harvester shown in Figure 5.

Variety	Regression model	R^2
(a) Height of cut = 4 cm:		
Ife Brown,	L=16.86 EXP(0.25S)	0.50
TVX 3236	L=11.84 EXP(0.59S)	0.64
IT84E-124	L=2.34 EXP (2.30S)	0.97
(b) Height of cut = 10 cm:		
Ife Brown	L=16.12 EXP(0.50S)	0.50
TVX 3236	L=14.53 EXP(0.56S)	0.72
IT84E-124	L=13.63 EXP(0.63S)	0.74

4.3 Lodging loss

At 4 cm height of cut, the lodging loss increased linearly with increase in speed for IT84E-124 and Ife Brown but was slightly curvillinear for TVX 3236. The magnitude of the loss due to height of cut was not quite different for each speed level as shown in Figure 6. Table 1 shows that crop variety, machine forward speed and the interaction of variety with speed were significant at 95% confidence limit. Also, the Analysis of Variance for the lodging loss as percent of total loss was equally significant for crop variety, plant height, and, the variety – plant height and variety – plant height – machine speed interactions at 95% confidence limit. These effects are further illustrated in Figure 7 which shows that TVX 3236 was more susceptible to lodging at the confidence level investigated. The regression equations describing each curve in Figure 6 are presented in Table 5.

Table 5. Regression models for cowpea harvesting losses (L) at the three forward speeds (S) of harvester shown in Figure 6.

Variety	Regression model	R^2
(a) Height of cut = 4 cm:		
Ife Brown,	L=13.08 + 7.50S	0.96
TVX 3236	L=15.44 EXP(0.37S)	0.75
IT84E-124	L=13.92 + 2.50S	0.75
(b) Height of cut = 10 cm:		
Ife Brown	L=16.68 + 4.75S	0.91
TVX 3236	L=15.0 + 10.0S	0.98
IT84E-124	L=16.36 EXP(0.14S)	0.66

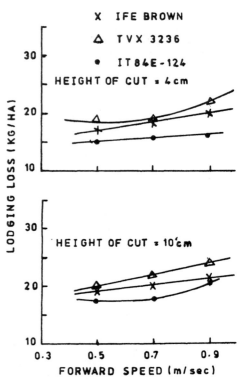

Fig. 6 Effect of machine speed on lodging loss of three cowpea varieties

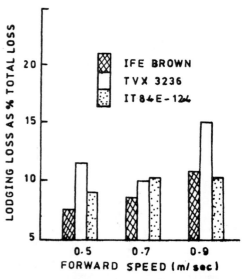

Fig. 7 Histogram of percent lodging loss versus machine speed for three cowpea varieties (Height of cut = 4 cm)

5 CONCLUSIONS

Effects of crop variety and the forward speeds of cowpea harvester were in varying magnitude for the three cowpea varieties under the conditions investigated. Observations have shown mechanical harvesting to be more expedient than manual harvesting when a large area is to be harvested. This investigation was conducted for two plant heights of cut and three machine forward speeds. It would therefore be instructive if greater variations in these crop and machine parameters and pod moisture content would minimize field loss. A delay in field removal of cowpea plants have shown to result in higher lodging tendencies. Lodging loss was particularly pronounced for TVX 3236. Additional field tests aimed at further improving the design and the performance of the prototype would be conducted.

6 REFERENCES

Anonymous. 1973. Annual Report: Grain Legume improvement programme. International Institute of Tropical Agriculture (IITA), Ibadan, Nigeria.

Anonymous. 1981. Annual Report: Mechanical harvesting of cowpea. IITA, Ibadan, Nigeria.

Daisy, E. K. 1979. Food Legumes. Tropical Products Institute. Crop and Product Digest 1 (3):89-90.

Dunn, W.R., Nave, W.R and Butler, B.J. 1973. Combine header component losses in soybean. Trans ASAE 16 (5):1032-1035.

Ichikawa, T. 1981. Performance of soybean reapers in Japan. Tropical Agriculture Research Center, JARJA9 14 (4):229-234.

Lamp, B.J., Johnson, W.H. and Harkness,K.A. 1961. Soybean harvesting losses - approaches to reduction. Trans ASAE 4(2): 203-205, 207.

Navasero, N.C. and Ajirenike, S.A. 1986. A modified lawn mower for cutting and windrowing cowpea and interrow weeding. Paper presented at the Annual Conference of the Nigerian Society of Agricultural Engineers, Ile-Ife, Nigeria.

Nave, W. R. and Yoerger, R. R. 1975. Use of air-jet guards to reduce soybean harvesting losses. Trans ASAE 18 (4): 626-629.

Nave, W. R. and Hoag, D. L. 1975. Relationship of sickle guard spacing and sickle frequency to soybean shatter loss. Trans ASAE 18 (4): 630-632, 637.

Quick, G. R. 1973. Laboratory Analysis of the combine header. Trans ASAE 16(1): 5-12.

Land and Water Use, Dodd & Grace (eds), © 1989 Balkema, Rotterdam. ISBN 90 6191 980 0

Future developments in stripper harvesting

J.S.Price
AFRC Institute of Engineering Research, Silsoe, Bedford, UK

ABSTRACT: In-situ stripping of crops is a very old concept. The Gallic Vallus (70AD) was a crude stationary comb stripper propelled by bullocks. The last century has seen many stripper developments but two problems remained - high grain losses and an inability to harvest laid crop.

The AFRC Engineering stripper uses flexible comb elements mounted on a rotor spinning on a horizontal axis, its upper half enclosed by a movable hood. In contrast to earlier machines, laid crops can be stripped and grain losses are low. Full patents are held by the National Research and Development Corporation's British Technology Group.

A retrofit stripper header can be attached to a combine in place of the conventional header. In cereal crops the reduction in MOG. (material other than grain) passing through the combine is approximately 75% in standing crops, and approximately 50% in badly laid crops. Over half the crop is threshed at the point of stripping. Combine grain throughput, at a given loss level, is increased by 50 - 100%.

The 1988 harvest saw the first commercial stripper headers in use on farms in Great Britain and abroad, manufactured by patent licencee Shelbourne Reynolds Engineering Ltd. Work at AFRC Engineering continues on further development of the stripping process and the design of new stripper-harvesters which will fully exploit the advantages of the stripping technique.

Développements dans l'égrenage

SOMMAIRE: L'égrenage des céréales au champ est un très vieux concept. Le Vallus gaulois (70 AD) était une égreneuse rudimentaire à peigne immobile poussée par des boeufs. Pendant le siècle passé il y a eu de nombreux développements dans l'égrenage mais il reste toujours deux problèmes - les pertes de grains importantes et l'impossibilité de récolter les céréales versées.

L'égreneuse développée à AFRC Engineering utilise des élements de peigne flexibles montés sur un rotor qui tourne sur un axe horizontal et dont la moitié supérieure est enfermée dans un boîtier mobile. Par contraste avec les égreneuses précédentes, cette machine est capable de récolter les céréales versées et les pertes de grains sont basses. Les brevets complets sont détenus par le British Technology Group de la National Research and Development Corporation.

Un bec égreneur rétromonté peut être installé dans une moissonneuse-batteuse à la place de la table de coupe classique. Dans les céréales on voit une réduction de 75% environ dans le volume de matériel autre que grains dans le batteur pour les céréales sur pied et de 50% dansles cultures versées. Plus de la motié de la récolte est battue au point d'égrenage. Le débit de grains dans la moissonneuse-batteuse, à un niveau de pertes donné, augmente par 50-100%.

Les premiers becs égreneurs commerciaux ont été utilisés dans des exploitations en Grande Bretagne et à l'étranger pendant la moisson de 1988. Ils sont fabriqués par le licencié du brevet, Shelbourne Reynolds Engineering Ltd. Le travail à AFRC Engineering continue avec le développement avancé des processus de l'égrenage et la conception de nouvelles égreneuses qui exploiteront en pleine mesure les avantages des techniques de l'égrenage.

Weitere Entwicklungen im Pflück-Drusch

ZUSAMMENFASSUNG: Das Pflücken von Getreide im Felde ist ein sehr alter Begriff. Das gallische Vallus (70 AD) war ein primitiver, unbeweglicher, ochsenbetriebener Kammpflücker. Im letzten Jahrhuyndert hat sich das Getreidepflücken weitgehend entwickelt, aber es bestehen noch zwei Probleme - hohe Körnerverluste und die Unfähigkeit, Lagergetreide zu ernten.

Der vom AFRC Engineering entwickelte Pflück-Drescher, deren obere Hälfte von einer beweglichen Haube eingeschlossen ist, besteht aus flexible, um eine horizontale Achse drehende Kammelemente. Im Gegensatz zu frü heren Maschinen, kann mit diesem Pflücker Lagergetreidemit kleinen Körnerverlusten geerntet werden. Volle Patente sind im Besitz der British Techology Group der National Research and Development Corporation.

Ein nachgerüsteter Pflückvorsatz kann anstatt des herkömmlichen Ährenköpfers in einen Mähdrescher eingebaut werden. In Getreidekulturen wird der Strohanteil im Drescher von ungefär 75% in stehenden Kulturen und 50% in Lagergetreiden verringert. Mehr als die Hälfte der Getreide wird an der Pflückstelle gedroschen. Der Getreidedurchsatz des Mähdreschers zu einem gegebenen Verlustniveau wird um 50 - 100% erhöht.

Während der 1988 er Ernte wurden die ersten kommerziellen, von Patentlizenzinhaber Shelbourne Reynolds Engineering Ltd hergestellten Pflückvorsätze in Betrieben in Grossbritannien und im Ausland benutzt. Bei AFRC Engineering wird die Arbeit mit der weiteren Entwicklung des Pflückverfahrens und der Konstruktion neuer Pflück-Drescher, die die Vorteile der Pflücktechnik völlig ausnützen, fortgesetzt.

1. INTRODUCTION

A harvesting technique in which only the most valuable parts of the crop are harvested is potentially fast and efficient. Some of the earliest attempts to mechanise harvesting involved simple apparatus to detach grain from standing straw. The AFRC Engineering stripper has succeeded where previous machines have failed. Grain losses are comparable with, and often lower than those of a cutterbar and its ability to recover lodged crops is good. The implications for increased harvesting efficiency become clear when two characteristics of the strippers' performance are considered; (a) intake of M.O.G. (material other than grain) is normally reduced to between 25 and 50% compared with a cutterbar, and (b) about 60% of the crop is threshed at the point of stripping.

2. HISTORIAL DEVELOPMENT OF STRIPPING

The Gallic Vallus (70 AD) was a simple wooden container with a forward projecting comb, mounted on wheels and pushed into the crop by bullocks. Stripped heads were raked back into the container by an attendant.

In the 19th Century, the stripping principle was developed further. In 1843, Australian John Bull patented a powered beater to move stripped heads away from the comb. Later, strippers using Bulls' principles were sold and used commercially.

The Australian comb stripper was equipped with a cutterbar and crop control rotor in 1913 (Taylor)[1]. This enabled heads to be cut from straw that was held by the rotor and comb. The Australian machine had evolved into a high cutting header, able to harvest standing and slightly lodged crops with a minimum of MOG.

The 20th Century saw the development of a different class of stripper, that with 'active' stripping elements usually attached to a rotor spinning on a horizontal axis. In the USA, Baldwin's 'Standing Grain Thresher' (1910)[1], employed a horizontal stripping rotor. Crop movement over the rotor was aided by an air blast which was subsequently replaced by a moving canvas in 1916. Britain saw the introduction, in the 1940's, of a stripper consisting of a number of parallel stripping plates, approximately 500 mm in diameter, mounted on a horizontal spindle. The Wild Harvest Thresher had little success, being effective only in standing crops.

Both the Baldwin and Wild machines were ultimately defeated by the two problems that have frustrated all of the pioneers of stripping. These are high grain shatter losses and an inability to recover severely lodged crops, common in wet European climates.

The AFRC Engineering stripper belongs to the second class of strippers. It consists of flexible arrow head elements mounted on a horizontal rotor of 540 mm diameter and rotating at 600 to 800 rev/min, giving a peripheral speed of 17 to 22.7 m/sec. The stripping element shape combines comb-like projections to penetrate the crop with a wide gap, or relief aperture between each individual element projection to ensure that stripped straw is released. The flexibility and smoothness of the polyurethane elements results in a gentle action which reduces straw intake.

3 DEVELOPMENT OF THE AFRC ENGINEERING STRIPPER

Initial investigations into stripping began at AFRC Engineering in 1984. Arrow-head shaped stripping elements of polypropylene were used. The stripping rotor was driven in the 'overshot' mode, i.e. so that the stripping elements passed forward and upward through the crop. The upper half of the rotor was enclosed by an adjustable hood. Two experimental rigs, a tractor mounted field rig 1.5 m wide, and a 300 mm wide laboratory rig were constructed. Crop cut by a binder was stored for winter use on the laboratory rig. It was clamped between wooden boards and passed through the rig by means of a variable speed carriage. With this rig, shown in Fig.1, close analysis of the stripping process was possible using high speed ciné photography.

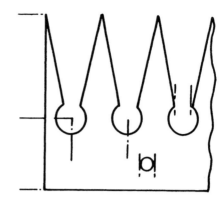

Fig.2. Design of the stripping elements

aperture, shown in Fig.2, emerged as the most effective and gave the lowest grain losses. Later elements were moulded from softer and more flexible polyurethane which had two advantages: reduced straw intake and reduced risk of breakage.

World patents[3] on the design of the stripping rotor and stripping elements are held by the National Research and Development Corporation's British Technology Group.

For the 1985 season, a 3.6 m wide stripper header was constructed, based on the shell of a standard combine pick-up header. The stripping rotor and hood were mounted ahead of a standard centre delivery auger. The stripper header was quickly interchangeable with the standard combine cutterbar header, allowing easy comparison between cutting and stripping techniques in the same crop conditions. Fig.3. shows a diagram of the combine mounted stripper header.

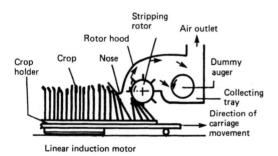

Fig.1. Cross section of the laboratory stripping rig.

The tractor mounted rig consisted of a stripping rotor with a crop collecting box behind. It allowed investigation of the stripping process in real crop and field conditions.

As a result of the early laboratory and field experiments[2] a stripping rotor of 540 mm diameter, with 8 rows of stripping elements. was selected. A range of stripping element shapes were tested, cut from 9 mm thick polypropylene sheet. An arrow-head element with a round relief

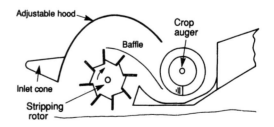

Fig.3. Cross section of the grain stripping header

4 GRAIN LOSS MEASUREMENTS

Early experiments with the 1.5 m wide field rig, had indicated that grain losses were comparable with those of a cutterbar and the basic machine settings were established. During 1985 and 1986 a considerable number of header loss measurements[4,5] were made in a range of crops, including cereals, linseed, peas and beans. Cutterbar loss measurements were also made. Performance of the stripper header was particularly good in lodged barley and in linseed. Table 1 shows the minimum header losses of the stripper and cutterbar headers in cereals in 1986.

Table 1. Minimum grain losses (kg/ha) from cutterbar and stripper headers in 1986.

Crop	Stripping	Cutting
Winter barley	72	390
Spring barley	100	228
Winter wheat	32	12
Spring wheat	38	5
Winter oats	96	45

Extensive header loss measurements enabled the effects of the five settings of the stripper to be investigated:

(1) Vertical hood height was shown to be the principal machine setting affecting grain loss. To minimise the grain losses in a typical wheat crop it was essential that the lower edge of the hood was maintained 100 to 150 mm below the crop height. In crops of varying height, the hood position needed frequent readjustment by the operator. Fig.4 illustrates the effect of hood height on grain loss in a crop of wheat.

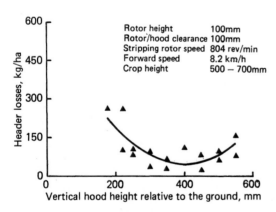

Fig.4. The effect of hood height on grain losses in wheat.

(2) Forward speed also had a significant effect on grain loss. Below a speed of around 8 km/h, losses rose quickly with decreasing speed. Above 8 km/h grain losses continued to fall slightly.

(3) Rotor speeds could have a significant effect on performance. At very low speeds some grains remained unstripped, while very high speeds increased MOG intake. In most cereal crops a speed of 700 rev/min was satisfactory.

(4) Rotor height did not have a major effect on grain losses. In standing wheat a height of 100 to 150 mm was satisfactory, however, lodged crops required the rotor to be set lower.

(5) The horizontal clearance between the rotor and hood had the least effect on grain losses. A clearance of around 100 mm was satisfactory.

5.0 COMBINE PERFORMANCE INCREASES

Grain separation efficiency and therefore combine harvester workrate is directly proportional to MOG throughput rate. The use of a stripper header with its low MOG intake and a high proportion of threshing at the point of stripping, can therefore be expected to increase workrate considerably. Comparative tests[5,6] were undertaken in 1986 and 1987, exchanging cutterbar and stripper headers on a single combine. A minimum of 3 runs with each header, at increasing rates of grain throughput, were carried out. On each 45 m run, grain throughput was measured and the effluxe from the combine sieves and straw walkers collected on sheets. The grain separation losses they contained were quantified using a stationary rethresher[7]. No modifications to the combine were necessary when changing to the stripper header, except to reduce the concave clearance to a narrower setting.

With a straw walker combine, the grain throughput was increased by as much as 91% in winter barley, at a separation loss of 100 kg/ha. A typical result in wheat, giving an increase of 55% is shown in Fig.5.

In 1987, similar comparative tests were done using a combine with multi-cylinder separation. The more aggressive rotary separation was appropriate for the material taken in by the stripper which is predominantly flag leaf. Measured performance increases were therefore higher than those of the straw walker combine. Fig.6 shows the result of a comparative test in spring wheat, where the performance increase at 100 kg/ha separation loss, was 74%.

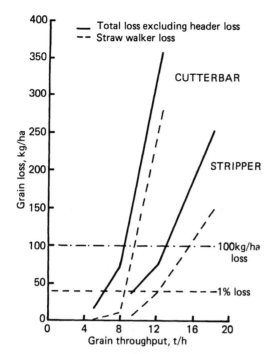

Fig.5. Performance comparison for a
conventional combine equipped
with cutterbar and stripper headers
in spring wheat.

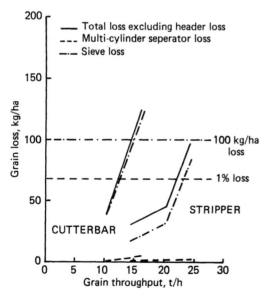

Fig.6. Performance comparison for a multi-
cylinder separation combine equipped
with cutterbar and stripper headers
in spring wheat.

6.0 COMMERCIAL MANUFACTURE OF THE STRIPPER HEADER

The performance advantages of the stripper
header became widely recognised as results
from the development programme were
released. A British company, Shelbourne
Reynolds Engineering Ltd., was licenced by
the British Technology Group to manufacture
the stripper header. The Shelbourne
Reynolds' machine is basically similar to
the AFRC experimental machine, but has
detail changes in hood design and includes a
short conveyor between the stripping rotor
and the crop delivery auger. The 'retrofit'
header can be fitted to all leading makes of
combine.

Twenty-four stripper headers were
manufactured and sold for the 1988 season in
widths up to 6m.

7.0 CONTINUING RESEARCH AND DEVELOPMENT

Research and development on stripping
continues at AFRC Engineering to further
refine and improve the technology. The
design of the rotor and the assessment of
new stripping elements is being carried out
and work is being done on sensing grain
losses and to add automatic control to the
important hood height setting. Work is also
being done to adapt the machine to crops
such as beans, in which losses are at
present too high. Development is in
progress to strip dry rice crops. The
implications of stripper harvesting on the
management of cultivations and cropping are
being examined. Field trials are comparing
the breakdown of incorporated stripped straw
and incorporated chopped straw. The
widespread use of strippers may change the
farmers cropping decisions. Certain crops,
previously unattractive because of
difficulties in harvesting, (for example
linseed), become more attractive as they can
be stripper harvested with ease.

Alongside the continuing development of
the stripping process, work has begun to
develop a new generation of stripper
harvesters. These will be designed
to fully exploit all the advantages of the
stripping technique and produce cheaper,
lighter and more efficient harvesters for
the future.

8.0 A NEW GENERATION OF STRIPPER HARVESTERS

The potential for improved harvesting
efficiency comes from two basic
characteristics of the stripper's
performance.

(a) Low intake of MOG, rarely over 50%
that taken by a cutterbar. In certain crops
in early season, it can be as low as 10%.

(b) A large proportion of the crop is threshed at the point of stripping. Field tests in 1988 showed results from 52 to 73% in wheat, and 62 to 86% in barley.

When a stripper header is attached to a combine these advantages are only partly exploited. They can be fully exploited in a new design of harvester.

Reduced MOG throughput improves the final separation efficiency, but all combine separation systems are designed for the type of MOG taken in by the cutterbar. A separation system designed for MOG which is predominantly flag leaf with little stiff straw should, with development, have increased efficiency.

Grain threshed by the stripping rotor enters the combine and much of it is quickly separated in the forward part of the concave. If, however, this free grain were removed immediately after the stripping rotor, then the forward section of the concave would be free to allow separation of ears threshed by the drum. Also any fears of free grains being damaged or passing through to the straw walkers, would be removed. This early separation or 'pre-separation' is seen as a vital part of a stripper harvester of the future.

Stripper harvester development at AFRC Engineering is approached by looking, in turn, at each of the stages required for the complete machine, and attempting to optimise each stage. Experimental rigs have been built to look at pre-separation and the final grain/straw separation.

Pre-separation is achieved using a simple separator which is being patented[8]. A perforated sheet steel sieve replaces the trough in which the conventional crop auger rotates. At this stage of development between 40 and 50% of the total grain throughput can be removed, (Fig.7).

Final separation will be by a rotary separator. The rig will allow optimisation of speeds, sieve types etc.

Threshing and cleaning stages of the machine are likely to be very similar to those on existing machines, but development work may show that detail changes can be made.

A 2 m wide experimental rig is also in use to aid development. This comprises stripping, pre-separation, threshing and final grain/straw separation stages but at present has no cleaning. None of these are fully optimised but the rig has a grain throughput of over 10 t/h at 100 kg/ha separation loss. This indicates the enormous potential for a fully optimised new generation stripper harvester.

The work described in Section 7.0 to improve and adapt the stripping process, will further increase the capabilities of the stripper harvester and its potential economic benefits to the farmer.

REFERENCES

1. QUICK, G.R.; BUCHELE, W.F. The Grain Harvesters. A.S.A.E.

2. KLINNER, W.E.; NEALE, M.A.; HOBSON, R.N.; GEIKIE, A. Feasibility assessment of an in-situ grain and seed stripping system. Divisional Note DN 1315, natl Inst. agric. Engng, Silsoe.

3. UK PATENT APPLICATIONS. 8424395, 8608585.

4. KLINNER, W.E.; NEALE, M.A.; ARNOLD, R.E.; GEIKIE, A.A.; HOBSON, R.N. Development and first evaluations of an experimental grain stripping header for combine-harvesters. Divisional Note DN 1316, natl Inst. agric. Engng, Silsoe.

5. HOBSON, R.N.; PRICE, J.S.; TUCK, C.R.; NEALE, M.A.; HALE, O.D. Evaluation and development of the grain stripping system – 1986. Divisional Note DN 1443, AFRC Institute of Engineering Research, Silsoe.

6. PRICE, J.S. Performance comparison of a combine with multi cylinder separation using cutterbar and stripper headers in 1987. Divisional Note DN 1449, AFRC Institute of Engineering Research, Silsoe.

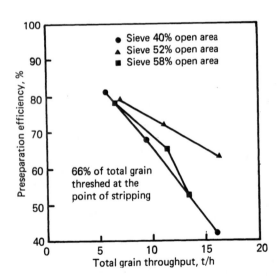

Fig.7. The effect of total grain throughput rate on pre-separation efficiency using three different 12 mm round hole sieves in wheat.

7. BURGESS, L.R.; HALE, O.D. The construction of a small rethresher for measuring combine-harvester grain losses. Divisional Note DN 1455, AFRC Institute of Engineering Research, Silsoe.

8. UK Patent Application 8801740.

Land and Water Use, Dodd & Grace (eds), © 1989 Balkema, Rotterdam. ISBN 90 6191 980 0

Usefulness of different representations of the material in the pressing chamber of a baler to dynamic analysis of a pressing system

J.Uziak
Academy of Agriculture, Lublin, Poland

ABSTRACT: The analysis of the behaviour of the pressed material in the pressing chamber of balers is presented. Four different representations of the reaction of the pressed material on the piston of the pressing system is described. The model of the pressing system and the motion equation of this system is presented. The equation of motion is solved for different working conditions. For each case of working conditions and formula describing force acting on the piston the coefficient of speed fluctuation of the system is calculated.

RÉSUMÉ: On a présenté l'analyse du comportement de produits en tiges dans la presse à haute densité. On a donné quatre modes différents de la description de l'influence du matérial pressé sur le piston du système pressant. On a presenté le modèle du système pressant et l'équation du mouvement de ce système. L'équation du mouvement à été resolue pour les conditions différentes de travail et pour les modes différents de description de la force agissante sur le piston. On a calculé le coefficient de l'irrégularité de mouvement.

ZUSAMMENFASSUNG: In dieser Arbeit wurde die Analyse des Verhaltens von dem gepressten Material in einer Presskammer der Hochdruckpresse dargestellt. Es wurden drei verschiedene Beschreibungsmethoden der Wirkung des gepressten Materials auf den Kolben des Presssystems angegeben. Es wurden ein Modell des Presssystems und die Bewegungsgleich-ung dieses Systems dargestellt. Die Bewegungsgleichung wurde für verschiedene Arbeitsbedingungen und fur verschiedene Beschreibungsmetho-den der Kraftwirkung auf den Kolben gelost. Es wurde auch der Laufunruhe-koeffizient berechnet.

1 INTRODUCTION

Balers have been used for collecting post-combined straw and hay for years. The improvement in the construction of these machines is conected with calculations of the load applied to the machine elements. The pressing system is one of the fundamental mechanisms used in balers. It is subjected to the highest load of a high variability. It is also characterized by a high energy consumption up to 60% of the power applied to the baler. In order to give some ideas regarding the design of the baler it is neccessary to determine the performance of the pressing system. The knowledge of the forces acting on this mechanism is essential to describe its motion. Besides the forces of gravity, the most important force in this system is the force coming from the pressed material. It is also the most difficult quantity to describe. Despite the fact that this problem was investigated many times and that several formulae describing the material in the pressing chamber are available, it is still difficult to decide which expression should be used in the specific problem. These expressions were derived by analytical methods or obtained experimentally. They differ from one another in the mathematical form, in complexity, and in the number of experimental coefficients used. Therefore, the problem arises which of the expressions should be choosen in order to perform calculations.

In this paper four different expressions describing the force coming from the pressed material have been used. For these four expressions a dynamic analysis of the pressing system have been performed. In this analysis for different working conditions the coefficient of speed fluctuation have been determined.

2 THE EQUATION OF MOTION OF THE PRESSING SYSTEM

The typical pressing system used in balers is the classical slider-crank mechanism (Fig. 1).

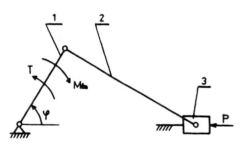

Fig.1 A physical model of the pressing system.

The equation of motion of this mechanism has the following form:

$$J(\varphi)\frac{d^2\varphi}{dt^2} + \frac{1}{2}\frac{dJ}{d\varphi}(\varphi)(\frac{d\varphi}{dt})^2 = T+M_b \quad (1)$$

where φ is the angle of the crank rotation, $J(\varphi)$ is the total moment of inertia of the system reduced to the crank (depending on the masses of links, their lengths, and on the moment of inertia of the flywheel), T and M are the driving and braking torque, respectively, also reduced to the crank.

The braking torque is given by the expression:

$$M_b(\varphi)= \frac{1}{2}l_1(G_1+ G_2)\cos\varphi + Pl_1(\sin\varphi+$$

$$+ \frac{\lambda \ \sin2\varphi}{2\sqrt{1-\lambda^2\sin^2\varphi}} \ ) \quad (2)$$

where G_i is the link weight (i=1,2), l_i is its length (i=1,2),

$$\lambda = \frac{l_1}{l_2} \ ,$$

and P is the force acting on the ram coming from the pressed material.

The equation of motion (1) has been solved using the method introduced by Uziak (1987). This method requires the use of a digital computer and allows a multiple solution of the equation of motion. The angular velocity $\omega = \dot\varphi$ has been assumed as the unknown value in the equation. This quantity has been calculated for the full rotation of the crank from 0 to 2π rad. in the steady motion. In each cycle extremal values and the mean value of angular velocity have been determined. Thus, the coefficient of speed fluctuation can be calculated. The coefficient of speed fluctuation is defined as:

$$C_s = \frac{\omega_{max} - \omega_{min}}{\omega_o} \quad (3)$$

where ω_{max} and ω_{min} are maximal and minimal angular velocity during the working cycle, and ω_o is the mean angular velocity of the crank.

The method introduced by Uziak (1987) can be used on assuming that the driving torque T is constant and that the braking torque M does not depend on the angular velocity ω, but is only a function of the angle of crank rotation φ.

3 ANALYSIS OF THE FORCE ACTING ON THE RAM

The force P in formula (2) results from the reaction between the compressed material on the ram. This reaction depends on the parameters of the material and the type of the chamber in which pressing takes place. In balers, pressing takes place in an open chamber. In such a chamber, each portion of the material fed through a side inlet is further pressed by the ram to the portions previously pressed. The pressure increases with the growing density of the material until the displacement resistance of the whole block of the material in the chamber is overpowered.

The force P reaches relatively high values. It gives in turn a high load applied to the pressing system. Moreover, this load is also characterized by high variability in its values. It is due to the fact that the reaction between the material and ram does not last over the whole working cycle. The pressing begins when the crank takes the position determined by the angle φ_p:

$$\varphi_p = 2\pi - \arccos \frac{1_1^2 - 1_2^2 + (1_1 + 1_2 - h)^2}{21_1 (1_1 + 1_2 - h)} \quad (4)$$

where h is the initial width of the portion of the pressed material which depends on the mass of this portion (G), the initial density of the material γ_o, and

the cross-section of the pressing chamber (F):

$$h = \frac{G}{\gamma_o F} \quad (5)$$

The process of pressing ends when the ram takes the dead position, i.e. when the angle of the crank position is equal 2π.

The value of force P acting on the ram can be expressed as:

$$P = p F \quad (6)$$

where F is the cross-section area of the pressing chamber, p is the unit pressure on the ram.

4 RELATIONSHIP BETWEEN PRESSURE AND COMPRESSED DENSITY

The behaviour of the agricultural materials during pressing is believed to depend on their physical and biochemical properties. Owing to the fact that the compression of agricultural materials is a fundamental part of such processing operations as baling, wafering, and briquetting, several researches were carried out in order to understand the mechanism of compression. This mechanism was investigated by using dimensional analysis (Rehkugler and Buchele (1969), Abd-Elrahim, Huzayyin and Taha (1981)), mathematical models (Vinogradev and Dimitriev (1969)), and rheological models (Mohsenin and Zaske (1976), Osobov (1967)).

The design of the compression device requires the knowledge of the pressure needed to obtain the desired compressed density. This necessitates a relation to predict the pressure in terms of the compressed density. Useful empirical relations have been obtained in this regard (Osobov (1967), Mewes (1958) and (1959), O'Dogherty, Wheeler (1984)) and some of the theoretical analyses are of particular interest (Osobov (1967), Mewes (1958) and (1959).

Skalweit (1938) proposed the following relation between pressure (p) and the current density of material (γ):

$$p = c \, \gamma^m \qquad (7)$$

where c and m are the empirical constants. Mewes (1958) modified expression (7) to obtain:

$$p = b \, (\gamma - \gamma_o)^n \qquad (8)$$

where b and n are, again, the empirical constants and γ and γ_o are the current and initial densities, respectively. The analyses of Mewes were limited to a low pressure (< 1 MPa).

Osobov (1967) assumed that the pressure rise on any section of the material during compression depends on the degree of compression, which he further assumed to be a linear function of the compression pressure. He finally obtained an expression of the form:

$$p = \frac{C}{a}\left\{\exp\left[a\,(\gamma - \gamma_o)\right] - 1\right\} \qquad (9)$$

where C and a are experimental constants, γ and γ_o are densities.

Experimental investigations carried out by Chrapacz (1958) resulted in the following expression:

$$p = 1.92 \cdot 10^{-5} \, \gamma^{2.178} \, \alpha \, \beta \, K \qquad (10)$$

where,

γ – the current density of the material,

$\alpha = 1 - 0.02(W - 15.3)$ the coefficient depending on the moisture content W [%],

$\beta = (0.084 \, v)^{2.75}$ the coefficient depending on the velocity of of the compression,

K – the dimensionless coefficient depending on the state of the material.

5 THE CALCULATIONS OF THE SPEED FLUCTUATION COEFFICIENT FOR DIFFERENT REPRESENTATIONS OF COMPRESSED MATERIAL

Expressions (7),(8),(9), and (10) have been used to calculate force P acting on the ram accordingly to the formula (6). The current density of the material has been determined comparing the mass of the material in the pressing chamber before and after compression. This leads to the formula:

$$\gamma = \gamma_o \, \frac{h}{h-x} \qquad (11)$$

where h is the initial width of the portion of the material (expressed by (5)), x is the position of the ram measured from the origin of the pressing chamber, γ and γ_o are current and initial densities, respectively. The position of the ram x depends on the geometrical parameters of the mechanism and the angle of crank's rotation:

$$x = l_1(\cos\varphi - 1) + l_2(\sqrt{1 - \lambda^2 \sin^2\varphi} - 1) + h_k \qquad (12)$$

where h_k is the width of the material corresponded to the final density of material γ_k, i.e.

$$h_k = \frac{G}{\gamma_k \, F} \qquad (13)$$

In order to analyse the usefulness of the pressure-density relations to investigate the dynamics of the pressing system, the speed fluctuation coefficient for each relation has been calculated. The calculations concern the coefficient of speed fluctuation (defined by formula (3)), because this quantity in a simple and compact form, by using only one number, characterized the motion of the mechanism. It creates the basis for a simple and convenient comparison between the results obtained by using different formulae discribing the pressure-density relation.

The coefficients of speed fluctuation were calculated by using the method presented in paper (Uziak (1987)). The pressed material was post-combined wheat straw, and calculations were done for a baler Z-225 (produced by "Agromet" in Lublin, Poland).

The calculations were performed for the following combinations of the material parameters:

1. moisture content (W) of the pressed material: 15%, 20%, and 25%;

2. mass (G) of the pressed material: 2 kg, 2.5 kg, and 3 kg;

3. initial density ($_p$) of the material: 35 kg/m^3, 40 kg/m^3, and 45 kg/m^3;

4. final density ($_k$) of the material: 105 kg/m^3, 115 kg/m^3, and 125 kg/m^3.

The constant coefficients needed for calculations of the values of the force acting on the ram (expressions (7)-(10)) have been assumed on the basis of the papers (Alfierov (1957), Busse (1963) and (1967), Matthies and Busse (1966), Zelcerman (1958)).

The results show that the value of the coefficient of speed fluctuation is practically independent of the form of the pressure-density relation. The maximal difference of the calculated coefficient is 0.018 for the same parameters of the material. However, a regularity should be noticed that, independently of the parameters of the material, the coefficient of the speed fluctuation is the biggest for the formula given by Skalweit, and the smallest for the formula given by Osobov. Still the differences are so small that for calculations of the dynamics any of the four formulae can be used. Because of the simplest mathematical form the formula given in 1938 by Skalwei is recommended. The formula given by Mewes, which in fact is the modified Skalwiet's formula, can

also be useful. An additional argument for these two expressions is the fact that the constant coefficients used in them are well tabulated.

6 CONLCUSIONS

The effect of different formulae describing the pressure-density relation on the coefficient of speed fluctuation has been examined. It has been found that the form of these formulae minimally affects the value of the coefficient. The biggest value of the coefficient of speed fluctuation occurs for the formula given by Skalweit, the smallest value occurs for the formula given by Osobov. Because the value of the examined coefficient is practically independent of the form of the pressure-density relation, it is recommended to use a formula of the simplest mathematical form, i.e. formulae given by Skalweit and by Mewes.

REFERENCES

Abd-Elrahim, Y.M.; A.S. Huzayyin, & I.S. Taha 1981. Dimensional analysis and wafering of cotton stalk. Trans. ASAE 24(4): 829-832.

Alfierov, S.A. 1957. Zakonomiernosti pri szatji solomy. Sel'Khozmash. 3: 6-10.

Busse, W. 1963. Untersuchungeg uber das brikettieren von halmgut. Grundlagen der Landtechnik 18: 50-57.

Busse,W. 1967. Die theorie auf dem gebiet des verdichtens landwirscheftlicher halmgut. Landtechnische Forschung 14: 6-15.

Chrapacz, E.I. 1958. Rasczot na procznost priessovalnoj kamiery siennych priessov. Trakt. Sel'kozmash. 1: 19-22.

Matthies, H.J. & W.Busse 1966. Neue erkentnisse auf dem gebiete des verdichtens von halmgut mit hohen normaldruck. Grundlagen der Landtechnik 3: 10-16.

Mewes, E. 1958. Zum verhalten von pressgutern in prestopfen. Landtechnische Forschung 8(6): 154-164.

Mewes, E. 1959. Berechung der druckverteilung an stroh-und heu pressen. Landtechnische Forschung 9(6): 160-170.

Mohsenin, N. & J. Zaske 1976. Stress relaxation and energy requirement in compaction of unconsolidated materials. J. of Agric.Engng.Res. 21: 193-205.

Osobov, V.I. 1967. Theoretical principles of compressing fibrous plant material. Trudy Viskhom, 55: 221-265.

O'Dogherty, M.J. & J.A. Wheeler 1984. Compression of straw to high densities in closed cylindrical dies. J. of Agric.Engn.Res. 29: 61-72.

Rehkugler, G.E. & W.F. Buchele 1969. Biomechanics of forage wafering. Trans. ASAE 12(1): 1-8.

Skalweit, H. 1938. Krafte und beanspruchungen in strohpressen. In: 4 Konsruckteurkursus. RKTL-Schriften (88). Berlin: Beuth-verreib: 30-35.

Uziak, J. 1987. Determination of speed fluctuation coefficient of a baler. ASAE International Winter Meeting Paper No. 87-1547.

Vinogradov V.I. & G.N. Dimitriev Modelling of the process of compressing straw. Zeml. Mekh. 12 Akad. Sel'khozmashini Nauk 14(2): 62-74.

Land and Water Use, Dodd & Grace (eds), © 1989 Balkema, Rotterdam. ISBN 90 6191 980 0

Apple firmness and sensory quality using contact acoustic transmission

H.A.Affeldt Jr. & J.A.Abbott
USDA Beltsville Agricultural Research Center, Beltsville, Md., USA

ABSTRACT: A rapid, nondestructive method for evaluating selected sensory attributes of apples was developed employing sonic transmission. 'Red Delicious' variety apples from packinghouses in four states across the U.S. were tested. Sonic measurements were correlated with state of origin, Magness-Taylor firmness, soluble solids (sugar content), titratable acidity and maturity indices generated from a panel of trained inspectors. Correlations were highest between sonic resonance and either Magness-Taylor firmness or maturity indices. However, variations in transmission profiles resulted from measurement geometry and coupling media. Various materials were tested for transmissibility and the optimum was chosen for development.

1 INTRODUCTION

Apple production in the U.S. exceeded 192 billion bushels in 1988 with an estimated value of $686 million (Int'l Apple Inst. 1988). Nearly 40% of these apples are of the 'Delicious' variety. Effective worldwide marketing of this fruit with minimal loss requires efficient standardized grading and consistent quality evaluation. Barring appearance, consumer acceptance of firm, fleshy fruits relies heavily on texture as a quality factor (Lott 1965).

Texture is a multifaceted term encompassing mechanical properties and sensory reactions associated with firmness, crispness, juiciness, tenderness and mealiness. Firmness is directly related to fruit development and ripening after maturity. United States standards for grading apple maturity therefore rely heavily on the textural attribute of firmness as an index of fruit quality (USDA 1978a, 1978b).

Many techniques have been developed to measure fruit texture. Most, however, are destructive in nature, inflicting cuts, bruises or other damage during testing. Examples include the Magness-Taylor firmness tester (Magness and Taylor 1925), the penetrometer (Mitchell et al. 1961), the shear press (Kramer et al. 1951), the extrusion cell (Lanza and Kramer 1967), the compression device (Mohsenin 1963) and the Instron texture profile (Abbott et al. 1984). Furthermore, these methods are time

consuming and neglect to account for variations prevalent within the whole intact apple. They do, however, provide a repeatable, specified standard by which to sample and judge a fruit lot.

A nondestructive technique for evaluating fruit texture was patented as early as 1942 when Clark and Michelson observed changes in the natural frequency of intact fruit vibration during the ripening process. Recent investigations (Abbott et al. 1968a, 1968b; Finney and Norris 1968; Finney 1970; Cooke 1970) have reported that resonances in apples are significantly influenced by changes in fruit texture. When combined with the apple's mass in the form f^2m, direct relations develop with the dynamic elastic properties of the fruit flesh (Abbott et al. 1968b). The development of these nondestructive techniques was a significant achievement, yet these methods remained time consuming, were not practical for commercial application and were limited in accuracy by the amount of information the data acquisition system could gather.

Renewed interest by the apple industry in the development of high speed, nondestructive objective measurements for use in grading and segregating fruit for storage and market has prompted this research. This paper examines a technique of this nature and the relationships which exist with sensory evaluation and other textual attributes.

2 METHODS AND TECHNIQUES

2.1 The samples

'Red Delicious' apples, grown, harvested and stored under commercial conditions, were collected in four states across the U.S.: Washington, New York, Virginia and Michigan. Experienced, Federally licensed Market Inspectors from each state supervised the sample selection with the premise that a large maturity range was desired for testing. These mixed, unspecified 'Delicious' sports from air and controlled atmosphere storages were packed in pulp trays in apple cartons and shipped express to the Beltsville Agricultural Research Center. Cartons were received January 11-15, 1988.

Individual apples were numbered sequentially as received, with an identifying digit for the supplying state. Each apple was weighed to the nearest gram. Fruit were then stored in air up to 11 days at 0° C and 95% relative humidity until all cartons arrived.

2.2 Intact fruit sonic tests

Apples were allowed to equilibrate to room temperature (22° C) for sonic resonance tests. Whole, intact apples were individually mounted on an MB Electronics[*] Model EA 1500 electromagnetic vibrator (frequency range 5-10,000 Hz, force vector 50 lb.; MB Electronics, New Haven, CT) by placing them horizontally onto a coupling medium resting on a vibration pedestal. A variety of coupling media and mounting configurations for those media were tested. The amplitude and frequency of the first longitudinal mode of vibration (Af_{n1}, f_{n1}) and the first flexural mode of vibration (Af_{n2}, f_{n2}) were recorded for each mounting configuration and compared to that developed with floral clay (Vogue Tip Floral Clay; Beagle Manuf. Co. Inc., El Monte, CA) on a plexiglass pedestal. Floral clay has been used by previous

researchers (Finney 1970; Finney et al. 1978) because it provides a nearly flat transmission spectrum with minimum attenuation. A medium which approximates the floral clay characteristic without requiring the adhesion that floral clay exhibits was then selected to facilitate higher speed applications.

Nondestructive sonic resonance tests were conducted on each intact apple prior to sensory evaluations. Sonic resonance techniques were similar to that of Finney and Norris (1968) and Finney (1970). For these laboratory tests, apples were anchored horizontally on the vibrator with floral clay. A time-compressed electronic signal, equivalent to a 5-2000 Hz sinusoidal scan, was fed to an Electrodyne Model N100 power amplifier (250 va, frequency range 5-20,000 Hz; Electrodyne Inc., Alexandria, VA) for the vibrator's electromagnet. The apple vibration from the resulting acoustic excitation was detected directly opposite the accelerator with an Endevco model 2222 accelerometer having a mass of 0.5 grams and a frequency response flat from 10-5000 Hz ± 10%. Detector signals were analyzed to produce the apple's frequency response curve over the excitation range. A complete response curve could be obtained in less than 1 second. From the mass and the resonant frequencies identified, the nondestructive "stiffness coefficient", f^2m, was calculated for each apple.

2.3 Firmness and soluble solids

Tests were made with an 11 mm (7/16 in) dia Magness-Taylor probe (Magness and Taylor 1925). One of six inspectors tested all apples within a tray (20-23 fruit per tray). Inspectors rotated assignments between trays. Puncture tests were made on both the blush and blush-opposite sides of each fruit.

Juice exuded from the punctures was caught on a hand refractometer. The soluble solids was measured and recorded by an assisting inspector.

2.4 Maturity scores

Six Federally licensed, senior apple inspectors from various regions, including the four states supplying apples, convened on January 20-22. Inspectors were selected on the basis of their knowledge and experience in grading apples for maturity. Inspectors were briefed on the purpose of

[*] Mention of trade names is solely for the purpose of providing specific information and does not constitute endorsement by the U.S. Department of Agriculture over others of a similar nature not mentioned.

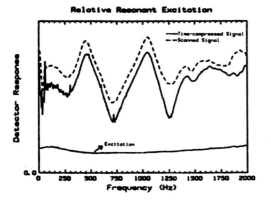

Figure 1. Excitation and frequency transmission profile through floral clay for scanned and time-compressed sonic resonance techniques on 'Red Delicious' apples.

the experiment and a review of the shipping point and market inspection instructions for judging apple maturity (USDA 1978a, 1978b) was given. Inspectors were told to judge maturity as they normally judged during inspection in accordance with the Agricultural Marketing Service (AMS) standards. Other characteristics normally scored were to be disregarded.

In a separate room, the apple lots were randomly distributed under fluorescent lights. One at a time, the inspectors examined the apples and assigned a maturity score to each. The inspectors' AMS classifications were converted to a numerical value as follows: 0, severely overripe or too injured to evaluate; 1, overripe; 2, ripe; 3, firm ripe; 4, firm; and 5, hard. Apples which were assigned a 0 by any inspector were eliminated from further analyses.

Inspectors were observed to use visual inspection, hand pressure, thumb pressure, cutting, biting and chewing to make a maturity judgement. Their manual, oral and auditory senses appeared to be the major determinants in assigning scores.

2.5 Titratable Acids

After scoring by all six inspectors, a slice of the remaining apple (about 1/4 of the intact apple) was juiced in an electric fruit juicer. A 5 ml aliquot of juice was mixed with 20 ml of water and 5 drops of hematoxylin dye. This was titrated with 0.01 NaOH until a distinct color change occurred. Volume of NaOH was recorded and percent titratable acids was calculated.

2.6 Analysis

Unequal numbers of apples were received from the four states. Apples were randomly grouped into lots of about 20 within sources. Lots were completely randomized within tests.

Simple (r) and multiple (R) correlation coefficients are presented. A total of 702 apples were used for classification. Composite samples were generated by averaging 5 and 10 consecutive apple records sorted by mass or sample number. Composite samples were also generated by sorting over mean maturity class and averaging sample records which fell into each of 22 distinct classes.

3 RESULTS

3.1 Coupling media and mounting

Floral clay anchored in an aluminum form provided the greatest transmissibility of vibration energy to the apple. The frequency response through the clay was nearly flat from 0 - 2000 Hz (Fig. 1). Over 50 combinations of other vibration coupling materials were tested to find a suitable substitute (Table 1). The rigid form mounted to the pedestal to hold the sample and the material placed in the form were varied. Variations in the rigid form material had a significant effect on the frequency transmission. A plexiglass pedestal or concave wooden semi-sphere minimized amplitude attenuation and shifts in detected resonant frequencies.

Floral clay, used by previous researchers in vibration, exhibits a viscous, adhesive trait which tends to stick and conform to objects at room temperature. This characteristic makes it a useful acoustic coupler at low frequency. However, this coupling method is impractical for higher speed operations where test material may be conveyed mechanically past the measurement point.

Alternative coupling media placed in the forms minimized frequency shifts when made up of fine, granular material. Commercially gleaned sand, 60/80 micron glass beads (Wilkens Instr. and Res. Inc., Walnut Creek, CA) and powdered sugar induced well defined resonance peaks, particularly f_{n2}, the first flexural frequency, which has been identified with changes in apple texture (Abbott et al. 1968b). These materials will conform to the contour of the apple when excited with a 3-5 kHz sonic

2039

Table 1. Resonant frequencies and associated attenuation detected in 'Red Delicious' apples through various acoustic coupling media.

Form	Medium	Af_{n1} (dB)	f_{n1} (Hz)	Af_{n2} (dB)	f_{n2} (Hz)
Plexiglass Pedestal					
REFERENCE MEASUREMENT::floral clay		-17.23	360	-17.94	920
	powdered sugar	-20.25	145	-27.34	835
	commercial sand	-19.59	255	-21.60	885
	60/80 glass beads	-23.23	335	-23.92	915
floral clay ring in	surgical rubber	-18.53	335	-20.02	915
Wood Semi-sphere					
	powdered sugar	-18.12	265	-21.24	895
	commercial sand	-17.73	275	-21.41	890
	60/80 glass beads	-20.61	305	-20.32	900
Plexiglass Semi-sphere					
	powdered sugar	-19.93	275	-20.56	880
	commercial sand	-19.03	275	-20.71	865
	60/80 glass beads	-21.43	265	-21.40	875

burst, providing a natural mold to each sample.

Alternatively, a floral clay ring sealed in a thin film of surgical rubber provided approximately the same efficiency as the exposed clay. This configuration cradled the apple, while the surgical rubber resisted slippage without adhesion.

Response curves from apple samples excited and measured at discrete sinusoidal frequencies between 0 and 2 kHz agreed well with those reported by Finney (1970) and Finney et. al. (1978, Fig. 1). This discrete scan typically consumed 20 minutes which allowed for computer controlled frequency changes and instrument settling at each frequency. In comparison, sample excitation with a time-compressed acoustic signal enhanced character at higher frequencies. Noise at lower frequencies (<300 Hz) generated with the time-compressed technique did not influence measurements of f_{n1} or f_{n2} due to the large signal-to-noise ratio. The time-compressed scan, requiring less than 1 second per sample, provides a justifiable alternative with commercial practicability.

Apples were rotated eight times 45° each to examine the variation in frequency response around the apple's equator directly opposite the accelerator. The first longitudinal and flexural frequencies shifted less than 15 Hz for uniform, intact specimens (Fig. 2). Variation in flexural harmonics was evident beyond f_{n2} as the specimen was rotated. Large distortions in transmission occurred where surface defects (scarring, callus development) were present. Small bruises were evident by probing in the local vicinity of the bruise for frequency distortion. A large bruise caused global modal distortion and could be detected by a measurement at almost any point around the equator. Equatorial measurements at a particular site were repeatable within ± 5 Hz providing the temperature and condition of the apple were stable.

3.2 Firmness

Ranges of apple firmness varied among states (Table 2). Washington apples were

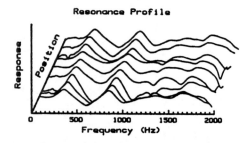

Figure 2. Profile of sonic wave transmission around the equator of a 'Red Delicious' apple. Each curve represents a 45° rotation of the fruit.

Table 2. Summary of Magness-Taylor (MT) firmness values for 'Red Delicious' apples from four states (All values reported in Newtons).

	No. Apples	Mean	Std. Dev.	Min	Max	Diff
Michigan	318	54.4	8.3	33.6	76.2	42.7
New York	171	51.3	6.2	36.4	67.8	31.4
Virginia	138	50.2	13.2	26.9	79.5	52.6
Washington	75	43.6	6.7	32.5	59.4	26.9
Total	702	51.6	9.5	26.9	79.5	

uniformly soft in comparison to Michigan apples which were generally more firm. The firmness of Virginia apples covered a broader range than other states. The largest apples by dia were provided by Washington, while the smallest arrived from Michigan. 'Red Delicious' sports were not identified by suppliers but appeared to differ between and within states.

The difference in firmness ranges may have resulted from different sports, from individual inspectors' maturity selection or from handling prior to shipment and testing. Preliminary regressions on variables separated by state differed significantly from regressions with all samples together (data not presented). Clark and Shackelford (1973) reported similar results where correlations with firmness tests differed among cultivars of peaches. Since most packinghouses will process various sports of the same cultivar in a lot, further regressions assumed all sports were lumped as one cultivar.

Magness-Taylor (MT) values for the blush side, the opposite side and the mean were regressed on the means of six inspectors' maturity scores and on the mean of the best three inspectors' maturity scores for each apple (Table 3). The three inspector mean was determined by regressing each inspector's maturity score against the mean of the remaining five. The best three were chosen to generate a mean maturity score for each apple. The r^2 values were highest for the MT means, .70 and .72, with the 6-inspector and 3-inspector means, respectively. Firmness has generally been identified as an unreliable index of maturity (Haller and Magness 1944), in agreement with our results. Therefore, both the MT mean and the three inspector mean maturity score will be reported in further analysis.

3.3 Correlations with resonant frequencies

Features of the apple frequency response curve including the amplitude and frequency of the first longitudinal resonant mode (Af_{n1}, f_{n1}), the first flexural resonant mode (Af_{n2}, f_{n2}) and the absorption peak between f_{n1} and f_{n2} (Af_{a1}, f_{a1}) were investigated. Correlation coefficients were highest between f_{n1} and f_{n2} (.72, Table 4) F_{n1} was correlated with f_{a1} (.67) while f_{n2} was significantly related to both Af_{a1} and

Table 3. Correlation coefficients for Magness-Taylor (MT) firmness with inspectors' maturity rating of 'Red Delicious' apple quality (n=702).

	MT firmness		
	Blush	Opposite	Mean
Inspectors' Maturity Scores			
Mean of all 6	0.80	0.82	0.83
Mean of best 3	0.82	0.84	0.85

Table 4. Correlation coefficients between properties of the acoustic transmission profile of intact 'Red Delicious' apples (n=702).

	Af_{n1}	Af_{n2}	Af_{a1}	f_{n1}	f_{n2}	f_{a1}
Af_{n1}	1	.26	.10	.49	.36	.30
Af_{n2}		1	.00	-.04	-.14	-.23
Af_{a1}			1	.35	.55	.23
f_{n1}				1	.72	.67
f_{n2}					1	.50
f_{a1}						1

Table 5. Correlation coefficients for sonic properties of 'Red Delicious' apples with apple properties and state of origin.

All samples, n=702.

	State	Mass	MT Mean	Maturity
f_{n1}	.00	-.59	.49	.46
f_{n2}	.38	-.78	.59	.54
$(f_{n1})^2_m$	-.26	.02	.31	.32
$(f_{n2})^2_m$	.20	-.20	.59	.57
f_{al}	.09	-.38	.38	.35

Composite samples by mass, n=140.

f_{n1}	.26	-.86	.72	.68
f_{n2}	.45	-.95	.75	.68
$(f_{n1})^2_m$	-.34	.03	.33	.36
$(f_{n2})^2_m$	.19	-.43	.71	.67
f_{al}	.33	-.67	.56	.50

Composite samples by mass, n=70.

f_{n1}	.42	-.90	.81	.76
f_{n2}	.58	-.97	.84	.78
$(f_{n1})^2_m$	-.40	.05	.26	.28
$(f_{n2})^2_m$	.14	-.50	.76	.74
f_{al}	.50	-.74	.60	.55

f_{al} (.55, .50 respectively). Abbott et al. (1968b) concluded that f_{n2} exhibited greater repeatibility than f_{n1}. Our results support the use of f_{n2} when using a single resonant frequency predictor. Correlations for all samples were highest between f_{n2} and mass (-0.78, Table 5). The negative coefficient suggests an inverse relationship where f_{n2} increases as mass decreases. Correlation coefficients between sonics or MT and soluble solids and titratable acids were too low to be of practical significance and are not reported. Both f_{n1} and f_{n2} were highly correlated with MT mean (0.49, 0.59 respectively) and maturity score (0.46, 0.54 respectively). However, correlations with maturity score were lower than with MT mean for all resonant properties tested. As recommended by Abbott et al. (1968a, 1968b), f^2m values were generated using both f_{n1} and f_{n2}. Correlations with MT mean and maturity decreased when using f_{n1} while those using f_{n2} increased insignificantly. Composite scores reduced variability between texture measurements and increased correlation between properties. The small, nondestructive cyclic strains (deformations) imposed by the sonic measurement are testing different components of apple texture or firmness (Bourne 1969) than are the puncture and shear action of the MT probe which deforms, ruptures and destroys cellular tissues. Falk et al. (1958) report the modulus of elasticity and resonant frequency of plant tissues are significantly related to turgor pressure within the cells. Thus, turgor may influence the measurement of f^2m and the textural ratings of fruit. The correlation coefficients presented quantify the relationship between the two types of firmness being measured.

3.4 Correlations with acoustic transmission profiles

Single and multiple frequency classification models were developed to compare with the use of f_{n1} and f_{n2} in maturity prediction. Models employed the amplitude at select frequencies from the acoustic transmission profile to predict maturity or firmness and were based on the forms:

Single frequency:
$$TS = k_1 * Af_1 + c$$
Two term frequency:
$$TS = k_1 * Af_1 + k_2 * Af_2 + c$$
Ratio:
$$TS = k_1 * (Af_1/Af_2) + c$$
Difference:
$$TS = k_1 * (Af_1 - Af_2) + c$$
Single derivative:
$$TS = k_1 * D(Af_1) + c$$
Two term derivative:
$$TS = k_1 * D(Af_1) + k_2 * D(Af_2) + c$$

Where: TS = 6-inspector mean or measured textural attribute
k_i = ith regression coefficient
c = intercept of regression
Af_i = Amplitude at ith select frequency
$D(Af_i)$ = Slope at ith select frequency calculated using frequency values 25 Hz to either side of f_i

Implementation of these fixed frequency models would be easier than using resonant frequencies since f_{n1} and f_{n2} float and must be located over the scanned regime. Individual sample correlations were slightly lower for all models than the resonant frequency models except for the two term frequency model, which was

Table 6. Prediction model regression based on amplitude of vibration transmission (Multiple regression R value presented, selected frequencies (Hz) used in the model are shown in parentheses).

All samples, n=702.

		MT Mean		Maturity
f_{n2} - Comparative Model		.59		.54
Af_1	(850)	-.58	(845)	-.52
Af_1,Af_2	(860,165)	-.63	(350,100)	-.59
Af_1/Af_2	(1205,285)	.57	(545,850)	.53
$Af_1 - Af_2$	(1250,820)	.58	(605,820)	.53
$D(Af_1)$	(695)	-.50	(470)	.46
$D(Af_1),D(Af_2)$	(680,755)	-.57	(680,770)	-.52

Composite by mass, n=140.

Af_1	(835)	-.75	(835)	-.68
Af_1,Af_2	(845,125)	-.78	(350,100)	-.75
Af_1/Af_2	(155,545)	-.76	(135,545)	-.70
$Af_1 - Af_2$	(530,845)	.76	(545,835)	.70
$D(Af_1)$	(685)	-.74	(680)	-.69
$D(Af_1),D(Af_2)$	(685,35)	-.77	(435,35)	-.73

Composite by mass, n=70.

Af_1	(820)	-.85	(825)	-.79
Af_1,Af_2	(815,105)	-.88	(815,105)	-.83
Af_1/Af_2	(870,1960)	-.86	(885,1965)	-.82
$Af_1 - Af_2$	(835,1960)	-.87	(845,1960)	-.82
$D(Af_1)$	(685)	-.82	(685)	-.77
$D(Af_1),D(Af_2)$	(675,1380)	-.85	(680,1575)	-.80

slightly higher (Table 6). The two term frequency model, employing the frequencies near 165 and 860 Hz for MT prediction, resulted in the highest correlations. These frequencies provide the most reliable predictor when employing fixed-frequency excitation to determine maturity or firmness.

The ratio model which normalizes the amplitude at one frequency with that at another, did not perform as well as the non-normalized model. Non-linear amplitude effects in the frequency response curve proportional to the mode of coupling were observed and may indicate the need for a standardized coupling media. Furthermore, in order to excite the same vibrational modes in each apple without restriction from the coupling media, a point source exciter may be required.

Using means of apples grouped into composites for 22 distinct maturity categories based on the mean score from all six in-spectors, composites were highly correlated with the MT mean, f_{n1} and f_{n2} (0.98, 0.92 and 0.94, respectively; Table 7). Correlations were slightly lower for composites based on apple mass, Af_{a1} and f_{a1} (0.81, 0.86 and 0.85, respectively) and significantly lower for Af_{n1} and Af_{n2} (0.65, -0.73). State of origin was not related to any of the sensory or instrument ratings indicating a diverse selection of apple firmness and maturity from each state. The strong relationship between firmness,

Table 7. Correlation coefficients between 22 score classes identified from the six inspectors' mean maturity score for each apple and other apple attributes.

Mass	-.81	f_{n1}	.92	Af_{n1}	.65
MT Mean	.98	f_{n2}	.94	Af_{n2}	-.73
State	.38	f_{a1}	.85	Af_{a1}	.86

maturity, f_{n1} and f_{n2} justifies the need to develop objective measurements which can distinguish properties involved in the different textural attributes sensed by the sonic probing technique.

4 SUMMARY

A rapid, non-destructive instrument for evaluating firmness and maturity in apples was developed. Sonic measurements were found to vary insignificantly around the equator of uniform, intact apples. Resonance frequencies excited are related to the acoustic coupling media; floral clay has been superior. Micro-sized glass beads, however, were found to be an adequate substitute where adhesion to the apple's surface is undesirable, for example, in high-speed applications.

Based on the acoustic transmission profiles of whole intact apples, mathematical models were developed for prediction of apple properties. Correlations with sonic measurements were too low to be significant for predicting state of origin, soluble solids and titratable acidity. Correlation coefficients were highest between sonic measurements and MT mean or maturity score. For the 'Red Delicious' cultivar tested, regression coefficients were comparable between models based on frequencies of resonance and models based on amplitudes of vibration transmission at select frequencies. In agreement with other reports (Finney 1971), maturity of 'Delicious' apples was found to be correlated more highly with MT pressures than with sonic measurements, partly because the two instruments seem to be measuring different mechanical properties of the fruit. A significant relationship existed between sonic measurements and mass, though mass did not aid in sonic prediction of maturity.

5 CONCLUSION

A high-speed, nondestructive technique has been developed capable of measuring acoustic transmission profiles through apples in less than 1 second; significantly faster than previous methods requiring as much as 20 minutes per fruit. Models employing resonant frequencies and amplitudes of select frequencies from these profiles have been used to predict Magness-Taylor firmness and maturity with correlations in agreement with other published techniques. Further research needs to be conducted, however, to identify and differentiate the biomechanical properties being sensed when using a destructive measurement from those being sensed with a nondestructive measurement.

REFERENCES

Abbott,J.A.,A.E.Watada & D.R.Massie 1984. Sensory and instrument measurement of apple texture. J.Amer.Soc.Hort. Sci.109(2):221-228.

Abbott,J.A.,G.S.Bachman,N.F.Childers,J.V. Fitzgerald & F.J.Matusik 1968a. Sonic techniques for measuring texture of fruits and vegetables. Food Tech.22(5): 101-112.

Abbott,J.A.,N.F.Childers,G.S.Bachman,J.V. Fitzgerald & F.J.Matusik 1968b. Acoustic vibration for detecting textural quality of apples. Proc.Amer.Soc.Hort.Sci.93: 725-737.

Bourne,M.C 1969. Two kinds of firmness in apples. Food Tech.23(3):59-60.

Clark,H.L. & W.Mikelson 1942. Fruit ripeness tester. U.S. Patent 2,277,037.

Clark,R.L. & P.S.Shackelford 1973. Resonance and optical properties of peaches as related to flesh firmness. Trans.Amer.Soc.Agr.Eng.16:1140-1142.

Cooke,J.R. 1970. A theoretical analysis of the resonance of intact apples. ASAE Paper No.70-345. ASAE, St. Joseph, MI 49085.

Falk,S.,C.H.Hertz & H.I.Virgin 1958. On the relation between turgor pressure and tissue rigidity. Physiol.Plant.11:802-817.

Finney,E.E. & K.H.Norris 1968. Instrumenttion for investigating dynamic mechanical properties of fruits and vegetables. Trans.ASAE 11(1):94-97.

Finney,E.E. 1970. Mechanical resonance within Red Delicious apples and its relation to fruit texture. Trans.Amer.Soc. Agr.Eng.13:177-180.

Finney,E.E. 1971. Dynamic elastic properties and sensory quality of apple fruit. J.Texture Studies 2:62-74.

Finney,E.E.Jr.,J.A.Abbott,A.E.Watada &
D.R.Massie 1978. Nondestructive sonic
resonance and the texture of apples.
J.Amer.Soc.Hort.Sci. 103(2):158-162.

Haller,M.H. & J.R.Magness 1944. Picking
maturity of apples. U.S. Dept. Agr. Cir.
711.

International Apple Institute 1988.
National and International Apple News
19(3).

Kramer,A.,K.Aamlid,R.B.Guyer & H.Rogers
1951. New shear-press predicts quality
of canned limas. Food Engr.23(4):112,
187.

Lanza,J. & A.Kramer 1967. Objective
measurement of graininess in applesauce.
Proc.Amer.Soc.Hort.Sci. 90:491.

Lott,R.V 1965. The relationship between
stage of maturation and degree of
quality in apple and peach fruits.
Trans.Ill.Hort.Soc. 99:131-138.

Magness,J.R. & G.F.Taylor 1925. An improved
type of pressure tester for the
determination of fruit maturity. U.S.
Dept.Agr.Cir. 350.

Mitchell,R.S.,D.J.Casimir & L.J.Lynch 1961.
The maturometer--instrumental test and
redesign. Food Tech. 15:415.

Mohsenin,N. 1963. A testing machine for
determination of mechanical and
rheological properties of agricultural
products. Pa.Agr.Exp.Sta.Bul. 701.

U.S.D.A. 1978a. Market Inspection
Instructions. Agr. Marketing Service. pp.
140.

U.S.D.A. 1978b. Shipping point inspection
instructions. Agr. Marketing Service. pp.
53.

Land and Water Use, Dodd & Grace (eds), © 1989 Balkema, Rotterdam. ISBN 90 6191 980 0

Handled apples as calibrated sensors of apple handling gentleness

O.J.Berentsen
Agricultural University of Norway, Norway

ABSTRACT: When handling gentleness is assessed through the extent of damage to the handled produce, this produce serves as a kind of measuring instrument, and this instrument obviously needs calibration. The ability of apples to withstand impacts without damage being caused depends on their strength, i.e. the magnitude of the stress that will produce the critical strain needed to cause bruises. In the author's tests this strength varied between 0.2 and 0.8 MN/m^2, while the critical strain was reached when a flat plate was pressed into the rounded apple cheek to a depth of 2.5 - 3 mm.

This paper is a report on some findings on the mechanical strength and susceptibility to damage in apples.

In the testing of any harvesting or handling machine the harvested or handled produce serves as a measuring instrument. Hence calibration of this "instrument" is necessary to obtain true or comparable test results in respect of the quality-of-work performance of the machine or system. Also, if strain is the critical factor in the determination of susceptibility to damage, the measured value of the stress that can produce that critical strain will be an important parameter in the functional relationships of the machine or system performance, and will provide the correction factor needed for calibration.

In some studies it has been attempted to relate the bruising of apple flesh to either the applied load or the amount of energy absorbed during the loading process. In other studies it has been concluded that the shear stress is the more significant failure parameter, as measured on cylindrical samples of apple flesh. And some researchers maintain that strain is the more significant failure parameter.

MATERIALS AND METHODS

In the investigations apples were subjected to compression tests as well as to flat-plate "impression" tests. The object was not to find typical or variety-related values of bruise-causing strains and stresses in apples, but to try and find some defined parameter for the susceptibility to bruising damages. For this it was thought that tests on whole apples would be the more meaningful.

Through a cage provided, the "JJ Lloyd Tensile Testing Machine", which was used in the investigations, could be used for compression tests. External forces on the whole apple were applied in two ways: Impression of a flat steel plate into the round cheek of a whole apple, and compression of the whole apple embedded in plaster of Paris at the two sides, or the two ends, on which forces were applied. See Figs. 1 and 2. The measuring instrument, the JJ machine, made possible the measurement and recording of very small forces and displacements. The rate of deformation, or the crosshead speed, was very low, 6 mm/min.

Fig. 1. Flat plate impression   Fig. 2. Whole apple compression

RESULTS

The compressive strengths of whole apples compressed "on top" and "on side" were found not to be significantly different.

When the sectional areas of the embeddings were used for the calculations of normal stresses and not the "girth" sectional area of the apple, the stresses calculated were very similar to those obtained with the flat steel plate impression, both at the "points of failure". With the impression method, "failure" meant bruising or brown spots being produced, while with the compression method failure meant cracking occurring. The cracking pattern typically showed shear failure planes or zones at 45^0 angles to the normal force direction.

With the apples tested the average compressive strength at failure was 0.35 MN/m^2, and the average "impressive" yield strength was 0.33 MN/m^2, both varying between 0.2 and 0.8 MN/m^2 with different stages of ripeness and different lengths of storage period for the apples.

The critical impression depth or deformation was around 2.5 - 3 mm. Whole apples of around 50 - 80 mm of girth diameter were usually compressed between 4 and 7 mm before they reached the yield or failure point. The modulus of elasticity for the apples measured was found to be around 4 MN/m^2.

A typical example of the force-displacement curves for impression can be seen on Fig. 3, and one for strength of whole apple in compression on Fig. 4. It is remarkable that the critical impression depth for bruising or discolouring to occur was usually somewhere around the second yield point on the curve, and always above the unexplained first step of it. The depths of the brown spots or bruises themselves were usually only slightly larger than the depths of impression. When bruising was produced with the impression method, there was usually a permanent flattening left on the otherwise rounded apple cheek surface.

While bruise-causing deformations showed little variation, the corresponding stresses varied widely, in these experiments mostly with the ripeness of the apples or with the length of storage period. The main result, therefore, was that bruising is primarily related to strain. Consequently, the measured value of the stress that produces the critical strain should be a useful measure of the susceptibility to bruising damage, and so be the parameter desired for "calibration" of the crop that is being harvested or handled.

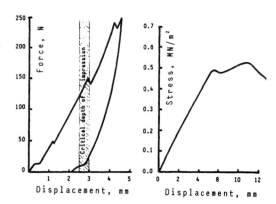

Fig. 3. Example of force-displacement relationship for apple impression

Fig. 4. Example of strength of whole apple in compression

Land and Water Use, Dodd & Grace (eds), © 1989 Balkema, Rotterdam. ISBN 90 6191 980 0

Halmfutter-Mähaufbereitung mit Intensivaufschluß und Mattenformung

Th.Bischoff, H.Wandel, S.Öztekin & K.Walther
Institut für Agrartechnik, Universität Hohenheim, Bundesrepublik Deutschland

RESUME: On développe et éprouve une nouvelle methode pour la préparation intensive et la production de mattes de plantes d'herbage pour la situation de l'Europe centrale. Les recherches sont realisées comme essai de laboratoire et expérience au champ avec de la luzerne et de l'herbe. Ils traitent la récolte journalière, l'ensilage et le séchage sur le champ et sous abri de la matière préparée et pressée en mattes. Les résultas montrent, ce qu'on peut secher des plantes d'herbage, qui etaient fauchées fraichement et preparées intensivement en un jour jusque à une humidité de 20 % et moins. L'indice pH de la luzerne ensilée etait influence positivement par la préparation intensive et par adjonction d'herbe. Le temps necessaire pour l'aération du foin des plantes préparées intensivement etait diminué en moyenne de la moitié. En comparaison avec les méthodes ordinaires de préparation on peut souligner un risque minime de mauvais temps des pertes minimes et des dépenses d'énergie minimes.

KURZFASSUNG: Ein neues Verfahren zur Mähaufbereitung mit Intensivaufschluß und Mattenformung von Halmfutter wird für mitteleuropäische Verhältnisse entwickelt und erprobt. Die Untersuchungen werden mit Wiesengras und Luzerne im Labor- und halbtechnischen Maßstab durchgeführt. Sie behandeln die eintägige Feldtrocknung, die Silierung sowie die Unterdachtrocknung des aufbereiteten und zu Matten gepreßten Gutes. Die Ergebnisse zeigen, daß frisch gemähtes und intensiv aufbereitetes Halmfutter an einem Tag auf einen Feuchtegehalt von 20 % und darunter auf dem Feld getrocknet werden kann. Bei der Silierung von Luzerne wurde der pH-Wert-Verlauf durch das Aufbereiten und Beimischen von Wiesengras positiv beeinflußt. Die Belüftungszeit bei der Unterdachtrocknung wird durch die intensive Aufbereitung im Durchschnitt auf weniger als die Hälfte reduziert. Im Vergleich zur herkömmlichen Mähaufbereitung sind geringes Wetterrisiko, geringe Verluste und geringer Leistungsbedarf sowie Wegfall jeglicher Feldwerbung hervorzuheben.

EINLEITUNG

In den Grünlandregionen Mitteleuropas ist für die Halmfutterkonservierung die Heuwerbung mit Bodentrocknung noch immer stark verbreitet. Je nach Klimabedingungen werden hierzu 4 bis 8 Tage benötigt (Olfe). Eine Verbesserung ergibt sich durch schnellere Wasserabgabe der Pflanzen mittels Aufbereiten. Mit der bisherigen konventionellen Mähaufbereitung kann eine Verkürzung der Trocknungszeit um 25 bis 30 % erreicht werden (Höhn). In den USA ist eine Technik der intensiven Aufbereitung mit Mattentrocknung für Luzerne entwickelt worden, mit welcher es möglich ist, vormittags gemähte und aufbereitete Luzerne bereits am Nachmittag trocken

einzufahren. Das Ziel der "Eintagesfutterernte" ist damit nahegerückt (Wandel und Koegel). Aufgabe dieser Untersuchungen ist es, die intensive Mähaufbereitung auf die Verhältnisse, wie sie im mitteleuropäischen Raum, insbesondere durch Klima und Futterart gekennzeichnet sind, zu übertragen bzw. anzupassen.

VERFAHRENSTECHNIK DER INTENSIVAUFBEREITUNG

Die in den USA entwickelte Mähaufbereitung für Luzerne gliedert sich in 3 Teilschritte:
a) Mähen
b) Intensivaufbereitung

c) Formgebung (Mattenpressung des auf
bereiteten Halmgutes)

zu a: Die Mahd erfolgt mit einem Messer-
mähwerk

zu b: Die hierzu eingesetzte Technik
besteht aus einer zentralen Schneidkantenwalze, um welche halbkreisförmig kleinere Riffelwalzen, an geordnet sind (Abb.
1). Das Halmgut wird zwischen der Schneidkantenwalze und den Riffelwalzen durchgeführt. Die Aufbereitung erfolgt durch die
Differenzgeschwindigkeit zwischen Riffelwalzen und Schneidkantenwalze und dem
geringen Abstand der beiden Walzentypen
zueinander. Dabei wird die Wachs- und
Cutinschicht der Blatt- und Stengelober-
flächen aufgebrochen und abgerieben, die
äußere Zellschicht (Epidermis) aufgerissen
und der Stengel mehrfach
längsgeschlitzt, d.h. zerfasert.

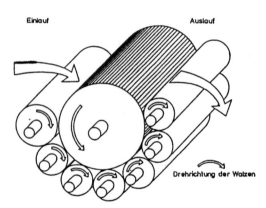

Abb.1:   Schema Aufbereiter

zu c: Die intensiv aufbereitete Luzerne
passiert, gleichmäßig dosiert, eine Walzenpresse. Dadurch entsteht, je nach Gut-
menge, eine faserige, formstabile Matte
mit einer Stärke von 6 bis 10 mm. Der beim
Aufbereiten an die Oberfläche tretende
Zellsaft wirkt dabei als Bindemittel für
die feineren Teile an die Matte. Die
Preßkraft beträgt zwischen 2 und 5 kN,
d.h. die Linienlast variiert zwischen 40
und 100 N pro cm Walzenlänge.

EIGENE UNTERSUCHUNGEN

Nach ersten Voruntersuchungen mit einem in
den USA für Luzerne konzipierten Intensiv-

aufbereiter mit Wiesengras, zeigte sich
der Bedarf einer Neukonstruktion mit noch
stärkerem Aufschlußeffekt. Außer diesem
Aufbereiter besteht die Versuchseinrich-
tung aus einem Zuführband, einem Antriebs-
aggregat (Elektromotor mit Meßwelle) sowie
einer stationären und einer mobilen Walzenpresse, mit der die Matten auf das Feld
ausgelegt werden können. Das Zuführband
simuliert das Mähen, und mit der Meßwelle
wird der Kraftbedarf des Aufbereiters
gemessen. Die Preßkraft der Walzenpresse
wird mit einer Kraftmeßdose registriert.

Die Versuche mit Wiesengras und Luzerne
im Labor- und halbtechnischen Maßstab umfassen die Feldtrocknung, Silierung und
Unterdachtrocknung. Zur Feldtrocknung ist
morgens um 9.30 gemäht, aufbereitet, zu
Matten gepreßt und um 10.30 Uhr auf die
gemähte Wiese ausgelegt worden. Das Mattengewicht beträgt 1,4 bzw. 1,8 kg/m². Zum
Vergleich dient nicht aufbereitetes Gut,
die Null-Variante mit nur 1,2 kg/m²
Frischgut. Die TS-Bestimmung erfolgt
stündlich. Entsprechend der herkömmlichen
Heuwerbung wird die Nullvariante alle 2
Stunden gewendet.

Silierversuche sollen die Wirkung auf
die Silierung zeigen. Bekanntlich ist beim
Silieren auf gute Verdichtung und Luftabschluß zu achten. Um gleichzeitig genügend
Wiederholungen einer Versuchspartie zu ermöglichen, werden 1,5-Liter-Gebinde gewählt. Die Umgebungstemperatur der Silier-
behälter beträgt zwischen 20 und 24 °C.
Die Silierung der Aufbereitungsvarianten
ist hier zunächst nach dem pH-Wert beurteilt. Die Gesamtproben der Silage setzen
sich aus Teilproben aus dem unteren, mittleren und oberen Bereich der Behälter zusammen. Folgende Varianten der Silierung
kamen zunächst in Betracht:
  0-Variante: Luzerne-Gras-Gemisch, ange-
                  welkt, nicht aufbereitet
  Variante 1: Luzerne-Matte
  Variante 2: Luzerne-Gras-Matte
  Variante 3: Gras-Matte
Bei den gemischten Varianten beträgt das
Verhältnis Luzerne zu Gras jeweils 50 % zu
50 % (Frischsubstanz).

Die Unterdachtrocknungsanlage besteht
aus Gutbehälter (0,2 m³), Luftverteilraum,
Luftkanal und Radialgebläse. Im Bereich
der Luftzuführung werden die Daten der
Trocknungsluft gemessen. Als spezifische
Trocknungsluftmenge wurden 0,15 m³/m³ s
angestrebt. Die Unterdachtrocknung mit
kalter und teils erwärmter Trocknungsluft
betrifft die Gutvarianten:
  a) Luzerne, angewelkt, nicht aufbereitet
  b) Frisch aufbereitete Luzerne
  c) Aufbereitete und angewelkte Luzerne
mit Anfangsfeuchtegehalten von 60, 50, 40
und 30 % mit jeweils 3 Wiederholungen

ERGEBNISSE DER FELDTROCKNUNG

Die Versuche zeigen, daß Halmfutter mit 80 % Feuchtegehalt, intensiv aufbereitet, als Matten am Tag des Mähens auf 20 % und darunter trocknet ohne irgendwelche Arbeitsgänge der Heuwerbung. In Abbildung 2 ist der Trocknungsverlauf beispielhaft für Wiesengras dargestellt.

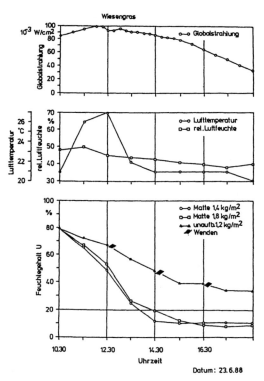

Abb.2:   Mattentrocknung auf dem Feld

Die Halmfuttermatte mit 1,4 kg je m² war bereits nach 3,5 Stunden auf einen Feuchtegehalt von 20 % abgetrokknet. Die Halmfuttermatte mit 1,8 kg je m² benötigte bis zu diesem Restfeuchtegehalt 4 Stunden Trocknungszeit. Das nicht aufbereitete Halmgut, die Null-Variante, hatte zu diesem Zeitpunkt nach einmaligem Wenden noch einen Feuchtegehalt von 49 %. Nach weiteren Trocknungsstunden und zweimaligem Wenden sank der Feuchtegehalt der Null-Variante noch um 15 auf 34 %. Die Halmfuttermatte (mit 1,4 bzw. 1,8 kg/m² erreichte dagegen schon nach zusätzlichen 1,5 Trocknungsstunden einen konstanten Feuchtegehalt von ca. 10 %.

Von großem Interesse ist der Einfluß der Formgebung (Mattenpressung) des aufberei-

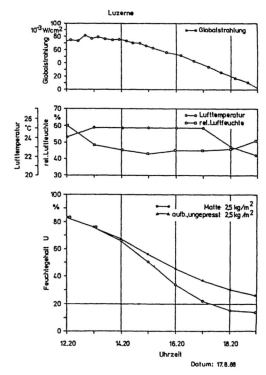

Abb.3:   Mattentrocknung auf dem Rost

teten Gutes auf den Trocknungsverlauf (Abb. 3).

Zu diesem Zweck wurde aufbereitete Luzerne lose und als Matte jeweils mit 2,5 kg/m² auf Rosten auf dem Feld ausgelegt. Bei der Mattenformung ist mit einer Linienlast von 40 N je cm Walzenlänge gearbeitet worden. Keine der Varianten wurde gewendet. Als Ergebnis zeigt sich deutlich die Abhängigkeit der Trocknungsgeschwindigkeit von der Kompaktierung des zerfaserten Gutes zur Matte. Während die Matte nach 5,5 Stunden einen Feuchtegehalt von 20 % erreichte, ist zu diesem Zeitpunkt beim losen Gut der Feuchtegehalt noch um 15 Prozentpunkte höher.

ERGEBNISSE DER SILIERUNG

Der Einfluß der intensiven Aufbereitung auf eine verbesserte Silierung kann durch Zellaufschluß, Vermischung und erhöhte Lagerdichte der Futterpflanzen erklärt werden (Zimmer). Für die Bewertung des Gärfutters ist der pH-Wert ein anerkannter Maßstab. Je niedriger der pH-Wert, um so höher ist der Gärsäuregehalt. Er ist in Abbildung 4 dargestellt.

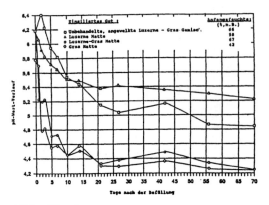

Abb. 4: Silierung

er ab dem 20. Tag nach der Befüllung unter 4,4. Die Wirkung des aufbereiteten Grasanteiles entspricht hier geradezu einem Silierhilfsmittel.

ERGEBNISSE DER UNTERDACHTROCKNUNG

Der Verlauf der Unterdachtrocknung ist stark vom Zustand der Trocknungsluft, aber ebenso von der Wasserabgabefähigkeit des Trocknungsgutes abhängig. Bei nicht aufbereitetem, angewelktem Gut (Abb. 5) richtet sich die Trocknungsgeschwindigkeit bis etwa 35 % Gutfeuchte nach der Zustandsänderung der Trocknungsluft.

Der pH-Wert beträgt bei der Befüllung bei der Null-Variante (Luzerne-Gras-Gemisch, angewelkt), der Variante 1 (Luzerne-Matte) und der Variante 2 (Luzerne-Gras-Matte) ca. 6,2. Bei der Variante 3 (Gras-Matte) ist der pH-Wert mit ca. 5,8 niedriger als bei den ersten drei Varianten. Nach einiger Gärzeit (unter Luftabschluß) unterschied sich der pH-Wert der einzelnen Varianten deutlich voneinander. Am 10. Tag nach der Befüllung liegt er bei der Null-Variante (Luzerne-Gras-Gemisch, angewelkt) noch über dem der Variante 1 (Luzerne-Matte). Ab dem 10. Gärtag sinkt er aber bei der Null-Variante noch schneller ab als bei der Variante 1, obwohl die Null-Variante nicht aufbereitet, sondern nur angewelkt ist. Bereits durch das Beimischen von Gras verbessert sich der pH-Wert Verlauf der nicht aufbereiteten Luzerne deutlich. Gras vergärt wegen seines Hauptkohlenhydrates Fructosan besser (Mc Donald). Der durchschnittliche pH-Wert der Null-Variante lag am Ende der Silierungsperiode um 4,8.

Die intensive Aufbereitung scheint wenig Einfluß auf die Silierung der Luzerne-Matte auszuüben, da der pH-Wert noch nach 70 Tagen 5,2 beträgt. Dies liegt wahrscheinlich in den ungünstigen Siliereigenschaften der Luzerne begründet. Gegen pH-Veränderungen weist sie einen höheren Widerstand auf als Gräser, bedingt durch ihre höheren organischen Säuregehalte (Mc Donald). Zu diesen Substanzen gehört beispielsweise das Roheiweiß. Intensive Aufbereitung vermag also die negativen Siliereigenschaften der Luzerne allein nicht zu kompensieren. Bei den Varianten 2 (Luzerne-Gras-Matte) und 3 (Gras-Matte) sank er dagegen innerhalb der ersten 5 Tage stark ab. Nach 20 Tagen zeigten beide Varianten pH 4,3. Bei der Variante 3 sinkt

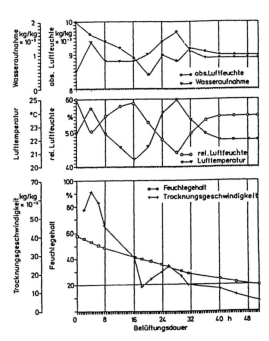

Abb. 5: Unterdachtrocknung : Luzerne, nicht aufbereitet

Ab diesem Punkt stagniert die Trocknung, obwohl die Trocknungsfähigkeit ein Maximum erreicht. Beim Trocknen unter 30 % Gutfeuchte verlangsamt sich gewöhnlich die Trocknungsgeschwindigkeit; man spricht vom energieaufwendigen Trocknungsende. Günstiger verhält es sich bei intensiv aufbereitetem Trocknungsgut (Abb. 6). Hier kommt es bei relativ ungünstigem Trocknungsluftzustand zu einer absolut höheren Trocknungsleistung aufgrund einer überwiegend höheren Sättigung der Trocknungsluft. Zunächst mag dies mit dem extrem hohen Anfangsfeuchtegehalt für ein Trocknungsgut (74 %) zusammenhängen, jedoch

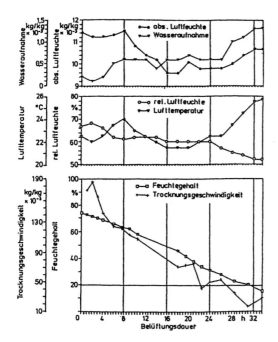

Abb.6: Unterdachtrocknung: Luzerne, aufbereitet

Tab. 1: Zusammenhang zwischen Feuchteabgabe und Belüftungsdauer bei den unterschiedlichen Belüftungsphasen; Luzerne, nicht aufbereitet

| Belüftungs-phase | Feuchte-gehalt (%) | Differenzfeuchte | | Belüftungsdauer | | Verhältniswert: rel.Belüftungsdauer/ rel.Differenzfeuchte |
		absolut (%)	relativ (%)	absolut (Stunden)	relativ (%)	
Anfang	37,80					
1.Knickpunkt	40,00	17,80	47,08	18,50	36,27	0,770
2.Knickpunkt	30,00	10,00	26,46	12,50	24,51	0,926
Ende	20,00	10,00	26,46	20,00	39,22	1,482
Gesamt	37,80		100,00	51,00	100,00	

Tab. 2: Zusammenhang zwischen Feuchteabgabe und Belüftungsdauer bei den unterschiedlichen Belüftungsphasen; Luzerne, aufbereitet

| Belüftungs-phase | Feuchte-gehalt (%) | Differenzfeuchte | | Belüftungsdauer | | Verhältniswert: rel.Belüftungsdauer/ rel.Differenzfeuchte |
		absolut (%)	relativ (%)	absolut (Stunden)	relativ (%)	
Anfang	74,00					
1.Knickpunkt	47,00	27,00	50,00	18,00	60,00	1,200
2.Knickpunkt	35,00	12,00	22,62	5,00	16,66	0,736
Ende	20,00	15,00	27,38	7,00	23,34	0,852
Gesamt		54,00	100,00	30,00	100,00	

Der Einfluß des Trocknungsgutzustandes auf
die Belüftungs-Trocknungsdauer ist in
Tabelle 3 dargestellt. Den höchsten Zeitbedarf benötigt die Kaltbelüftung von
nicht aufbereiteter Luzerne mit 55 bis 60
Stunden bei 55 bis 65 % Einlagerungsfeuch-
te. Diese Belüftungszeit ist hier gleich
100 % gesetzt.

Tab.: 3: Der Einfluß des Anfangsfeuchtegehaltes auf das Belüftungsverfahren

| Anfangs-feuchte-gehalt (%) | Belüftungsdauer | | | | | | | |
	absolut (Stunden)	relativ (%)	absolut (Stunden)	relativ (%)	absolut (Stunden)	relativ (%)	absolut (Stunden)	relativ (%)
25 - 35	2,33	4,19						
35 - 45	3,68	6,63						
45 - 55	4,91	8,84						
55 - 65	11,16	20,10	55,50	100,00	4,00	7,20	24,00	43,24
65 - 75	31,00	55,85			9,75	17,56		
75 - 85	31,50	56,75			28,00	50,45	29,00	52,25
Verfahren	aufbereitet		nicht aufbereitet		aufbereitet		nicht aufbereitet	
	Kaltbelüftung				Warmbelüftung			

spätestens beim Restfeuchtegehalt von 30 %
trifft dies nicht mehr zu.

Selbst ab dieser kritischen Restfeuchte
zeigt hier die Trocknungsgeschwindigkeit
noch deutliche Erhöhungen. Das Ergebnis
schlägt sich in einer höheren Trocknungs-
rate in der letzten Trocknungsphase nie-
der.
In Tabelle 1 und 2 sind diese Zusammenhänge aus Abbildung 5 und 6 verdeutlicht.
Der Zusammenhang zwischen relativerBelüf-
tungsdauer und relativer Differenzfeuchte
ist im Verhältniswert ausgedrückt. Der
ansteigende Verhältniswert in Tabelle 1
bedeutet einen zunehmenden Zeitbedarf mit
fortschreitender Belüftungsdauer bzw.
Trockenheit des Gutes. Tabelle 2 zeigt den
abnehmenden Verhältniswert für aufbereitete, nicht angewelkte Luzerne als Trocknungsgut.
Beispiel: Um die ersten 50 % der Gutfeuchte abzuführen, sind 60 % der Belüftungszeit notwendig (Tab. 2). Die restli-
chen 50 % konnten aber in 40 % der Belüftungszeit abgeführt werden. Diese Zeit-
verschiebung und die Verkürzung der Trocknungszeit dürfte mit dem Aufschlußeffekt
zu begründen sein.

ÖKONOMISCHE GESICHTSPUNKTE

Ohne Versuche im technischen Maßstab ist
eine abschließende Verfahrensbewertung
nicht möglich, wohl aber eine überschlä-
gige Einschätzung. Gegenüber der konventionellen Mähaufbereitung kennzeichnet das
untersuchte Grundprinzip verfahrenstech-
nisch eine sehr kurze Feldphase, damit
geringes Wetterrisiko, geringe Verluste
sowie Wegfall jeglicher Feldwerbung wie
Zetten, Wenden, Schwaden und Breitstreuen.
Somit wird auch die Ernte von Halmfutter an

einzelnen Schönwettertagen zum optimalen Schnittzeitpunkt ermöglicht. Bezüglich Energiebedarf für Mähen und Aufbereiten einschließlich Mattenformung ist der voraussichtliche Bedarf von 17 kWh/tTM relativ gering im Vergleich zu herkömmlichen, ähnlich leistungsfähigen Aufbereitern mit 15 kWh/t TM. Der Aufbereiter mit Intensivaufschluß verzichtet auf feststehende Schikanen und Leiteinrichtungen mit hohem Energieverbrauch. Alle Aufbereitungswerkzeuge, die Haupt-, Planeten- und Formgebungswalzen wirken fördernd auf den Gutstrom.

Der Preis für einen Intensivaufbereiter mit Mähwerk wird sich nach der Einsatzfläche richten. Nach überschlägiger Kalkulation bewegt er sich im oberen Bereich konventioneller Aufbereiter neuerer Bauart. In jedem Fall kann dem Intensiv Aufbereiter die Einsparung der Heu- bzw. Silagefeldwerbung angerechnet werden. Als Richtwert für die Flächenleistung wird 0,8 bis 1,5 ha/h je nach Ertrag (bei 2m Arbeitsbreite) angenommen. Die täglich zu mähende Fläche beim Aufbereiten richtet sich nach der benötigten Trocknungszeit, z.B. 4 oder 6 Stunden, abhängig vom Futteranfall bzw. Mattengewicht je Flächeneinheit und der verfügbaren Schlagkraft beim Bergen der Matten.

LITERATUR

Höhn, E.,1987. Quetschen und Schlagen, zahlt sich das aus ? DLZ 5/87, S. 661-667

Mc Donald, P., 1981. The Biochemistry of Silage. John Wiley and Sons, p. 36/37

Olfe, H. - C., 1971. Untersuchungen über die Korrespondenz der meteorologischen Daten mit den thermodynamischen Trocknungsmöglichkeiten von Halmfutter. KTBL-Berichte über Landtechnik, Nr. 141

Wandel, H., R. Koegel, 1986. Ein neuer Weg der Halmfutteraufbereitung. Landtechnik 4, 41. Jahrgang

Zimmer, E., 1967. Der Einfluß der mechanischen Aufbereitung auf die Silierfähigkeit von Halmfutter. Grundlagen der Landtechnik 17(6), S. 197-202

ABSTRACT: A new treatment of mower-conditioning with intensive maceration and mat pressing has been developped and tested for middle European conditions. The research took place in the laboratory and in the field. It concerns the field-drying in one day, the ensiling and the barn drying of macerated and mat—pressed forage of grass and alfalfa. The results show, that freshly cut macerated and mat—pressed forage can achieve a moisture content of 20 % and less in one day. The mixture of macerated grass and alfalfa improves the pH-value during ensiling process. The barn-drying time of macerated forage crops can decrease on average to less than half of the time of untreated forage crops. Compared to conventional conditioners it lowers the risk of bad weather, reduces losses and energy requirements and enables the omission of some post harvest operations.

Land and Water Use, Dodd & Grace (eds), © 1989 Balkema, Rotterdam. ISBN 90 6191 980 0

Ensilability of mat-processed alfalfa

R.E.Muck & R.G.Koegel
USDA/ARS Dairy Forage Research Center, Madison, Wis., USA

K.J.Shinners & R.J.Straub
Agricultural Engineering Department, University of Wisconsin, Madison, Wis., USA

ABSTRACT: An alternative machine to the mower-conditioner, which increases forage drying rates by macerating and pressing the forage into a 6 to 8 mm thick mat for wilting, was tested with regard to its effects on ensiling. In two harvests, mat-processed and normally-harvested alfalfa were wilted to approximately the same dry matter level, harvested and ensiled in mini-silos with and without a commercial inoculant. In both harvests, the uninoculated, mat-processed alfalfa reached final pH twice as quickly as the uninoculated, normally-harvested alfalfa. The drop in pH in the uninoculated, mat-processed alfalfa during the initial 2 to 4 days of fermentation was not statistically different from that of the inoculated silages.

RÉSUMÉ: Caractéristiques fermentaires de la luzerne récoltée au moyen de la technologie des "nattes".
Les caractéristiques fermentaires de la luzerne ont été comparées pour deux techniques de récoltes: (1) faucheuse conditionneuse normale et (2) technologie des "nattes" (conditionnement intensif suivi par la formation d'une mince couche de 6 à 8 mm d'épaisseur). Deux coupes successives de luzerne, préfanées au champ jusqu'aux même teneurs en matière sèche, ont été récoltées par les deux techniques et ensilées dans des mini-silos avec et sans inoculant commercial. Pour les deux coupes de luzerne non-inoculées, le pH minimal a été atteint avec une vitesse double pour l'ensilage natte (2) comparé à l'ensilage normal (1). La diminution du pH pendant les 2 à 4 premiers jours de fermentation n'a pas été significativement différente entre l'ensilage natte non-inoculé et l'ensilages inoculés.

ZUSAMMENFASSUNG: Gärungsfähigkeit von Luzernesilage geerntet mit dem Mattenverfahren (Intensive Aufbereitung).
Die Gärungsfähigkeit von intensiv aufbereitetem Grünfutter wurde untersucht. Die normale Mäher-Aufbereitung wurde durch das sogenannte Mattenverfahren ersetzt, wodurch die Feldtrocknung schneller wurde. Das Verfahren verbindet intensive Aufbereitung mit anschliessendem Pressen zu 6 bis 8 mm dicken Matten. In zwei Schnitten wurde normale und mattengeerntete Luzerne auf den gleichen Wassergehalt vorgewelkt und in Mini-Silos, mit- und ohne handelsübliche Impfung einsiliert. Die nicht geimpfte mattengetrocknete Luzerne erreichte den End-pH Wert in beiden Schnitten doppelt so schnell wie die nicht geimpfte normal geerntete Luzerne. Die Abnahme des pH Wertes in der nicht geimpften mattengetrockneten Silage unterschied sich in den ersten zwei bis vier Tagen statistisch nicht von der Abnahme in der geimpften Silage.

1 INTRODUCTION

The losses in harvesting alfalfa with current machinery are substantial. Handling alfalfa as hay results in dry matter (DM) losses of 15 to 22% (Moser 1980; Koegel, et al. 1985) under typical U.S. conditions. Rainfall can increase those losses by leaching nutrients, shattering leaves and prolonging plant respiration.

Harvesting wilted alfalfa for silage reduces the mechanical and plant respiration losses and reduces the weather risk. However, DM losses are still significant. Chopping and blowing of wilted alfalfa has been reported to cause DM losses of 5% (Kjelgaard 1979). This is in addition to plant respiration losses over 1 to 2 days of wilting.

In order to reduce these losses, a radically different machine was proposed to replace the standard mower-conditioner (Koegel, et al. 1988, Shinners, et al. 1989). This machine carries out four processes: mows, macerates (severely conditions) the crop, presses the crop into a continuous mat 6 to 8 mm thick and deposits the mat onto the stubble for drying. The maceration process splits the stems into long fibrous pieces while mashing or pureeing the leaves. The combination of severe conditioning, the thinness of the mat and the darkened color of the forage cause drying rates to be several times faster than rates from normal mowing and conditioning. Typically, mat-processed alfalfa dries to 80% DM in 6 h or less (Shinners, et al. 1987). Even under poor drying conditions, wilting to normal silage dry matters (35 to 50% DM) should occur in a few hours, creating a wilted silage system with little or no weather risk.

Several studies have looked at the potential of ensiling mat-processed alfalfa. Macerated alfalfa has been shown to consolidate at a faster rate under both static and dynamic loading than normally harvested alfalfa although final silage bulk densities under a constant static loading were not significantly different (Shinners, et al. 1988). Such characteristics would make the mat-processed alfalfa more desirable for ensiling in a clamp or bunker silo where packing is done by driving a tractor over the forage.

Two studies have looked at the feeding potential of macerated alfalfa (Lu, et al. 1980; Hong, et al. 1988). Macerated alfalfa had increased digestibility and feed intake as compared with normally-harvested alfalfa.

The objectives of this study were to compare the rate and extent of fermentation in mat-processed alfalfa with normally-harvested alfalfa with and without the use of a commercial inoculant.

2 MATERIALS AND METHODS

Two similar trials were performed on first (June 9) and second (July 14) cutting alfalfa in 1988. Alternate swaths were mowed with a mower-conditioner and a prototype mat-making machine, which was the same as that described by Koegel, et al. (1988) with some minor modifications. Three hours after mowing the mat-processed alfalfa was loaded using a belt and tine pickup and taken to a stationary chopper. In first cutting, the mat-processed alfalfa was chopped to a 20 mm theoretical length of cut (TLC). In second cutting, a smaller chopper with a TLC of 10 mm was used. The normally-mowed (or control) alfalfa was wilted until the moisture content was approximately equal to that of the mat-processed alfalfa. In both control cuttings, wilting was approximately 24 h. In first cutting, the normally-mowed alfalfa was harvested with a precision chop harvester with a TLC of 6 mm. In second cutting, the alfalfa was picked up manually and chopped with the same small stationary chopper as the mat-processed alfalfa.

The alfalfa was ensiled in mini-silos (100 ml centrifuge tubes) as per Muck (1987a). In the first trial, there were three treatments: mat-processed, control and control inoculated with lactic acid bacteria. In the second trial, a fourth treatment, mat-processed with inoculation, was added. Inoculation was with a commercial inoculant containing Lactobacillus plantarum and Pediococcus cerevisiae. One ml of diluted inoculant containing 2.5×10^7 colony forming units (CFU)/ml was added to 50.0 g alfalfa whereas the uninoculated alfalfa received 1.0 ml distilled water.

For each treatment, 18 mini-silos were filled, sealed and incubated at 30°C in a water bath. At 1, 2, 4, 8, 16 and 32 days, 3 mini-silos of each treatment were removed from the water bath and immediately frozen at -18°C until analyzed. In trial 2, the procedure was the same except the final silages were taken at 25 days.

The initial, unensiled alfalfa samples were analyzed for moisture content by oven drying at 60°C for 72 h (ASAE 1983), pH, lactic acid bacteria (LAB) via counting on Rogosa SL agar (Muck and O'Connor 1985), total nonstructural carbohydrates (TNC) (Smith 1981), reducing sugars (Smith 1981), buffering capacity (Muck and Walgenbach 1985), total Kjeldahl nitrogen (TKN) (AOAC 1980), soluble non-protein nitrogen (NPN) (Muck 1987a), free amino acids (FAA) and ammonia (NH_3) (Broderick and Kang 1980), and fermentation endproducts (lactate, acetate, succinate, formate, propionate, butyrate and ethanol) via HPLC (Muck and Dickerson, 1988). The silage samples were analyzed for the same constituents as the initial alfalfa with the exception of LAB, TNC, reducing sugars and buffering capacity.

Statistical analyses were performed using one-way analysis of variance and separation of means was by least significant difference.

Table 1. Characteristics of alfalfa en-siled in Trial 1			
	Mat	Control	Inoc. Control
DM, %	54.8c	60.7a	56.8b
Buffer Cap., meq/kg DM	504	549	480
Sugar, % DM	4.1	4.2	4.3
TNC, % DM	6.6	5.9	5.7
LAB, Log(CFU/g)	4.82a	2.38b	3.10b
Inoculant LAB			5.62
pH	5.92b	6.14a	6.10a
TKN, % DM	2.80	2.81	2.78
NPN, % TKN	14.01	13.78	18.40
FAA, % TKN	6.35	5.03	5.55
NH3, % TKN	0.50	0.42	0.47
Succinic, % DM	0.05	0.12	0.09
Formic, % DM	0.19	0.17	0.08
Acetic, % DM	0.12	0.22	0.12

abc Concentrations followed by different letters are different at P<0.05.

Table 2. Characteristics of alfalfa en-siled in Trial 2		
	Mat	Control
DM, %	46.2a	42.4b
Buffer Capacity, meq/kg DM	561	569
Sugars, % DM	3.0	3.0
TNC, % DM	5.5	5.3
LAB, Log(CFU/g)	7.11a	4.81b
Inoculant LAB	5.74	5.59
pH	5.81b	6.08a
TKN, % DM	3.58b	3.95a
NPN, % TKN	16.49	13.49
FAA, % TKN	7.67	8.52
NH3, %TKN	0.38	0.36
Succinic, % DM	0.06	0.09
Formic, % DM	0.13	0.19
Acetic, % DM	0.10	0.14

ab Concentrations followed by different letters are different at P<0.05.

3 RESULTS AND DISCUSSION

3.1 Alfalfa characteristics

The characteristics of the alfalfa en-siled in the two trials are shown in Tables 1 and 2. The major differences between the mat and control alfalfa were in the number of epiphytic LAB, pH and DM. In both trials, the LAB on the mat-processed alfalfa were significantly higher than on the control forage. Counts immediately before and after mak-ing the mats have shown a thousand-fold or more increase in LAB numbers (Muck, unpublished data). Apparently, the macer-ating and/or the pressing operations are sources of inoculation as plant juices and particles accumulate on the equip-ment. The juices provide sugars for LAB growth, and LAB are removed as new forage passes through the machine. This inoc-ulation is much greater than that which has been measured with normal mower-conditioners (Muck 1987b) and was most likely the cause of the higher numbers on the mat-processed alfalfa.

The pH of the mat-processed alfalfa was consistently 0.2 pH unit less than the control alfalfa. Even though this was significant, the effect on the amount of acids required to reach a given pH was probably small. Buffering around pH 6 in alfalfa has been reported to be less than half that at pH 5.0 and less than a third that at pH 4.5 (Greenhill 1964).

In both trials, there were significant differences in the initial DM level among the treatments. The wettest treatment should have had a somewhat faster fermen-tation. However, the final DM contents were not significantly different (Tables 3 and 4) so that DM variation among treat-ments was probably not an important fac-tor.

Sugar and TNC contents were similar for all treatments within a trial so that substrate levels for the LAB were not significantly different (P>0.10). In trial 2, TKN levels in the control alfal-fa were significantly higher than for the mat-processed alfalfa. No reason for this difference was evident.

3.2 Final silage characteristics

Silage quality at 32 days in trial 1 is given in Table 3. The inoculated control had the lowest final pH, but that of the mat was similar although significantly higher. The differences in pH were large-ly due to the ratio of fermentation prod-ucts. Acetic acid content in the mat treatment was more than four times that in the control treatments. With a pK of 4.75 (Weast, et al. 1964), the acetic acid would have added additional buffer-ing to the mat silage and prevented as low a pH. The lactic-to-acetic acid ratios for the mat and uninoculated con-trol were similar (3.2 and 2.5, respec-tively) whereas the inoculated control produced a ratio of 15.2. Such a differ-ence would be expected when a good inocu-lant is applied and dominates fermenta-

Table 3. Characteristics of the final silage (32 d) in Trial 1[*]

	Mat	Control	Inoc. Control
DM, %	56.2	58.1	57.2
pH	4.68b	5.66a	4.52c
TKN	2.82a	2.81a	2.49b
NPN, % TKN	40.49b	50.83a	46.56ab
FAA, % TKN	33.96	28.66	35.18
NH3, % TKN	3.35	2.39	3.07
Succinic	0.22a	0.14b	0.12b
Lactic	6.17a	1.13b	5.64a
Formic	0.19a	0.15a	0.00b
Acetic	1.91a	0.45b	0.37b
Propionic	0.00	0.00	0.00
Ethanol	0.12a	0.08ab	0.04b
Butyric	0.00	0.00	0.00

[*] All concentrations as % DM except as noted.
abc Concentrations followed by different letters are different at P<0.05.

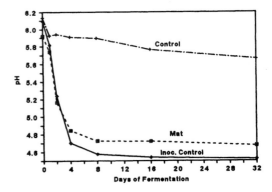

Figure 1. Trial 1 pH time course.

Table 4. Characteristics of the final silage (25 d) in Trial 2[*]

	Mat	Inoc. Mat	Control	Inoc. Control
DM, %	41.5	40.3	41.7	41.7
pH	5.04a	5.17a	5.17a	4.82b
TKN	3.65c	3.78bc	3.94ab	4.02a
NPN, %TKN	50.50	50.29	46.71	43.96
FAA, %TKN	41.62	38.18	41.52	41.33
NH3, %TKN	8.80a	8.78a	6.13b	3.93c
Succinic	0.68	0.62	0.68	0.63
Lactic	5.47a	4.78b	3.92c	5.14ab
Formic	0.19a	0.28a	0.17ab	0.05b
Acetic	3.67a	3.89a	2.97b	1.98c
Propionic	0.01	0.02	0.04	0.00
Ethanol	0.21b	0.31a	0.36a	0.17b
Butyric	0.00	0.00	0.01	0.01

[*] All concentrations as % DM except as noted.
abc Concentrations followed by different letters are different at P<0.05.

tion. Considering this variation in fermentation products, the mat treatment had a surprisingly similar final pH to the inoculated control.

Little fermentation occurred in the uninoculated control. No clostridial fermentation was evident as noted by the lack of butyric acid, and none would be expected at high DM levels (Muck 1988). The silage appearance and smell were typical of good silage. As indicated in Figure 1, the pH declined slowly but steadily and ostensibly was still dropping at 32 days.

There was no difference in the soluble nitrogen fractions between the mat treatment and the inoculated control. The higher NPN in the uninoculated control was due most likely to the long duration of elevated pH (McKersie 1985).

Final silage characteristics in trial 2 are shown in Table 4. All silages finished at a rather high pH although there were no signs of clostridial activity. All the fermentations apparently were stopped by a lack of fermentable substrate.

As in trial 1, the inoculated control had the lowest final pH, and both mat treatments had elevated acetic acid levels as compared with the controls. Based on the pH and acid contents, it would appear that the inoculant had little effect on the mat-processed alfalfa.

The total amount of fermentation products was higher in the mat treatments than for either of the control treatments. The same was true for the first

trial; however, in the earlier trial, pH rather than lack of substrate may have stopped fermentation due to the high DM level. These results suggest that the mat process may cause enhanced plant enzyme breakdown of complex carbohydrates providing additional substrate for fermentation. If further studies would prove this true, the mat process would be valuable for ensiling low sugar crops like alfalfa.

The soluble nitrogen fractions in trial 2 were unaffected by treatment with the exception of NH3. Ammonia levels were highest in the mat treatments, and the uninoculated control was higher than the inoculated control. It appears that the effect on NH3 in this trial is related to the high epiphytic LAB numbers in both

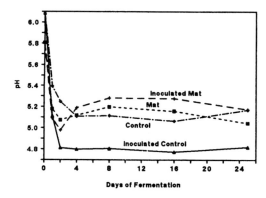

Figure 2. Trial 2 pH time course.

the mat and control alfalfas. Certain
LAB species can deaminate amino acids
(Ohshima and McDonald 1978) plus the high
epiphytic LAB level may indicate the
presence of other proteolytic microorganisms.

The charactistics of the mat-processed
silages in both trials show that these
silages ensiled as well as or better than
the uninoculated control silages. The
inoculated control silages had slightly
improved pH's in both trials and lower
NH3 in one case. However, the amount of
fermentation acids produced in both trials was highest in the mat treatments,
suggesting that mat-processed alfalfa may
provide more substrate for fermentation.

3.3 Fermentation rates

The rate of fermentation is most clearly
shown in the pH time courses (Figures 1
and 2). In the first trial, the pH in
the mat-processed alfalfa paralleled that
of the inoculated control. The pH's at
days 1 and 2 for these two treatments
were not significantly different (P>0.10)
from each other. By day 4, the inoculated control had a significantly lower pH.
Both treatments reached final pH at approximately 8 days, indicating that the
natural LAB in the mat-processed alfalfa
fermented the available substrate at
approximately the same rate as the inoculant LAB on the control alfalfa. This
was unexpected. Earlier work (Muck and
O'Connor 1988) with normally-harvested
alfalfa indicated that a good inoculant,
supplying LAB at 10% of the natural LAB
population, would dominate the fermentation and improve fermentation rate. In
our trial, the epiphytic population on
the mat-processed alfalfa was approximately 10% of the inoculant LAB applied to

the control alfalfa, but the epiphytic
population still fermented the available
substrate as quickly as the inoculant LAB
on the control alfalfa. The most likely
explanation of these results is that the
mat-processing made sugars and other substrate more immediately available to the
LAB, allowing faster growth.

As indicated earlier, the uninoculated
control had a much higher final pH.
Figure 1 shows that the decline in pH was
a gradual one and appeared to be continuing even after 32 days. This treatment
had a significantly higher (P<0.05) DM
content initially, but DM was not significantly different in the final silages.
The higher DM would have slowed the fermentation rate with respect to the other
treatments. However, much of the differences between this treatment and the
others should have been due to the very
low epiphytic LAB population and the
slower availability of the substrate.

In the second trial, there were no
significant differences in pH at any time
point between the inoculated and uninoculated mat-processed alfalfa. Considering
that the epiphytic LAB population on the
mat-processed alfalfa was more than 10
times that applied by the inoculant,
these results agree with those from
normally-harvested alfalfa (Muck and
O'Connor 1988). The inoculant LAB probably had a very limited effect on fermentation.

Inoculation of the control alfalfa
greatly increased the rate of pH decline. Final pH was reached in 2 days
with inoculation and in 4 days in the
uninoculated control. Evidently, the
inoculant LAB dominated on the control
alfalfa.

Overall in trial 2, pH declined fastest
in the inoculated control during the
first two days. However, the rate of
accumulation of fermentation products in
that time was significantly higher
(P<0.05) in the mat treatments. With
more acetic acid being produced, the pH
in the mat treatments did not fall quite
as fast as the inoculated control. The
uninoculated control was the only treatment that did not reach final pH by day
2. Thus, the LAB grew fastest in the mat
treatments, but the increased acetic acid
production buffered the mat silages and
prevented a pH decline as fast and as
great as the inoculated control.

4 CONCLUSIONS

Mat-processed alfalfa had characteristics
which caused it to ensile as well as or

better than normally-harvested alfalfa. Uninoculated, mat-processed alfalfa fermented quickly, reaching final pH in half the time, or less, of the uninoculated, normally-harvested alfalfa. This was in part due to higher levels of epiphytic LAB on the mat-processed alfalfa at ensiling. However, comparisons between the mat-processed and inoculated control treatments suggest that the mat-making process causes the sugars and other substrates used by the LAB to be more available, increasing fermentation rates.

In both trials, the mat-processed silages had the highest concentrations of fermentation products. Since the levels of sugars and non-structural carbohydrates were similar between the mat-processed and normally-harvested alfalfa at ensiling, these results suggest that mat processing of the alfalfa increases the breakdown of complex carbohydrates in the plant during ensiling. This could enhance the ensiling of many forages like alfalfa which may have inadequate sugar levels.

In contrast, mat-processed alfalfa was not necessarily more prone to plant proteolytic activity. In the first trial, the mat-processed silage had the lowest NPN level. In the second, the NPN contents were higher in the mat-processed silages; however, there were no statistically significant differences among the four treatments.

Ammonia contents were unaffected in the first trial by treatment. In the second trial, mat-processed silage had significantly higher NH3 levels. This may have been related to the high initial LAB numbers on the mat-processed alfalfa in that trial. Effects on this parameter should be monitored in future studies because of its negative impact on feed value.

Finally, it appears that inoculation of mat-processed alfalfa is of less value than with normally-harvested alfalfa. Fermentation rates without inoculation were similar to inoculated, normally-harvested alfalfa. Also, the high epiphytic LAB levels suggest that current inoculants would supply insufficient levels of LAB to dominate many of the fermentations. More research is needed to investigate LAB numbers on mat-processed alfalfa and to test various inoculants before conclusive statements regarding their effectiveness on mat-processed alfalfa can be made.

Overall, the mat-processed alfalfa ensiled well, and this study suggests that it has advantages over alfalfa harvested with current machinery.

REFERENCES

AOAC 1980. Official methods of analysis of the AOAC, 13th ed. Arlington, VA: AOAC.

ASAE 1983. ASAE standard S 358.1; moisture measurement-forages. In Agricultural engineers yearbook of standards, 30th ed., p.327. St. Joseph, MI:ASAE.

Broderick, G.A. & J.H.Kang 1980. Automated simultaneous determination of ammonia and total amino acid in ruminal fluid and in vitro media. J.Dairy Sci. 63:64-75.

Greenhill, W.L. 1964. The buffering capacity of pasture plants with special reference to ensilage. Aust.J.Agric. Res. 15:511-519.

Hong, B.J., G.A.Broderick, R.G.Koegel, K.J.Shinners & R.J.Straub 1988. The effect of shredding alfalfa on cellulolytic activity, digestibility, rate of passage and milk production. J. Dairy Sci. 71:1546-1555.

Kjelgaard, W.L. 1979. Energy and time needs in forage systems. Trans.ASAE 22:464-469.

Koegel, R.G., R.J.Straub & R.P. Walgenbach 1985. Quantification of mechanical losses in forage harvesting. Trans.ASAE 28:1047-1051.

Koegel, R.G., K.J.Shinners, F.J.Fronczak & R.J.Straub 1988. Prototype for production of fast-drying forage mats. Appl. Engr.Agric. 4:126-129.

Lu, C.D., N.A.Jorgensen & G.P.Barrington 1980. Intake, digestability, and rate of passages of silages and hays from wet fractionation of alfalfa. J.Dairy Sci. 63:2051-2059.

McKersie, B.D. 1985. Effect of pH on proteolysis in ensiled legume forage. Agron.J. 77:81-86.

Moser, L.E. 1980. Quality of forage as affected by post-harvest storage and processing. In C.S.Hoveland (ed.), Crop quality, storage and utilization, p.227-260. Madison,WI: Am.Soc.Agron.

Muck, R.E. 1987a. Dry matter level effects on alfalfa silage quality I. nitrogen transformations. Trans.ASAE 30:7-14.

Muck, R.E. 1987b. Factors affecting numbers of lactic acid bacteria on lucerne prior to ensiling. In Eighth silage conference: summary of papers, p.3-4. Hurley,UK:Agric.Food Res.Counc.

Muck, R.E. 1988. Factors influencing silage quality and their implications for management. J.Dairy Sci. 71:2992-3002.

Muck, R.E. & J.T.Dickerson 1988. Storage temperature effects on proteolysis in alfalfa silage. Trans.ASAE 31:1005-1009.

Muck, R.E. & P.L.O'Connor 1985. Initial bacterial numbers on alfalfa prior to ensiling. ASAE Paper No. 85-1538. St. Joseph,MI: ASAE.

Muck, R.E. & P.L.O'Connor 1988. Effect of inoculation level on alfalfa silage quality. ASAE Paper No.88-1070. St. Joseph,MI: ASAE.

Muck, R.E. & R.P.Walgenbach 1985. Variations in alfalfa buffering capacity. ASAE Paper No. 85-1535. St. Joseph,MI: ASAE.

Ohshima, M. & P.McDonald 1978. A review of the changes in nitrogenous compounds of herbage during ensilage. J.Sci.Food Agric. 29:497-505.

Shinners, K.J., R.G.Koegel & R.J.Straub 1987. Drying rates of macerated alfalfa mats. Trans.ASAE 30:909-912.

Shinners, K.J., R.G.Koegel & R.J.Straub 1988. Consolidation and compaction characteristics of macerated alfalfa used for silage production. Trans.ASAE 31:1020-1026.

Shinners, K.J., R.G.Koegel & R.J.Straub 1989. Rapid field drying of forages utilizing shredded mat technology. In Proceedings of the 11th International Congress on Agricultural Engineering. Rotterdam: Balkema.

Smith, D. 1981. Removing and analyzing total nonstructural carbohydrates from plant tissue. Agric.Extens.Publ. R2107. Madison: Univ.Wisconsin-Madison.

Weast, R.C., S.M.Selby & C.D.Hodgman (eds.) 1964. Handbook of chemistry and physics. Cleveland: Chemical Rubber Co.

Land and Water Use, Dodd & Grace (eds), © 1989 Balkema, Rotterdam. ISBN 90 6191 980 0

Rapid field drying of forages utilizing shredded mat technology

K.J.Shinners
Agricultural Engineering Department, University of Wisconsin, Madison, Wis., USA

R.G.Koegel
USDA/ARS Dairy Forage Research Center, Madison, Wis., USA

R.J.Straub
Agricultural Engineering Department, University of Wisconsin, Madison, Wis., USA

ABSTRACT: If forages are shredded and then properly formed into a continuous mat, they can be field-dried from cutting moisture to safe baling moisture in a single day, thereby reducing the risks of weather damage. Shredding could also increase forage intake and rate and extent of digestability. An economic analysis has shown that the shredded forage mat technology can be economically attractive.

ZUSAMMERFASSUNG: Wenn Grünfutterpflanzen intensiv aufbereitet und anschliessend in kohäsiver Matten gepresst werden, können sie auf dem Stoppelfeld innerhalb eines Tages trocknen, sodass sie sofort mit einer Ballenpresse geerntet werden können. Dieses Verfahren vermindert weitgehend Wetterverluste. Die intensive Aufbereitung kann auch den Verbrauch und die Verdaulichkeit des Futters steigern. In einer ökonomisohen Analyse hat sich die Mattentechnik als lohnend erwiesen.

RESUMÉ: Lorsque les fourrages sont macérés au moment de coupe et pressés en une mince couche, appelée "natte", ils peuvent en un seul jour sécher jusqúa un niveau qui permette le ramassage en balles. Ce procédé réduit beaucoup les risques de dommages du à la pluie. La macération peut aussi accroitre les quantités de fourrages ingérés ainsi que la digestibilité. Une analyse économique a montré que la technique "natte" à fourrage macéré peut être rentable.

1. INTRODUCTION

Forages contain large quantities of proteins and other valuable nutrients per unit land area, help conserve soil, require relatively low energy inputs and are an important source of feeds for ruminant animals. Therefore, forages have been an important agronomic crop for centuries. For as long as forages have been cut, dried, and placed in storage for future use, there has been a desire to field dry the crop as quickly as possible to avoid the large losses which occur when the crop is exposed to adverse weather. In the 1930's it was suggested that fracturing the plant stem by passing it through a set of rolls after cutting would reduce the field drying time because moisture would have a less restrictive route out of the plant (Zink, 1933). This practice did not gain widespread acceptance in the US until the 1950's, by which time many of the technical aspects of this process had been improved (Bruhn, 1955). Today, the use of intermeshing rubber roll conditioners are the most common method to mechanically improve forage drying in the US, while brush and flail type conditioners are used in Europe (Hale and Klinner, 1981).

The above system has limitations. As the extent of mechanical conditioning is increased, mechanical losses also increase due to the many small plant parts (mainly leaves) removed from the stem. In the case of alfalfa (Medicago sativa), these leaves contain a large percentage of the plants protein and their loss results in significant quality reduction. Furthermore, the limp material resulting from extreme mechanical treatment forms dense layers in which the drying rate may actually be lower than with less severely conditioned material.

Measurements of solar radiation in the north central US indicate that the energy available is normally sufficient to evaporate enough water to dry forage to baling moisture in less than 1 day. Since two or three days are typically required, it appears that conventional field drying of forages uses the available solar energy very inefficiently. This is to be expected, since in order for plants to prosper they must have the ability to limit water loss to the environment. This ability is, at least partially, retained after the plant has been cut. Dernedde (1985) has proposed three means for radically decreasing field drying time : (a) increase the absorption of solar energy by the cut forage in order to vaporize the plant moisture more quickly; (b) decrease resistance of water vapor movement from the plant into the atmosphere; and (c) prevent any rewetting of the plant material from the environment.

Research and development has been carried out at the University of Wisconsin on a rapid-drying forage process that incorporates the three goals proposed by Dernedde. This process, known as shredded forage mat technology, consists of four basic steps: (a) mowing; (b) maceration (also known as shredding or fibrizing); (c) formation of the shredded forage into a thin, cohesive mat; and (d) depositing the formed mat onto the field stubble for drying. In this paper, maceration and shredding are used interchangeably. Ajibola et al. (1980) showed that a maceration process, in which alfalfa was both shredded and bruised, which darkens the forage color, increased the solar absorption of the material. If shredded forage mats are spread over a larger fraction of the field than is typical for conventional methods then solar absorption will be further improved. Since maceration destroys the plant structure, the resistance to moisture movement out of the material is minimized because the length of the flow path is reduced and the specific surface area is increased. Research has shown (Shinners et al., 1987a) that shredded forage mats dry to baling moisture (20% wet basis) in 2 1/2 to 6 h under good drying conditions. Thus, rewetting by dew at night can normally be eliminated, and the probability of being rained upon can be greatly reduced.

In order to retain the many small plant parts that are created during the maceration process, it is necessary to form the forage into a thin, contenuous, cohesive "mat" immediatley after maceration. This mat is then placed on, and supported off the ground by, the crop stubble. Previous research (Risser et al., 1985; Shinners et al., 1985) had defined some of the requirements for maceration and mat formation and certain machine configurations were developed. Based on this expe-

rience a complete field-going machine was designed and constructed which incorporated the above four basic functions (Koegel et al., 1988). The prototype was smaller than farm-scale, having a cutting width of 120 cm. Fig. 1 shows a schematic representation of a shredded mat machine.

2. MACHINE COMPONENTS

2.1 Cutting Mechanism

The cutting mechanism used on the prototype mat formation machine was a reciprocating knife type that was not significantly different from those in general use. This type of cutting mechanism was chosen because of its simplicity and its low power requirements per unit width. A parallel-tine reel was used to sweep the cut material onto a belt type conveyor for transport to the macerator. Re-positioning of the macerator could eliminate the need for the conveyor.

2.2 Macerator

A schematic of two configurations used for the macerator is shown in Fig. 2. The maceration device must : (a) severely condition (macerate, shred, fibrize) the forage to increase its specific surface area and reduce its resistance to moisture removal; and (b) perform this severe conditioning without unduly reducing the fiber length of the material. Previous research (Shinners et al., 1987b) has determined that long fibers are required for the subsequently formed mat to have sufficient strength to be supported by the forage stubble.

The process by which forages are severely conditioned while still maintaining fiber length is to repeatedly shear the plant stem along its longitudinal axis. The theoretical energy to shear an alfalfa stem longitudinally was found to be approximately 1/10 the energy required to shear it transversely per unit area (Shinners et al., 1987c). Therefore, a macerator which incorporated this type of processing was hypothesized to have relatively low power needs.

Both configurations shown in Fig. 2 operate on the same principal. If forage stems are fed approximately parallel to the shafts of the rotating elements, and there is a difference in surface speeds of opposing elements, the material will be sheared along the stems longitudinal axis. The degree of maceration can be controlled by the speed ratio between opposing rotating elements, the clearance between the elements, and the number of active nip points.

It has been found (Shinners et al., 1988a) that a straight coin knurl surface configuration provided a sufficiently aggressive surface for shredding and feeding and was also self cleaning. The knurled surface was easy to fabricate and should lend itself to renewal without removing the elements from the machine. All elements were successfully driven using a serpentine double V-belt. Speed ratios between the drum and the rolls of 1.3-1.5:1 were recommended for alfalfa and 2.0-2.5:1 for grass. The number of active nip points should be 6 for alfalfa and 8 for grass. Minimum surface speeds should be 1000 m/min.

2.3 Press

The macerated forage must be formed into a mat for two reasons: (a) to facilitate handling and prevent losses and (b) to improve the forage drying rate. Because many small particles are created during the maceration process, it is necessary to form the resulting material into cohesive mats to prevent excessive losses when the forage is placed on the stubble for drying. Continuous mats that are supported by the forage stubble when dry are also easily picked-up with a minimum of material loss. It has been shown (Shinners et al., 1987a) that mats that are not supported by the stubble and come in contact with the soil will dry at significantly reduced rates.

The most important physical parameter of the macerated forage that effects the wet strength of the formed mat was found to be the quantity of long fibers remaining after maceration (Shinners et al., 1987b). Wet strength of the mats was also affected by the pressure of formation and the length of time under the formation pressure. Risser et al. (1985) determined that mats of acceptable wet strength could be produced with pressures of 690 kPa and time of formation as short as 0.25 s. As the pressure of formation is increased, the time of formation can be reduced.

Two press configurations used to form mats are shown in Fig. 3. The belt and drum press utilized a long arc of contact in order to ensure a long time of formation and a reduced formation pressure. The 120 deg arc of contact results in a dwell time of pressure of 0.4 sec at forward speed of 6.4 km/h. Moderate belt tension (2.6 kN/m) is used to pre-compress the material before pneumatic roll pressure is applied. Pneumatic rolls were used to increase the effective area over which the roll force was applied. Roll force can range from 4 kN/m to 20 kN/m and is usually increased from first to third rolls.

The drum and release belt are perforated to allow the escape of excess plant juice expressed during the mat formation. This juice should be collected and placed back onto the top surface of the mat to prevent nutrient loss.

The roller-to-roller contact press utilized a set of rollers forced together under high force to press the material into a mat in a short formation time, thus reducing the complexity of the press. A problem often encountered with roller-to-roller contact presses was the difficulty of uniform feeding of wet material through the press because the material would "flow" back from the roller nip point rather than feed through the press. A pressing belt was included in this configuration which by means of its tension held the material by friction between the input and pressing belts.

Both rolls should be driven to provide satisfactory performance and at least one belt should be porous to allow a free escape for moisture expressed during formation.

The press should be configured in such a way that it can receive the shredded material directly from the macerator outlet and place the mat continuously onto the stubble. The mat should not drop more than approximately 5 cm onto the stubble in order to prevent mat damage. Since the press peripheral speed must be equal to ground speed in order to prevent mat damage, the press should be ground driven from the machine wheels. The press should be allowed to "float" over irregularities in the field and be raised for transport.

2.4 Functional Performance of Components

2.4.1 Macerator

Both macerator configurations have been used extensively. The drum macerator with a 40 cm diameter, 71 cm width drum has produced capacities in the range of 7.8 to 11.3 Mg wet matter /h at specific energy requirements of 1.9 to 1.6 kWh/Mg wet matter. Both capacity and specific energy requirements of the macerator were dependent on the element speed ratio and the number of active elements (Shinners et al., 1988a). The roll type macerator has performed satisfactorily although no performance data has been quantified.

2.4.2. Press

The belt and drum type press has been used extensively and performance has been satisfactory. Belt tracking has been a problem, but the addition of rolling edge guides has eliminated this. The pressed mat occasionally has difficulty releasing from the press belt. A "doctor blade" that forces a small volume of air between the mat and the press belt has helped solve this problem. The mat release belt was run over a small diameter idler roll near the point of release onto the stubble, so that belt flexure aided in separating the mat from the release belt. The specific energy for processing in this press ranged from 0.4 to 0.6 kWh/Mg wet matter (Koegel et al., 1988). The roller-to-roller press has been used only in minor laboratory tests, although with satisfactory results.

Initially, difficulty was experienced with obtaining a uniform flow of macerated material onto the press belt. There was also a tendency for fine material to be placed on the press belt first and for coarse material to be placed on top of the fines. This appeared to be due to a difference in trajectories between fine and course particles as they moved from the macerator outlet to the press belt. If material enters the press in a non-uniform manner, a non-continuous mat is formed, which will not be supported by the stubble. The non-uniform flow will also produce thick spots of mat which will dry with great difficulty. A continuous mat will more easily release from the press belts. When material segregation occurs, the fines on the bottom of the mat causes an impervious layer which slows the drying rate of the mat.

A cylindrical spinner with radial fins was placed at the outlet of the macerator to help allieviate the above problems. This spinner has equally spaced, full width radial rubber flights with a maximum diameter of 21 cm. The spinner rotates at 800 rpm. Material coming out of the macerator impinges on the spinner, and is then thrown out in "rooster tail" fashion over the exposed part of the press belt. This performed two functions. By slowing the velocity of the material out of the macerator and mixing it as it is thrown onto the press belt, the tendency for the material to segregate was greatly reduced. By spreading the material over the exposed length of the press belt, the tendency for uneven flow of material to cause lumps of material is reduced because the material flow is "averaged" over the belt length.

3. FUNCTIONAL PERFORMANCE OF PROCESS

3.1 Drying Rate of Mats

The ultimate goal of the mat harvesting process is to ensure that the forage crop can be harvested as dry hay within one daylight period to assure consistent harvest with little or no loss due to adverse weather. A properly formed alfalfa mat can be placed on the stubble and dried to 20% moisture (wet basis) within one day. Drying rate of an alfalfa mat was found to be inversely proportional to mat thickness, but a properly formed and supported mat of 6-7 mm thickness was found to dry under good weather conditions in times ranging from less than 3 h to no greater than 6 h (Shinners et al., 1987a). Mats greater than approximately 7 mm thickness did not dry within the required single daylight period (Shinners et al., 1985). A dense, thick mat apparently restricts air movement to the lower, moist layers which reduces the drying rate. For a given mass, mats that have more surface area (a thin mat) also adsorb more radiant solar energy, which tends to speed the drying process. To achieve the desired single day drying, the mat should have a maximum thickness of 7 mm, which for a dense crop will cover a minimum of 65 to 70% of the area from which the forage was cut (Shinners et al., 1985).

Most drying rate trials thus far have been conducted using alfalfa. Some tests have recently been conducted using grass. Experiments conducted in a controlled environment in Quebec, Canada (Savoie et al., 1989) have shown that mats formed from grass dried 2-3 times as fast as equal mass per unit area normally conditioned swaths and that single day drying of grass mats is possible. Similar results were found in limited trials conducted at Kiel, West Germany (Lüusse, 1986) using grass to form mats.

3.2 Rain and Respiration Loss

Losses from respiration and rain damage were determined from experimental tests conducted in East Lansing, MI, (Rotz et al., 1988). Alfalfa mats and normally conditioned alfalfa swaths were field cured on trays and the loss of dry matter between the fresh crop and dry crop measured. A portion of the samples were exposed to controlled rain from an overhead sprinkler system. Rain amounts of 0, 5 and 17.5 mm were used on various samples.

The rate of dry matter loss due to rain damage was approximately 6 times the rate found with normally conditioned swaths. This high rate of loss was to be expected because the plant structure is so disrupted by the maceration process that soluble plant nutrients should be easily leached from the plant by the rainfall. However, because forage mats dry so quickly compared to normally conditioned swaths, the risk of rain damage is considerably reduced. The rate of loss due to respiration was very small and similar to that which occured in the normally conditioned swaths.

3.3 Harvesting and Storing

A modified windrow pick-up has been developed to lift the mats from the stubble. This pick-up consisted of a wide belt with many plastic fingers attached. The belt was driven from a ground support wheel such that its speed was matched to ground speed. This pick-up was very similar to those used currently by combines to lift windrowed small grains from the stubble. The pick-up has been used to successfully lift both wilted mats for silage, and dry mats for hay. Continuous mats were more easily lifted than broken mats, but the pick-up was also successful in lifting small mat sections created by a non-continuous mat formation.

The pick-up was mounted on a small rectangular baler, replacing the normal tine type pick-up. Mats were successfully baled with this baler with only one minor modification. Because of the compliant nature of the plant structure, the formed bales tended to be quite limp. Wedges were added to the bale chamber to further restrict the flow of hay through the chamber, thereby increasing bale density. This modification was moderately successful, although further flow restriction modifications are planned. Although neither pick-up or bale chamber losses were quantified, it was felt that these losses would be about the same as those experienced with normal harvesting practices.

A machine has also been developed to lift wilted mats from the stubble for silage production. This machine was a modified double-chop, flail type forage harvester. The flail rotor was removed, the front housing removed, and the above mentioned belt type pick-up mounted to deposit the mat directly into the machine cross-auger. The action of the cross-auger, blower knives and blower paddles effectively broke or chopped the mat into handleable pieces. The material was transported pneumatically into a self-unloading wagon.

Consolidation and compaction characteristics of macerated alfalfa used for silage production have been determined using laboratory sized silos (Shinners et al., 1988b). Macerated alfalfa consolidated at a faster rate than conventionally chopped alfalfa but maceration did not affect the final bulk density. This would imply that tower silos could be filled more quickly and silage quality might be im-

proved due to faster oxygen exclusion, but that silo capacity would not be increased. Maceration improved the dynamic compaction characteristics of alfalfa compared to chopped alfalfa, especially at higher moistures. This could mean reduced number of compaction trips required when placing forage in horizontal storage structures.

The ensiling characteristics of shredded alfalfa mats have been determined in laboratory mini-silos (Muck et al., 1989). The shredded alfalfa mat silage reached final pH twice as quickly as normally harvested alfalfa. Other silage parameters also indicated that the alfalfa mat silage ensiled better than the control silage.

3.4 Feed Utilization

Maceration of alfalfa has been found to have a positive effect on the feeding value of the forage to the ruminant animal (Hong et al., 1988a). The neutral detergent fiber (NDF) constituent of macerated alfalfa was found to be 12% more digestible than conventional alfalfa hay in a sheep feeding trial. The increased digestibility allows the animal to receive more energy from the forage. Dry matter intake of alfalfa mats was increased 6% over conventional alfalfa hay in the same sheep trial. Daily production of 4% fat corrected milk (FCM) was increased 12% in goats fed macerated alfalfa. A scanning election microscope study (Hong et al., 1988b) revealed that maceration increased the specific surface area of alfalfa stems by damaging the structural integrity of the stems. This allowed greater bacterial colonization of the cell walls and explains the positive effect of maceration on fiber digestion.

The improvement in feed quality of alfalfa has not been confirmed through feeding trials with dairy cows, although this is planned for the near future. The effects of maceration of the feed value of grass has also not been determined.

4. ECONOMIC ANALYSIS

A simulation of the economics of maceration and mat drying of alfalfa on dairy farms in the North Central US has been performed (Rotz et al., 1988a). This simulation was conducted using 26 years of weather conditions in a model of the dairy forage system developed by Rotz et al. (1988b). Maceration and mat drying of alfalfa can be economically used in hay production on a dairy farm. The reduction in field losses decreases feed costs enough to more than offset the increased initial and operating costs of the mat making machine compared to a standard mower-conditioner. When maceration is assumed to improve the digestibility and intake of alfalfa, the increase in farm return is up to 5 times the value spent on increased machinery costs.

The economic benefit of implementing the maceration and mat drying process on a farm producing only silage was limited. The economic benefit of the mat system is very sensitive to the increase in digestibility obtained through maceration, the initial cost and power requirement of the mat making machine, and the drying rate of the forage mat. The rate of loss from the machine treatments and rain damage have a small to moderate impact on the economic benefit.

5. CONCLUSIONS

The shredded forage mat harvesting system has considerable promise as a method to increase forage quanity and quality by reducing field drying time to a single day. Many important considerations of the process, including mat drying rate, mat uniformity and strength, machine energy requirements, functional performance of components, some storage characteristics of the mats, and feed utilization, have been determined in the laboratory or using small scale field equipment. Economic simulation has shown that the process has considerable promise to improve dairy farm profitability.

Only limited work has been done as yet on picking up the dried mats and processing them for storage as dry hay or silage, although initial results are very promising. Field loss data and storage characteristics compared to conventional methods are important information yet to be obtained. Because feed utilization has been shown to be important in the economic benefit of the mat system, full scale feeding and lactation studies utilizing dairy cattle should be conducted.

Research on the shredded forage mat technology has been initiated at several North American and European agricultural research institutions, in addition to the work that continues at the University of Wisconsin. One institution is developing a farm scale (2.2 m width) mat making machine scheduled to be ready for field trials during the summer of 1989. Several North American and European agricultural equipment manufacturers have expressed interest in developing a farm scale prototype. It appears that shredded forage mat harvesting has the potential to become the forage harvesting system of the 1990's and beyond.

REFERENCES

Ajibola, O. O., R. G. Koegel, and H. D. Bruhn. 1980. Radiant energy and its relation to forage drying. Transactions of the ASAE 23(5):1297-1300.

Bruhn, H. D. 1955. Status of hay crusher development. Agricultural Engineering 36(3):165-170.

Dernedde, W. 1985. Vorwelken Von Gras. Halmgulernetchnik. VDI - KolloquimLandtechnik Heft 2.

Hale, O. D. and W. E. Klinner. 1981. An initial performance comparison of experimental brush conditioning system, using a new method of assessment. Divisional Note 1053. AFRC Engineering, Silsoe, Bedford, England.

Hong, B. J., G. A. Broderick, R. G. Koegel, K. J. Shinners and R. J. Straub. 1988a. Effect of shredding alfalfa on cellulolytic activity, digestibility, rate of passage, and milk production. J. Dairy Sci. 71:1546-1555.

Hong, B. J., G. A. Broderick, M. P. Panciera, R. G. Koegel and K. J. Shinners. 1988b. Effect of shredding alfalfa stems on fiber digestion determined by in vitro procedures and scanning electron microscopy. J. Dairy Sci. 71:1536-1545.

Koegel, R. G., K. J. Shinners, F. J. Fronczak and R. J. Straub. 1988. Prototype for production of fast-drying forage mats. Applied Eng. in Agric. 4(2):126-129.

Lüusse, T. 1986. Drying of forage in mats. Senior thesis, University of Kiel, Department of Agricultural Engineering. Kiel, West Germany.

Muck, R. E., R. G. Koegel, K. J. Shinners and R. J. Straub. 1989. Ensilability of mat-processes alfalfa. In Proceedings of the 11th International Conference on Agricultural Engineering. Dublin, Ireland.

Risser, P. E., R. G. Koegel, K. J. Shinners and G. P. Barrington. 1985. Factors affecting the wet strength of macerated forage mats. Transactions of the ASAE 28(3):711-715, 721.

Rotz, C. A., R. G. Koegel, K. J. Shinners and R. J. Staub. 1988a. Economics of maceration and mat drying of alfalfa on dairy farms. ASAE Paper No. 88-1550, ASAE, St. Joseph, MI, USA, 49085.

Rotz, C. A., J. R. Black, D. R. Mertens, and D. R. Buckmaster. 1988b. DAFOSYM: A model of the dairy forage system. J. Prod. Agric. In Press.

Savioe, P., S. Beauregard, A. Abbar. 1989. Maceration of grass hay. To be presented at the ASAE Summer Meeting, Quebec City, Quebec, Canada. ASAE, St. Joseph, MI, USA, 49085.

Shinners, K. J., G. P. Barrington, R. J. Straub and R. G. Koegel. 1985. Forming mats from macerated alfalfa to increase drying rates. Transactions of the ASAE 28(2):371-374.

Shinners, K. J., R. G. Koegel, G. P. Barrington and R. J. Straub. 1987b. Physical parameters of macerated alfalfa related to wet strength of forage mats. Transactions of the ASAE 30(1)22-27.

Shinners, K. J., R. G. Koegel, G. P. Barrington and R. J. Straub. 1987c. Evaluating longitudinal shear as a forage maceration technique. Transactions of the ASAE 30(1):18-22.

Shinners, K. J., R. G. Koegel and R. J. Straub. 1987a. Drying rates of macerated alfalfa mats. Transactions of the ASAE 30(4):909-912.

Shinners, K. J., R. G. Koegel and R. J. Straub. 1988a. Design considerations and performance of a forage maceration device. Applied Engr. in Agric. 4(1):13-18.

Shinners, K. J., R. G. Koegel and R. J. Straub. 1988b. Consolidation and compaction characteristics of macerated alfalfa used for silage production. Transactions of the ASAE 31(4):1020-1026.

Zink, F. J. 1933. The mower-crusher in hay making. Agricultural Engineering 14(3):71-73.

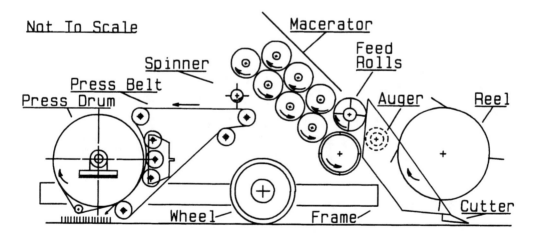

Fig. 1 - Schematic of mat making machine.

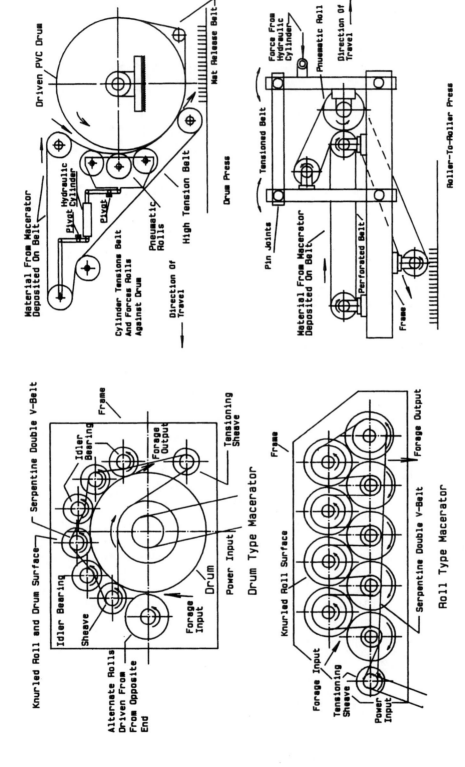

Fig. 2 – Schematic of macerator.

Fig. 3 – Schematic of mat press.

3.3 Application of plant nutrients and pesticides

Land and Water Use, Dodd & Grace (eds), © 1989 Balkema, Rotterdam. ISBN 90 6191 980 0

Keynote paper:

Application and control of the distribution of plant nutrients and pesticides

A.Nordby
Norwegian Institute of Agricultural Engineering, Norway

Abstract: The essential features of the equipment and the development of equipment for applying fertilizers and plant protection chemicals are emphasized. Problems with and the demands on precision of discharge, evenness of distribution, effectiveness of deposit on the target, and the minimization of the loss of chemicals drifting outside the treated fields are discussed. The choice of methods and equipment is discussed and comparisons between them are made. Finally the state of the art is suggested.

In general we need an improvement of both spreading and spraying. The agriculture is accused of incorrect use of fertilizer and plant protection chemicals:

a. Using too much
b. Uneven spreading of fertilizers and chemicals
c. Too low deposit on the target
d. Drift
e. Losing chemicals to the soil, brooks, rivers and lakes outside the treated field

The government in many countries try to improve the situation. In a resolution made by our government and the government in Sweden the aim is to reduce the use of plant protection chemicals with 50 %.

- Laws, rules and directives about the use and spreading of fertilizer and plant protection chemicals
- Better training of the crews
- Certification of the crews
- Keep the equipment in good order (voluntary tests and inspections)
- Better labels on the pesticide packages
  Landers (1988), Jonsson (1989) and Stubsjøen (1989)

1. SPREADING OF FERTILIZERS

1.1 Main methods for spreading fertilizers

Good surveys of various methods and equipment for fertilizers are given by Kepner, Baines and Barger (1978) and by Culpin (1987).

Some of the main methods of the spreading of fertilizers:

a. Broadcast before ploughing
b. Deep placement with chisel type cultivator
c. Broadcast and mixed into the soil after ploughing and before planting
d. Surface distribution to a growing crop
e. Drilled into established pastures and other sods with special equipment
f. Apply fertilizers as solutions, for instance a solution of NH_3 in water. Anhydrous ammonia is used as a fertilizer in USA and on a small scale in some European countries

Dry fertilizers, granulated or prilled, can be distributed from the air, by fixed wing planes or by helicopters.

1.2 Some parameters for the designing of fertilizer spreaders

a. Discharge rate, proportional to the forward speed
b. Discharge rate, adjustable in small increments
c. Definite relation to the reference provided on the unit
d. Accurately made
e. Easy to empty and disassemble for thorough cleaning
f. Corrosion-resistant materials, especially for the working parts

1.3 The fertilizer

Our main fertilizers are granulated or prilled. The main method used is spreading from the ground. In some countries and for certain purposes fertilizers are spread as solutions both of single constituents and of N-K-P.

The properties of the fertilizers affect both discharge, distribution, working conditions for the crew and drift to the neighbouring areas.

The particle size, evenness, drillability and hygroscopicity are not the same for the various fertilizers. The spreader has to be set for each fertilizer used and the discharge rate has to be controlled.

1.4 Fertilizer distributors

Spreaders with spinning discs are widely used. They are light and relatively cheap. Under good weather conditions (wind and humidity) and with good control of the working width the distribution can be sufficiently good.

Bulk hoppers with air delivery have come into common use during the last ten years. Working width up to 25-30 m. If one takes care of the working width, the machines can do a good job. The machines are expensive and therefore used on big farm and by contractors.

The combined grain and fertilizer drills were developed for placing the fertilizer in a favourable position in the soil relative to the seeds and plants. Due to the high weight of both seed and fertilizer in the hoppers of such machines their working widths are normally small, from 2.5-5 m. The machines suit the small and medium size farms.

When it is impractical or impossible to do the job from the ground, for instance in forests, in rough terrain, or in flooded rice fields, fixed-wing aircraft and helicopters can be used.

1.5 The distribution of fertilizer

The exact aim of the evenness of distribution is not easy to define. Therefore it seems not to be exactly defined, so far. Culpin (1987) reckoned a CV of 10-15 % as a reasonable aim. A test report from Statens jordbrugstekniske Forsøg, Denmark, Report no 284 (1983), reports a coefficient of variation of 9.2-12.8 %. This is characterized as fairly good (in Danish, "ret

god"). Statens Maskinprovningar, Sweden, presents the following table in Report no. 3118 (1987):

Table 1. Estimation of the evenness of distribution of fertilizers

CV, %	Distribution of the fertilizer
0- 5	Good
6-10	Sufficient
11-20	Not completely sufficient
>20	Poor

Most of the test reports today present the evenness of distribution as the CV in percent. This makes it possible to compare results. But of course, the testing methods and the equipment used play an important role.

In Table 2 the distribution results are presented for the most common spreader types. If these results could be transferred to practice the situation would look promising. But this is not the case. Culpin (1987) presents some results from an ADAS study in 1980. The distribution of the fertilizer was measured on 177 machines of 18 different makes. With most of the machines it was possible to achieve a coefficient of variation of 20 % or less. But half of the machines were above the barely satisfactory performance measure, CV 10-15 %, which is reckoned as reasonable.

In experimental field plots it is also important to spread the fertilizer at the wanted rate and well distributed. A lot of different kinds of equipment have been designed, manufactured and used in field plot work. Most of them discharge the fertilizer to the wanted rate with a sufficiently good distribution.

A high deposit on the target well distributed is very desirable. Kepner, Baines and Barger (1978) state that the effects of variations in distribution have not been established. It is quite certain that application rates could be reduced, thereby reducing costs and residues problems, if a better distribution than today could be achieved.

2. SPREADING PLANT PROTECTION CHEMICALS

2.1. Developing methods and equipment

Many researchers have presented their points of view on methods and equipment, such as Combellack (1984), Hagenvall

(1985), Hislop (1987) and Nordby (1979).
Miller (1987) has described the state of
the art of spray application technology.

The most important kinds of equipment for
the farmers is developed and made by
specialized manufacturers. If the manufacturers have competent people in their
developmental departments and they keep in
contact with the research work, I believe
this is the main way to plant protection
equipment also in the future.

It seems to be almost impossible to cooperate well in developing plant protection
equipment. Manufacturers of plant protection chemicals and equipment, governmental
institutes, organizations and private
persons have very differing interests. When
profit goes into it, then the problems
start. A link between the manufacture and
sale of chemicals and that of equipment may
also be difficult to handle in the long
run.

If you look around and see what equipment
they are using in the different countries,
you can notice some differences. What is
the reason?

- The crops cultivated
- Sizes of the farms
- Diseases and needs to fight them
- Development of equipment
- Equipment manufactured in the country
- Marketing of equipment
- Advisory set-up
- Research work and results

2.2 Where goes the spray ?

The agriculture has been accused of uncritical use of plant protection chemicals.
A great deal of the criticism is fair. But
as elsewhere in the society, one is now
going too far in the other direction.
Everybody now seems to be experts on dose
and equipment. The agriculture should meet
this challenging situation with knowledge
and safe use of the chemicals. If not,
somebody else will tell us how to use
chemicals in agriculture.

Where is the target for the spray ? That
differs with crop, disease, chemical, time
of treatment and other factors. Everyone
spraying should know where the target is,
and set up and use the equipment accordingly. As much as possible of the spray should
be deposited on the target. Today the
deposit is too uneven and its percentage
too low.

Too small an amount of the spray is deposi-

ted on the target. For crop sprayers, the
deposit is found to be from 40 to 75 % of
the total spray delivered (Nordby &
Jørgensen 1985 - Robinson & Garnet 1984).
In practical spraying it is quite certain
that it is less.

By mist spraying of fruit trees, the result
is also unsatisfactory. Before the leaves
have broken only 10-20 % is deposited on
the trees, and with fully grown leaves,
still only 25-60 % of the total spray is
deposited. (Herrington, Mapother & Stringer
1980 - Vittrup 1965).

Spray drift has to be considered by every
spreading of chemicals. Increases in wind
speed, boom height and working pressure
increase the drift by crop spraying. By
uncritical use of mistsprayers you can get
a lot of spray drift. The driving direction
in relation to the wind direction is also
important. Some countries have implemented
rules that, for instance, permit no
spraying if the wind speed is more than
5 m/s measured two metres above the ground.

The spray drift is a main problem and has
to be reduced. A low spray deposit rate on
the target also means deposit on the soil
and drift of chemicals outside the field to
be treated, into lakes, rivers, forests and
pastures.

2.3. Research work on pesticide
     application.

Every country has some pesticide application research. The work can be carried out
by:

- Governmental institutions
- Organizations
- Manufacturers of plant protection
  chemical
- Manufacturers of plant protection
  equipment
- And others

If we want the results to be transferred to
practical use and to give basic information
for further work, then we need good,
documentary reports on the work. This could
for instance cover:

- Droplet size
- Liquid distribution from the sprayer and
  of the deposit on the plants
- Driving speed
- Boom height
- Nozzle type and size
- Working pressure
- Volume rate

- Quantity of the spray measured as deposit on the target
- Quality of the spray measured as deposit on the target
- Percentage of the spray deposit on the target
- Weather conditions
- Spray drift

So far there are few papers which present such documentation. If data on for instance distribution, deposit, driving speed, and working pressure are missing, then it is impossible to make good use of the results. There are examples in which the documentation is good, for instance in papers from Long Ashton Research Station in England. They also measure where the spray goes, and they make correlations of these measurements with biological results. (Hislop 1987).

In Norway we are also trying to get results that can be used in practice and that can form a basis for further work. (Skuterud, Nordby and Tyldum 1988). Investigations that give useful answers to problems are expensive. You need scientists and assistants, equipment in labs and in the field, grants to cover the expenses. In many countries there is only one way: Give the grant and the possibilities to people who have a chance to cope with the problems.

2.4 The distribution of plant protection chemicals

The distribution of spray from spray booms are often measured with the boom stationary and with standardized equipment and methods. Norsk standard, NS 1170 (1968). The liquid is collected in 5 or 10 cm wide channels. The results of such measurements can give a CV of 5-12 %, at pressures of 2-10 bar, a boom height of 40 cm, and with 110^0 fan nozzles spaced at 50 cm. Here again it is difficult to find what is an acceptable CV. A CV less than 10-12 % is characterized as very good. With the nozzles on the market, for instance Albuz, Delavan, Hardi, Lechler, Lurmark and Teejet, it is possible to obtain such results. In Germany they earlier used the deviation from the mean. The deviation should not exceed ±15 %. In "Merkmale Spritz- und Sprühgeräte für Flächenkulturen" 1.1.1.1.(1988) is a CV of 7-9 % a demand. This is based on 10 cm wide channels.

What is happening with the distribution in the field? Hislop (1987) presents some CV percentages for the spray deposit. These are valuable figures for our evaluating of the results.

2.5 Main methods and equipment used

Table 3 gives a short survey of methods.

There are few granular chemicals on the market. Some of the granule spreaders are difficult to calibrate. The distribution of the granules is unsatisfactory.

The crop sprayer has in principle been the same the last hundred years, with nozzles spaced along a pipe. The crop sprayer is robust, simple and relatively cheap. It has been discussed and criticized since the Second World War. But still spraying is the main method for applying plant protection chemicals. In Scandinavia and on the Continent I guess 85-90 % of the plant protection chemicals is applied by spray booms.

In agriculture we have many different targets for the spray. It is still quite uncertain what is optimal in each case considering deposit on target, distribution, penetration and loss of spray. The fan nozzle produces a range of droplets. To a certain extent the droplets can be controlled by discharge rate and working pressure. In Scandinavia we have reduced the pressure during the last ten years. In weed spraying with 110^0 fan nozzles spaced at 50 cm, 2 bar is used. For spraying against potato blight the working pressure is from 5-8 bar. Bjugstad (1987).

The volume rate in field spraying may vary from 60-400 l/ha. Cooke et al. (1988) and Skuterud et al. (1988) conclude that the crop sprayer is still the best and most versatile piece of plant protection equipment.

For UK Hislop (1987) advises that in large scale arable farming the existing CDA equipment should be more used. The big farmers could have one piece of CDA equipment and one ordinary crop sprayer. If they were commercially available, the spinning-disc electrostatic and the electrodynamic type of equipment should be more used. The crop sprayer is still the main type of equipment in Scandinavia and on the Continent. In some countries CDA is hardly seen in practical use. 5-10 years ago there was a great deal of interest in CDA. Due to overoptimistic advertisements and problems with the equipment during work the interest has rapidly waned. Results from trials are variable and do not stimulate the interest.

Air assisted nozzle systems are a quite new development. Results published by Hardi, Denmark (1988) look quite promising. The penetration and the drift of spray seem to be controlled quite well.

Mistspraying is one of the methods that have brought the spreading of plant protection chemicals in discredit. A big part of the spray is lost. The working situation for the crew is also bad, especially with the knapsack-mistsprayers. On new types of mist sprayers the air and the liquid are directed more horizontally against the trees. With low trees and a better adjustment of the air and the liquid the deposit percentage on the target should increase and the loss of spray decrease.

Aerosol (fog) is used in glasshouses. There is discussion about the choice of chemicals and the use of the equipment. For the crew it seems safer to place the equipment outside the houses.

Electrostatics has been investigated carefully during the last 10 years. But some of the results presented are non-consistent. Law (1987) gives a survey of basic phenomena that are active in electrostatic pesticide spraying. He concludes with optimism, in the area of electrostratic crop spraying technology, that sufficient fundamental understanding and engineering design know-how now exist internationally to support its routine usage in production of food and fibre for mankind. Hislop (1987) states that spinning disc electrostatic spraying or electrosta-tic dynamic spraying would be preferred if they became commercially available. The situation is not clear at the moment. I doubt we will see the method in common use during the first five years.

2.6 State of art

- The crop sprayer with certain nozzles such as large drop fan nozzles or three-hole nozzles, can be used for fertilizer spreading.

- Applying plant protection chemicals as dry substances has little interest.

- Generally we need separate equipment for fertilizer and for plant protection chemicals.

- Too much attention is paid to developing large and sophisticated equipment.

- Too little attention seems to be concentrated on medium and small equipment.

- The cleaning and maintenance of the equipment are often difficult and time-consuming. This has to be improved by the manufacturers through better materials and better designs.

- In practice the users do not make the most of their equipment. That means the equipment is not in the shape to do a good job.

The situation can be improved by:

   Training of the crew (certificate)
   Testing and inspection of the equipment (voluntary inspection)

- The delivery rate should be exactly controlled.

- The distribution has to be improved (design - accuracy of manufacture - research - advice - practice)

- The loss of chemicals outside the treated area has to be reduced

- Safety and precision
  Check and calibrate the sprayer
  Use multiple nozzles, bayonet fittings, clean water tanks, good filters and foam markers
  Follow the instructions for the equipment and on the label

- Monitors and controls can help improve the spraying quality

Remember: The most important factor is the driving speed.

Table 2. Fertilizer distributors and distribution results based on testing in Denmark, Norway and Sweden

Type	Fertilizer	Kg/ha	Distribution CV in %	Working width m
CENTRIFUGAL AMAZONE ZA-E 602, WEST Germany Melding nr. 451 LTI 1)	Compound granules	350-1800	3.5-7.1	7.0-9.0
RAUCH-KOMET ZS 600 N, WEST GERMANY Melding nr. 600 LTI 1)	Compound granules NPK 21-4-10	200-1000	6.9-7.9	9.5-10
	Nitrogen f. granules	120-770	3.07-7.9	6
BULK HOPPERS AIR DELIVERY ÖVERUM TIVE JET 1218 Med. 3171 Statens M.provningar 2)	Nitrogen f. prilled	150-750	8.8-13	18
	PK 5-16 granules	150-750	7-6-8	18
	Nitrogen f. granules N28	150-600	7-5-6	18
AERO 1112 WEST GERMANY Med. 3024 Statens M.provningar 2)	Nitrogen f. granules	200-900	9-10-13	12
	Compound granules PK 713	200-900	8-7-10	12
Nordsten Air-O-Matic, type KSF Denmark, Meddelelse nr. 284 Statens jordbr.tekn. Forsøg 3)	Spheroidal NPK 21-4-10	215-630	11.2-11.1-11.4	12
	Prilled NPK 21-4-10	200-580	11.3-11.9-9.2	12
COMBINED GRAIN AND FERTILIZER DRILL (fluted wheel/TUME KL 2500, Finland Melding nr. 554 LTI 1)	Compound granules 21.4.10	500	2.4 horizontal position •	2.5
NORDSTEN COMBI-MATIC CKF, DENMARK Melding nr. 564 LTI 1)	Compound granules 21-4-10	530	4.1 horizontal position	2.5

1) Landbruksteknisk institutt, 1432 Ås-NLH, NORWAY

2) Statens Maskinprovningar, 750 07 Uppsala, SWEDEN

3) Statens jordbrugstekniske Forsøg, 8700 Horsens, DENMARK

Table 3. Spreading and distribution of plant protection chemicals

Methods	Remarks
Dusting	Only for certain purposes Risk of drift
Spreading granules	Against weeds and insects Difficult to calibrate Unsatisfactory distribution
Spraying	
Hydraulic nozzle	Most common and used all over the world
Fan type	Good distribution
LP	" "
XR	" "
Cone type	Problems with boomheight and distribution
Gaseous energy nozzle ..... (twin fluid)	Various types. One or two are in practical use.
CDA	Overestimated, over-optimistically adver- tised. Used mainly in developing countries and in England.
Air assisted nozzle systems	May increase penetration and reduce drift
Mistspraying	Mainly in trees and scrubs Difficult conditions for the spraying crew Spray drift is a problem
Fog (aerosol)	Glasshouse and stores. Problems with approval of chemicals Difficult conditions for the crew
Contact applicator (Weed wiper)	In a small scale for weed control Only minor interest
Electrostatics	Non-consistent results Not in common use

REFERENCES

Bjugstad, N., 1987: Equipment for spraying against fungus and insect pests in row crops. Norwegian Agricultural Research. Supplement no. 1. 201 p.

Bundesblatt Teil I, 1986: Gesetz zum Schutz der Kulturpflanzen. Nr. 49, 19 September, p. 1505-1519.

Combellack, J.H. (1984). Herbicide application techniques. Crop Protection 3, 9-34.

Cooke, B.K., Hislop, E.C, Herrington, P.J. Estern, N.M., Jones, K.G., Woodley, S.E. and Chapple, A.C., 1986: Physical, chemical and biological appraisal of alternative spray techniques in cereals. Crop Protection 5 (3), 155-164.

Culpin, C., 1987: Farm Machinery. London. 450 p.

Hagenwall, H., 1985: Framtidens växt-skyddsteknik - något nygammalt. Lant-mannen 7, p. 11-13.

Hardi, 1989: Hardi Twin Systems, Copen-hagen, 9 p.

Herrington, P.J., Mapother, H.R. & Stringer, A., 1981: Spray Retention and Distribution on Apple Trees. Pestic. Sci. 12. p. 515-520.

Hislop, E.C., 1987: Requirements for effective and efficient pesticide application. Rational pesticide use. Cambridge University Press, p. 53-71.

Jonsson, E., 1988: Lägesrapport om åt-gärder enligt handlingsprogrammet för att minska hälso- och miljöriskerna vid användning av bekämpningsmedel. Fra Lantbruksstyrelsen 17.07.1988, 26 p. Bilag fra kemikalieinspektionen. Lågdos-medel ved Vibeke Bernson, 2 p.

Kepner, R.A., Bainer, R. and Barger, E.L., 1978: Principles of Farm Machinery, AVI, USA, 527 p.

Landers, A.J., 1988: Spraying systems and legislation. The Agricultural Engineer. Vol. 43, 2, p. 51-52.

Law, S.E., 1987: Basic phenomena active in electrostatic pesticide spraying. Rational pesticide use, Cambridge University Press, p. 81-105.

Miller, P.C.H., 1987: Spray application technology. Agricultural Engineer. Autumn, p 77-84.

Nordby, A., 1979: Nordisk prosjekt. Utredning og undersøkelser med plante-vernutstyr. Del 2. Generelt om metoder og utstyr. Landbruksteknisk institutt. Serie A, L.nr. 11/79, nr. 597, 83 p.

Nordby, A. & Jørgensen, L., 1985: Gjennom-trengelighet av sprøytevæske for ulike arbeidstrykk, kjørehastigheter og dyse-størrelser. 2. Danske Planteværnkon-ference, Nyborg, p. 85-98.

Richtlinien für die amtliche Prüfung von Pflanzenschutzmitteln, 1988: Merkmale Spritz- und Sprühgeräte für Flächenkul-turen. Biologische Bundesanstalt für Land- und Forstwirtschaft. Bundesrep. Deutschland, 18 p.

Robinson, T.H. & Garnet, R.P., 1984: The influence of electrostatic charging, drop size, and volume of application on the deposition of propiconazole and its re-sultant control of cereal diseases. British Crop Production Conference - Pests and Diseases. vol. 2, p. 1057-1065.

Skuterud, A., Nordby, A., and Tyldum, A., 1988: Effects of application methods, spray volumes, pressures and herbicide rates on weed control in spring cereals. Crop Protection Vol. 7, p. 303-308.

Stubsjøen, M., 1989: Tiltak for redusert bruk av kjemiske plantevernmidler i landbruket. Aktuelt fra Statens Fag-tjeneste for Landbruket nr. 3, p. 27-35.

Land and Water Use, Dodd & Grace (eds), © 1989 Balkema, Rotterdam. ISBN 90 6191 980 0

Pesticide deposit and activity as a function of spray atomizers and liquid formulation

N.B.Akesson & W.E.Yates
University of California, Davis, Calif., USA

ABSTRACT: The use of the highly effective ruby laser spray drop analyzer has enabled a much more detailed study of spray atomization than was heretofore possible. Studies conducted over a 10 year period have gathered a significant amount of data on drop size and size range produced by various atomizers as well as effects on drop size by injection of the spray into an airstream and alteration of the drop size by the spray formulation.

As a result of our ability to more quickly and accurately measure the drop size of released sprays, we are able to more clearly identify the effects of drop size on deposit and biological activity of the spray in relation to a specific crop pest and also to measure the losses of spray materials to airborne transport out of the treatment area. This information is not only of value to pesticide researchers but also to farmers and other applicators and will assist them with the decisions required whenever a pesticide application is to be made.

1 INTRODUCTION AND REVIEW OF LITERATURE

The importance of agricultural spray drop size to overall safety and application efficacy cannot be overstated and should be given first consideration whenever an application is to be made. The calibration of the sprayer and choice of nozzle flow rate for the dosage rate of active ingredient and total volume to be applied to the crop is generally understood and has received considerable attention, particularly as regards under- and over-dosage (Rider and Dickey, 1982; Editors (1) 1976). But very little consideration among pesticide applicators has been given for selection of the most effective and safe pesticide spray drop size for a given application. A very broad generalization of coarse, medium and fine drops has recently been adopted by the British Crop Protection Council (Editors (2) 1986) and selected nozzles (ground machine only) from six manufacturers are listed for drop size characteristics along with flow rate and pressure to complete the necessary information to properly select the general drop size and total volume required to do the job. A selection data-bank of atomizers, drop size and flow rates, for aircraft and air-carrier spraying has been developed by Akesson and Gibbs (1988) with data storage on computer disk which in-

cludes as well pertinent information on spray deposit and losses from movement outside the treated area.

The Agricultural Engineering Department, University of California at Davis has had on-going projects on spray drop size evaluation for many years. These have been in relation to biological efficacy and minimizing of undesirable contamination of, and damages to nearby non-target crops; and of contact with workers and human and domestic animal habitat as well as the general wild life environment in or near sprayed areas. This has involved aircraft and ground application systems for agriculture and forestry and also for disease vector control, mosquitoes and flies.

Drop size studies prior to about 1975 were conducted using various substrate surfaces on which to collect drops. Classic among these are the manganese dioxide coated glass slides, the heavily clay-impregnated kromkote and printflex papers and various other films, papers and plastics. Also air filters and impactors are used for extracting drops from the air (Akesson and Yates, 1986-3). It was obvious to the users of these materials, which had to actually collect and hold the spray drops, that severe limitations on accuracy as well as a tedious and expensive data analysis operation was involved. With the advent in the late 1970's of the in situ particle size-frequency analyzers using the ruby laser (columnated light source) system, an entirely new technique has been made available. Now rapid scanning and analysis of a wide selection of atomizers, formulations and operational conditions can be easily made. The Agricultural Engineering Department at Davis acquired their Particle Measuring Systems 2-D and Foward Light Scatter probes and data analysis equipment in 1979 and have now had about 10 years of experience on collection of data on which to base a selected but still somewhat limited evaluation of spray drop characteristics of atomizers and spray formulations. Because of the high usage of aircraft and air carrier sprayers in California, the emphasis of research has been directed toward these machines and the spray drop spectrum they produce and disperse.

In order to evaluate various atomizers and formulations discharged into an airstream a wind tunnel was designed and built by the Agricultural Engineering Department. Fig. 1 shows a schematic diagram of the installation. A 91 cm radial fan driven directly from the shaft of a 300 kW internal combustion engine is used to draw air through a 60 x 60 cm tunnel throat section in which the probe from the PMS analyzer is placed. The spray from the nozzle under test is directed into the probe laser beam which crosses between the two extended horns of the probe. The probe interfaces with a small computer which stores the drop size and frequency counts and calculates the various statistical data on the test. Between 10 and 15 thousand drops are counted for each test run; 2 or 3 replicate runs are made and averaged for the final report. An x-y scanner is used to automatically scan the spray nozzle across the laser sensor.

2 SPRAY ATOMIZERS

By far the most commonly used atomizers for agricultural spraying, both ground and aircraft, are the hydraulic pressure types shown in Fig. 2. On the left is a simple disc-orifice nozzle producing a jet or solid stream of relatively large spray drops. Next to the right is a hollow cone nozzle showing the whirl core and disc-orifice combination. The slots in the core spin the liquid and along with pressure alterations control the drop size produced. Following to the right is another hollow cone-producing nozzle utilizing a tangential lead tube to the spin cavity and orifice to produce the hollow discharge pattern. If a center hole is made in the whirl disc of type two, the discharged spray produces a solid pattern as shown from nozzle 4. The far right nozzle produces a fan-shaped spray, the discharge volume may be tapered on the edges to permit matching with adjacent nozzles, or may be uniform across a given spray pattern with specific spray angle at a stated pressure. All of these nozzles may be obtained with a range of discharge volumes and drop sizes. An example of the drop sizes produced are shown in Fig. 3 for a typical hollow cone nozzle such as one with a 0.24 cm orifice and Number 46 core. Here (upper histogram) the nmd or $D_{N.5}$ is 258 μm and the number of drops under 100 μm is 10%. The vol. median is 327 μm and 0.1% of the vol. of spray discharged is under 100 μm dia.

Fig. 2 Hydraulic pressure nozzles

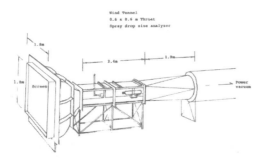

Wind Tunnel
0.6 x 0.6 m Throat
Spray drop size analyzer

Fig. 1 Wind tunnel and spray drop analyzer

Another measure of spray distribution or size range is the RS or relative span. This is defined as:

$$RS = \frac{D_{V.9} - D_{V.1}}{D_{V.5}}$$

where an RS of 1 would indicate a normal or Gaussian distribution of drops and

values below or above 1 would indicate
narrower range and broader range of sizes
respectively. Note that the RS of the
hollow cone nozzle (typical of smaller
drop size producing atomizers) was 0.71.

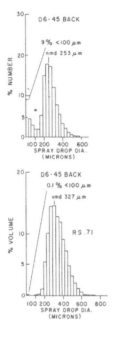

Fig. 3   Hollow cone spray drop distribution

Fig. 4 shows a typical rotary screen
type atomizer driven by an electric motor.
The drop size produced by these devices is
governed by (1) the peripheral (surface)
speed of the screen and (2) the rate of
flow (pressure and hole size) through the
screen. When operated in an airstream
smaller drops result as air velocity
increases. Fig. 5 shows a drop size
distribution by number and volume from one
such type atomizer. The nmd at 180 μm is
smaller than that of the hollow cone
nozzle in Fig. 3, as is also the vmd at
240 μm. This is because the vmd is
smaller for the rotary and the % of drop
volume under 100 μm dia. is higher at
around 5% compared with the hollow cone
at 0.1%. There is no evidence showing
that a rotary type screen, disc, cup or
plate device in the airstream of an
aircraft or an air-carrier sprayer produces a significantly narrower range of
drops, or any lesser drop vol. under 100

μm than a comparable drop size hollow cone
nozzle. A fan nozzle, for example, at the
same flow rate as the rotary screen produced a vmd ($D_{V.5}$) of 610 μm, an RS of
1.25 and showed 1.67% of drop vol. in
drops under 100 μm. Thus the rotary atomizers are better able to handle a larger
quantity of liquid and produce smaller
drops than will a pressure nozzle. But at
a given drop size range the two produce
about the same skewed Gaussian distribution. Without the disruptive effects of
the airstream (static operation) various
rotary and other monodisperse atomizers
can be operated in such a way as to significantly narrow the drop size range
placing a large portion of the drops into
2 or 3 of the PMS class sizes.

Fig. 4   Rotary screen atomizer

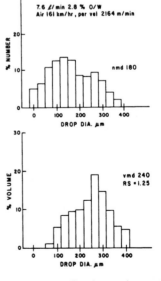

Fig. 5   Spray distribution rotary atomizer

3 DROP SIZE AND FLOW RATES

Graphical material has been prepared to indicate the correlation found between throughput or flow rate from various cone, fan and rotary nozzles.

3.1 Hollow cone nozzles

Graphics for drop size vs. flow rate for various hollow cone (core-disc type) nozzles are shown in Fig. 6 (Yates et al., 1984). Water sprayed from two directions, 0° and 90° to the airstream, are given as well as one set for a cottonseed oil and solvent formulation. The effect of the oil base liquid was to reduce surface tension by 50% and increase viscosity to 19 mPa·s. Liquid density remained approximately the same. As the curves show, the oil base spray increased the drop size at all flow rates for the 0° position but decreased the drop size for both water and oil to about the same level at the 90° position.

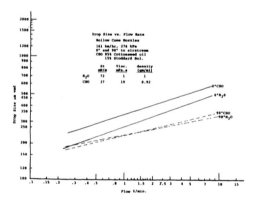

Fig. 6 Drop size vs flow rate, hollow cone

3.2 Fan nozzles

Fig. 7 (Yates and Akesson, 1985-2) indicates 3 positions of the nozzle discharge to the airstream with water and 4 with cottonseed oil and solvent formulation. As to be expected the water curves show largest drop size for the 0° position followed by the 90° and smallest for the 135°, with the nozzle discharge directed into the 161 km/hr air. The oil base data produced an unexpected response. The 45° position produced drops much the same as for water at 0° but the oil base 0° position dropped to below that of the 90° for both oil and water. There is no logical explanation for this apparent aberration, logic would have placed the oil data for 0° at about the position that the 45° CSO is graphed. The 135° curves appear quite logical. The lower surface tension reduces the drop size of the oil base sprays and thus reduces overall drop size. Of interest to note is the rapid increase in drop size for all the runs as the flow rate is increased.

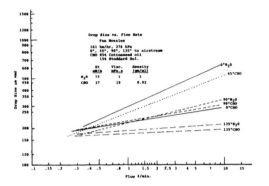

Fig. 7 Drop size vs flow rate, fan nozzle.

3.3 Rotary screen atomizers

This graph (Fig. 8) presents drop size vs. peripheral velocity from 3 different rotary screen type atomizers operated in a 161 km/hr airstream, but at different peripheral or screen outer surface velocity. The runs were made with water with the exception of 4 tests marked by asterisks(*) for a cottonseed oil and solvent base run. As would be expected the reduced surface tension of the oil base liquid reduced the drop size delivered by the rotary atomizer.

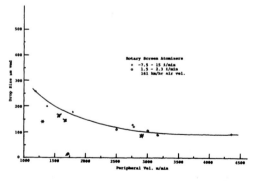

Fig. 8 Drop size from rotary screen atomizers

Drop size from all types of atomizers can be affected by changes in the physical properties of the liquid, basically as viscosity surface tension and density. Of these the surface tension reduction from that of water at 72 mN/m to around 35 mN/m for oils, or for emulsifiers and other adjuvants and solvents appears to be the most significant change taking place with the addition of pesticides to a water base. Viscosity for most finished tank mix formulations does not change greatly, perhaps increasing to 2 or 3 mPa·s with the addition of water soluable thickening agents, cellulose, sugar, syrups and others or with addition of non-soluable flowables and wettable powders. Various vegetable oils have been popularized as adjuvants and carriers for pesticide materials and the two most used, cotton-seed and soybean oils, have viscosities of 75 to 100 mPa·s. Density of water at 1 g/cc is usually reduced a small amount by addition of various petroleum and vege-table oils and aside from a few limited use pesticide materials does not vary greatly from water (Akesson et al., 1985).

Haq et al. (1983) did a study on the effect of surface tension and viscosity on cone and jet nozzles as well as an attempt to relate the effects of various water soluable polymers to drop size. The latter appear to cause a viscoelastic response when added to water which is not wholly reflected in normal viscosity mea-sure with a rotating element viscometer such as the Brookfield. Haq did not have access to the PMS drop size analyzer but used kromkote paper cards to collect drops and a Quntamet drop size scanner to make size-frequency analysis. Fig. 9 shows effects on drop size from hollow cone nozzles, orifice size from 0.8 to 4 mm and with Number 46 core. There is evidence of a small identifiable increase in drop size as the viscosity is increased 1 to 30 mPa·s.

Fig. 10 shows the response of the hollow cone nozzles (same sizes as Fig. 9) to reduction of surface tension from water at 72 mN/m to 25 mN/m. The 2.4 mm orifice and Number 46 core shows a decrease of 30% in vmd when ST is reduced from 72 to 25 mN/m.

Response of drop size (to addition of polymers) with two nozzles (1) an 8006 fan nozzle operated at 276 kPa and 135° to the 161 km/hr airstream, and (2) a D8-46 (3 mm orifice, 46 core) hollow cone nozzle at 0° to the airstream are shown on Fig. 11

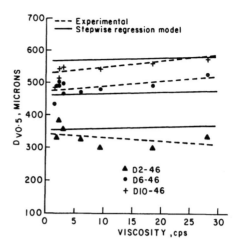

Fig. 9 Viscosity vs median drop size

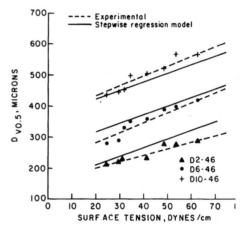

Fig. 10 Surface tension vs median drop

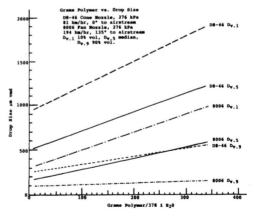

Fig. 11 Polymer concentration vs drop size

(Yates et al., 1985-2). The drop size vs. the grams per 378 l of water are plotted with curves for (solid lines) the vmd or $D_{V.5}$, dashed curves for the small (10% by vol.) drop end of the spectrum $D_{V.1}$ and the dash-dot curves for the large drop end representing 90% by volume of the spray drops. The increase with increased polymer in the $D_{V.1}$, or small drop end of the spectrum, shows the positive effect of adding the polymer for both fan and cone nozzles. The effect on the median or $D_{V.5}$ is also positive, but there is a barely discernable effect on the $D_{V.9}$ or large end of the spectrum for the high shear rate 8006 fan nozzle. The D8-46 $D_{V.9}$ shows a greater response which leads to the suggestion that a high shear rate tends to break down the viscoelastic effect of the polymer adjuvant. Thus these materials have greatest response from atomizers already producing large drops. There is no question but what the addition of polymers will increase the drop size, shifting the spectrum upward, except in the case of small drop producing atomizers or in a high velocity high air-shear situation. The use of water soluble polymers to provide thickening effect on pesticide sprays has enjoyed considerable success over the past 15 years. However as Bouse et al. (1988) in a very comprehensive study points out, there is considerable difference in various of the polymers used and their ability to increase drop size in the sprays. The effect on percentage of spray vol. in drops smaller than 200 μm dia. was used by Bouse as an indicator for small drop size reduction by polymers. Bouse found a linear relationship with concentration of polymers, for most of the materials tested, with increase in the median or $D_{V.5}$ and an inverse linear relationship with vol. in drops below 200 μm. This is consistent with any type atomizer and points out that (1) use of polymers does not significantly alter the skewed Gaussian distribution of drop spectra as occurs with all atomizers; (2) increasing the drop size by adding polymers merely moves the entire spectrum upward; and (3) larger drop size sprays may be achieved by using larger throughput nozzles or nozzles such as jets and the RD type obviating the need for the use of sticky and awkward polymer formulations.

Additives and adjuvants placed in the spray tank for whatever purpose should follow manufacturer's label instructions for specific pesticide materials. Applicators should be cautioned against undiscriminate use of these materials which will not only affect the drop size produced by a given nozzle system but may have far reaching biological responses on crop plants.

5 DROP SIZE AND EFFICACY OF INSECTICIDES AND FUNGICIDES

There are widespread differences of opinion as to the proper drop size to use with a given insecticide or fungicide application. There is little doubt that chemical activity is increased as small drops of 50 and as small as 10 microns diameter are applied to a pest. Fumigation by vapors of pesticides of gas molecules of as much as 1/1000 of a micron diameter may be effective on certain pests if these are confined to the treatment area in a closed space. Thus, while biologists can demonstrate highly effective control of insects and fungi with these aerosol type applications, the physical requirements of generating and delivering these to target crops and pest habitat generally requires larger drops of more predictable movement and deposit characteristics.

Forest insect sprays have been widely investigated for most effective drop size. The early work has been summed up by Barry (1984) and Ekblad and Barry (1984). Field tests indicated that 87% of the drops found on Douglas fir needles were very small arerosol size of 1 to 16 μm dia. Also on Eastern spruce budworm larvae drops of 6 to 15 μm dia. were dominant at 76%. In other tests using sprays of around 300 μm vmd, 86 to 94% of the drops on the needles (Douglas fir) were less than 61 μm and 71 to 40% were less than 16 μm dia. Since the initial drops released by the aircraft are so much larger (200-300 μm vmd) than what is caught by the fir needles it raises the question of whether larger drops are shattering, evaporating or just how are sufficient small drops generated to produce the coverage found in the trees (Wedding et al., 1978). It is also to be noted that deposit in the fir trees is largely confined to the upper crown or to the upwind side of the tree where sprays are rapidly filtered out by the highly efficient needles of the tree (Armstrong, 1984; Kristmanson et al., 1984).

In agricultural spraying by ground equipment Burt et al. (1970) and others summed up the considerable field test work that had been done with ULV (ultra-low-volume) sprays on various crops, cotton, soybean, and corn. Burt, using narrow size range drops from a spinning disc atomizer of 140, 200 and 300 μm vmd found satisfactory insect control for the smaller drops but inadequate coverage gave poor boll weevil control, on cotton for example, with drops of 300 μm vmd. The implication here is that while volume of toxicant deposited may be the same, the difference in coverage of drops/unit area reflects the comment that drops/unit area are more important than size of the drops (Behrens, 1957). This could be even more important for insect and fungi control than for weed and brush control. Burt concluded that drift studies showed sprays of 140 μm vmd or smaller produced significant drift-loss and suggested that sprays even by ground equipment should be greater than 140 μm vmd.

Probably the most widely accepted aerosol applications are those used for mosquito and fly control. Akesson et al. (1986) applying Scourge® (a pyrethroid) by aircraft found sprays in the range of 10-15 μm vmd to be equally as effective as applications of 25-30 μm vmd for adulticiding under temperature inversion conditions. But a temperature inversion is essential to hold the aerosol down near to the ground so that flying mosquitoes can contact the drops in the aerosol cloud. Due to difficulties in obtaining the 10-15 μm vmd size by aircraft equipment it was concluded that a 30 μm vmd aerosol was as small as was practical for use by the California Mosquito Control Districts for aircraft operations. It is interesting to note that spraying for Tsetse fly control as reported by Coutts (1980) indicates a 30 μm vmd aerosol has been proven in field tests to be more effective than larger drop size ranges, while still being manageable with aircraft. But he notes it is also best performed under inversion type weather.

6 HERBICIDE EFFECTIVENESS AND DROP SIZE

Biological effectiveness of herbicides has been shown by many observers to be affected by the drop size and active chemical concentration of the applied sprays. Ennis and Williamson (1963) showed that plants (soybeans, flax, sugar beet and wheat) sprayed with 2,4-D, a translocated herbicide, had more stem curvature and showed reduced yield as drop size was reduced from 500-600 μm to 100 μm dia. using a monodisperse applicator. Solution concentrations of both 3% and 10% for the different size drops displayed similar yield reduction with the 100 μm drops. Behrens (1957) reported that a minimum of 10-12 drops/cm^2 was required for maximum plant response and he felt drop coverage (spacing) was more important than size of the drops within the range of 100 to 500 μm dia. Ennis (1965) suggests that even smaller drops would have greater effectiveness and felt that the better coverage plus low drop concentration prevented rapid "burn" of the plants and facilitated entry of the 2,4-D and 2,4,5-T to plant tissue. But he also acknowledged that for field applications consideration for spray drift would require that larger drop sizes be used.

Similarly, McKinlay et al. (1972) found the response of sunflower to paraquat was significantly enhanced with 100 μm drops having greater phototoxicity effects (tissue damage) than 350 μm drops. Tests with two levels of concentration showed the higher concentration such as used with reduced volume application produced greater toxic response.

Prasad (1987) conducted a detailed study in the greenhouse with troclopyr, glyphosate and hexazinone. He sprayed 4 month old seedlings of birch, alder and aspen using a monodisperse atomizer for drops of 155, 335, 465 and 665 μm dia. Evaluations after 3 weeks showed the 155-335 drop applications were more biologically effective than the larger sizes. As with Ennis's earlier work with 2,4-D, Prasad attributed these results primarily to biological response of small amounts of active herbicide uniformly applied to the individual leaf surface. Troclopyr and glyphosate are translocated herbicides while hexazinone is a contact type material. As would be expected the latter showed greater response to the smaller drops than did the translocated materials. Note that Prasad applied the same dosage of ai but not in the same numbers of drops. Thus coverage was different for the different size drops. The use of surface active adjuvants with herbicides has been shown to improve spreading of drops and penetration of plant cuticle and tissue (McWhorter, 1962). The use of a surfactant can in part improve the spreading and redistribution of chemicals applied in large drops, thus making possible use of the safer large drop sprays. An additional function of surface active materials is to reduce the rebounding of drops from plant surfaces and increase the potential for break-up of large drops striking leaves, twigs and needles. Hess et al. (1974) using a scanning electron microscope showed that large drops over 500 μm dia. shattered on impact with surfaces and hairs on leaves forming many more small drops that give additional coverage of the leaf. Smaller drops (<200 μm dia.) were less prone to shatter and tended to cling to leaf hairs without reaching the active leaf surface. This observation could indicate that larger drops traveling at greater velocities could be helpful in penetrating brush canopy and with shattering produce a significant increase in coverage through generation of many small drops.

In California restricted use herbicides were identified in the early 1950's (later reclassified as restricted materials) and large drop size producing spray systems were required by state regulation for their application. These regulations initially included the phenoxy herbicides, later amended to include propanil, paraquat, picloram, dicamba, phenols and cresols and Molinate. The restricted materials require county level permits for their use and for all liquid sprays specific equipment and operation conditions are required for both aerial and ground application. For fixed wing aircraft the jet nozzle (simple orifice, no whirl core) of 0.159 cm (1/16 in.) dia. or larger directed with the airstream and operated at not over 276 kPa (40 lb/in^2) is specified. For helicopters operating at 96 km/h or less, the 0.159 cm orifice may have a #46 or larger whirl core added to provide better spray atomization at the lower air speed. Spray drop size for the jet nozzles is in the range of 800 to 950 μm vmd, while the addition of the #46 whirl core reduces that to 450-550 μm vmd at reduced airspeed. These nozzle systems have been used for about 35 years for California restricted herbicides, while there is little doubt that wider more uniform swaths and better control efficacy could be achieved with smaller drops the potential for damages under California multi-crop agriculture are too great to permit the use of smaller drop size sprays for highly active herbicides.

Deposit on plant surfaces by various size drops is a function of impaction by drop motion and sedimentation by action of gravity (Elliott and Wilson, 1983). The rate of mass deposition by impaction is directly a function of drop velocity (or wind velocity for drops less than about 50 μm) drop dia. and an inverse function of object frontal width. But again the rate of number deposition will be increased by the total numbers of drops available, a function of drop size. Limitations on minimum drop size for impaction on plants and insects is such that drop size and object frontal width should not go below a

ratio of unity. Thus drops smaller than the object tend to flow around the object instead of impacting plant targets. Object size varies from perhaps one to several thousand mm, from plant stems to large leaves (Cramer and Boyle, 1976; Dumbauld, 1984).

Obviously, the functions of sedimentation and impaction will have various degrees of influence on a given spray usage. For pesticides where plant targets are aimed for, drops of 300 to 400 μm, primarily affected by sedimentation and not impaction would appear to be most advantageous to use for forest sprays from the viewpoint of physical application. However, it is to be noted that biological response favors the use of smaller drops and coverage may dominate the effective result of a given spray application and not the drop size per se. Although applications of sprays in drop size ranges less than 300 μm vmd may be found more effective, the choice of a size range for a given application job returns to the problem of drift loss particularly for herbicides from the application area (Akesson and Yates, 1976).

7 SUMMARY

Biologists have clearly shown aerosol size drops (down to 10 μm) can be highly successful for insect and fungi control, and also quite clearly that drops of 100 to 200 μm dia. (at 1 to 1.5 drops/cm^2) are most effective for control of brush and other plant species. The problem of loss of small drops, generated by all of the presently available aircraft systems, which move out of the spray area as airborne drift, must be given priority consideration in any pesticide spray work. This means that minimally the drop size vmd should be above 400-500 μm vmd for herbicides and where any nearby sensitive crops, gardens or waterways exist even larger drops would be advisable. The application with giant size drops, over 1 mm dia., is partially offset by increasing total volume applied in order to maintain drop frequency necessary for effective coverage. This starts at 90 to 100 ℓ/ha and may require double this amount for very heavy growth situations.

The future of all spray activity rests to a great extent with development of more sophisticated, and likely to be expensive monodisperse spray application systems. Such devices for aircraft have been demonstrated experimentally but have not attracted development funding to make them a reality. Granular pesticides are only effective through the soil and do not hold very much promise for future use, although several chemical companies continue to look at these as no-drift application.

But applications, both aircraft and ground, are finding increasing resistance from the public and from growers to the use of pesticides. While these materials are essential to our high productivity and blemish-free produce today, increasing care and precision of application must be the primary goal for continued pesticide use. Greater attention and utilization of drop size information will aid in achieving this goal.

REFERENCES

Akesson, N.B. & Yates, W.E. 1976. Application parameters relating to pesticide application. In Pesticide Spray Application, Behavior and Assessment: Workshop Proceedings, USDA FS General Tech. Report PSW-15/1976.

Akesson, N.B., Raubach, N. & Gibbs, R. 1986. Optimizing applications conditions for mosquito control with Scourge. Proceedings 54th Annual Conference CMAVCA Inc., Sacramento, p. 47-52.

Akesson, N.B., Bayer, D.E. & Yates, W.E. 1986-2. Application effects of vegetable oil additives and carriers on agricultural sprays. 1st Int. Congress Adjuvants and Agrochemicals, CRC Press Vol. II.

Akesson, Norman B. & Yates, Wesley E. 1986-3. Development of drop size-frequency analysis of sprays used for pesticide application. In Pesticide Formulations and Application Systems, Fifth Volume, ASTM STP 915, p. 83-93.

Akesson, Norman B. & Gibbs, Richard 1988. Precision and safety with PARIS: Some basic challenges for aerial application. AA-ASAE Paper 88-008, St. Joseph, MI.

Armstrong, J.A. 1984. Fate of chemical insecticides in foliage and forest litter. Chapter 17, Chemical and Biological Controls in Forestry, American Chemical Society Symposium Series #238, Washington, DC, p. 241-251.

Barry, J.W. 1984. Deposition of chemical and biological agents in conifers. Chapter 10, Chemical and Biological Controls in Forestry, American Chemical Society Symposium Series #238, Washington, DC, p. 118-151.

Behrens, R. 1957. Influence of various components on the effectiveness of 2,4,5-T sprays. Weeds 5:183-196.

Bouse, L.F., Carlton, J.B. & Jank, P.C. 1988. Effect of water soluable polymers on spray droplet sizes. Trans. ASAE 31(6).

Burt, E.C., Lloyd, E.P, Smith, D.B., Scott, W.P., McCoy, J.R. & Tingle, F.C 1970. Boll weevil control with insecticide applied in sprays with narrow-spectrum droplet sizes. Journal Econ. Ent. 53(2):365-370.

Coutts, H.H. 1980. Aerial applications of insecticides for the control of Tsetse flies. VI Int. Agr. Avia. Congress Proceedings, Int. Centre for App. of Pesticides, Cranfield, England, p. 179-185.

Cramer, H.E. & Boyle, D.G. 1976. The micrometeorology and physics of spray particle behavior. In Pesticide Spray Application, Behavior and Assessment Workshop Proceedings, p. 27-39.

Dumbauld, R.K. 1984. Modeling of aerial spray drift and canopy penetration. Chapter 12, Chemical and Biological Controls in Forestry, American Chemical Society Symposium Series #238, Washington, DC, p. 153-174.

Editors (1) Agricultural Development and Advisory Service. 1976. The utilization and performance of field crop sprayers. Ministry of Agr. Fisheries and Food, Middlesex, England Farm Mech. Studies 29.

Editors (2) British Crop Protection Council 1986. Nozzle selection handbook. BCPC 20 Bridgeport Rd. Thornton Heath CR4 7QG.

Ekblad, R.B. & Barry, J.W. 1984. Technological progress in aerial application of pesticides. Chapter 8, Chemical and Biological Controls in Forestry. American Chemical Society Symposium Series #238, Washington, DC, p. 80-94.

Elliott, J.G. & Wilson, B.J., editors 1983. The influence of weather on the efficiency and safety of pesticide application. British Crop Protection Council Publication #3, Croydon, England.

Ennis, W.B., Jr. & Williamson, R. 1965. Influence of droplet size on effectiveness of low-volume herbicidal sprays. Weeds 11(1):67-72.

Haq, Kondaker, Akesson, Norman B. & Yates, Wesley E. 1983. Analysis of droplet spectra and spray recovery as a function of atomizer type and fluid physical properties. In Pesticide

Formulations and Application Systems, Third Symposium ASTM STP 828, p. 67-82.

Hess, F.G., Bayer, D.E. & Falk, R.H. 1974. Herbicidal dispersal patterns: I, A function of leaf surface. Weed Science 22(4):394-401.

Kristmanson, D.D., Picot, J.J.C. & Henderson, G.W. 1984. The inertial impact of droplets on foliage of some conifers. Poster Paper #7. In Proceedings Symposium of the Future of Aviation in Agriculture, Nat. Res. Council Canada AFA-TN-17, NRC 23504, p. 401-409.

McKinlay, K.S., Brandt, S.A., Morse, P. and Ashford, R. 1972. Droplet size and phytotoxicity of herbicides. Weed Science 20(5):450-452.

McWhorter, C.G. 1962. The use of adjuvants. Chapter 2, p. 10-25, Champaign, IL.

Prasad, R. 1987. A study of droplet size and density in relation to efficacy of herbicides. WSSA Abstracts, p. 98-99, Urbana, IL.

Rider, A.R. & Dickey, E.C. 1982. Field Evaluation of calibration accuracy for pesticide application equipment. Trans. ASAE 25(2):258-260.

Wedding, J.B., Carney, T.C., Ekblad, R.B. Montgomery, M.E. & J.E. Cermak 1978. Particle deposition in a Douglas fir canopy: A wind tunnel simulation of aerial spraying. Trans. ASAE 21:253-260, 266, St. Joseph, MI.

Yates, W.E., Akesson, N.B. & Cowden, R.E. 1984. Measurement of drop size frequency from nozzles used for aerial application of pesticides in forests. USDA, FS EOC 8434-2804, Davis, CA.

Yates, W.E. & Akesson, N.B. 1985. Characteristics of atomization systems for reduced volume application of pesticides from agricultural aircraft. Int. Conference on Liquid Atomization and Spray Systems, Imperial Coll., London, England, Vol. 2 IVB/3/1-3/12.

Yates, W.E., Cowden, R.E. & Akesson, N.B. 1985-2. Effects of Nalco-Trol on atomization. USDA, FS. FPM 85-2, Davis, CA.

ABRIβ: Die Anwendung eines hochwirksamen Analysierers von Rubinlasersprühtropfen erlaubt eine wesentlich ausführlichere Untersuchung von Sprühzerstäubern, als jezuvor möglich. Untersuchungen, in einer Zeitspanne von über 10 Jahren, resultierten in einer beduetenden Ansammlung von Werten im Bezug auf die Größe der Tropfen, auf die Größenspanne, die von den verschiedenen Zerstäubern erzeugt wird, sowie auf die Wirkung auf die Größe der Tropfen bei der Einführung von dem Schaum in einen Luftstrom und bei Veränderung von der Tropfengröße durch die Sprühformulierung. Als ein Resultat unserer Fähigkeit schneller und genauer die Tropfengröße, die von Sprühern abgegeben wird, zu messen, sind wir in der Lage die Wirkungen von Tropfen auf Ablagerung und biologische Aktivität des Sprühers, in Bezug auf eine spezifische Pflanzenpest, eindeutiger zu identifizieren, und ferner den Verlust von Sprühmitteln an Luftbeförderung aus dem Behandlungsgebiet zu ermitteln. Diese Information ist nicht nur wertvoll für die Schädlingsbekämpfungsmittelforscher, sondern auch für die Landwirte und andere Anwender, und zusätzlich ist es auch hilfreich im Bezug auf die notwendigen Entscheidungen, die gefällt werden müßen, wann immer Schädlingsbekämpfungsmittel angewendet werden sollen.

RESUME: L'utilisation du ruby laser "spray drop analyser" (analyseur de gouttes de pulvérisation) a permis, par sa haute efficacité, une étude beaucoup plus détaillée de l'atomisation-pulvérisation qu'il ne l'était possible jusqu'alors. Des recherches menées sur une période de dix ans ont fourni une quantité significative de données sur la taille des gouttes et dus à l'injection de "spray" (la pulvérisation) dans un courant d'air et l'altération de la taille des gouttes due à la formulation de la pulvérisation. Par suite de notre capacité à mesurer plus rapidement et plus précisément la taille des gouttes dans le dégagement de pulvérisations, nous pouvons identifier plus clairement les effets de la taille des gouttes sur le dépôt et l'activité biologique de la pulvérisation par rapport à un parasite de plantes spécifique et nous pouvons aussi mesurer les pertes de matières pulvérisées dans le transport aérien en dehors de la zone de traitement. Ces renseignements sont non seulement inestimables pour les chercheurs, mais aussi pour les fermiers et autres utilisateurs, en ce qu'ils les aideront à prendre leurs décisions chaque fois qu'une pulvérisation pesticide devra être accomplie.

Land and Water Use, Dodd & Grace (eds), © 1989 Balkema, Rotterdam. ISBN 90 6191 980 0

Testing of herbicide application equipment in the vineyard

E.Cini, D.Vannucci & M.Vieri
Institute of Agricultural Mechanization, Florence, Italy

ABSTRACT: Some of the most useful machinery for chemical weeding on the row in vineyards of Tuscany are considered. The results of qualitative field tests and quantitative laboratory tests are reported. The computer elaboration of the different regulation parameters shows the importance of some operative characteristics for the use of weeding machines in the field.

RESUME: Essais d'un certain nombre d'equipements pour le desherbage chimique de la file de vigne en Toscane ont fait l'objet de notre étude. Nous reportons les résultats qualitatifs d'une recherche sur le terrain, ainsi que les résultats quantitatifs obtenu dans essais de laboratoire.
L'analyse du functionnement des gicleurs avec différents réglefes, simulé sur l'ordinateur, montre les caractéristiques qu'ont obtient "sur le terrain", tout en confirmant l'importance de ces paramètres lors del l'utilisation d'équipements pour le désherbage de la file.

ZUSAMMENFASSUNG: Analyse von Geräten für die chemische Unkrautvertilgung im Weinberg. Gegenstand der Betrachtung sind einige der meistgebrauchten Geräte für die chemische Unkrautvertilgung in toskanichen Weinbergen. Die qualitativen Ergebnisse einer Untersuchung in den Weinbergen und die in Labors ermittelten quantitativen Resultate werden vorgelegt.
Die Betriebsanalyse der verstellbaren Verteiler, die durch Simulation mit den Computer elektronisch durchgeführt wurde, verdeutlichte die operativen Eigenschaften im praktischen Einsatz und hob die Bedeutung dieser Werte die der Verwendung von Geräten für die Unkrautvertilgung im Weinberg hervor.

1 INTRODUCTION

The application of health standards in agriculture has now become a major issue. This is the result of numerous ecological accidents as well as an increased awareness of the dangers inherent in products, especially chemicals, that damage or modify biological cycles. In fact, new field application techniques are being developed to reduce chemical pollution in agriculture. On the other hand, the lack of interest on the part of the farmer in the consequences of pesticide use results in a striking ignorance both of the machinery required for the application of pesticides and herbicides and of the field parameters regulating application for the protection of health and the environment.

The herbicide application machinery currently available for field crops is quite adequate, but regarding hill tree-crops the market could be significantly improved. In fact, if we observe the equipment for hill vineyard herbicide application in operation, we can see its imperfections in the characteristics of the soil along the row boundary strips: steep lateral slopes, irregular soil profile, etc..
All of these factors influence negatively the distribution performance of the equipment, and the farmer is forced to use more chemicals to obtain a good result that is not easy to calculate in monetary terms, but the global cost is certainly high.

Denomination	Distribution diagram	V.C. %	Flow [l/min]	Press [kPa]	Size [μm]	Width (AB) [m]	Nozzle Position (A [m]
Double fan nozzle		70	2.25	450	3-400	1	0.15
Lateral fan nozzle		45	1.70	450	3-400	1	0.0
Mirror nozzle		94	1.70	450	4-600	0.9	0.4
Rotary nozzle (horiz.axis)		20	0.30	50	2-300	1	0.4
Rotary nozzle (vertic.axis)		60	0.25	100	2-300	0.7	0.0

Fig.1 Synthesis of the nozzle working parameters tested.

2 MATERIALS AND METHODS

The goal of our experimentation is to determine the basic parameters regulating the quality of work of a herbicide sprayer in the vineyard. We shall then offer some solutions geared to obtain an improved performance.

The field tests consisted in the evaluation of a distinct set of parameters regarding a variety of equipment, all of which conforms to the classical configuration - three point hitch, tank, pump, pressure regulator, spraying boom.

The sprayers tested are those most frequently used for application in vineyard row:
- nozzle mirror;
- nozzle double fan;
- nozzle lateral fan;
- obstacle avoiding dribble bar; spinning rotary nozzle (vertical axis);
- spinning rotary nozzle (horizontal axis).

To perform the tests we adopted a lateral bar on the market in France, which avoids trunks and obstacles in the row. Moreover, we obtained a spray recovery system consisting of a shield on the row side and a device with a "Venturi" tube to convey the liquid to the tank. Some qualitative tests were also performed with an experimental ultra low volume device of our own design which uses an aerosol spray under a plastic bell. This machine uses an air compressor carried and powered by the tractor, and it consists of a small tank and a pneumatic spraying system which emits a mist into the vegetation between the plants on the row. There is also a recovery system.

To obtain the full range of data required, we performed field and laboratory tests adopting the manufacturer's working parameters (pressure and volume in particular). The parameters taken into account were: uniformity of spray distribution, spray width, theoretical and practical volume distibution, and product loss with respect to distance from the row centre (0.0m, 0.3m, 0.5m). To make an overall evaluation of the equipment in question we also took into account the costruction and safety of operation. The former consideration did not include the vertical and horizontal variation angle, even though it is an important parameter bearing upon distribution quality. In this case we performed field tests only where the variation of this parameter is clearly random in relation to the advancement of the tractor along the vineyard row.

Table 1. Synthesis of the other tested equipment parameters.

Equipment	Working Press. [kPa]	Volume [1/ha]	Velocity [km/h]	Flow [l/m]	Width [m]
Dribble bar	100	315	3	1.1	0.9
Weed wiping brush	100	100	3	0.3	1
Weed wiping strip	100	60	3	0.2	0.7
Fog chamber	350	20	3	0.08	0.7

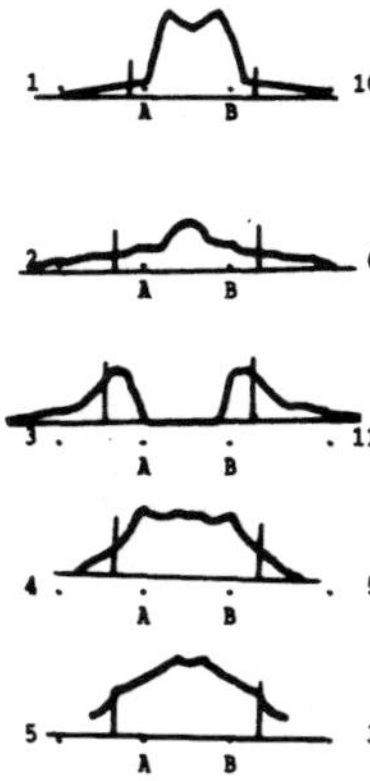

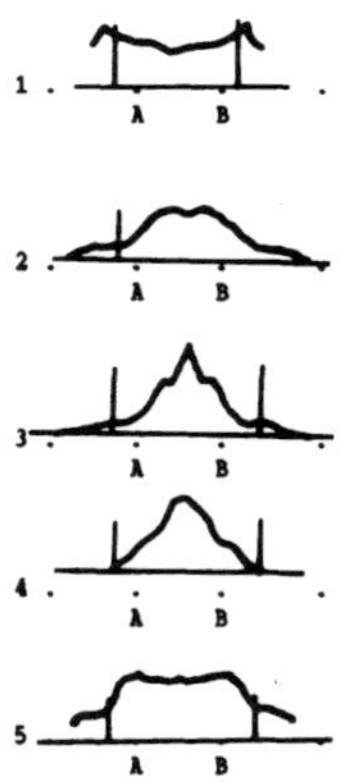

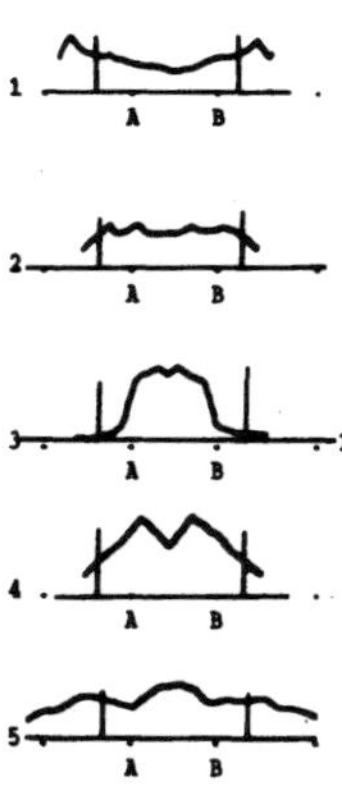

Nozzle working at 0.1m from the row Nozzle working at 0.3m from the row Nozzle working at 0.5m from the row

Nozzle type	Diagram	V.C. tot. %	Vol. theo. [1/ha]	V.C. AB %	Loss %
1		101	591	69	11
2		64	336	41.6	37
3		117	318	112	60
4		59	76	30	16
5		37	103	20	21

Nozzle type	Diagram	V.C. tot. %	Vol. theo. [1/ha]	V.C. AB %	Loss %
1		18	874	19	16
2		74	470	45	18
3		108	429	64	27
4		70	112	78	0
5		40	92	21	2

Nozzle type	Diagram	V.C. tot. %	Vol. theo. [1/ha]	V.C. AB %	Loss %
1		27	690	18.7	41
2		22	660	12	7
3		144	581	79	12
4		37	97	28	5
5		29	70	20	49

Fig.2 Distribution diagram and measure values of the nozzle working parameters tested on double side of the row. 1- double fan nozzle; 2- lateral fan nozzle; 3- mirror nozzle; 4- spinning rotary nozzle (vertical axis); 5- spinning rotary nozzle (horizontal axis).

3 RESULTS

The following results evaluate aspects of the performance of the equipment and the quality of herbicide distribution, including the spray path around the obstacles along the row.

Currently, hydraulic pression nozzles are most commonly used for vineyard herbicide application in Tuscany.

In fact, as we have seen in the field, it is possible to adopt these sprayers (mirror, double and lateral fan) mounted on a strong bar which is able to avoid obstacles on the row with a simple and effective obstacle avoiding mechanism. Moreover, we did not encounter any problems with vine branches near earth level.

The performance of the rotary sprayers is comparable to that of the nozzles mentioned above, but they present some problems with vine vegetation on the row: they avoid obstacles less easily, but they are clearly able to distribute very small volumes of liquid with a good level of uniformity.

Moreover, the power, and the precision, necessary are small (for example the working pressure is 0.2 bar for rotary spayers with horizontal axes, and 1.5 bar for rotary sprayers with vertical axes). Obviously, the size of the drops varies with rotation velocity.

The drop polverization is obtained by means of a centrifugal rotary system driven by a D.C. electric motor powered by tractor battery. There is also a small electric pump to catch excess product.

The product loss of the rotary nozzle with vertical axis is greater due to the trailing of the tractor in the field. Fig.1 reports the working parameters of the sprayers mentioned above: the distribution diagram and the relative variability coefficient, working pressure, flow per minute, mean drop diameter, spray width and distance of nozzle from centre.

Fig.2 shows the distribution diagram and the respective variability coefficients; the volume theoretically distributed, increasing with a velocity of 0.8 m/s; the herbicide path variation on the working area (0.9m), measured from the centre of the row; and the product losses, hypothesizing 3 distribution distances with the nozzles at the same height (0.4 m). The comparison between Fig.1 and Fig.2 furnishes the evaluation The "fog chamber" (see specifications in Tab.1) needs a great deal of energy because it requires a tractor-mounted air compressor which must deliver a large of the equipment tested.

The results obtained from the present experimentation with the dribble and weed-wiper bar, and the experimental prototype of "fog chamber" are reported in Tab.1.

Fig.4 expresses in graphic terms the results of a computer elaboration of the avoiding bars at work. The different degrees darkness reflect the differing concentrations of herbicide distribution on the soil. To perform this simulation

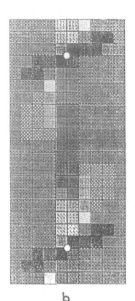

a

b

Fig.3 Scheme of the covering degree on the vineyard row, obtained using obstacle selecting equipment (a- single side; b- double side).

it has been necessary to test the avoidance velocity of the bar around trunks or similar obstacles in the vineyard rows.

Obviously this kind of equipment does not have drift problems, in the case of weed-wiper. However, there are some problems due to the irregular soil profile in the vineyards which causes an irregular motion of the tractor so that the distribution can be less than optimal. Moreover, there are also some problems in cleaning the distribution apparatus due to contact with the soil. Furthermore, with the weed-wiper distribution device, soil can in some cases limit the chemical effect of the active components because of a neutralization process caused by soil colloidal components.

volume of air in order to obtain the same effect as other equipment, but at the moment our prototype also presents some problems concerning drop penetration into the vegetaion. In fact, the drop size (<100μm) is such that the drops are affected by laminar motion and do not cover the vegetation. In other words, after the equipment passed we saw a great quantity of tiny drops carried away upon the thermal air currents, thus drifting into the atmosphere without covering the vegetation.

4 CONCLUSIONS

Our work has shown that it is necessary to select herbicide equipment much more carefully in order to maximize the effect, while minimizing the quantity, of the product distributed. Consequently, it hypothesizes a diminished environmental impact and optimal distribution costs. In Fig.2, an example of equipment selection is shown, based on the analysis of the characteristics of the machinery available on the market, according the performance required.

The simulation of Fig.2 was performed adopting software present on market, and is thus easy for anyone who can work with a personal computer.

We therefore believe that in the near future it will be necessary for manufacturers to furnish working specifications that farmers will be able use for simple simulations performed on a personal computer to verify a choice of equipment. This philosophy could be important in the development of a new agricultural management technique, reducing environmental problems and optimizing the cost of herbicide and pesticide spraying.

REFERENCES

Miele,S. 1982. Nuove attrezzature per il diserbo chimico. Macchine e Motori Agricoli. 4:37-40.

Baraldi,G. 1982. Irroratrici per la difesas delle piante: efficienza ed impiego. Macchine e Motori Agricoli. 11:131-136.

Baraldi,G. 1987. Aspetti meccanici nella corretta distribuzione dei fitofarmaci in agricoltura. Informatore Fitopatologico. 5:5-10.

Baraldi,G. 1987. Metodologie di distribuzione dei diserbanti. Proceedings "Diserbo chimico e ambiente" Padova, 7-10 June.

Musillami,S. 1982. Les trattaments par pulverization et les polverizateur an agricolture. CEMAGREF.

Land and Water Use, Dodd & Grace (eds), © 1989 Balkema, Rotterdam. ISBN 90 6191 980 0

Etude des possibilités d'amélioration du pulvérisateur à dos pour le traitement des mauvaises herbes dans la culture des céréales

K.Houmy & M.Bouhache
Institut Agronomique et Vétérinaire Hassan II, Rabat, Maroc

RESUME : Les essais sur le pulvérisateur à dos testé avec un banc d'essai de répartition ont montré que la répartition la plus homogène a été obtenue avec la buse à miroir suivie de la rampe à trois buses à fente, de la rampe avec quatre buses à chambre de turbulence et enfin de la buse à chambre de turbulence. Au champ, la mauvaise herbe Ranunculus sardous Grantz a été bien contrôlée avec aussi bien le 2,4D qu'avec le bentazone. De même, aucune différence d'efficacité entre les quatres dispositifs étudiés n'a été relevée. Cependant, Cichrium endivia L, a été moins contrôlée par les mêmes-herbicides et la buse à chambre de turbulence a été la moins efficace comparativement aux autres dispositifs testés dans le cas de l'utilisation de 2,4-D.

ABSTRACT : Laboratory tests on a lever operated knapsack sprayer showed that the most uniform spray was obtained with one flooding nozzle spray design followed by three flat spray nozzles, four hollow cone nozzles and by one hollow cone nozzle. field experiments showed that the weed, Ranunculus sardous Grantz was better controlled by both 2, 4-D and bentazon herbicides. However no difference in efficiency was noted between the four nozzle designs. In contrast to the first weed , Cichorium endivia L. was less controlled by 2,4-D and bentazon. In addition in one test one hollow cone nozzle had a lower efficiency than the other nozzles.

ZUSAMMENFASSUNG : Die versuche an-einem Rückenspritzgerät haben an einem Versuchsstand gezeigt, dass die regelmässigste Verteilung bei einer Pralledüse ist; an der 2.Reihe kommt das Spritzgestänge mit 3 Flachstrachldüsen und an der 3.Reihe das Spritzgestänge mit 4 Dralldüsen und anschliessend das spritzgestänge mit einer Dralldüse. Auf dem Feld wurde das Unkraut Ranunculus Sardous Grantz gut kontrolliert sowohl mit 2,4-D als auch mit Bentazone. Es wurde auch keinen wesentlichen Unterschied bezüglich der wirknug bei den Veschiedenen Systemen festgestellt . Im Gegensatz wurde das Umkraut Cichorium endivia L.wenig mit den oben genannten Herbiziden kontrolliert. Die Dralldüse hat sich im Vergleich zu den anderen Düsen mit 2,4-D am schlechtesten gezeigt.

1 INTRODUCTION

Jusqu'à présent, le pulvérisateur à dos représente l'un du matériel le plus utilisé pour le traitement des céréales au Maroc. En effet, il occupe 85 % du parc national du matériel de traitement phytosanitaire (Yadini 1984). Cette dominance s'explique essentiellement par son prix d'achat abordable, sa simplicité et aussi par le fait que la majorité des agriculteurs ne disposent pas de tracteurs agricoles.

Quant à l'utilisation de cet appareil, elle reste encore loin d'être bien maitrisée. En plus des problèmes, de dosage qui est généralement non respecté, le jalonnement qui est rarement effectué, l'intérêt du choix de type de buse n'est pas encore pris en considération pour la réussite d'un traitement.

Dans la plupart des cas, la lance avec une seule buse à chambre de turbulence est utilisée. En outre, la largeur de traitement est déterminée par le mouvement de va et vient de la main de l'opérateur. Par conséquent, la répartition des produits appliqués est loin d'être homogène, ce qui implique en général une diminution de l'efficacité des traitements herbicides.

L'objectif poursuivi par ce travail est d'étudier les différents types de buses adaptées sur le pulvérisateur à dos et de voir l'effet de chaque type de buses sur l'efficacité de deux herbicides, de contact et systémique, dans la culture des céréales.

2 MATERIEL ET METHODES

Nous avons utilisé un pulvérisateur à dos à pression entretenue, équipé d'une pompe à piston et dont la capacité de la cuve est de 20l. A ce pulvérisateur, nous avons adapté un manomètre qui permet de contrôler la pression de liquide au cours des essais. Ainsi quatre dispositifs de pulvérisateur ont été étudiés :
- La rampe "conventionnelle", c'est une lance portant une seule buse à chambre de turbulence. La largeur de travail est obtenue par un mouvement de va et vient de la lance.
- Une lance munie d'une seule buse à miroir.
- Une rampe pourvue de trois buses à fente.
- Une rampe munie de quatre buses à chambre de turbulence.

La largeur de travail de chaque dispositif est de 1,5m. Pour les deux derniers dispositifs un support a été confectionné pour fixer la rampe et pour régler la hauteur des buses.

a LES ESSAIS AU LABORATOIRE

Ces essais avaient pour l'objectif d'étudier la répartition du liquide de chaque dispositif étudié. Pour cela, nous avons utilisé un banc de répartition avec un profil en V et une distance entre les gouttières de 9,8cm. Quinze éprouvettes ont été utilisées pour receuillir ainsi les volumes qui s'écoulent le long des gouttières. Le dispositif "conventionnel" n'a pas été testé du fait que le banc de répartition ne lui est pas adapté. Le cœfficient de répartition (kv) a été utilisé pour comparer la répartition de chaque dispositif de pulvérisation. Ce cœfficient a été calculé selon la formule suivante :

$$kv \ \% = \frac{100}{\bar{v}} \sqrt{\frac{(vi - \bar{v})^2}{n}}$$

n : Nombre de volumes élémentaires.

$\bar{v}$: Volume moyen et égal à $\frac{\sum vi}{n}$

b LES ESSAIS AU CHAMP

Les essais au champ ont été réalisés sur une parcelle de blé tendre (Triticum carstirum L) avec une dose de semis de 170 kg/ha. Le désherbage chimique a été effectué au stade fin tallage du blé. Nous avons utilisé deux herbicides, un de contact (bentazone) et un systémique (2,4-D) avec un volume de bouillie de 300l/ha.

Les essais ont été conduits sur des parcelles élémentaires de 30m² de superficie (20 x 1,5m).

Au niveau de chaque parcelle élémentaire, nous avons choisi au hasard sur la médiane trois placettes de 0,5 x 0,5m. La densité des adventices a été noté 45 jours après le traitement. Les résultats sont exprimés en pourcentage de réduction de la densité par rapport au témoin non traité et jugés selon l'échelle de la commission des essais Biologique (CEB) de la Société Française de phytiatrie et Phytopharmacie :
100 à 95 % de réduction : très bonne efficacité
95 à 80 % de réduction : bonne efficacité
80 à 60 % de réduction : efficacité moyenne
60 à 40 % de réduction : efficacité faible
Moins de 40 % de réduction : sans intérêt pratique.

3 RESULTATS

a AU LABORATOIRE

Les différents volumes receuillis sont reportés sur les figures 1. 2 et 3. La buse à miroir et la rampe

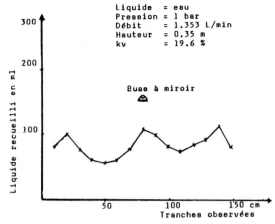

Fig. 1 : Répartition transversale du liquide
(Buse à miroir)

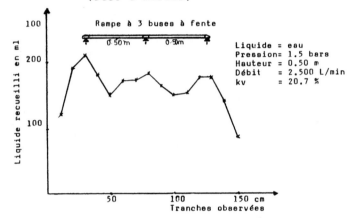

Fig. 2 : Répartition transversale du liquide
(Rampe à 3 buses à fente)

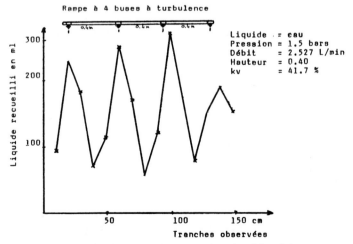

Fig. 3 : Répartition transversale du liquide
(rampe à 4 buses à turbulence)

à trois buses à fente ont donné une répartition relativement homogène (fig. 1 et 2) avec les coefficients de répartition respectivement de 19,6 et 20,7 %. Cependant, la répartition obtenue avec la rampe à quatre buses à chambre de turbulence est hétérogène (fig. 3) avec un coefficient de répartition de 41,7 %.

b AU CHAMP

La flore adventice rencontrée sur les parcelles de notre essai est composée d'une quinzaine d'espèces apparetenant à quatorze genres différents et à huit différentes familles botaniques. Cependant, l'étude de l'efficacité de différents dispositifs de pulvérisation a été jugées uniquement sur la base de deux espèces abandantes et dominantes, Cichorium endivice L et Ranunculus sardous Grantz.

Les résultats obtenus ont permis de montrer qu'avec les herbicides utilisés R. sardous est bien contrôlée. Avec l'utilisation de 2,4-D, les quatres dispositifs ont parmis une bonne efficacité avec un pourcentage de réduction de 100% pour la buse à miroir, 100,0 % pour la rampe à trois buses à fente, 100,0 % pour la rampe à quatre buses à chambre de turbulence et 96,9 % pour une seule buse à chambre de turbulence (fig. 4).

De même, nous avons noté une bonne efficacité du bentazone pour les quatres dispositifs avec un pourcentage de réduction de 100,0 % pour la buse à miroir, 95,0 % pour la rampe à trois buses à fente, 85,0 % pour la rampe à quatre buses à chambre de turbulence et 95,0 % pour une seule buse à chambre de turbulence (fig.5).

Comparativement à la première espèce, C. endivia est moins contrôlée par les mêmes herbicides. Sous l'effet de 2,4-D, une efficacité moyenne est obtenue avec la buse à chambre de turbulence avec un pourcentage de réduction de 69,6 %. En revanche, une bonne efficacité est obtenue avec la buse à miroir (94,9 %) suivie de la rampe à trois buses à fente (86,4%) et de la rampe à quatre buses à chambre de turbulence (86,1%) (Fig. 6). L'utilisation du bentazone a permi une efficacité moyen-

ne pour les quatres dispositifs avec un pourcentage de réduction de 78,9 % pour la buse à miroir, 77,0 % pour la rampe à trois buses à fente, 68,6 % pour la buse à chambre de turbulence et 61,7 % pour la rampe portant quatres buses à chambre de turbulence (Fig.7).

DISCUSSIONS

Concernant le Ranancule, aussibien pour l'herbicide de contact que l'herbicide systémique, aucune différence n'a été relevée entre les différents dispositifs expérimentés. Ceci pourrait être expliquer par la petite taille de feuilles et par la faiblesse de la masse foliaire exposée par cette espèce. Par conséquent, les chances d'atteindre la plante par chaque dispositif sont très grandes. Ces considérations semblent être à l'origine de la bonne efficacité des deux herbicides. En opposition, pour C. endivia qui présente une masse foliaire assez importante et des feuilles assez grandes, le recouvrement obtenu par les quatres dispositifs n'est pas suffisant pour mettre en évidence la bonne efficacité du produit de contact (bentazone). L'efficacité moyenne observée dans le cas de 2,4-D appliqué sur la chicorée avec la buse à chambre de turbulence explique par l'hetérogénéité de la répartition de la bouillie engendrée par le mouvement de va et vient de la main de l'opérateur. Cependant, la différence de répartition observée au laboratoire entre la buse à miroir, trois buses à fente et quatre buses à chambre de turbulence n'avait pas de répercussions sur l'efficacité du 2,4-D sur la chicorée.

Bien qu'une amélioration de l'efficacité du produit pourrait être obtenue avec l'addition des buses, il convient de signaler que ceci demande plus d'effort comparativement à une seule buse.

En tenant compte de ces considérations, le dispositif de pulvérisation utilisant une seule buse à miroir pourrait présenter un intérêt pratique qui permet un meilleur recouvrement et par conséquent un meilleur contrôle des adventices.

Toutefois, toute augmentation de

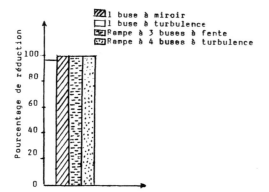

Fig. 4 : Pourcentage de réduction
de la ranoncule traitée
avec 2,4-D.

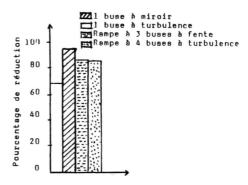

Fig. 6 : Pourcentage de réduction
de la chicorée traitée
avec 2,4-D.

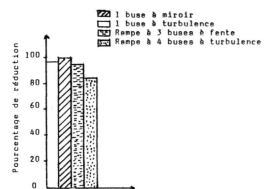

Fig. 5 : Pourcentage du réduction
de la ranoncule traitée
avec la bentazone.

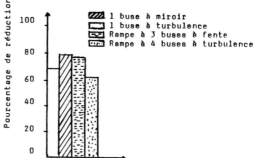

Fig. 7 : Pourcentage de réduction
de la chicorée traitée
avec le bentazone.

l'angle du jet de ce type de buse
et par conséquent sa largeur du
travail ne pourrait qu'améliorer
les performances de celle-ci.

REFERENCES

Gentour, H. 1987. Matériel de traitement contre les mauvaises herbes dans la culture des céréales
(région de Settat, Maroc). Mémoire de fin d'étude, Département
de Machinisme Agricole, I.A.V.
Hassan II, Rabat, Maroc.

Musillami, S. 1973. Buses à fente
en pulvérisation agricole. Etude du CNEEMA, Nº 376.
Triyeh, F. 1988. Comparaison de
quelques possibilités d'amélioration du pulvérisateur à dos
pour le traitement des mauvaises
herbes dans une céréale. Mémoire
de fin d'étude, Département de
Machinisme Agricole, I.A.V. Hassan II, Rabat, Maroc.
Yadini, D. 1984. Matériels de traitement au Maroc. Collogue ANAPPAV,
I.A.V. Hassan II, Rabat, Maroc.

Land and Water Use, Dodd & Grace (eds), © 1989 Balkema, Rotterdam. ISBN 90 6191 980 0

Closed system sprayers – The design and development of the DOSE 2000

A.J.Landers
Royal Agricultural College, Cirencester, Glos., UK

ABSTRACT: The need for closed system spraying is discussed along with recent legislation affecting the use of pesticides in Europe and the USA. The design requirements for the DOSE 2000 closed system are outlined along with the study of closed systems at the Royal Agricultural College. The paper concludes that as legislative requirements to control pesticide use within Europe become tighter, it will be necessary to use crop sprayers which ensure the safety of the operator and avoid environmental pollution. Closed system sprayers will satisfy these demands and provide considerable benefits to farmers and growers.

RESUME: Les avantages du système clos de pulverisation sont examinés en vue de la législation récente touchant l'utilisation des pesticides en Europe et aux Etats-Unis. Les exigences techniques pour le système clos DOSE 2000 sont indiqués, aussi bien que le programme d'études des systèmes clos au Royal Agricultural College, Cirencester. A mesure que la législation sur les pesticides devient plus stricte en Europe, il sera nécessaire d'utiliser seulement ces pulverisateurs qui sauvegardent l'opérateur et éliminent la pollution de l'environnement. Les pulverisateurs en système clos satisferont à ces demandes tout en apportant des avantages aux agriculteurs et aux cultivateurs.

ZUSAMMENFASSUNG: Die Vorteile des geschlossenen Spritzsystems wird besprochen, und zwar in Zusammenhang mit neuer Gesetzgebung über die Verwendung von Pflanzenbehandlungsmitteln in Europa und in den Vereinigten Staaten. Die Bauanforderungen für das geschlossene System DOSE 2000 werden beschrieben, sowie die Forschung über geschlossene Spritzsysteme im Royal Agricultural College, Cirencester. Während die gesetzliche Forderungen strenger werden, die Verwendung von Pflanzenbehandlungsmitteln in Europa zu regulieren, wird es nötig werden, Spritzgeräte zu benutzen, die die Sicherheit des Anwenders versichern und Boden – und Umweltverschmutzung vermeiden. Geschlossene Spritzsysteme werden diese Forderungen genügen, und Bauern und Züchtern grosse Vorteile anbieten.

1 INTRODUCTION

During recent years there has been a considerable tightening of the legislation governing the application of pesticides. In the United Kingdom, the introduction of the Food and Environment Protection Act, Part II, 1985 along with the Draft Code of Practice (MAFF 1988) has resulted in the need to train operators in the safe use and handling of pesticides along with the safe operation of application equipment. Training is being carried out under the auspices of the Agricultural Training Board. Certificates of competence are issued to candidates who are successful in tests organised by the National Proficiency Tests Council.

The introduction of the Control of Substances Hazardous to Health (HSC 1989) will ensure a move towards a safer working environment. Employers will be required to make an assessment of the risks to health which arise from hazardous substance exposure, and to take precautions to protect people. The Code also states that technical or engineering methods, for example, closed systems should be considered.

Two main areas of concern with pesticide use where closed system sprayers can help, are the minimising of operator contamination and the reduction in environmental pollution.

1.1 Operator Contamination

1 Figure 1 shows areas of potential operator contamination. Splashes can be a great risk to the operator – when decanting or measuring concentrate pesticide from a manufacturer's container into a sprayer tank. The pesticide container label states the protective clothing that should be worn. (Abbott et al 1987) noted that mixing and loading the pesticide concentrate into sprayers was potentially more contaminating to the operator than the subsequent spraying of the diluted material. The extent of potential dermal exposure is quite high.

2 Nozzle blockages

Nozzle blockages require immediate attention from the operator,

resulting in potential contamination. Blockages should be kept to a minimum by correct filtration. Good sprayer design should reduce this problem. Hydraulic boom folding will reduce contamination when compared with manual folding.

3 Spray drift

When spraying, drift can be a problem if the correct procedure is not followed. The correct nozzles at the correct pressure in suitable weather conditions is most important if drift is to be minimised. The use of closed tractor cabs and carbon filtration systems result in a better environment for the sprayer operator.

4 Washing out

Washing out sprayers after use is another potential hazard due to the risk of splashes generated by the use of pressure water hoses.

Good operator training and machine design will reduce operator contamination. Closed systems allow the concentrate pesticide container to be connected to the sprayer without the need for decanting or measuring. The water tank on the sprayer does not need to be rinsed as it contains clean water. (Rutz 1987) observed a decrease in mixer/loader illness with the use of closed systems in the USA.

1.2 Environmental pollution

In recent months, widespread public concern and controversy has arisen due to nitrate residues in drinking water in the United Kingdom. Public attention is being focused on farmers' attitudes to environmental pollution.

Concentrate pesticide spillage whilst filling crop sprayers and tank residues

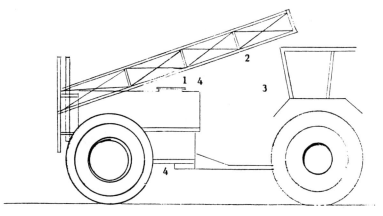

Figure 1   Areas of potential operator contamination

    1 Decanting/measuring concentrate pesticide
    2 Nozzle blockages/boom folding
    3 Spray drift
    4 Washing out

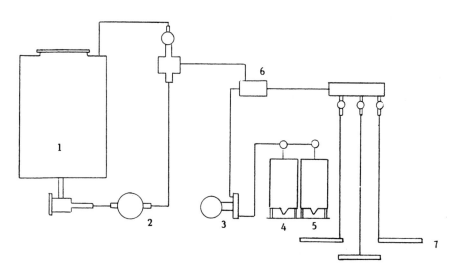

Figure 2   Schematic diagram of the DOSE 2000

    1 Water tank   5 Water container
    2 Water pump   6 Mixing chamber
    3 Pesticide pump   7 Booms
    4 Pesticide container

along with tank washings after spraying can result in environmental pollution. Care and attention to detail are required and are outlined in the Code of Practice (MAFF 1988). The planned disposal of tank residues and washings on an unsprayed area of crop should result in a reduction in pollution.

Washing out tank residues thoroughly is very important; failure to do so can result in disastrous effects to crops during subsequent spraying. The degree of cleaning required will depend upon the product just applied. (Taylor et al 1988) observed that thorough cleaning, using the method proposed in the Code of Practice (MAFF 1988) took in excess of one hour, resulting in 5.2ml of pesticide remaining from a 600 litre crop sprayer. Timeliness is of the essence whilst spraying, the pesticide must be applied during the correct weather conditions and at the correct growth stage. The need to spend in excess of one hour thoroughly cleaning a sprayer is expensive in terms of labour and lost production.

The closed system spraying technique results in only clean water remaining in the sprayer water tank. This water may be released safely on to the ground without any adverse effect on the environment.

2 LEGISLATION

Legislation in many countries is affecting the application of pesticides. In Germany, for example, Federal regulations govern correct professional conduct in plant protection. The testing of application equipment is required, as it is in Sweden.

In Holland, pesticide containers must be rinsed efficiently to a standard 0.01 per cent pesticide residue.

The Ministry of Environment in Denmark aims at a 25 per cent reduction in pesticide use in agriculture before 1990 with a further reduction of 25 per cent before 1997, based upon the average use from 1981 to 1985.

In the State of California, legislation for closed systems for category one liquid pesticide was introduced in 1973. (Rutz 1987) observed that it was not until 1983 that the Environment Protection Agency agreed that enforcement action would not take place provided conditions regarding protective clothing, operator training and inspection of equipment was adhered to.

(Brazelton and Akesson 1987) outlined various principles required for handling pesticides within a closed system. They observed the need to bring pressure upon regulating agencies and pesticide operators to improve working conditions.

3 DESIGN REQUIREMENTS FOR THE DOSE 2000 CLOSED SYSTEM SPRAYER:-

1 To help the operator carry out the spraying operation safely and efficiently without incurring any risk of operator contamination.

The pesticide containers should be capable of being handled without risk of contamination and be connected via non-drip connectors. Pesticide container size should match the water tank capacity to ensure efficient sprayer operation.

2 To allow thorough emptying and cleaning without contaminating the environment.

The water tank can be drained of clean water on to the ground without fear of polluting the environment. The use of the distance remaining function allows the operator to rinse the pipelines to the boom before leaving the field.

3 To be constructed of durable materials and to withstand the effects of pesticides and solvents.

Stainless steel and ceramics can withstand such attacks. The correct choice of 'plastic' pipes and Viton or PTFE seals should be used.

4 To inject a wide range of pesticides with varying viscosities, from water soluble pesticides through dispersable granules to wettable powders.

The operator should be able to adjust the control box according to the viscosity of the pesticide, via a simple calibration procedure.

5 To apply several pesticides/additives at the same time without any pre-mixing.

This will allow up to three products approved for tank-mixing to be used. Many farmers use up to three pesticides at any one time.

6 To mix the pesticide and water thoroughly.

It is imperative that a thorough mix occurs to avoid scorching the crop or under-applying pesticide, resulting in crop loss. Vigorous mixing should be the aim.

7 To function accurately at the range of dose rates found in practice.

Arable farmers may need to apply between 0.5 litre to 10 litres per hectare of concentrated pesticide. Consideration must be given to the concentrate-pesticide viscosity in arriving at the correct dose rate via a simple calibration procedure.

8 To change dose rates quickly and accurately due to changes in operating parameters.

Examples such as changes in forward speed, engine speed and boom width. Electronic monitors should detect any changes and relay them to the control box.

9 To be easy to use and understand.

It is imperative that the operator is able to use the control box to his advantage. To obtain the maximum benefit from such a device a clear, concise manual must be provided along with a thorough training course.

10 To be capable of being fitted to most existing sprayers.

The modern crop sprayer varies considerably in its design and complexity. The injection system needs to be easily adapted to fit most machines.

11 To be commercially viable.

4 DESCRIPTION OF THE DOSE 2000

The DOSE 2000 was developed in Sweden by Agri Futura AB International. It is a kit which can be fitted to any conventional field sprayer.

The system works in the following way. The operator can select the required dose level on the control box mounted in the tractor cab. The microprocessor controlled ceramic piston pump will then pump the pesticide concentrate from the container into the mixing chamber where it mixes thoroughly with the water coming from the sprayer water pump. The solution then goes to the boom nozzles via the boom manifold valves.

Figure 2 shows a schematic diagram of the DOSE 2000 components.

4.1 The containers

The containers used have a capacity of 30 litres, with a wide opening to allow powders to be poured in. The size of the container is important to ensure that it matches the sprayer water tank capacity so that this may be refilled at the same time. The weight of the full container is an important consideration in relation to access to the container rack for lifting, and the extra weight imposed upon the sprayer. The containers must empty completely, and so a probe is fitted into the container reaching down into a well formed in the base. An alternative considered was to invert any containers, but problems could arise due to incorrect sealing particularly in a farm environment. The standardisation of container openings will allow easier connection to the suction probe.

The present situation of many small containers being required for use with large capacity tank sprayers all needing to be rinsed and disposed of is quite a problem. Larger capacity containers allowing improved logistics have been used in the past for seed treatment products. Perhaps the time is now right to consider returnable containers, as in the USA.

The 30 litre containers can be filled or rinsed out by the use of a small filling station. The electrically operated filling station comprises a suction probe, pump and filling probe. The small manufacturers' containers can be emptied into the larger container on the sprayer, and the rinsing water can also be drawn out into a separate container. This diluted waste can then be applied at a higher rate or disposed of according to the Codes of Practice, or taken away by a professional contractor.

4.2 The pump

The piston pump needs to be very durable to withstand the action of the pesticide concentrate. It comprises a ceramic piston in a stainless steel cylinder; the seals are PTFE. The piston pump is driven via the pto input to the sprayer. It is switched on only when water flow occurs through the sprayer (this is detected via a flow meter situated near the main control valve). The piston pump stroke length is adjusted according to certain parameters; forward speed (sensed by a wheel sensor or a radar speed sensor), the number of boom sections open, the dose level set by the operator, and the pto speed into the pump.

The stroke length of the pump (and thus the output) is adjusted by means of an electric stepper motor, controlled by the in-cab controller. Output is directly proportional to the dose setting.

The pump needs to be very durable to withstand the concentrate pesticide and a well-proven pump was chosen. This is used in many industrial applications, eg fluoride and chlorine dosing for water authorities. Other simpler, cheaper pumps such as peristaltic ones are available. However they are not suitable because of their lack of precision, reliability and longevity.

The control box on the DOSE 2000 displays the following features:

1 Dose

One of four dosing levels can be selected and the dose chosen is shown on a display; the others can be programmed in advance. During operation the dose is altered simply by switching between the various settings.

2 Calibration

Changes in parameters such as viscosity can lead to small changes in the dosing level; a simple test can be run. New settings can be inserted by press-button.

3 Remaining distance

To avoid having unused pesticide remaining in the sprayer system, the pesticide can be switched off and replaced by water as the driver nears the end of the field. The display indicates at what distance to switch-over to water, thus rinsing the pipes of pesticide.

4 Tractor Speed

5 The area sprayed with each pump

6 The current flow rate of water and the total volume of water used.

In addition audio visual indicators will function when:
- pesticide container is empty
- water tank is empty
- hose leaks
- system error

Pumps 1, 2, 3
The operator can select any one pump or a combination of up to 3 pumps, if fitted. The spot treatment of patches of weed is a useful possibility with this system.

The controls are easy to use, and the use of a simple magnetic card allows a service engineer to fault-find in the sprayer system.

4.3 The mixing chamber

The mixing chamber is connected by a hose between the main control valve and the boom section valves. The water flow enters at the side of the chamber, and the pesticide is injected at one end of the chamber. Up to three approved products may be injected concurrently. The resulting diluent is delivered via the boom valves to the nozzles.

The mixing chamber needs to a) mix the concentrate with the water thoroughly; b) be of a large enough capacity to even out the pulsing of the piston pump (along with the rubber hoses) and, c) have a small enough capacity so as not to create a long time delay when changing dose levels. To keep time delay to a minimum the ideal position of the chamber would be at the rear of the sprayer next to the booms. Consideration must be given to the physical constraints of mounting a closed system on to existing sprayers.

5 DISCUSSION: THE ADVANTAGES OF THE DOSE 2000 CLOSED SYSTEM

(Landers 1988) outlined the following advantages of the DOSE 2000:-

1 The disposal of surplus spray liquid has been a financial and environmental problem. This is minimised since the tank contains only clean water and the chemical can be returned to the store.

2 Operator and environmental contamination due to splashes when pouring, rinsing or measuring concentrates is reduced.

3 Accuracy of metering due to sophisticated electronics and precisely calibrated pumps results in financial benefits.

4 Spot treatment of weeds is readily possible, resulting in lower pesticide use on farms and its resulting benefits.

5 The logistics of spraying are improved due to less time spent at the headland mixing and calculating.

6 Nozzle throughput can be checked by using clean water.

7 Dose rate adjustments can be made whilst on the move so, for example, heavy or light infestation on headlands

can be sprayed at appropriate rates.

8 The use of a distance meter allows the operator to rinse out the sprayer lines before leaving the field. The display indicates at what distance to switch off the pesticide and switch on the rinsing water.

9 A conventional spraying system is still used, which complies with existing legislation and likely future developments.

10 Water may be withdrawn from a source without fear of suck-back resulting in contamination of ponds, streams, etc.

6 EVALUATION TRIALS AT THE ROYAL AGRICULTURAL COLLEGE FARMS

The Royal Agricultural College farms extend to 740ha, where emphasis is placed on commercial farming systems. Trials were carried out in conjunction with the Farms Manager and staff from the Engineering and Biology Departments of the College. The DOSE 2000 has been evaluated on different crops using pesticides at various application rates during 1988. The biological performance has been assessed along with the following mechanical evaluation:-

6.1 Ease of installation

The attachment of the dosing system to the conventional sprayers was studied. The dosing system could be installed within a few hours; the amount of plumbing required is quite small but the power take-off and pulley needs an extra guard in some cases.

6.2 Ease of operation of the control box

There is a need for a comprehensive operator manual. The ease of use was evaluated in respect of the controls and the display. The dose rate was easy to set and adjust; the water flow rates, forward speed and area were easily monitored via digital display.

6.3 Field operation

The filling of the pesticide container required the correct protective clothing and skills as the filling of a conventional sprayer. (This operation would be simpler if larger pesticide containers were commercially available which could be fitted directly on to the sprayer).

The couplings connecting the pesticide containers to the dosing unit pipes operated without problems. The accuracy of the dosing pump has been observed. The pump is extremely accurate with water and aqueous solutions, but further research is needed into the viscosities of various pesticides. The dose level can easily be adjusted by means of a press-button.

Calibration was straightforward using the calibration vessel and the area meter. Numerous calibrations trials with water have been carried out to ensure accuracy before applying any pesticides.

Spot spraying response times were recorded using paraquat on grassland. The response time to a change in dose rate or the introduction of a pesticide was clearly observed. Modification of the sprayer could result in a response time of five seconds on the 18 metre sprayer applying 200 litres/ha.

6.4 Mechanical reliability

The DOSE 2000 was used with numerous pesticides throughout the course of the spraying season. The only problem has been the need to replace the dosing pump seals of Viton with seals made from PTFE.

6.5 Further studies

Further studies are to be carried out on viscosities, and the effect on timelines of the spraying operation due to faster turnround on the headlands (due to not having to mix or measure concentrated pesticides or rinse out the water tank).

7 CONCLUSIONS

1 Current United Kingdom and European Community legislation has resulted in a significant move towards safer practices when storing, handling and applying pesticides.

Future legislation, brought about by Governments and public awareness, can only improve safety standards. As more emphasis is placed upon safer use of pesticides such as the Control of Substances Hazardous to Health, so there will be a greater need for closed system sprayers to avoid operator contamination and environmental pollution.

2 The advantages of closed system techniques in avoiding operator contamination and environmental pollution, are likely to outweigh the disadvantage of the extra capital cost. The advantages of not having to measure or decant pesticide at the headland, coupled with not needing to wash or rinse the main tank, will result in a saving of time, leading to improved work rates.

3 The advantages of being able to spot treat as and when required, along with the ability to change dose rates as levels of infestation alter, will be of considerable benefit to farmers during this period of lower profit margins.

4 There is a need for the agriculture-related industries to respond to innovative techniques to reduce environmental pollution and operator contamination. There is also a need for standardisation of container openings to allow universal couplings to be connected along with a larger choice of pack size to match sprayer tank capacity.

REFERENCES

Abbott, J.M., Bonsall, J.L., Chester, G., Bernard Hart, T., Turnbull, J. 1987. Worker exposure to a herbicide applied with ground sprayers in the United Kingdom. American Industrial Hygiene Association. 48(2) 167-175.

Brazleton, R.W. and Akesson, N.B. 1987. Principles of closed systems for handling of agricultural pesticides in The Proceedings of the American Society of Testing Materials 7th Symposium on pesticide formulations and application systems, pp 16-27. Eds. G.B. Beestman and D.I.B. Van der Hooven, ASTM, Philadelphia.

HSC 1989. Draft approved Code of Practice for the control of substances hazardous to health: control of exposure to pesticides at work. London. Health and Safety Commission.

Landers, A.J. 1988. Closed System Spraying - the DOSE 2000. Aspects of Applied Biology 18: 361-369.

MAFF 1988. Revised Draft Code of Practice for the Agricultural and Commercial Horticultural Use of Pesticides. London. Ministry of Agriculture, Fisheries and Food.

Rutz, R. 1987. Closed system acceptance and use in California. In The Proceedings of the American Society of Testing Materials, 7th Symposium on pesticide formulations and application systems, 29-34. Eds. G.B. Beestman and D.I.B. Van der Hooven. ASTM Philadelphia.

Taylor, W.A., Pretty, S., Oliver, R.W.,
1988. Some observations quantifying and
locating spray remnants within an
agricultural crop sprayer. Aspects of
Applied Biology 18: 385-393.

Land and Water Use, Dodd & Grace (eds), © 1989 Balkema, Rotterdam. ISBN 90 6191 980 0

Modélisation de la répartition transversale du volume dispersé par une rampe de pulvérisateur

P.Leunda & C.Debouche
Faculté des Sciences Agronomiques de Gembloux, Gembloux, Belgique

R.Caussin
Comité de Recherches pour l'Amélioration des Techniques de Traitements Phytosanitaires (IRSIA), Gembloux, Belgique

ABSTRACT: The reduction of volumes spread over an hectare, which is the most significant trend in field spraying practice, requires a more accurate control of the homogeneity of volume distribution under the spraying boom. This homogeneity admits two components, a longitudinal and a transversal one, the former being ensured by the recent developments of electronical equipment controlling the flow according to the speed of advancement. The transversal component is linked to the distribution of each isolated nozzle on the boom and depends on parameters such as nozzle spacing and spraying height and pressure, whose values are fixed by the users.
A possible approach for optimizing the values given to those parameters consists in establishing a model of the distribution of an isolated nozzle, taking into account nozzle calibre, spraying pressure and boom height. This individual model can then be used to predict the transversal repartition under a boom and improve it by varying the spraying parameters and nozzle spacing on the boom.

RESUME: La réduction des volumes de bouillie épandus à l'hectare, tendance dominante dans l'évolution actuelle de la pulvérisation agricole, requiert une précision accrue du contrôle de l'homogénéité de la répartition au sol. Cette homogénéité admet deux composantes, longitudinale et transversale, dont la première est assurée par le développement récent des appareils électroniques d'asservissement du débit à la vitesse d'avancement. La composante transversale est liée à la répartition de chaque buse isolée et dépend de paramètres comme l'écartement des buses, la hauteur et la pression de pulvérisation, dont les valeurs sont fixées par les utilisateurs.
Une voie d'approche pour optimiser les valeurs à donner à ces paramètres consiste à établir un modèle de la répartition propre à une buse isolée, modèle où interviennent le calibre de la buse, la pression et la hauteur de pulvérisation. Ce modèle individuel peut alors être utilisé pour prévoir et optimiser la répartition transversale obtenue sous une rampe en faisant varier les paramètres de la pulvérisation et l'écartement des buses sur la rampe.

1 INTRODUCTION

La variabilité du volume épandu au sol par une rampe de pulvérisation a deux composantes, transversale et longitudinale. La composante longitudinale est conditionnée par la régularité de la relation entre le débit et la vitesse d'avancement. La composante transversale est liée à la variabilité des caractéristiques d'un buse à l'autre, à la répartition individuelle de chaque buse qui compose la rampe, ainsi qu'à l'écartement auquel sont montées les buses. La répartition individuelle des buses dépend, elle, de leur facture et des hauteur et pression de pulvérisation.

Réputées tant pour leurs propriétés de répartition que pour la composition de leur spectre des gouttes, les buses à fente sont actuellement les plus utilisées dans le domaine de la pulvérisation agricole. Ce sont des "buses de pulvérisation hydraulique ayant un orifice en forme de fente et produisant un jet plat" (définition ISO). Ces buses fournissent une répartition transversale en forme de cloche, traditionnellement assimilée à une courbe de Gauss.

Pour former un jet plat en éventail, les buses à fente emploient souvent un écoulement conique convergent de liquide, émergeant dans l'atmosphère à travers un orifice de forme plus ou moins elliptique. La manière exacte dont l'orifice est façonné pour fournir l'angle désiré est propre au fabricant. Approximativement, il est généré par l'intersection d'un prisme triangulaire avec un cône ou une calotte sphérique, la limite de l'orifice résultant étant décrite par deux arcs elliptiques inclinés (Liljedahl 1971). Il faut remarquer que, si des études existent sur certaines propriétés des buses à fente, telles leur spectre des gouttes, il n'y a toujours pas d'analyse de l'écoulement du fluide dans et au travers de ces buses (Nation 1985).

L'objectif poursuivi par ce travail est d'établir à partir de données expérimentales un modèle mathématique de la répartition transversale d'une buse à fente et d'utiliser les résultats de cette modélisation pour prévoir la répartition sous une rampe composée de plusieurs buses et en établir un modèle mathématique. Ce modèle devrait permettre d'optimiser la répartition sous rampe, pour différents écartements des buses, en faisant varier les paramètres hauteur de rampe et pression de pulvérisation.

Seront testées des buses de trois marques courantes, présentant des angles de jet de 80° et 110°. Pour chacune des 6 variantes seront choisis trois calibres usuels. Les buses Lurmark LP sont moulées dans un plastique récent, le Kematal, et ne sont utilisables qu'à basse pression. Elles seront testées à 1 et 2 bars. Les buses XRTeejet sont usinées dans un acier durci; elles

seront soumises aux pressions de 1, 2, 3 et 4 bars pour les modèles 110° et à 2, 3 et 4 bars pour les autres. Enfin, les buses Albuz fabriquées en alumine frittée seront essayées à 2, 3 et 4 bars. Chacune des 51 variantes ainsi définies sera utilisée à 3 hauteurs différentes (40, 60 et 80 cm).

Le montage expérimental se compose d'un répartiteur normalisé, surmonté d'une rampe en position horizontale. La rampe porte une buse à laquelle parvient de l'eau mise sous pression par l'action de l'air dans un réservoir fermé ou par une pompe. La pression de l'eau et son débit sont mesurés en permanence et sont maintenus constants pendant la durée de l'essai au moyen d'un détendeur ou d'un dispositif électronique de contrôle de vannes.

Le répartiteur est un plan incliné limité en aval par une rangée de 53 éprouvettes graduées formant l'axe des abscisses. La rampe est placée parallèlement à cet axe, et la buse est fixée à l'aplomb de l'éprouvette centrale portant l'abscisse 0. Le plan est divisé perpendiculairement à cet axe en 53 gouttières de 5 cm de largeur, les éprouvettes recueillant les volumes qui s'écoulent le long des gouttières.

2 MODELISATION DE LA REPARTITION D'UNE BUSE A FENTE

2.1 Le modèle théorique

Si on considère l'abscisse du point de chute de chaque unité de volume (la gouttelette) comme la valeur observée d'une variable aléatoire, on assimile les volumes recueillis dans les éprouvettes aux fréquences observées, groupées en classes, de cette variable aléatoire. L'unité de mesure de ces fréquences de classe sera le cm^3, qui correspond au plus petit volume décelable dans les éprouvettes. On peut alors interpréter la répartition transversale des volumes épandus comme une distribution de probabilité observée, et chercher à la représenter par une distribution théorique de probabilité qui présenterait des caractéristiques semblables. Ce problème présente deux volets: le choix du modèle ou fonction de densité de probabilité et l'estimation de ses paramètres à partir de ceux de la répartition transversale mesurée, considérée comme un échantillon.

De manière cohérente avec le modèle théorique choisi, le processus de modélisation utilisé sera donc l'ajustement d'une distribution théorique de probabilité sur les répartitions observées, distribution dont les paramètres seront estimés au mieux pour qu'elle corresponde à la répartition observée.

Dans un second temps, les paramètres estimés de la distribution théorique seront exprimés en fonction des facteurs qui conditionnent l'aspersion: pression, hauteur et calibre de la buse.

Cependant, d'autres facteurs, incontrôlés ceux-là, peuvent également influencer la répartition: la géométrie du montage expérimental, la variabilité des buses, enfin l'ensemble de facteurs aléatoires qui se manifestent par des différences d'une répétition à l'autre d'un même essai de répartition. L'influence de ces trois types de facteurs a été étudiée par des méthodes de l'analyse statistique, qui permettent de conclure que leur influence n'est pas significative. Ainsi, une seule répétition de l'essai de répartition suffit-elle pour donner une représentation acceptable de la distribution de la population-parent, et ce d'autant plus qu'on s'éloigne des extrêmes de la répartition et que les volumes recueillis dans les éprouvettes sont plus importants.

2.2 L'ajustement à la distribution normale tronquée

Les répartitions transversales observées sont des courbes approximativement en forme de cloche, unimodales, symétriques et à marges finies, toutes caractéristiques que possède la distribution normale tronquée.

Une variable aléatoire X possède une distribution normale doublement tronquée lorsque sa fonction de densité de probabilité est (Johnson et Kotz 1970):

$$f(x) = \frac{1}{\sigma\sqrt{2\pi}} e^{-\frac{1}{2\sigma^2}(x-\xi)^2} \left[\frac{1}{\sigma\sqrt{2\pi}} \int_A^B e^{-\frac{1}{2\sigma^2}(t-\xi)^2} dt \right]^{-1}$$

$$= \sigma^{-1} Z\left(\frac{x-\xi}{\sigma}\right) \left[\phi\left(\frac{B-\xi}{\sigma}\right) - \phi\left(\frac{A-\xi}{\sigma}\right) \right]^{-1},$$

pour $A \leq X \leq B$,

où σ^2 est la variance de la distribution normale, $Z(x)$ est la densité de probabilité de la distribution normale réduite relative à la valeur x, et $\phi(x)$ est la valeur de la fonction de répartition de la même distribution pour la valeur x.

Les degrés de troncature sont $\phi\left(\frac{A-\xi}{\sigma}\right)$ et $\left(1-\phi\left(\frac{B-\xi}{\sigma}\right)\right)$.

On notera que si $\frac{A-\xi}{\sigma} = -\left(\frac{B-\xi}{\sigma}\right) = -\delta$, alors $E(X) = \xi$ et

$$var(X) = \left[1 - \frac{2\,\delta\,Z(\delta)}{2\,\phi(\delta) - 1}\right] \sigma^2 .$$

La distribution normale doublement tronquée est entièrement définie par 4 paramètres: ξ, σ^2 et les deux points de troncature A et B. Si on suppose que la répartition transversale des débits épandus par une tête de jet placée à la verticale du point central de la classe (éprouvette) centrale, ou classe 0, est symétrique, alors on aura effectivement: A = -B et $\xi = 0$. Cela signifie que la largeur de travail d'une buse est symétrique par rapport à l'orifice et comprise entre les deux points de troncature. Vu les faibles degrés de troncature et l'importance de l'effectif, on estimera la variance de la distribution ajustée par la variance de l'échantillon.

Deux exemples de l'ajustement sont présentés au paragraphe suivant, ainsi qu'un tableau reprenant les paramètres de l'ajustement pour les buses XRTeejet 110°.

2.3 Critique de l'ajustement

Deux accidents se rencontrent simultanément ou non dans la plupart des répartitions observées: la présence de deux épaulements latéraux plus ou moins symétriques par rapport à la classe centrale et d'une troncature du sommet de la répartition. La part du volume total représentée par ces accidents est faible (de 0 à 5%) et variable selon

la buse et les conditions de
pulvérisation. L'importance de ces
accidents diminue avec la pression;
une hauteur croissante augmente la
troncature du sommet et diminue
l'importance des épaulements.

D'une manière générale, la forme
de la répartition est assez bien
rendue par l'ajustement, mais le
degré de ressemblance devra être
quantifié de manière plus
rigoureuse, par le test chi carré
d'ajustement (Dagnelie 1975).

Le tableau 1 ci-dessous reprend
les caractéristiques des buses
XRTeejet 110° (calibre et débit Q,
en l/min), les valeurs des
paramètres de l'aspersion (hauteur
H, en cm et pression P, en bars) et
les valeurs des paramètres
déterminant la distribution
théorique (nombre de classes No et
écart-type σ). L'écart-type est
exprimé en nombre de classes de 5 cm
(ou d'éprouvettes). La dernière
colonne du tableau reprend les
valeurs calculées pour le rapport
$\chi^2_{obs}/\chi^2_{1-\alpha}$, issu du test chi carré,
désigné par Chirap et calculé avec α
égal à 0.05. Ce rapport doit être
inférieur à un pour pouvoir admettre
la validité de l'ajustement.

Les figures 1 et 2 représentent
deux de ces ajustements relatifs aux
buses XRTeejet 110° de calibre 02:
l'un est acceptable à l'issue du
test chi carré (chirap = 0,83);
l'autre ne l'est pas (chirap =
3,53).

Tableau 1. Paramètres de
l'ajustement des buses XRTeejet 110°

Calibre	H	P	No	σ	Q	Chirap
02	40	1	21	4.30	0.42	3.53
		2	29	5.58	0.63	1.75
		3	31	5.80	0.78	1.70
		4	31	6.01	0.90	0.83
	60	1	31	6.11	0.42	4.20
		2	39	7.23	0.63	1.53
		3	41	7.69	0.78	1.52
		4	43	7.92	0.90	1.63
	80	1	39	7.54	0.42	3.40
		2	47	8.70	0.63	1.53
		3	49	9.15	0.78	2.23
		4	49	9.38	0.90	3.79
04	40	1	23	4.52	0.87	4.35
		2	29	5.57	1.27	2.15
		3	31	5.87	1.55	1.63
		4	31	6.05	1.79	1.24
	60	1	33	6.64	0.87	4.50
		2	39	7.79	1.27	2.13
		3	41	8.15	1.55	1.70
		4	43	8.33	1.79	1.35
	80	1	41	8.34	0.87	3.82
		2	51	9.56	1.27	2.17
		3	51	9.96	1.55	2.15
		4	51	10.12	1.79	2.62
06	40	1	23	4.42	1.35	3.90
		2	29	5.45	1.91	3.63
		3	31	5.76	2.36	3.67
		4	33	5.92	2.72	4.54
	60	1	33	6.51	1.35	3.98
		2	41	7.58	1.91	3.57
		3	43	7.98	2.36	3.68
		4	45	8.16	2.72	3.51
	80	1	43	8.34	1.35	3.63
		2	51	9.54	1.91	2.74
		3	51	9.89	2.36	3.05
		4	51	9.93	2.72	2.98

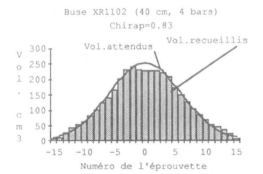

Fig.1 Ajustement acceptable

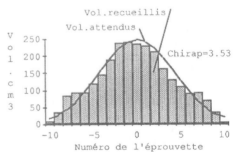

Fig.2 Ajustement inacceptable

2.4 Conclusions

Sur les 153 essais de répartition effectués avec les trois marques de buses, seuls six ont donné un rapport des chi carré inférieur à 1, et 46 un rapport inférieur à 2. Il faut en conclure que l'ajustement est inacceptable au sens statistique. En effet, les différences entre les distributions observée et ajustée, exprimées par le rapport des chi carré, sont trop importantes pour être dues au hasard. Toutefois, cette conclusion est à nuancer en fonction des marques, des calibres et des conditions de pulvérisation.

D'autre part, d'un point de vue plus pragmatique, la ressemblance entre répartition observée et distribution ajustée peut être très bonne; les différences dues aux accidents observés sont relativement faibles. De plus, comme le montrent les graphiques ci-dessus, il n'apparaît aucune différence qualitative entre les répartitions qui s'ajustent bien à la distribution théorique et les autres. Au vu des résultats graphiques, il paraît difficile de trouver une forme de distribution théorique qui s'ajuste mieux, par une prise en compte des épaulements observés, tout en conservant un nombre raisonnable de paramètres.

3 RELATIONS ENTRE LES PARAMETRES DE L'AJUSTEMENT ET CEUX DE LA PULVERISATION

3.1 Introduction

D'après les hypothèses suivantes:

$$\xi = 0$$
$$A = -B,$$

deux paramètres du modèle, l'écart-type et la largeur de travail, doivent encore être estimés à partir des variables explicatives choisies: la pression et la hauteur de pulvérisation et le calibre de la buse.

La méthode utilisée pour mettre en évidence les relations entre ces deux ensembles de paramètres est celle de la régression multiple.

3.2 Régression de l'écart-type en fonction de la hauteur, de la pression et du débit

Une observation rapide du tableau 1 montre que l'écart-type de la répartition observée est une fonction croissante de la pression d'utilisation et de la hauteur de l'orifice. Il apparaît que l'équation de régression recherchée exprime l'écart-type en fonction des logarithmes népériens de la hauteur, de la pression et du débit. Pour l'ensemble des essais, le calcul du coefficient de corrélation et la réalisation des tests de signification ont montré que les relations obtenues sont très étroites et significatives, comme l'indique l'exemple ci-dessous, relatif aux buses XRTeejet 110°, où la régression a porté sur 36 points (voir tableau 1). L'écart-type σ et la hauteur H sont en cm, la pression P en bars, R^2 représente le coefficient de détermination, en pourcent; Q_2 est le débit à deux bars, en litres par minute.

L'équation de régression est:

$$\sigma = -78.0^{***} + 27.1^{***} \ln(H) + 6.15^{***} \ln(P) + 1.72^{**} \ln(Q_2)$$

et $R^2 = 0.980$.

L'écart-type résiduel est égal 1.263 pour un écart-type total de 8.576.

Les résultats de l'analyse de la variance sont consignés dans le tableau suivant, où Ddl signifie le nombre de degrés de liberté, SCE la somme des carrés des écarts et CM le carré moyen.

Source	Ddl	SCE	CM
Régression	3	2523.34	841.11
Ln (H)	1	2131.61	
Ln (P)	1	369.44	
Ln (Q2)	1	22.28	
Erreur	32	51.03	1.59
Total	35	2574.37	

On constate donc que la presque totalité de la variation de l'écart-type est expliquée par la régression, de manière très hautement significative; la hauteur de pulvérisation justifie 80 % de la variation, le reste étant dû à la pression et, dans une moindre mesure, au débit.

Les corrélations et les degrés de signification sont tout aussi élevés pour les autres buses essayées.

3.3 Régression de la largeur de travail

La largeur de travail est, dans notre modèle, le nombre d'éprouvettes contenant un volume significatif, nombre auquel on ajoute un s'il est pair, de manière à avoir une largeur identique de part et d'autre de l'éprouvette centrale et être cohérent avec l'hypothèse de symétrie. La caractéristique du dispositif expérimental consistant à regrouper les volumes en classes de largeur égale à 5 cm va augmenter l'imprécision de la relation recherchée entre ce paramètre et ceux qui le conditionnent.

Cette relation exprime la largeur de travail LT (en cm) en fonction des logarithmes népériens de la hauteur et de la pression, mais le débit standard peut y intervenir sous forme logarithmique ou non, avec des valeurs comparables des coefficients de corrélation et du degré de signification.

A titre d'exemple, la régression logarithmique est présentée ci-dessous pour les buses XRTeejet 110°.

L'équation de régression est:

$$LT = - 401*** +1.39*** \ln (H)$$
$$+ 36.5*** \ln(P) + 8.94** \ln(Q2)$$

et $R^2 = 0.976$.

L'écart-type résiduel est égal à 7.286 cm pour un écart-type total de 45.237.

Les conclusions qu'on peut tirer sont les mêmes que lors de l'étude de l'écart-type; celui-ci est d'ailleurs lié à la largeur de travail par une corrélation de l'ordre de 0.95.

3.4 Conclusions

L'ajustement à la distribution normale tronquée de la répartition transversale des volumes épandus par une tête de jet est d'une qualité généralement inacceptable d'un point de vue statistique. D'un point de vue pratique, la ressemblance de la distribution ajustée et de la répartition observée est satisfaisante, d'autant plus qu'il paraît difficile de trouver une autre distribution théorique qui s'ajuste mieux. Cette ressemblance varie selon la marque de buses, et selon le type (angle au sommet) au sein d'une même marque.

Les paramètres de l'ajustement (points de troncature et écart-type) présentent une très nette corrélation avec les paramètres de la pulvérisation. Ainsi, deux équations par type de buse permettent de retrouver les paramètres caractérisant complètement la distribution théorique à partir de la connaissance de la hauteur et de la pression d'utilisation, ainsi que du débit nominal de la buse.

4 MODELISATION DE LA REPARTITION SOUS RAMPE

4.1 Introduction

Lorsque des buses à jet en éventail sont associées sur une rampe, il se produit un recouvrement transversal des répartitions des jets voisins, qui conduit à une meilleure homogénéité de la répartition résultante. Selon les valeurs des paramètres de la pulvérisation, qui conditionnent la largeur de travail d'une buse isolée, et selon l'écartement des buses sur la rampe, le recouvrement peut être double ou triple, voire même quadruple. Comme l'écartement est constant et que les angles des jets sont approximativement les mêmes, il apparaît une périodicité dans les séquences de recouvrement. Le plus souvent, chaque répartition individuelle est différente, du fait de l'usure des buses et de leurs différentes provenances. A l'inverse, la simulation de la répartition sous rampe à partir de la répartition individuelle présuppose que toutes les buses possèdent une répartition identique. Cette supposition est cohérente avec les conclusions de l'étude des facteurs non-contrôlés, mais ne peut s'appliquer qu'à des buses neuves de même série de fabrication, montées correctement.

D'autre part, dans la suite du texte, on appellera répartition sous rampe observée celle obtenue par simulation à partir des répartitions

individuelles observées, et répartition théorique celle obtenue par les
répartitions individuelles ajustées.

4.2 La simulation de la répartition sous rampe

La longueur usuelle des rampes est
de 12 à 20 mètres, et une valeur
habituelle de l'écartement des buses
est de 50 cm. Mais comme les résultats de la simulation doivent
montrer la répétition périodique
d'une même séquence, une simulation
portant sur un plus petit nombre de
buses suffira à décrire la répartition sous une rampe plus longue. La
répartition simulée sera produite
par sommation des répartitions
individuelles observées et ajustées
de 7 buses écartées de 50 cm, et ce
pour un passage du pulvérisateur
partiellement recouvert par les
passages latéraux (précédent et
suivant). La figure 3 ci-dessous
montre la répartition simulée
obtenue, pour un essai de répartition correctement ajusté à la
distribution normale.

Buse XR11002 (40 cm, 4 bars)

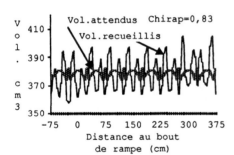

Fig.3 Simulation de la répartition
sous rampe

On constate que le recouvrement
des répartitions individuelles
voisines identiques, décalées d'une
distance constante, engendre une
résultante dont les motifs se repro-
duisent périodiquement, cette périodicité se manifestant aussi bien sur
la répartition sous rampe observée
que sur la répartition ajustée.
Mais là s'arrête la ressemblance
des deux répartitions simulées: la
marge de variation de la répartition
observée est nettement plus
importante, et ses pics (ou ses

minima) ne correspondent pas à ceux
de la répartition théorique. De
plus, on constate que la qualité de
l'ajustement n'influe pas qualitati-
vement sur ces dissemblances. Il
faut donc en conclure que le processus de sommation des répartitions
individuelles amplifie leurs irrégularités par rapport à la distribution ajustée. La répartition
théorique résultante est généralement plus homogène que la répartition dite observée.

4.3 Les paramètres de la répartition sous rampe

Deux paramètres décriront le
résultat de la simulation de la
répartition résultante: ils quanti-
fieront l'homogénéité des répartitions théorique et observée. Ces
coefficients de variation relatifs
aux répartitions observée et
théorique seront désignés respectivement par CVO et CVT et se
calculeront par la relation:

$$CV = \frac{100}{\bar{x}} \sqrt{\frac{1}{n} \sum_{i=1}^{n} (x_i - \bar{x})^2} \ ,$$

où $\bar{x}$ est le volume moyen de la
répartition sous rampe, x_i est le
volume de la répartition sous rampe
relatif à la classe i, n est le
nombre total de classes touchées par
les répartitions.
Ces deux paramètres ont été calculés
pour une buse et des conditions de
pulvérisation données, en faisant
varier l'écartement des buses sur la
rampe, pour l'ensemble des
conditions d'utilisation.
La figure 4 ci-dessous montre pour
une buse XRTeejet 110° dans des
conditions données de pulvérisation
l'évolution des coefficients de
variation en fonction de
l'écartement. Il faut noter qu'il
s'agit dans cet exemple d'un essai
dont l'ajustement est acceptable
(rapport des chi carré égal à 0.83).

4.4 Discussion des résultats

L'examen des répartitions sous rampe
révélait déjà la dissemblance des
répartitions théorique et observée,
même lorsque l'ajustement de la

répartition individuelle est acceptable au sens statistique. La figure 4 le confirme, qui met en évidence l'absence de relation commune entre les deux répartitions et l'écartement.

Par conséquent, le modèle théorique individuel n'est pas utilisable pour décrire la répartition observée sous rampe, et a fortiori pour l'optimiser. On constate toutefois que l'homogénéité est plus grande pour les répartitions théoriques que pour les observées, et que les coefficients de variation de ces dernières sont plus bas lorsque l'ajustement est bon, c'est-à-dire lorsque leur forme est plus régulière.

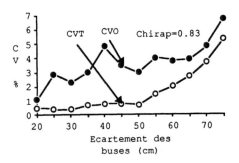

Buse XR11002 (40 cm, 4 bars)

Fig.4 Coefficients de variation en fonction de l'écartement des buses

Vu que le modèle théorique est inutilisable, seuls seront discutés dans la suite les répartitions sous rampe observées et leurs coefficients de variation.

Les coefficients de variation diminuent en général quand l'écartement des buses diminue. Cette tendance présente toutefois des exceptions localisées telles que les coefficients peuvent diminuer de moitié en augmentant l'écartement de 10 cm. La relation entre les coefficients de variation et l'écartement est différente pour tous les essais: elle est imprévisible. Ce caractère imprévisible de l'évolution des coefficients de variation se manifeste aussi dans leur relation avec la pression et la hauteur. Aussi les observations générales qu'on pourra tirer sur l'influence de ces deux paramètres

sont-elles susceptibles de nombreuses exceptions.

D'une manière générale, le coefficient de variation de la répartition sous rampe diminue nettement lorsque la hauteur de pulvérisation passe de 40 à 60 cm. Il se maintient ou diminue de manière négligeable quand la hauteur passe de 60 à 80 cm. Toutefois, ces observations doivent être nuancées en fonction de la pression. A 40 cm de hauteur, l'effet d'une pression croissante est d'augmenter de manière sensible l'homogénéité de la répartition sous rampe. A 60 cm de hauteur, l'effet de la pression sur le coefficient de variation est faible et ne présente pas de caractère systématique. A 80 cm, cet effet devient négligeable.

Cela signifie qu'il y a un intervalle de hauteurs dans lequel le coefficient de variation est généralement peu affecté par une variation de la pression; pour des buses écartées de 50 cm, cet intervalle va de 60 à 80 cm au moins. L'intérêt de cette observation est lié au système de régulation du débit d'un pulvérisateur. Rappelons que ce système vise à obtenir une répartition longitudinale homogène en compensant la variation de la vitesse du pulvérisateur par des variations proportionnelles du débit, et donc de la pression.

Buse XR11006, espac. 50 cm

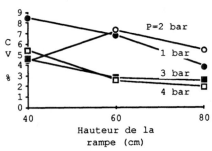

Fig.6 Coefficients de variation en fonction de la hauteur et de la pression

5 CONCLUSIONS

La répartition transversale des volumes épandus par une tête de jet se reproduit avec de faibles différences d'une répétition à l'autre d'un essai de répartition et se montre peu sensible à de petites erreurs de montage, ce qui permet de limiter l'expérimentation à une seule observation pour une buse et des conditions de pulvérisation données.

Cette répartition a été ajustée à la distribution normale à troncature symétrique, courbe de mêmes caractéristiques que la répartition. La qualité de cet ajustement est essentiellement limitée par la présence quasi constante d'accidents dans les répartitions observées: troncature au sommet, épaulements latéraux. Malgré ces accidents, l'ajustement est parfois acceptable en toute rigueur, et souvent assez proche des répartitions observées. Les paramètres de ces dernières, ainsi que ceux des distributions ajustées, dépendent de manière étroite des paramètres de la pulvérisation: hauteur, pression et calibre de la buse. Ainsi, l'écart-type et la largeur de travail s'expriment en fonction des logarithmes de la hauteur, de la pression et du débit nominal. Ces deux relations permettent de décrire complètement le modèle théorique.

Lorsque des buses sont montées sur une rampe, les répartitions transversales des jets voisins se recouvrent et engendrent ainsi une répartition sous rampe constituée d'une succession périodique de pics et de creux, d'amplitude généralement moindre pour les répartitions sous rampe obtenues à partir des distributions ajustées que pour celles obtenues à partir des répartitions individuelles observées. Les différences entre les répartitions sous rampe théorique et observée existent quel que soit la qualité de l'ajustement: le recouvrement des répartitions individuelles voisines amplifie toutes les irrégularités de ces répartitions par rapport à leurs distributions ajustées.

Le modèle théorique individuel est donc inutilisable pour simuler et optimiser la répartition réelle sous rampe. Cependant, le procédé de simulation, appliqué aux répartitions individuelles observées, a permis de dégager quelques observations quant à l'influence sur l'homogénéité de la répartition sous rampe des paramètres de la pulvérisation: pression, hauteur de rampe, écartement des buses. Ainsi le fait qu'un écartement croissant diminue généralement l'homogénéité de la répartition, qu'une hauteur ou une pression croissantes l'augmentent, et qu'il existe un intervalle de hauteurs où l'homogénéité est peu affectée par une variation de pression.

Il reste que la comparaison des répartitions théorique et observée montre l'influence importante des accidents des répartitions individuelles sur l'homogénéité des répartitions sous rampe. Il appartient finalement aux fabricants de buses de poursuivre leurs efforts pour atténuer ces accidents et améliorer ainsi la précision et la régularité des rampes de pulvérisation.

BIBLIOGRAPHIE

Dagnelie, P. 1975. Théorie et méthodes statistiques. Gembloux: Presses agronomiques de Gembloux.
Johnson, N.L.and Kotz, S. 1970. Distributions in Statistics, Continuous Univariate Distributions-1. Boston: Houghton Mifflin Co.
Leunda, P. 1987. Contribution à la modélisation du fonctionnement des organes d'épandage d'un pulvérisateur (TFE), Faculté.des Sciences Agronomiques de l'Etat à Gembloux, Belgique.
Liljedahl, L.A. 1971. Effect of fluid properties and nozzle parameters on drop size distribution from fan spray nozzles. PhD.thesis, Iowa State University, Ames, Iowa, USA.
Nation, H.J. October 1985. A selective review of literature on hydraulic pressure nozzle design and performance. Div.Note DN1300, National Institute of Agricultural Engineering, Silsoe, UK.

Land and Water Use, Dodd & Grace (eds), © 1989 Balkema, Rotterdam. ISBN 90 6191 980 0

A method for determination of theoretical characteristics of centrifugal pumps for field sprayers

M.Miszczak
Institute of Agricultural Engineering, Agricultural University, Warsaw, Poland

ABSTRACT: Centrifugal pumps have been widely used in agriculture of many countries. Very often the required shape of the pump characteristics should be determined at the early stage of pump designing by its manufacturer. This paper presents a new method for predicting the characteristics of centrifugal pump of $n_{SQ} < 45$, basing on its constructional parameters. An algorithm for determination of these parameters is presented also. The hydraulic losses were assumed to be proportional to the full trinomian square of dimensionless output of the pump. The paper was based on empirical and analytical investigations carried out at the Warsaw Agricultural University.

Die Kreiselpumpen haben eine breite Anwendung in der Landwirtschaft vieler Länder gefunden. In vielen Fällen soll der erwünschte Verlauf der Pumpenkennlinie bereits auf der Stufe der Projektierung durch den Hersteller bestimmt werden. In vorliegender Arbeit wurde eine neue Methode der Prognostizierung von Kennlinien zentrifugaler Kreiselpumpen von $n_{SQ} < 45$ auf Grund von ihren Konstruktionsparametern dargestellt Diese Methode wurde ausgearbeitet bei Annahme, dass die Strömungsverluste proportional zum vollständigen quadratischen Trinum der dimensionslosen Pumpenleistung sind. Der Arbeit liegen die, in der Landwirtschaftlichen Universität durchgeführten empirische und analytische Untersuchungen zugrunde.

Les pompes centrifuges sont largement utilisées dans l'agriculture de beaucoup de pays. Dans nombreux cas la forme désirée de la caractéristique de la pompe doit être déterminée déjà à l'étape de l'établissement du projet de la pompe par le constructeur. On a présenté dans ce travail une méthode récente de la prévision scientifique de la caractéristique de la pompe centrifuge à $n_{SQ} < 45$, établie à partir de ses paramètres de construction. La méthode a été élaborée en supposant que les pertes du débit sont proportionnelles au trinôme total du second dégré du débit adimensionné. Le travail est basé sur les essais empiriques et analytiques effectués à l'Université Agronomique.

To relate the characteristics of centrifugal pump to its constructional parameters the empirical investigations were carried out, enabling to verify the methods used for this purpose. The investigations on characteristics of the pumps were accomplished with the use of special stand at the Warsaw Pump Factory. For further analysis the mean characteristics of several pumps of the same type were taken. The obtained in such way material was related to three types of pumps: PM, PJM and A types produced in Poland. This material includes most of the single-stage centrifugal pumps with volute collecting casing produced in Poland. The pumps of PM type are single-stage centrifugal pumps with the single-stream closed impeller and volute collecting casing, suitable for monoblock connecting to the driving engine.

The pumps of PJM types are like PM type, but with excential position of impeller axis to the volute center. As a result, the part of impeller's circumference is placed inside the volute. The pumps of A types are similar to that of PM type, but with the impeller's vanes drawn in towards the inlet, even in pumps of small shape number $n_{SQ} < 25$ This feature should enable to reduce the cavitation and improve conditions of flow on the vanes at outputs different from the calculated ones.

All experimental characteristics were elaborated with the use of regression method

and equations relating the pressure head
of the pump to its output were determined.
The optimal equations were searched with
the use of power, exponential and polynomial dependences.

The highest values of correlation and
determination coefficients with the least
errors of estimation were obtained for regression equations in the second degree
polynomial form:

$$H = AQ^2 - BQ - C \qquad (1)$$

where: H - pressure head, m
       Q - output of pump, m^3/s

Considering the obtained results one can
observe that the regression coefficients
determined are quite accurate, for their
correlation coefficients exceed 0.0050.

Then, using constructional parameters
of the pumps presented, the A, B and C
parameters of characteristics equations
for the pumps with impellers of basic
types were calculated according to: Borowski's, Bajbakow'-Matwiejew's, Gusev's,
Pfleiderer's and Wajda's methods. The
recommended values of coefficients were
taken. At the further stage, the characteristics obtained with the methods mentioned above were compared with the real
characteristics. It was found that within
the range of pumps output from 0.4 Q_n to
1.2 Q_n (recommended with respect to
their efficiency) the highest deviations
from the real characteristics, amounted
in average to 40%, showed characteristics
obtained by Borowski's method, while characteristics obtained by Gusev's method
had the mean value of deviation equal to
33%. The least deviations were found for
the methods of: Bajbakow-Matwiejew - 16%,
Pfleiderer - 13% and Wajda - 15%. The
exemplary characteristics obtained by different methods are presented in Fig. 1
against the real characteristic. Substantial discrepancies between the obtained
analytical characteristics and the real
ones point at uselessness of the hitherto
methods for predicting characteristics of
the pumps presented. In that case an
attempt was undertaken towards working
out of a new method for predicting the
characteristic with greater accuracy. As
a starting point there was assumed that
equation of characteristic, similarly to
Gusev's, Pfleiderer's and Wajda's methods,
can be obtained by substracting the friction and shock losses from theoretical
characteristic:

$$H = H_{th} - \cdot \Delta h_{St1} - \Delta h_{St2} \qquad (2)$$

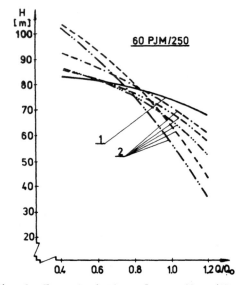

Fig. 1 Characteristics of pump 60PJM/250.
1 - real, 2 - according to methods hitherto
used.

It was assumed, that the theoretical characteristic equation is based on Pfleiderer's theory of passing from the theoretical pressure head at infinite large number
of vanes to theoretical pressure head of
impeller:

$$H_{th} = \frac{\pi \cdot d_2 \cdot n}{60g(1+p)} \left[\frac{\pi \cdot d_2 \cdot n}{60} - Q \frac{\varphi_2 \cdot ctg\,_2}{d_2 \cdot b_2} \right] \qquad (3)$$

where: p - Pfleiderer's amendment defined
by equation:

$$p = \frac{\mathcal{H} \cdot d_2^2}{4z \cdot M_{St}} \qquad (4)$$

The experimental coefficient   for the
pumps considered was determined from
dependence:

$$\mathcal{H} = 0,65 \left(1 + \frac{\beta_2}{60} \right) \qquad (5)$$

The shock losses can be determined similarly to the Gusev's and Pfleiderer's
methods as proportional to the square of
difference between the output at non-shock
inflow and actual output from equation:

$$\Delta h_{St1} = \frac{\xi \cdot \pi^2 \cdot d_1^2 \cdot n^2}{2g \cdot 60} \left(1 - \frac{Q}{Q_o} \right)^2 \qquad (6)$$

where: Q_o - output at non-shock inflow

2122

calculated as follows:

$$Q_o = \frac{\pi^2 . d_1^2 . b_1 . n . tg\, \beta_1}{60 . \varphi_1} \qquad (7)$$

As opposed to Gusev's and Pfleiderer's methods, where the value of proportionality factor is defined only by certain range of possible numbers, in the proposed methods this factor can be determined exactly for given types of pumps produced at given technology.

The friction losses were assumed as proportional to trinimial square of ratio between actual output and the output at non-shock inflow. Such assumption resulted from consideration to additional losses caused by fluid swirling between the impeller's vanes, apart from the flow losses usually assumed. Thus, the friction losses can be calculated from the equation:

$$\Delta h_{St2} = (1 - \eta_h) . H_{th} . (k_1 (\frac{Q}{Q_o})^2 + k_2 . Q . n + k_3) \qquad (8)$$

where: H_{th} - theoretical pressure head at output corresponding to non-shock inflow determined by equation (3).

η_h - hydraulic efficiency at output corresponding to non-shock inflow.

k_1, k_2, k_3 - empirical coefficients determined for a given pump and for its fabricating technology.

Taking into consideration the above assumptions and dependences as well as geometrical and kinematic relationships prevailing in centrifugal pump impeller, the equation of characteristic according to method proposed takes the form:

$$H = AQ^2 - BQn - Cn^2 \qquad (9)$$

where: $A = \dfrac{-\varphi_1}{\pi^2 . g . d_1^2 . b_1 . tg\, \beta_1 (1+p)} \cdot$

$$\left[\frac{\xi \varphi_1 (1+p)}{2 . b_1 . tg\, \beta_1} + k_1 (1 - \eta_h) . \left(\frac{d_2^2 . \varphi_1}{d_1^2 . b_1 . tg\, \beta_1} + \right.\right.$$

$$\left.\left. - \frac{\varphi_2}{b_2 . tg\, \beta_2} \right) \right]$$

$$B = \frac{1}{60g(1+p)} \left[\frac{\xi . \varphi_1 (1+p)}{b_1 . tg\, \beta_1} \right.$$

$$- \frac{\varphi_2}{b_2 . tg\, \beta_2} - k_2 (1 - \eta_h) .$$

$$\left. \left(\frac{d_2^2 . \varphi_1}{d_1^2 . b_1 . tg\, \beta_1} - \frac{\varphi_2}{b_2 . tg\, \beta_2} \right) \right]$$

$$C = \frac{\pi^2}{60^2 g(1+p)} \left[d_2^2 - \frac{d_1^2 . \xi . (1+p)}{2} + \right.$$

$$\left. - k_3 (1 - \eta_h) \left(d_2^2 - \frac{\varphi_2 . d_1^2 . b_1 . tg\, \beta_1}{\varphi_1 . b_2 . tg\, \beta_2} \right) \right]$$

The presented method enables to determine predicted characteristic of the pump basing on its constructional parameters. The proportionality factors: k_1, k_2, k_3 and ξ were determined basing on 12 selected at random pumps of each type (PM, PJM, A), considering the range of outputs from 0.4 to 1.2 of the output at non-shock inflow, thus, the entire range of admissible work of pump. Then, taking the series of values of shock coefficient ξ, the shock losses Δh_{St1} were calculated for each considered pumps by equation (1) for 10 values of pump output within the above range. At the same points the flow losses Δh_{St2} were calculated by equation (2), taking the pressure head of pump H from the mean real characteristics. Using the dependence (8) the coefficients k_1, k_2 and k_3 were calculated by quadratic polynomial regression method. From the obtained results there were chosen these values of coefficients ξ, k_1, k_2, k_3, which best corresponded to the real losses. The obtained values of proportionality factors are presented in Table 1.

To verify the method presented above there were determined, on the basis of the obtained coefficients and constructional parameters, the predicted characteristics for all sizes of pumps produced of PM, PJM and A types. The calculated characteristics were compared with the real ones, determining for this purpose the maximal relative error within considered pump output. The exemplary characteristics obtained by the presented method, against the real characteristics for the pumps with turned impellers are illustrated in Fig. 2.

Table 1

Type of pump	k_1	k_2	k_3	Coefficient of correlation	Coefficient of determination	Standard error of estimation
PM	1,0	1,0268	−0,9212	0,8944	0,9145	0,8344	0,1113
PJM	1,0	1,4749	−1,6396	1,1647	0,9099	0,8279	0,1412
A	0,6	0,3655	0,1176	0,5168	0,9236	0,8530	0,0894

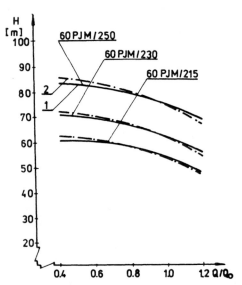

Fig. 2 Characteristics of pumps 60PJM/250-215. 1 - real, 2 - according to proposed method

REFERENCES

1. Bajbakow O.W., Matwiejew I.W.: Progno-zirowanie charakteristic centrobieznych nasosov. Wiestnik Maszinostrojenia, nr 10 1973.
2. Borowskij B.I.: Analiticzeskij metod rasczota energeticzeskich charakteris-tik centrobieznych nasosov so spiralnym otwodom. Energomaszinostrojenie, nr 2, 1970.
3. Gusiev W.W.: K rasczotu energeticzes-kich charakteristik centrobieznego na-sosa. IWUZ Gornyj Zurnal, nr 1, 1973.
4. Pfeliderer C., Peterman H.: Strömungs-maschinen. Berlin-Heidelberg-New York-Springer-Verlag, 1972.
5. Stepanoff A.J.: Centrifugal and axial flow pumps. Wyd. 2. J. Wiley and Sons Inc. New York, 1957.

NOTATION

A, B, C - coefficients of pump characteristics
b_1, b_2 - width of rotor flow channel at inlet and outlet, respectively
d_1, d_2 - inner and outer diameter of rotor vane, respectively
d_3 - diameter of pump scroll beginning
d_r - reduced diameter
H - pressure head
n - rotational speed of pump shaft
p - Pfleiderer amendment
s - thickness of vane
z - number of vanes
Q - output of pump
β_1, β_2 - angle of vane inclination at inlet and outlet of rotor
φ_1, φ_2 - coefficient of blanking off of rotor's inlet and outlet
η_h - hydraulic efficiency

Considering the obtained results one can observe, that the mean values of devia-tion of characteristics calculated by the method presented from the real ones amount to: for pumps of PM type - 5.48%, for PJM pumps - 5.90%, and for A pumps - 4.43%. Thus, the deviations are substan-tially smaller than those obtained with the use of other methods and do not exceed the values of measurement error.

The method proposed can be then used for predicting the characteristics of new centrifugal pumps of described types and also of other types when the proportiona-lity factors are determined. This method enables also to correct the characteristic depending on requirements at the early stage of pump designing without fabrica-tion of expensive prototypes.

Land and Water Use, Dodd & Grace (eds), © 1989 Balkema, Rotterdam. ISBN 90 6191 980 0

Plant protection equipment – Use and misuse

A.Nordby
Norwegian Institute of Agricultural Engineering, Norway

ABSTRACT: The problems of spray application are emphasized. The development of spraying equipment during the last 10-15 years is discussed and the number produced estimated. The ordinary type crop sprayer, in good order, with correct settings and used according to recommendations still is the best equipment for field spraying. However, only approximately 50 % of the sprayers are working satisfactorily, due to bad maintenance. One way of improval is the voluntary inspections of spraying equipment.

The use of plant protection chemicals for controlling weeds, fungi and insects has played an important role in the increase of plant production. But there are risks in too much depending on chemicals. The misuse is a threat against human health and against the environment. Brundtland (1987) reports that 10 000 people died in developing countries from poisoning by plant protection chemicals in 1983, and 400 000 more were seriously ill from the same reason.

Investigations in the USA and in Europe have revealed residues of pesticides in lakes and rivers, especially shortly after the spraying has taken place. At some places plant protection chemicals are also found in the ground water. The situation has alarmed the authorities, especially in several European countries (Statens plantevern - Gefo 1987). The aim is: Reduce the use of plant protection chemicals in agriculture. This means reduced dose of active ingredients and reduced number of sprayings per season. Directions, rules and laws to this effect have been presented.

PLANT PROTECTION EQUIPMENT

Lack of reliable statistics makes it difficult to discuss the production of plant production equipment in Europe. The industry estimate the production of crop sprayers to approximately 70 000 per year.

The mostly sold type is a tractor mounted crop sprayer with 400-600 l tank size and a spray boom with working width 10-12 m. There are a few large producers of crop sprayers in Europe, for instance in Denmark, West Germany, France and Italy. One factory report a production volume of 10 000 booms of 12 m working width per year.

Some countries do have statistical data of sprayers in practical use. But here again, the available information is obviously incomplete. In West Germany there are approximately 170 000 crop sprayers in use and in Norway 27 000.

The lifetime of a crop sprayer is usually 10-12 years.

In Scandinavia, approximately 75 % of the plant protection chemicals are weed killers. Some 90 % of the plant protection chemicals are spread by hydraulic nozzles placed on spray booms.

SOME MAIN PROBLEMS IN SPRAY APPLICATION

Too small amount of the spray is deposited on the target. For crop sprayers, the deposit is found to be 40-75 % of the total spray delivered (Nordby & Jørgensen 1985 - Robinson & Garnet 1984).

Also by mist spraying of fruit trees, the result is not quite satisfactory. Before breaking into leaf, only 10-20 % can be deposited on the trees, and with fully grown leaves, still only 25-60 % of the total spray was deposited. (Herrington,

Mapother & Stringer 1980 - Vittrup 1965).

Drift of spray

The problem of drift of spray has been thoroughly investigated and one can easily find a hundred references about drift measurements. Increasing wind speed, boom height and working pressure increase the drift of spray. The driving direction in relation to the wind direction is also important. Some countries have implemented rules that, for instance, permit no spraying if the wind speed is more than 5 m/sec measured 2 m above ground. The spray drift is a main problem and has to be reduced. The drift is an inherent hazard for:
    the natural environment
    neighbouring crops
    contamination of the spraying crew

Uneven deposit on the crop results in:
    less than optimal pest control
    excessive amounts of chemicals on crops
    for sale

Low spray deposit rate also means drift of chemicals outside the field to be treated, into lakes, rivers, forests and pastures. Spraying against bushwood in the forestry, mainly from helicopters or other aircraft, greatly impairs the recreational value of berry picking, if not necessarily the quality of the berries themselves.

In the last several years, the justification of using plant protection chemicals has been continuously discussed. Non-chemical methods may be an alternative for weed control. However, for insect and fungus control non-chemical measures seldom are an alternative to spraying but a supplement and in some cases a compensation.

DEVELOPMENT OF SPRAYING EQUIPMENT THE LAST 10-15 YEARS

Crop sprayers

Flat spray nozzles with 110^0 degree fan angle more or less have become a standard in Europe, and at a spacing of 50 cm.

Some manufacturers are: Albuz, Delavan, Hardi, Lechler, Lurmark, Teejet and Tecnoma

There is a good choice in both flow rate and in materials.

In Europe, and perhaps especially in Scandinavia, 2 bar is the usual working pressure in weed spraying. If in some cases one is afraid of drift one might reduce the pressure to one bar only. When reducing the working pressure to one bar, it is important to keep the boom height at not less than 40 cm to ensure proper spray distribution.

A certain type of 110^0 flat nozzle, the XR Teejet, gives reasonably good distribution at 1 bar and 40 cm boom heigth.

Tab. 1. Liquid distribution from a spray boom. (CV, coefficient of variation) Nozzle spacing 50 cm. Boom heigth 40 cm [1])

Nozzle		Working pressure, bar			
Size	Type	1	2	3	4
11002	X R common type Teejet	12,9 17,5	5,9 9,7	8,9 6,1	9,0 4,9
11003	X R common type Teejet	13,7 19,4	5,6 10,7	5,4 13,5	6,0 10,3
11004	X R common type Teejet	6,3 15,9	7,2 8,2	5,7 8,3	6,0 8,1

[1]) Measurements from Norwegian Institute of Agricultural Engineering 1989

To avoid having to stop the spraying by nozzle clogging, double or triple holders can be installed. The snap fit nozzle, Hardi - Lurmark - Spraying Systems, automaticly repositions the spray tip.

Clean water tank

When dealing with chemicals, it is wise to have plenty of clean water available for washing the hands etc. It is also necessary for rinsing sprayer parts. A clean water tank should be compulsory equipment on sprayers.

Filter

When using volume rates down to 100 l/ha, a good filtering system is a necessity. On some crop sprayers the filter is placed for easy control and maintenance. Some filters are self cleaning.

For many sprayers, emptying the tank is complicated and time consuming. Some makes have the emptying valve handle on top of the tank. This is convenient and reduces the risks of contaminating the spraying crew.

Filling the sprayer tank

Filling chemical on the tank is often done in a difficult working position with risks of contamination. With a filling device mounted on the sprayer one can conveniently and without risks fill the exact amount of chemicals into the sprayer tank.

Foam markers help avoiding gaps and double spraying.

On sprayers with working width 12 m and more hydraulic lowering and lifting of the boom is a practical feature. Hydraulic folding and unfolding is also convenient.

Operating unit and control of spraying

The control boxes are placed in the tractor cab. The components are completely sealed off from chemical exposure. The different makes and models may offer one or more of the following:
  Constant application rate (by adjusting the working pressure in relation to driving speed
  Constant pressure control
  On and off of the spray delivery from the cabin
  Individual boom sections shut off
  Manual pressure adjustment
  Folding and unfolding spray boom, left and right hand side individually

Electrostatic spraying, spinning disc, and rotary cage sprayers

In field spraying, electrostatic distribution and deposit of the liquid can not compete with the ordinary crop sprayer. Cooke et al (1986) state that 200 l/ha distributed with 110^0 flat nozzle produced the most even spray within the crop and gave the lowest coefficient of variation (CV) of mean deposits both on cereal plants and on weeds.

Cooke also found that the spinning disc and the rotary cage sprayers either yielded too high CV values or lacked penetration ability.

Skuterud et al (1988) found that the flat spray nozzle gave better results than spinning discs for both systemic and contact herbicides.

Weed wipers

Weed wipers have been used to some little extent in row crops. In the main spraying crop, the cereals, it is of no interest at all.

Air assisted sprayer

For all well known spraying methods and spraying equipment there are at least three weak points:
1. Drift of spray
2. Insufficient penetration
3. Difficulties with applying low spray volumes

The air assistance is working on the whole boom width. On the Hardi Twin System sprayer, the air is assisting the transport of the droplets produced by an ordinary spray boom. The booklet Hardi Twin System (1989) presents a few results for the combinations:

Flat sprayer nozzle	150 l/ha
Giro jet	41 l/ha
Danfoil	45 l/ha
Hardi Twin prototype	50 l/ha

The results by weed spraying are nearly the same for all equipment and for the doses 2,5-1,25-0,63 1 Faneron per ha.

When spraying with Tilt Turbo against rust in wheat, doses 1,0-0,5 and 0,25 l/ha, the air assisted spray gave the best results.

So far, just a few results are available. However, the system of air assistance seems promising. Beside Hardi Twin there is also another Danish make, Danfoil. In the Netherlands, van den Munkhof has presented a prototype of air assisted sprayer. (Anon (1987)

Enclosed systems - direct injection

This system employs two tanks, one for water and one for the plant protection chemical. By varying the amount of chemicals injected into the water before the spray boom the dose of active ingredients may be altered. In West Germany, metering systems are now being used by spraying along the railways (Jacobi & Kramer 1984).

Metering pumps for application of chemicals injection, widely used in some industrial plants, are on sale in West Germany. Schmidt (1983) gives a survey of closed system, direct feeding of liquid plant protection chemicals.

So far, there has been gathered very little experience with direct injection in practical spraying. However, research and development work is going on. Perhaps it could be benefital to look to the industry instead of starting from scratch? One obvious advantage with direct injection systems is that there will be no diluted chemicals to dispose of.

Miller (1987) emphasizes that the system offers considerable improvements in terms of operator contamination and the risk of environmental pollution. The system is, however, not without problems. A leakage in the system with the concentrates line or pump can lead to very serious consequences. The concentrates part also have to be cleaned every day.

Crop tilter

In order to improve the penetration of the spray liquid into the crop one may make openings in the crop. Ciba Geigy has introduced a crop tilter boom. It can be used by insect and fungus spraying in for instance cereals.

Spraying in a chamber

For certain purposes, chamber spraying can be employed. This method ensures a one hundred per cent utilization of the spray liquid, and there will be no drift. The Norwegian Institute of Agricultural Engineering has developed an equipment whereby trays with 2/0 forest tree seedlings (peg plants) are conveyed through a spraying chamber (Nordby 1983). On the conveyer the underside of the trays are cleaned with one non-rotating and one rotating brush before the trays enter the spraying zone. In order to keep spray drift away from the crew, the chamber has three sets of doors. There are six nozzles spraying from each side and four spraying in the direction of travel. These latter four nozzles are placed two and two at 45^0 on a "row-opener". Liquid not deposited on the plants is filtered and led back to the tank. About 25 % of the liquid sprayed is deposited on the plants. The spray distribution is satisfactory. The equipment has been in production in some years. In 1988, some 50 to 60 million plants in 16 nurseries were treated with this chamber spraying equipment.

Aerial spraying

Aerial spraying is still in use. However, no big improvements have been reported during the last few years.

Mist spraying

Mist spraying is mainly used for spraying bushes and trees. The main problems have been loss of spray and drift. With new types (Holder/Platz) it is possible to send air and liquid more horisontal against the trees. This will give a higher percentage of chemical deposited on the trees.

RESEARCH AND INVESTIGATIONS. WHERE ARE WE?

Today we have all sorts of remedies, conferences, abstracts, good surveys from the libraries and exchange of students and scientists. If you read any English or German publication and take a look at the references, what do you often discover? You will notice that most of the references also are in either English or German, respectively. Is it still a closed border in the canal? I think we have to use the available knowledge and information better than we have done so far.

Strangely enough, you can sometimes come about experimental plot spraying equipment unable to do as good a job as a crop sprayer in practical use. Distribution of liquid is perhaps not good enough. Driving speed too low to be compared with usual practice. These imperfections may lead to: Many publications without reliable documentation about the spray applied on the plots. The information given should at least cover:
  Droplet size
  Liquid distribution
  Driving speed
  Boom height
  Nozzle type
  Working pressure
  Quantity of spray
  Quality of spray

STATE-OF-THE-ART OF FIELD SPRAYING 1989

Cooke et al (1986), Skuterud et al (1988) and Permin (1983) conclude that crop spraying with 110^0 degree nozzles on a spray boom, and 100-200 l/ha is still the best choice for field weed spraying.

Probably the ordinary crop sprayer will still be the most common equipment for the next 10 years. With some additionals or modifications, the crop sprayer can also be made to good use in for instance strawberries and brassica crops.

HOW IS THE SITUATION CONCERNING THE CROP SPRAYERS IN PRACTICE?

Half of the sprayers in use are not performing in accordance with our requirements, (Bengtsson 1989 & Osteroth 1988).

Errors are frequently found on:
Nozzles
Pumps
Manometers
Spray booms

What, then, can be done to improve the results we get from the crop sprayers?

EDUCATE THE SPRAYING CREW

Some countries require a certification of sprayer operators. Upon attending a training course, the operator is awarded a certificate. So far, the courses have dealt mainly with plant protection chemicals and the danger of contaminating the crew. Today, some courses also have the spraying equipment on the program: Maintenance, setting, use and so on. The passing of a training course therefore should ensure a properly qualified spraying crew.

HOW TO KEEP THE SPRAYERS IN PRACTICAL WORK IN GOOD ORDER. VOLUNTARY INSPECTIONS OF SPRAYERS IN PRACTICAL USE

To ensure that sprayers in practical use are properly maintained is a difficult task. One of the best ways may be the voluntary inspection of sprayers. This has been carried out in West Germany for the last ten years. Sweden is making a good start in 1988. Norway is also preparing voluntary inspection.

In order to get this voluntary inspection working one will need:
Good organization
Educated testers
Good, transportable testing equipment
Economic support from the authorities

The testing equipment has to be taken out to selected places in the countryside, to where the farmers can then bring their equipment.

The test should include measuring the following:
Pump capacity
Distribution of the spray
Delivery from each of the nozzles
Accuracy of the manometer

The farmer should participate directly in the test measuring. This procedure is a very valuable demonstration of sprayers. The sprayer can be approved on the spot. If it is not approved, the farmer will get a written statement of what has to be done in terms of adjustments, replacements or other

repairing before approval. In West Germany approximately 27 000 crop sprayers were tested in 1987 (Osteroth 1988).

TYPE APPROVAL

In West Germany new sprayers are tested for type approval (Merkmale Spritz- und Spruhgeräte für Flächenkulturen 1988). In Sweden they are also working with directions and rules for type approvals. The testing will start on 01.01.1991.

Tests of this kind need to be based on very clear and carefully prepared directions and legal provisions. What, then, can one hope to attain?

For good types of sprayers and conscientious manufacturers the effect will be small. For bad sprayers the positive effect might be considerable.
1. Improve the sprayers
2. Get bad sprayers away from the market
3. Improve the working conditions for the spraying crew

If not revised in view of new knowledge of spraying practices and advances in materials technology, parts design, and so on, the directions and demanded reqirements for type approval may become an obstacle for further development of plant protection equipment. This is not different from any standardization. Design progress can be complicated, costly and will require a lot of work by the manufacturer. Stimulating such work should be a major concern of the authorities. Adjustment and renewal of the directions, for instance every 5 years, may help.

The type approval will be expensive. Therefore the directions and rules ought to be standardized. It might not be necessary for every country to do such testing. By assigning the task to a few institutes and countries the testing could be carried out both better and cheaper.

REFERENCES

Anon, 1987: Van den Munkhof bringt nieve luchtondersteunde spuit. Weekblad Groenten & Fruit. 8 mei, p. 52.

Bengtsson, A., 1989: Funktionstest av lantbrukssprutor. Erfarenheter från 1988. 30:e svenska ogräskonferensen, Uppsala.

Biologische Bundesanstalt für Land- und Forstwirtschaft Bundesrepublik Deutschland, 1988: Merkmale Spritz- und Sprühgeräte für Flächenkulturen. Braunschweig, 18 p.

Brundtland, G. H. 1987: Our common future. World Commission on Environment and development, Geneva.

Cooke, B. K., Hislop, E.C, Herrington, P.J. Estern, N.M., Jones, K.G., Woodley, S.E. and Chapple, A.C., 1986: Physical, chemical and biological appraisal of alternative spray techniques in cereals. Crop protection 5 (3), 155-164.

Hardi, 1989: Hardi Twin Systems, Copenhagen, 9 p.

Herrington, P.J., Mapother, H.R. & Stringer, A., 1981: Spray Retention and Distribution on Apple Trees. Pestic. Sci. 12. p. 515-520.

Jakobi, F. and Kramer, M., 1984: Die neuen Spritzzüge für chemische Aufwuchs-bekämpfung bei der Deutschen Bundesbahn. Verkehr und Technik Berlin, Heft 1, 5 p.

Miller, P.C.H., 1987: Spray application technology. Agricultural Engineers. Autumn, p 77-84.

Nordby, A., 1983: Plant Protection Equipment Developed and used in Norway. EPPO Bull, 13 (3): 509-511.

Nordby, A. & Jørgensen, L., 1985: Gjennomtrengelighet av sprøytevæske for ulike arbeidstrykk, kjørehastigheter og dysestørrelser. 2. Danske Planteværnkonference. Nyborg. p. 85-98.

Osteroth, H.J., 1988: Freiwillige Kontrolle von Pflanzenschutzgeräten. Ergebnisse für das Jahr 1987. Nachrichtenbl. Deut. Pflanzenschutzd. 40. p. 76-79.

Permin, O., 1983: Fordeling af bladherbicide med hydrauliske dysetyper. Tidskrift for planteavl, 87 p, 69-96.

Robinson, T.H. & Garnet, R.P., 1984: The influence of electrostatic charging, drop size, and volume of application on the deposition of propiconazole and its resultant control of cereal diseases. British Crop Production Conference -Pests and Diseases. Vol. 2, p. 1057-1065.

Skuterud, A., Nordby, A., and Tyldum, A., 1988: Effects of application methods, spray volumes, pressures and herbicide rates on weed control in spring cereals. Crop protection Vol. 7 October 303-308.

Schmidt, M., 1983: Direkteinspeisung von Flüssigen pflanzenbehandlungsmitteln. Forschungsbereicht Agrartechnik. Berlin.

Statens plantevern - Gefo, 1987: Plantevernmidler i overflatevann og grunnvann. 39 p.

Vittrup Christensen, J., 1965: Beskjæringens og frugtbæringens indflydelse på løvmengde og frugtstørrelse på æbler. Tidsskr. Planteavl. 69. p. 93-97.

Land and Water Use, Dodd & Grace (eds), © 1989 Balkema, Rotterdam. ISBN 90 6191 980 0

Recent developments in the use of aircraft in the control of locust plagues in subsaharan Africa

J.D.Parker
Cranfield Institute of Technology, Bedford, UK

ABSTRACT: Plagues of locusts and grasshoppers are once again ravaging African agriculture. National pest control authorities are not equipped to deal with this problem and international agencies, based in Africa, which used to have responsibility for the control of migratory pests, have either been disbanded altogether or are operating at a level which is too low to prevent the situation deteriorating. Operational research has identified the logistic and material requirements of the countries of sub-saharan Africa and proposals, based on the use of aircraft, are made for a far-reaching reorganization of facilities for the control of migratory pests.

RESUME: Des nuées de criquets et de sautériaux sont une fois de plus en train de ravager l'agriculture africaine. Les autorités nationales pour la lutte contre les insectes nuisibles n'ont pas les compétances nécessaires pour faire face à ce problème et les agences internationales qui sont basées en Afrique et qui avaient la responsabilité du contrôle des insectes nuisibles migrateurs ont été entièrement dispercées ou elles fonctionnent à un niveau qui est trop réduit pour empêcher une détérioration de la situation. Des recherches opérationnelles ont pû identifier les exigences logistiques et matérielles des pays africains sub-sahariens. Des plans fondés sur l'usage des aéronefs sont proposés pour une réorganisation compréhensive pour la lutte contre les insectes nuisibles migrateurs.

ZUSAMMENFASSUNG: Heuschrecken- und Grashüpferschwärme verwüsten erneut die afrikanische Landwirtschaft. Nationalstaatliche Ämter zur Schädlingsbekämpfung sind zur Behandlung dieses Problems nicht entsprechend ausgerüstet, und internationale Behörden mit Standort in Afrika, die ehemals für die Bekämpfung von Wanderschädlingen zuständig waren, sind entweder ganz aufgelöst worden oder nur in so eingeschränktem Masse tätig, dass sie eine Verschlimmerung der Situation nicht verhindern können. Mit Hilfe der Unternehmensforschung wurden Logistik- und Materialbedarf der afrikanischen Länder südlich der Sahara identifiziert. Vorschläge, die sich auf den Flugzeugeinsatz stützen, werden unterbreitet mit dem Ziel weitreichender Reorganisation der Einrichtungen zur Bekämpfung von Wanderschädlingen.

1. INTRODUCTION

For many years in the 1940s and 1950s locust plagues ravaged the crops of sub-Saharan Africa. Reported damage caused by the Desert Locust (Schistocerca gregaria) alone was estimated at between £.Stg.0.5 and 5 million per year over the period and the actual cost was probably considerably greater. (COPR 1982).

Locusts and grasshoppers may, under favourable rainfall conditions and with sufficient vegetation for food, pass through several breeding cycles in a single season, multiplying their numbers by a factor of 100 or more with each cycle. Thus a relatively small swarm of the insects can, if left unchecked, cause considerable damage to crops in the course of one growing season. Furthermore, eggs deposited in the soil at the end of the rainy season may remain viable for a number of years, hatching to form new swarms when favorable conditions return. The situation can deteriorate each year if control measures are not carried out or are ineffective.

The adults of some species of locust and grasshopper can travel considerable distances, flying with the wind several hundred kilometers in a single day. Thus if control of the flightless juvenile stages is ineffective or incomplete, what started as a local problem, capable of being dealt with by national authorities using ground-based equipment, becomes international in scale, requiring the use of aerial spraying to provide the necessary control measures.

Highly effective air-to-air spraying techniques were developed in the 1950s for the control of flying locust swarms. (Rainey & Sayer 1953). This technique consisted of forming a dense cloud of small, concentrated pesticides droplets above, and down wind of, the swarm, which then flew into the cloud. Small droplets had to be used, so that the spray cloud remained airborne for an appreciable length of time, and it was found that the spraying method could be very efficient. (Sayer 1959). Flying insects are very effective collectors of airborne droplets and it was calculated that up to 6 per cent of the lethal doses of insecticide applied in this way could be credited with a "kill". (MacCuaig 1962, Joyce 1979). This compares with an efficiency which may be less than 0.1 per cent in a conventional pesticide application.

In the late 1960s, weather patterns in Africa became drier and less favorable for both the growth of crops and the development of locusts. With the exception of a few years, and in restricted areas, locusts have not presented a major problem anywhere in Africa for some 25 years, until the mid '80s when the return of more favorable rainfall patterns has been beneficial to both crops and locust and grasshoppers.

2. CONTROL ORGANIZATIONS

Several international organizations were established in the 1960s to deal with the problems caused by migratory pests such as locusts, grasshoppers, armyworm, quelea birds and tsetse fly, and these invariably were given authority to operate across national boarders. The Desert Locust Control Organization for Eastern Africa was set up in 1962 and now operates in Djibouti, Ethiopia, Kenya, Somalia, Sudan, Tanzania and Uganda. DLCO-EA's original remit was to cooperate with National pest control authorities of the member countries in the control of the Desert Locust. However, with the onset of drier condi-

tions in the 1970's, the Organization's governing council agreed to extend its area of interest to other migratory pests, including tsetse fly, quelea and armyworm, but with Desert Locust problems having priority.

In this way the DLCO-EA maintained its viability and operational capability and is now, in 1989, well equipped to treat initial locust and grasshopper outbreaks in the member countries. The Organization is staffed entirely by nationals of member countries, who contribute the whole of its recurrent budget in US dollars. However, DLCO-EA is not able to deal with major outbreaks with its present resources and it does not normally operate outside the boarders of its member countries. Since locusts, grasshoppers and other migratory pests can travel great distances across national boundaries, these restrictions impose severe limitations on the effectiveness of DLCO-EA in a Continent-wide context.

The Organisation Commune de la Lutte Antiacridienne et de la Lutte Antiaviaire, OCLALAV, was established in the 1960s to control locusts and bird pests in 10 francophone countries in West Africa, with considerable financial and administrative assistance from France and other donors. After a period of some years of effective action, this organization has steadily declined until today it is so deprived of resources that it is largely ineffective in most of its member countries. Proceedings are now in motion to carry out the formalities of winding up OCLALAV's affairs.

A further international organization in West Africa, the Organisation Internationale Contre le Criquet Migrateur (OICMA), was also established in the 1960s specifically for the control of the African migratory locust (Locusta migratoria migratorioides). During the period of locust recession, OICMA had little to occupy its manpower and resources, which were in consequence dispersed. Now that its expertise could be used to good effect in West Africa, it is no longer available as the organization has ceased to exist.

In Central and Southern Africa, The International Red Locust Control Organization (IRLCO-CSA), whose remit is mainly the control of the Red Locust (Nomadacris septemfasciata), has no less that 12 member countries, many of them also members of DLCO-EA. This organization is operational and effective, as far as its very limited resources permit, but suffers from chronic lack of funds and equipment which severely restrict its ability to respond to demands for its services. IRLCO-CSA has, in recent years, demonstrated the considerable value of helicopters in locust survey and control. There are also several national organizations in Africa, for example ADMAS in Ethiopia, Kilimo Anga in Tanzania and UTAVA (Unité des Traitements Agricoles par Voie Aérienne) in Cameroon, which are supposed to have some responsibility for the control of locusts and grasshoppers. All these are non-operational to a greater or lesser extent, despite the fact that their services are badly needed for the control of locusts, grasshoppers and other migratory pests.

As a result of the deficiencies described above, locust and grasshopper control using aircraft has been undertaken on a piecemeal basis and, because of the inevitable delays involved in setting up aerial spraying operations, control measures have usually been undertaken to late to be fully effective. Furthermore, many of the skills and techniques developed some 30 years ago for the aerial spraying of locusts have either been forgotten or overlooked. Thus the author is aware of no recent air-to-air spraying of locust or grasshopper swarms, with the exception of limited work carried out by DLCO-EA in East Africa in 1987 and 1989, and most of the aircraft provided are not equipped with the most suitable rotary atomizers required for the ULV spraying techniques developed in the 1950s. (Sayer 1959). Flying swarms are therefore allowed to escape from their breeding areas to invade other areas, possibly in another country.

3. PROPOSED SOLUTIONS

Before considering what can be done about the deplorable state of affairs regarding the control of locusts, grasshoppers and other migratory pests in sub-Saharan Africa, it is worth analyzing what common factors were present in the various international locust control organizations which led to their initial, and in some cases continuing, success. These are:
1. The organizations had the political and financial support of their member governments, who recognized that their financial contributions were to be regarded as insurance premiums against pest attack, and that the organizations had to be maintained in a state of readiness, even during periods of locust recession.
2. Staff of the organizations possessed a considerable degree of expertise, discipline and motivation in their various fields, based largely on in-service training, good management and security of employment.
3. Control operations were based largely on the use of aircraft, which were used for surveys, transport and, where necessary, aerial spraying. This demanded the establishment and maintenance of a team of highly trained and well motivated professionals who were provided with the facilities needed to operate aircraft safely and effectively.
4. Inexpensive and long-lasting pesticides were available for the control of flightless juvenile stages and of flying swarms of the pest insects.

Deficiencies in all of these factors have been responsible for the decline and terminal malaise seen today in so many of Africa's aerial spraying operations.

It is clear that only an international approach will be sufficient to reverse the declining capability of African countries to control migratory pests. There have been many international conferences on the subject in recent years, none of which has done more than highlight the difficulties and consider technical methods for tackling the problem. In many respects the problem is less technical than political. Since the 'golden' days of locust control, many African countries have gained independence and now jealously guard their national identity and borders. This does not create a suitable climate for international cooperation in the operation of aircraft. National borders which were once crossed without formality by pilots of light aircraft now required two weeks of planning, and then there is every chance that permission will be rescinded at short notice. Even DLCO-EA, which has been operating in East Africa for 26 years, cannot now overfly the national boundaries of its member states without several days of delay.

A guide to a possible solution to the problem lies in the experience of the Onchocerciasis Control Programme, now operating in 11 countries of West Africa. The OCP was set up in 1974, with the World Health Organization as executing agency, to control the disease River Blindness. In the initial stages, some 12 000 kilometers of river in the programme area were treated with a larvicide at regular intervals to control the aquatic stages of the small fly which carries the disease and aircraft were found to be the only means by which this could be achieved. Since its inception, the OCP has expanded and nearly 50 000 km of river are now protected by the aircraft fleet. (Parker 1978). Nineteen different countries, funds and international agencies contribute to the programme's cost and an Onchocerciasis Committee has been formed in each member country to provide the political framework for international cooperation.

As always, technical problems have been less important than political ones but a mechanism has been developed whereby OCP aircraft can overfly all national boundaries within the area with a minimum of formality. Technically, the OCP has evolved effective control strategies through a process of continuing research and development, for which adequate provision has been made in the Programme's budget. After experimenting with various means of larvicide application, aircraft, and in particular helicopters, were recognized as being the only feasible means of undertaking a large-scale international pest control programme.

A major impediment to the successful control of locusts and grasshoppers today is the lack of suitable pesticides. The problem is not that such pesticides are no longer available or are ineffective but that the most suitable products, dieldrin and HCH (BHC), have been banned by the international community, under pressure from environmental groups, because of their persistence and possibly harmful effects on the environment. The alternative chemicals now proposed, such as fenitrothion, are both more expensive and less persistent than dieldrin and, because they have to be used more frequently and over larger areas, may also be more damaging to the environment as well. FAO has strongly recommended the reintroduction of dieldrin to control the present locust and grasshopper emergency and has estimated that costs will be some twelve times higher if the more expensive, less effective pesticides are used. (FAO 1988). Many African countries are unable to afford the newer pesticides proposed and control operations are either curtailed or abandoned altogether.

Thus the solution to the present locust and grasshopper problem appears to the author to be as follows:

1. Establish an International Locust and Grasshopper Control Programme in sub-Saharan Africa, with the UN Food and Agriculture Organization as the executing agency. The Programme should incorporate all existing national and international organizations with a similar remit and would operate a fleet of spray aircraft equipped for all the various techniques of controlling migratory pests.

2. Base control measures on the use of pesticides which have been shown to have no long-term harmful effects, under African conditions and applied appropriately. This would probably allow both dieldrin and HCH to be used.

3. Be prepared to carry out control measures against other migratory pests, such as armyworm, quelea and tsetse fly, as required.

Modern techniques of remote sensing can considerably ease the difficulty of detecting locust and grasshopper breeding and newly developed ULV pesticides application equipment for aircraft should enable control activities to be undertaken with a considerable degree of success. What is needed now is the political will on the part of both African countries and donor nations and agencies, to provide the necessary international cooperation. The above proposals may appear ambitions but it is salutary to consider what the locust and grasshopper problem in Africa would be if plagues are not effectively controlled. Experts cannot foresee an end to the problem, and it will probably deteriorate further, unless rainfall patterns in sub-Saharan Africa revert to become unfavourable once again. Then of course, there will not be sufficient crops to support large numbers of pests, or indeed of people.

REFERENCES

COPR (1982) The locust and grasshopper agricultural manual. London: Centre for Overseas Pest Research. 690pp.
FAO (1988) UN Food and Agriculture Organization, Rome. Unpublished Reports.
Joyce, R.J.V. (1979) The evolution of an aerial application system for the control of Desert Locusts. Phil.Trans.R. Soc.Lond. B 287:305-314.
MacCuaig, R.D. (1962) The collection of spray droplets by flying locust swarms. Bull.ent.Res. 53:111-123.
Parker, J.D. (1978) The use of aircraft in the WHO Onchocerciasis Control Programme. Proc. V int.Agric. Aviat. Congr.:127-136.
Rainey, R.C. & Sayer, H.J. (1953) Some recent developments in the use of aircraft against flying locust swarms. Nature, Lond. 172:224-228.
Sayer, H.J. (1959) An ultra-low volume spraying technique for the control of Desert Locusts. Bull.ent.Res.50:371-386.

Land and Water Use, Dodd & Grace (eds), © 1989 Balkema, Rotterdam. ISBN 90 6191 980 0

Manual and automatic systems of controlling the application rate of tractor-mounted ground-crop sprayers

B.Rice & J.Croke
Agricultural Engineering Department, Teagasc, Oak Park Research Centre, Carlow, Ireland
B.McKenna
Department of Agricultural and Food Engineering, University College, Dublin, Ireland

ABSTRACT: In a survey of the on-farm operation of tractor-mounted sprayers, only 8 out of 20 were applying within 10% of the intended application rate. Mechanical control systems based on a fixed-orifice type regulator could maintain a linear relationship between boom output and pump speed in controlled tests, but had no apparent effect on calibration errors on farms. Radar and fifth-wheel forward speed transducers and turbine flow-meters were found to be adequate for instantaneous application rate measurement.

RESUME: Dans le cadre d'une enquête sur le fonctionnement dans une exploitation agricole d'un vaporisateur de récoltes monté sur un tracteur, 8 sur 20 exploitants seulement utilisaient, à 10% près, le taux d'application recommandé. Des systèmes de contrôle basés sur un régulateur de type orifice fixe pouvaient maintenir un rapport linéaire entre le débit à la potence et la vitesse de la pompe dans le cadre d'essais sous contrôle mais n'avaient apparemment aucun effet sur les erreurs de calibrage chez l'agriculteur. Un radar et des transducteurs de vitesse a cercle horizontal et des fluxmètres à turbine se sont avérés adéquats pour une mesure instantanée du taux d'application.

AUSZUG: Bei einer Untersuchung uber den Hofeinsatz von auf Treckern aufgebauten Bodenpflanzenspritzen zeigte sich, daß nur 8 von 20 Geräten einen Ausstoß hatten, der innerhalb einer Toleranzgröße von 10% der vorgesehenen Menge lag. Zwar konnten bei Versuchstests mechanische Steuersysteme, die auf einem Festmundstück-Regler basierten, eine lineare Beziehung zwischen Gestängeausstoß und Pumpengeschwindigkeit aufrechterhalten, sie hatten jedoch keine offensichtlichen Auswirkungen auf Eichfehler, die auf Bauerhöfen entstehen. Es zeigte sich, daß Radar und Antriebsgeschwindigkeits-Signalumsetzer an der Aufsatteleinrichtung sowie Turbinenfluβzähler fur eine augenblickliche Messung der Anwendungsrate ausreichten.

1 INTRODUCTION

The accuracy with which spray chemicals are applied to agricultural crops is coming under increasing scrutiny in many countries. The need for farmers to reduce input costs, consumer demands to reduce residues in food crops, and pressures to minimize environmental damage, all make it necessary to achieve control of pests and diseases with reduced amounts of spray chemical and to avoid over-dosing of the crop through incorrect sprayer setting or operation.

The first requirement for accurate spraying is the correct setting of forward speed and pressure to give the desired application rate. The second requirement is uniform distribution, by accurate matching of successive runs of the sprayer, maintenance of the correct boom height, and a set of jets with consistent outputs which can give an even spray pattern.

Past surveys of the on-farm operation of sprayers on farms, e.g. those by MAFF/ADAS (1976) in the U.K. and Rice (1971) in Ireland have shown that substantial calibration errors frequently occur.

To reduce or eliminate calibration errors, improved monitoring and control systems have been introduced by sprayer and instrumentation manufacturers. Givelet (1981) has reviewed the control systems that have reached the market. On many tractor-mounted sprayers with constant-displacement pumps, the spring-loaded pressure regulator has been replaced by a valve in which the orifice is not affected by pressure. This is intended to allow the sprayer to maintain a constant application rate if the forward speed is adjusted within a tractor gear. Alnajjar and Schmidt (1979) studied the design of this system in detail, but there is little information on its contribution to more accurate spraying in practice.

A number of electronic control systems have also been introduced, in which the signals from forward speed and spray flow transducers are used to compute application rate; this is compared with the pre-set intended rate and a correcting signal sent to a motorized valve in the flow or return line.

In recent times, numerous studies have been made of forward speed measuring devices. The sensitivity of radar devices to surface condition, pulse frequency and mounting position has been investigated by Fichtel and Balcarek (1986), Tsuha et al. (1982), and Thelen and Neher (1985). In general, radar has been found to give an accurate measure of forward speed, with some worries about its sensitivity to height and angular mounting position and its accuracy when travelling in wind-blown crops. The use of undriven and fifth-wheels for speed measurement has also been investigated and compared with radar by Tompkins et al. (1988), Bell and Isensee (1987), Richardson et al. (1984), Makin and Reid (1985), and Speckmann and Jahns (1986). All wheel-based systems are affected by variation in rolling resistance, which is difficult to control in the case of undriven vehicle wheels. The performance of a fifth-wheel is very much affected by its design; in particular its diameter, width, ground pressure and suspension system.

Hughes and Frost (1985) have reviewed the performance of flow-meters that might be considered for use on agricultural sprayers, and favour the turbine type as being relatively cheap, small and light, easy to maintain and reasonably reliable. There is little published information on their performance on sprayers on farms.

A number of complete electronic control systems have been marketed for several years by sprayer and instrumentation companies. They have been based mainly on radar or undriven wheel magnetic pick-up for speed measurement, turbine flow-meter

or inference from pressure for flow measurement, and a motorized valve in the boom flow or return line for correction of application rate. There is very little published information on the performance of these systems, either as new or on farms. Swedish test reports on five control systems suggest that over large areas they held the application rate within 5-7% of the intended value when 200 l/ha or more was being applied. The reports contain no information about accuracy at lower application rates, or about local variations in application rate over small areas (Statens Maskinprovningar 1983).

The work described in this paper has three objectives: 1. To examine the extent to which sprayer calibration is still a problem on Irish farms, 2. To examine the performance of mechanical fixed-orifice type control systems in controlled tests and on farms, and 3. To test the components of an electronic control system (speed and flow transducers, motorized valve and computing system), as the first phase in evaluating the suitability of electronic control systems for use on tractor-mounted sprayers.

2 MATERIALS AND METHODS

Over the period May-July 1988, a survey was carried out of the operation of 25 sprayers on farms within 20 km of Oak Park. One objective of the survey was to establish the level of precision being achieved in the setting of application rates, and to identify the main sources of inaccuracy.

An instrumentation package was developed consisting of a radar speed-measuring unit, four turbine flow-meters, a logging unit to which could be attached a solid-state memory, and a battery pack. Brackets and connectors were constructed to allow the package to be connected to all the widely used sprayers in the area. Flow to each boom section and forward speed were logged once every three seconds while one tankful of spray was being applied.

To measure the accuracy with which sprayers with fixed-orifice type controls maintained constant application rate at varying forward speed, the instrumentation described above was successively attached to two new sprayers from different manufacturers, and the sprayers were operated at forward speeds which produced a range of pressures from 0.1 to 0.6 MPa. Application rate was logged at each pressure.

The performance of three forward speed-measuring devices was measured; a magnetic pick-up on an undriven wheel, a similar device on a fifth-wheel, and a radar unit (TRW). The radar and fifth wheel were compared on a range of surfaces from tarmac to soft tilled ground and high wind-blown crops. The fifth-wheel was examined separately on tilled ground only.

Six turbine flow-meters were tested for linearity and repeatability over the range of throughputs normally encountered on sprayers. Three mounting positions were compared - vertical, horizontal and inclined at 45°.

A motorized butterfly valve (Spraying Systems No. A244-3/4) was located alternatively in the flow line to the boom and in the return to the tank. The valve setting was adjusted from fully open to fully closed and the boom flow and pressure were measured. This test was repeated for two nozzle sizes - No. 4110-20 and No. 1553-8.

3 TESTS AND RESULTS

3.1 On-farm survey of sprayer calibration

All the sprayers in the survey were tractor-mounted, with boom widths from 9.6 to 12.7 m. Tank sizes were 600-800 litres. Nozzle spacings on the boom were 0.33 or 0.5 m; 11 sprayers had fan nozzles, 14 had hollow-cone. Nine had fixed-orifice type regulators, 16 had conventional spring-loaded relief valves. Sprayers from four manufacturers were represented in the survey; Allman (6), Berthoud (9), Evrard (2), and Hardi (8).

Application rates measured were from 118 to 258 l/ha, and forward speeds from 6.9 to 11.7 km/h. Of 20 operator estimates of forward speed, 6 were within 5% of the measured value, a further 2 were within 10%, and the remaining 12 were in error by more than 10%. Of 18 estimates of application rate, 6 were within 5% of the measured value, 10 were within 10%, and 8 were in error by more than 10% (Table 1).

Two operators (8 and 22) were achieving within 5% of the desired application rate in spite of big discrepancies in their forward speed estimates. This suggests that they had arrived at their chosen settings by adjusting their speed in the field rather than by static calibration or from handbooks. On the other hand, eight operators would have achieved much more accurate application rates if they had been driving at their target speed, i.e. they had chosen the correct nozzle size,

Table 1. Actual and intended speeds and application rates

Sprayer no.	Application rate			Forward speed			Boom output (% of rated)
	Intended	Actual		Intended	Actual		
	l/ha	l/ha	%	km/h	km/h	%	
1	157	165	105	7.0	7.0	100	137
3							98
4	201	200	100	8.8	8.8	100	99
5	213	225	106	8.1	8.1	100	91
7	–	185		8.5	9.6	113	105
8	201	197	98	9.7	8.0	82	102
9	168	118	70	8.1	11.0	136	100
10	179	156	87	9.7	9.7	100	120
11	–	176		9.7	11.1	114	126
12	206	194	94	8.9	9.4	106	102
13	118	167	142	4.8	8.8	183	117
14	196	133	68	9.7	11.7	121	97
15	224	168	75	11.5	9.5	83	101
16	134	126	94	8.1	8.5	105	119
17	201	158	79	12.9	10.3	80	114
18	224	151	74	6.4	11.8	184	96
20	207	190	92	6.4	6.9	108	104
21	213	191	90	8.1	9.6	119	104
22	257	258	100	9.7	7.5	77	97
23	224	198	88	8.1	10.0	123	90
24	224	213	95	8.9	8.6	97	95

pressure and speed to achieve close to the desired application rate, but had failed to set the forward speed correctly.

Seven sprayers had outputs 5% or more above the rated value for the nozzles at the pressure used. This difference could be attributed mainly to nozzle wear.

Clearly, serious errors are being made in sprayer calibration on farms. Calibration must include forward speed measurement, since this is contributing to application rate setting errors. Nozzle wear can lead to over-application errors if re-calibration is not carried out frequently.

3.2 Tests on fixed-orifice pressure control systems

These systems are based on two features: a pump whose output is independent of pressure, and a by-pass system which maintains constant the proportion of spray passing to the boom and returning to the tank.

On the first sprayer tested, a linear relationship was found between pump speed and boom output (Fig. 1). For comparison, the performance of a spring-loaded regulator on the same sprayer is also shown. Initial tests on the second sprayer showed a highly non-linear relationship (Fig. 2). When an alternative pump and larger suction hose were fitted, a linear relationship between pump speed and boom output was established.

These results show that fixed-orifice control systems can maintain a constant application rate at varying forward speed, provided their design accords with the

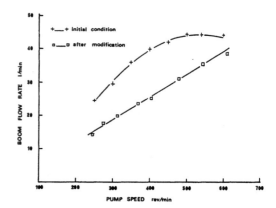

Fig. 2 Boom flow rate vs. pump speed before and after modification, sprayer 2

principles set out by Alnajjar and Schmidt (1979).

In the on-farm survey, no difference in calibration could be detected between the sprayers with spring-loaded and fixed-orifice regulators. Neither was there any indication from the forward speed records that operators availed of the facility to change forward speed while spraying, or had any difficulty maintaining a constant forward speed. A typical record of forward speed for one sprayer in the survey is shown in Table 2.

3.3 Tests on components of electronic control system

3.3.1 Undriven wheel

A Hardi LX-600 sprayer was mounted on a MF-148 two-wheel drive tractor with

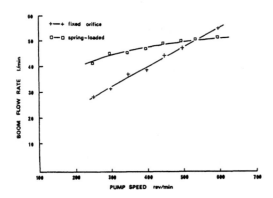

Fig. 1 Boom flow rate vs pump speed with fixed-orifice and spring-loaded regulator, sprayer 1

Table 2. Speeds and application rates in successive sprayer runs (sprayer no. 24)

Run no.	Mean speed (km/h)	Application rate (1/ha)
1	8.8	210
2	8.6	214
3	8.7	212
4	8.8	212
5	8.9	206
6	8.9	209
7	9.0	204
8	9.0	204
9	9.0	204
10	9.0	207

6.00-19 front tyres inflated to 0.13 MPa, and 11-32 rear tyres inflated to 0.08 MPa. On three typical soil conditions the time to travel 100 m was measured and the number of front wheel revolutions recorded. The test was repeated three times in each set of wheel-tracks. This complete test was carried out with the sprayer tank full and empty.

The difference between the first pass on uncompacted ground and subsequent passes in the same wheel-tracks is a good indication of the error that can be introduced by calibrating on a hard surface and subsequently using the sprayer on softer ground. The results show that errors of up to 12% could be introduced (Table 3). Tompkins et al. (1988) recorded errors of up to 11% in similar tests. The effect on indicated speed of the change in weight distribution as the tank empties appears to have varied from zero to 5%.

3.3.2 Tests on fifth-wheel and radar systems

Fifth-wheel and radar systems were attached to a tractor and compared on four terrains (Fig. 3). The speed indicated by both units was logged at 3-second intervals over a distance of 100 m, and the time to cover 100 m was measured and calculated. These tests were repeated at

Fig. 3 Radar and fifth-wheel speed sensors mounted on tractor

15 speeds between 1 and 12 km/h.

The radar output was affected by tractor vibration which introduced a slight error at speeds below 1 km/h. Between 1 and 12 km/h, the average error in 61 tests on tarmac and three typical spraying surfaces was 0.9%. In two high crops in conditions too windy for spraying, the average error increased to about 2%, and the maximum error in an individual run was 4.6% (Table 4).

The average error of the fifth-wheel (0.7%) was even lower than that of the radar on the four surfaces on which they were compared. The scatter in the

Table 3. Undriven wheel calibration on successive passes over 3 terrains with sprayer tank full and empty

Terrain	Pass	Tank full		Tank empty	
		Distance travelled	Error compared with final pass	Distance travelled	Error compared with final pass
		(m/rev)	(%)	(m/rev)	(%)
Stubble (firm)	1	2.50	1.0	2.49	1.6
	2	2.48	0.0	2.46	0.4
	3	2.48	0.0	2.45	0.0
	4	2.48	0.0	2.45	0.0
After potato harvest	1	2.57	2.8	2.66	6.8
	2	2.50	0.0	2.49	0.0
	3	2.50	0.0	2.49	0.0
	4	2.50	0.0	2.49	0.0
Freshly tilled	1	2.83	11.4	2.85	11.6
	2	2.51	1.2	2.60	6.1
	3	2.49	0.4	2.46	0.4
	4	2.48	0.0	2.45	0.0

Table 4. Radar and fifth-wheel calibration on 6 terrains

Terrain	Transducer	Error (%)			COV of individual readings per run (%)
		Mean	Min.	Max.	
Tarmac	Radar	0.63	-0.6	+1.4	0.017 - 0.052
	Fifth-wheel	0.16	-0.2	+1.8	0.019 - 0.060
Short grass	Radar	0.87	-0.4	+1.4	0.004 - 0.036
	Fifth-wheel	0.82	-2.5	+1.3	0.008 - 0.049
Bare, firm soil	Radar	1.20	-0.9	+2.9	0.017 - 0.052
	Fifth-wheel	1.65	-2.9	+0.33	0.019 - 0.060
Long grass (15 cm)	Radar	0.98	-0.1	+2.1	0.007 - 0.040
	Fifth-wheel	0.34	-0.6	+0.6	0.007 - 0.034
Barley (1 m) wind-blown	Radar	1.59	-3.6	+4.6	0.009 - 0.076
Weeds (0.6 m)	Radar	2.47	-3.9	-1.2	0.020 - 0.064

individual readings taken at 3-second intervals was slightly greater with the fifth-wheel; this was probably due to the lower pulse frequency of its output (10 pulses/metre compared with 129 pulses/metre for the radar) and could be improved by changing the design of the toothed wheel.

While both devices gave an acceptable level of accuracy, the radar system was chosen for the remaining tests because of the greater damage susceptibility of the fifth-wheel on rough ground, on corners, and at high forward speeds.

3.3.3 Tests on flow-measuring devices

In the on-farm survey, nozzle outputs were on average 5.4% greater than the rated values taken from instruction manuals and calibration charts. If the flow were to be inferred from a pressure measurement and a discharge coefficient, frequent re-calibration would be needed to maintain an accurate relationship. Since the main purpose of an electronic control system is to maintain the desired application rate without calibration, which the survey shows not to be carried out with sufficient frequency on farms, flow-measuring systems based on pressure measurement could hardly be expected to improve the situation.

Thirteen tests on turbine flow-meters included six different meters at three orientations; vertical downward, horizontal and downward at 45 deg. Each test included about eight flow measurements; from linear regression, a calibration constant and coefficient of determination were calculated.

Turbine outputs showed a high degree of linearity (Table 5). Calibration constants varied slightly, but not consistently, with orientation, and they also varied slightly from one meter to the next (Table 5). These results indicate that the performance of turbine meters is adequate for their use in sprayer control systems, if they are individually calibrated after installation in their working position on the sprayer.

3.3.4 Tests on motorised control valve

There are three possible situations in which the motorised valve could operate:
1. In the line to the spray boom, with a spring-loaded valve in the return to the tank.
2. In the line to the boom, with a constant-orifice valve in the return line.
3. In the return line.
A 19 mm motorised valve (Spraying Systems No. A244-3/4) was operated in these three situations on a Hardi LK sprayer fitted alternatively with No. 4110-20 fan and No. 1553-8 cone nozzles, and with the pump driven at 540 rev/min. Boom flow-rate was measured at valve

Table 5. Tests on turbine flow-meters

Flow-meter no.	Orientation	Output (1/pulse)	Coefficient of determination (r^2)
1	Horizontal	1.130	0.9969
	Vertical	1.105	0.9996
	45°	1.146	0.9992
2	Horizontal	1.068	0.9998
	Vertical	1.067	0.9997
	45°	1.040	0.9997
3	Horizontal	1.057	0.9998
	Vertical	1.039	0.9998
4	Horizontal	1.052	0.9990
	Vertical	1.056	0.9994
5	Horizontal	1.068	0.9999
	Vertical	1.076	0.9996
6	Horizontal	1.104	0.9996
	Vertical	1.068	0.9999

positions from fully open to fully closed. Initially the pressure was adjusted to mid-position.

When the motorised valve was located in the boom line with No. 4110-20 nozzles fitted, the boom flow-rate varied from 33 to 40 1/min with the constant-orifice control, and from 27 to 43 1/min with a spring-loaded control valve over the full range of valve movement (Fig. 4). This corresponds to a pressure range from 2.2 to 3.2 bar and from 1.5 to 3.7 bar with the respective control systems, a somewhat narrow range, especially in the case of the constant-orifice system. Also the first 20 deg. of valve movement from the fully open position has very little effect on flow. A start-up procedure would have to be devised so that the motorised valve would be near to the middle of its effective movement at the pressure required to give the intended application rate.

When No. 1553-8 nozzles were fitted, adjustment of the motorised valve had very little effect on pressure or boom output. In its fully closed position the valve had a discharge coefficient about 8 times greater than the nozzles (33 times greater when fully open), so clearly its adjustment could have little effect on flow. Fitting an alternative valve with a lower discharge coefficient would overcome this problem, but its use would entail generating a pressure considerably higher

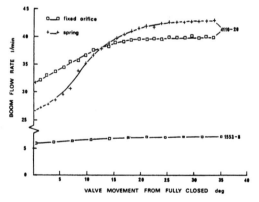

Fig. 4. Boom flow rates with motorized valve in line to boom

than the spraying pressure in order to achieve control.

When the motorised valve was in the return line, it achieved a much wider range of adjustment of boom output; 17 to 43 1/min (0.07 to 0.37 MPa) with the No. 4110-20 nozzles and 3 to 10 1/min (0.07 to 1.0 MPa) with No. 1553-8 nozzles (Fig. 5). Settings outside this range could be achieved by adjusting the pump speed.

It is therefore considered that the motorised control unit should be located in the return to the tank rather than the flow line to the boom.

2141

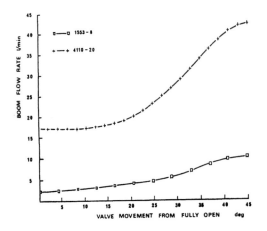

Fig. 5 Boom flow rates with motorized vale
in return to tank

4 CONCLUSIONS

Calibration errors·are a major source of
inaccuracy in spray application.
Fixed-orifice mechanical control systems
can keep sprayer application rate constant
with varying speed, but do not reduce
calibration errors.

A good closed-loop electronic control
system should be based on radar or fifth
wheel forward speed measurement, a turbine
flow-meter for boom output measurement and
a motorized control valve in the return
line to the tank. Systems based on speed
measurement by undriven tractor wheel,
flow inference from pressure measurement,
or a motorized valve located in the boom
flow line are unlikely to be fully
satisfactory. The performance of sprayer
control systems cannot be easily assessed
on the farm, so independent evaluation of
their performance is desirable to make
them more acceptable to farmers.

REFERENCES

Alnajjar, B. and M. Schmidt 1979.
Konstantes Aufwandvolumen durch
Blendensteuereinrichtungen. Grundl.
Landtechnik 29(6): 192-195.
Bell, E. and E. Isensee 1987.
Eignung von Geschwindigkeitsgebern.
Landtechnik 42(2): 57-59.
Fichtel, H. and O. Balcarek 1986.
Vergleichende Untersuchnngen von
Radargeraten zur
Geschwindigkeits-messung an
Landmaschinen. Grundl. Landtechnik
36(3): 68-73.

Givelet, M.P. 1981. Electronic control
systems in pesticide application.
Outlook on agriculture 10(1): 357-360.
Hughes, K.L. and A.R. Frost 1985. A review
of agricultural spray metering.
J. agric. Engng. Res. 32: 197-207.
MAFF/ADAS 1976. The utilization and
performance of field crop sprayers. Farm
Mechanization Studies No. 29.
Makin, A.W. and P.J. Reid 1985. Some
factors affecting the use of non-powered
wheels for speed and wheelslip
measurement. Agric. Engng. Austr. 14(1):
8-17.
Rice, B 1971. A survey of the operation
and maintenance of ground-crop sprayers.
Ir. J. agric. Res. 10: 71-79.
Richardson, N.A., R.L. Lanning, K.A. Kopp
and E.J. Carnegie 1984. Getting a true
measure of ground speed. Agric. engng.
65(6): 14-20.
Speckmann, H. and G. Jahns 1986.
Grundlagen zur Geschwindigkeitsmessung
auf fahrenden landwirtschaftlichen
Arbeitsmaschinen. Grundl. Landtechnik
36(3): 78-86.
Statens Maskinprovningar 1983. Meddelande
2789-94. Group test of control equipment
for farm sprayers. Statens
maskinprovingar, Box 7035, 750 07
Uppsala 7.
Thelen, W and A. Neher 1985. Weg-und
Geschwindigkeitsmessung mittels
Ultraschall-Dopplerverschiebung
vorzugsweise fur landwirschaftliche
Fahrzeuge. Grundl. Landtechnik 35(2):
40-42.
Tompkins, F., W. Hart, R. Freeland,
L. Wilkerson and L. Wilhelm 1988.
Comparison of tractor ground speed
measurement techniques. Trans. Am. Soc.
agric. Engrs. 31(2): 369-374.
Tsuha, W., A. McConnell and P. Witt 1982.
Radar true ground speed sensor for
agricultural and off-road equipment.
Paper No. 821059, SAE, 400 Commonwealth
Drive, Warrendale PA 15096.

Land and Water Use, Dodd & Grace (eds), © 1989 Balkema, Rotterdam. ISBN 90 6191 980 0

Gewichtsermittlung und Wegmessung am Traktor für eine verbesserte Applikationstechnik bei Dünger und Pflanzenschutz

H.Auernhammer, M.Demmel & J.Rottmeier
Institut für Landtechnik, TU Weihenstephan, Bundesrepublik Deutschland

ZUSAMMENFASSUNG: Bei Verteilarbeiten können in der Praxis große Abweichungen vom Soll-wert auftreten. Deshalb werden Überprüfungsmöglichkeiten für die ausgebrachten Mengen auf dem Feld benötigt und es muß die schlupffreie Vorfahrt in die Verteilregelung ein-bezogen werden. Zur Überwachung der Ausbringmengen eignen sich Wiegemöglichkeiten im Schlepperheckkraftheber. Bei der Nachrüstung erbringen Sensoren in den Hubarmen die besten Ergebnisse. Bei Schleppern mit EHR erbringt die Druckmessung in der Hydraulik gleich gute Ergebnisse wie die Messung in den Hubarmen. Generell sind die geringsten Abweichungen jedoch bei speziellen Wiegerahmen zu erwarten. Radarsensoren sollten in der Fahrspur angebracht werden. In Spurschächten liegt die damit erreichbare Genauigkeit bei 97 bis 98 %. Bei kurzen Meßwegabschnitten kann der Fehler auf 5 % ansteigen. Unter ungünstigen Bedingungen zeigten die untersuchten Sensoren große Unterschiede.

ABSTRACT: During application of fertilizer and pestizides big differences to the planned value are usual. Only inspection possibilities on the field can avoid these errors. In addition to it the slip free motion of the vehicle has to be included into the distribu-tion control. Monitoring of distribution material can be done in specific weighing units in the three point linkage of tractors. As an add on device sensors in the lift arms produce good results. In tractors with electronic draft control the system pressure offers as good results as the sensor in the lift arms. Generally a weighing frame has the lowest deviations from the true weight. Radar sensors should be mounted between front and rear wheel. In fields with trafic lanes the accurace is in the range from 97 to 98 %. Only under bad circumstances the tested sensors show big differences.

RESUME: En épandage et en pulvérisation, qui sont des travaux exemplaires de distribu-tion, il y a sonvent des grandes différences entre le dosage théorique et le dosage vrai. Ainsi on a besoin des systèmes effectifs pour contrôler en champ les quantités distribuées. De plus, pour régler la distribution en dépendence de la vitesse vraie, il est nécessaire de respecter le patinage des rous motrices. Pour surveiller les quantites à distribuer des mécanismes, différents au relevage trois points du tracteur s'offrent. En cas d'équipement supplémentaire les capteurs de force dans les deux bras supérieurs de relevage garantissent le memsurage le plus précis. Pour les tracteurs équipés du reve-lage avec régulation hydroélectronique (Système BOSCH) les capteurs de pression four-missent la même précision. Généralement on excompte le pesage le plus exact par une balance en forme de cadre accouplée au revelage trois points. Il est avantageux d'installer des détecteurs radars dans les traces du véhicule. Ainsi, la precision du memsurage de vitesse respectivement de voie est entre 97 % et 98 %. L'erreur pent angmenter à 5 % en maximum quand les distances à mesurer sont très courtes. A des con-ditiones défavorables, les capteurs examinés out révélé des grandes différences.

1. EINFÜHRUNG

Eine hochproduktive Landwirtschaft muß zur Erzeugung guter Ernten eine nicht unbe-trächtliche Menge an Dünger und Pflanzen-schutzmitteln je ha aufwenden. Obwohl heute dafür eine ausgereifte Technik zur Verfü-gung steht, kann von einer gleichmäßigen Ausbringung je Flächeneinheit bei wechseln-den Verhältnissen nicht gesprochen werden. Insbesondere durch den Schlupf der An-triebsräder verändert sich bei wechselnden

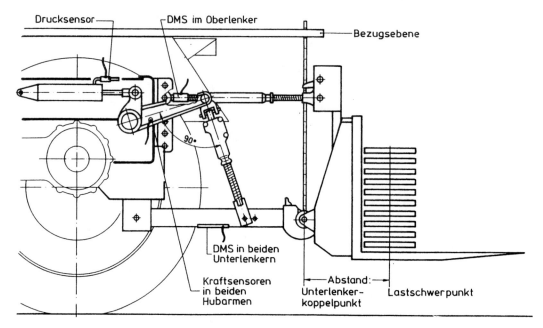

Abbildung 1: Applikationsstellen für Sensoren zur Gewichtsermittlung in der Dreipunkthydraulik

Bodenarten oder in hängigem Gelände die Ausbringmenge lokal sehr stark. Die Einbeziehung der wahren Vorfahrt in einen Regelkreis kann dabei zwar Abhilfe schaffen, jedoch liegen über die Genauigkeit dafür infrage kommender Sensoren unterschiedliche Ergebnisse vor. Zudem fehlt insbesondere bei der Ausbringung von Mineraldünger in loser Form eine geeignete Überprüfungsmöglichkeit, um Fehler während des Ausbringvorganges schnell erkennen und darauf gezielt reagieren zu können.

Aus diesem Grunde sollte nach Möglichkeiten einer mobilen Wägung im Schlepperheckkraftheber gesucht werden. Für die infrage kommenden Systeme war dann die mögliche Genauigkeit unter Praxisbedingungen zu analysieren. Ergänzend dazu sollten dann die auf dem Markt verfügbaren Radarsensoren auf ihre mögliche Genauigkeit hin untersucht werden.

2. GEWICHTSERMITTLUNG IM SCHLEPPERHECK-KRAFTHEBER

Zur Gewichtsermittlung im Schlepperheckkraftheber bieten sich mehrere Möglichkeiten an. Zum einen kann der Druck im Hydrauliksystem für eine Gewichtsermittlung herangezogen werden. Daneben sind Sensoren am

oder im Dreipunktgestänge denkbar und letztendlich können eigens für diesen Zweck konzipierte Wiegerahmen zum Einsatz gelangen.

Alle diese Möglichkeiten basieren auf spezifischen Sensoren. Diese werden bei den Möglichkeiten am Schlepper nachträglich an den entsprechenden Stellen appliziert. Integraler Bestandteil sind sie dagegen beim Wiegerahmen.

2.1 Vorversuche mit unterschiedlichen Sensoren am Schlepper

Um einen ersten Überblick über die generellen Möglichkeiten zu erhalten, wurden in Zusammenarbeit mit einem Schlepperhersteller an einem Testschlepper vier verschiedene Meßstellen mit entsprechenden Sensoren versehen (Abb. 1).

Am einfachsten gestaltete sich die Anbringung eines Drucksensors unmittelbar an der Ölzuführung vor dem Hydraulikzylindern. Dagegen mußten zur Anbringung von Dehnungsmeßstreifen an den Unterlenkern und am Oberlenker entsprechende Schwächungen vorgenommen und nach der Applikation die Sensorstellen dauerhaft dicht verschlossen werden. Die größten Schwierigkeiten ergaben sich bei der Applikation der Sensoren in den Hubarmen. Dafür wurde vom Sensor-hersteller eine äußerst hohe Genauigkeit der Bohrung vorgegeben. Zudem mußte der günstigste Anbringungsort mittels "finiter Elementmethode" bestimmt werden.

Für die Untersuchung wurde der Schlepper mit einer Ladegabel im Heckkraftheber versehen. Als Belastung wurden dann Einzelgewichte mit jeweils 90 kg nacheinander auf die Gabel aufgebracht. Nach jeder Einzellast erfolgte ein vollständiges Anheben des Dreipunktgestänges und danach ein Absenken auf eine Position, in welcher Unter- und Oberlenker parallel zueinander standen.

An dieser Position wurden dann alle Sensorsignale nach einer entsprechenden Aufbereitung und notfalls erforderlicher Verstärkung auf einem Mehrkanalmeßschreiber aufgezeichnet. Für jede Einzellast erfolgten nach einer jeweiligen Absenkung auf den Boden vier Wiederholungsmessungen. Die maximal aufgebrachte Last betrug 975 kg und entspricht damit den Verhältnissen eines Düngerstreuers. Nach der Maximallast wurden die Einzelgewichte wieder einzeln abgenommen und dabei wiederum vier Messungen durchgeführt. Alle Messungen erfolgten bei laufendem Motor, um die Verhältnisse der Praxis einzubeziehen. Gesonderte Untersu-

chungsvarianten bezogen neben der waagrechten Stellung des Schleppers auch Neigungen nach links, nach rechts, nach vorne und nach hinten von jeweils 3 und 5 % mit ein.

Generell ergaben sich bei Neigungen bis 3 % keine Beeinflussungen der Ergebnisse. Stärkere Neigungen bis 5 % können ebenso vernachlässigt werden. Die Ergebnisse der erreichten Genauigkeit bei ebenem Boden sind in Abbildung 2 dargestellt.

Alle Sensoren zeigen für sie typische Ergebnisse. Weitgehend linear sind die Meßfehler über der aufgebrachten Last bei den Sensoren in den Hubarmen. Der konstante Vertrauensbereich für den Mittelwert ist sehr eng und verspricht eine sehr exakte Gewichtsermittlung.

Hingegen zeigen die Sensoren in den Unterlenkern eine stärkere Abweichung vom Nullwert mit größerem Vertrauensbereich. Hinzu kamen bei diesen Sensoren schon nach kurzer Zeit Genauigkeitsprobleme durch eindringende Feuchtigkeit nach Einsätzen des Schleppers zur Spätdüngung.

Einen überraschend engen Vertrauensbereich erreichte beim Testschlepper auch der Drucksensor im Hydrauliksystem. Allerdings

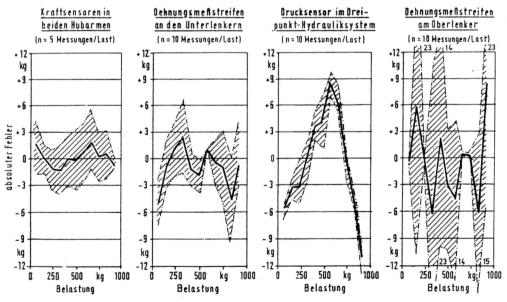

(Schlepper mit 63 kW; Messungen im Stand bei laufendem Motor; Unterlenkerposition parallel zur Bezugsebene; konstanter Schwerpunktabstand)

Abbildung 2: Mittlere Meßfehler und deren Vertrauensbereiche für verschiedene Sensoren zur Gewichtsermittlung im Schlepperheckkraftheber

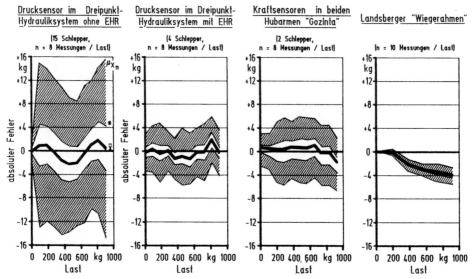

Abbildung 3: Untersuchungsergebnisse für den Hydraulikdruck, die Hubarmsensoren und den Wiegerahmen

weicht dabei das Signal sehr stark von der Nullinie ab und müßte deshalb gesondert elektronisch nachbearbeitet werden.

Als weitgehend unbrauchbar erwies sich bei diesen Untersuchungen der Sensor im Oberlenker.

2.2 Vergleichsuntersuchungen mit ausgewählten Sensoren

Aufbauend auf diese Voruntersuchungen wurden dann gezielte weitere Untersuchungen durchgeführt. In diese Untersuchungsreihe wurden insgesamt 20 Schlepper für die Hydraulikdruckmessung einbezogen. Bewußt wurden auch ältere Schlepper getestet, um eventuell zu erwartende Alterungserscheinungen feststellen zu können. Zudem wurde ein weiterer Schlepper mit Sensoren in den Hubarmen ausgestattet und schließlich wurde auch ein Wiegerahmen berücksichtigt. Für alle Schlepper wurde der oben geschilderte Meßablauf durchgeführt. Folgende Ergebnisse wurden ermittelt (Abb. 3).

Beim Hydraulikdruck wurden Schlepper mit und ohne EHR berücksichtigt. Schlepper ohne EHR zeigen dabei sehr große Streubreiten im Vertrauensintervall für den mittleren Meßfehler (innerer Bereich) und eine sehr große Streuung für die insgesamt zu erwartenden zufälligen Ergebnisse (äußerer Be-

reich). Hingegen erreichten Schlepper mit EHR in etwa die gleichen Ergebnisse wie die beiden Schlepper mit den Sensoren in den Hubarmen. Deutlich aber hebt sich von allen der Wiegerahmen ab. Er weist die geringste Streuung auf und kann deshalb ohne Einschränkung empfohlen werden.

3. RADARMESSUNGEN

Auch die Untersuchungen an den Radarsensoren sollten die Verhältnisse der Praxis sehr stark berücksichtigen. Deshalb wurde als Anbringungsort die rechte Fahrspur zwischen dem Vorder- und Hinterrad gewählt. Die Erfassung der Meßgenauigkeit erfolgte in einem Weizenbestand etwa drei Wochen vor der Ernte in den vorhandenen Spurschächten, auf Getreidestoppel und in Senf (Abb. 4).

Erfaßt wurden die vom Sensor abgegebenen Signale für einen über Lichtschranken gemessenen Weg von 4 m mit Geschwindigkeiten von 0,7, 1,7, 2,5 und 3,5 m/s. In einem 80 bis 120 cm hohen Senfbestand wurden zudem Messungen über Meßweglängen von 4 m, 1 m und 0,5 m bei jeweils 2,5 m/s Geschwindigkeit durchgeführt. Die ermittelten Ergebnisse sind in Abbildung 5 und 6 dargestellt.

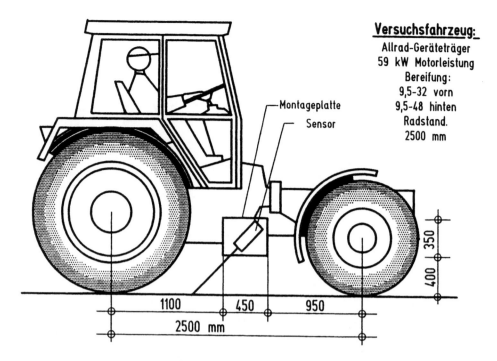

Versuchsfahrzeug:
Allrad-Geräteträger
59 kW Motorleistung
Bereifung:
9,5-32 vorn
9,5-48 hinten
Radstand.
2500 mm

Abbildung 4: Sensoranbringung am Testschlepper und Referenzmeßanordnung

Beide Radarsensoren zeigen unterschiedliche Genauigkeiten. Im Weizenbestand liegt der ermittelte Fehler zwischen 1,4 und 3,5 %. Sensor A bringt das beste Ergebnis im Hauptarbeitsbereich, Sensor B dagegen bei 12 km/h.

Auf Getreidestoppel verringern sich die Meßfehler beider Sensoren. Sensor B ist dabei überlegen. Überraschenderweise zeigt nun Sensor A den größeren Fehler im Hauptarbeitsbereich.

Wird die Meßweglänge reduziert, dann erhöht sich der Fehler bei den beiden Sensoren. Generell bringt dann Sensor B ungünstigere Ergebnisse, wobei der Fehler bei 0,5 m Meßweglänge dann in der Praxis nicht mehr akzeptiert werden kann. Allerdings ist zu bedenken, daß die gewählte Testumgebung sicher die in der Landwirtschaft ungünstigsten Bedingungen darstellt und somit eigentlich der Maximalfehler ermittelt wurde.

4. EINORDNUNG UND DISKUSSION DER ERGEBNISSE

Die durchgeführten Untersuchungen mit den ermittelten Ergebnissen lassen eine klare Einordnung zu.

Für die Gewichtsermittlung im Schlepperheckkraftheber sind derzeit drei wesentliche Systeme geeignet. Als Nachrüstung für Altschlepper und als Grundausrüstung für neue Schlepper mit und ohne EHR eignen sich die Sensoren in den Hubarmen. Dabei ist allerdings höchste Präzision bei der Bohrung erforderlich, weshalb die entsprechende Applikation ausschließlich vom Spezialisten durchgeführt werden kann.

In Schleppern mit EHR sollte künftig aufgrund der guten Möglichkeiten eine Gewichtsermittlung - nach Möglichkeit elektronisch gesteuert - integriert werden.

Hingegen kann der Wiegerahmen aufgrund seiner hohen Genauigkeit für alle Anwendungsfälle empfohlen werden. Allerdings muß dazu der derzeit noch sehr hohe Preis reduziert werden. Dies könnte durch die Einbeziehung der Signale und der Anzeige im Schlepperbordcomputer erfolgen. Generell hat aber dieses Wägeprinzip keinerlei Einfluß durch den unterschiedlichen Schwerpunktabstand zu befürchten und zudem kann

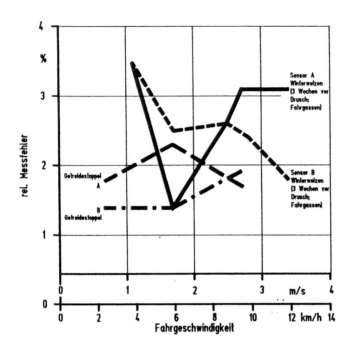

Abbildung 5: Genauigkeit zweier Radarsensoren bei unterschiedlicher Vorfahrt in Weizen und auf Getreidestoppel (Meßweglänge 4 m)

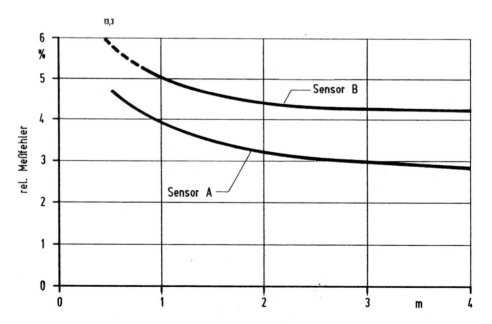

Abbildung 6: Genauigkeit zweier Radarsensoren bei unterschiedlicher Meßweglänge (v = 2,5 m/s; Senf 80 bis 120 cm hoch; keine Spurschächte)

diese Einheit problemlos an unterschiedlichen Schleppern eingesetzt werden. Allerdings können sich durch den Geräteversatz Einsatzschwierigkeiten bei leichteren Schleppern ergeben.

Radarsensoren erbringen je nach Vorfahrtgeschwindigkeit und Bodenbewuchs unterschiedliche Meßgenauigkeiten. Für die typischen Anwendungsfälle in Getreide mit Spurschächten und Anbringung der Sensoren in der Fahrspur sind Fehler zwischen 1,5 und 4 % zu erwarten, wenn die Signale über eine Meßweglänge von 4 m integriert werden. Auf Getreidestoppel reduzieren sich diese Fehler auf etwa 2 bis 2,5 %.

Unter ungünstigen Verhältnissen ohne Spurschächte und Meßweglängen von nur 0,5 m ergeben sich dagegen maximale Fehler bis zu 13 %. Dies sollte aber nicht darüber hinwegtäuschen, daß damit der Meßfehler bei üblichen Arbeitsgeschwindigkeiten zwischen 0,2 und 0,3 km/h liegt. Somit darf bei Einbeziehung der schlupffreien Vorfahrt in die Regelvorgänge bei Verteilgeräten immer von einer ausreichenden Genauigkeit ausgegangen werden.

5. LITERATUR

Auernhammer, H., M. Demmel und H. Stanzel 1988. Wiegemöglichkeiten in der Schlepperdreipunkthydraulik. Landtechnik 10: 414-418

Auernhammer, H. 1988. Wiegen in der Schlepperdreipunkthydraulik - In den Hubarmen stecken die Sensoren. Agrar-Übersicht 10: 26-27 und 33

Boll, E. 1988. Elektronik an Drillmaschinen, Steuern und Regeln von Geräten für die Bestellung. Darmstadt: KTBL (Nr. 322)

Knechtges, H. 1988. Behälterwägung im Dreipunktanbau. Landtechnik 5: 218-219

Isensee, E. 1988. Ansprüche der Produktionstechnik an die elektronische Steuerung/Regelung. Düsseldorf: VDI (Landtechnik 88)

Land and Water Use, Dodd & Grace (eds), © 1989 Balkema, Rotterdam. ISBN 90 6191 980 0

Investigations into thermal techniques for weed control

S.Parish
West of Scotland College, Auchincruive, Ayr, Scotland, UK

ABSTRACT: Mustard and Italian ryegrass plants were exposed to flame and infrared treatments in laboratory tests. It was found that the effect of the gas burner depended on its design and angle to the horizontal, and that the height of the burner above the ground controlled the effect on ryegrass, but not on mustard. Other results also contribute ideas to refine the design of flame weeding equipment.

ABSTRACT: Senf- und italienische Raygraspflanzen wurden in Laborversuchen Flammen- und Infrarotbehandlungen ausgesetzt. Es wurde herausgefunden, dass die Wirkung des Gasbrenners von seinem Design und von dem Winkel zum Horizontalen abhaengt, waehrend die Hoehe des Brenners ueber dem Boden die Wirkung auf Raygras, nicht aber auf Senf beeinflusste. Weitere Ergebnisse steuern auch Ideen zur Vervollkommnung des Designs von Abflammgeraeten bei.

ABSTRACT: Des plantes de moutarde et de fromental italien ont été exposées dans le laboratoire aux traitements de flammes et de rayons infrarouge. On a trouve que l'effet du bruleur à gaz dependait de son dessin et de l'angle au plan horizontal, et aussi que l'hauteur du bruleur au-dessus du sol controlait l'effet sur le fromental, mais pas sur la moutarde. D'autres resultats contribuent des idées pour raffiner le dessin des desherbeuses par le feu.

1 INTRODUCTION

The use of flame cultivation for weed control, stubble burning and land clearance can be considered as well established agricultural practices. The mechanisation of flame weeding, for crops such as cotton, continued into the 1960's, until the oil crisis, and the development of more effective herbicides, made it a less favourable operation (Kepner et al 1978). Stubble burning equipment using liquefied petroleum gas (LPG) to volatise the crop residues was developed more recently to overcome the hazards associated with traditional straw and stubble burning (Vernon 1985).

In recent years there has been a renewed interest in flame weeding for non-chemical weed control, due to the increasing consumer demand for organically grown vegetables (Hill 1986). Concern over the effect of herbicides on the environment may also encourage a wider use of integrated weed control systems, including thermal treatments.

Two categories of equipment have been developed to use heat for weed control.

Flame equipment relies on the contact of the flame with the plant, whereas the effect of infrared equipment is from the heat radiated from a hot surface close to, but not touching, the plants. Both types are normally used for weed control pre-emergence of a row-crop, although some crops can tolerate post-emergence treatment at specific growth stages (Vester 1984). For flame equipment Vester reports the effect on the weeds depends on the type of burner, the angle of the burner, the height above the ground, the gas pressure, and the travel speed. According to Hoffmann (1980), the plant temperature must reach 100°C for at least 0.1 seconds to cause sufficient damage for the plant to subsequently wilt, although some species, particularly grasses, are more resistant to thermal treatments. Infrared equipment requires an exposure time of 2 seconds to be effective (Van't Rood 1984), but is claimed to be more fuel efficient.

There are examples of both commercial and home-made, infrared and flame equipment currently in use in the United Kingdom. Most of the commercial equipment is imported, and trials with different types have been carried out in other European countries. Results from some of this work, together with a review of a range of commercial and experimental equipment, are also presented by Hoffman.

The majority of results published are based on the effect of thermal treatments in field trials, and there appears to be a lack of basic data on the effect of flames and infrared radiation on plants. This paper reports results from laboratory based investigations into the effects of flame treatments on two plant species.

2 OBJECTIVES

The main aim of the work was to investigate the effects of varying the design of an LPG flame gun on the dry matter yield of white mustard and Italian ryegrass. These were chosen as examples of monocotyledon and dicotyledon plants. Infrared treatments were also carried out, and are included here for comparative purposes. An electric infrared element was used, which emitted the same range of wavelengths as that identified in previous work as being the most effective (Parish 1989).

The experimental approach was to attempt to identify important factors associated with the plants, and those associated with the flame. To identify the former, trays of plants were prepared to be treated at separate growth stages, and after treatment were cut at different time intervals.

Table 1. Details of plants, growth stages when treated, and cutting intervals (in weeks).

Species	Early stage	Late stage	When cut
Ryegrass	Shoot only	2 or 3 leaf	1,2,3
Mustard	Cotyledons	2 or 4 leaf	1,2,4

To assess flame performance, variations in burner height, angle, burner type, gas pressure, and travel speed were made. The round burner was 25 mm diameter and the fishtail attachment 60 x 8 mm.

The combinations of the variables meant that 48 treatments were carried out on each species at each growth stage, for all three cutting intervals. There was no replication of treatments.

Table 2. Details of flame variables.

Burner	Pressure (bar)	Height (mm)	Angle °	Speed (m/s)
Round	1	75	30	0.5
Fishtail	2	100	60	0.25
		125		

3 EXPERIMENTAL METHOD

3.1 Plant production

The aim was to produce 20 plants in a narrow band, grown in potting compost in polystyrene trays, measuring 160 x 210 mm. To avoid the possibility of effects on plants from hot polystyrene at the edge of the trays, both ends of the trays were cut away to prevent flame contact. Each of these trays were placed inside an unmodified tray, and only removed when treated.

The trays were partly filled with compost, the seeds sown randomly into bands 55 mm wide and 170 mm long, and covered with a thin layer of compost. The trays were numbered, using a table of random digits, and transferred into an unheated glasshouse, with automatic ventilation control. The compost was kept moist, before and after treatment.

For treatment the trays were transferred in batches to the laboratory. Each tray was checked to ensure that the plant leaf surfaces were dry, any droplets of water being carefully removed using tissue paper. After treatment the batch was returned to the glasshouse. The positions of the trays in the glasshouse were not monitored, as it was assumed that even conditions existed.

The plants were cropped by cutting the stems at the surface of the compost, and the fresh weight of plant material per tray was recorded. After drying in an oven overnight at 100°C, the dry weights per tray were obtained. Dry weights of plants from untreated control trays for each species, growth stage and cutting date were also recorded.

3.2 Experimental rig

The burner was connected to a 3.9 kg propane cylinder, via a 0.5 m extension tube, a gas torch handle with a control valve, flexible hose, a 0 to 2 bar pressure gauge, and the cylinder pressure control valve. This equip-

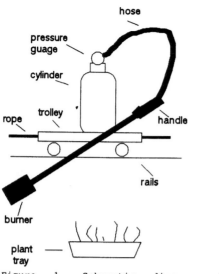

Figure 1. Schematic diagram of the experimental rig.

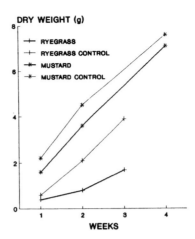

Figure 2. Dry weight versus cutting date, using all data.

ment was mounted on a trolley, which ran on rails above a shallow soil tank (see Figure 1).

The trolley was pulled along by a rope from a hydraulically powered winch, and the speed of travel controlled by a pressure compensated flow control valve. Constant speed, and braking, were achieved by attaching the rear of the trolley to a counterbalancing mass, via rope and pulleys. The extension tube was held in position by adjustable clamps, which allowed both the angle and height of the burner to be set. The gas pressure was set by adjusting the pressure control valve.

Three trays with plants at the same growth stage were treated in each pass of the flame, one for each cutting date.

For infrared treatment of plants, the trolley was exchanged for one with the emmitter attached.

4 DISCUSSION OF RESULTS

4.1 Flame data

An analysis of variance was carried out on the complete set of data. This showed that all the variables, except the burner angle, had a significant effect on dry weight. The trends appeared to be as might have been predicted, that slower, closer application of a flame at a higher gas pressure was more effective than a faster, less close application at a lower pressure. Figure 2 shows

that in general the plants did manage to grow on after treatment.

The round burner was more effective than the fishtail burner, but the effect of the burner angle depended on the burner type. The round burner was more effective at a 30° angle to the horizontal, than at a 60° angle, whereas for the fishtail burner the opposite was the case (see Figure 3). At the shallower angle the flat flame was observed to deflect upwards, possibly due to the convection from its own heat.

Partioning the data by crop, growth stage and cutting date showed differences between the effects of treatments on the plant spe-

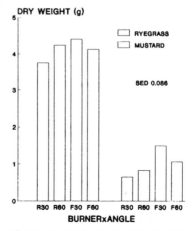

Figure 3. Effect of burner and angle on dry weight (all data).

cies. Increasing the gas pressure, which gives a more intense flame, was significantly more effective on mustard, particularly at the early cotyledon leaf stage. Increasing the pressure did not have such a significant effect on ryegrass (see Figure 4). However, the height of the burner above the plant was significant on ryegrass, but not on mustard, as shown in Figure 5. Therefore, in grasses it appears important to position the burner closer to the low growing point of the plants.

The significance of the main variables are summarised in Tables 3a and 3b, and details from the ryegrass treated at the later stage, and the mustard treated at the early stage, both cut after 2 weeks, are shown in Figures 6 to 9.

Table 3a. The effect of flame treatments on ryegrass.

Variable	Early treatment			Later treatment		
	1	2	3	1	2	3
Pressure					**	
Speed	***	*			***	***
Height		*			*	**
Burner x angle		***	***		***	***

Table 3b. The effect of flame treatments on mustard.

Variable	Early treatment			Later treatment		
	1	2	4	1	2	4
Pressure	*	**				
Speed	*	**		*		
Height						
Burner x angle	*	***		**		

* significant difference at 5% level
** significant difference at 1% level
*** significant difference at 0.1% level

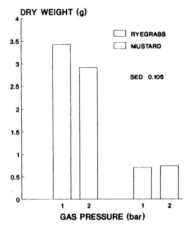

Figure 4. Effect of pressure on dry weight (all data).

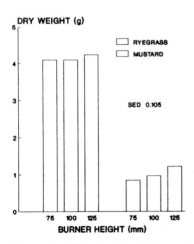

Figure 5. Effect of height on dry weight (all data).

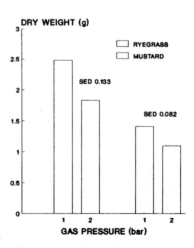

Figure 6. Effect of pressure on dry weight.

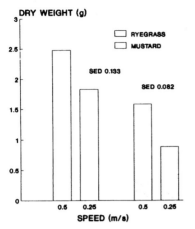

Figure 7. Effect of speed on dry weight.

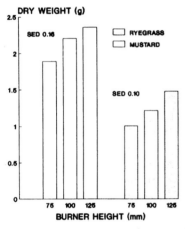

Figure 8. Effect of height on dry weight.

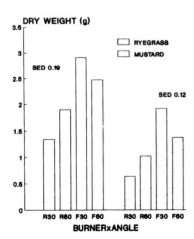

Figure 9. Effect of burner and angle on dry weight.

From these results it is possible to make some tentative suggestions to refine the design of flame weeders. The suggestions need to be assessed by further trials, but are made as follows:

1. Plants are best treated at an early growth stage, and less energy may be used as a faster speed is possible for the same effect.

2. The most effective angle of the burner depends on its design, with a flat flame needing a steeper angle.

3. The height of the burner above the ground should be lower if grass weeds are a major problem.

4. A higher gas pressure may be more effective, to control dicotyledonous weeds.

4.2 Comparison with infrared

The infrared emmitter had an input power of 9 kW in an area 1 x 0.225 m, and was run at 0.125 m/s. The results from this treatment are shown in Figure 10 in comparison with those from the round burner at 2 bar, and 0.5 m/s, on ryegrass. To make a similar comparison for mustard (see Figure 11), the infra red emitter results at the faster speed of 0.2 m/s are used. The increased effect of the infrared treatment on mustard is most probably due to the vulnerable leaf area directly facing the incoming radiation, and therefore more easily damaged.

A more detailed comparison between the flame and infrared treatments is being prepared, but it is interesting to compare the energy inputs to achieve the effects described

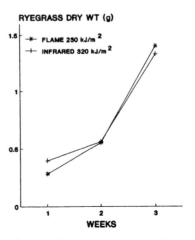

Figure 10. Ryegrass dry weight versus cutting date, showing similar effects from gas and electricity inputs.

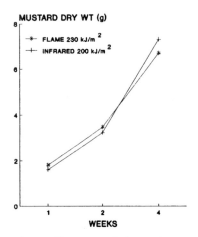

Figure 11. Mustard dry weight versus cutting date, showing similar effects from gas and electricity inputs.

above. On mustard the inputs were at similar levels, 230 kJ/m^2 of LPG and 200 kJ/m^2 of electricity, whereas on ryegrass the electricity input was 320 kJ/m^2, and LPG input 230 kJ/m^2.

5 CONCLUSIONS

The general pattern of the results from the flame treatments is in line with those previously reported.

The angle at which the burners were set controlled their effect, the round burner performing better at a shallow angle, and the fishtail burner at a steeper angle. Setting the burner lower was more effective on ryegrass, but not on mustard. It is hoped to confirm these effects in field trials.

In the infrared treatments, mustard appears more susceptible than ryegrass. To achieve a similar effect on ryegrass from the two treatments, a higher energy input was required for the infrared treatment than for the flame treatments.

ACKNOWLEDGEMENTS

The author would like to thank the Leverhulme Trust for funding this work, and Calor Gas Ltd for the loan of equipment.

REFERENCES

Hill, C.E. 1986. The future for organically grown food. London: Food from Britain.

Hoffmann, M. 1980. Abflammtechnik. Darmstadt: K.T.B.L.

Kepner, R.A., Bainer, R. & Barger, E.L. 1978. Principles of farm machinery, p.247-251. Connecticut: A.V.I.

Parish, S. 1989. The effects of infrared radiation on plant survival. Journal of the Institution of Agricultural Engineers (accepted for publication).

Van't Rood, A.C. 1984. Weed control by infrared heat radiation. In Flame cultivation for weed control. Opprebais: C.R.A.B.E.

Vernon, G. 1985. Recent developments in flame cultivation in the United Kingdom. In Proceedings of the symposium on the LPG situation in the EEC region 1985-2000. Madrid: United Nations Economic Commission for Europe.

Vester, J. 1984. New experience with flame cultivation for weed control. In Flame cultivation for weed control. Opprebais: C.R.A.B.E.

Land and Water Use, Dodd & Grace (eds), © 1989 Balkema, Rotterdam. ISBN 90 6191 980 0

Fertilizer application for small farmers in the Mexican humid tropics

B.G.Sims
AFRC Institute of Engineering Research, Silsoe, UK

R.Jiménez R. & A.Aragón R.
Instituto Nacional de Investigaciones Forestales y Agropecuarias, Mexico

ABSTRACT: Increase in fertilizer use in Mexico over the last thirty years has been one of the most rapid in the world. The bimodal nature of Mexican agricultural development means that fertilizer use is concentrated in the irrigated areas of the North and Central regions of the country, however use by small farmers in other regions is common and increasing. The paper describes the work done by the UK/Mexico Agricultural Engineering Project within the National Agricultural and Forestry Research Institute (INIFAP), on the development and evaluation of low cost machinery for fertilizer application. The work was done in the humid tropical south-eastern state of Veracruz where six animal drawn fertilizer distributors were subjected to a detailed test procedure which included laboratory and field evaluation. The results reveal the difficulty of applying accurate rates of fertilizers in the humid tropics as their physical characteristics vary with changes in the relative humidity of the atmosphere. The ease of calibration and thorough cleaning of fertilizer residues and the choice of construction materials to reduce corrosion are important selection factors. The best machine tested was a design from the International Institute for Tropical Agriculture (IITA) in Nigeria which incorporates an open wire auger distribution mechanism which also acts as an agitator and prevents bridging of moist fertilizers.

RESUME: L'augmentation de l'utilisation d'engrais au Méxique pendant les dernières trente années a été une des plus rapide du monde. A cause du caractère bimodal du développement agricole Mexicain, l'utilisation d'engrais a été concentrée dans les zônes irriguées du Nord et du Centre du pays; cependant, l'utilisation par les petits agriculteurs dans d'autres régions est courant et augmente. Cet article décrit les travaux du UK/Méxique Project Machinisme Agricole, au sein de l'Institut National de Recherches sur l'Agriculture et les Forêts (INIFAP), sur le développement et l'évaluation de machines peu coûteuses pour la distribution d'engrais. Le travail a eu lieu dans l'état sud-est humide et tropical de Veracruz, où six distributeurs d'engrais à traction animale ont subit des essais détaillés avec évaluation dans le laboratoire et sur le terrain. Les résultats démontrent les problèmes de l'application de volumes précis d'engrais dans les tropiques humides parce que le caractère physique de l'engrais change avec les variations de l'humidité rélative de l'atmosphère. La possibilité de calibration et de nettoyer les résidus d'engrais et le choix de materiaux de construction qui limitent la corrosion sont des facteurs de sélection importants. La meilleure machine testée a été développée par l'IITA (Institut International de l'Agriculture Tropicale) au Nigéria et contient un mécanisme de distribution basé sur un vérin ouvert en fil que sert aussi comme agitateur et empêche le bouchement d'engrais humide.

ZUSAMMENFASSUNG: Die Zunahme an Düngermittelverbrauch in Mexiko während der letzten dreissig Jahre war eine der schnellsten der Welt. Der Verbrauch von Düngermitteln ist wegen des doppelten Charakters der landwirtschaftlichen Entwicklung in Mexiko in den bewässerten Zonen der nördlichen und zentralen Gebiete konzentriert, obwhol der Verbrauch durch Landwirten in anderen Gebieten verbreitet ist und zunimmt. Dieser Artikel beschreibt die Arbeit des UK/Mexiko Landtechnischen Projektes innerhalb vom Staatlichen mexikanischen land- und forstrwirtschaftlichen Forschungsinstitut (INIFAP) über die Entwicklung und Auswertung preiswerter Düngungsgeräte. Die Arbeit fand im feuchten tropischen südöstlichen Staat Veracruz statt, wo sechs tiergezogene Düngerstreuer in Laboratorium- und Feldtests ausführlich ausgewertet wurden. Die Ergebnisse weisen die Schwierigkeiten davon auf, Düngermittel in den feuchten Tropen in präzisen Mengen zu verteilen, weil sich die physikalischen Eigenschaften mit Änderungen der relativen Feuchte der Luft auch ändern. Einfaches Kalibrieren, gründliche Reinigung von Düngerresten und Benützung von korrosionsbeschränkende Materialien sind wichtige Auswahlsfaktore. Das Beste von den getesteten Geräten wurde am IITA (Internationales Institut für Tropische Landwirtschaft), Nigeria, entwickelt und besteht aus einer offenen Drahtförderschnecke, die als Verteilsystem und Rührapparat dient und auch das Verstopfen von feuchten Düngermitteln verhindert.

1. INTRODUCTION

According to Villanueva (8) in 1983, the annual increase in fertilizer use in Mexico over the previous 30 years has been one of the most rapid in the world, although it is also true that her starting point was very low. The bimodal character of Mexican agriculture has led to a concentration of input use, including fertilizers, in the modernised areas of the Pacific North and Central regions. In 1969, for example, Sonora and Sinaloa states accounted for more that 50 percent of national fertilizer consumption.

Fertilizer use by small farmers is widespread, especially in basic grains production (maize, beans, rice and wheat) for which the Rural Credit Bank (BANRURAL) extends credit lines for inputs. To date, small farm fertilizer application has been practically 100 percent manual, the fertilizer being carried in a bucket or large can and applied by hand either in a band along the crop row, or at the base of each plant.

In response to the lack of mechanical distributors the Agricultural Mechanisation Unit (UIMA) of the Mexican National Institute of Forestry and Agricultural Research (INIFAP), established a programme of development and evaluation of fertiliser and seeder-fertilizer distributors with the aim of promoting the commercial manufacture of promising machines. This paper describes the results of the programme which was carried out in Veracruz state in the humid tropical Gulf region of south-eastern Mexico.

2. FERTILIZER AND SEEDER-FERTILIZER DISTRIBUTORS EVALUATED

In this section all the machines evaluated in the UIMA programme between 1981 and 1987 are described. Only two have reached commercial manufacture (the VACSA vertical disc seeder/fertiliser distributor, and the International Crops Research Institute for the Semi-Arid Tropics (ICRISAT) fertilizer distributor. The principle of the International Institute of Tropical Agriculture (IITA) distributor is incorporated in a pre-production prototype machine.

The majority of the devices described are designed to be attached to one of the multi-purpose animal drawn tools developed by the UIMA, the Multibarra pedestrian controlled tool frame and the Yunticultor wheeled tool carrier (Sims (8), 1987).

2.1. Seeder-fertiliser for wide beds.

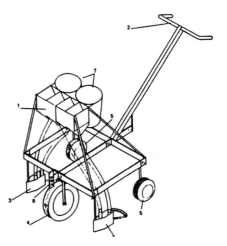

Fig.1 The "Araña" seeder-fertilizer for wide beds.

1. Double hopper 2. Handle 3. Furrow openers with seed covers 4. Depth control wheel 5. Furrow press wheels 6. Hitch point 7. Buckets for seed and fertilizer.

The Araña was developed to relay sow between standing crops grown on wide beds, a system being evaluated in ICRISAT (Krantz (4)),1980 and in Mexico (Camacho (2)),1983.

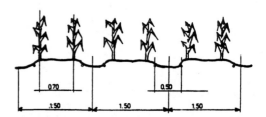

Fig.2 1.5m wide beds each with two rows of maize. The second crop is relay sown on the edges of the beds. Dimensions in m.

Figure 2 shows 1.5m wide beds each with two rows of maize. The Araña passes down the furrow between beds, hauled by a single animal while two operators hand feed the seed and fertilizer from the buckets to the hopper. A third operator controls the maize with the handle-bar.

2.2. Reciprocating plate seeder-fertilizer distributor

This machine (Figure 3) has a double hopper (for seed and fertilizer) which discharges on to a reciprocating perforated plate. The figure shows a transmission system developed in Mexico for the machine which is based on Alan Stokes' (Project Equipment, Oswestry, UK) design.

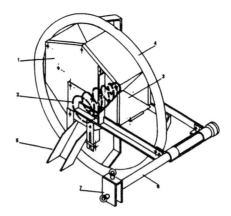

Fig 4 IITA surface band fertilizer distributor

1. Wooden hopper 2. Tapered feed auger 3. Sliding shutter 4. External drive wheel 5. Adjustable outlet 6. Floating linkage 7. Attachment to the IITA rolling jab planter

2.4. UIMA seeder-fertilizer distributor

The seeder unit incorporated in this machine is based on a Project Equipment design, the fertilizer distributor is designed by the UIMA in Mexico (Figure 5).

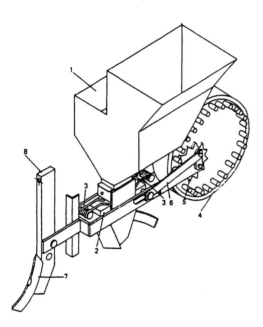

Fig 3. Reciprocating plate seeder-fertilizer distributor.

1. Double hopper for seed and fertilizer 2. Reciprocating plate 3. Spring seed ejectors 4. Peg gear wheel 5. Secondary gear 6. Connecting rod 7. Furrow opener 8. "Multibarra" attachment stalk

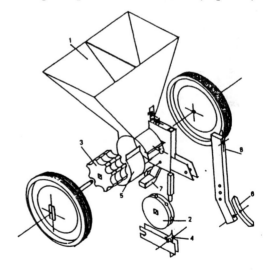

Fig 5 UIMA seeder-fertilizer distributor

2.3. Surface band fertilizer distributor

This machine was designed at IITA in Nigeria (Baryeh et al. (1) 1984), to permit band application of fertilizer in no-till crops (Figure 4).

1. Double hopper 2. Seed metering disc 3. Fertilizer metering wheel 4. Seed ejector 5. Adjustable fertilizer outlet 6. Furrow opener 7. Seed cover 8. "Multibarra" attachment stalk.

2.5. VACSA vertical disc seeder-fertilizer distributor

VACSA, a commercial farm machinery manufacturer in Mexico has based the seeder of their machine on a Project Equipment design, while the fertilizer distributor is their own (Figure 6).

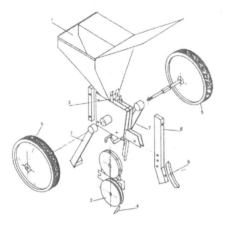

Fig 6 VASCA vertical disc seeder-fertlizer distributor.
1. Double hopper 2. Fertilizer metering disc 3. Disc cleaner (top) 4. Disc cleaner (bottom) 5. Drive wheels 6. Furrow opener 7. Seed coverer 8. "Multibarra" attachment stalk

2.6. ICRISAT fertilizer distributor
This oscillating fertilizer distributor was developed at ICRISAT centre for use with a wheeled tool carrier (Figure 7) (India. ICRISAT (3), 1980).

Fig 7 The ICRISAT fertilizer distributor mounted on the Yunticultor wheeled tool carrier

The hopper discharges into a primary outlet with a central keel. Beneath this is an oscillating concave which delivers the fertilizer to the secondary outlet from where it falls to the furrow openers through flexible plastic tubes (Figure 8).

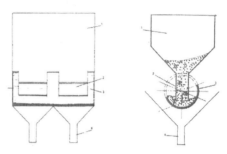

Fig 8 Metering mechanism of the ICRISAT fertilizer distributor

1. Hopper 2. Primary outlet 3. Oscillating concave 4. Secondary outlet connected by plastic tubing to the furrow openers

3.   TEST PROCEDURE

The small farmer fertilizer distributor test procedure developed has two main categories:

- workshop tests (durability, metering, longitudinal distribution of fertilizer, effect of vibration).

- Field tests (application rates, wheel slip, construction quality, ease of operation).

Key principles of the procedures are based on Patterson et al. (5) 1964.

3.1.   Detailed description of the machine and its operation including:

- description of mechanisms
- principle dimensions
- hopper capacity
- weight
- adjustments

3.2.   Fertilizers

The fertilizers most commonly used by farmers in the region will be used in the test. If possible granular, powdered and crystalline fertilizers will be included.

3.3. Workshop tests

For all the workshop tests (except durability) there should be at least three repetitions.

3.3.1. Durability
The machine will be run on a test rig (or in the field), for 20 hours. Breakages and wear will be reported.

3.3.2. Application rate
The machine will be run without furrow openers, and with the hopper half full over a distance of 10m on a smooth surface. Fertilizer delivered will be collected in bags attached to each outlet for subsequent weighing. Application rates (in kg/ha assuming an inter-row distance of 1m) will be reported for the following application rate adjustments:
- minimum
- maximum
- approximately mid way between maximum and minimum (median position).

3.3.3. Longitudinal distribution
The machine, set at its median application rate and with the hopper half full, will be run over a distance of 10m on a smooth surface. Fertilizer delivered in each 0.5m segment will be collected and weighed. The results will be presented graphically.

3.3.4. Effect of vibration on application rate and longitudinal distribution
The aim of this test is to simulate the effect of uneven terrain. Standard bumps (20mm x 20mm) will be placed on the 10m test track at 1m intervals and workshop tests 3.3.2 and 3.3.3 will be repeated. An alternative would be to attach bumps to the drive wheel of the machine.

3.4. Field tests

3.4.1. Application rate
With the same settings used in workshop test 3.3.2 (if appropriate for the fertilizer being used) the machine will be used over 0.1 ha starting with the hopper completely full. The fertilizer required to refill the hopper will be weighed and the application rates per hectare calculated. These will be compared with the results of Workshop Test 3.3.2.

3.4.2. Wheel skid
The machine will be slowly pulled (hopper empty) over a smooth surface and the distance (l_0) covered by 10 wheel revolutions will be measured. In field test 3.4.1 the distance (l_1) covered by 10 revolutions will be measured.

$$\text{Wheel skid} = \frac{(l_1 - l_0)}{l_0} \times 100\%$$

Using the results of field application and wheel skid tests, and comparing them with the workshop application rate results, the effects of vibration and wheel skid in field conditions on application rate will be discussed.

3.4.3. Construction
At the end of the field test the machine will be inspected and comments made on the quality of construction and signs of wear. Special attention will be given to evidence of corrosion.

3.4.4. Operation
Ease of operation adjustment and cleaning will be reported on.

4. TEST RESULTS FOR THE IITA DISTRIBUTOR

Detailed results of the tests of the fertilizer distributors described are given elsewhere (Sims et al. (6), 1987), here the results of the test procedure applied to the IITA fertilizer distributor are given as an example.

4.1. Description

The IITA fertilizer distributor (Figure 4) has a wooden hopper separated from the tapered feed auger by a sliding stainless steel flow control plate. An external drive wheel rotates the auger as the machine advances and the fertilizer falls down an adjustable outlet. The original IITA design is manually operated and single purpose, but the UIMA version is designed to be attached to a rolling jab planter by means of a floating linkage. The combination planter/fertilizer distributor is coupled to a multi-purpose tool carrier pulled by draft animals.
The machine standing alone has the following dimensions:
-length: 0.65m
-width: 0.25m
-height: 0.54m
-weight: 12.7kg
-hopper capacity: 3.1kg urea fertilizer
The distributor has two adjustments:
- the sliding flow control plate
- the angle of the fertilizer outlet which allows the fertilizer band to be displaced laterally by about 50mm.

4.2. Fertilizers

The three fertilizers used were:
- Granular: diammonium phosphate
- Powdered: ammonium sulphate and super phosphate mixture
- Crystalline: urea

4.3. Workshop tests

4.3.1. Durability
The tapered feed auger drive shaft (12.7mm diameter mild steel) broke after six hours work on a test rig simulating a forward speed of 1.1m/s.

4.3.2. Application rate
Results of tests with and without vibration are given in Table 1. Vibration was produced by welding two bumps to the drive wheel.

Comparing the rates it can be seen that vibration had a large effect on the granular fertilizer (diammonium phosphate), increasing rates by an average of 40 percent. The effect on powdered fertilizer was insignificant with an average difference of 2 percent.

Table 1 IITA fertilizer distributor application rates

App.rate Setting	Fertilizer	Application rate (kg/ha) Normal	with vib.	Mean Diff. (vib. vs normal) (%)
Minimum)	ammonium sulphate	38.0	36.9	
Mean)	+	188.1	196.0	+2
Maximum)	superphosphate	401.3	425.7	
Minimum)		97.0	NR	
Mean)	urea	406.2	NR	-
Maximum)		686.3	NR	
Minimum)	diammonium	39.0	66.7	
Mean)	phosphate	236.0	250.6	+38
Maximum)		332.7	450.7	

4.3.3. Longitudinal distribution
Figure 9 shows the longitudinal distribution of granular diammonium phosphate with and without vibration and with the feed rate plate in its middle position. Variation about the mean is approximately ± 20 percent without vibration, which is acceptable, but rises to ± 38 percent with vibration.

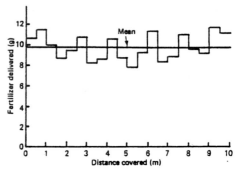

a) Without vibration. Mean rate/0.5m:9.8g (Maximum 111.8g, minimum 7.8g).

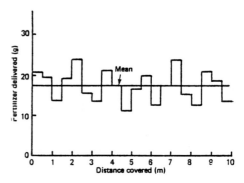

b) With vibration. Mean rate/0.5m:17.3g (Maximum 23.9g, minimum 11.0g)

Fig 9 IITA fertilizer distributor. Longitudinal distribution of granular fertilizer.

4.4. Field tests

4.4.1.& 4.4.2. Application rate and wheel skid
The application rate of powdered fertilizer achieved in the field with the control plate in an intermediate position between minimum and mean was 100.4kg/ha with 6.8 percent wheel skid in a disc harrowed dry sandy loam soil.

The application rate on a level surface with the same calibration was 93.0kg/ha, which suggests that the effects of vibration are negligible and that wheel skid directly reduces application rate. Results for field tests with other fertilizers are unavailable but would be needed for a complete report.

4.4.3. Construction
The machine is of robust and simple construction. The failure of the drive

shaft in the durability test could have
been due to faulty material. Slight
leakage of fertilizer occurred between the
rear wall of the hopper and the feed
control plate, exacerbated by shrinkage of
the wooden hopper walls. The fertilizer
outlet, auger and shaft are all susceptible
to corrosion and require meticulous
cleaning after each period of use.

4.4.4. Operation
During field use two problems were
observed:
-the hopper is too small reducing field
efficiency due to frequent re-filling.
-the feed control plate does not maintain
its' setting and had to be tied to the
chassis to maintain the application rate.

5. CONCLUSIONS

Although only the test results for the
IITA machine have been reported here, all
the distributors described were tested in
Mexico and some general conclusions can be
drawn.
   The tests revealed the difficulty of
applying fertilizers in the humid tropics.
Physical characteristics of the fertilisers
change with variations in the relative
humidity of the atmosphere and at times it
is impossible to continue field application
of hygroscopic products even though on
opening the bag the fertilizer is free
flowing.
   The results indicated the advisability of
calibrating the distributors each time
field conditions, fertilizer or atmospheric
relative humidity change. This is not an
attractive proposition for the small
farmer.
   The problem of serious and rapid
corrosion in the humid tropics reinforces
the resistance of small farmers to adopt
combined seeder/fertilizer distributors as
they lose the whole investment when the
distributor corrodes.
   The IITA design is the most promising as
the auger acts as an agitator and reduces
bridging of moist fertilizer in the hopper.
The problems observed (hopper too small and
feed plate not securely fastened) would be
easy to overcome.
   Nevertheless the conclusion as a result
of the test programme is that small farmers
in the Mexican humid tropics are likely to
continue to apply fertilizer by hand until
a simple, reliable and financially
justifiable machine becomes available.

REFERENCES

1. Baryeh, E. Navasero, Garman, C and Kang
   B. T. 1984. Fertilizer band applicator.
   In: Fertilizer application equipment for
   small farmers. Rome, FAO. p 35-42.

2. Camacho Castro, R. 1983 Desarrollo de un
   prototipo de explotación agropecuaria
   para pequeños productores en una región
   del trópico sub-húmedo de México. Tésis de
   MC. Chapingo, México. Colegio de
   Postgraduados. 348p.

3. India. International Crop Research for
   the Semi-Arid Tropics (ICRISAT), 1980.
   Annual Report 1979/80. Patanchera P.O.
   India p 198-199

4. Krantz, B. A. 1980.
   Rainfall collection and utilization in the
   semi-arid tropics. Papel presentado al
   EEUU-México taller de recursos sobre
   colección de la lluvia la agricultura de
   regiones arias y aridas y semi-aridas.
   Tucson Arizona. Septiembre de 1980.

5. Patterson, D. E. Hebblethwaite, P.,
   Phillipson, A. and Bull K. G. 1964. A
   detailed test procedure for seed drills.
   In: Stevens G. N. Ed. Equipment testing
   and evaluation. Bedford, England,
   National Institute of Agricultural
   Engineering. p67-76.

6. Sims, B. G., Jiménez Regalado, R. and
   Aragón Ramírez A. 1987. Mise au point et
   essais de'expandeurs d'engrais destines
   aux petites agriculteurs mexicains des
   regiones tropicales humides. Machinisme
   Agricole Tropical, France (99): 22-40.

7. Sims, B. G. 1987
   Mecanización para el pequeño agricultor.
   Mexico. Secretaría de Agricultura y
   Recusos Naturales, Instituto Nacional de
   Investigaciones Forestales y
   Agropecuarias. 384p

8. Villanueva Martínez, M 1983.
   La producción de granos básicos en
   México. Estudio de sus tendencias
   recientes, sus causas y perspectivas.
   Econotecnia Agrícola (México) 7(12): 25-
   27.

Land and Water Use, Dodd & Grace (eds), © 1989 Balkema, Rotterdam. ISBN 90 6191 980 0

NPK content of cotton plants as a diagnostic means for fertilization practice

I.Levin & M.Meron
ARO, Institute of Soils and Water, Bet Dagan, Israel

ABSTRACT: NPK fertilization experiments in cotton, under sprinkler and drip irrigation methods were carried out in the Hula Valley, in Israel. The purpose of these experiments was to study the response of cotton to different NPK quantities and methods of application. In addition, the experiments aimed to examine the relationship between the yields and the NPK content of the petioles during the season, in intensively irrigated and high-yielding SJ-2 cotton.

Frequent fertigation gave significantly higher yields than zero fertilizing or soil-applied NPK fertilizer. In the optimal treatments, the $N-NO_3$ content of the petioles, during the course of the season, followed and closely reflected the effects of N treatment. The P content of the petioles was inversely proportional to the $N-NO_3$ content. There was no significant correlation between the yields and the P content of the petioles in experiments in which P was applied. The K content of the petioles, however, showed a clear correlation to yields in all cases. A drop in the K content (under 4%), one month after flowering, indicated potassium deficiency.

Our conclusions are that N and K content levels in the cotton petioles, during the main growing season, may be of practical diagnostic value to cotton growers.

1 INTRODUCTION

Determining quantities and the timing of fertilizer application in cotton by chemical analyses is an accepted approach, known for a long time (Olson and Bledsoe, 1942; Joham, 1951). However advances in cotton cultivars, agrotechnology, irrigation methods and especially the continual increase in yields have necessitated new diagnostic approaches in the effective use of fertilizer.

NPK uptake was studied in Acala cotton under furrow cultivation (Basset et al., 1970) and sprinkler irrigation (Halevy, 1976). The NPK content of the petioles under these conditions was published for California cotton by Basset and McKenzie (1976) and by Grimes et al. (1974).

High yields of seed cotton (6 - 8.2 t/ha) were obtained in drip and sprinkler irrigation experiments in the Hula Valley of northern Israel (Levin and Meron, 1980; Meron and Levin, 1985; Ben-Horin and Ronen, 1983).

In order to study the possibility of using the NPK content of the petioles during the main growing season, as an interactive diagnostic tool for fertilizer requirements of such high yielding cotton long-term fertilization experiments were carried out.

The results and their implications regarding fertilization management in commercial cotton fields are presented here.

2 MATERIALS AND METHODS

Fertilization experiments under sprinkler and drip irrigation were performed on cotton (cv Acala SJ-2) in the Hula Valley of Israel. The climate is typically Mediterranean, with rainless summers and an average temperature range during the peak season (July, August) of 34°C maximum to 18°C minimum. The soil is alluvial vertisol; 56% clay, 28% silt. The experimental plots were part of large commercial cotton fields and, apart from the application of fertilizer, were treated alike, agrotechnically. The sprinkler-irrigated experiment lasted two years and the drip-irrigated experiment four years, continuing on the same plots and with the same fertilization treatments. In the sprinkler-irrigated experiment, the response of the cotton to the quantities of nitrogen, applied in two different ways, was studied. P and K were supplied equally in all treatments. The treatments were: 1) zero-N (control); 2) 100 kg N/ha as solid urea, applied before planting; 3) 180 kg N/ha, applied as in Treatment 2; 4) 100 kg N/ha fertigated in two portions during the growing season; 5) 180 kg N/ha, applied in two or three portions during the season. Available N in the soil before treatment was about 140 t/ha at the 1.2 m depth profile. The treatments are therefore given in terms of total nitrogen.

In the drip-irrigated experiments, the response of cotton to 50 kg/ha P and 250 kg/ha K, applied as solid fertilizer, incorporated into the soil under the laterals (P_S and K_S), as against fertigation throughout the season (P_W and K_W) was studied. In all these treatments, nitrogen fertilizer was applied in equal quantities. The experiment was factorially designed, including zero levels (K_O and P_O) as control treatments, in five replicates.

In addition to recording the routine experimental data (not reported here), 30 petioles from every plot were sampled each week. The petioles were taken from the youngest fully developed mature leaves. They were dried and ground, and their $N-NO_3$, P and K contents were determined in 1% acetic acid extract, by standard colorimetric and flame emission procedure.

The yields of seed-cotton obtained in the second year of the sprinkler-irrigation experiments, and the fourth year of the drip-irrigation experiments are presented in Table 1, being more meaningful than earlier results because of the depletion of soil nutrients.

3 RESULTS AND DISCUSSION

In the nitrogen application experiment, the highly significant decrease in seed-cotton yield under zero-N application, and the relatively high yields in the rest of the treatments (Table 1) correlated to a high degree with the $N-NO_3$ contents of the petioles (Fig. 1). The level of $N-NO_3$ content in all treatments, apart from the control, were fairly close to the proposed reference line. In the zero-N treatment, $N-NO_3$ in the petioles decreased sharply immediately after the beginning of flowering. These findings open up the possibility of using the $N-NO_3$ content of the petioles for diagnostic purposes in N fertilization management, and for correcting N application rates at

an early stage of plant growth. It is of interest to point out the high P content of the petioles in the zero-N treatment (Fig. 2). This phenomenon means that in well irrigated but N deficient plants, the P content is higher than in the healthy ones, independently of the quantity of P applied.

Table 1. Seed cotton yields

Sprinkler Irrigation Experiment

Application Method	Total N (kg/ha)*	Yield (t/ha)
Control	140	3.8 a.
Soil dressing	240	5.5 b.
Soil dressing	320	5.6 b.
Fertigation	240	5.3 b.
Fertigation	320	5.5 b.

Drip Irrigation Experiment

Application Method	Total PK (kg/ha)		Yield (t/ha)**
P_o Control	P	0	5.6 a.
K_o Control	K	0	5.5 a.
P_s Soil incorporated	P	50	5.7 ab.
K_s Soil incorporated	K	250	5.6 a.
P_w Fertigated	P	50	5.9 b.
K_w Fertigated	K	250	6.0 b.

* 140 kg/ha of available N was found in the 0-120 cm soil profile before fertilizing.

** Each treatment of P and K presented here is an average of the three combinations of variables (e.g. P_o = average of $P_o K_o$; $P_o K_s$; $P_o K_w$).

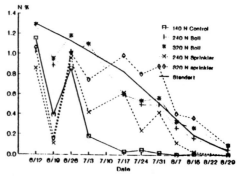

Fig. 1 $N-NO_3$ content of cotton petioles in sprinkler fertigation experiment. Treatments notated in terms of total available nitrogen.

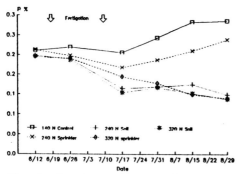

Fig. 2 Phosphorus content of cotton petioles in sprinkler fertigation experiment

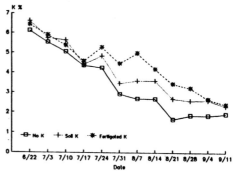

Fig. 4 Potassium content of cotton petioles in drip fertigation experiment

In the experiments of P and K application, the yields of seed-cotton were significantly lower in the zero-application (control) experiments than in the P and K fertigated treatments (Table 1). This significant decrease in yield is reflected in the P content of the petioles during the season only to a minor degree (Fig. 3). These findings do not warrant the use of the P content of the petioles for P fertilization control. In contrast to the P content, the K content of the petioles reflects the yield differences clearly (Fig. 4). The steep drop in the K content of the petiole (down to less than 5%) after 4 weeks from the beginning of flowering, under zero-K (K_O) and soil-applied treatment (K_S) correlates well to the lower yields in these treatments (Fig. 4 and Table 1). Such a drop may indicate potassium deficiency, therefore.

Summing up this study, therefore, we conclude that cotton yields respond more favorably to timely NPK fertigation treatments than to the application of solid fertilizers into the soil. The nitrogen and potassium content of the petioles may be used as a diagnostic tool for the control and management of cotton fertilization.

REFERENCES

Basset, D.M., Andersen, W.D. and Werkhoven, C.H.E. 1970. Dry matter production and nutrient uptake in irrigated cotton. Agr. J. 62: 299-303.

Basset, D.M. and McKenzie, A.J. 1976. Plant analyses as a guide to cotton fertilization. In: Reisentauer, H.M. (ed.) Soil and plant tissue testing in California. U.C. Div. of Agr. Sci. Bull., No. 1879.

Ben-Horin, D. and Ronen, Z. 1983. How we exceed yields of 8 t/ha seed cotton and 2.7 t/ha lint. Hassadeh 63:890-892 (in Hebrew).

Halevy, J. 1976. Growth rate and nutrient uptake of two cotton cultivars grown under irrigation. Agr. J. 68:701-708.

Joham, H.E. 1951. The nutritional status of the cotton plant as indicated by tissue tests. Plant Physiol. 26:76-89.

Levin, I. and Meron, M. 1980. Cotton response to trickler irrigation regimes in the Hula Valley of Israel (1974-1978). ASAE Winter Meeting Paper, no. 80, 2081.

Meron, M. and Levin, I. 1985. Nitrogen, phosphorus and potassium uptake of drip-irrigated cotton. Proc. Third Drip/Trickle Irrigation Congress. Fresno, CA. pp. 382-386.

Olson, L.C. and Bledsoe, R.P. 1942. The chemical composition of the cotton plant and the uptake of nutrients at different stages of growth. Georgia Exp. Sta. Bull. 222.

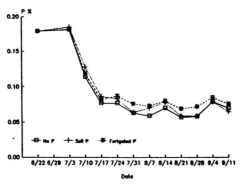

Fig. 3 Phosphorus content of cotton petioles in drip fertigation experiment

Land and Water Use, Dodd & Grace (eds), © 1989 Balkema, Rotterdam. ISBN 90 6191 980 0

Evolution de la teneur des feuilles de blé dur en azote, phosphate et potassium au cours de la croissance sous conditions naturelles et en plein champ: Temps d'échantillonnage

M.Mechergui & S.Lazaar
Institut National Agronomique de Tunis, Tunisie

RESUME: Nous avons étudié l'évolution de la concentration de l'azote (N), du phosphore (P) et du potassium (K) dans les feuilles de blé dur en fonction de l'âge de la plante et de la dose d'engrais.
Le niveau de N et K dans les feuilles diminue rapidement avec l'âge de la plante et est affecté par la dose apportée en ces éléments. Le taux de P augmente pendant la montaison par rapport au tallage et décroit à l'épiaison. L'absorption de P se poursuit pendant la montaison et ceci quelque soit le niveau de fertilisation.
Les rapports P/N et N/K ont montré une variation significative rendant dificile l'utilisation de la méthode de DRIS pour la détermination des normes de fertilisation. Une analyse de regression a montré que le rendement est lié à la concentration de N et P dans les feuilles pendant le tallage et à celle du K pendant l'épiaison. Ces résultats pourront servir dans l'utilisation de la méthode de DRIS.

ABSTRACT: This work studies possible interactions between the plant and its environment in order to better understand fertilisation problems of rainfed cereals in semi-arid regions. Results showed that nitrate and Potassium levels in Durum wheat leaves are maximum at tillering and then decrease rapidly with âge. Phosphate on the other hand showed a small increase and decrease suggesting that the absorption will continue until the heading.
Such a study reveals that sampling cereal leaves at tillering would be the best indicator of nutrient status and the best predictor of crop yield level. Those results will help using the DRIS technique for determining the norms of fertilisation.

ZUSAMMEN FASSONG: Die Entwicklung der Stickstoff (N)-, Phosphor (P)-und Kalium (K)-Konzentrationen in den Hartweizenblättern wurde untersucht in Besug auf das Alter der Pflanze und der Düngungsmenge. Das Stickstoff-und Kalium-Niveau in den Blättern nimmt ab mit dem Alter der Pflanze und wird von den Düngungsmengen dieser Flemente beeinflusst.
Die Phosphorrate steigt nach der Bestockung während des Halmwuchses an und nimmt während der Ahrenbildung ab. Die Phosphorabsorption setzt sich während des Aufwachsens unabhängig von der Düngungsmenge fort.
Die N/K-und P/N-Verhältnisse zeigten eine signifikante Schwankung, sodass die Drissmethode zur Bestimmung der Düngungsnormen schwer anwendbar ist. Eine Regressionsanalyse zeigte, dass der Ertrag mit - der N-und P-Konzentration in den Blättern während der Bestockung und mit - der K-Knozentration während der Ahrenbildung verbunden ist.
Diese Resultate können bei der Verwendung der Driss-Methode benutzt werden.

1. INTRODUCTION

L'étude du mode de variation de la composition minérale des feuilles de la plante est d'une importance capitale. Elle permet de suivre la distribution de l'absorption des nutriments en fonction du temps et la détermination des périodes de nutrition critique.
L'absorption des éléments minéraux par la plante varie considérablement en fonction des facteurs édaphiques et climatiques. Elle varie également en fonction des espèces végétales et même au sein de l'espèce en fonction des variétés.
Des travaux de recherche antérieurs ont montré que la concentration des éléments nutritifs varie au cours de la croissance de la plante en fonction de l'organe et les conditions de nutrition Heller (1969).

Heller (1969) a montré que les jeunes ♂
feuilles de blé absorbent N,P et K alors
que les feuilles sénéscentes accumulent
plutôt du calcium.
Les taux d'azote et de phosphore dans plusieurs organes de blé diminuent avec l'âge
de la plante (Boatwright et Hass, 1960),
Prevel (1978).
Beaufils (1954) a conclu que la concentration de N, P et K décroit avec l'âge, alors
que celle du calcium augmente.
Beaufils (1971) a étudié l'évolution de N,
P et K dans les feuilles de canne à sucre
avec l'âge et a constaté une diminution de
ces trois éléments au cours de la croissance.
Par ailleurs, Racz et al(1965) ont montré
que la teneur de la plante de blé en potassium baisse avec l'âge. Il ont trouvé par
ailleurs, dans leur étude sur l'effet
de la fertilisation sur la composition minérale, une augmentation de la teneur des
feuilles du blé en azote suite à un apport
d'engrais azoté.
Une étude faite sur l'avoine par Nielson et
al(1960) a montré que l'application d'une
fertilisation potassique engendre une augmentation de la concentration de potassium
dans son tissu pendant les stades de 7
feuilles et l'épiaison.
L'objectif de cette étude est de déterminer
l'effet de l'âge de la plante et de la dose
d'engrais apporté sur le niveau des éléments
N, P et K dans les feuilles. Ceci aura pour
rôle de faire une analyse de choix d'une
méthode d'analyse pour pouvoir déterminer
les normes de fertilisation par la méthode
de DRIS.

2. MATERIEL ET METHODES

L'étude est basée sur deux essais de fertilisation en plein champ installés respectivement dans le domaine de l'Ecole Supérieure
d'Agriculture du Kef (ESAK) et dans la zone
de Lorbous comme le montre la Figure.1

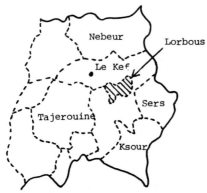

Fig.1.Situation générale de la zone d'étude.

La zone est caractérisée par un climat semi-
aride supérieur avec une pluviométrie moyenne de 490 mm.

2.1. Essai ESAK : (Ecole Supérieure d'Agri-culture du Kef).

Ce protocole est installé sur un sol peu
évolué, d'apport alluvial dont les caractéristiques physico-chimiques figurent sur
le tableau 1.

Tableau 1. Propriétés physico-chimiques du
sol.

Z	%A	%L	%S	pH	%MO	CE	K	CEC
1	21,3	14	64,7	8,3	1,2	1,4	1,1	17,0
2	21,3	14	58,7	8,3	1,2	1,0	0,8	12,8
3	29,3	16	54,7	8,3	0,6	0,8	1,0	10,9

1 = 0-35, 2=35-65, 3=65-115 cm
+=Moyenne des trois échantillons.

Le dispositif expérimental est un bloc alé-
toire complet ayant trois répétitions formant chacune un bloc à part. Chaque bloc
groupe neuf parcelles élémentaires dont la
superficie unitaire est de l'ordre de $25m^2$
(2,5x10). Chaque parcelle a reçu au hasard
un traitement qui consite en la combinaison
d'une dose d'azote avec une dose de phosphate.
L'essai est un factoriel N3 qui consiste en
l'apport de trois doses de nitrates d'ammonium (17;33, et 67kgN/ha) et de trois
doses de superphosphate (50,100 et 200 kg
de P_2O_5/ha).

2.2. Essai Lorbous

Ce protocole est aussi installé sur un sol
peu évolué, d'apport alluvial dont les caractéristiques physico-chimiques figurent
sur le tableau 2.

Tableau 2. Propriétés physico-chimiques du
sol de Lorbous

Z	%A	%L	%S	pH	%MO	CE	K	CES
1	39	25	36	8,1	2,6	1,1	1,1	7,0
2	41	23	36	8,2	1,6	0,8	0,8	10
3	51	23	26	8,3	1,4	0,6	0,8	8

1=0-25, 2=25-65, 3=65-110

C'est un essai factoriel $N^3 P^3 K^3$. Les
unités parcellaires ont une surface de
$25m^2$. Chaque unité a reçu au hasard un
traitement qui consiste en la combinaison
de trois dose d'engrais azotés, phosphatés
et potassiques. Ces différentes doses
d'engrais utilisées figurent sur le
tableau 3.

Tableau 3. Doses d'engrais utilisées

Nitrate d'ammonium		Superphosphate triple		Sulfate de potassium	
dose	kg/ha	dose	kg/ha	dose	kg/ha
NO	O	PO	O	KO	O
N1	100	P1	60	K1	60
N2	150	P2	100	K2	100

2.3. Considération générales

L'épandage de P a été réalisé entièrement avant semis. L'azote est apporté au semis et au tallage. Il n'ya pas eu d'apport d'eau vu qu'on voulait étudier la fertilisation dans les conditions naturelles.
Les parcelles de l'essai ESAK ont été semées en blé dur Badri "D56-3" alors que celles de Lorbous ont été semées en blé dur "INRAT 69".
Le prélèvement des échantillons a été réalisé à trois reprises : pendant la formation des talles, la montaison et l'épiaison. La partie aérienne de la plante a été prélevée minitieusement sur chacune des unités expérimentales. Les feuilles collectées ont été lavées à l'eau potable pour éliminer toute trace de produits chimiques et notamment les produits de désherbage. Elles ont été séchées dans l'étuve pendant 24h à une température de 80°C et enfin broyées.
La méthode de Microkjeldahl a été utilisée pour l'analyse de l'azote total dans la plante. Le phosphore a été déterminé après calcination de l'échantillon de la plante.
Le phosphore a été déterminé après calcination de l'échantillon de la plante dans un four à moufle à une température de 550°C pendant 24h. Le potassium a été mesuré au flamme photomètre après calcination de la poudre végétale par voie sèche.

3. RESULTATS ET DISCUSSION

3.1. Evolution de la concentration de N.P et K dans les feuilles

3.1.1. Essai ESAK
- L'azote
Les données du tableau 4 montrent que quelque soit le traitement d'engrais appliqué, la concentration de l'azote dans les feuilles diminue avec l'âge de la plante (P=0,05). Ce résultat confirme les conclusions de Boatwright et Hass (1960), Prével (1978).
La vitesse de diminution d'azote est rapide entre le tallage et la montaizon et lente entre la montaison et l'épiaison.
Cette diminution de taux d'azote entre le tallage et la montaison est dûe à l'effet de dilution imposé par la production de la matière sèche par la plante. Par contre la diminution du taux d'azote entre la montaison et l'épiaison s'explique par la migration de cet élément vers l'épi (Salisbury et Ross, 1978). La concentration d'azote dans la plante semble ne pas être affectée par l'augmentation des doses d'azote et de phosphate dans n'importe quel stade végétatif du blé. Il y a une légère augmentation en fonction de la dose pendant le tallage mais elle n'est pas significative.
- Le phosphore dans les feuilles de blé dur est maximale pendant le stade tallage, néanmoins, elle chute pendant les stades les plus avancés de la croissance de la plante (p=0,05). La diminution enregistrée entre le tallage et la montaison ne peut être dûe qu'à l'effet de dilution produite à la suite de l'accumulation par la plante de matière sèche. Le déclinoté entre la montaison et l'épiaison est le résultat de la migration de phosphore vers l'épi. La faible tendance de dimunition du taux d'azote dans les feuilles après le tallage montre que la plante continue à absorber le phosphore pendant la montaison.
- Le potassium
Le taux de potassium dans le tissu des feuilles de blé est maximum pendant le stade de formation des talles. Il décroît rapidement au fur et à mesure que la plante s'approche de la maturité. Ceci confirme les conclusions de Miller (1939), de Klebesadle (1969). La diminution enregistrée entre le tallage et la montaison est expliquée par la dilution produite par l'accumulation de la matière sèche, alors que la migration de K vers l'épi justifie la diminution de la concentration entre la montaison et l'épiaison. La vitesse de diminution de la concentration est plus faible pendant la première période que pendant la deuxième. La concentration de K dans la plante est indifférente aux apports phosphatés. Par ailleurs une augmentation de la dose d'azote apportée engendre une augmentation non significative de la concentration du potassium dans les feuilles de blé

3.1.2. Essai Lorbous
- Evolution de la concentration de N dans les feuilles.
Les données du tableau 5 montrent que quelque soit le traitement d'engrais appliqué, la concentration de l'azote dans les feuilles diminue avec l'âge de la plante (P=0,05). La chute de l'azote dans les feuilles est lente entre le tallage et la montaison et plus rapide entre la montaison et l'épiaison.
Une analyse de la variation relative aux taux d'azote dans les feuilles du blé a montré que :

Tableau 4 : Evolution de la concentration de N,P et K dans les feuilles de blé dur et pour l'essai ESAK.

Traitement kg/h		Tallage Montaison %			Stade de croissance Montaison %			Epiaison			Sol à la montaison (ppm)		
N	P	N	P	K	N	P	K	N	P	K	N	P	K
50	50	4,52	0,37	2,84	1,91	0,26	2,13	1,41	0,20	1,43	27,1	31,9	435,3
50	100	4,72	0,34	2,74	1,90	2,27	2,19	1,32	0,20	1,39	27,2	25,7	469
50	200	4,11	0,32	2,76	2,01	0,25	2,09	1,17	0,18	0,97	27,8	24,2	417
100	50	4,67	0,29	2,57	2,02	0,26	2,38	1,45	0,18	1,64	32	24,1	431
100	100	5,05	0,32	2,87	1,90	0,27	2,07	1,34	0,21	1,31	32,4	28,7	443
100	200	4,72	0,33	2,77	1,96	0,27	2,15	1,29	0,24	1,36	31,7	32	421
200	50	4,88	0,30	2,92	1,98	0,27	2,29	1,61	0,20	1,73	43,7	29,7	459
200	100	4,70	0,29	2,90	1,97	0,28	2,42	1,32	0,24	1,37	39,29	29,3	411
200	100	4,82	0,34	2,64	1,90	0,28	2,22	1,56	0,21	1,15	30,5	27,3	430

Tableau 5 : Evolution de concentration de N,P et K dans les feuilles du blé dur et pour l'essai Lorbous.

	Tallage %			Montaison %			Epiaison %			Sol à la montaison (ppm)			Rdt q/ha
	N	P	K	N	P	K	N	P	K	N	P	K	
$N_0P_0K_0$	3,21	0,15	2,03	1,27	0,16	1,22	0,85	0,12	0,45	29,23	18,32	199,0	23
$N_0P_0K_1$	3,42	0,17	2,64	1,44	0,20	1,87	1,19	0,18	0,94	34,90	27,16	326,3	14
$N_0P_0K_2$	3,50	0,24	2,85	1,51	0,23	2,00	1,20	0,20	1,15	39,81	25,81	375,7	23
$N_0P_1K_0$	3,35	0,17	2,33	1,34	0,22	1,44	0,96	0,19	0,57	35,31	24,24	311,1	23
$N_0P_1K_1$	3,68	0,20	2,92	1,71	0,23	2,02	1,39	0,18	0,86	36,33	21,80	336,0	22
$N_0P_1K_2$	3,85	0,25	3,50	2,04	0,27	2,13	1,62	0,20	1,16	40,43	24,17	420,9	18
$N_0P_2K_0$	3,25	0,21	2,43	1,30	0,26	1,56	0,89	0,23	0,63	34,90	33,54	255,3	25
$N_0P_2K_1$	3,24	0,14	2,14	1,28	0,20	1,29	0,86	0,14	0,89	29,67	13,68	199,3	16
$N_0P_2K_2$	3,20	0,16	2,13	1,24	0,14	1,31	0,87	0,12	1,15	29,64	11,97	198,4	22
$N_1P_0K_0$	3,49	0,20	2,38	1,62	0,22	1,40	1,31	0,19	0,76	34,25	23,26	304,0	22
$N_1P_0K_1$	3,69	0,21	3,01	1,68	0,24	1,93	1,38	0,21	0,86	34,21	26,74	340	20
$N_1P_0K_2$	3,79	0,24	3,14	1,83	0,27	2,20	1,45	0,26	1,01	40,98	26,74	412,4	38
$N_1P_1K_0$	3,48	0,20	3,01	1,51	0,20	1,71	1,19	0,20	0,63	37,70	23,03	312,2	18
$N_1P_1K_1$	3,52	0,19	3,00	1,71	0,21	1,81	1,28	0,16	0,82	35,16	29,52	336,9	22
$N_1P_1K_2$	3,48	0,22	2,96	1,48	0,26	1,79	1,22	0,22	0,97	39,68	31,06	300,4	20
$N_1P_2K_0$	3,42	0,27	3,12	1,38	0,28	1,84	1,11	0,24	0,78	37,35	27,36	265,7	17
$N_1P_2K_1$	3,37	0,28	3,07	1,37	0,28	1,91	1,09	0,23	0,82	38,18	29,15	280,2	28
$N_1P_2K_2$	3,76	0,23	3,29	1,96	0,23	1,86	1,69	0,17	0,94	37,76	26,95	400,8	12
$N_2P_0K_0$	3,69	0,25	2,50	1,91	0,28	1,54	1,60	0,24	0,82	37,76	26,43	379,6	18
$N_2P_0K_1$	3,80	0,23	3,32	1,95	0,22	1,90	1,64	0,19	0,98	34,25	25,17	392,9	25
$N_2P_0K_2$	3,92	0,25	3,50	2,02	0,23	2,17	1,66	0,20	1,15	37,70	25,81	426,4	25
$N_2P_1K_0$	3,63	0,20	2,78	1,83	0,24	1,91	1,42	0,17	0,52	42,14	23,10	327,8	24
$N_2P_1K_1$	3,72	0,24	3,13	1,81	0,27	2,14	1,45	0,21	1,06	38,93	24,58	318,8	4
$N_2P_1K_2$	3,53	0,21	2,87	1,68	0,23	1,96	1,37	0,21	1,23	35,85	28,60	382,9	22
$N_2P_2K_0$	3,54	0,27	2,90	1,67	0,24	1,98	1,33	0,19	0,69	39,06	31,99	305,2	22
$N_2P_2K_1$	3,61	0,20	2,74	1,79	0,26	1,81	1,42	0,19	0,97	36,47	24,58	325,8	21
$N_2P_2K_2$	3,65	0,23	3,12	1,80	0,26	1,79	1,40	0,19	1,12	37,49	27,98	379,6	23

Lorsque la dose N augmente la concentration du même élément dans les feuilles augmente.
+ Suivant la même allure que l'azote, la dose de potassium influence la concentration de l'azote dans les feuilles.
+ La dose de phosphate n'a pas montré d'effet significatif sur la concentration d'azote dans les feuilles.
- Evolution de la concentration de phosphate dans les feuilles.
Nous remarquons que d'après l'analyse de variance, la concentration de phosphate dans les feuilles n'est pas affectée ni par la dose de phosphate dans le sol, ni par la dose de potassium ou de l'azote dans le sol. Par contre, il y a une légère augmentation de phosphate à la montaison par rapport au tallage. Cette variation n'est pas significative. Ces résultats sont dûs au fait que le sol est initialement riche en phosphate tableau 5.
- Evolution de la concentration de potassium dans les feuilles
Les données du tableau 5 montrent que :
+ Une augmentation de la dose d'azote dans le sol engendre une augmentation de la concentration de K dans la plante.
+ De même, une augmentation de K dans le sol engendre une augmentation de K dans la plante.
+ La chute de concentration en fonction du temps est impotante mais presque constante.

3.1.3. Conclusion
Le blé absorbe le maximum de ses besoins en azote et en potassium pendant le tallage, ce qui laisse entrendre que l'apport d'une quantité suffisante de ces éléments pendant cette période critique est indispensable. La dose a un effet positif sur la concentration de la plante sauf pour le cas de phosphate qui peut être expliqué par un ecxès de ce même élément dans le sol. L'absorption de P commence dès le jeune âge et se poursuit jusqu'à la montaison.

3.2. Evolution de P/N, K/P et N/K dans les feuilles :

3.2.1.
Le calcul de ces rapports a été fait mais il n'est pas présenté ici. Seules les conclusions sont présentées.
Le rapport P/N augmente avec l'âge. Cette variation temporelle est plus importante dans l'essai Lorbous que celui de l'ENSAK.

3.2.2. K/P
Dans le cas de l'essai Lorbous le rapport diminue avec l'âge. Le cas de l'ENSAK illustre une légère diminution mais qui n'est pas généralisée pour tous les traitements.

3.2.3. N/K
Le rapport N/K diminue du tallage au montaison. Ceci est général pour les deux essais.
Nous remarquons que ces résultats ne confirment pas en totalité ce qui est venu dans les travaux de Beaufils (1959,1971,1973) ou ceux de (Beaufils et Sumner 1976) ou Sumner (1977). En effet pour eux ces rapports sont constants ou peu variables avec l'âge vu qu'ils sont basés sur leur observation qui consiste à ce que les courbes N,P,K en fonction du temps sont parallèles. Ceci n'est pas observé et surtout pour le cas de P où la teneur en cet élément a augmenté et ensuite diminué.
Quelle concentration pourrons-nous considérer dans l'application de la méthode DRIS.

3.2. Pour résoudre ce problème les corrélations entre le rendement et la concentration de chaque élément dans la plante sont étudiées.

3.3. Choix du temps d'échantillonnage .
L'évolution de la concentration des éléments nutritifs en fonction du stade de croissance de la plante limite l'utilisation de l'analyse de la plante en tant que guide de fertilisation.
Une analyse de regression linéaire a été faite et qui a pour objectif de choisir parmis les concentrations les mieux corrélées avec le rendement.
Les résultats sont regroupés dans le tableau 6.

Tableau 6. Les coefficients de corrélation (R^2) en %

Culture	Azote			Phosphate			Potassium		
	T	M	E	T	M	E	T	M	E
Blé dur(Badri)	25	18	10	29	18	12	05	20	38
Blé dur(INRAT) (69)	23	10	10	38	22	13	12	15	25
Blé tendre(FXA)	28	25	23	35	24	06	24	16	36
Orge (Taj)	28	10	08	50	18	11	15	22	34

T=Tallage, M=Montaison,E=Epiaison

Nous remarquons que :
- Le coefficient de corrélation du rendement avec la concentration en azote et phosphate dans la plante est maximum pendant le tallage. Ce même coefficient diminue avec l'âge.
- Le coefficient de corrélation du rendement avec le potassium est maximum pendant l'épiaison. Il est croissant du tallage à l'épiaison.
Ces résultats sont identiques pour les deux variétés de blé dur utilisées.

Des essais similaires ont été installés dans la zone sur l'orge et le blé tendre.

Dans le même objectif une analyse de ces données a aboutie aux mêmes résultats que précédement comme le montre le tableau6.

Par conséquent dans le cas des sols de la région d'étude du Kef où le potassium est disponible à la plante en quantité suffisante et où l'azote et le phosphore doivent être apportés chaque année, l'analyse de la plante pendant le tallage donne une meilleure indication sur le niveau du rendement attendu. Ces concentrations peuvent être utilisées par conséquent dans la méthode de Dris pour la détermination des normes de fertilisation.

4. REFERENCES BIBLIOGRAPHIQUES

Beaufils,E.R.1954 Minéral diagnosis of Hevea Brasiliensis Arch Rubercult.32,1.

Beaufils, E.R.1959 Physiological Diagnosis III Method of interpreting the data.Rev. Gen Caoutch, 36, 225.

Beaufils, E.R.1971 Physiological Diagnosis a guide on principles developped for rubber trees.F.S.S.A. Journal 1,1-30.

Beaufils,E.R.1973. Diagnosis and Recommandation Integrated system(Dris).A general sheme for experimentation an calibration based on principles developed from research in plant nutrition. Soil Sci. Ful.1, University of Natal.

Beaufils,E.R. and Sumner,M.E., 1976. Application of the Dris approch for calibrating the soil and plant yield/quality factors of sugarcane. Proc.S.A.Sug.Tech Assoc.

Boatwright,G.C. and H.J.Hass 1960. Developpement and composition of spring cubeat as influenced by nitrogen and phosphorus fertilization. Agron.J 53:33-36.

Heller,R. 1969. Biologie végétale, Nutrition et matabolisme tome II.ed.Masson et Cie. Paris.

Heller,R..1978.Abrégé de physiologie végétale.Tome I et II.ed.Masson.Paris.

Klebesadel,L.J.1969.Chemical composition and yield of oast and peas separated from a forage at saccessive stage of growth.J. 61 : 713 - 716.

Miller, E.C.1939. A physiological study of the winter wheat plant at different stages of its development. Kansas Agr.Exp.Sta. Tech.Ball.47.

Nielson,K.F.,R.L.Holstead,A.J.Machlean,R.M. and S.H.Bourget.1960. The influence of soil temperature on the growth and mineral composition of oats.Can.J.Soil.Sci.40:255.

Racz,G.L.,M.B.Webber,R.J.and Hedlin,R.A.1965 Phosphorus and Nitrogen Utilization by rape,Flox,and wheat.Agro.J.57 : 335-337.

Prevel,P.M.1978. Role des éléments minéraux, fruits.Volume 33 (7-8).

Salisbury,F.B.and C.W.Ross.1978. Plant Physiology.2D.ed.Wads Worth.Belmont, California.

Sumner,M.E.1977.Effect of Corn Leaf Sampled on N,P,K,Ca an mg content and calculated Dris indices Plant Anal.,8(3), 269-280.

Sumner, M.E.and E.R.Beaufils.1977.Application to Diagnosis and Recommandation Integrated System(Dris) to sugar cane. Final Report on research project Sponsored by the South Africa Sugar Association. Department of soil Science and Agrometeorology. University of Natal - Pietermaritz burg.79p.

Land and Water Use, Dodd & Grace (eds), © 1989 Balkema, Rotterdam. ISBN 90 6191 980 0

Testing of spreaders for granular fertilizer

P.Spugnoli, M.Vieri & M.Zoli
Institute of Agricultural Mechanization, Florence, Italy

ABSTRACT: The autors examine some kind of fertilizer spreaders
using rotary and pneumatic system. Urea and bulkblending (10.10.21) are employed in
tests. Spreading analisys shows the best quality of pneumatic spreaders and the hight
variability in the rotary spreaders. Simulation with computer of different scheme of
work in spreading shows optimum operating breadth.

RESUME: Les auteurs analysent les résultats d'essais comparatifs de l'éspandange
d'éngrais granulés (urée et complexe 15.15.21) avec des éspandeuses centrifuges et des
éspandeuses pneumatiques. Les diagrammes de répartition de l'engrais montrent la bonne
qualité du travail pour ce qui est du type pneumatique et son énorme variabilité
lorsqu'il s'agit du type centrifuge. La simulation avec un ordinateur a permis de
déterminer les largeurs de travail optimales pour améliorer la qualité de la
répartition sous différentes conditions d'exercice.

ZUSAMMENFASSUNG: Vergleichende Versuche über die Verteilung fester Düngemittel.
Es werden die Ergebnisse von vergleichenden Versuchen über die Ausstreuung von
Granulat-Dünger (Urea und Komplex 15.15.21) mit pneumatisch betriebenen und
Zentrifugal-Düngerstreuern geprüft und besprochen. Die Verteiler-Diagramme zeigen gute
Arbeitsergebnisse bei pneumatischen Geräten und beachtliche Unterschiede bei
Düngerstreuern mit Reaktionszentrifuge. Die Simulation mit Computer zeigte ein
optimales Arbeitsspektrum auf, um die Streuungsqualität unter verschiedenen
Einsatzbedingungen zu verbessern.

1 FOREWORD

A return in agriculture, as in other
areas, to production techniques designed
to minimize energy losses has encouraged
thoughtful decision-making regarding the
use of raw materials and equipment, but
also concerning those operational
procedures most effective in optimizing
the input/output relationship.
Fertilization represents one of those
phases with the greatest expenditure of
energy, because of the high energy level
of the fertilizers themselves, sometimes
higher than 70 MJ/kg. With some
intensive crops, such as maize,
fertilizer expenses represent about 50%
of overall costs (4).
While the type of fertilizer, its
composition and the quantity applied
represent important choices in
determining the relationship of raw
materials used to product obtained, the
aptness of the spreader and above all the
organization of work constitute an index
of the effectiveness of that
relationship, inasmuch as they have a
direct bearing upon agricultural
productivity. Yields very often fall
short of their potential because of
irregular fertilization. This is clearly
legible in strips of differential
vegetation growth; the non-uniform
bunching and the looging of the product.
In practise, solid fertilizers are
applied using distribution systems which
are either localized or "full field"
(broadcast), and the equipment most
commonly used relies on centrifugal
propulsion or pneumatic action.
Equipment with full field distribution is
still the most widespread, despite the
introduction of precision mechanisms,
which, however, are penalized by a higher

Rotary spreader

Pneumatic spreader

Fig.1 Scheme of testing procedure

Table 1. Physical features of employed fertilizers

Fertilizer	Mass dens. of sample [kg/dm3]	Mass dens. of solid [kg/dm3]	Sample granulometric analysis				
			<1mm	1-2mm	3-4mm	4-5mm	>5mm
Urea	1.3	0.7	2.8	71.0	25.8	0.4	0.0
10.10.21	1.8	1.1	0.0	2.3	26.9	45.8	0.6

purchase price.

Full field spreaders, both pneumatic and centrifuge, can nonetheless create problems because of their irregular transverse distribution, which is accentuated and complicated when fertilizers differing in grain size are distributed.

With these problems in mind, tests were conducted on three spreaders, different models produced by the same company, each for the full field distribution of various granular fertilizers, in order to compare both systems for transporting the product, those based on centrifugal and pneumatic propulsion.

2 MATERIALS AND METHODS

As noted above, three types of spreader were tested, two using centrifugal propulsion, differing, however, in dimensions and hopper capacity, and one using pneumatic action.

The machines, each tractor-drawn, have: conoid hoppers with capacities, respectively, of 150 and 500 lt for the centifuge models and 250 lt for the pneumatic model; devices for moving the mass towards the exit aperture; dosing devices; distribution apparatus.

The transport and distribution devices are propelled by the tractor's power take off and inserting a multiplier with a transmission ratio of 1:1.25 for the centrifuge model and of 1:8 for the pneumatic type.

Dosage is controlled by widening or narrowing a rectangular slit set at the bottom of the hopper, its maximum aperture being 50 x 60 mm.

In both centrifugal models the distributor consists of a rotating disc with a vertical axis (velocity $\approx$ 500 rpm) and slightly concave (inclination 5°), and with three radial scoops for throwing the granules.

In the pneumatic model a fan produces an air flow of 600÷800 m /h, with a velocity of c. 80 m/s. The fertilizer alloted by the metering device falls into an inclined conduit and is drawn by depression into the air flow. Distribution is lateral, unlike that of the centrifugal system, with its posterior, fan-shaped projection.

The fertilizers used were: urea and a fertilizer complex (15.15.21), whose chemical characteristics are presented in Tab.1.

The aim of the testing was to determine the range and uniformity of distribution of the aforementioned fertilizers achieved by each of the spreaders.

For testing range, the setting was that most commonly used, with the apertures of the metering at 25 x 50 mm and 45 x 50 mm. The centrifuge types were used with the scoops removed from the distibutor disc and collecting the fertilizer for a set period of time. For the pneumatic type the hopper was loaded with a given amount of fertilizer, and the time taken to empty it was measured.

To determine uniformity of distribution we used a series of boxes measuring 500 x 360 mm and 30 mm high (0.18 m²) set on the ground at 1 m intervals, as shown in Fig.1, and repeating each trial three times in order to verify the longitudinal progression, which generally varies as the hopper is emptied.

From the resulting figures diagrams were then constructed of the distribution characteristics of each piece of equipment and for the two fertilizers employed, their quality revealed by the coefficient of variability (C.V.).

Finally, a computer simulation was set up to accentuate variations in the uniformity of distribution in the field. Strips were fertilized both moving in a single direction and in alternating directions varying the width of the area

Table 2. Flows of fertilizer in tested spreaders.

Implements	Urea		Bulk blending	
	index 10 [kg/min]	index 15 [kg/min]	index 10 [kg/min]	index 15 [kg/min]
1° Rotary spreader	26.2	76.5	27.2	80.9
2° Rotary spreader	43.4	99.6	64.3	125.5
Pneumatic spreader	62.5	----	69.5	-----

First rotary spreader

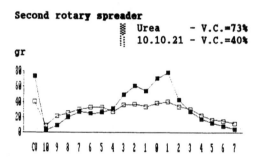

Width [m]

Second rotary spreader

Width [m]

Pneumatic spreader

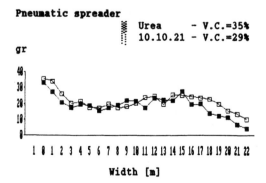

Width [m]

Fig.2 Spreading analisys in different tests.

worked. To determine the ideal width, that which permits the largest spread possible combined with the most uniform distribution, an acceptable limit was set at V.C. = 45%, following the judgements of a number of authors (3).

3 RESULTS

Table 2 presents the ranges determined for the various devices and for the two types of fertilizer.
In the distribution tests, there was no significant difference between the three repetitions, and Fig.2 illustrates the median distribution curves according to the thesis adopted and in relation to the width and the coefficient of variability for each curve.
The results indicate the distinctive behavior of urea with respect to the complex, which is more highly concentrated near the distributor. One also notes higher V.C.'s and diminished width of coverage. This distinction is not apparent with the pneumatic model.
The differing behavior of the two fertilizers can be explained by considering the relative dimensions and density of the granules, as indicated in Table 1: the greater density of the complex makes for greater kinetic energy and thus a wider projection when subjected to centrifugal force. With pneumatic action the particle's behavior is linked to the initial thrust and then to the varying intensity and uniformity of the free flow. Those particles with a relatively great mass and surface area draw the greatest force from the initial thrust, and the distance reached is not strictly conditioned by the flow of air. The smaller particles, however, are more strongly affected by the flow, and their projection is directly related to it. This purely qualitative consideration may be tested more rigorously in a subsequent study.
Fig.2 also indicates the greater uniformity of the pneumatic action as compared to the centrifugal drive. The distibution curve reaches a minimum at the end of the jet and a maximum at the distributor, probably due to defective entry of the fertilizer into the air stream, so that many granules get a lateral push and are thus sent sideways, falling close to the machine.
With the computer simulation of distribution in the field the difficulties in effecting effecient fertilization using this sort of

2177

A - single pattern
   l=13m; V.C.=130%

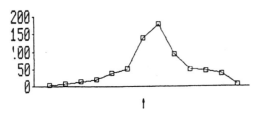

B - alternate pattern with single breadth
   l=7m;   V.C.=55%

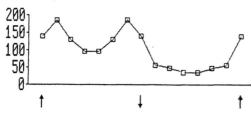

C - alternate passage with double breadth
   l=7+5m; V.C.=42%

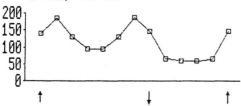

D - homologous passage
   l=7m;   V.C.=48%

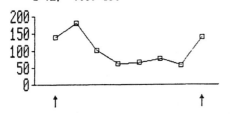

Fig.3 Operative scheme in first rotary spreader (spreading Urea).

equipment became manifest, expecially as regards the centrifugal models.
The irregular broadcasting of these models, which is basically triangular in shape, makes it necessary that each lap overlap its predecessor if one is to achieve acceptable results. This further complicates the problem first, because various fertilizers determine distinctive distribution patterns and second, because the pattern itself is very often asymmetrical.
This is most evident with the centrifugal type spreader (Fig.3) in the distribution of urea, where the distribution pattern (in a single passage) is clearly triangular and off-axis with respect to the tractor, and distribution is highly irregular, with a V.C. of 103%.
Adjacent and alternating laps yield a good distribution, with a width of 7 m for a V.C. of 55%. Better still, but complicated in practise, are alternating laps of 7 and 5 m, where the C.V. falls to 42%.
It becomes much simpler to determine the optimal width with adjacent laps in the same direction. A 7 m width gives an acceptable V.C. of 48%, with 6.5 m the V.C. drops to 40%. However, this approach is not efficient, owing to the time wasted in returning to the other side of the field.
Table 3 presents the optimal operation of the machines according how the work is performed. For alternate laps, both even and dual widths are considered. The figures in parentheses indicate those widths which permit optimal uniformity. In certain cases one notes that only a slight variation in width greatly improves the uniformity of application. In Table 3 one sees that with the pneumatic spreader overlapping applications lead to no significant improvement, except when the distal portions are superimposed. With the spreading of urea superimposition even worsens the situation.

4 CONCLUSIONS

The results indicate extreme variability in the operation of the equipment tested. To this one may add the divergent physical behavior of the fertilizers, which prescribes a variety of working widths.
Comparison of the equipment favors the spreader using pneumatic action, which combines an impressive working width with a uniform distribution of both fertilizers.
A computer simulation designed to determine optimal working situations has defined different working widths for each thesis according to how the work is organized, and in some cases an acceptable fertilization is possible only where the applications alternate in width.
These considerations indicate the need to

Table 3. Operative features of fertilizer spreaders with different scheme of work.

Implement	Theorical width [m]	Single pattern Width [m]	V.C. %	Homologous passage Width [m]	V.C. %	Alternate passage 1 width [m]	V.C. %	2 widths [m]	V.C. %
		....		Urea		....			
1° Rotary Spreader	20	13	103	6.5	43	7	55	7+5	42
2° Rotary Spreader	30	18	73.7	13 / 12	41 / 34	15	59.7	9+17	42
Pneumatic Spreader	28	24	36	24 / 23	36 / 37	24 / 23	36 / 49.4	24+22	33

Implement	Theorical width [m]	Single pattern Width [m]	V.C. %	Homologous passage Width [m]	V.C. %	Alternate passage 1 width [m]	V.C. %	2 widths [m]	V.C. %
		....		Bulk Blending(10.10.21)		....			
1° Rotary Spreader	20	14	58	12 / 10	41 / 30	12 / 11	46 / 37	13+11 / 13+7	45 / 21
2° Rotary Spreader	30	19	42	19 / 15	40 / 16	17.5 / 17	45 / 38	10+18	12
Pneumatic Spreader	28	24	32	24 / 23	32 / 30	24 / 23	32 / 27	24+22	23

furnish farmers with detailed explanations of the operation of spreaders according to the various fertilizers adopted. It would be especially useful to include with each piece of equipment an owner's manual that is detailed enough to respond to the full range of working conditions, but also clearly set forth and easy to consult. Finally, as regards fertilizers and their divergent behavior as they are spread, it might be useful to favor products with granules whose density and mass are as uniform as possible, and whose irregularities are compensated by inert additives.

REFERENCES

Casella, A. 1956. Prove comparative di laboratorio di uno spandiconcime a spaglio con tubo di lancio oscillante. Atti del Centro Nazionale Meccanico Agricolo, Vol II.Torino, Italy.

Ade, G. 1984. Spandiconcime centrifughi. Macchine e Motori Agricoli, n°4.

Cavalli, R. 1984. Analici della distribuzione di una miscela di fertilizzanti solidi mediante spandiconcime di tipo centrifugo. Macchine e Motori Agricoli. n°10.

Balsari, P. Provolo, G. Airoldi, G. 1987. Concimazione minerale con impiego di spandiconcime pneumatico. Macchine e Motori Agricoli. n°3.

O.E.C.D. 1975. Draft revised standard testing procedure for fertilizer distributors.

Parish Chaney 1985. Evaluation of a new lawn fertilizer application technology. Transaction of the ASAE, Vol.28(1).

Parish 1986. Evaluation of two methods of fertilizer spreader pattern correction. Transaction of the ASAE, Vol.29(2).